AUTOMOTIVE TRANSMISSIONS AND POWER TRAINS

Other Books and Instructional Materials
by William H. Crouse and * Donald L. Anglin

Automotive Chassis and Body*
  Workbook for Automotive Chassis and Body*
Automotive Electrical Equipment
  Workbook for Automotive Electrical Equipment*
Automotive Engines*
  Workbook for Automotive Engines*
Automotive Fuel, Lubricating, and Cooling Systems*
  Workbook for Automotive Fuel, Lubricating, and Cooling Systems*
Automotive Transmissions and Power Trains*
  Workbook for Automotive Transmissions and Power Trains*
Automotive Service Business: Operation and Management
Automotive Emission Control
Automotive Engine Design
Workbook for Automotive Service and Trouble Diagnosis
Workbook for Automotive Tools
Automotive Mechanics
  Study Guide for Automotive Mechanics
  Testbook for Automotive Mechanics*
  Workbook for Automotive Mechanics*
  Automotive Troubleshooting Cards
The Auto Book
  Auto Shop Workbook*
  Auto Study Guide
  Auto Test Book*
  Auto Cassette Series
General Power Mechanics (with Robert Worthington and Morton Margules*)
Small Engines: Operation and Maintenance
  Workbook for Small Engines: Operation and Maintenance

Automotive Transparencies
by William H. Crouse and Jay D. Helsel

Automotive Brakes
Automotive Electrical Systems
Automotive Engine Systems
Automotive Transmissions and Power Trains
Automotive Steering Systems
Automotive Suspension Systems
Engines and Fuel Systems

AUTOMOTIVE TRANSMISSIONS AND POWER TRAINS

CONSTRUCTION, OPERATION, AND MAINTENANCE

William H. Crouse
Donald L. Anglin

Fifth Edition

McGRAW-HILL BOOK COMPANY
GREGG DIVISION
NEW YORK
ST. LOUIS
DALLAS
SAN FRANCISCO
AUCKLAND
DÜSSELDORF
JOHANNESBURG
KUALA LUMPUR
LONDON
MEXICO
MONTREAL
NEW DELHI
PANAMA
PARIS
SÃO PAULO
SINGAPORE
SYDNEY
TOKYO
TORONTO

ABOUT THE AUTHORS

William H. Crouse

Behind William H. Crouse's clear technical writing is a background of sound mechanical engineering training as well as a variety of practical industrial experience. After finishing high school, he spent a year working in a tinplate mill. Summers, while still in school, he worked in General Motors plants, and for three years he worked in the Delco-Remy Division shops. Later he became Director of Field Education in the Delco-Remy Division of General Motors Corporation, which gave him an opportunity to develop and use his natural writing talent in the preparation of service bulletins and educational literature.

During the war years, he wrote a number of technical manuals for the Armed Forces. After the war, he became Editor of Technical Education Books for the McGraw-Hill Book Company. He has contributed numerous articles to automotive and engineering magazines and has written many outstanding books. He was the first Editor-in-Chief of the 15-volume McGraw-Hill Encyclopedia of Science and Technology.

William H. Crouse's outstanding work in the automotive field has earned for him membership in the Society of Automotive Engineers and in the American Society of Engineering Education.

Donald L. Anglin

Trained in the automotive and diesel service field, Donald L. Anglin has worked both as a mechanic and as a service manager. He has taught automotive courses and has also worked as curriculum supervisor and school administrator for an automotive trade school. Interested in all types of vehicle performance, he has served as a racing-car mechanic and as a consultant to truck fleets on maintenance problems.

Currently he serves as editorial assistant to William H. Crouse, visiting automotive instructors and service shops. Together they have coauthored magazine articles on automotive education and several books in the McGraw-Hill Automotive Technology Series.

Donald L. Anglin is a certified general automotive mechanic and holds many other licenses and certificates in automotive education, service, and related areas. His work in the automotive service field has earned for him membership in the American Society of Mechanical Engineers and the Society of Automotive Engineers. He is also an automotive instructor at Piedmont Virginia Community College, Charlottesville, Virginia.

Library of Congress Cataloging in Publication Data

Crouse, William Harry, (date)
Automotive transmissions and power trains.

(McGraw-Hill automotive technology series)
Includes index.
1. Automobiles—Transmission devices. I. Anglin, Donald L., joint author. II. Title.
TL262.C76 1976 629.2′44 75-15517
ISBN 0-07-014637-3

AUTOMOTIVE TRANSMISSIONS AND POWER TRAINS

7 8 9 0 WCWC 8 3 2 1

The editors for this book were Ardelle Cleverdon and Susan Berkowitz, the designer was Dennis G. Purdy, and the production supervisor was Rena Shindelman. The cover illustrator was Shelley Freshman. It was set in Melior by York Graphic Services, Inc.
Printed and bound by Webcrafters, Inc.

CONTENTS

PREFACE

This is the fifth edition of *Automotive Transmissions and Power Trains*. The book has undergone many changes in its five editions. Now, in this new, larger edition, the latest developments in manual and automatic transmissions are covered. The basic principles of clutches, transmissions, and power trains remain unchanged. However, many refinements have been made in transmissions and power trains in recent years. Automatic transmissions, especially, have reached a degree of sophistication unknown in earlier years. Upshifts and downshifts are managed so smoothly, with such ready and accurate response to operating conditions, that the driver is hardly aware of them. This smoothness has been obtained by the tailoring of the automatic transmission controls to the engine and the vehicle, as well as by the addition of new and more subtle controlling devices.

New information on the five-forward-speed manual transmission has been added in this edition. This transmission provides overdrive in fifth. Also, the three-speed Borg-Warner automatic transmission, which uses a compound planetary gearset to get three forward speeds, is covered in detail. Elsewhere in the book, explanations and service procedures have been rewritten as necessary to clarify the material and add the latest details on construction, operation, and service procedures.

The new, larger format of this edition permits larger illustrations, which make study easier for the student. The new format is designed to make the book more attractive and thus more readily accepted by the student. Hundreds of new illustrations have been added to show new models and developments and also to provide greater clarity of details. A feature of the new edition is the introduction of the metric system of measurements. When a United States Customary measurement is used, it is usually followed by its metric equivalent in brackets, for example, 0.002 inch [0.0508 mm].

The 32-page color insert showing hydraulic circuits of the most popular automatic transmissions is retained. The new Borg-Warner automatic transmission, mentioned previously, has been added to this insert. Also, other changes and additions have been made in the insert to further help the student understand this difficult subject matter.

A new *Workbook for Automotive Transmissions and Power Trains* has been developed specifically to accompany this basic textbook. It includes the basic automotive transmission and power-train service jobs that are the latest recommendations of the Motor Vehicle Manufacturers Association-American Vocational Association Industry Planning Council. Together, the textbook and the workbook provide the background information and hands-on experience needed to prepare a student to become a qualified and certified automotive transmission and power-train expert.

To assist the automotive instructor, the *Instructor's Planning Guide for Automotive Transmissions and Power Trains* is available from McGraw-Hill. The instructor's guide was prepared to help the automotive instructor do the best possible job of teaching by most effectively utilizing the textbook, workbook, and other related instructional materials. The instructor's guide contains suggestions on student motivation, classroom instruction and related shop activities, the automotive curriculum, and much more. In addition, it includes the answer key for the tests at the end of each jobsheet in the *Workbook for Automotive Transmissions and Power Trains*.

Also in the instructor's guide is a list of various related textbooks and ancillary instructional materials available from McGraw-Hill. Used singly or together, these items form a comprehensive student learning and activity package. They provide the student with meaningful learning experiences and help the student develop job competencies in automotive transmissions and power trains and related fields. The instructor's guide explains how the various available materials can be used, either singly or in combination, to satisfy any teaching requirement.

WILLIAM H. CROUSE
DONALD L. ANGLIN

TO THE STUDENT

Automotive Transmissions and Power Trains is one of eight books which are included in the McGraw-Hill Automotive Technology Series. These books cover in detail the construction, operation, and maintenance of automotive vehicles. They are designed to give you all the information you need to become successful in the automotive service business. The books satisfy the recommendations of the Motor Vehicle Manufacturers Association–American Vocational Association Industry Planning Council. The books also meet the requirements for automotive mechanics certification and state vocational educational programs, as well as the recommendations for automotive trade apprenticeship training. Furthermore, the comprehensive coverage of the subject matter in these books make them valuable additions to the library of anyone interested in any aspect of automotive engineering, manufacturing, sales, service, and operation.

Meeting the Standards

The eight books in the McGraw-Hill Automotive Technology Series meet the standards set by the Motor Vehicle Manufacturers Association (MVMA) for an associate degree in automotive servicing and in automotive service management. These standards are described in the MVMA booklet "Community College Guide for Associate Degree Programs in Auto and Truck Service and Management." The books also cover the subjects recommended by the American National Standards Institute in their detailed standard D18.1-1972, "American National Standard for Training of Automotive Mechanics for Passenger Cars and Light Trucks."

In addition, the books cover in depth the subject matter tested by the National Institute for Automotive Service Excellence (NIASE). The tests given by NIASE are used for certifying general automotive mechanics and automotive technicians working in specific areas of specialization under the NIASE voluntary mechanic testing and certification program.

Getting Practical Experience

At the same time that you study the books, you should be getting practical experience in the shop. That is, you should handle automotive parts, automotive tools, and automotive servicing equipment, and you should perform actual servicing jobs. To assist you in your shop work, there is a workbook for each book in the Automotive Technology Series. For example, the *Workbook for Automotive Transmissions and Power Trains* includes the jobs which cover every basic procedure for servicing automotive transmissions and power trains. If you do every job in the workbook, you will have hands-on experience with all automotive-transmission and power-train work.

If you are taking an automotive mechanics course in school, you will have an instructor to guide you in your classroom and shop activities. But even if you are not taking a course, the workbook can act as an instructor. It tells you, step by step, how to do the various servicing jobs. Perhaps you can meet others who are taking a school course in automotive mechanics. You can talk over any problems you have with them. A local garage or service station is a good source of practical information. If you can get acquainted with the automotive mechanics there, you will find they have a great deal of practical information. Watch them at their work if you can. Make notes of important points for filing in a notebook.

Service Publications

While you are in the service shop, study the various publications received at the shop. Automobile manufacturers, as well as suppliers of parts, accessories, and tools, publish shop manuals, service bulletins, and parts catalogs. All these help service personnel do a better job. In addition, numerous automotive magazines are published which deal with problems and methods of automotive service. All these publications will be of great value to you; study them carefully.

These activities will help you get practical experience in automotive mechanics. Sooner or later this experience, plus the knowledge that you have gained in studying the books in the McGraw-Hill Automotive Technology Series, will permit you to step into the automotive shop on a full-time basis. Or, if you are already in the shop, you will be equipped to step up to a more responsible job.

Checking Up on Yourself

You can check up on your progress in your studies by answering the questions given every few pages in the book. There are two types of tests, progress quizzes and chapter checkups, the answers to which are given at the back of the book. Each progress quiz should be taken just after you have completed the pages preceding it. The quizzes allow you to check yourself as you finish a lesson.

On the other hand, the chapter checkups may cover several lessons since they are review tests of entire chapters. Because they are review tests, you should review the entire chapter by rereading it or at least paging through it to check important points before trying the test. If any questions stump you, reread the pages in the book that will give you the answers. This sort of review is valuable and will help you to remember the information you need when you work in an automotive shop.

Keeping a Notebook

Keeping a notebook is a valuable part of your training. Start it now, at the beginning of your studies of automotive transmissions and power trains. Your notebook will help you in many ways: It will be a record of your progress. It will become a storehouse of valuable information you will refer to time after time. It will help you learn. And it will help you organize your training program so that it will do you the most good.

When you study a lesson in the book, have your notebook open in front of you. Start with a fresh notebook page at the beginning of each lesson. Write the lesson (or textbook page number) and date at the top of the page. As you read your lesson, jot down the important points.

In the shop, use a small scratch pad or index cards to jot down important points. You can transfer your notes to your notebook later.

You can also make sketches in your notebook showing wiring or fluid diagrams, power flow, and so on. Save articles and illustrations from technical and hot-rod magazines and file them in your notebook.

Your notebook will become a valued possession—a permanent record of what you learned about automotive transmissions and power trains.

Glossary and Index

There is a glossary (a definition list) of automotive terms in the back of the book. Whenever you have any doubt about the meaning of a term or the purpose of an automotive part, you can refer to this glossary. Also, there is an index at the back of the book. This index will steer you to the page in the book where you will find the information you are seeking.

And now, good luck to you. You are studying a fascinating, complex, and admirable machines—the automobile. Your studies can lead you to success in the automotive field, a field where opportunities are nearly unlimited.

ACKNOWLEDGMENTS

During the preparation of this new edition of *Automotive Transmissions and Power Trains,* the authors were given invaluable aid and inspiration by many people in the automotive industry and in the field of education. The authors gratefully acknowledge their indebtedness and offer their sincere thanks to these people. All cooperated with the aim of providing accurate and complete information that would be useful in training automotive mechanics.

Special thanks are owed to the following organizations for information and illustrations that they supplied: American Motors Corporation; Buick Motor Division of General Motors Corporation; Cadillac Motor Car Division of General Motors Corporation; Chevrolet Motor Division of General Motors Corporation; Chrysler Corporation; Ford Motor Company; General Motors Corporation; Oldsmobile Division of General Motors Corporation; Pontiac Motor Division of General Motors Corporation; Simca; Society of Automotive Engineers; Sun Electric Corporation; Toyota Motor Sales Co., Ltd., Volkswagen; and Young and Rubican. To all these organizations and the people who represent them, sincere thanks.

WILLIAM H. CROUSE
DONALD L. ANGLIN

chapter 1

POWER-TRAIN COMPONENTS

This book has been prepared to give you an understanding of the construction, operation, maintenance, and servicing of automotive power trains. By the time you have finished this book, you should have a good acquaintance with the various components of the automotive power train, including the clutch, transmission, drive shaft, and differential. Also, if you are able to visit or work in an automotive repair shop or school automotive shop during the time you study this book, you should ultimately be able to service these various power-train components yourself. Naturally, you should not expect to become an expert in this work overnight because that takes long practice and experience. However, this book, combined with necessary practical shopwork, is designed to give you the basic information you need to become an expert.

⊘ 1-1 Components of the Automobile Before we begin our studies of the power train, let us first take a quick look at the complete automobile. This will show us how the power train operates in relation to the other parts of the automobile. The automobile might be said to consist of five basic parts, or components. These are:

1. The power plant, or engine, which is the source of power
2. The frame-and-wheel assembly, which supports the engine and body
3. The power train, which carries the power from the engine to the car wheels and which consists of the clutch (on vehicles with nonautomatic transmissions), transmission, drive shaft, differential, and axles
4. The car body
5. The car-body accessories, which include the heater, lights, radio, windshield wiper, convertible-top raiser, and air conditioner

Many of these components can be seen in the illustration of the automotive chassis (Fig. 1-1). The engine (Figs. 1-2 and 1-3) produces power by burning a mixture of gasoline vapor and air in the engine combustion chambers. Combustion creates high pressure, which forces the engine pistons downward. The downward thrusts on the pistons are carried through connecting rods to cranks on the engine crankshaft. These thrusts on the cranks cause the crankshaft (Fig. 1-4) to rotate. The engine flywheel is attached to the rear end of the crankshaft. The rear face of the flywheel is flat and smooth and serves as the driving member of the clutch (on vehicles with nonautomatic transmissions). When the clutch is engaged, the rotary motion of the engine crankshaft is carried through the flywheel and the clutch to the other members of the power train (transmission, drive shaft, and differential).

⊘ 1-2 Power Train The power train (Figs. 1-5 to 1-7) carries the power from the engine to the car wheels. The power train can provide several different *gear ratios* (explained in ⊘ 1-5) between the engine crankshaft and the wheels. In the standard arrangement the gear ratio can be changed so that the engine crankshaft will rotate approximately four, eight, or twelve times to cause the wheels to rotate once. On cars equipped with automatic transmissions, these approximate gear ratios may not hold true. Automatic transmissions with torque converters provide a very large number of gear ratios. All this is explained in detail in later chapters. For the moment, let us concentrate on the standard power-train arrangement. An understanding of this arrangement will make it easier to understand all automatic transmissions. The standard power train consists of a series of gears and shafts which mechanically connect the engine crankshaft and car wheels. As mentioned previously, it contains the clutch, a transmission, or gear changer, a drive shaft, and the final drive, or the differential and wheel axles. Let us look at each of these in detail.

⊘ 1-3 Clutch The clutch (Fig. 1-8, p. 5) permits the driver to connect the crankshaft to, or disconnect it from, the power train. A clutch, or disconnecting

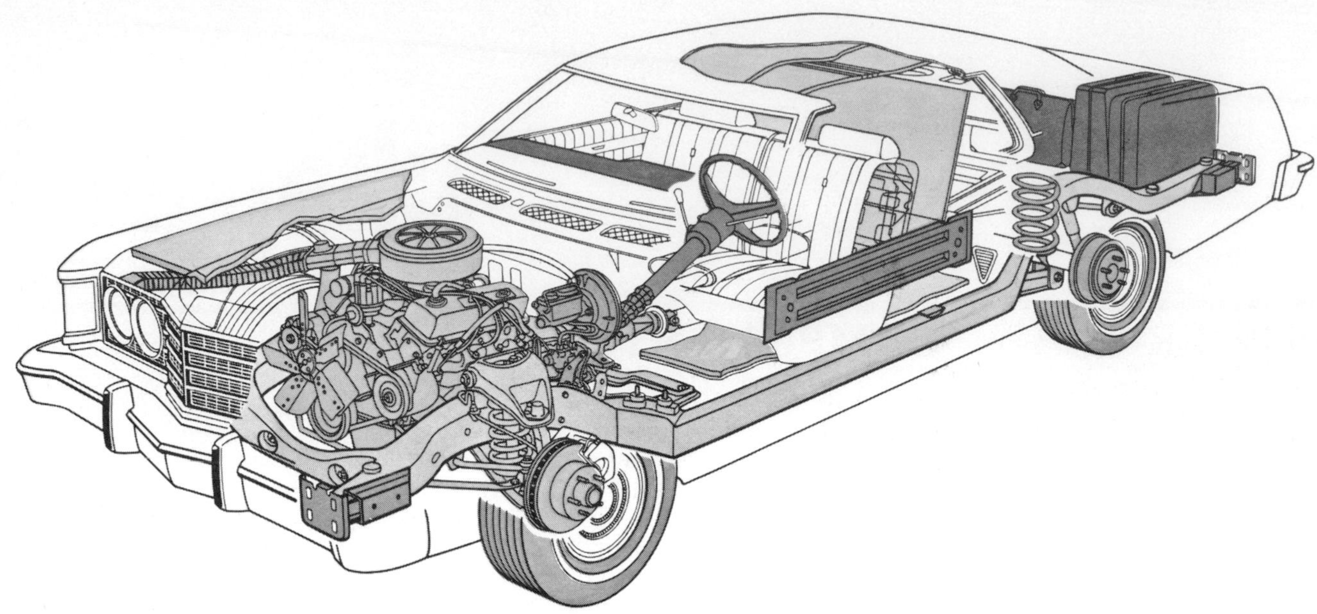

Fig. 1-1. Chassis of a passenger car. The chassis contains the source of power, or engine; the frame, which supports the engine, wheels, and body; the power train, which carries the engine power to the rear wheels; and the steering and braking systems. (*Ford Motor Company*)

device, is necessary since the automobile engine must be started without load. That is, it must not be required to deliver any power during the starting period. In order for the engine to deliver power, the crankshaft must be rotating at a reasonable speed of several hundred revolutions per minute (rpm). The engine will start at speeds below 100 rpm, but it would not continue to operate at this low speed if a load were immediately thrown on it. Consequently, a clutch is placed in the power train between the crankshaft and the transmission. The clutch permits the engine to run freely without delivering power to the power train. It also permits operation of the transmission so that the various gear ratios between the engine crankshaft and wheels can be obtained.

The clutch contains a double-faced friction disk (driven plate) about 1 foot [30.5 millimeters (mm)] in diameter. The friction disk is splined to the clutch shaft, and there is a spring arrangement for forcing this disk tightly against the smooth face of the engine flywheel. The splines are internal teeth on the friction-disk hub and matching external teeth on the clutch shaft. They permit the disk to move back and forth along the shaft but cause the disk and shaft to rotate together. External splines can be seen on the shaft in Fig. 1-8. The flywheel is attached to the end of the engine crankshaft. When the clutch is

Fig. 1-2. Typical six-cylinder engine, partly cut away so that the internal construction can be seen. This engine is known as an *in-line six* because its six cylinders and pistons are located one behind the other. Note the cylindrical piston and its connecting rod shown in the cutaway section. (*Ford Motor Company*)

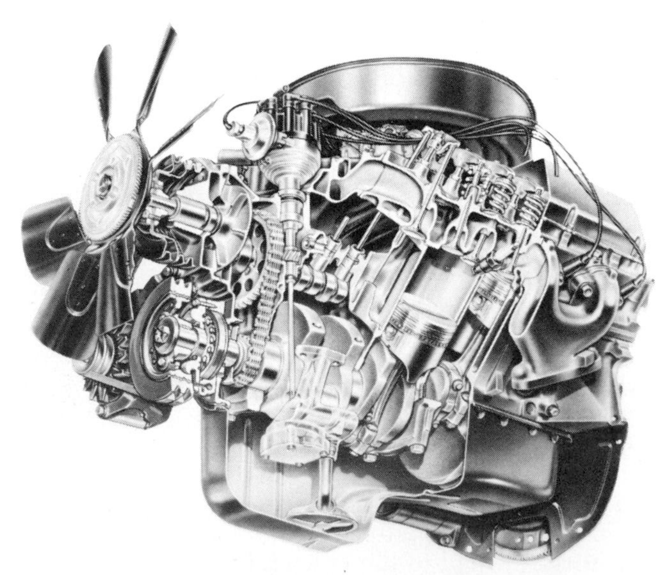

Fig. 1-3. Cutaway view of an eight-cylinder V-type overhead-valve engine. (*Ford Motor Company*)

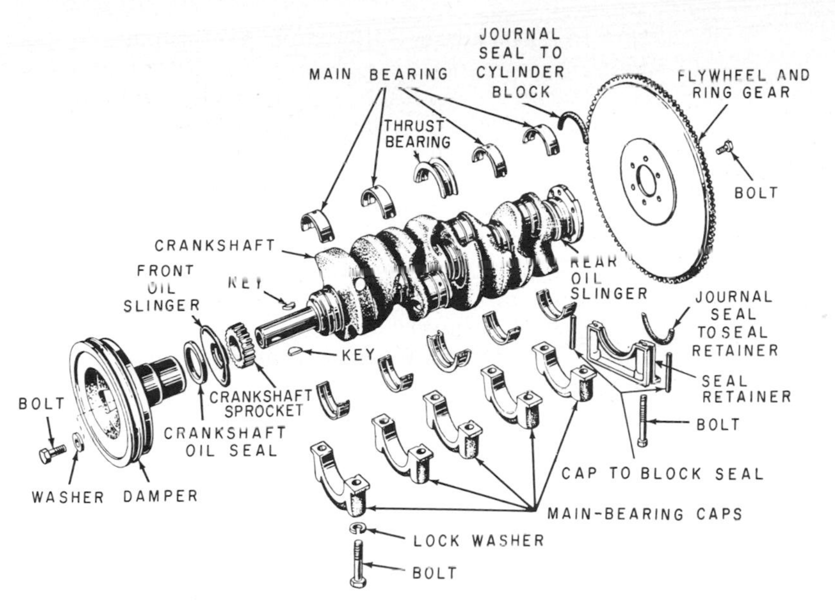

Fig. 1-4. Crankshaft and related parts for an eight-cylinder V-type engine. (*Ford Motor Company*)

engaged, the friction disk is held against the flywheel so that it revolves with the flywheel. This rotary motion is carried through the clutch and clutch shaft to the transmission and from there to the car wheels.

When the clutch pedal is pushed down, the clutch fork moves against the clutch throw-out bearing, forcing the bearing inward. This operates release levers that take up the spring pressure so that pressure against the friction disk is relieved and the disk can move away from the flywheel face. When this happens, the friction disk and the shaft stop revolving. When pressure on the clutch pedal is removed, the springs again cause the pressure plate to force the friction disk against the flywheel face so that it rotates with the flywheel once more.

⊘ 1-4 Transmission The transmission provides a means of varying the gear ratio. Thus the engine crankshaft may turn four, eight, or twelve times to

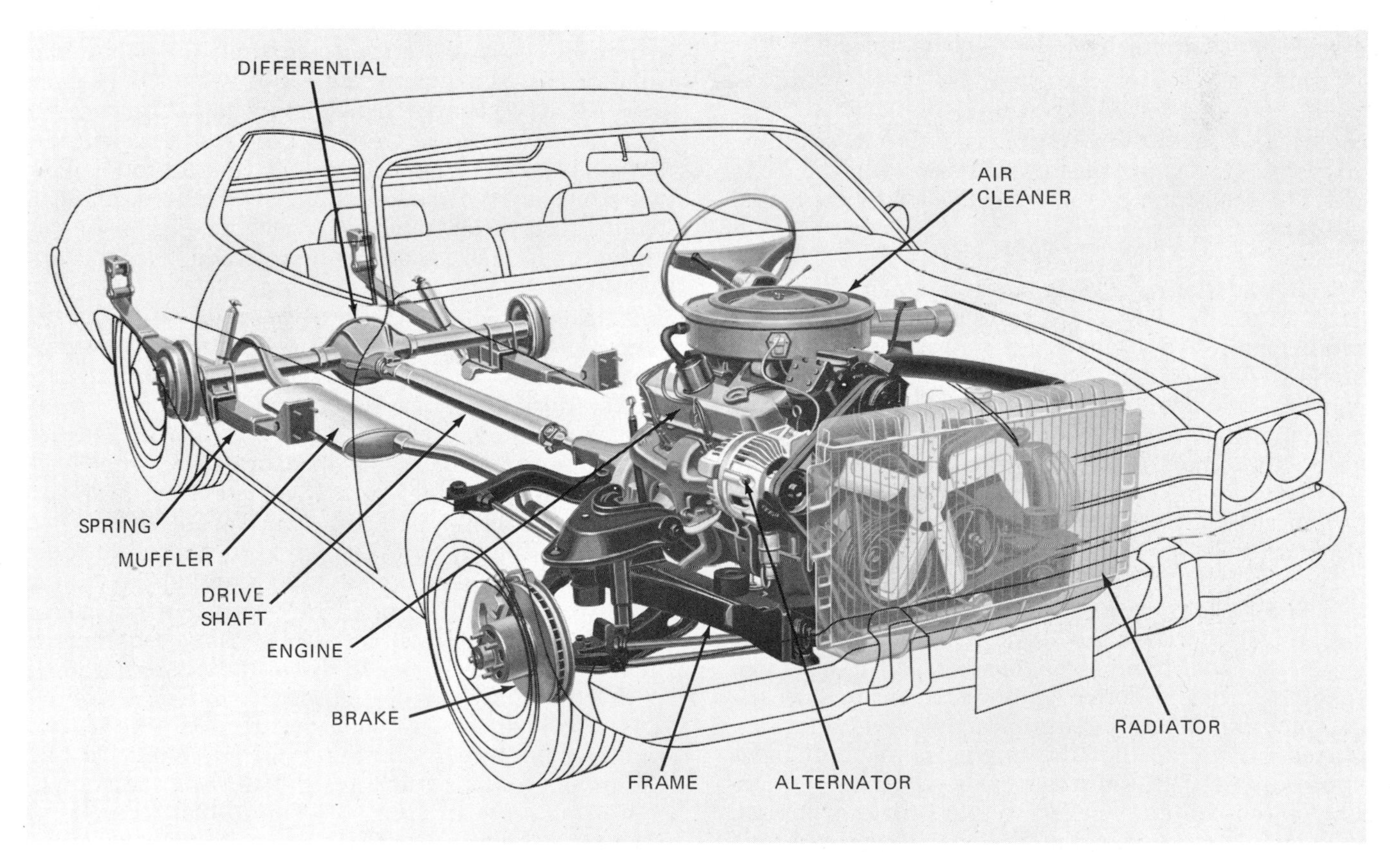

Fig. 1-5. Location of major components in the automobile. (*Young and Rubican*)

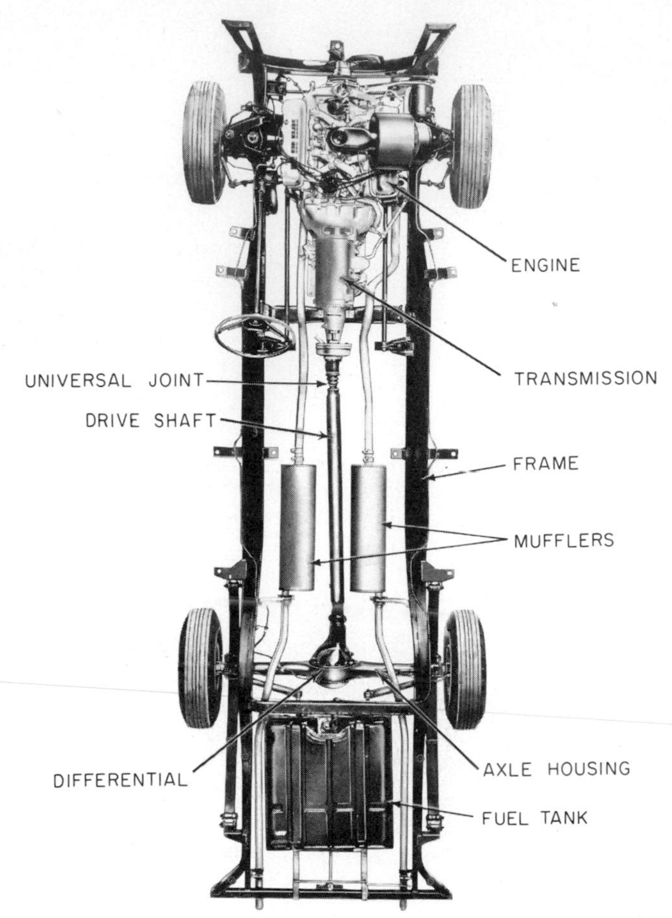

Fig. 1-6. View from the top of a typical automotive chassis showing the locations of the engine, transmission, drive shaft, differential, and rear-axle housing. (*Chrysler Corporation*)

each wheel revolution (approximately). In addition, a reverse gear is provided that permits backing the car. The operation of the transmission is explained in ⊘ 1-6.

NOTE: We don't want to complicate matters this early in the book, but we should mention that the ratios mentioned so far are for a three-speed transmission. This is a transmission that has three forward speeds: low, second, and high. However, today most manual transmissions in cars are four-speed units. They have four forward speeds: low, second, third, and high. When a four-speed transmission has a floor-shift lever, the combination is often called "four on the floor."

The various gear ratios are necessary since the gasoline engine does not develop much power at low engine speeds. It must be turning at a fairly high speed in order to deliver enough power to start the car moving. Thus, on first starting, the gears are placed in *low* so that the engine crankshaft turns approximately 12 times for each wheel revolution. The clutch is then engaged so that power is applied to the wheels. Car speed increases with engine speed until the car is moving 5 to 10 mph (miles per hour) [8.05 to 16.09 km/h (kilometers per hour)]. At this time the engine crankshaft may be turning as many as 2,000 rpm. The clutch is then disengaged and the engine crankshaft speed reduced to permit gear changing. The gears are shifted into *second*, and the clutch is again engaged. Since the ratio is now about 8:1, a higher car speed is obtained as engine speed is again increased. The gears are then shifted into *high*, the clutch being disengaged and engaged for this operation, and the ratio between the engine and wheels will be approximately 4:1. In other words, the engine crankshaft will turn four times to cause the wheels to turn once.

In automatic transmissions, the various ratios between the engine crankshaft and wheels are achieved by automatic means. That is, the driver does not need to shift gears because the automatic controls in the automatic transmission supply the proper ratio to suit the driving conditions. Such transmissions make use of a torque converter as well as mechanical, hydraulic, or electric controls. All these are discussed in detail in later chapters.

⊘ 1-5 Gears and Torque Before considering the transmission further, we shall take a closer look at gears (Fig. 1-9). Let us find out what takes place when power is transmitted from one to another meshing gear. The relative rotation between two meshing gears (the gear ratio) is determined by the number of teeth in the gears. For instance, when two meshing gears have the same number of teeth, they will both turn at the same speed (Fig. 1-10). However, when one gear has more teeth than the other, the smaller gear will turn more rapidly than the larger one. Thus, a gear with 12 teeth will turn twice as fast as a gear with 24 teeth (Fig. 1-11). The gear ratio between the two gears is 2:1. If the 12-tooth gear were meshed with a 36-tooth gear, the 12-tooth gear would turn three times for every revolution of the larger gear. The gear ratio between these gears would be 3:1.

1. TORQUE Not only does the gear ratio change with the relative number of teeth in the meshing gears, but the *torque* also changes. Torque is twisting, or turning, effort. When you loosen the lid on a jar, you apply a twisting force, or torque, to the jar (Fig. 1-12). Torque is measured in pound-feet (abbreviated lb-ft) or, in the metric system of measurement, kilogram-meters [kg-m]. Torque should not be confused with work, which is measured in foot-pounds (ft-lb) or meter-kilograms [m-kg].

To calculate torque, multiply the push (in pounds or kilograms) by the distance (in feet or meters) from the center to the point where the push is exerted. For example, suppose you had a wrench 1 foot [0.31 m (meter)] long and that you used it to tighten a nut (Fig. 1-13). If you put 10 pounds [4.54 kg (kilograms)] of pressure on the wrench, as shown, you would be applying 10 pound-feet [1.38 kg-m] torque to the nut. If you put 20 pounds [9.07 kg] of pressure on the wrench, you would be applying 20

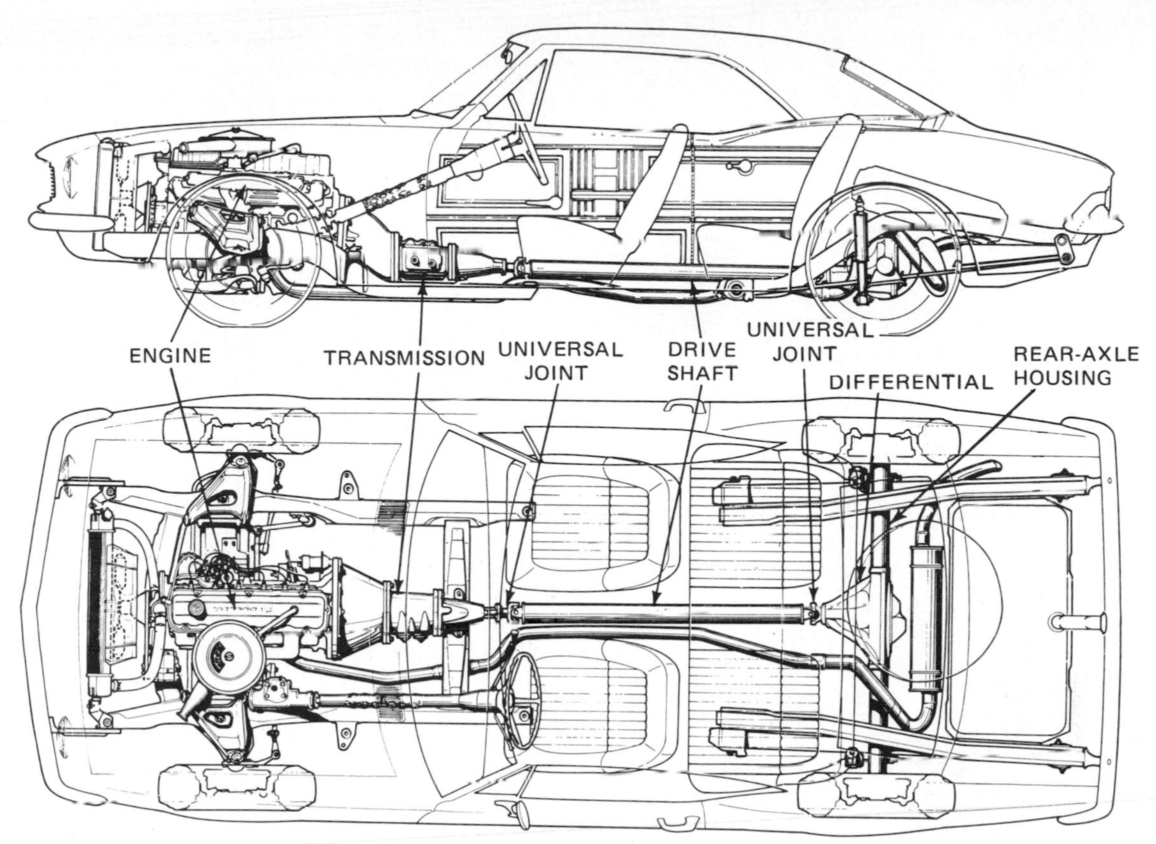

Fig. 1-7. Top and side outline views of a passenger car showing the power train. (*Chevrolet Motor Division of General Motors Corporation*)

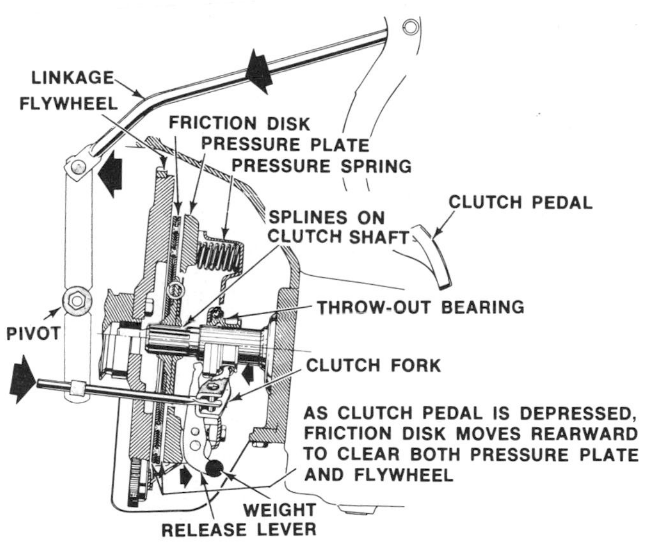

Fig. 1-8. Sectional view of a clutch, with linkage to the clutch pedal shown schematically. (*Buick Motor Division of General Motors Corporation*)

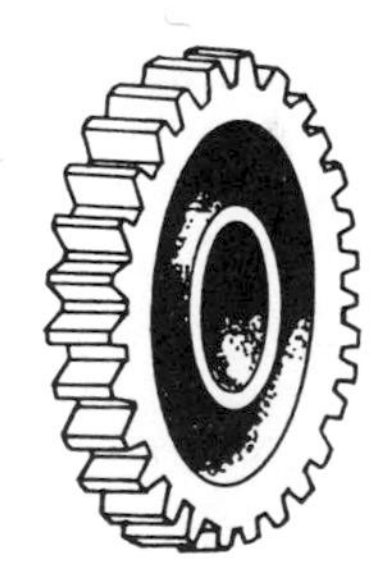

Fig. 1-9. Spur gear.

Fig. 1-10. Two meshing spur gears with the same number of teeth.

pound-feet [2.76 kg-m] torque. If the wrench were 2 feet [0.62 m] long and applied 10 pounds [4.54 kg] of pressure, the torque on the nut would be 20 pound-feet [2.76 kg-m].

Any shaft or gear that is turned has torque applied to it. The engine pistons and connecting rods push on the cranks on the crankshaft, thereby applying torque to the crankshaft and causing it to turn. The crankshaft applies torque to the gears in the transmission, and so the gears turn. This turning effort, or torque, is carried through the power train to the rear wheels so that they turn.

2. TORQUE IN GEARS Torque on shafts or gears is measured as a straight-line force at a distance

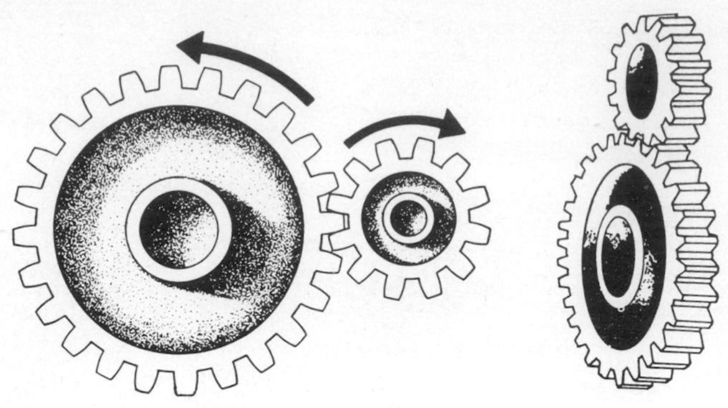

Fig. 1-11. Two views of a pair of meshing spur gears. The smaller gear turns faster than the larger gear.

Fig. 1-12. Torque, or twisting effort, must be applied to the jar lid to loosen and remove it.

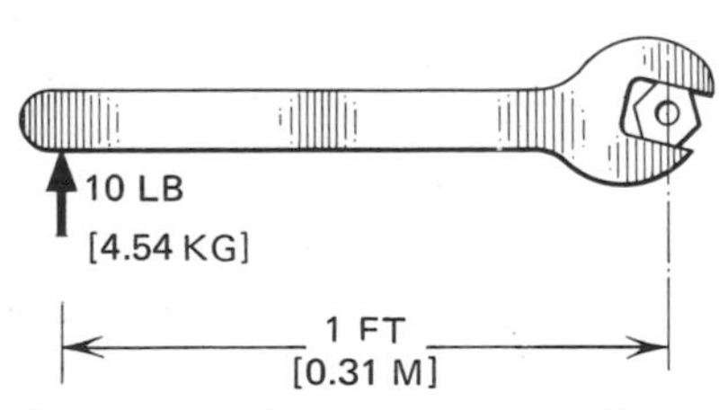

Fig. 1-13. If you are tightening a nut with a wrench, you can figure the torque you apply by multiplying the amount of push times the distance from the center of the nut.

from the center of the shaft or gear. For instance, suppose we want to measure the torque in the gears shown in Fig. 1-11. If we could hook a spring scale to the gear teeth and get a measurement of the pull, we could determine the torque. (Actually, a spring scale could not be used, although devices exist that measure torque of rotating parts.) Suppose, for example, we found that the tooth of the driving gear is pushing against the tooth of the driven gear with a 25 pound [4.34 kg] force (Fig. 1-14). This force, at a distance of 1 foot [0.31 m] (the radius, or distance from the center of the driving gear), means there is a torque of 25 pound-feet [3.46 kg-m]. That is, the smaller (driving) gear is delivering a torque of 25 pound-feet [3.46 kg-m].

The 25-pound [11.34 kg] push from the gear teeth of the smaller gear is applied to the gear teeth of the larger gear. But it is applied at a distance of 2 feet [0.61 m] from the center. Therefore, the torque on the larger gear is 50 pound-feet (25×2) [6.91 kg-m (3.455×2)]. The same force is acting on the teeth of the larger gear, but it is acting at twice the distance from the shaft center.

3. TORQUE AND GEAR RATIO Now, the important point of all this is that if the smaller gear is driving the larger gear, the gear ratio will be 2:1. But the torque ratio will be 1:2. The larger gear will turn only half as fast as the smaller gear. But the larger gear will have twice the torque of the smaller gear. In gear systems, *speed reduction means torque increase*. For example, in ⊘ 1-4 we mentioned that when the transmission is in low gear, there is a speed reduction (or gear reduction) of 12:1 from the engine to the wheels. That is, the crankshaft turns 12 times to turn the rear wheels once. This means that the torque *increases* 12 times (ignoring losses due to friction). In other words, if the engine produced a torque of 100 pound-feet [13.82 kg-m], then 1,200 pound-feet [165.8 kg-m] torque would be applied to the rear wheels.

To see how this torque produces the forward thrust, or push, on the car, refer to Fig. 1-15. In the example shown, we assume that the torque delivered by the engine is 100 pound-feet [13.82 kg-m]. We assume also that the gear reduction from the engine to the rear wheels is 12:1, with a torque increase of 1:12. Wheel radius is assumed to be 1 foot [0.31 m] (for ease of figuring). With the torque acting on the ground at a distance of 1 foot [0.31 m] (radius of wheel), the push of the tire on the ground is 1,200 pounds [544.31 kg]. Consequently, the push on the wheel axle, and thus on the car, is 1,200 pounds [544.31 kg].

NOTE: Actually, the torque is split between the two rear wheels. Thus, the torque on each rear wheel is 600 pound-feet [82.92 kg-m], and so each tire

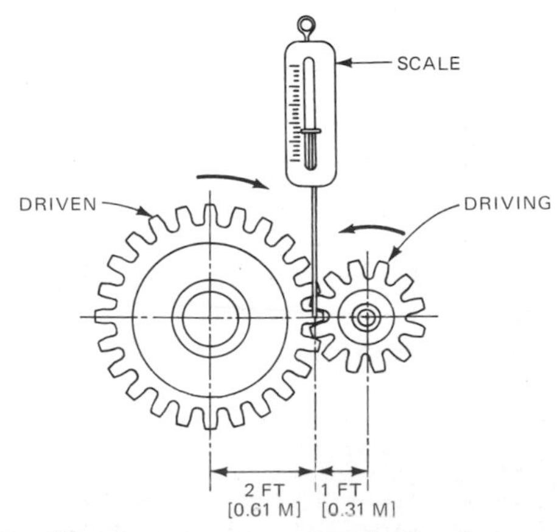

Fig. 1-14. The torque on a gear is the force on a gear tooth times the distance from the center of the shaft to the point on the tooth where the force is applied.

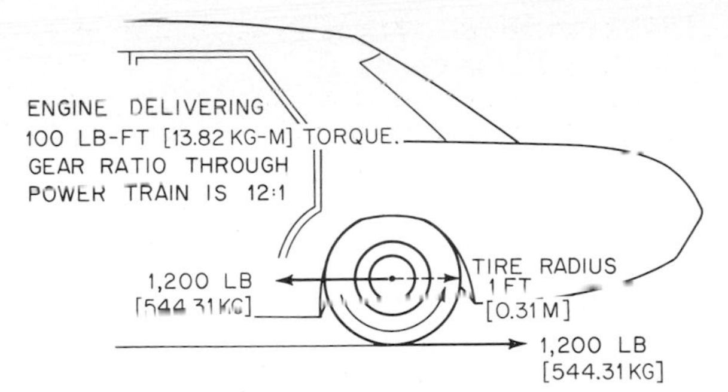

Fig. 1-15. How torque at the rear wheels is translated into a forward push on the car. The tire is turned with a torque of 1,200 pound-feet [165.8 kg-m]. Since the tire radius is 1 foot [0.31 m], the push of the tire on the ground will be 1,200 pounds [544.31 kg]. That is, the car is pushed forward with a 1,200-pound [544.31 kg] force.

pushes on the ground with a force of 600 pounds [272.16 kg]. Both tires together push with a force of 1,200 pounds [544.31 kg], giving the car a forward thrust of 1,200 pounds [544.31 kg].

4. OTHER GEARS The gears discussed so far are *spur gears.* The teeth are parallel to and align with the center line of the gear. Many types of gears are used in the automobile. They differ from the spur gear mainly in the shape and alignment of the gear teeth. Thus, *helical gears* are like spur gears except that, in effect, the teeth have been twisted at an angle to the gear center line. *Bevel gears* are shaped like cones with the tops cut off; the teeth point inward toward the apex of the cone. Bevel gears are used to transmit motion through angles. Some gears have their teeth pointing inward; these are *internal gears.* Several typical gears are shown in Fig. 1-16.

Check Your Progress[1]

Progress Quiz 1-1 In almost any job you are doing, it is a good idea to pause occasionally and check the progress you are making. For instance, the battery technician checks a battery being charged periodically to see how the battery is taking the charge. In a somewhat similar way, the Progress Quizzes you find every few pages in this book will help you check yourself to find out how you are taking the "charge" of new information. The questions that follow give you a chance to determine quickly how well you remember the important facts you have just read in the past few pages in the book. If you have difficulty answering any of the questions, don't be discouraged. Just reread the pages that will give you the answer. Most students make a practice of reading and rereading their lessons several times. This helps them understand and remember the facts covered in the lessons.

You are asked to write the answers to the questions in your notebook. Writing the answers, plus reviewing the pages in the book covered by the quiz, will help fix the essential facts firmly in your mind. As you continue to do these things during your progress through the book, you will find it easier and easier to remember the important points covered in the book. In other words, you will become an expert student. And from being an expert student to being an expert automotive technician is only a step.

Correcting Parts Lists The purpose of this exercise is to enable you to spot the unrelated item in a list. For example, in the list "animal: dog, cat, horse, chair, bird," you can see that "chair" does not belong because it is the only item that is not an animal. In each of the following lists, you will find that *one* item is included that does not belong. Write each list in your notebook, but *do not write* the item that does not belong.

1. Heater, radio, windshield wiper, hoist, air-conditioner
2. Power train: clutch, transmission, drive shaft, steering knuckle, differential, axle
3. Clutch: friction disk, rear axle, throw-out bearing, pressure springs, release levers

[1] Answers to questions in the progress quizzes and chapter checkups are given at the back of the book.

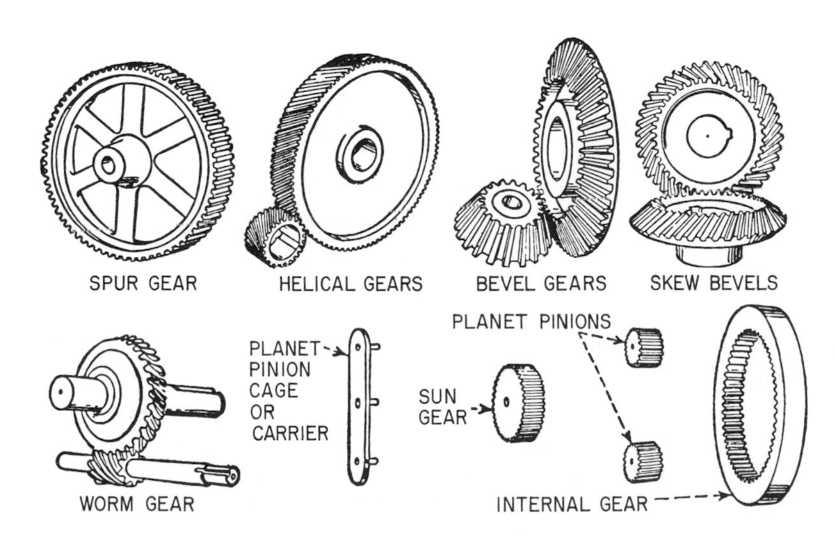

Fig. 1-16. Types of gears. The bottom gearset is a disassembled view of a planetary-gear system. Planetary gears are used in overdrives and automatic transmissions.

4. Gear ratios in power train with standard transmission: 4:1, 8:1, 12:1, 16:1

Completing the Sentences The sentences that follow are incomplete. After each sentence there are several words or phrases, only one of which will correctly complete the sentence. Write each sentence in your notebook, selecting the proper word or phrase to complete it correctly.

1. The power train includes the clutch, drive shaft, differential, and: (*a*) chassis, (*b*) transmission, (*c*) front axles, (*d*) steering gear.
2. There is a double-faced friction disk splined to a shaft in the: (*a*) transmission, (*b*) differential, (*c*) engine, (*d*) clutch.
3. Two meshed gears have a gear ratio of 3:1. Every time the larger gear turns once, the smaller gear will turn: (*a*) one-third time, (*b*) once, (*c*) three times.
4. If two meshing gears have a 4:1 gear ratio and the smaller gear has 12 teeth, the larger gear will have: (*a*) 12 teeth, (*b*) 24 teeth, (*c*) 36 teeth, (*d*) 48 teeth.
5. The device that produces different gear ratios in the power train is called a: (*a*) differential, (*b*) transmission, (*c*) speed changer.
6. In gear systems, speed reduction means torque: (*a*) reduction, (*b*) stabilization, (*c*) increase.
7. In the example shown in Fig. 1-15, if the gear reduction were 8:1, the push on the car would be: (a) 400 lb [181.44 kg], (*b*) 800 lb [362.87 kg], (*c*) 1,200 lb [544.31 kg].
8. In the example shown in Fig. 1-15, if the wheel radius were 18 in [457.2 mm], the push on the car would be: (*a*) 400 lb [181.44 kg], (*b*) 800 lb [362.87 kg], (*c*) 1,200 lb [544.31 kg], (*d*) 1,600 lb [725.75 kg].

⊘ 1-6 Operation of the Transmission There are many kinds of transmission. Some of the simpler types are found on passenger cars. More complex transmissions are used on trucks and buses. However, all manually shifted transmissions are quite similar in operation even if they are different in construction. Let us look at a simple transmission and find out how it works. Later, in another chapter, we shall see that the more complex transmissions work on the same principles. The transmission we shall discuss consists essentially of three shafts and eight gears of various sizes (Fig. 1-17). In the illustration, only the moving parts are shown. The transmission housing and bearings are not shown.

Actually, no modern transmission shifts gears by moving the gears, as shown in Figs. 1-18 to 1-21. But showing the gears being shifted into and out of mesh with other gears is the simplest way to introduce you to transmission action. Later, we shall learn that in most transmissions, the gears are in constant mesh. Shifting is accomplished by locking gears to each other or to shafts. The end effect is the same with either system.

Four of the gears are rigidly connected to the countershaft (Fig. 1-17). These are the drive gear,

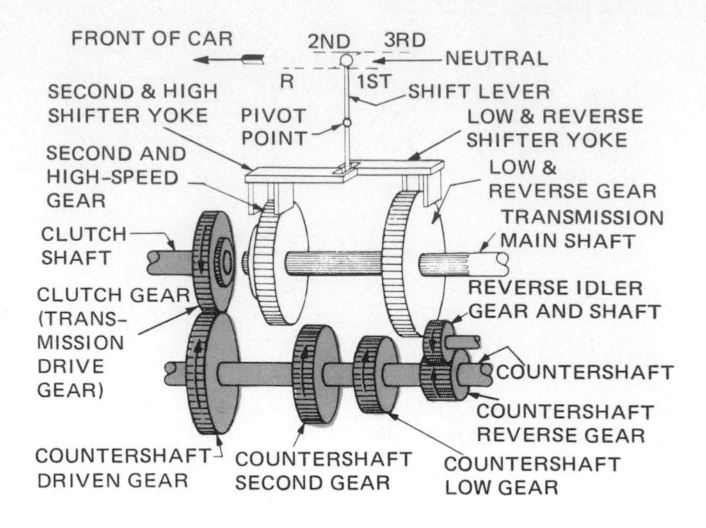

Fig. 1-17. Transmission with gears in neutral.

second gear, low gear, and reverse gear. When the clutch is engaged and the engine is running, the clutch-shaft gear drives the countershaft drive gear. This turns the countershaft and the other gears on the countershaft. The countershaft rotates in a direction opposite, or counter, to the rotation of the clutch-shaft gear. With the gears in neutral, as shown in Fig. 1-17, and the car stationary, the transmission main shaft is not turning. The transmission main shaft is mechanically connected by shafts and gears in the final drive to the car wheels. The two gears on the transmission main shaft may be shifted back and forth along the splines on the shaft by operation of the gearshift lever in the driving compartment. The splines are matching internal and external teeth that permit endwise (axial) movement of the gears but cause the gears and shaft to rotate together. Note, in Figs. 1-18 to 1-21, that a floorboard-shift lever is shown. This type of lever is shown because it illustrates more clearly the lever action in shifting gears. The transmission action is the same, regardless of whether a floorboard type

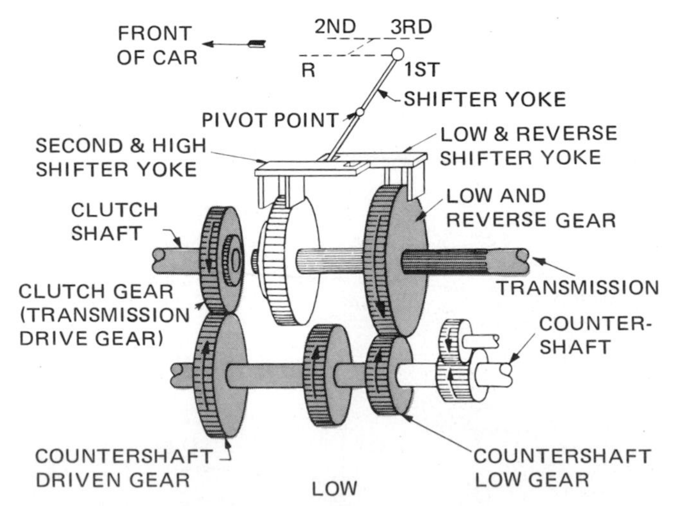

Fig. 1-18. Transmission with gears in low.

of shift lever or a steering-column lever is used.

1. *LOW GEAR* When the gearshift lever is operated to place the gears in low (Fig. 1-18), the large gear on the transmission main shaft is moved along the shaft until it meshes with the small gear on the countershaft. The clutch is disengaged for this operation, and so the clutch shaft and countershaft stop rotating. When the clutch is again engaged, the transmission main shaft rotates as the driving gear on the clutch shaft drives it through the countershaft. Since the countershaft turns more slowly than the clutch shaft and since the small countershaft is engaged with the large transmission main-shaft gear, a gear reduction of approximately 3:1 is achieved; that is, the clutch shaft turns three times for each revolution of the transmission main shaft. Further gear reduction in the final drive at the rear wheels produces a still higher gear ratio (approximately 12:1) between the engine crankshaft and wheels.

2. *SECOND GEAR* When the clutch is operated and the gearshift lever moved to second (Fig. 1-19), the large gear on the transmission main shaft demeshes from the small countershaft gear. The smaller transmission main-shaft gear is slid into mesh with the large countershaft gear. This action provides a somewhat reduced gear ratio, so the engine crankshaft turns only about twice while the transmission main shaft turns once. The final drive gear reduction increases this gear ratio to approximately 8:1.

3. *HIGH GEAR* When the gears are shifted into high (Fig. 1-20), the two gears on the transmission main shaft are de-meshed from the countershaft gears and the smaller transmission-shaft gear is forced axially against the driving gear. Teeth on the ends of the two gears mesh so that the transmission main shaft turns with the clutch shaft and a ratio of 1:1 is obtained. The final drive gear reduction produces a gear ratio of about 4:1 between engine crankshaft and wheels.

4. *REVERSE GEAR* When the gears are placed in reverse (Fig. 1-21), the larger of the transmission main-shaft gears is meshed with the reverse idler gear. This reverse idler gear is always in mesh with the small gear on the end of the countershaft. Interposing the idler gear between the countershaft gear and the transmission main-shaft gear causes the transmission shaft to be rotated in the opposite direction, or in the same direction as the countershaft. This action reverses the rotation of the wheels so that the car backs up.

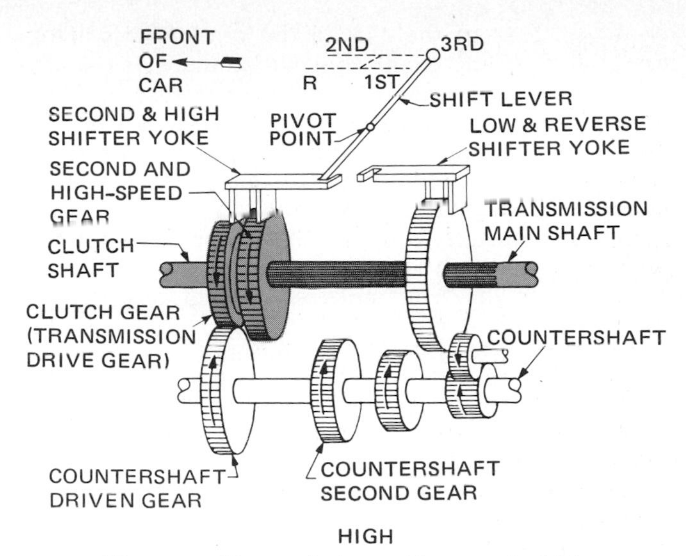

Fig. 1-20. Transmission with gears in high.

Although the preceding description outlines the basic principles of all transmissions, somewhat more complex transmissions are used on modern cars. These include helical or herringbone gears and gear shifting in conjunction with synchromesh devices that synchronize the rotation of gears that are about to be meshed. These devices eliminate clashing of gears and facilitate gear shifting.

⊘ 1-7 Final Drive In most cars, the *final drive* transmits the power from the transmission, which

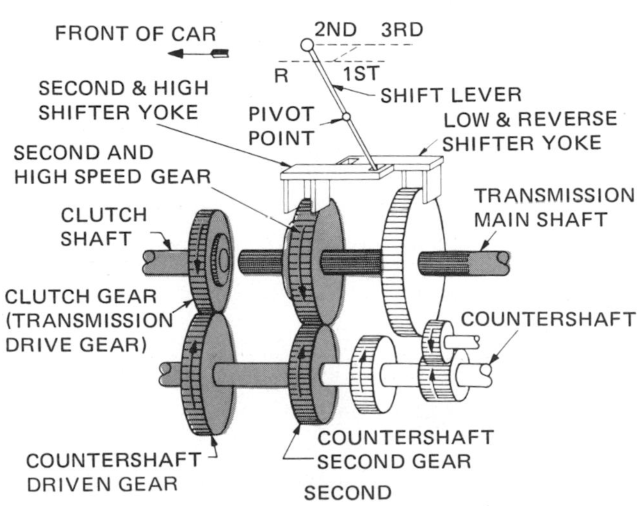

Fig. 1-19. Transmission with gears in second.

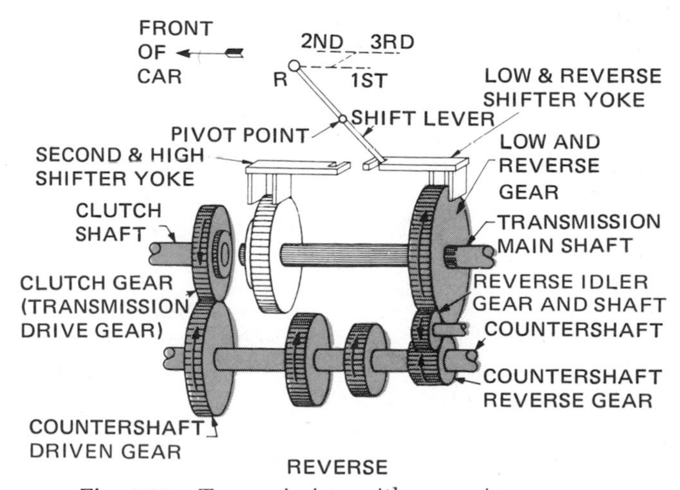

Fig. 1-21. Transmission with gears in reverse.

is at the front of the car, to the rear wheels. This final-drive arrangement consists of the drive shaft, differential, and rear axles. Other cars use a front-wheel drive in which the transmission is directly connected to the differential. Still other cars have the engines in the rear, and on these the transmission is also directly connected to the differential. Some vehicles, especially the type that must travel off the road and in rough terrain, have four-wheel drive; that is, both the front and rear wheels are driven by the engine.

The drive shaft used in the standard arrangement (engine at front, rear wheels driven) connects the transmission shaft and the rear-wheel driving mechanism, or differential (Figs. 1-6, 1-7, and 1-22). The drive shaft is more than a simple line shaft since it is connected at one end to the rigidly mounted transmission and at the other end to the wheel axles that move up and down with the wheel-spring movement. Two separate effects are produced by this movement. First, the distance between the transmission and wheel axles changes as the springs compress and expand and the axles move toward and away from the car frame. Second, the driving angle changes with the spring movement.

1. *UNIVERSAL JOINTS* To take care of the differences in the angle of drive as the axle moves up and down, the drive shaft incorporates one or more *universal joints.* A universal joint (see Fig. 1-23) is essentially a double-hinged joint through which the driving shaft can transmit power to the driven shaft even though the two shafts are several degrees out of line with each other.

Each of the two shafts has a Y-shaped yoke on the end, and between these yokes there is a center member that is shaped like a cross. The four arms of the center member are assembled in bearings in the ends of the shaft yokes. The bearings can turn on the crossarms to take care of any angularity between the shafts as the shafts rotate. The driving shaft causes the center member to rotate by its pressure on two of the crossarms. The other two crossarms cause the driven shaft to rotate. Many types of universal joints have been designed. The universal joint actually used on modern automobiles is more complicated than this, but the principle is the same.

2. *SLIP JOINT* Since the drive shaft tends to shorten and lengthen with the wheel-axle movement, it is necessary that some device be incorporated that will permit this action. The device used is the *slip joint,* located at either the front or the rear end of the drive shaft (Fig. 1-24). The slip joint is merely an externally splined shaft and a matching internally splined shaft. The two shafts can slide back and forth with respect to each other and still transmit driving power. The slip joint shown in Fig. 1-24 consists of the externally splined slip yoke and the internally splined front shaft.

⊘ 1-8 Differential If a car were driven in a straight line only, no differential would be necessary. However, when a car rounds a turn, the outer wheel must travel farther than the inner wheel. If a right-angle turn is made with the inner wheel turning on a 20-foot [6.096 m] radius, this wheel travels about 31 feet [9.449 m] (Fig. 1-25). The outer wheel, being nearly 5 feet [1.5 m] from the inner wheel (56 inches), turns on a 24⅔-foot [7.518 m] radius and travels nearly 39 feet [11.887 m] (Fig. 1-25).

If the drive shaft were geared rigidly to both rear wheels so that they would both have to rotate together, then each wheel would have to skid an average of 4 feet [1.219 m] in making the turn just discussed. On this basis, tires would not last long. In addition, the skidding would make the car hard to control around turns. The differential eliminates these troubles because it allows the wheels to rotate by different amounts when turns are made.

To study the construction and action of the differential, let us build up, in effect, a simple differential (Fig. 1-26). The two rear wheels are attached, through the axles, to two small bevel gears called *differential side gears* (Fig. 1-26*a*). There is a differential case assembled around the left axle (Fig. 1-26*b*). The case has a bearing that permits it to turn independently of the left axle. Inside the case is a shaft that supports a third bevel gear (Fig. 1-26*c*). This third bevel gear, called the *differential pinion*

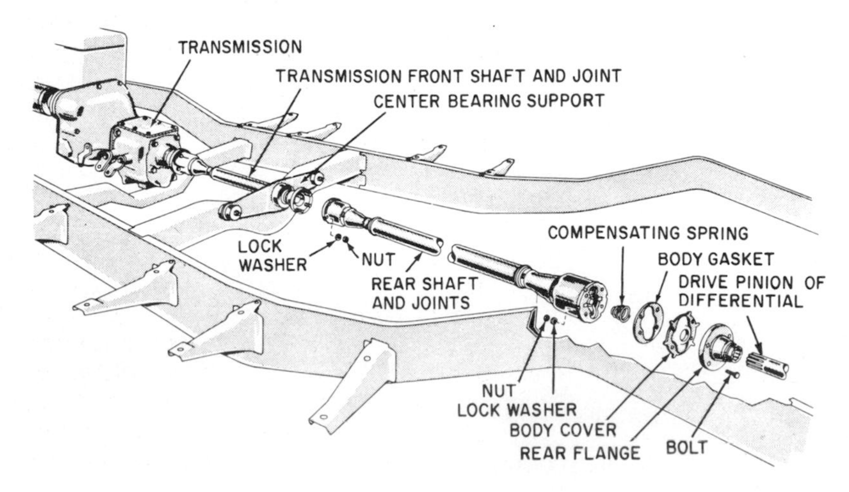

Fig. 1-22. One type of drive shaft, partly disassembled so that the component parts can be seen.

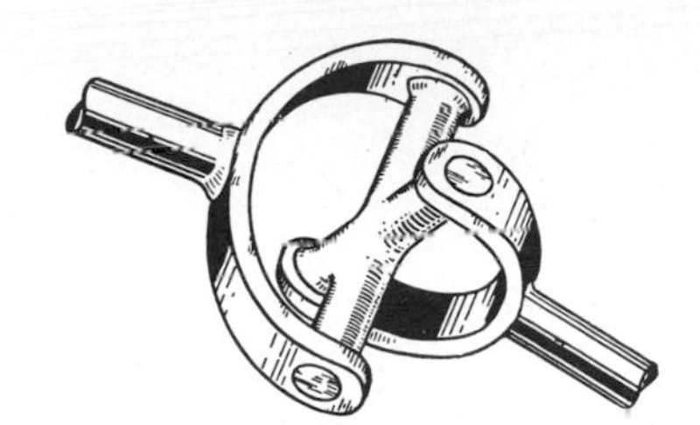

Fig. 1-23. Simple universal joint.

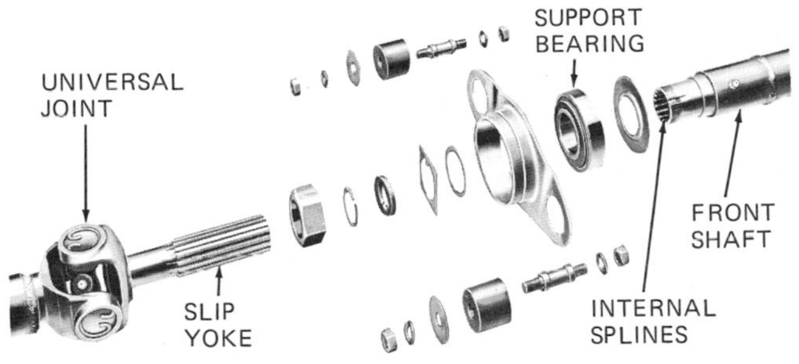

Fig. 1-24. Drive shaft and support bearing partly disassembled so that the slip joint can be seen. External splines are on the universal-joint yoke, and internal splines are in the shaft.

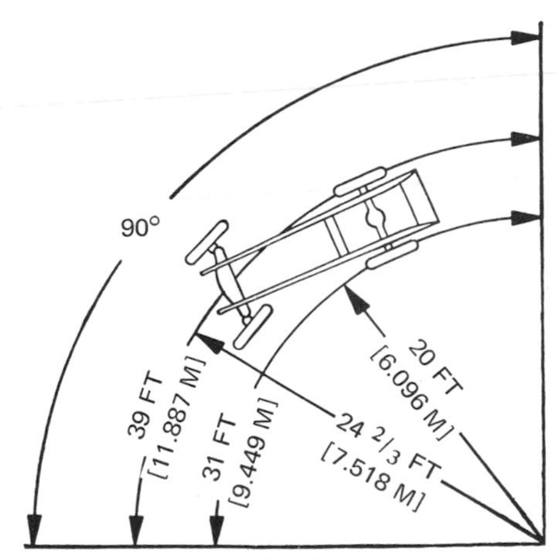

Fig. 1-25. Difference of wheel travel as a car makes a 90-degree turn with the inner rear wheel on a 20-foot [6.096 m] radius.

gear, is meshed with the two axle bevel gears. Thus, when the differential case is rotated, both axle bevel gears rotate and thus both wheels turn. However, suppose that one wheel is held stationary. Then, when the differential case is rotated, the differential pinion gear will also rotate as it "runs round" on the stationary axle bevel gear. As it rotates in this manner, the differential pinion gear carries rotary motion to the other axle bevel gear, causing it and the wheel to rotate.

It can be seen that when one rear wheel turns more rapidly than the other, the differential pinion gear spins on its shaft, transmitting more rotary motion to one rear wheel than to the other. When both rear wheels turn at the same speed, the differential pinion gear does not rotate on its shaft.

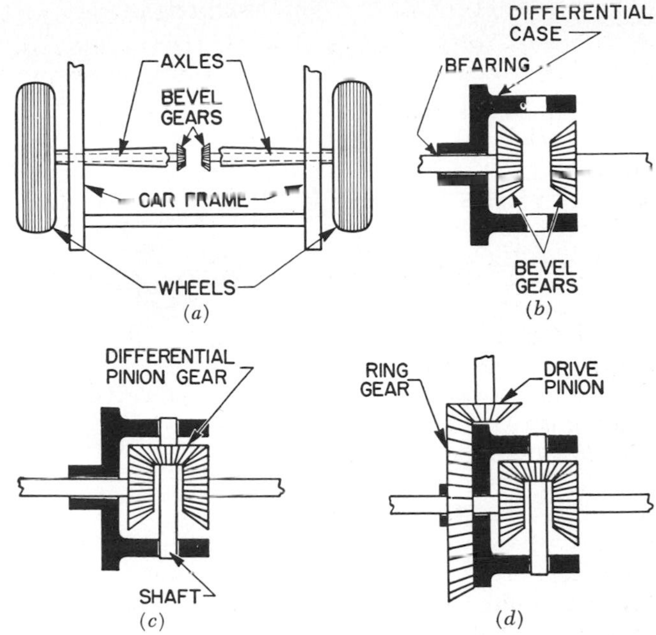

Fig. 1-26. Rear axles and differential: (*a*) The rear axles are attached to the wheels and have bevel gears on their inner ends. (*b*) The differential case is assembled on the left axle but can rotate on a bearing independently of the axle. (*c*) The differential case supports the differential pinion gear on a shaft, and this gear meshes with the two bevel gears. (*d*) The ring gear is attached to the differential case so that the case rotates with the ring gear when the latter is driven by the drive pinion.

The differential case is rotated by means of a ring gear attached to it. This ring gear is meshed with a drive pinion on the end of the drive shaft (Fig. 1-26*d*). When the car is on a straight road, the ring gear, differential case, differential pinion gear, and two axle bevel gears all turn as a unit without any relative motion. However, when the car begins to round a curve, the differential pinion gear rotates on its shaft to permit the outer rear wheel to turn more rapidly than the inner rear wheel.

The actual differential is somewhat more complicated than the one shown in Fig. 1-26. An actual differential, partially cut away to show the parts, is illustrated in Fig. 1-27. The driving power enters the differential through the drive pinion on the end of the drive shaft. The drive pinion is meshed with a large ring gear so that the ring gear revolves with the pinion. Attached to the ring gear (through the differential case) is a differential pinion shaft on which are assembled two differential pinion gears. Each rear car wheel has a separate axle, and there are two side gears splined to the inner ends of the two wheel axles. The two differential pinion gears mesh with these two side gears. When the car is on a straight road, the two differential pinion gears do not rotate on the pinion shaft; but they do exert pressure on the two side gears so that the side gears turn at the same speed as the ring gear, causing both rear wheels to turn at the same speed also. When the car rounds a curve, the outer wheel must turn

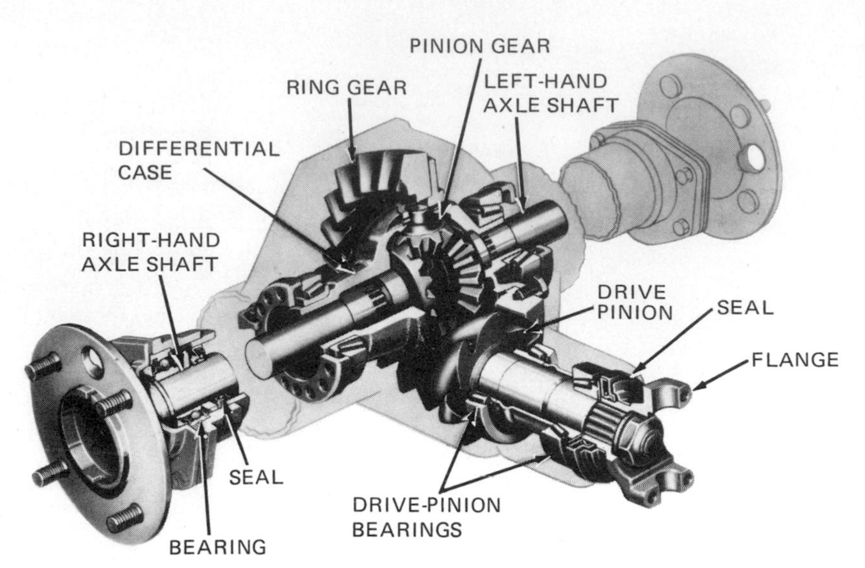

Fig. 1-27. Cutaway view of a differential and rear axle. (*Ford Motor Company*)

faster than the inner wheel. To permit this, the two pinion gears rotate on their pinion shaft, transmitting more turning movement to the outer side gear than to the inner side gear. Thus the side gear on the outer wheel axle turns more rapidly than the side gear on the inner wheel axle, permitting the outer wheel to turn more rapidly while the car is rounding the curve.

⊘ 1-9 Dynamometer Tests of the Power Train Many modern automotive shops have chassis dynamometers to check the power train (Figs. 1-28 and 1-29). The chassis dynamometer can duplicate any kind of road test at any load or speed. The part of the dynamometer you see consists of two heavy rollers mounted at or a little above floor level. The car is driven onto these rollers, as shown in Fig. 1-29, so that the car can drive the wheels. Next, the engine is started and the transmission put into gear. The car is then operated as though it were actually out on the highway.

Under the floor there is a device that can place varying loads on the engine and power train. This device allows the technician to check power-train action as well as engine performance. Instruments are hooked up to measure action during acceleration, cruising, coasting, and so on. Thus, the shift points of automatic transmissions can be determined under varying load conditions. Also, faulty conditions of the clutch, manual transmission, and drive shafts can be detected.

Fig. 1-28. Chassis dynamometer of the flush-floor type. The rollers are set at floor level. (*Sun Electric Corporation*)

Fig. 1-29. Automobile in place on a chassis dynamometer. The rear wheels drive the dynamometer rollers. At the same time, instruments on the test panel measure engine speed, car speed, engine-power output, and so on. (*Sun Electric Corporation*)

Check Your Progress

Progress Quiz 1-2 The following questions will help you determine how well you remember what you have been reading. If you have difficulty answering any of the questions, you should reread the past few pages. Most students make a practice of reading and rereading their lessons several times in order to make sure they understand them. Do not be discouraged if the questions stump you the first time. Just reread the past few pages and try the questions again. As you do this, you will begin to learn how to pick out the important facts you should remember. After you have repeated this procedure in the next few sections of the book, you will begin to find it increasingly easy to read and retain the essential facts. By the time you have finished the book, if you continue the practice, you will have become an expert student. And from being an expert student to being an expert automotive technician is only a step.

Correcting Parts Lists The purpose of this exercise is to enable you to spot the unrelated part in a list. For example, in the list "clutch: friction disk, rear axle, throw-out bearing, pressure springs, shaft," you can see that "rear axle" does not belong because it is the only item named that is not part of the clutch.

In each of the following lists, you will find one item that does not belong. Write each list in your notebook, but *do not write* the item that does not belong.

1. Transmission: clutch shaft, countershaft, transmission main shaft, motor, reverse idler gear, second-and-high-speed gear, low-and-reverse gear
2. Differential: bevel gears, differential pinion gear, ring gear, crankshaft, drive pinion
3. Drive shaft: universal joint, knee joint, slip joint
4. Countershaft drive gear, countershaft second gear, countershaft low gear, countershaft reverse gear, countershaft bevel gear

Completing the Sentences The sentences that follow are incomplete. After each sentence there are several words or phrases, only one of which will correctly complete the sentence. Write each sentence in your notebook, selecting the proper word or phrase to complete it correctly.

1. Essentially, the transmission described in the chapter consists of three shafts and: (*a*) three gears, (*b*) five gears, (*c*) eight gears, (*d*) ten gears.
2. In shifting into low, a gear on the transmission main shaft is moved into mesh with the: (*a*) countershaft low gear, (*b*) countershaft idler gear, (*c*) clutch gear, (*d*) output gear.
3. In the transmission, the countershaft drive gear is meshed with a gear on the: (*a*) output shaft, (*b*) main shaft, (*c*) clutch shaft.
4. In high gear, the transmission main shaft turns at the same speed as the: (*a*) countershaft, (*b*) clutch shaft, (*c*) idler shaft.
5. To take care of the difference in driving angle as the rear axle moves up and down, the drive shaft has one or more: (*a*) slip joints, (*b*) elbow joints, (*c*) release joints, (*d*) universal joints.
6. To take care of the lengthening and shortening of the drive shaft with rear-axle movement, the drive shaft has a: (*a*) slip joint, (*b*) elbow joint, (*c*) release joint, (*d*) universal joint.
7. In the differential, the drive pinion meshes with the: (*a*) bevel gear, (*b*) pinion gear, (*c*) ring gear.
8. The power train includes the clutch (on some cars), transmission, differential, and: (*a*) carburetor, (*b*) drive shaft, (*c*) axle housing.

CHAPTER 1 CHECKUP

NOTE: Since the following is a chapter review test, you should review the chapter before taking it.

Now that you have completed a chapter in the book, you will want to test your knowledge of the subjects covered in the chapter. The questions that follow have two purposes: One is to test your knowledge; the other is to help you review the chapter. It may be that you will not be able to answer all the questions offhand. If this happens, turn back into the chapter and reread the pages that will give you the answer. For instance, under "Listing Parts" you are asked to list the parts, besides bearings, that are in motion when the car is moving. If you cannot remember them all, turn back to the illustrations of the differential in the chapter and refer to them when writing your list. The act of writing the names of the parts will help you remember them.

NOTE: Write your answers in your notebook. Then later, when you have finished the book, you will find your notebook filled with valuable information to which you can refer quickly.

Completing the Sentences The sentences that follow are incomplete. After each sentence there are several words or phrases, only one of which will correctly complete the sentence. Write each sentence in your notebook, selecting the proper word or phrase to complete it correctly.

1. In most cars the power train transmits power from the engine to the: (*a*) crankshaft, (*b*) steering gear, (*c*) front wheels, (*d*) rear wheels.
2. The clutch part that is between the pressure plate and the engine flywheel is called the: (*a*) throw-out bearing, (*b*) clutch fork, (*c*) friction disk, (*d*) clutch pedal.
3. The gear in the transmission that is always in mesh with the clutch gear is called the: (*a*) second-and-high-speed gear, (*b*) low-speed gear, (*c*) idler gear, (*d*) countershaft drive gear.

4. The drive shaft has one or more: (*a*) elbow joints, (*b*) universal joints, (*c*) fluid couplings, (*d*) spur gears.
5. In the standard differential described in this chapter, the total number of gears and pinions is (*a*) two, (*b*) three, (*c*) six, (*d*) eleven.
6. In the differential, the ring gear is attached to the: (*a*) bevel gear, (*b*) drive gear, (*c*) differential case, (*d*) drive shaft.
7. The device in the drive shaft that permits changes in shaft length is called the: (*a*) length joint, (*b*) change joint, (*c*) slip joint, (*d*) shaft joint.
8. In the transmission, the reverse idler gear is always in mesh with the: (*a*) low gear, (*b*) second gear, (*c*) countershaft reverse gear, (*d*) main-shaft reverse gear.

Listing Parts In the following, you are asked to list parts that make up the various automotive components discussed in the chapter. Write these lists in your notebook.

1. List four major components of the automobile.
2. List four major components of the power train.
3. List five parts in a clutch.
4. List six gears used in a typical standard transmission.
5. List two types of joint used in a drive shaft.
6. List the major parts in the differential, aside from bearings, that are in motion when the car is moving.
7. List the parts through which power is flowing in the transmission when the gears are in first; in second; in reverse.

Purpose and Operation of Components In the following, you are asked to write the purpose and operation of certain components of the automobile discussed in the chapter. If you have any difficulty in writing your explanation, turn back in the chapter and reread the pages that will give you the answer. Then write your explanation. Don't copy; try to tell it in your own words just as you would explain it to a friend. This is a good way to fix the explanation more firmly in your mind. Write in your notebook.

1. What is the purpose of the clutch?
2. How does the clutch operate?
3. What is the purpose of the transmission?
4. Explain what takes place when the transmission gears are shifted into low, second, high, and reverse.
5. What is the purpose of the universal joints in the drive shaft?
6. What is the purpose of the slip joint in the drive shaft?
7. Explain what happens in the differential when the car turns a corner.

SUGGESTIONS FOR FURTHER STUDY

If you would like to study further the various components of the automobile, aside from the power-train components discussed in this book, then you will want to study the other books in the McGraw-Hill Automotive Technology Series. From these books you will learn how the engine; the engine fuel, lubrication, and cooling systems; the electric system; and the various chassis units operate. In addition, you can spend some time in your local school automotive shop or a friendly service shop where repair work on various automotive and power-train parts is done. By watching what goes on, you can learn a great deal about how the various automotive parts are constructed and how they are put together and serviced.

You may be able to obtain shop repair manuals from your local school automotive shop or service shop. These manuals are very interesting since they contain many illustrations and detailed explanations of automotive components, including clutches, transmissions, drive shafts, and differentials. A careful study of these manuals will prove very rewarding. Be sure to write in your notebook any important facts that you learn.

chapter 2

CLUTCHES

This chapter discusses the purpose, construction, and operation of automotive clutches. The clutch is located in the power train between the engine and the transmission (Fig. 2-1). At one time, all cars had clutches. But today most cars are equipped with automatic transmissions, which do not require clutches. However, there are still many cars on the road with clutches, and many new cars are so equipped. The car with a manually shifted transmission requires a clutch to facilitate shifting.

⊘ 2-1 Function of the Clutch The clutch is a form of coupling that permits the driver to couple or uncouple the engine and the transmission. With the engine coupled, power is delivered to the transmission; with the engine uncoupled, by operation of the clutch, the engine runs free and does not deliver power to the transmission. The clutch temporarily uncouples the engine and the transmission so that the transmission gears can be shifted. Without temporarily interrupting the flow of power between the two, it would be difficult to de-mesh and mesh the transmission gears. The pressure between gear teeth in a set of gears through which power is flowing makes it hard to shift the gears out of mesh. Also, without a clutch, shifting gears into mesh would be a hazardous procedure. The driving gears and the driven gears would probably be running at different speeds, and broken gear teeth would result as meshing was attempted. The clutch, when operated, interrupts the flow of power so gear-tooth pressure is relieved for de-meshing. With the gears de-meshed and the engine uncoupled by operation of the clutch, the transmission driving gear runs free so that it can attain synchronous speed with other transmission gears. This is accomplished by synchronizing devices in the transmission. Meshing can thus be accomplished without clashing of gears.

At times during car operation, the clutch is operated to permit shifting of the transmission into neutral. This action de-meshes the transmission gears so that, even when the clutch is again engaged, engine power cannot be transmitted through the transmission. The engine can thus be started and brought to speed without delivering power through the transmission and power train to the car wheels. The amount of power that a gasoline engine can deliver during starting and at speeds below idle is small—too small to put the car into motion or keep it in motion. The engine must be rotating at several hundred revolutions per minute before it is able to deliver any appreciable amount of power. After the engine has been started, the clutch is operated to permit shifting of gears through the various speed ratios (Chap. 4) so that the car can be set into motion and increased in speed.

Figures 2-1 to 2-7 show various clutch parts and their relationship to the clutch pedal in the driving compartment of the car. Figure 2-1 shows the clutch partly cut away. Figure 2-2 shows, in disassembled view, the clutch and linkage parts between the clutch and the clutch pedal. Figure 2-3 shows the clutch-linkage parts in assembled view on an application that is somewhat different from the others. As you study different clutches on various cars, you will notice that several linkage systems are used. The arrangements of the linkages may differ, but the end result in the clutch is the same: A throw-out bearing in the clutch is forced inward to cause declutching when the clutch pedal is pushed down. This action is explained in detail later.

⊘ 2-2 Types of Clutches The type of clutch used on an automobile depends upon several factors, including the maximum engine torque developed, the type of transmission, and the nature of the service in which the clutch will operate. Although numerous types of clutches have been and are being used, the one most widely used today is the single-dry-plate type. This type of clutch includes a disk that is faced on both sides with friction material. The disk hub is attached to a shaft (called the *drive pinion, main drive gear,* or *clutch shaft*) by means of splines. The attachment is such that the disk hub can move endwise along the shaft but is forced to rotate with the shaft at the same speed as the shaft.

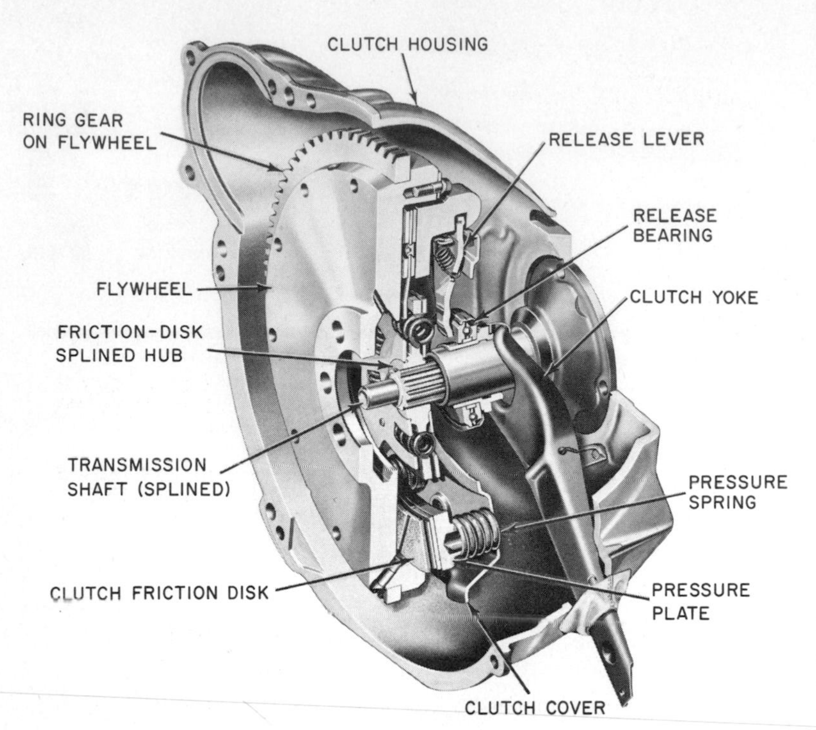

Fig. 2-1. Partial cutaway view of a typical clutch. (*Ford Motor Company*)

When the clutch is engaged, the disk is clamped tightly between two parallel metal faces by springs. In this position the friction between the disk facing and the two metal faces causes the disk and metal faces to rotate as a unit. Figure 1-8 illustrates an automotive clutch of this type, and Fig. 2-4 is an exploded view of this clutch. Figure 2-5 is a cutaway view of a somewhat similar clutch.

⊘ 2-3 Operation of the Clutch The operation of the clutch is based on the frictional contact between two smooth metallic driving surfaces and the facings riveted to the friction disk (also called the *driven plate,* or *clutch disk*). See Fig. 2-6. One of the metallic surfaces is on the flywheel, and the other is on the clutch pressure plate (Figs. 1-8 and 2-1 to 2-5). When the clutch is in the engaged position, spring pressure between the clutch cover and pressure plate clamps the disk tightly between the pressure plate and flywheel face (Fig. 2-7). The friction between these surfaces causes the disk to rotate with the flywheel and the pressure plate when the engine is running and the clutch is engaged. Since the hub of the disk is splined to the clutch shaft, the shaft rotates with the disk.

To uncouple the engine and the transmission, the clutch pedal is depressed. This action, in turn, through a series of levers, operates the clutch-yoke or fork assembly, causing the yoke or fork to move a throw-out bearing (also called the *release bearing*) in toward the flywheel. Movement of the throw-out bearing in this direction releases the spring pressure that holds the flywheel, friction disk, and pressure plate together, causing the pressure plate to be moved away from the friction disk (Fig. 2-8). This action permits the flywheel and pressure plate to rotate independently of the friction disk and clutch shaft. Figures 2-2 and 2-3 show typical linkages between the clutch pedal and fork.

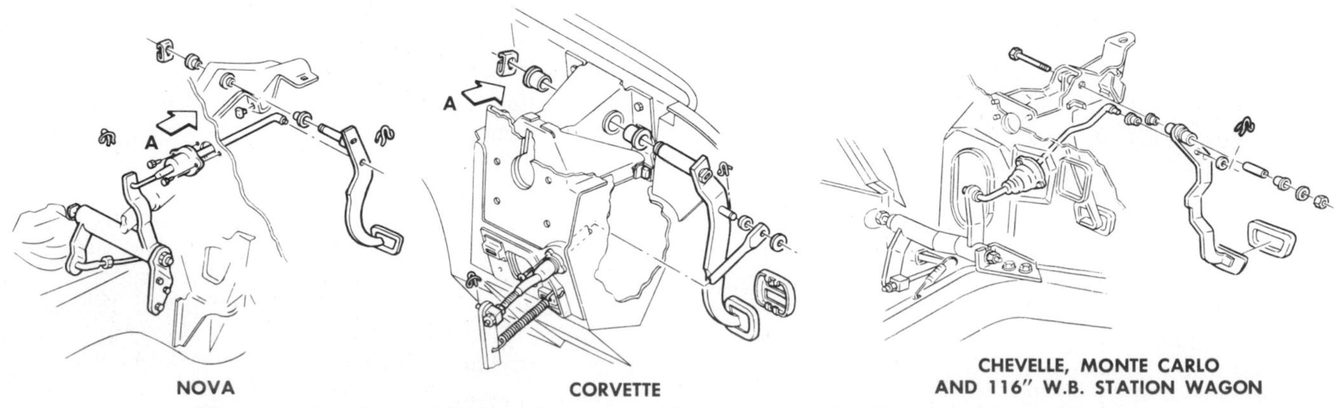

Fig. 2-2. Clutch-pedal linkage for several Chevrolet models. (*Chevrolet Motor Division of General Motors Corporation*)

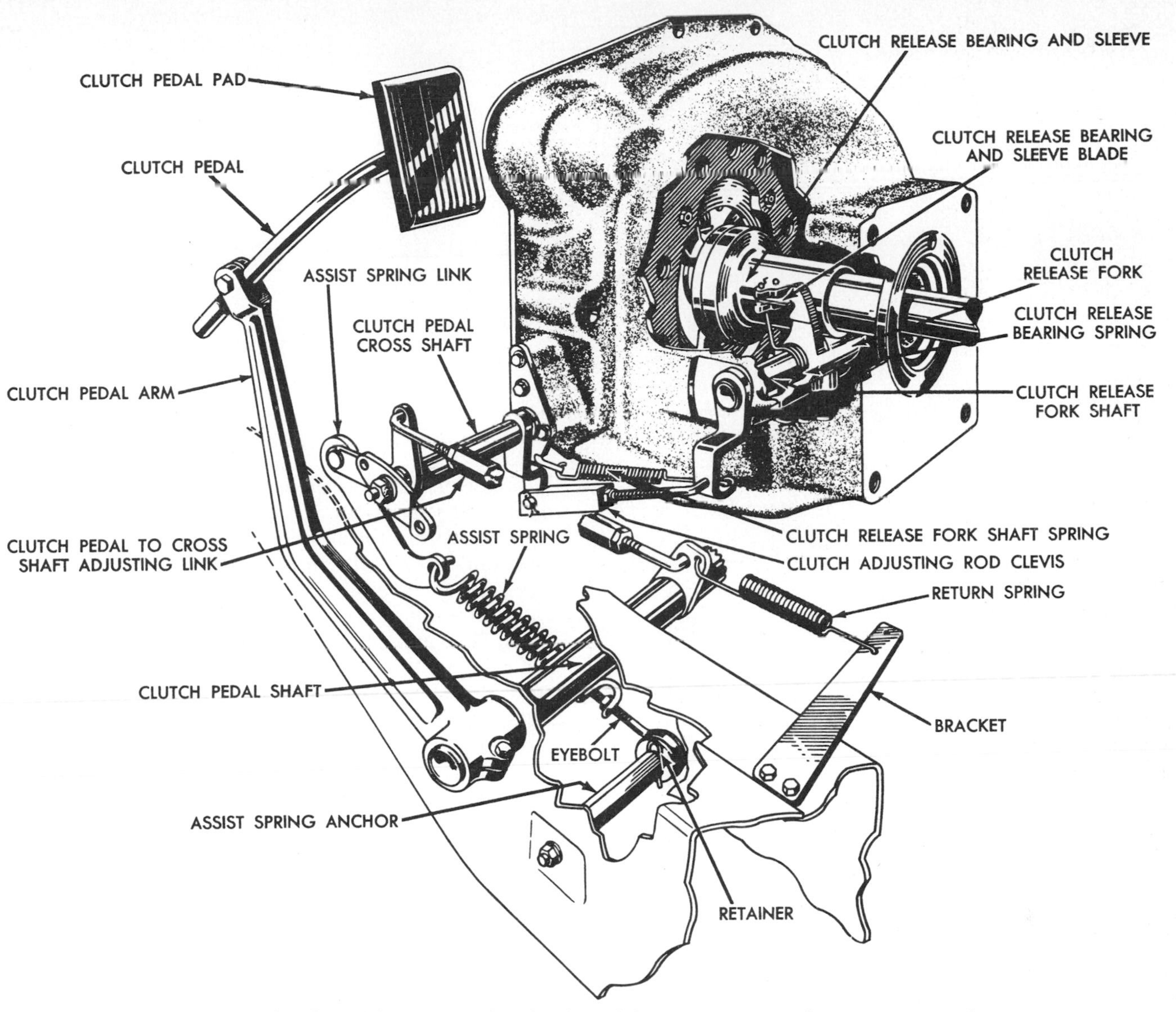

Fig. 2-3. Clutch and linkage to the clutch pedal. Housing is partly cut away so that clutch parts may be seen.

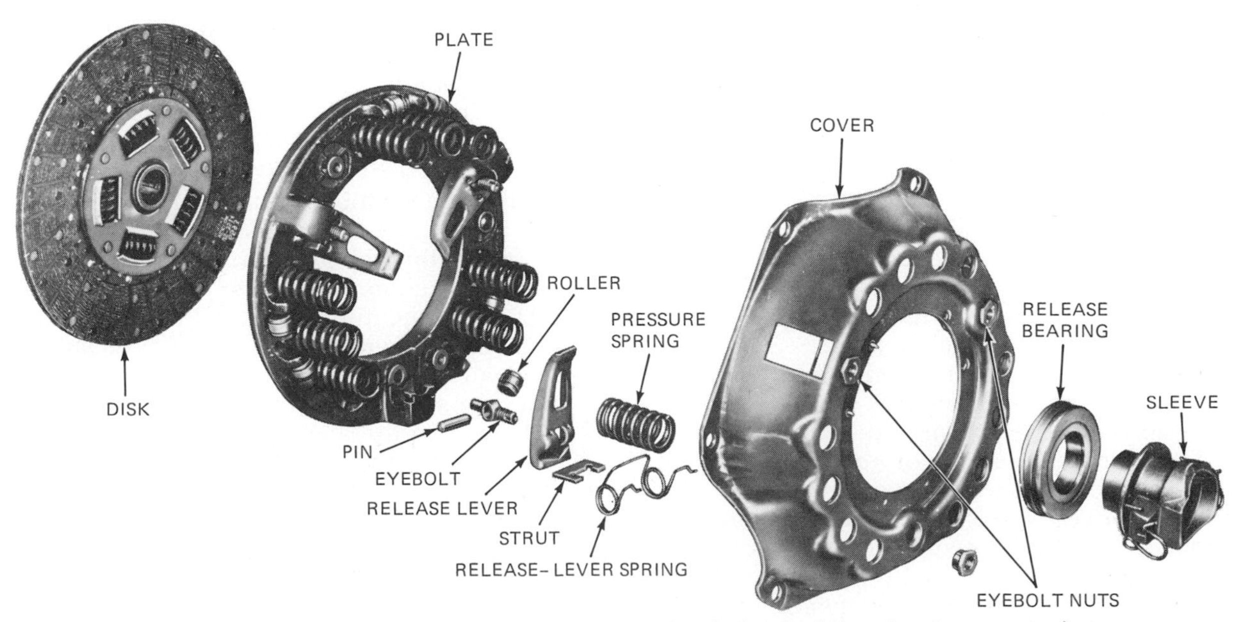

Fig. 2-4. Disassembled view of a semicentrifugal clutch. (*Chrysler Corporation*)

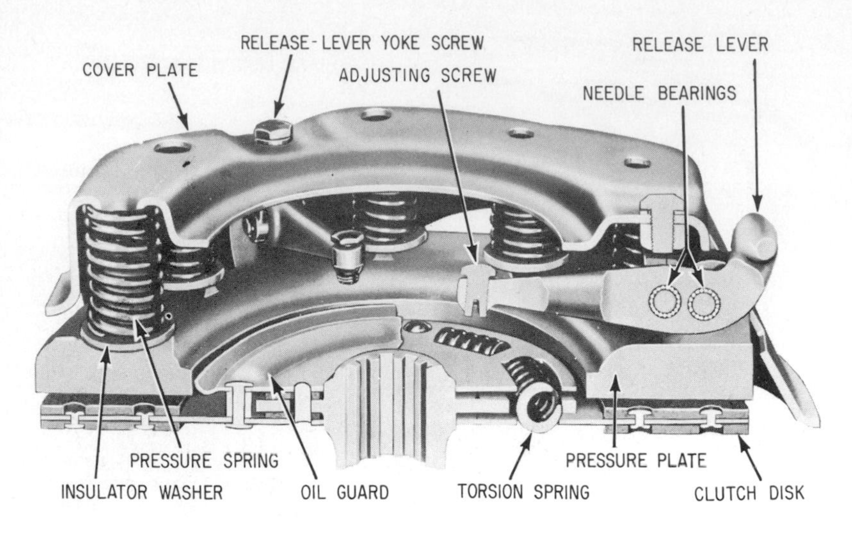

Fig. 2-5. Cutaway view of a clutch. (*Oldsmobile Division of General Motors Corporation*)

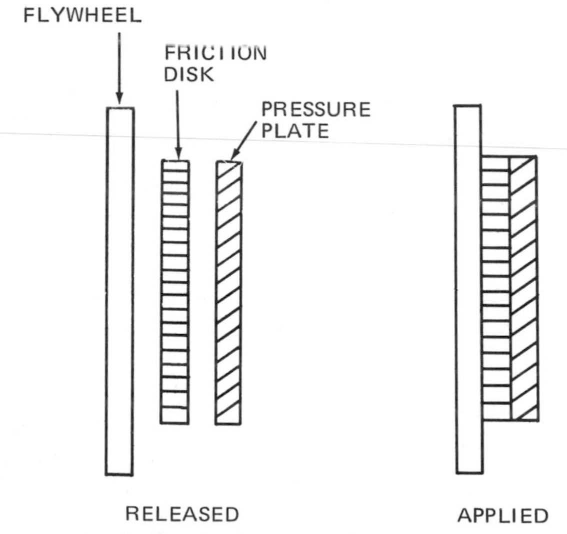

Fig. 2-6. Basic clutch elements, showing clutch action. *Left,* clutch released. The pressure plate and friction disk have moved away from the flywheel. *Right,* clutch applied. The pressure plate clamps the friction disk to the flywheel so that all parts must rotate together.

Figure 2-9 illustrates one type of clutch-yoke assembly. The yoke protrudes through the side of the clutch housing and, on the installation illustrated, includes a spring-loaded dust seal around the clutch-housing hole. Inside the housing, a clutch ball stud (stud with a round end) provides a ball point around which the yoke can pivot when its outer end is moved back and forth by operation of the clutch pedal. This movement causes the throw-out bearing, assembled on the forked end of the yoke, to be slid back and forth along the clutch shaft.

As the throw-out bearing is slid in toward the clutch, it comes up against a series of three or more release levers that are equally spaced around the clutch. The release levers pivot about pins, taking up the spring pressure and causing the pressure plate to be moved away from the friction disk. Figure 2-10 illustrates the two positions of a release lever: the engaged and released positions.

The pressure of the throw-out bearing against the inner end of the release lever causes it to move to the left (Fig. 2-10). This action causes the lever to pivot around the floating pin in the eyebolt. The outer end of the release lever is moved to the right, exerting pressure through the strut against the pressure plate and causing the plate to be moved to the right. This movement compresses the coil springs and relieves the spring pressure between the pressure plate, friction disk, and flywheel. Also, as the pressure plate moves to the right, clearances appear between the pressure plate and friction disk and between the friction disk and flywheel. These clearances allow the flywheel and the pressure plate to rotate independently of the friction disk.

⊘ 2-4 Friction Disk The friction disk includes a cushioning device to provide a cushion effect between the facings when the clutch is engaged. It also has a dampening device to prevent torsional vibration of the engine from being transmitted through the clutch. Figure 2-11 illustrates a typical friction disk. See also Fig. 2-5 for construction details of a friction disk.

The cushioning device consists of waved cushion springs to which the friction facings are riveted. These waves are compressed slightly as the clutch engages, providing the cushioning effect. The dampening device embodies a series of heavy coil springs placed between the drive washers, which are riveted to the cushion springs, and the hub flange, which is attached to the disk hub. The disk hub is thus driven through the springs, and they absorb torsional vibration. Stop pins limit the relative motion between the hub flange and the drive washers. A molded friction ring, compressed between the hub flange and the drive washers, provides frictional dampening that prevents oscillation between the hub flange and the drive washers.

Some special-duty clutches have metallic-ceramic friction facings (Fig. 2-12) instead of the asbestos or organic facings shown in Fig. 2-11.

⊘ 2-5 Other Types of Clutches The type of clutch we have described and illustrated in Figs. 2-1 to 2-5

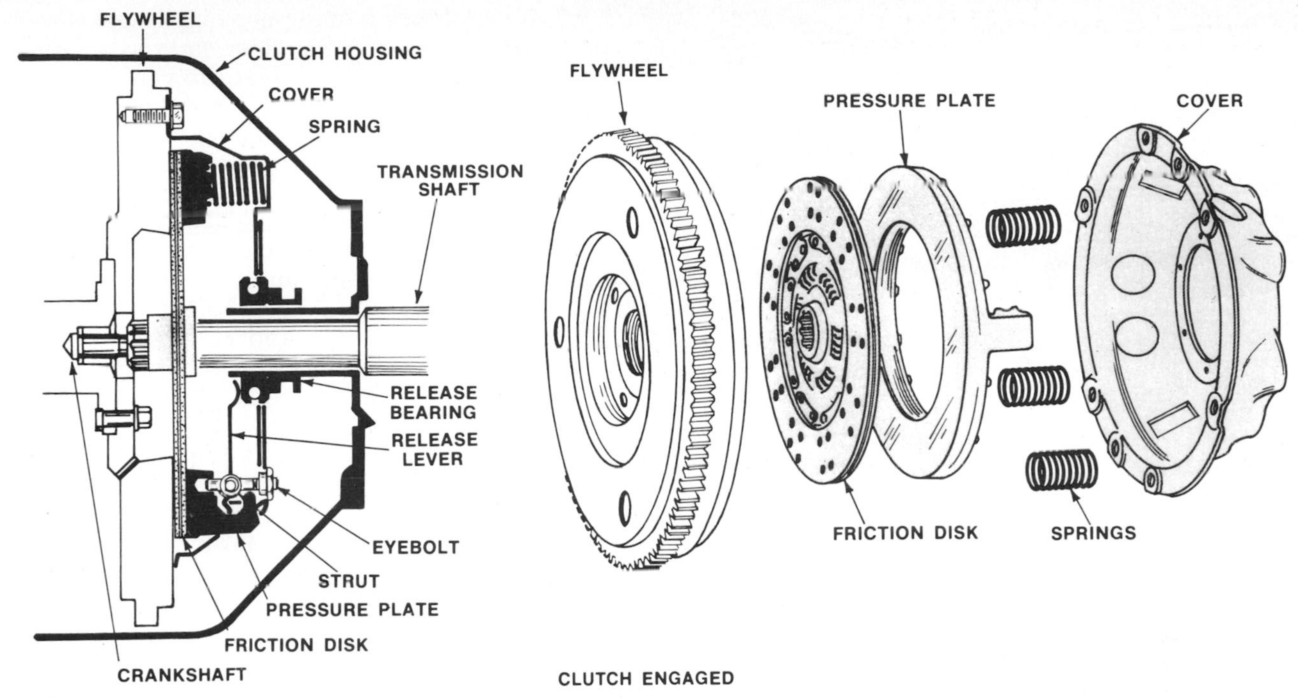

Fig. 2-7. Sectional view of the clutch in the engaged position (*left*). Major clutch parts are shown to the right.

is a *nine-spring clutch*. It has nine coil springs which furnish the pressure that holds the pressure plate against the flywheel facing when the clutch is applied. There are other types of clutches, including the three-spring clutch, diaphragm-spring clutch, two-friction-disk clutch, centrifugal clutch, and hydraulic clutch. We shall discuss these in the sections that follow.

⊘ 2-6 Three-Spring Clutch The three-spring clutch, shown in Fig. 2-13, uses three pressure

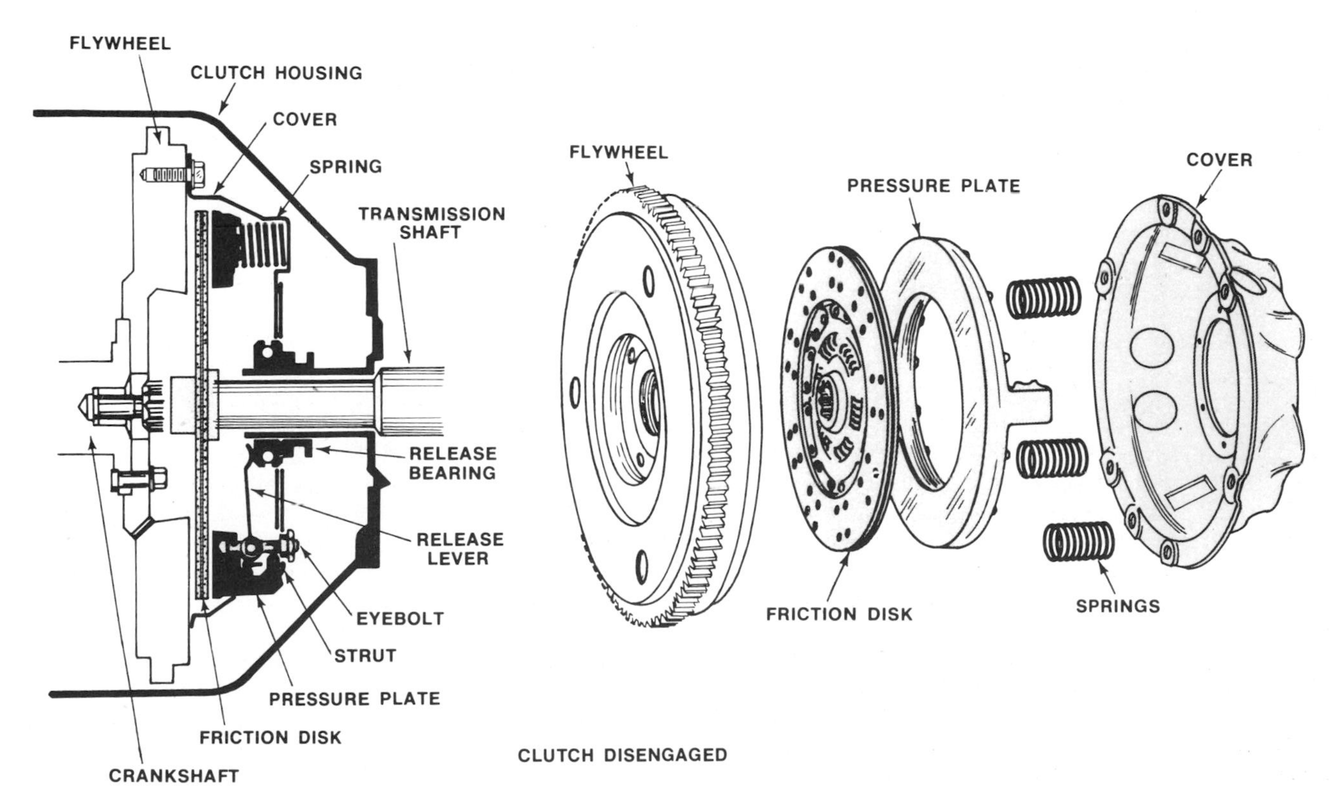

Fig. 2-8. Sectional view of the clutch in the released, or disengaged, position (*left*). Major clutch parts are shown to the right.

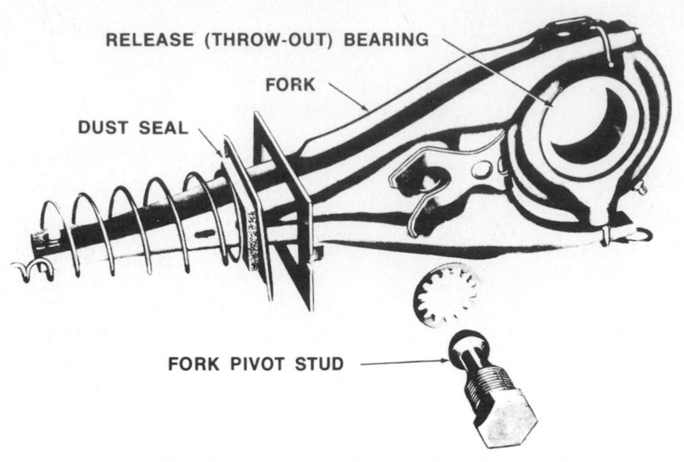

Fig. 2-9. Clutch-fork assembly with release (throw-out) bearing. (*Buick Motor Division of General Motors Corporation*)

springs instead of nine. Note how they are arranged at angles instead of perpendicular to the pressure plate as in the nine-spring clutch described previously. The operation of the clutch is the same however.

⊘ 2-7 Diaphragm-Spring Clutch This type of clutch incorporates a diaphragm spring that provides the spring pressure required to hold the friction disk against the flywheel. It also acts as the release levers that take up the spring pressure when the clutch is disengaged. Figures 2-14 and 2-15 show sectional and disassembled views of the diaphragm-spring clutch.

The clutch has a one-piece diaphragm that is a solid ring on the outer diameter and has a series of tapering fingers pointing inward toward the center of the clutch. The action of the clutch diaphragm is somewhat like the flexing action that takes place when the bottom of an oil can is depressed. When the throw-out bearing moves in against the ends of the fingers, the entire diaphragm is forced against a pivot ring, causing the diaphragm to dish inward. This raises the pressure plate from the friction disk.

Figures 2-16 and 2-17 illustrate the two positions of the diaphragm spring and clutch parts. In the engaged position (Fig. 2-16), the diaphragm spring is slightly dished, with the tapering fingers pointing slightly away from the flywheel. This position places spring pressure against the pressure plate around the entire circumference of the diaphragm spring. The diaphragm spring is naturally formed to exert this initial pressure. When the throw-out bearing is moved inward against the spring fingers (as the clutch pedal is depressed), the spring is forced to pivot round the inner pivot ring, dishing in the opposite direction. The outer circumference of the spring now lifts the pressure plate away through a series of retracting springs placed about the outer circumference of the pressure plate (Fig. 2-17).

Figure 2-18 is a sectional view of a diaphragm-spring clutch that is different in one important respect from the unit illustrated in Fig. 2-14. In the unit shown in Fig. 2-18, the inner ends of the tapered fingers are bent. However, both clutches function in a similar manner.

The diaphragm-spring clutch is also supplied in a two-friction-disk design (Fig. 2-19). The added

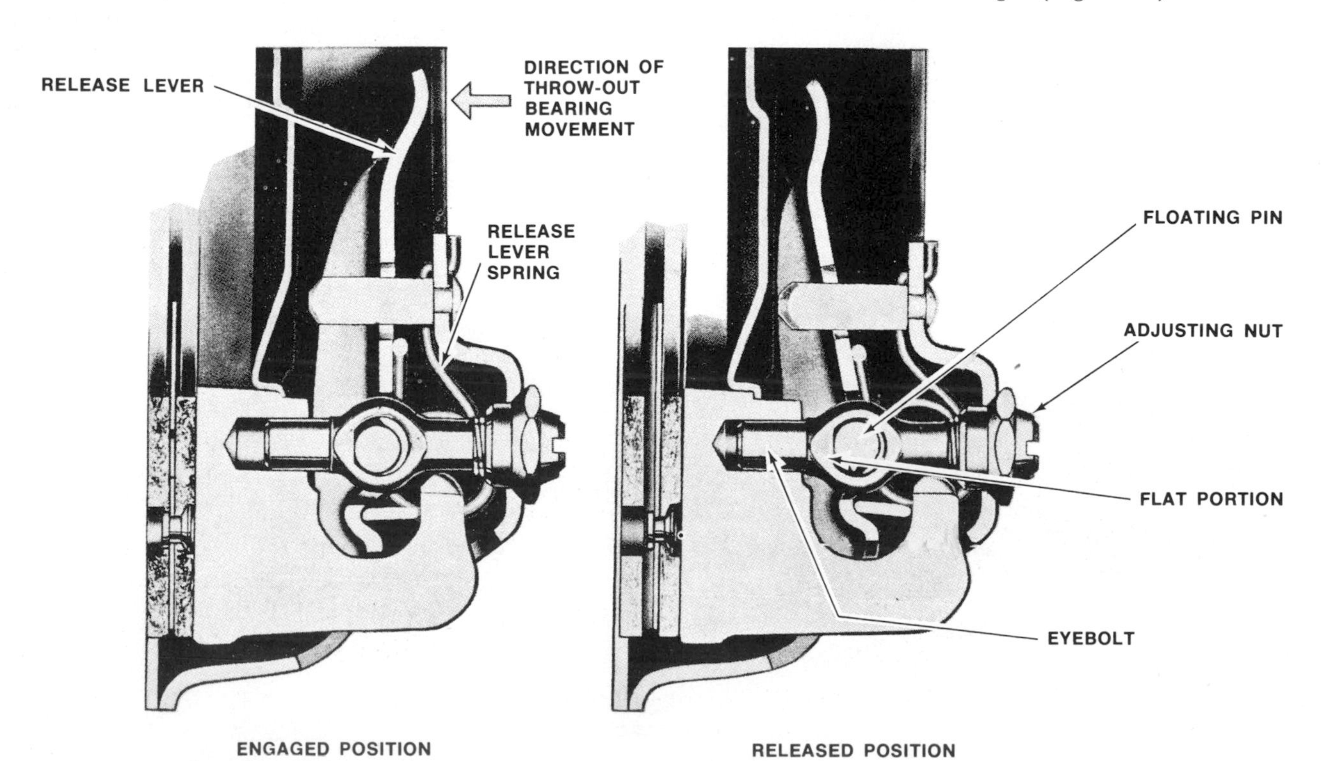

Fig. 2-10. The two limiting positions of the pressure plate and release lever. (*Oldsmobile Division of General Motors Corporation*)

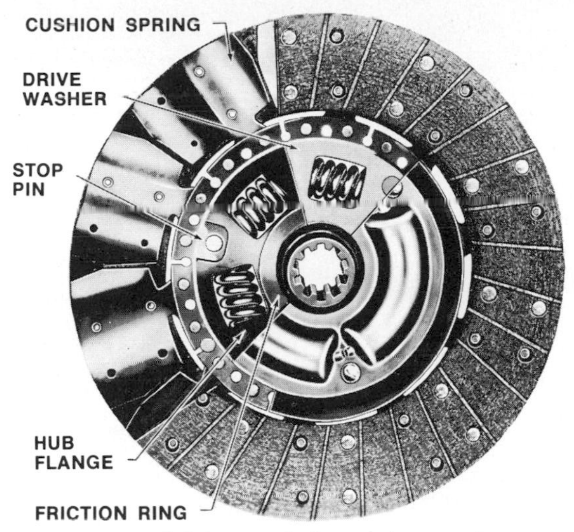

Fig. 2-11. Typical friction disk, or driven plate. Facings and drive washer are partly cut away to show springs. (*Buick Motor Division of General Motors Corporation*)

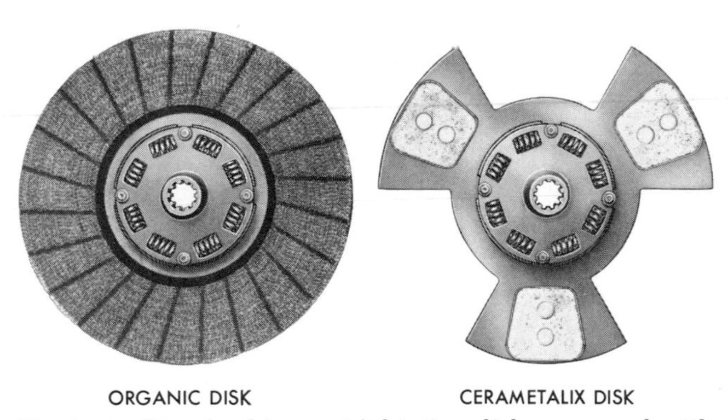

Fig. 2-12. Standard (organic) friction disk compared with a metallic-ceramic (Cerametalix) friction disk. (*Chevrolet Motor Division of General Motors Corporation*)

pressure plate and friction disk give the clutch greater holding power, which makes it suitable for use with higher-output engines.

⊘ 2-8 Two-Friction-Disk Clutch Where greater holding power is required, that is, where more power must go through the clutch, a two-friction-disk clutch is used. Figure 2-19 is a sectional view of a diaphragm-spring clutch using two friction disks and two pressure plates. Figure 2-20 is a sectional view of a coil-spring two-friction-disk clutch. These clutches work in the same way as the clutches described previously.

⊘ 2-9 Centrifugal Clutch Some clutches are designed to increase clutch holding power at high speeds. As clutch speed increases, weights on the clutch release levers are forced outward by centrifugal force. With this arrangement, fairly light pressure springs can be used so that the clutch operates without too much foot pressure. But at the same time, the centrifugal effect produces a higher pressure against the friction disk with increasing speed. This provides the extra holding power the clutch needs at high speeds.

Figure 2-21 shows one type of centrifugal clutch. The release levers have weights on their outer ends. As centrifugal force acts on the weights, the levers press down more tightly against the pressure plate. This movement adds to the coil-spring pressure holding the friction disk locked between the pressure plate and flywheel. Figure 2-4 shows a somewhat different arrangement. Here, the centrifugal effect acts on the rollers that are placed under the release levers. The centrifugal effect wedges the rollers between the cover and pressure plate to produce the increased pressure on the pressure plate.

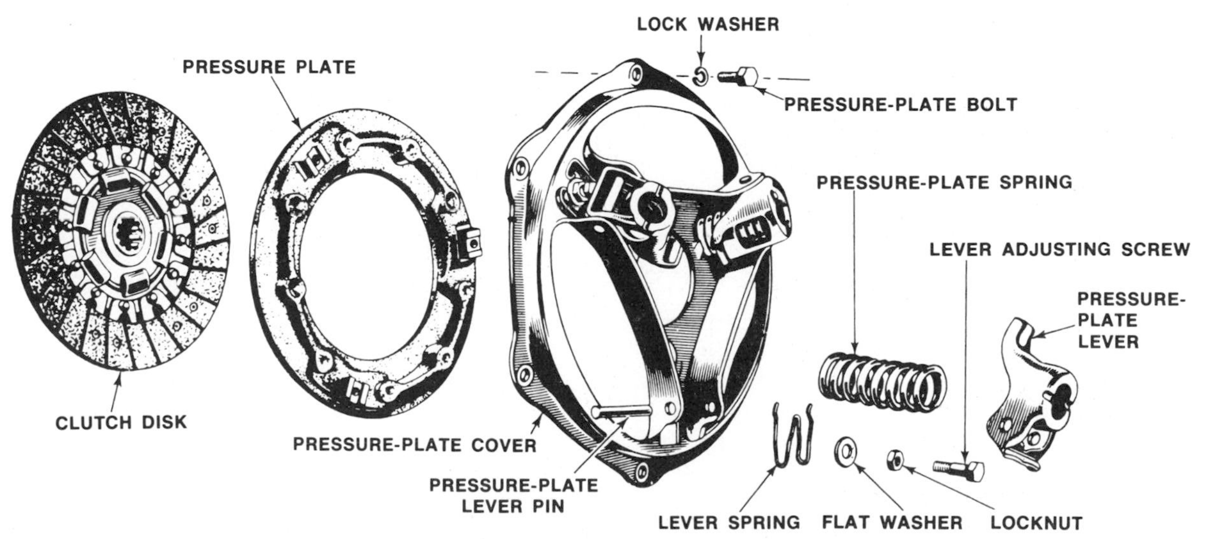

Fig. 2-13. Exploded view of a three-spring clutch.

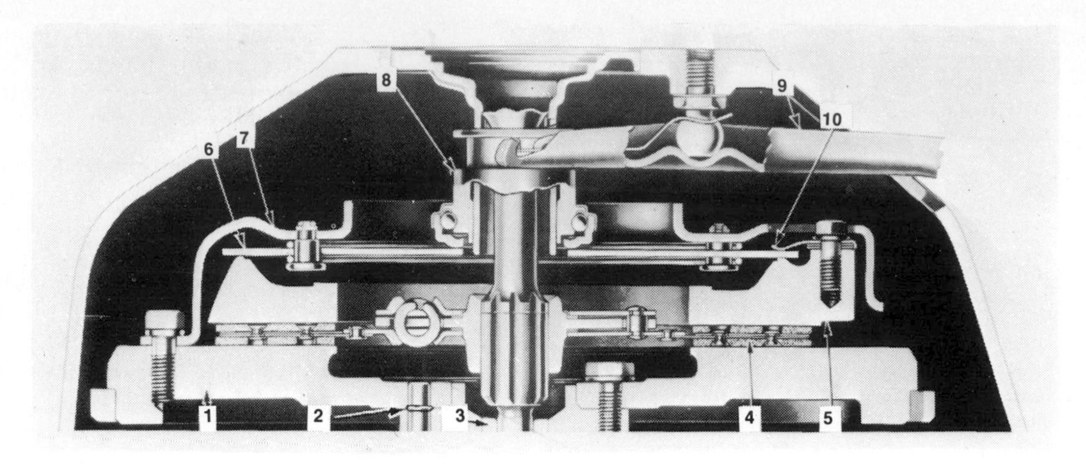

1. Flywheel
2. Dowel
3. Pilot bushing
4. Driven disk
5. Pressure plate
6. Diaphragm spring
7. Cover
8. Throw-out bearing
9. Fork
10. Retracting spring

Fig. 2-14. Flat-finger diaphragm-spring clutch in sectional view from the top. (*Chevrolet Motor Division of General Motors Corporation*)

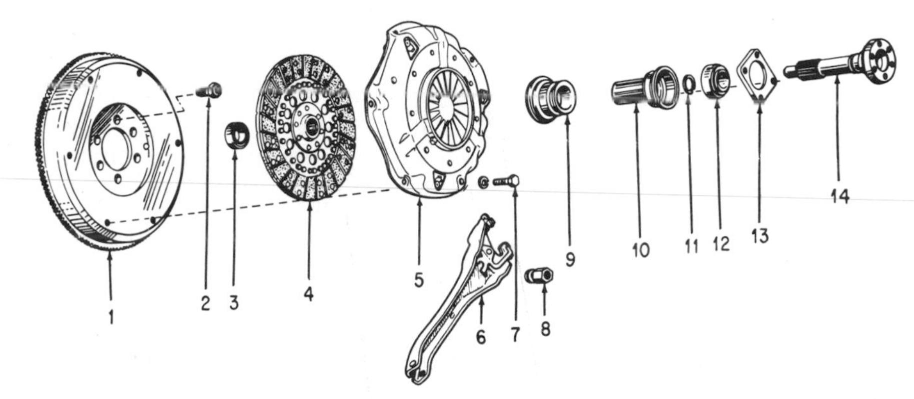

1. Flywheel
2. Flywheel attaching bolts
3. Clutch pilot bearing
4. Clutch driven plate
5. Clutch cover and pressure plate
6. Clutch-release fork
7. Clutch-cover attaching bolts
8. Clutch-release-fork ball nut
9. Clutch-release-bearing
10. Clutch-release-bearing support
11. Clutch-drive-shaft bearing retainer
12. Clutch-drive-shaft bearing
13. Retaining plate
14. Clutch drive shaft

Fig. 2-15. Disassembled view of a diaphragm-spring clutch. (*Pontiac Motor Division of General Motors Corporation*)

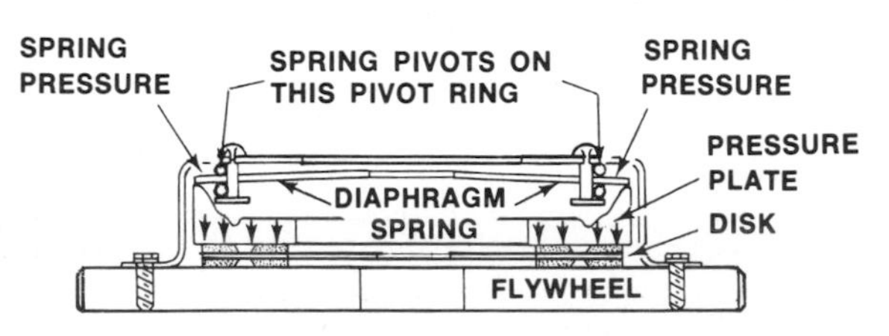

Fig. 2-16. Diaphragm-spring clutch in the engaged position. (*Chevrolet Motor Division of General Motors Corporation*)

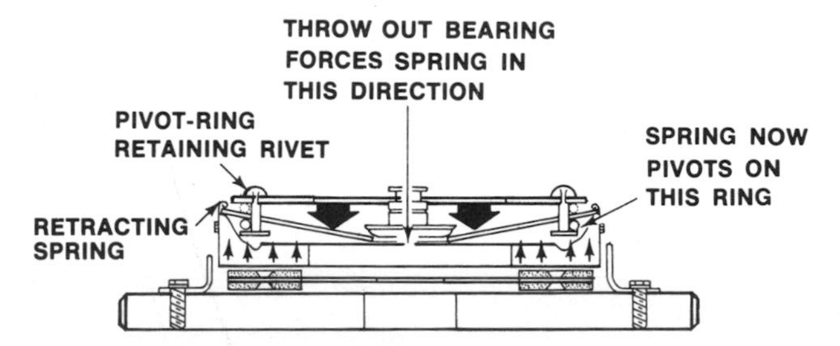

Fig. 2-17. Diaphragm-spring clutch in the disengaged position. (*Chevrolet Motor Division of General Motors Corporation*)

⊘ 2-10 Hydraulic Clutch The hydraulic clutch is used on vehicles where the clutch is remotely located so that it would be difficult to run linkages from the foot pedal to the clutch. Also it is used on high-output applications where heavy pressure-plate springs are required. When a clutch is designed to transmit high torques, the springs must be heavy in order to provide sufficient pressure on the friction disk. With insufficient spring pressure, the pressure plate and flywheel would slip on the friction disk, quickly ruining it.

The heavy spring pressure, however, increases the pressure that must be applied to the clutch release lever, or fork. This, in turn, increases the pressure that the driver must apply to the clutch pedal. To reduce the clutch-pedal pressure, a hydraulic system is used on the hydraulic clutch. Figure 2-22 shows one of these clutches in disassembled view.

1. Flywheel
2. Dowel hole
3. Pilot bushing
4. Driven disk
5. Pressure plate
6. Diaphragm spring
7. Cover
8. Throw-out bearing
9. Fork
10. Retracting spring

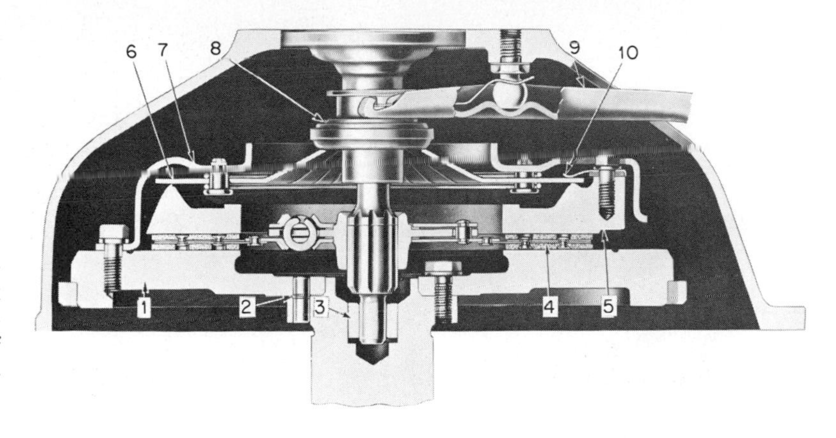

Fig. 2-18. Diaphragm-spring clutch with bent tapering fingers. (*Chevrolet Motor Division of General Motors Corporation*)

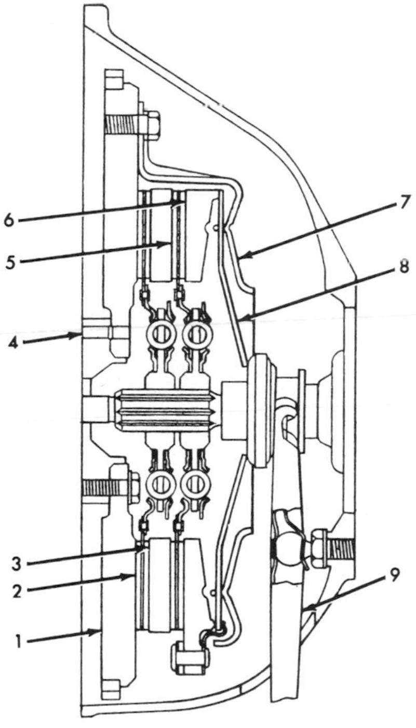

1. Flywheel
2. Front friction disk
3. Front pressure plate
4. Dowel hole
5. Rear friction disk
6. Rear pressure plate
7. Cover
8. Retracting spring
9. Fork

Fig. 2-19. Sectional view of a diaphragm-spring clutch using two pressure plates. (*Chevrolet Motor Division of General Motors Corporation*)

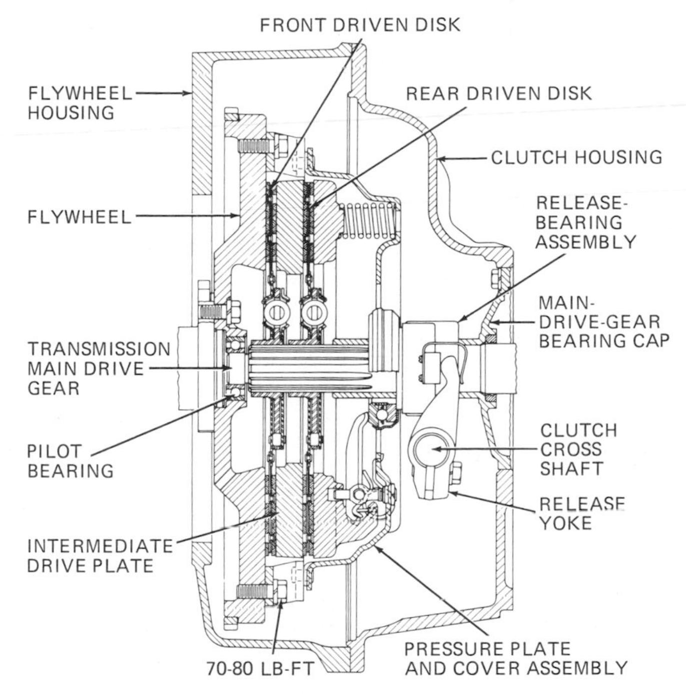

Fig. 2-20. Sectional view of a clutch with two friction disks and an intermediate drive plate. (*Chevrolet Motor Division of General Motors Corporation*)

Figure 2-23 shows the system schematically. The clutch pedal does not work directly to the release lever through linkages. Instead, when the driver pushes down on the clutch pedal, a pushrod is forced into a master cylinder. This action moves a piston and cup in the master cylinder, which forces hydraulic fluid out of the master cylinder, through a tube, and into a servo (slave) cylinder. (This action is much like that which takes place in a hydraulic brake system when the brakes are applied.)

As the fluid is forced into the servo cylinder, it moves a piston and pushrod. This movement causes the clutch release fork to move, thereby operating the pressure-plate release levers. The hydraulic system can be designed to multiply the driver's efforts so that a fairly light foot-pedal pressure produces a stronger push on the clutch release fork. There is the additional advantage that no mechanical linkage between the two is required. Thus, there is no problem of arranging long linkages if the engine is remotely located from the driver's compartment, for instance, at the rear of the vehicle.

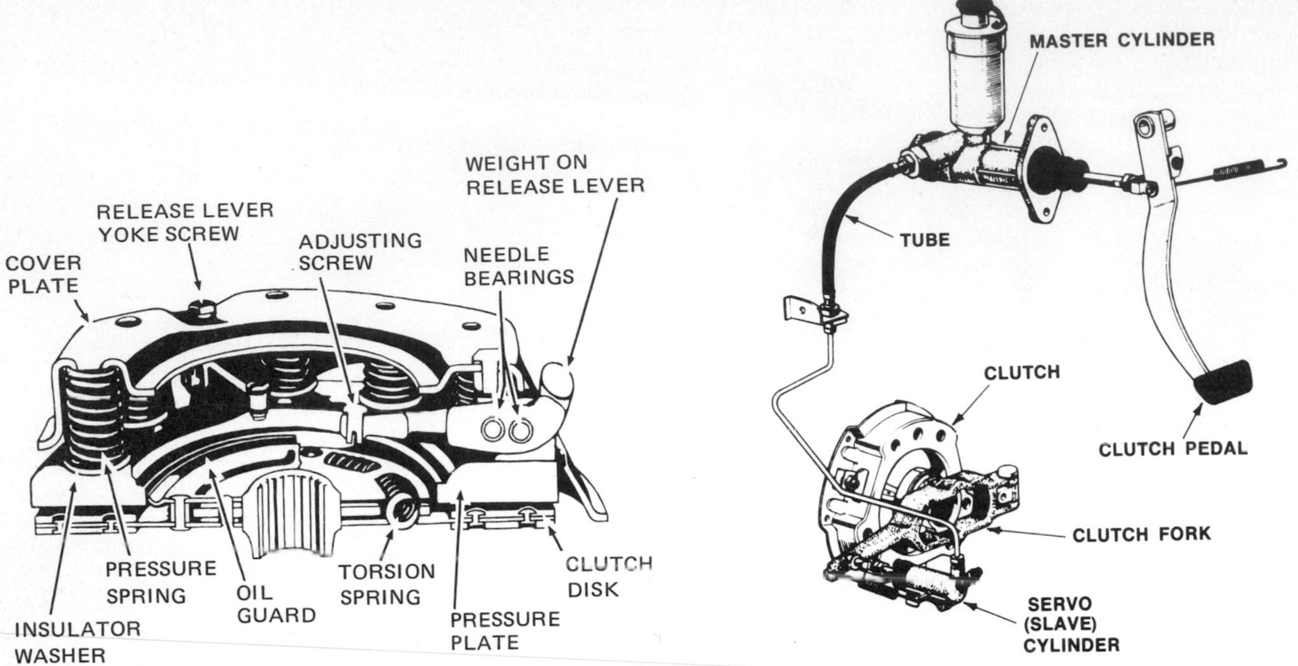

Fig. 2-21. Cutaway view of a centrifugal clutch.

Fig. 2-23. Hydraulically operated coil-spring clutch. (*Toyota Motor Sales, Limited*)

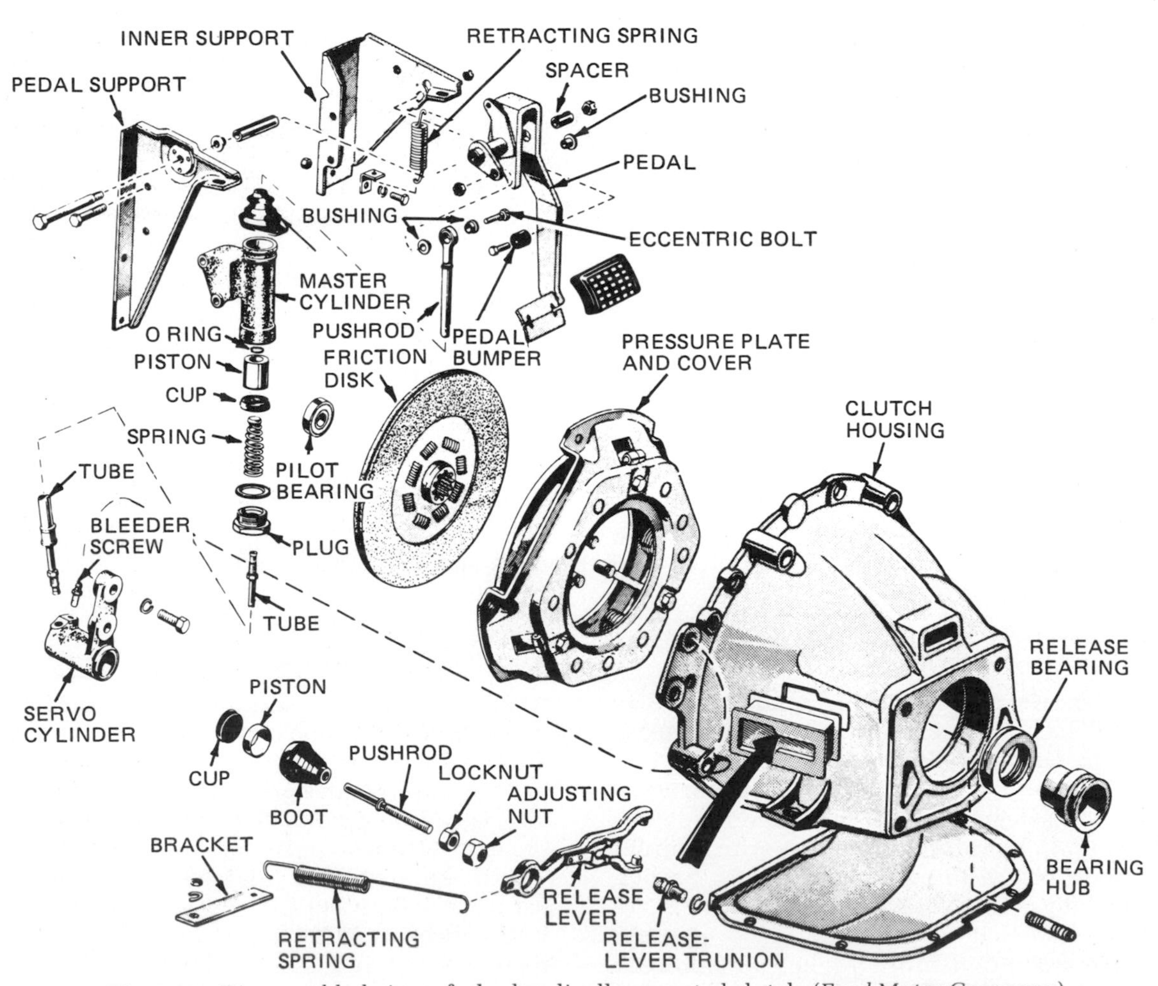

Fig. 2-22. Disassembled view of a hydraulically operated clutch. (*Ford Motor Company*)

⊘ **2-11 Clutch Safety Switch** Late-model cars have a clutch safety switch that prevents starting if the clutch is engaged. That is, the clutch pedal must be depressed at the same time that the ignition switch is turned to START. The movement of the clutch pedal closes the safety switch so that the circuit to the starting motor can be completed. The purpose of the switch is to prevent starting with the transmission in gear and the clutch engaged. If this happened, the engine might start and the car might move before the driver is ready. And that could lead to accidents.

CHAPTER 2 CHECKUP

NOTE: Since the following is a chapter review test, you should review the chapter before taking it.

Again, you will want to test your knowledge of the chapter you have just completed and find out how well you remember what you have been studying. The questions that follow have two purposes: One is to test your knowledge. The other is to help you review the chapter and fix the facts firmly in your mind. It may be that you will not be able to answer all the questions offhand. If this happens, turn back into the chapter and reread the pages that will give you the answers. For instance, under "Listing Parts" you are asked to list parts in clutches. If you cannot remember them all, turn back to the illustrations and refer to them as you make your list. The act of writing the names of the parts will help you to remember them.

NOTE: Write your answers in your notebook. Then later, when you finish the book, you will find your notebook filled with valuable information to which you can refer quickly.

Completing the Sentences The sentences that follow are incomplete. After each sentence there are several words or phrases, only one of which will correctly complete the sentence. Write each sentence in your notebook, selecting the proper word or phrase to complete it correctly.

1. The friction disk is splined to a shaft which extends into the: (*a*) differential, (*b*) drive shaft, (*c*) transmission, (*d*) engine.
2. The friction disk is positioned between the flywheel and the: (*a*) engine, (*b*) crankshaft, (*c*) pressure plate.
3. When the clutch is engaged, spring pressure clamps the friction disk between the pressure plate and the: (*a*) flywheel, (*b*) differential, (*c*) reaction plate, (*d*) clutch pedal.
4. Depressing the clutch pedal causes the throw-out bearing to move in and force the pressure plate to release its pressure on the: (*a*) throw-in bearing, (*b*) pressure springs, (*c*) friction disk.
5. The clutch cover is bolted to the: (*a*) friction disk, (*b*) flywheel, (*c*) car frame, (*d*) engine block.
6. In the typical friction disk, the cushioning device consists of a series of waved: (*a*) cushion pads, (*b*) cushion bolts, (*c*) cushion springs, (*d*) disks.
7. The release levers in the typical clutch pivot on: (*a*) springs, (*b*) levers, (*c*) threaded bolts, (*d*) pins.
8. In the typical friction disk, torsional vibration is absorbed by the use of a series of heavy: (*a*) cushion bolts, (*b*) waved pads, (*c*) coil springs.
9. In the diaphragm-spring clutch, inward movement of the throw-out bearing causes the diaphragm spring to: (*a*) dish inward, (*b*) expand, (*c*) contract.
10. In the semicentrifugal clutch, some pressure of the pressure plate against the friction disk increases with speed because of weights located on the: (*a*) pressure plate, (*b*) flywheel, (*c*) release levers, (*d*) clutch shaft.

Listing Parts In the following, you are asked to list the parts that go into various automotive clutches. Write these lists in your notebook.

1. List the parts used in typical clutch-pedal-to-clutch-linkage systems.
2. List the parts used in a nine-spring clutch.
3. List the parts used in a diaphragm-spring clutch.

Purpose and Operation of Components In the following, you are asked to write the purpose and operation of the clutches discussed in the chapter. If you have any difficulty in writing your explanation, turn back into the chapter and reread the pages that will give you the answer. Then write your explanation. Don't copy; try to tell it in your own words. This is a good way to fix the explanations more firmly in your mind. Write in your notebook.

1. What is the purpose of the clutch?
2. Explain how the three- or nine-spring clutch operates.
3. Explain how the diaphragm-spring clutch operates.
4. Explain how a semicentrifugal clutch operates.
5. Describe the construction of a typical friction disk and explain the purpose of the coil and the waved cushion springs.

SUGGESTIONS FOR FURTHER STUDY

In your school automotive shop or in a service garage, examine various clutches and clutch parts. Notice how these parts are constructed and how they go together. Observe the methods of linking the clutch pedal with the clutch on various cars.

Study whatever shop manuals you can find, and note carefully manufacturers' explanations of the operation and construction of various clutches. Be sure to write in your notebook any important facts that you learn.

chapter 3

CLUTCH SERVICE

This chapter describes the trouble-diagnosis, removal, overhaul, adjustment, reassembly, and installation procedures of various types of clutches used on passenger cars.

⊘ 3-1 Clutch Trouble-Diagnosis Chart Several types of clutch troubles may be experienced. Usually, the trouble itself is fairly obvious and falls into one of the following categories: slipping, chattering or grabbing when engaging, spinning or dragging when disengaged, clutch noises, clutch-pedal pulsations, and rapid friction-disk-facing wear. The chart that follows lists possible causes of each of these troubles and gives the numbers of the sections that explain more fully the ways to locate and eliminate the troubles.

NOTE: The complaints and possible causes are not listed in the chart in the order of frequency of occurrence. That is, item 1 (or item a) does not necessarily occur more often than item 2 (or item b).

CLUTCH TROUBLE-DIAGNOSIS CHART

(See ⊘ 3-2 to 3-9 for detailed explanations of the trouble causes and corrections listed below.)

COMPLAINT	POSSIBLE CAUSE	CHECK OR CORRECTION
1. Clutch slips while engaged (⊘ 3-2)	a. Incorrect pedal-linkage adjustment	Readjust
	b. Broken or weak pressure springs	Replace
	c. Binding in clutch-release linkage	Free, adjust
	d. Broken engine mount	Replace
	e. Worn friction-disk facings	Replace facings or disk
	f. Grease or oil on disk facings	Replace facings or disk
	g. Incorrectly adjusted release levers	Readjust
2. Clutch chatters or grabs when engaged (⊘ 3-3)	a. Binding in clutch-release linkage	Free; adjust
	b. Broken engine mount	Replace
	c. Oil or grease on disk facings or glazed or loose facings	Replace facings or disk
	d. Binding of friction-disk hub on clutch shaft	Clean and lubricate splines; replace defective parts
	e. Broken disk facings, springs, or pressure plate	Replace broken parts
3. Clutch spins or drags when disengaged (⊘ 3-4)	a. Incorrect pedal-linkage adjustment	Readjust
	b. Warped friction disk or pressure plate	Replace defective part
	c. Loose friction-disk facing	Replace defective part
	d. Improper release-lever adjustment	Readjust
	e. Friction-disk hub binding on clutch shaft	Clean and lubricate splines; replace defective parts
	f. Broken engine mount	Replace
4. Clutch noises with clutch engaged (⊘ 3-5)	a. Friction-disk hub loose on clutch shaft	Replace worn parts
	b. Friction-disk dampener springs broken or weak	Replace disk
	c. Misalignment of engine and transmission	Realign

CLUTCH TROUBLE-DIAGNOSIS CHART *(Continued)*

COMPLAINT	POSSIBLE CAUSE	CHECK OR CORRECTION
5. Clutch noises with clutch disengaged (⊘ 3-5)	a. Clutch throw-out bearing worn, binding, or out of lubricant	Lubricate or replace
	b. Release levers not properly adjusted	Readjust, replace assembly
	c. Pilot bearing in crankshaft worn or out of lubricant	Lubricate or replace
	d. Retracting spring (diaphragm-spring clutch) worn	Replace
6. Clutch-pedal pulsations (⊘ 3-6)	a. Engine and transmission not aligned	Realign
	b. Flywheel not seated on crankshaft flange or flange or flywheel bent (also causes engine vibration)	Seat properly, straighten, replace flywheel
	c. Clutch housing distorted	Realign or replace
	d. Release levers not evenly adjusted	Readjust or replace assembly
	e. Warped pressure plate or friction disk	Realign or replace
	f. Pressure-plate assembly misaligned	Realign
7. Rapid friction-disk-facing wear (⊘ 3-7)	a. Driver "rides" clutch	Keep foot off clutch except when necessary
	b. Excessive and incorrect use of clutch	Reduce use
	c. Cracks in flywheel or pressure-plate face	Replace
	d. Weak or broken pressure springs	Replace
	e. Warped pressure plate or friction disk	Replace defective part
	f. Improper pedal-linkage adjustment	Readjust
	g. Clutch-release linkage binding	Free; adjust
8. Clutch pedal stiff (⊘ 3-8)	a. Clutch linkage lacks lubricant	Lubricate
	b. Clutch-pedal shaft binds in floor mat	Free
	c. Misaligned linkage parts	Realign
	d. Overcenter spring out of adjustment	Readjust
9. Hydraulic-clutch troubles (⊘ 3-9)	a. Hydraulic clutches can have any of the troubles listed elsewhere in this chart	Inspect the hydraulic system; check for leakage
	b. Gear clashing and difficulty in shifting into or out of gear	

⊘ 3-2 Clutch Slips While Engaged Clutch slippage is extremely hard on the clutch facings and mating surfaces of the flywheel and pressure plate. The slipping clutch generates considerable heat. The clutch facings wear rapidly and may char and burn. The flywheel face and pressure plate wear. They may groove, crack, and score. The heat in the pressure plate can cause the springs to lose their tension, which makes the situation worse.

Clutch slippage is very noticeable during acceleration, especially from a standing start or in low gear. A rough test for clutch slippage can be made by starting the engine, setting the hand brake, and shifting into high gear. Then slowly release the clutch while accelerating the engine slowly. If the clutch is in good condition, it should hold so that the engine stalls immediately after clutch engagement is completed. The dynamometer can also be used to detect a slipping clutch. Connect a tachometer to read engine rpm. Run the vehicle at intermediate speed at part throttle. Note the engine rpm and speedometer reading. Then push the accelerator all the way down, using the dynamometer to load the engine while opening the throttle. Any increase in engine rpm at the same vehicle speed is clutch slippage.

Several conditions can cause clutch slippage. The pedal linkage may be incorrectly adjusted. If the incorrect adjustment reduces pedal lash too much, the throw-out bearing may be up against the release fingers even with a fully released pedal. This condition can take up part of the spring pressure, so the pressure plate is not locking the friction disk to the flywheel. The remedy for this problem is to readjust the linkage.

Binding linkage or a broken return spring may prevent full return of the linkage to the engaged position. Replace the spring if it is broken. Lubricate the linkage. Much of the clutch linkage is pivoted in nylon or neoprene bushings. These should be lubricated with silicone spray, SAE10 oil, or multipurpose grease, depending on the manufacturer's recommendation.

NOTE: If the linkage is not at fault, the slippage could be caused by a broken engine mount. This could allow the engine to shift enough to prevent good clutch engagement. The remedy here is to replace the mount.

If none of the above is causing slipping, then the clutch should be removed for service. Conditions in the clutch that could cause slipping include worn friction-disk facings, broken or weak pressure-plate or diaphragm springs, grease or oil on the disk facings, or incorrectly adjusted release levers.

The recommendation of most manufacturers is to replace the disk and pressure-plate assembly if there is internal wear or damage or weak springs. Pressure-plate assemblies can be rebuilt, but this is a job for the clutch-rebuilding shop.

NOTE: One clue to a slipping clutch is metal and facing material in the clutch housing. This condition can be detected by removing the inspection cover from under the clutch and flywheel.

CAUTION: If the clutch disk and pressure-plate assembly are replaced, the flywheel should be inspected carefully for damage—wear, cracks, grooves, and checks. Any of these conditions, if well advanced, will require replacement of the flywheel. Putting a new disk facing against a damaged flywheel will lead to rapid facing wear.

⊘ 3-3 Clutch Chatters or Grabs When Engaged The cause of clutch chattering is most likely inside the clutch. The clutch should be removed for service or replacement. Before this is done, however, check the clutch linkage to make sure it is not binding. If it binds, it could release suddenly to throw the clutch into quick engagement, with a resulting heavy jerk.

A broken engine mount can also cause the problem. The engine is free to move excessively, and this can cause the clutch to grab or chatter when engaged. The remedy is to replace the mount.

Inside the clutch, the trouble could be due to oil or grease on the disk facings or to glazed or loose facings. If this is the case, the facings or disk should be replaced. The trouble could also be due to binding of the friction-disk hub on the clutch shaft. This condition requires cleaning and lubrication of the splines in the hub and on the shaft.

NOTE: Clutch chatter after removal and installation of an engine may be caused by a misaligned clutch housing. Some clutch housings have small shims that can be lost during engine or clutch-housing removal. These shims must be replaced in the same positions to ensure housing alignment. It is also possible for dirt to get between the clutch housing and cylinder block, or either could be nicked or burred. Any of these conditions can throw off the housing alignment.

Other clutch problems—glazed or loose facings, oil or grease on the facings—require disk and pressure-plate replacement.

⊘ 3-4 Clutch Spins or Drags When Disengaged The clutch friction disk spins briefly after disengagement when the transmission is in neutral. This normal spinning should not be confused with a dragging clutch. When the clutch drags, the friction disk is not releasing fully from the flywheel or pressure plate as the clutch pedal is depressed. Therefore, the friction disk continues to rotate with or rub against the flywheel or pressure plate. The common complaint of drivers is that they have trouble shifting into gear without clashing; the dragging disk keeps the transmission rotating.

The first thing to check with this condition is the pedal-linkage adjustment. If there is excessive pedal lash, or free travel, even full movement of the pedal will not release the clutch fully. If linkage adjustment does not correct the problem, the trouble is in the clutch.

Internal clutch troubles could be due to a warped friction disk or pressure plate or loose friction-disk facing. One cause of loose friction-disk facings is abuse of the clutch. This abuse includes "popping" the clutch for a quick getaway (letting the clutch out suddenly with the engine turning at high rpm), slipping the clutch for drag-strip starts, and increasing engine power output ("souping up" the engine).

The release levers may be incorrectly adjusted, and so they do not fully disengage the clutch. Also, the friction-disk hub may be binding on the clutch shaft. This condition may be corrected by cleaning and lubricating the splines.

NOTE: A broken engine mount can also cause clutch spinning or dragging. The engine is free to move excessively, which can cause the clutch to spin or drag when disengaged. The remedy is to replace the mount.

⊘ 3-5 Clutch Noises Clutch noises are usually most noticeable when the engine is idling. To determine the cause, note whether the noise is heard when the clutch is engaged, when it is disengaged, or during pedal movement to engage or disengage the clutch.

Noises heard while the pedal is in motion are probably due to dry or dirty linkage pivot points. Clean and lubricate them as already noted in ⊘ 3-2.

Noises heard in neutral that disappear when the pedal is depressed are transmission noises. (These noises could also be due to a dry or worn pilot bushing in the crankshaft.) They are usually rough-bearing sounds. The cause is worn transmission bearings, sometimes caused by clutch-popping and shifting gears too fast. These conditions throw an extra load on the transmission bearings, as well as on the gears.

Noises heard while the clutch is engaged could be due to a friction-disk hub that is loose on the clutch shaft. This condition requires replacement of the disk or clutch shaft, or perhaps both if both are excessively worn. Friction-disk dampener springs that are broken or weak will cause noise. This condition requires replacement of the complete disk. Misalignment of the engine and transmission will cause a backward-and-forward movement of the friction disk on the clutch shaft. The alignment must be corrected.

Noises heard while the clutch is disengaged could be due to a clutch throw-out bearing that is worn, binding, or has lost its lubricant. Such a bearing squeals when the clutch pedal is depressed and the bearing comes into operation. The bearing should be lubricated or replaced. If the release levers are not properly adjusted, they will rub against the friction-disk hub when the clutch pedal is depressed. The release levers should be readjusted. If the pilot bearing in the crankshaft is worn or lacks lubricant, it will produce a high-pitched whine when the transmission is in gear, the clutch is disengaged, and the car is stationary. Under these conditions, the clutch shaft (which is piloted in the bearing in the crankshaft) is stationary, but the crankshaft and bearing are turning. The bearing should be lubricated or replaced.

In the diaphragm-spring clutch, worn or weak retracting springs will cause a rattling noise when the clutch is disengaged and the engine is idling. Eliminate the noise by replacing the springs without removing the clutch from the engine.

⊘ 3-6 Clutch-Pedal Pulsations Clutch-pedal pulsations are noticeable when a slight pressure is applied to the clutch pedal with the engine running. The pulsations can be felt by the foot as a series of slight pedal movements. As pedal pressure is increased, the pulsations cease. This condition often indicates trouble that must be corrected before serious damage to the clutch results.

One possible cause is misalignment of the engine and transmission. If the two are not in line, the friction disk or other clutch parts will move back and forth with every revolution. The result will be rapid wear of clutch parts. Correction is to detach the transmission, remove the clutch, and then check the housing alignment with the engine and crankshaft. At the same time, the flywheel can be checked for wobble. A flywheel that is not seated on the crankshaft flange will also produce clutch-pedal pulsations. The flywheel should be removed and remounted to make sure that it seats evenly.

If the clutch housing is distorted or shifted so that alignment between the engine and transmission has been lost, it is sometimes possible to restore alignment. This is done by installing shims between the housing and engine block and between the housing and transmission case. Otherwise, a new clutch housing will be required.

NOTE: These causes of clutch-pedal pulsation—bent flywheel, flywheel not seated on the crankshaft flange, and housing misalignment—are conditions that usually would not arise during normal operation. Most likely they would result from faulty reassembly after a service job.

Another cause of clutch-pedal pulsations is uneven release-lever adjustment (so that release levers do not meet the throw-out bearing and pressure plate together). Release levers of the adjustable type should be readjusted. Still another cause is a warped friction disk or pressure plate. A warped friction disk must be replaced. If the pressure plate is out of line because of a distorted clutch cover, the cover sometimes can be straightened to restore alignment.

In the diaphragm-spring clutch, a broken diaphragm will cause clutch-pedal pulsations. The clue here is that the pulsations develop suddenly, just as the diaphragm breaks.

⊘ 3-7 Rapid Friction-Disk Facing Wear Rapid wear of the friction-disk facings is caused by slippage between the facings and the flywheel or pressure plate. Thus, if the driver has the habit of "riding" the clutch (that is, keeping the foot resting on the clutch), part of the pressure-plate-spring pressure will be taken up so that slippage may take place. Likewise, frequent use of the clutch, incorrect clutching and declutching, overloading the clutch, and slow clutch release increase clutch-facing wear. Speed or "snap" gear shifting, increasing engine output ("souping up"), and drag-strip starts shorten clutch life. Also, the installation of wide oversize tires increases the clutch load. (Some manufacturers will not warranty the clutch if oversize tires are installed.)

Rapid facing wear after installation of a new friction disk can be caused by heat checks and cracks in the flywheel and pressure-plate faces. The sharp edges act like tiny knives. They shave off a little of the facing during each engagement. This is the reason why we mentioned, in ⊘ 3-2, that when a friction disk is replaced, the pressure-plate assembly should also be replaced. In addition, the flywheel face should be inspected, and if it is damaged, the flywheel should be replaced.

Several conditions in the clutch itself can cause rapid friction-disk-facing wear. For example, weak or broken pressure springs will cause slippage and facing wear. In this case, the springs must be replaced. If the pressure plate or friction disk is warped or out of line, it must be replaced or realigned. In addition, an improper pedal-linkage adjustment or binding of the linkage may prevent full spring pressure from being applied to the friction disk. With less than full spring pressure, slippage and wear may take place. The linkage must be readjusted and lubricated at all points of friction.

⊘ 3-8 Clutch Pedal Stiff A clutch pedal that is stiff or hard to depress is likely to result from lack of lubricant in the clutch linkage, from binding of the clutch-pedal shaft in the floor mat, or from misaligned linkage parts that are binding. In addition, the overcenter spring (on cars so equipped) may be out of adjustment or broken. Also, if the clutch pedal has been bent so that it rubs on the floorboard, it may not operate easily. The remedy for each of these troubles is obvious. Parts must be realigned, lubricated, or readjusted as necessary.

⊘ 3-9 Hydraulic-Clutch Troubles The hydraulic clutch can have any of the troubles described previously plus several in the hydraulic system. These special troubles include gear clashing and difficulty in shifting into or out of gear. The cause is usually loss of fluid from the hydraulic system. Such loss prevents the system from completely declutching for gear shifting. The hydraulic system should be checked and serviced in the same way as the hydraulic system in hydraulic brakes. Leaks may be in the master or servo cylinder or in the line or connections between the two. Hydraulic brakes and their possible troubles are discussed in *Automotive Chassis and Body*, another book in the McGraw-Hill Automotive Technology Series.

Check Your Progress

Progress Quiz 3-1 The material on clutch trouble-diagnosis that you have just been studying is not the easiest material to study. The listings of clutch troubles, along with the brief descriptions of them, supply a great many facts about clutch troubles, and all these facts are not easy to remember. At the same time, it is important to try to remember them because knowing these facts will help you when you go into the automotive shop. The following quiz will help you find out how well you remember the material you have just read on clutch trouble diagnosis. It will also help you review the material so that important points will be firmly fixed in your mind. If any of the questions seem hard to answer, reread the pages that will give you the answer.

Correcting Troubles Lists The purpose of this exercise is to help you spot related and unrelated troubles on a list. For example, in the list "clutch slips: incorrect pedal-linkage adjustment, broken pressure springs, engine misses, grease on facings," you can see that "engine misses" does not belong because it is the only condition that is not directly related to clutch slippage. In each of the following lists, you will find one item that does not belong. Write each list in your notebook, but do not write the item that does not belong.

1. Clutch pedal stiff: clutch linkage binding, misaligned linkage parts, overcenter spring out of adjustment, pressure springs weak
2. Rapid friction-disk-facing wear: driver rides clutch, warped pressure plate or friction disk, clutch-release linkage binding, weak or broken pressure springs, high-octane gas, improper pedal-linkage adjustment
3. Clutch-pedal pulsations: engine and transmission out of line, pressure plate out of line, warped pressure plate or friction disk, release levers not evenly adjusted, clutch housing distorted, differential out of line, flywheel not seated on crankshaft flange, crankshaft flange bent
4. Clutch noises with clutch disengaged: pilot bearing in crankshaft worn or out of lubricant, release levers not properly adjusted, gears worn, throw-out bearing worn or dry
5. Clutch noises with clutch engaged: friction-disk hub loose on clutch shaft, friction-disk dampener springs broken or weak, engine and transmission out of line, drive shaft out of line
6. Clutch spins when disengaged: incorrect pedal-linkage adjustment, warped friction disk or pressure plate, loose friction-disk facing, improper release-lever adjustment, spring weak, friction disk binding
7. Clutch chatters or grabs when engaged: oil or grease on disk facing, binding in clutch-release mechanism, friction-disk hub binding on clutch shaft, loose floating axle, broken disk facings or springs
8. Clutch slips while engaged: incorrect pedal linkage adjustment, broken or weak pressure springs, worn friction-disk facings, loose friction-disk facings, binding in release mechanism, grease or oil on facings, release levers incorrectly adjusted

⊘ 3-10 Clutch Service The major clutch services include clutch-linkage adjustment, clutch removal and replacement, and clutch disassembly, inspection, adjustment, and reassembly. It is important to remember that if a clutch defect develops, you must do more than just replace a worn part. You must determine what caused the worn part and fix the trouble so that the new part will not wear rapidly. For example, if you install a new friction disk between a cracked or worn flywheel and pressure plate, the new friction disk will not last long. In ⊘ 3-7 we explained what would happen to the new friction disk.

One of the most common causes of rapid disk-lining wear and clutch failure is improper pedal lash, or free play. If pedal lash is not sufficient, the clutch will not apply completely. It will slip and wear rapidly. In addition, the throw-out bearing will be operating continuously and will soon wear out.

⊘ 3-11 Clutch-Linkage Adjustment Clutch-linkage adjustment may be required from time to time to compensate for friction-disk-facing wear. In addition, certain points in the linkage or pedal support may require lubrication. The adjustment of the linkage changes the amount of clutch-pedal "free" travel (or pedal *lash*, as it is also called). The free travel of the pedal is the amount of travel that the pedal has before the throw-out bearing comes up against the release levers in the clutch. After this happens, there is a definite increase in the amount of pressure required for further pedal movement; pedal movement from this point on causes release-lever movement and contraction of the clutch-pressure-plate springs. In normal operation, free travel is lost; it is never gained. The test of pedal free travel should be made with a finger rather than a foot or hand. The finger can detect the increase of pressure more accurately than the hand or foot. Chevrolet suggests the following procedure to check for clutch slippage:

1. Drive in high gear at 20 to 25 mph [32.187 to 40.234 km/h].
2. Depress the clutch pedal to the floor and increase engine speed to 2,500 to 3,500 rpm.
3. Snap foot off the clutch pedal to engage the clutch quickly and at the same time push the throttle all the way down. Engine speed should drop noticeably for a moment as the clutch takes hold, but then the car should accelerate. If the engine speed increases when you simultaneously engage the clutch and open the throttle, the clutch is slipping.

CAUTION: Do not repeat this procedure more than once or you will overheat the clutch. Note that a clutch that has been slipping prior to readjusting the linkages may continue to slip after correct readjustment due to heat damage. Let the clutch cool for at least 12 hours and then repeat the test. If slippage still occurs, there may be enough damage inside the clutch to require removal and overhaul of the clutch.

1. *CHEVROLET* Clutch-linkage adjustments for late-model Chevrolets are shown in Figs. 3-1 and 3-2. To adjust the Corvette (Fig. 3-1), disconnect spring E. Loosen nuts A and B enough to cause the release bearing to touch the pressure-plate fingers. Then rotate the upper nut B until it is tight against the swivel C. Back the nut off $4\frac{1}{2}$ turns. Tighten lower nut A against the swivel. Install spring E. Check pedal free travel and compare it with specifications in the car shop manual.

For V-8 and L-6 engines (Fig. 3-2), disconnect the return spring at the clutch fork. Rotate the clutch lever and shaft until the clutch pedal is pressed firmly against the rubber bumper on the dash brace. Push the outer end of the clutch fork rearward until the throw-out bearing lightly touches the release levers. Then install the swivel or pushrod in the

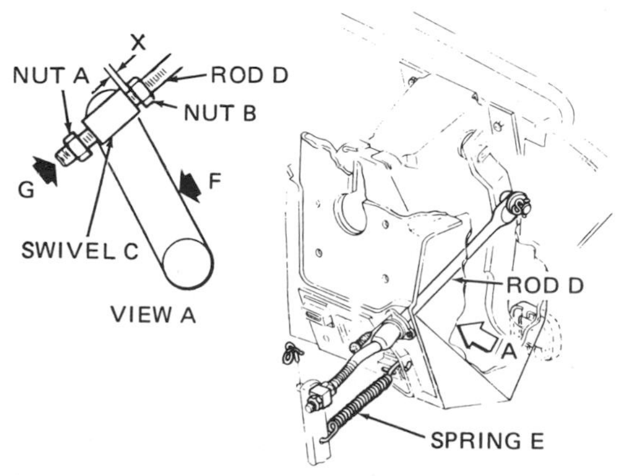

Fig. 3-1. Clutch-pedal free-travel (clutch-linkage) adjustment on the Corvette. (*Chevrolet Motor Division of General Motors Corporation*)

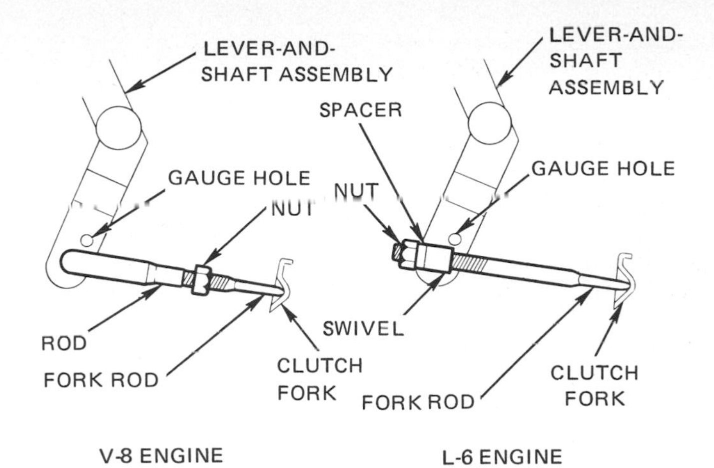

Fig. 3-2. Clutch-pedal free-travel (clutch-linkage) adjustment on Chevrolet cars. (*Chevrolet Motor Division of General Motors Corporation*)

gauge hole and increase its length until all lash is removed from the system. Remove the swivel or pushrod from the gauge hole and insert it into the lower hole on the lever. Install the retainer and tighten the locknut. Be careful not to change the adjustment of the rod length. Reinstall the return spring and check pedal free travel. Specifications vary on different Chevrolet models from a low of $\frac{3}{4}$ inch [19.05 mm (millimeters)] to a maximum of $1\frac{3}{8}$ inch [34.93 mm].

2. *FORD* Figure 3-3 shows the clutch-linkage arrangement for some late-model Ford cars. To adjust clutch-pedal free travel, disconnect the clutch-return spring from the release lever and loosen the release-lever-rod locknut and adjusting nut. Move the clutch yoke (release lever) to the rear until the throw-out (release) bearing lightly touches the clutch-release fingers (levers). Adjust the rod length until it seats in the yoke pocket.

Insert the specified feeler gauge between the adjusting nut and swivel sleeve. Then tighten the adjusting nut against the gauge, finger tight. Tighten the locknut against the adjusting nut. Be careful not to change the adjustment. Torque the locknut to specifications. Remove the feeler gauge and install the clutch-return spring. Repeatedly operate the clutch at least five times. Then check the clutch-pedal free travel and adjust again if necessary.

As a final check, measure the clutch-pedal free travel with the transmission in neutral and the engine turning at about 3,000 rpm. Free travel should be at least $\frac{1}{2}$ inch [12.7 mm]. If it is less, readjust.

3. *PLYMOUTH* Figure 3-4 shows assembled and disassembled views of some Plymouth-model clutch linkages. To adjust, loosen the clutch-rod-swivel-clamp screw (Fig. 3-5). Adjust the fork (yoke) rod by turning the self-locking adjustment nut (Fig. 3-6) to provide $\frac{5}{32}$ inch [3.97 mm] free movement at the end of the fork. This movement will provide the specified 1 inch [25.4 mm] of clutch-pedal free travel. Then adjust the interlock rod (Fig. 3-5) by first disconnecting the clutch-rod swivel from the interlock

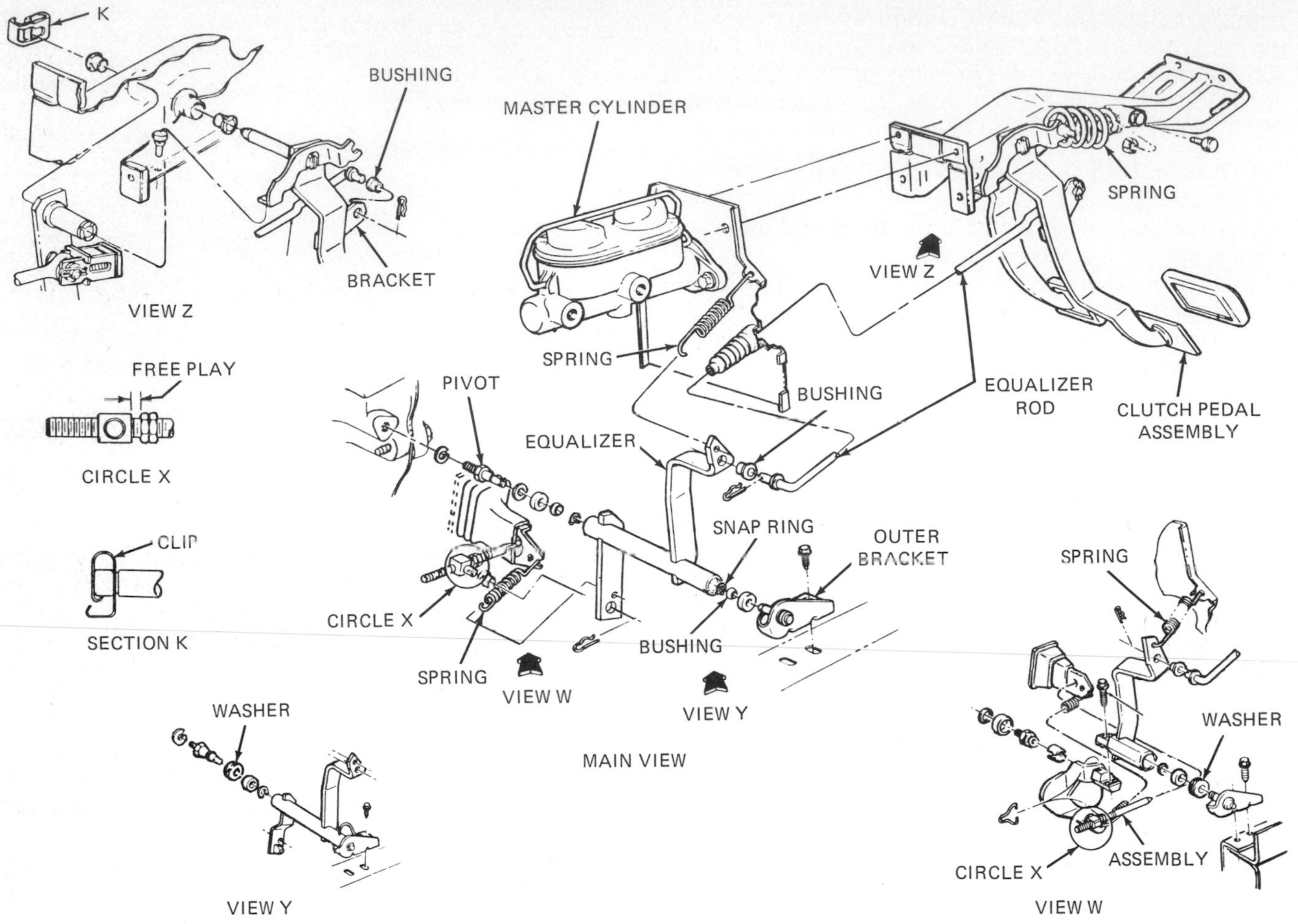

Fig. 3-3. Clutch-pedal and linkage adjustments on late-model Fords. (*Ford Motor Company*)

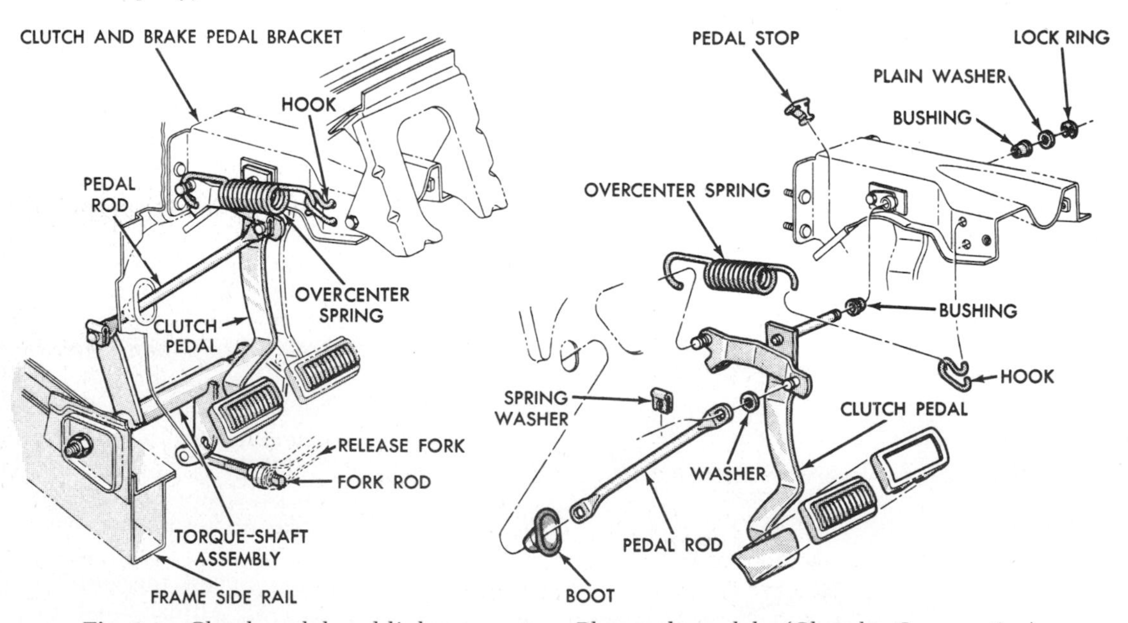

Fig. 3-4. Clutch pedal and linkage on some Plymouth models. (*Chrysler Corporation*)

pawl. Now, with the first-reverse lever on the transmission in neutral, the interlock pawl should enter the slot in the first-reverse lever. Slide the swivel onto the clutch rod so that the swivel enters the pawl. Hold the interlock pawl firmly into the first-reverse lever and tighten the swivel-clamp screw to specifications. The clutch pedal must be in the full up position.

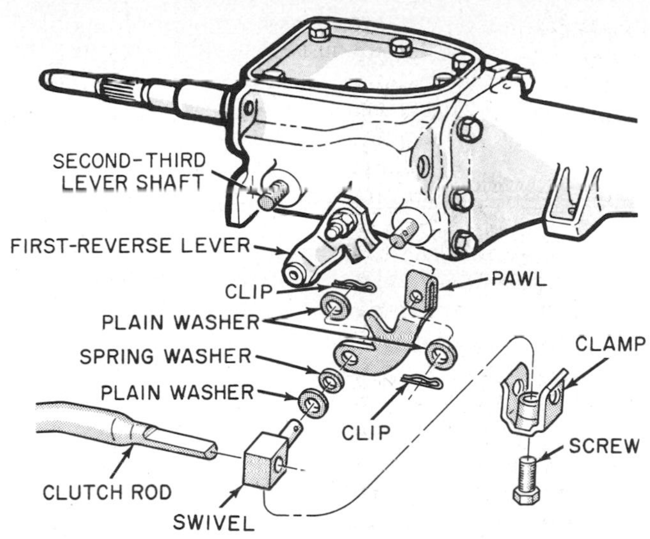

Fig. 3-5. Clutch-linkage adjustment. (*Chrysler Corporation*)

CAUTION: Do not pull the clutch rod rearward to cause the swivel to enter the pawl.

Try shifting the transmission, disengaging and engaging the clutch as you do so. Disengage the clutch and shift halfway to first or reverse. Interlock should hold the clutch pedal down to within 1 to 2 inches [25.4 to 50.8 mm] of the floor.

4. *LUBRICATION* Some specifications call for lubricating the clutch linkage at bushings or connections. These should be lubricated with silicone spray, SAE10 oil, or multipurpose grease, depending on the manufacturer's recommendation.

⊘ 3-12 Clutch Removal and Replacement Variations in construction and design make it necessary to use somewhat different procedures and tools when removing and replacing clutches on different cars. As a first step in clutch removal, the transmission must be removed. Transmission removal and replacement are covered in Chap. 6. One general caution to observe when the transmission is being removed is to pull it straight back from the clutch housing until the clutch shaft is clear of the friction-disk hub. Then the transmission can be lowered from the car. This procedure prevents distortion of and damage to the friction disk. Recommendations usually call for using special long pilot pins installed in place of two of the transmission bolts so that the transmission will maintain alignment as it is moved back.

1. *CLUTCH REMOVAL* On some cars, the engine must be supported by a special support bracket during transmission and clutch removal. With the transmission off, the clutch-housing pan or flywheel lower cover must be removed and the clutch-fork or -yoke linkage disconnected. On some cars, the brake-arm linkage also must be detached. Next, the throw-out bearing (where separate and detached) is taken out.

CAUTION: The cover must be reattached to the flywheel in exactly the same position as on the original assembly. If it is not, balance may be lost, and so vibration and damage will occur. To ensure correct realignment on reinstallation, both the flywheel and the cover are stamped with an X or some similar marking. These markings should align when the clutch is reinstalled on the engine. If you cannot locate the markings, you should carefully mark the clutch cover and flywheel with a hammer and punch (Fig. 3-7) before taking the clutch off so that you can restore the clutch to its original position.

On some cross-shaft-type clutches, the release fork and shaft must be pulled partly out of the clutch housing in order to provide room for the clutch assembly to pass the cross shaft. This can be done after the clutch-release-fork bracket is disconnected at the clutch housing and the release-fork-flange-cap screws are taken out. On other cross-shaft clutches, it is necessary only to detach the cross shaft so that

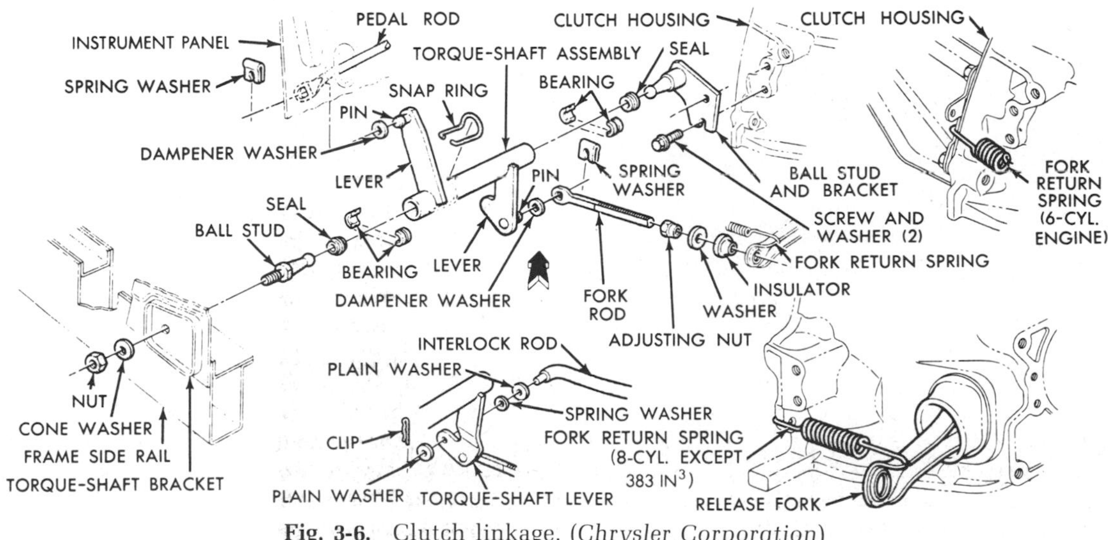

Fig. 3-6. Clutch linkage. (*Chrysler Corporation*)

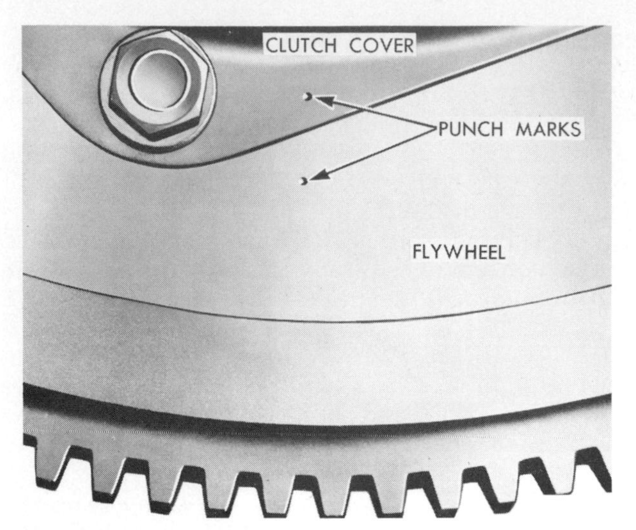

Fig. 3-7. Marking the clutch and flywheel. (*Chrysler Corporation*)

the fork can be swung up out of the way. On the diaphragm-spring clutch or the clutch using a ball stud on which the fork pivots, snap the fork off the ball stud with a screwdriver after removing the locknut from the adjustment link.

Loosen the clutch-attachment bolts one turn at a time so that the clutch cover will be evenly loosened and distortion will not occur. Loosen the bolts evenly until the spring pressure is relieved; then take the bolts out. A tool such as the one shown in Fig. 3-8 is very handy for turning the flywheel to get at the upper bolts. When the spring pressure is relieved and the bolts are out, the clutch can be lowered from the car.

2. CLUTCH REPLACEMENT In general, replacement of the clutch is the reverse of removal. Before the clutch is replaced, the condition of the pilot bearing in the end of the crankshaft should be noted and replacement made if necessary (⊘ 3-14). In addition, the condition of the throw-out bearing, as well as other clutch parts, should be checked and defective parts discarded (⊘ 3-14). Figure 3-9 shows a disassembled view of a clutch and housing, together with reassembly instructions.

FLYWHEEL

FLYWHEEL TURNING TOOL

Fig. 3-8. Using a special tool for turning a flywheel so that the upper-clutch attaching bolts can be reached.

NOTE: The clutch-housing alignment must be checked whenever the clutch is removed for service. A misaligned housing can cause improper clutch release, friction-disk failure, front-transmission-bearing failure, uneven wear of the pilot bushing in the crankshaft, clutch noise, and, in extreme cases, clutch vibration and jumping out of gear on deceleration. See ⊘ 3-15, which describes checking and adjusting clutch-housing alignment.

Turn the flywheel until the X or other marking is at the bottom. Then use the special clutch arbor (friction-disk-aligning tool), as shown in Fig. 3-10, to maintain alignment of the friction disk and pilot bearing in the crankshaft. Or you can use a spare transmission drive pinion (Fig. 3-11). Put the friction disk and clutch cover in place. Turn the cover until the X or other marking on it aligns with the similar marking on the flywheel. Install attachment bolts, turning them down a turn at a time to take up the spring tension gradually and evenly. Use a flywheel turning tool (Fig. 3-8) to make it easier to turn the flywheel and get at the upper bolts.

As a final step in the procedure, after the transmission has been reinstalled and the clutch linkages reattached, check the clutch-pedal free travel and make whatever adjustments are necessary (⊘ 3-11).

⊘ 3-13 Clutch Service Automotive manufacturers supply the various parts that go into the clutch assembly as service items. Thus, if the pressure springs lose tension because of overheating, or if the release-lever bearings wear enough to cause trouble, then the pressure-plate-and-cover assembly can be torn down so that these parts may be replaced. However, since about 1965, the instructions given in the manufacturers' shop manuals are to replace the old assembly with a complete new one. The shop manuals no longer give disassembly-assembly instructions. In today's shop manuals, the maximum disassembly covered is shown in Fig. 3-12.

There is one adjustment that manufacturers' shop manuals describe for coil-spring-type clutches: adjustment of the release levers. This adjustment requires a clutch-gauge plate and a clutch-lever-height gauge (Figs. 3-13 and 3-14). First, place the clutch-gauge plate on a flywheel as shown in Fig. 3-13. Then put the cover assembly on top with the release levers over the machined lands on the gauge. Next, attach the cover assembly to the flywheel. Draw the screws down a turn at a time in rotation to avoid distorting the cover. Depress the release levers several times to seat them. Measure their height with the height gauge as shown in Fig. 3-14.

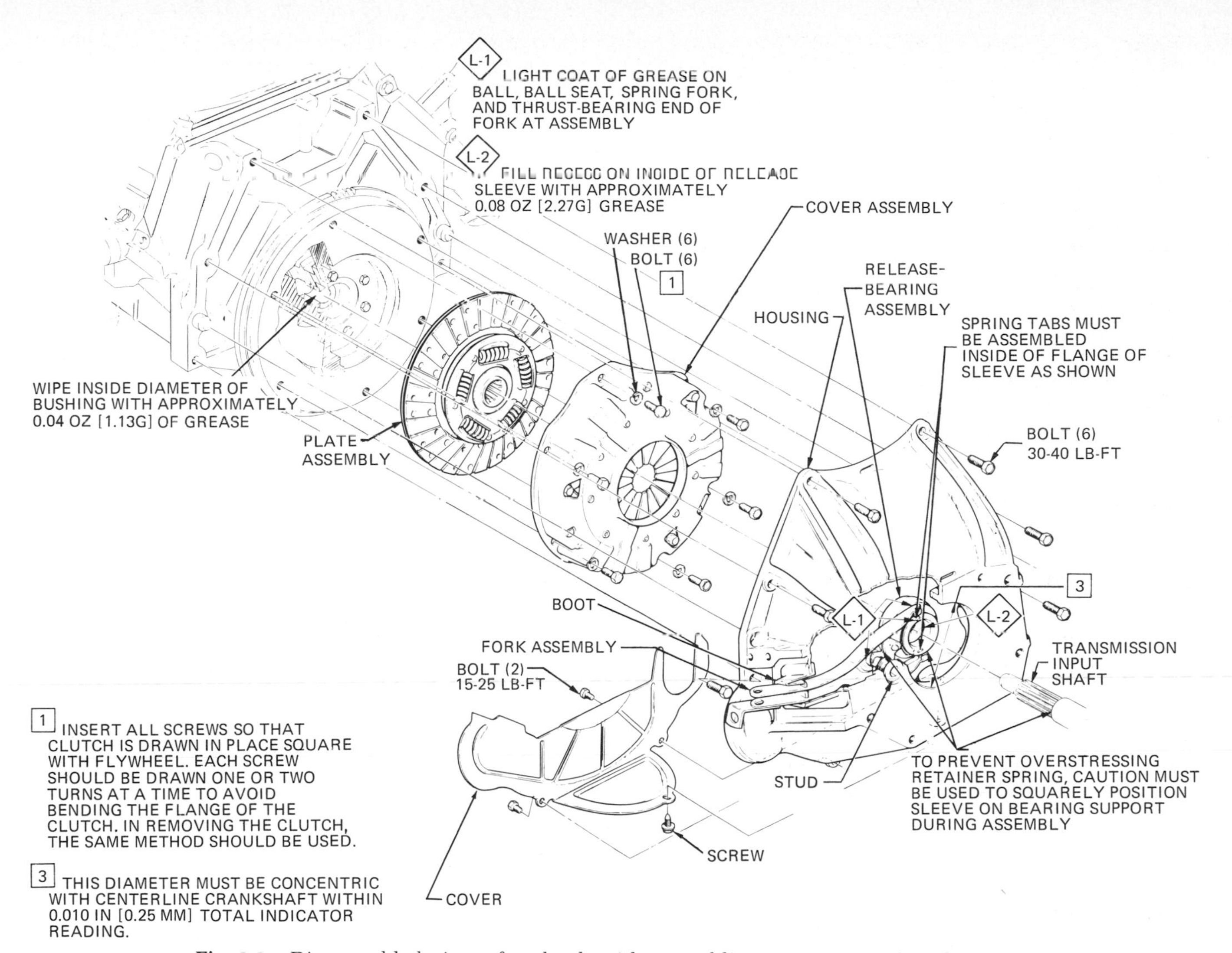

Fig. 3-9. Disassembled view of a clutch with assembling instructions. (*Buick Motor Division of General Motors Corporation*)

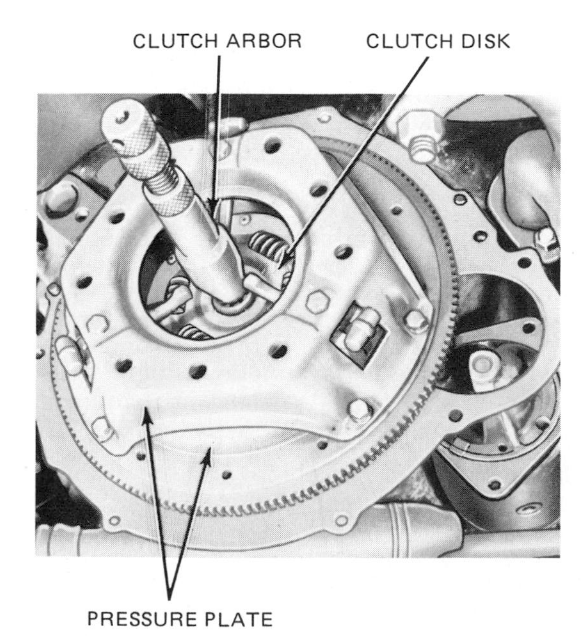

Fig. 3-10. Using a special tool (clutch arbor) to align the clutch during installation. (*Ford Motor Company*)

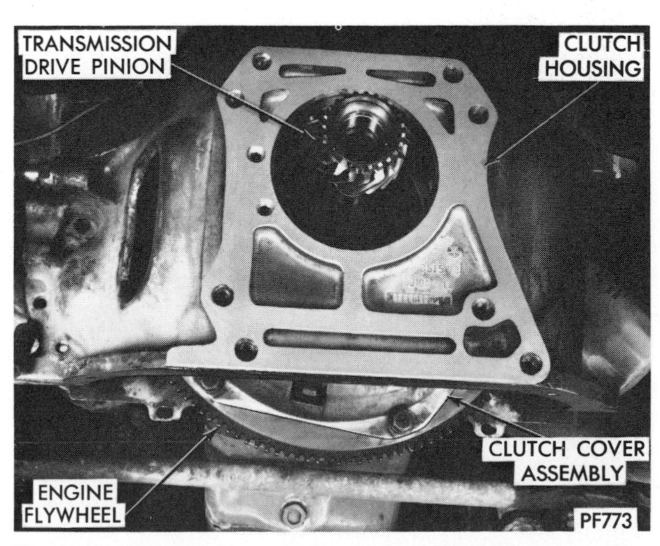

Fig. 3-11. Using a spare transmission drive pinion to align the clutch during installation. (*Chrysler Corporation*)

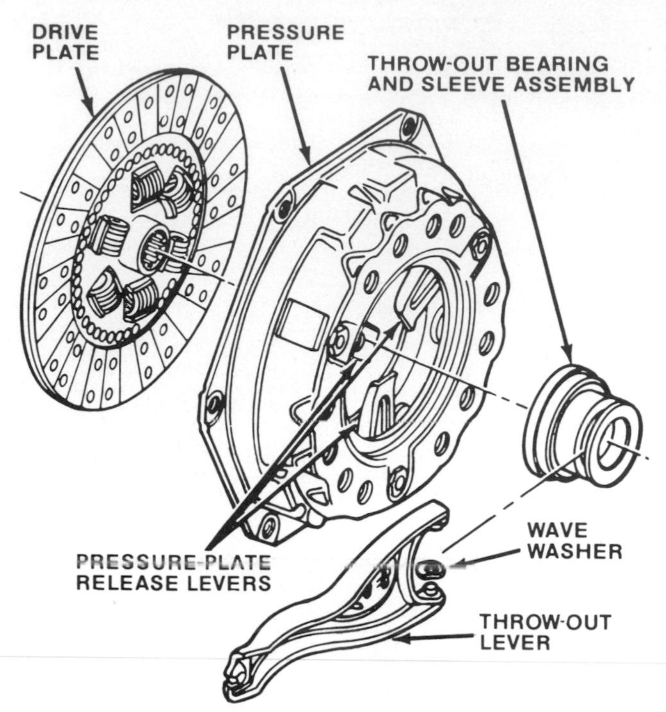

Fig. 3-12. Clutch assembly and related parts. (*American Motors Corporation*)

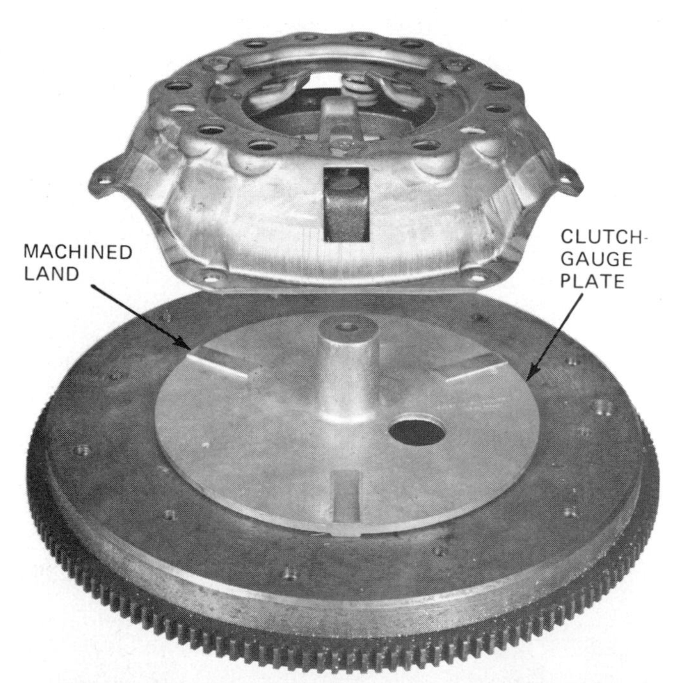

Fig. 3-13. Placing the clutch assembly on a flywheel and clutch-gauge plate to check release-lever adjustments. (*American Motors Corporation*)

Note that the height gauge has four settings that can be used for measuring above and below the hub. Figure 3-15 shows details of the adjustment.

On the indirect-spring pressure-type clutch, remove the release clip, loosen the locknut, and turn the adjusting screw until the lever is at the specified height. Tighten the locknut and recheck. If okay, install the release clip.

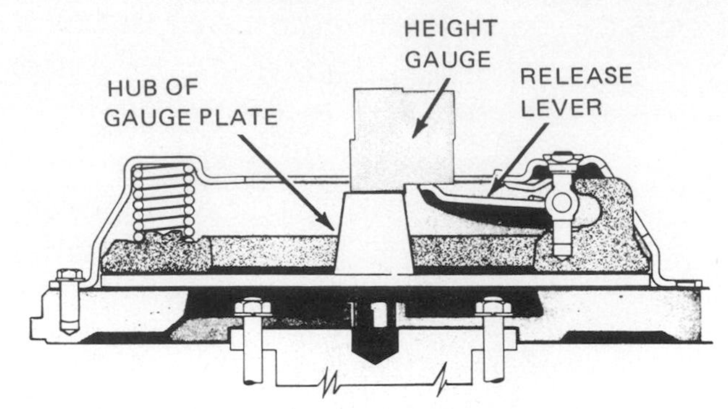

Fig. 3-14. Measuring release-lever adjustment with a height gauge. (*American Motors Corporation*)

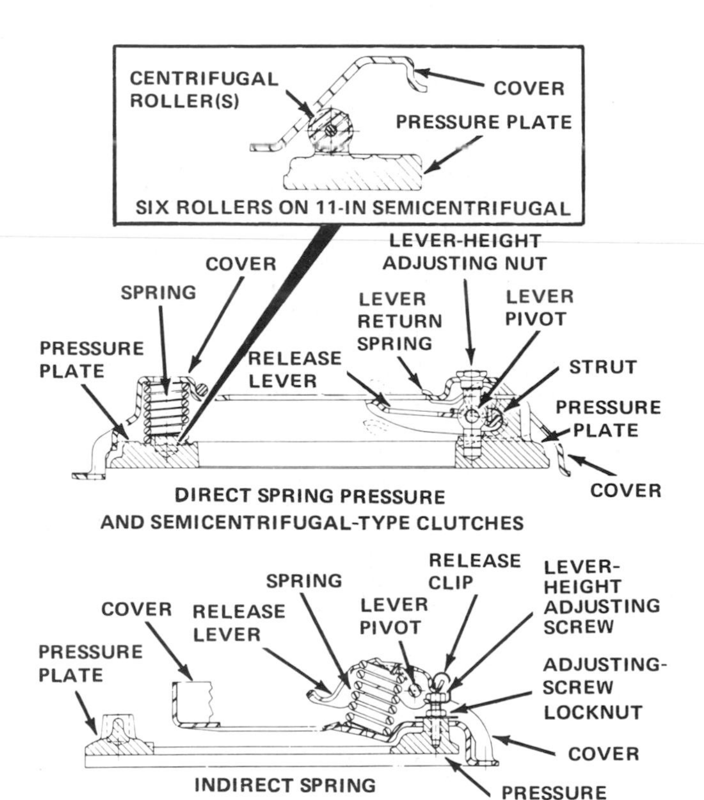

Fig. 3-15. Sectional views of clutches, showing adjustments of release levers. (*American Motors Corporation*)

On the direct-spring pressure-type clutch, turn the adjusting nuts until the lever is at the correct height. Work the lever several times. Recheck. If okay, stake the nut with a dull punch.

⊘ 3-14 Inspecting and Servicing Clutch Parts The various clutch parts can be checked as follows after the clutch is removed from the vehicle.

1. Use compressed air to blow dust out of the clutch housing. Check for oil leakage through the engine rear-main-bearing oil seal and transmission drive-pinion seal. If leakage is noted, replace the seal.
2. Check the friction face of the flywheel for uniform appearance and for cracks, grooves, and uneven wear. If there is uneven wear, check the flywheel runout with a dial indicator. A warped or otherwise

damaged flywheel should be replaced. See ⊘ 3-7 for the effect of heat checks and cracks on friction-disk-facing wear.

3. Check the pilot bushing in the end of the crankshaft. Replace it if it is worn. To remove it, fill the crankshaft cavity and bushing bore with heavy grease. Insert a clutch-aligning tool or spare transmission drive pinion (see Figs. 3-10 and 3-11) into the pilot bushing. Tap the end of the tool or drive pinion with a lead hammer. Pressure will force the bushing out. Install a new bushing with the aligning tool.

4. Check the journal on the end of the transmission input (clutch) shaft for wear. Replace it if it is worn.

5. Handle the disk with care. Do not touch the facings. Any trace of oil or grease will cause clutch slippage and rapid facing wear. Replace the disk if the facings show evidence of oil or grease, are worn to within 0.015 inch [0.381 mm] (Plymouth) of the rivet heads, or are loose. The disk should also be replaced if there is other damage, such as worn splines, loose rivets, or evidence of heat.

6. Wipe the pressure-plate face with solvent. Check the face for flatness with a straightedge. Check the face for burns, cracks, grooves, and ridges. See ⊘ 3-7 for the effect of heat checks and cracks on friction-disk-facing wear.

NOTE: If the friction disk is replaced, then as a rule the pressure-plate assembly should also be replaced.

7. Check the condition of the release levers. The inner ends should have a uniform wear pattern.

8. Test the cover for flatness on a surface plate.

9. If any of the pressure-plate parts are not up to specifications, replace the assembly. Also replace the friction disk.

10. Examine the throw-out bearing. The bearing should turn freely when held in the hand under a light thrust load. There should be no noise. The bearing should turn smoothly without roughness. Note the condition of the face where the release levers touch. Replace the bearing if it is not in good condition. Figure 3-16 shows lubrication points of release bearings. Light graphite grease is recommended.

CAUTION: Never clean the bearing in solvent or degreasing compound. It is prelubricated and sealed, and such cleaning would remove the lubricant and ruin the bearing.

11. Check the fork for wear on throw-out-bearing attachments or other damage. On reassembly, be sure that the dust seal or cover is in good condition to prevent dirt from entering.

NOTE: Steam cleaning can cause clutch trouble. Steam may enter and condense on the facings of the friction disk, pressure plate, and flywheel. The disk facings will absorb moisture. If the car is allowed to stand for a while with the facings wet, they may

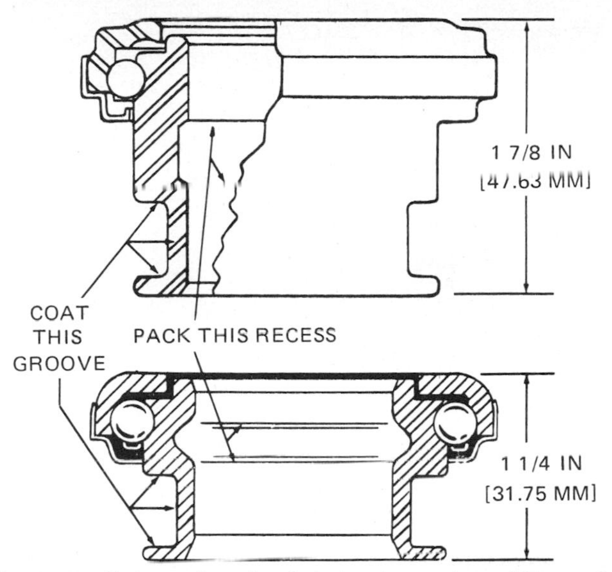

Fig. 3-16. Release-bearing lubrication points. (*Chevrolet Motor Division of General Motors Corporation*)

adhere to either the flywheel or the pressure plate. This means that the clutch would not disengage. To prevent this from happening, start the engine immediately after steam cleaning. Slip the clutch in order to heat and dry the facings.

12. Check the alignment of the clutch housing as explained in ⊘ 3-15.

⊘ 3-15 Checking the Clutch-Housing Alignment Whenever a clutch has been serviced, the clutch housing should be checked for alignment. This procedure includes checking the housing-bore runout and housing-face squareness (Figs. 3-17 and 3-18).

To check bore runout (Fig. 3-17), substitute a 3-inch [76.2 mm] bolt for one of the crankshaft bolts. Mount a dial indicator on this bolt with a C clamp. Center the dial indicator in the bore, as shown. Rotate the engine slowly clockwise to check runout. If the runout is excessive, it can be corrected by installing offset dowels (Fig. 3-19). Dowels come in various sizes, that is, with varying amounts of offset. To install the dowels, remove the clutch housing and

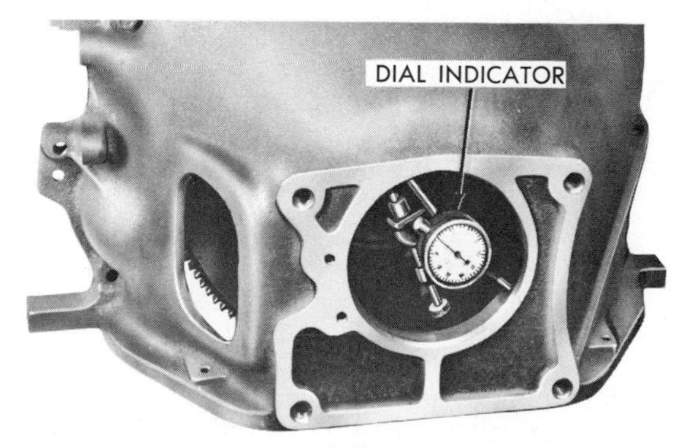

Fig. 3-17. Measuring clutch-housing-bore runout with a dial indicator. (*Chrysler Corporation*)

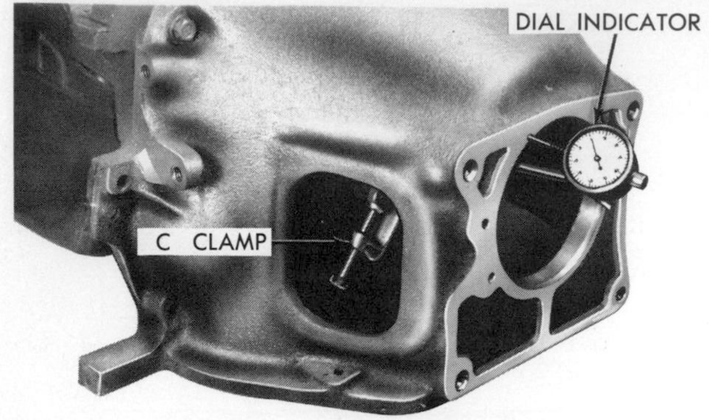

Fig. 3-18. Measuring clutch-housing face for squareness with a dial indicator. (*Chrysler Corporation*)

the old dowels. Your dial-indicator check has told you how much the bore is out of alignment and in which direction. These facts determine which pair of dowels you should select (the pair with the correct amount of offset). The slots in the dowels should align in the direction of maximum bore runout to correct the alignment.

To check housing-face squareness, reposition the dial indicator as shown in Fig. 3-18. Rotate the engine clockwise slowly to note how much the housing face is out of line. To correct alignment, place shim stock of the correct thickness in the proper positions between the clutch housing and engine block.

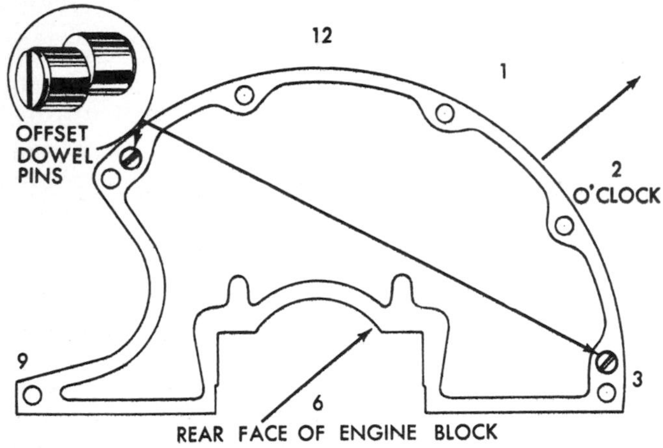

Fig. 3-19. Offset dowel diagram. (*Chrysler Corporation*)

CHAPTER 3 CHECKUP

Completing the Sentences The sentences that follow are incomplete. After each sentence there are several words or phrases, only one of which will correctly complete the sentence. Write each sentence in your notebook, selecting the proper word or phrase to complete it correctly.

1. Clutch slippage while the clutch is engaged is particularly noticeable: (*a*) during idle, (*b*) at low speed, (*c*) during acceleration, (*d*) when starting the engine.
2. Clutch chattering or grabbing is noticeable: (*a*) during idle, (*b*) at low speed, (*c*) when engaging the clutch, (*d*) when accelerating.
3. Clutch dragging is noticeable; (*a*) on acceleration, (*b*) at road speed, (*c*) when the clutch is disengaged, (*d*) at high speed.
4. Clutch noises are usually most noticeable when the engine is: (*a*) accelerating, (*b*) decelerating, (*c*) idling, (*d*) being started.
5. Clutch-pedal pulsation, or a nervous pedal, is noticeable when the engine is running and: (*a*) accelerating, (*b*) decelerating, (*c*) a slight pressure is applied to the pedal, (*d*) the car is moving at steady speed.
6. Slippage between the friction-disk facings and the flywheel or pressure plate will cause: (*a*) clutch-pedal pulsation, (*b*) rapid facing wear, (*c*) excessive acceleration.
7. The free travel of the pedal, or pedal lash, is the amount of travel the pedal has before the release bearing comes up against the: (*a*) release levers, (*b*) flywheel, (*c*) floorboard, (*d*) stop.
8. Heat checks or cracks on the flywheel and pressure-plate faces will cause: (*a*) excessive clutch slippage, (*b*) rapid flywheel and pressure-plate wear, (*c*) rapid friction-disk-facing wear.

Troubles and Service Procedures In the following you should write in your notebook the causes of the troubles and service procedures required. Do not copy the procedures from the book, but try to write them in your own words.

1. Name four causes of clutch slippage.
2. Name four causes of clutch chatter, or grabbing, when the clutch is engaged.
3. Name four causes of clutch spinning, or dragging, when the clutch is disengaged.
4. Into what two general groups can clutch noises be divided?
5. What could produce clutch noise when the clutch is engaged?
6. What could produce clutch noise when the clutch is disengaged?
7. Name two causes of clutch-pedal pulsations.
8. Name three causes of rapid friction-disk-facing wear.
9. Give some general instructions for making a clutch-linkage adjustment.
10. Give some general instructions for inspecting clutch conditions in the car.

SUGGESTIONS FOR FURTHER STUDY

Make a set of 3- by 5-inch trouble-diagnosis cards based on the clutch trouble-diagnosis chart given at the beginning of this chapter. Write complaints on one side of a card and possible causes on the other.

Write a page in your notebook for every clutch job you do. Include customer complaints, conditions found, repair performed, and special tools used.

chapter 4

MANUAL TRANSMISSIONS

With the introduction of different types of automatic transmissions in recent years, a great variety of transmissions are to be found on automobiles. Many cars have the standard manually operated gearshift. Many cars are equipped with completely automatic transmissions; these are so automatic that there is no clutch and thus no clutch pedal. The various automatic transmissions use rather complex controlling devices, as might be expected. These devices, however, function on very simple operating principles, as we shall see in later chapters.

In this book, several chapters are devoted to transmissions. The present chapter describes standard passenger-car and truck transmissions. Following chapters are devoted to transmission overdrives and automatic transmissions.

⊘ 4-1 Function of the Transmission As already explained (⊘ 1-6), the simplest manual passenger-car transmission provides three gear ratios between the engine crankshaft and the car wheels; the crankshaft is required to revolve approximately four, eight, or twelve times for each car-wheel revolution. In low gear the crankshaft revolves about twelve times per car-wheel revolution, thus permitting the engine to operate at fairly high speed when the car is first set in motion. Without this high gear ratio, the engine would turn so slowly that it could deliver little power to the wheels. The engine crankshaft must turn fairly rapidly before the engine develops sufficient power to start the car moving.

After the car has begun to move, a somewhat reduced gear ratio between the engine crankshaft and the car wheels is desirable. In the low-gear position, medium to maximum engine speed will cause the car to move only a few miles per hour. Shifting to the intermediate-gear position provides a gear ratio of approximately 8:1 between the engine crankshaft and car wheels; medium to maximum engine speed thus produces a car speed above 30 mph [48.3 km/h]. Shifting to the high-gear position provides a gear ratio of about 4:1 between the engine crankshaft and car wheels. This ratio permits higher car speeds at high engine speeds and also allows the engine to operate at lower and more efficient speeds when the car is traveling at fairly low or intermediate speeds.

While the car is in motion and in high gear, normally it is unnecessary to shift gears except when the car is brought to a stop or when additional power at low car speed is required. Such additional power is needed, for example, when climbing a steep hill or when rapid acceleration from low speed is desired. In addition to these three forward-gear positions, the reverse-gear position provides a means of reversing the direction of car motion so that the car can be moved backward.

NOTE: The fundamentals of transmission operation (⊘ 1-6) should now be reviewed because they will be helpful in understanding the modern transmissions which we shall now discuss.

⊘ 4-2 Four-Speed and Five-Speed Passenger-Car Manual Transmissions Today many more cars with manual transmissions are being equipped with four-speed transmissions rather than three-speed. For example, a recent tabulation indicated that about 7 percent of Ford Motor Company cars are equipped with four-speed transmissions. About 1 percent are equipped with three-speed transmissions. The other 92 percent are equipped with automatic transmissions. These figures are fairly representative of the automotive industry.

Some cars are now being equipped with five-speed manual transmissions. These transmissions are like the four-speed units but also have a fifth forward speed, which is actually an overdrive. See ⊘ 4-9. Overdrives are described in Chap. 5.

⊘ 4-3 Types of Manual Transmissions In addition to the three-speed and four-speed transmissions (with reverse) used on passenger cars and light trucks, there are other types of manual transmissions. For example, the manual transmissions used on some heavy-duty trucks and other equipment may have as many as ten forward speeds and two reverse speeds. These are combination units using a five-speed transmission with a two-speed auxiliary. Essentially, there is little difference between these various types of transmissions except that those providing more gear ratios, of course, have additional gears and shifting positions. Figures 4-10 to 4-38 illustrate various types of manually shifted transmissions in sectional and disassembled views. These include both passenger-car units with three forward speeds and one reverse speed and four-forward-and-one-reverse-speed transmissions. Descriptions of their operation follow.

Modern transmissions utilize synchromesh devices that make gear shifting easier. These devices ensure that gears that are about to be meshed are revolving at synchronized speeds. This means that the gear teeth that are about to mesh are moving at the same speed. As a result, the teeth mesh without any clashing of gears. Following sections describe the action of these devices.

⊘ 4-4 Steering-Column and Floor Shifts Many years ago the gearshift lever was located on the floor of the driver's compartment, the lower end being attached to shifting devices in the transmission case. This type of gearshift mechanism was used in ⊘ 1-6 to explain transmission action because the linkage to the shifter yokes or forks in the transmission is easier to understand than the linkages on many late-model cars. Today, most cars using standard manually shifted transmissions have the gearshift lever mounted on the steering column. With this mounting, the linkages between the lever and the transmission are more complex. However, the action of the transmission is similar; movement of the lever causes gears to move into or out of mesh.

Two separate motions of the gearshift lever are required in shifting gears. The first motion selects the gear assembly to be shifted; the second moves the gear assembly in the proper direction to complete the shift. A number of different types of selector and shifter devices are in use. Figure 4-1 shows the shifting patterns for steering-column and floor-shift levers. Let us look at the linkages for the two types of shifts.

1. *STEERING-COLUMN SHIFT* Figures 4-2 and 4-3 show one type of steering-column shift and its linkages to the transmission. To shift into first or reverse, the driver depresses the clutch pedal to momentarily disconnect the engine from the transmission. Then, the driver lifts the shift lever and moves it forward for reverse or back for low (first) gear. When the lever is lifted, it pivots on its mounting pin, which forces a tube, or rod, downward in the steering column. This downward movement pushes downward on a crossover blade at the bottom of the steering column. A slot in the blade engages a pin on the first-and-reverse shift lever (Fig. 4-4).

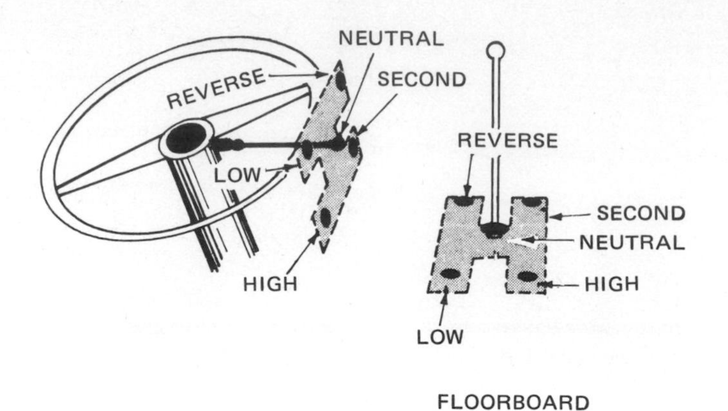

Fig. 4-1. Gearshift patterns for steering-column and floorboard shift levers.

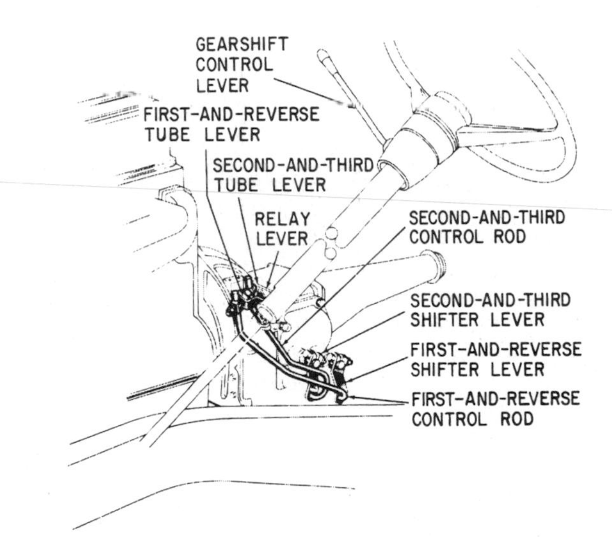

Fig. 4-2. Steering-column gearshift lever and linkage to the transmission. (*Chevrolet Motor Division of General Motors Corporation*)

Now, when the shift lever is moved, say, into first, the first-and-reverse lever is rotated. This movement is carried by linkage to the transmission (Fig. 4-5). At the transmission, the movement causes the first-and-reverse shift lever to move. This lever is on a shaft that extends through the transmission side cover (see Fig. 4-6). There is a lever on the inside end of this shaft, and a shifter fork is mounted on this lever. The arrangement does the following: When the shaft is rotated by movement of the first-and-reverse shift lever, it causes the shifter fork to move backward or forward inside the transmission. This motion shifts the first-and-reverse sliding gear into mesh. To shift into second or high, the driver moves the shift lever into the positions shown in Fig. 4-1. This movement causes the slot in the blade to engage the second-and-third shift lever at the bottom of the steering column. This lever is moved to actuate the second-and-third linkage to the transmission.

2. *FLOOR SHIFT* Many cars, especially sports-type cars with four-speed transmissions, have a floor-mounted shift lever, or "stick," as it is often called. A transmission using a floor-shift lever is shown in Fig. 4-7. Figure 4-8 shows the linkage on an automo-

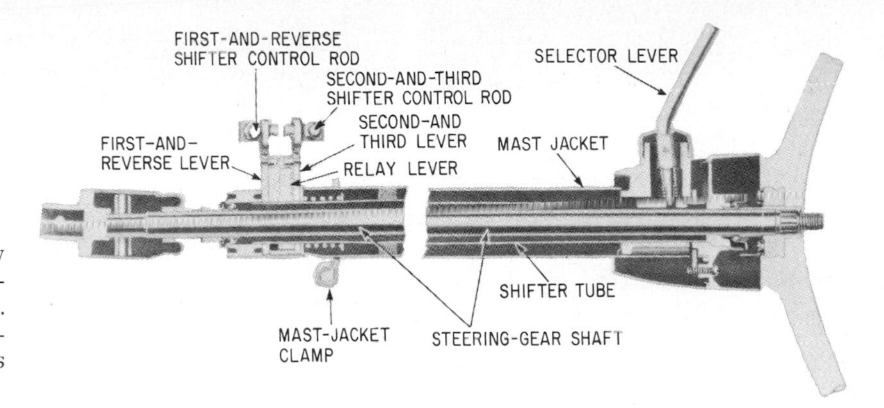

Fig. 4-3. Sectional view of a steering mast, showing gearshift controls. (*Chevrolet Motor Division of General Motors Corporation*)

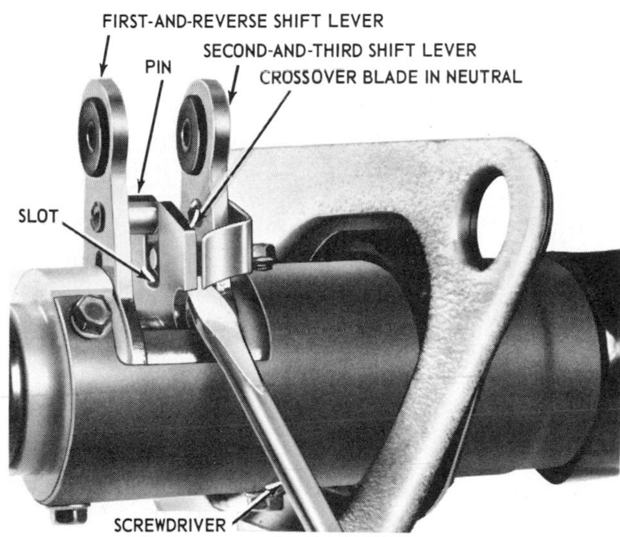

Fig. 4-4. Shift levers and crossover blade at the bottom of the steering column. The screwdriver holds the crossover blade in neutral for an adjustment check. (*Chrysler Corporation*)

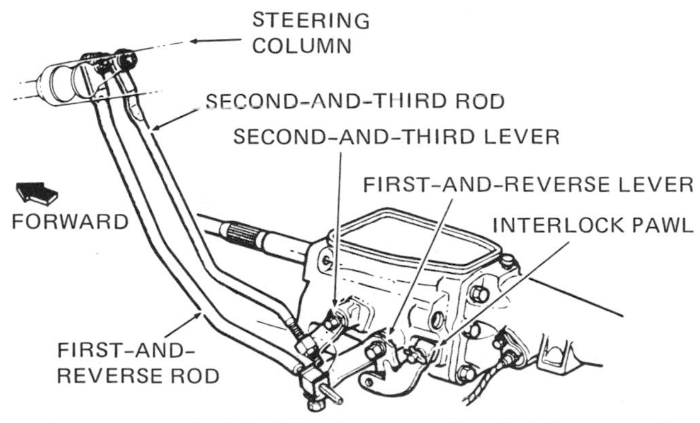

Fig. 4-5. Gearshift linkage between the shift levers at the bottom of the steering column and the transmission levers on the side of the transmission. (*Chrysler Corporation*)

bile that has a console. (You will remember that in automotive language a console is a small cabinet on the floor of the front compartment between the two front seats. It sometimes houses a glove compartment and various controls, such as a shift lever, electric window switches, and heater and air-conditioner controls.)

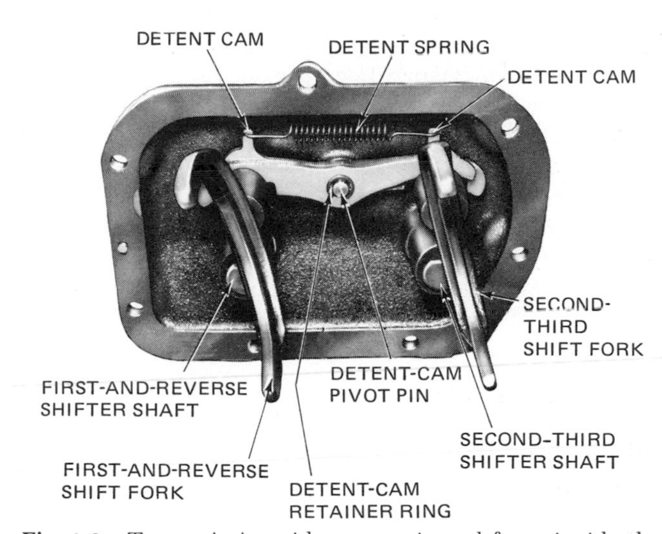

Fig. 4-6. Transmission side cover viewed from inside the transmission. The shift forks are mounted on the ends of levers attached to shafts. The shafts can rotate in the side cover. The detent cams and springs prevent more than one of the shift forks from moving at any one time. (*Chevrolet Motor Division of General Motors Corporation*)

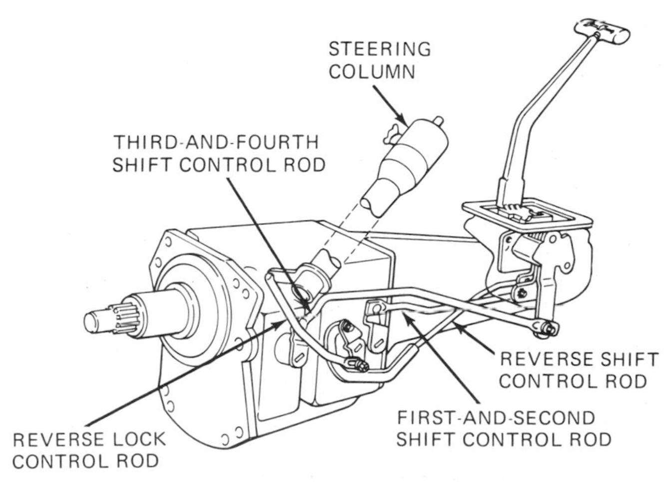

Fig. 4-7. External view of a four-speed floorshift transmission. This type of transmission is sometimes called *four on the floor*. (*Ford Motor Company*)

In the shift-lever support there are levers attached to each rod. These levers have slots that are selected by a tongue on the lower end of the shift lever as it is moved into the various gear positions.

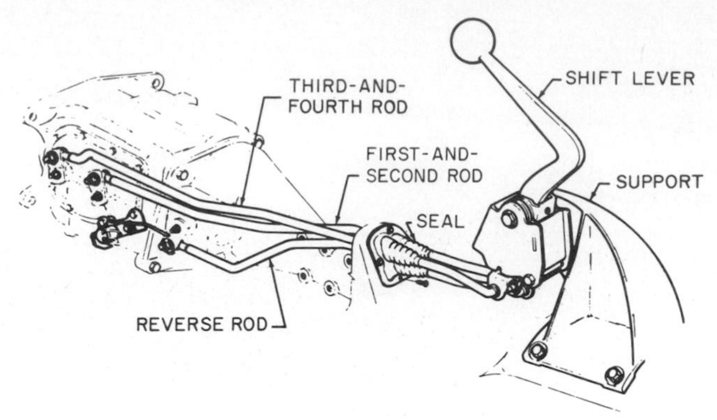

Fig. 4-8. Linkage between a four-speed transmission and a floor-mounted shift lever. (*Ford Motor Company*)

The first movement of the shift lever makes this selection. Then the second movement of the shift lever causes the selected lever and rod to move, thus causing the transmission lever to move and thereby shift the selected gear into the selected gear position.

Figure 4-9 is a disassembled view of a similar linkage for a floor-shift lever that mounts in the floor beside the front seats rather than in a console.

⊘ 4-5 Three-Speed Transmission Now let us look at an actual three-speed transmission (Figs. 4-10 and 4-11). An exploded view of a similar transmission is shown in Fig. 4-12. The transmission has the same type of countergear assembly as our simplified version (Fig. 1-17). The clutch shaft and gear are also similar. However, the transmission main shaft and the gears on it are a little different. We show the gears separately in Fig. 4-13 so that you can study them more carefully.

Note that the second-speed and first-speed gears do not move back and forth along the main shaft. They are in constant mesh with their respective gears on the countergear assembly. Also, these two gears are supported on bearings that allow them to rotate independently on the main shaft.

The hubs of the two synchronizer assemblies are splined to the main shaft and rotate with it. However, the synchronizer sleeves of the two assemblies can slide back and forth along the splines on the synchronizer hub. The forks shown in Fig. 4-6 fit in the grooves of the synchronizers (shown in Fig. 4-13). Now, you can see the connection between the gearshift lever and the synchronizers. When the gearshift lever is moved, the linkage selects one of the two synchronizers. It then moves this synchronizer.

1. *FIRST-SPEED GEAR* Now let us see what happens when the gearshift lever is moved into first. The clutch is disengaged for this operation. Movement of the gearshift lever causes the linkage to select the first-reverse synchronizer sleeve and move it to the left (in Fig. 4-14). As the synchronizer is moved to the left (in Fig. 4-14), the internal teeth in the synchronizer engage with the external teeth on the first-speed gear. This locks the first-speed gear, through the synchronizer, to the main shaft. Figure 4-15 shows a first-speed gear by itself so that you can see the external teeth. The synchronizer is

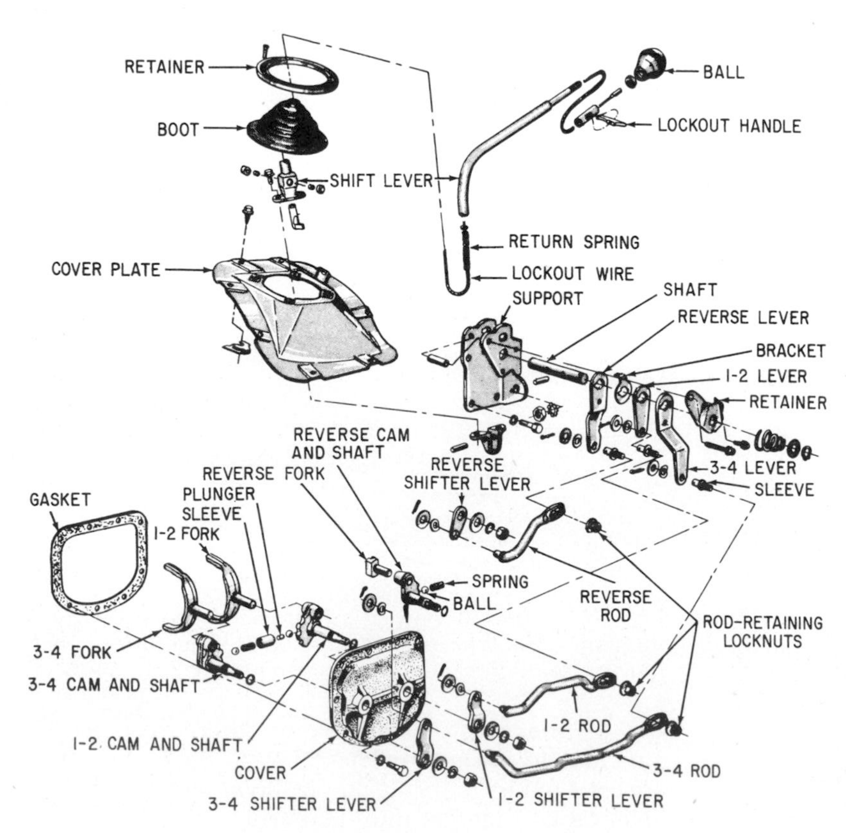

Fig. 4-9. Disassembled view, showing the linkage between a four-speed transmission and a floor-mounted shift lever. (*Ford Motor Company*)

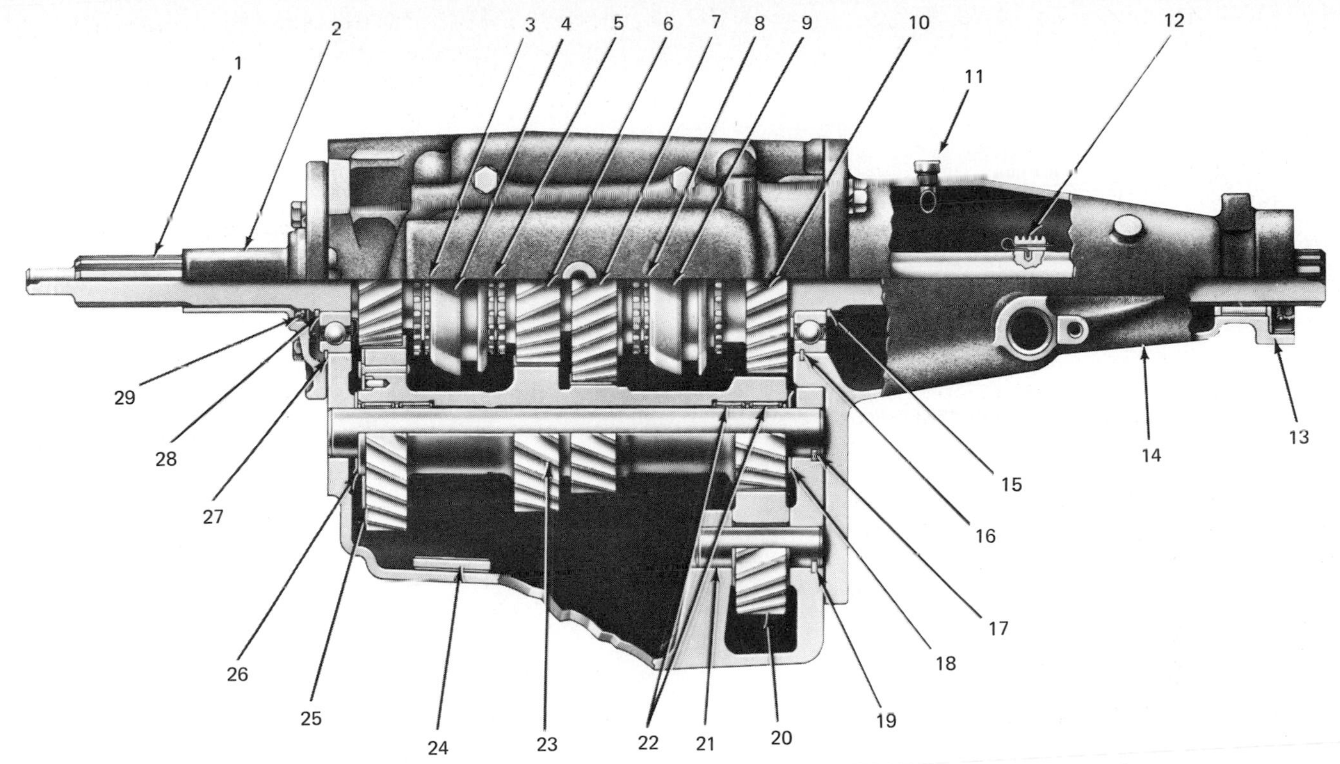

1. Clutch gear
2. Clutch-gear bearing retainer
3. Third-speed synchronizer ring
4. Second–third-speed-clutch assembly
5. Second-speed synchronizer ring
6. Second-speed gear
7. First-speed gear
8. First-speed synchronizer ring
9. First–reverse-clutch assembly
10. Reverse gear
11. Vent
12. Speedometer gear and clip
13. Rear-extension seal
14. Rear extension
15. Rear bearing-to-shaft snap ring
16. Rear bearing-to-extension snap ring
17. Countergear woodruff key
18. Thrust washer
19. Reverse-idler-shaft woodruff key
20. Reverse idler gear
21. Reverse idler shaft
22. Countergear bearings
23. Countergear
24. Case magnet
25. Antilash plate assembly
26. Thrust washer
27. Clutch-gear bearing
28. Snap ring
29. Clutch-gear-retainer lip seal

Fig. 4-10. Sectional view of a three-speed transmission. (*Chevrolet Motor Division of General Motors Corporation*)

shown by itself in Fig. 4-16 so that you can see the internal teeth.

Now, when the clutch is engaged, the power flow through the transmission is as shown in Fig. 4-14. Note that there is gear reduction as the small clutch gear drives the large gear on the countergear assembly. There is also gear reduction as the small gear on the countergear assembly drives the large first-speed gear. The total gear reduction in first-speed gear is a little less than 3:1 in most transmissions (2.636:1 in one model).

2. *SECOND-SPEED GEAR* Figure 4-17 shows the transmission in second-speed gear. The first-reverse synchronizer has been moved to its center position and out of mesh with the first-speed gear. The 2-3 synchronizer has been moved to the right so that its internal teeth engage the external teeth on the second-speed gear. The power flow is as shown by the arrow. The second-speed gear is smaller than the first-speed gear. And the countergear that meshes with the second-speed gear is larger than the countergear that meshes with the first-speed gear. Thus the gear reduction is less. It is a little less than 2:1 in most transmissions (1.605:1 in one model).

3. *THIRD-SPEED GEAR* In this gear, the power flow is straight through the transmission, as shown in Fig. 4-18. Third gear is achieved by moving the synchronizer to the left, as shown. Its internal teeth engage the external teeth of the clutch gear so that the drive is through these teeth and the synchronizer splines to the main shaft. The gear ratio is 1:1.

4. *REVERSE* To achieve reverse, an extra gear is inserted into the gear train. This extra gear is shown in Fig. 4-10 (item 20). It is in constant mesh with the

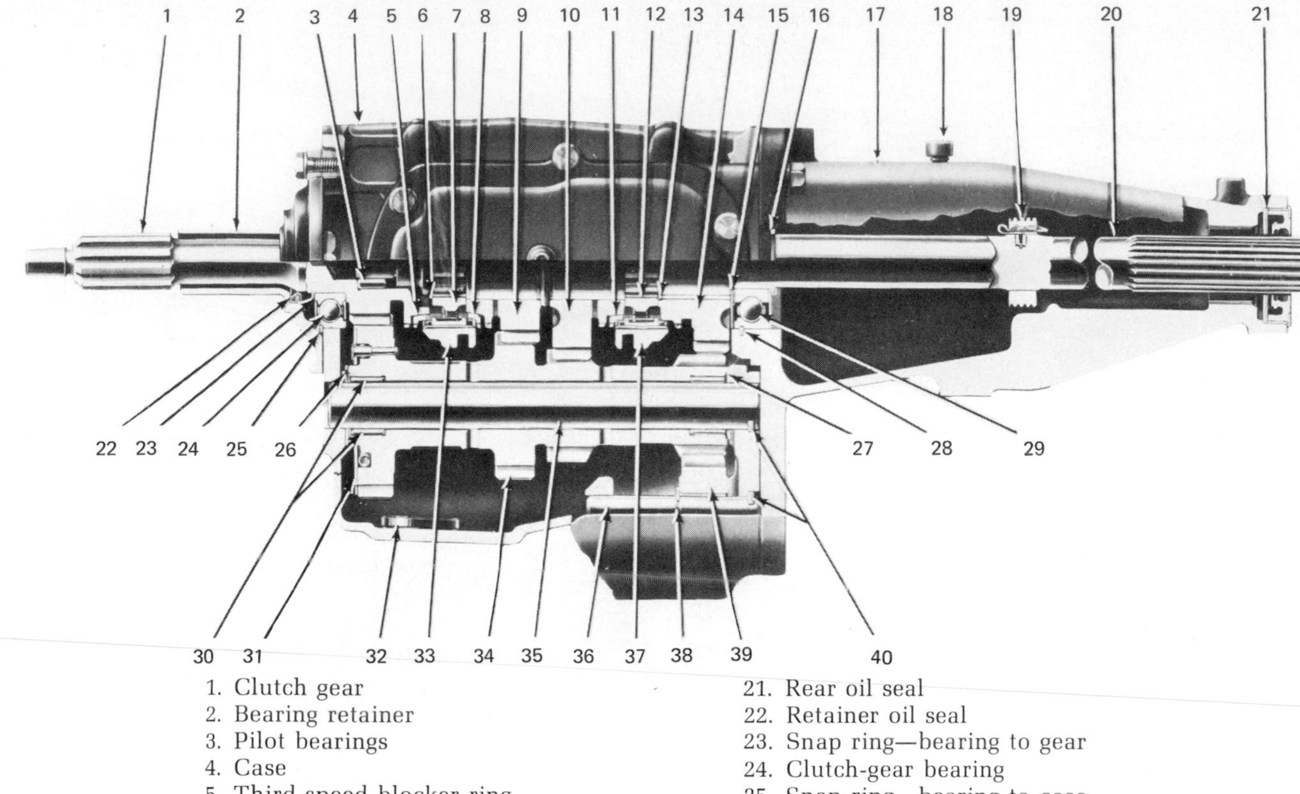

1. Clutch gear
2. Bearing retainer
3. Pilot bearings
4. Case
5. Third-speed blocker ring
6. 2–3 synch. snap ring
7. 2–3 synch. hub
8. Second-speed blocker ring
9. Second-speed gear
10. First-speed gear
11. First-speed blocker ring
12. First-speed synch. hub
13. First-speed synch. snap ring
14. Reverse gear
15. Reverse-gear thrust and spring washers
16. Snap ring—bearing to main shaft
17. Extension
18. Vent
19. Speedometer drive gear and clip
20. Main shaft
21. Rear oil seal
22. Retainer oil seal
23. Snap ring—bearing to gear
24. Clutch-gear bearing
25. Snap ring—bearing to case
26. Thrust washer (front)
27. Thrust washer (rear)
28. Snap ring—bearing to extension
29. Rear bearing
30. Countergear roller bearings
31. Antilash plate assembly
32. Magnet
33. 2–3 synch. sleeve
34. Countergear
35. Countershaft
36. Reverse idler shaft
37. First-speed synch. sleeve
38. E ring
39. Reverse idler gear
40. Woodruff key

Fig. 4-11. Sectional view of a three-speed transmission. (*Chevrolet Motor Division of General Motors Corporation*)

fourth gear on the countergear assembly, as shown in Fig. 4-19. When the shift lever is moved to reverse, the linkage moves the first-reverse synchronizer to the right, as shown in Fig. 4-19. The internal teeth in the synchronizer engage with the external teeth on the reverse gear so that the flow of power through the transmission is as shown in Fig. 4-19. Because of the extra gear in the gear train, the main shaft turns in the reverse direction. Therefore the wheels turn in the reverse direction, and the car moves backward. The gear ratio is a little less than 3:1 in most transmissions (2.636:1 in one model).

5. *INTERLOCK* To prevent the wrong shifter fork and gear in the transmission from moving, an interlocking mechanism is used. This mechanism locks the inoperative shifter fork as the other fork is moved into the shifted position. The mechanism consists of a spring-loaded plunger or shaft that is moved into a hole or slot in the inoperative shifter fork by movement of the active lever.

⊘ 4-6 Synchronizers To prevent the clashing of gears during shifting and to simplify the shifting action for the driver, synchronizing devices are used in transmissions. These devices ensure that gears that are about to mesh will be rotating at the same speed and thus will engage smoothly.

One type of synchronizer uses synchronizing cones on the gears and on the synchronizing drums (Fig. 4-20). In the neutral position, the sliding sleeve is held in place by spring-loaded balls resting in detents in the sliding sleeve (or ring gear). When a shift starts, the drum and ring gear, as an assembly,

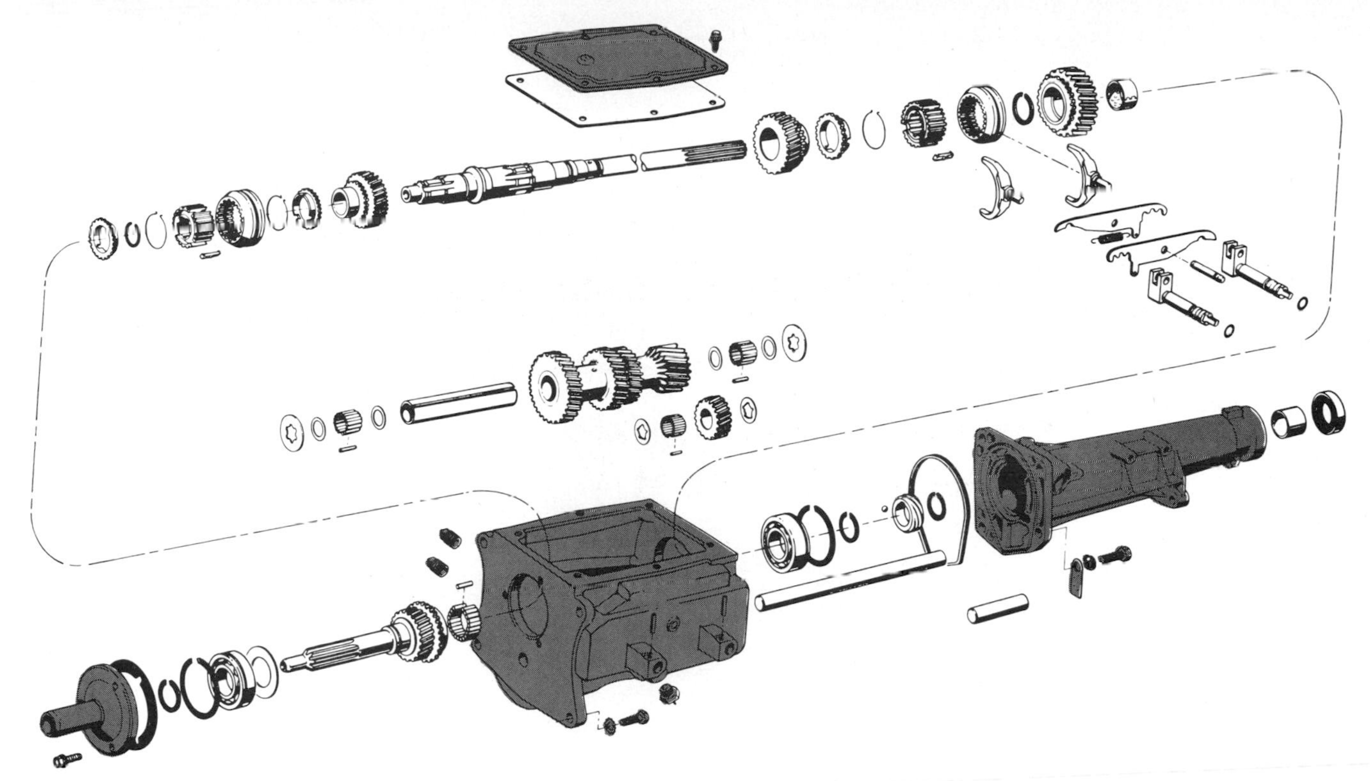

Fig. 4-12. Exploded view of a three-speed transmission. (*American Motors Corporation*)

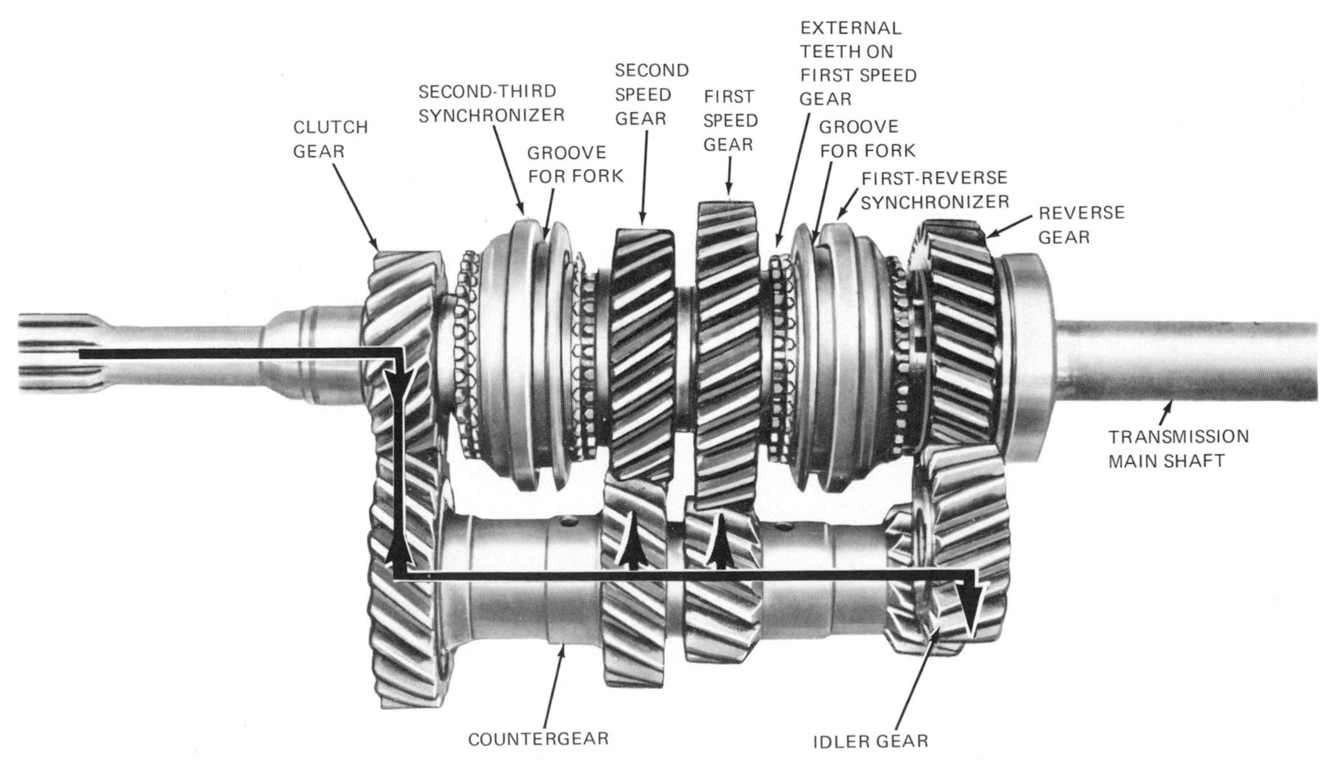

Fig. 4-13. Gear train and shafts in the three-speed transmission. (*Chevrolet Motor Division of General Motors Corporation*)

are moved toward the selected gear. The first contact is between the synchronizing cones on the selected gear and drum. This contact brings the two into synchronization. Both rotate at the same speed. Further movement of the shift fork forces the sliding sleeve on toward the selected gear. The internal teeth on the sliding sleeve match the external teeth on the selected gear. Now, the gears are locked, or

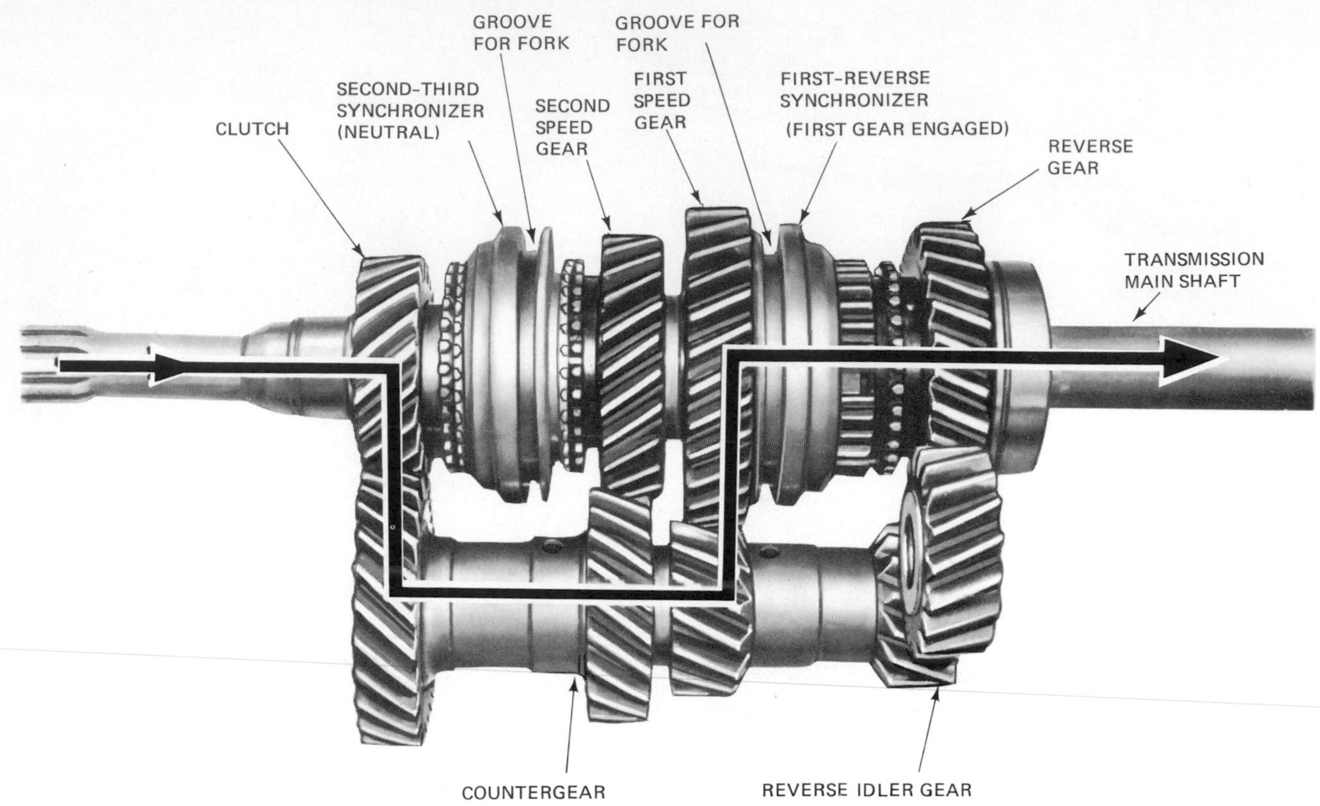

Fig. 4-14. Power flow through the gear train in first gear. (*Chevrolet Motor Division of General Motors Corporation*)

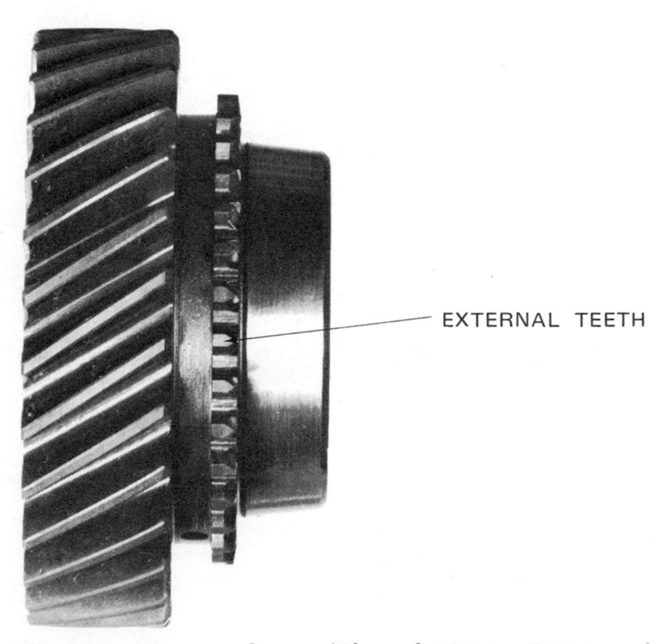

Fig. 4-15. First-speed gear. (*Chevrolet Motor Division of General Motors Corporation*)

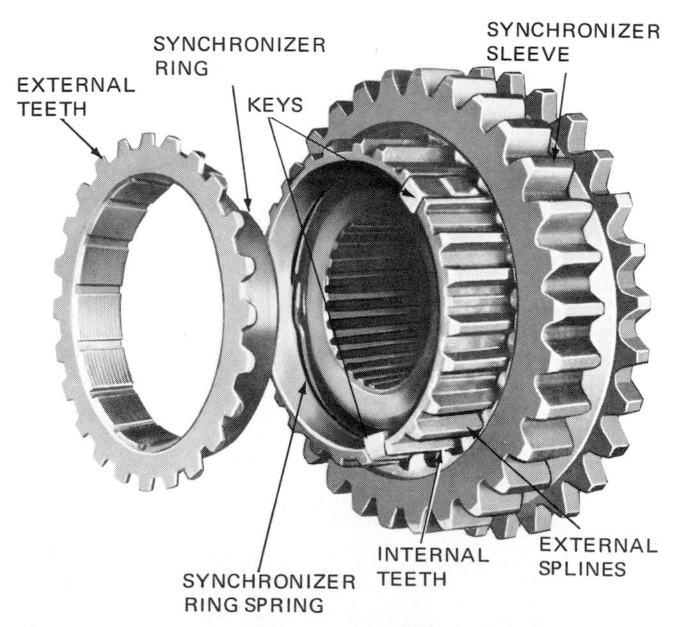

Fig. 4-16. Synchronizer assembly. (*Chevrolet Motor Division of General Motors Corporation*)

engaged, and the shift is completed. Note that the sliding sleeve moves off center from the drum for engagement. This pushes the balls down against the spring.

The pin-type synchronizer (Fig. 4-21) has a pair of stop rings. Each has three pins which pin it to the clutch-gear sleeve. The clutch gear is splined to the main shaft. External teeth on the clutch gear mesh with internal teeth on the clutch-gear sleeve. Thus, the clutch gear, clutch-gear sleeve, and two stop rings are always rotating with the main shaft. When a shift is made into second, for example, the main shaft and associated parts may be rotating at a different speed from the second-speed gear. How-

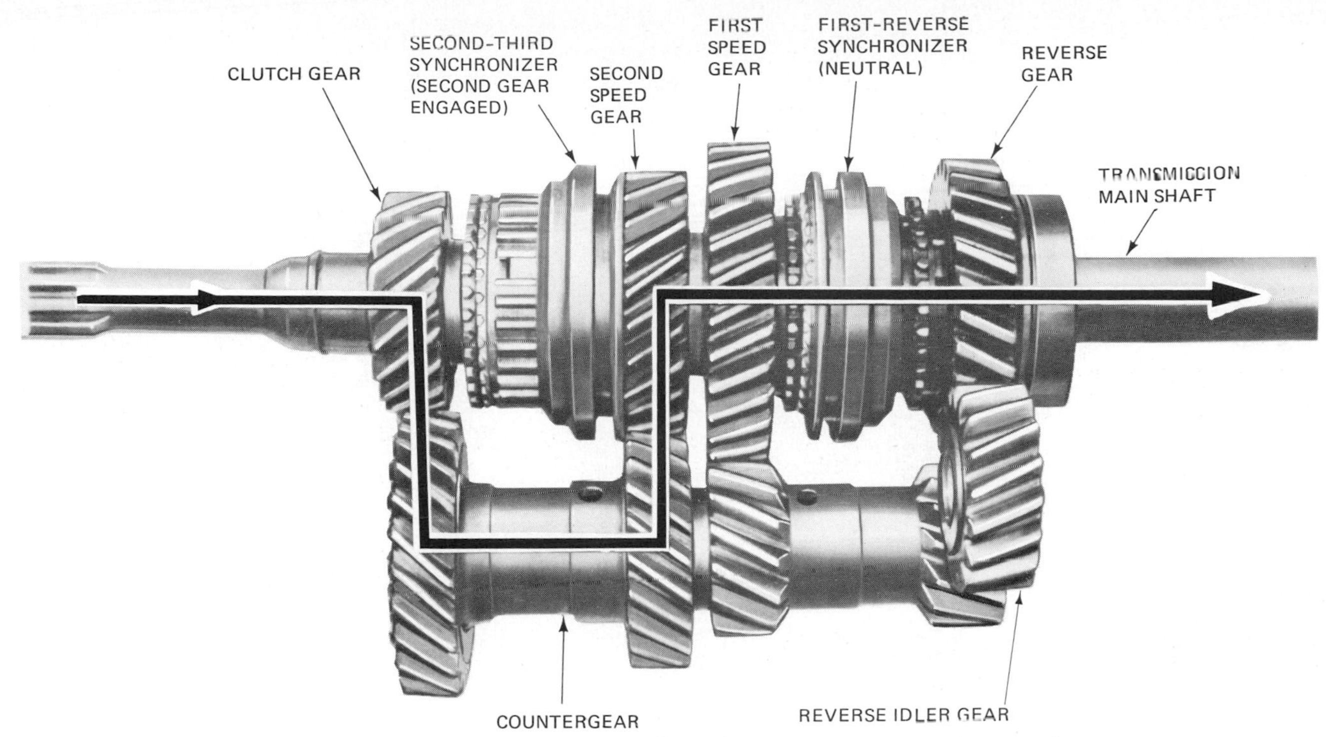

Fig. 4-17. Power flow in second gear. (*Chevrolet Motor Division of General Motors Corporation*)

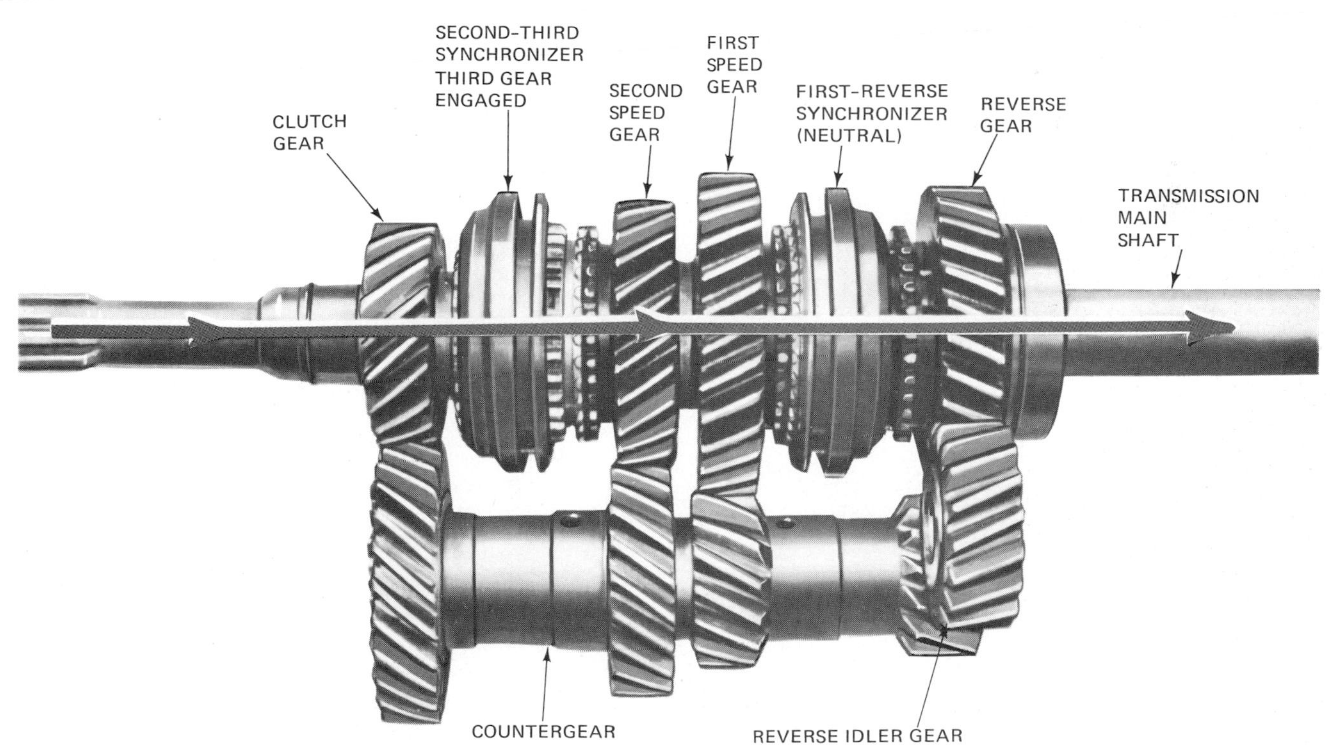

Fig. 4-18. Power flow in third gear. (*Chevrolet Motor Division of General Motors Corporation*)

ever, as the clutch-gear sleeve is moved toward the second-speed gear, the rear inner stop ring moves against the face of the second-speed gear. This brings the stop ring into synchronous rotation with the clutch-gear sleeve. It permits alignment of the external teeth on the clutch gear and the teeth on the small diameter of the second-speed gear. Now, the clutch-gear sleeve can slip over the teeth of the second-speed gear to couple with the second-speed gear and the clutch gear. Then, when the clutch is

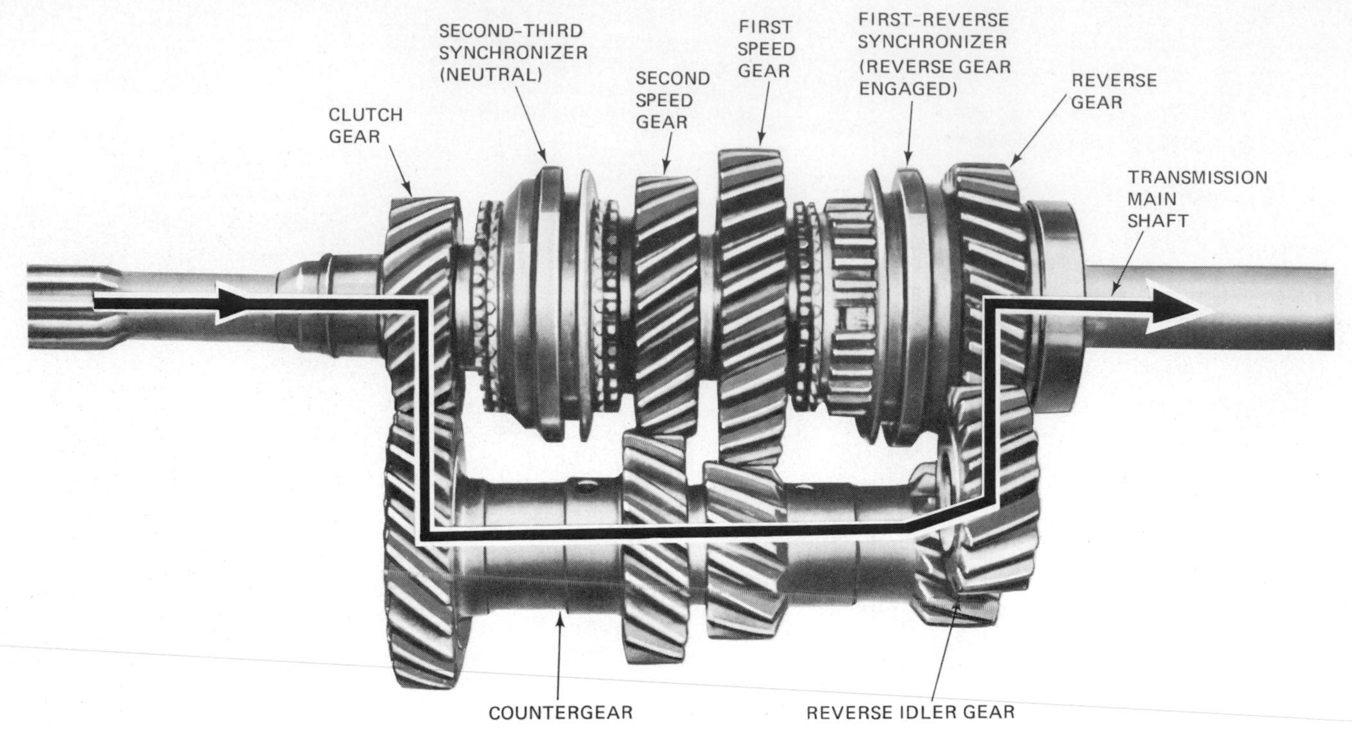

Fig. 4-19. Power flow in reverse. (*Chevrolet Motor Division of General Motors Corporation*)

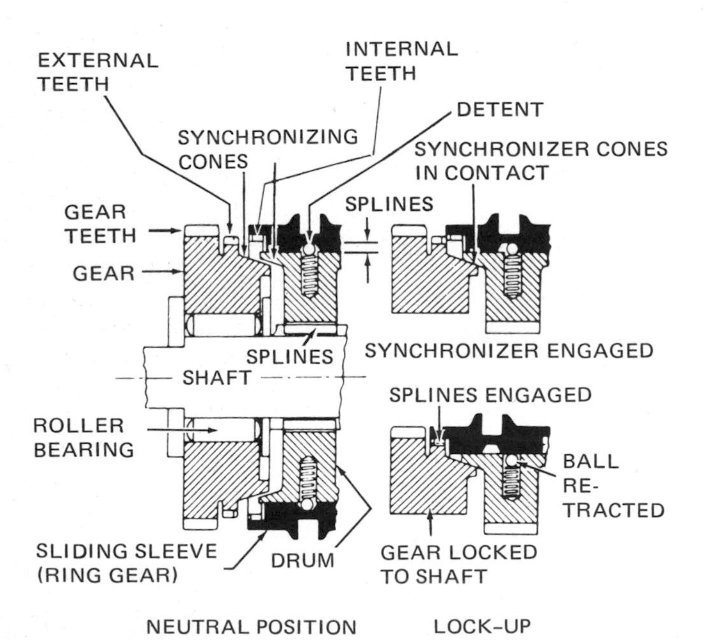

Fig. 4-20. Operation of transmission-synchronizing devices using cones.

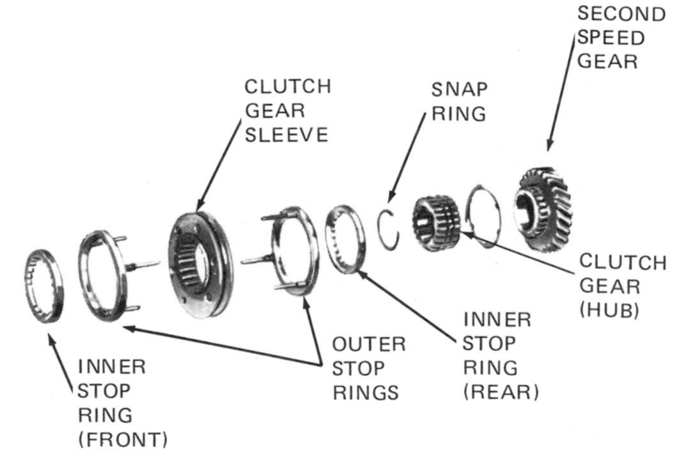

Fig. 4-21. Disassembled view of a pin-type synchronizing device used to ensure gear synchronization in shifting. (*Chrysler Corporation*)

engaged and the engine again delivers power through it, the second-speed gear drives the main shaft through the clutch gear and clutch-gear sleeve.

The action in a shift to third is very similar. The third-speed gear is supported on roller bearings.

Another type of synchronizer, used in the transmission illustrated in Figs. 4-10 to 4-12, is shown partly disassembled in Fig. 4-22. Instead of retracting balls, as in Fig. 4-20, this synchronizer has three keys and a pair of ring-shaped synchronizing springs. The keys are assembled in slots in the hub. The hub is splined to the main shaft. Assembled outside the hub is the synchronizing sleeve. The hub has external splines that fit the internal splines of the sleeve. The three keys have raised sections that fit in the annular groove of the sleeve.

NOTE: The sleeve shown in Fig. 4-22 has external teeth, but the synchronizing sleeves in the transmission shown in Figs. 4-10 to 4-12 do not. The action is the same, however.

Synchronizing is a three-stage action: First, the sleeve is moved toward the first-speed gear (when

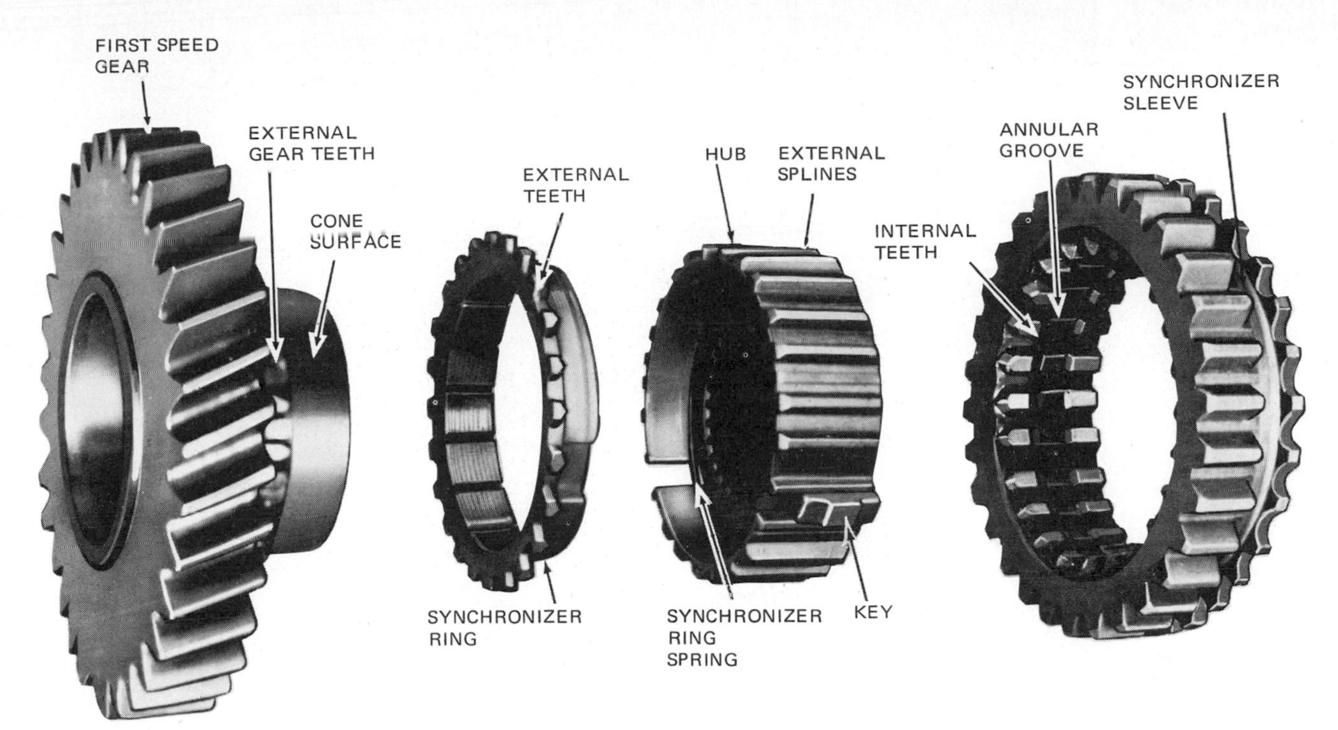

Fig. 4-22. Disassembled view of a synchronizing mechanism. (*Chevrolet Motor Division of General Motors Corporation*)

shifting to first). The sleeve slides on the hub splines and carries the three keys with it. Second, the keys move up against the synchronizer ring and push the ring toward the first-speed gear. The ring presses against the cone of the first-speed gear. Third, further sleeve movement causes the keys to be pressed out of the annular groove in the sleeve. The sleeve continues to move toward the first-speed gear. The friction between the synchronizing ring and the first-speed gear brings the two into synchronous rotation. Now, the final movement of the sleeve allows the internal teeth of the sleeve to engage the external teeth of the first-speed gear. Meshing is completed. Similar actions take place in the shifts to second and high.

⊘ 4-7 Four-Speed Transmission The four-speed transmission has four forward speeds and reverse. One type is shown in Fig. 4-23. The gears and shafts are shown by themselves in neutral in Fig. 4-24.

1. *FIRST GEAR* Figure 4-25 shows the power flow through the transmission in first gear. The 1-2 synchronizer has been moved to the right so that its internal teeth engage the external teeth of the first-speed gear.
2. *SECOND GEAR* Figure 4-26 shows the power flow through the transmission in second gear. The 1-2 synchronizer has been moved to the left so that its internal teeth engage the external teeth of the second-speed gear.
3. *THIRD GEAR* Figure 4-27 shows the power flow through the transmission in third gear. The 3-4 synchronizer has been moved to the right so that its internal teeth engage the external teeth of the third-speed gear.
4. *FOURTH GEAR* Figure 4-28 shows the power flow through the transmission in fourth gear. The 3-4 synchronizer has been moved to the left so that its internal teeth engage the external teeth of the clutch gear.
5. *REVERSE* In reverse, both synchronizers are in the neutral position. The reverse gear has been moved to the left, as shown in Fig. 4-29, so that it engages the reverse idler gear. Now, the extra gear in the train causes the main shaft to turn in the reverse direction, and so the car moves backward.
6. *GEARSHIFT LINKAGE* Figure 4-30 shows a typical linkage between the floor-mounted gearshift lever and the transmission. This is the "four-on-the-floor" arrangement. Note that there is an extra rod —the reverse rod—linking the shift lever and transmission. This rod carries the movement to the reverse fork that moves the reverse gear.

⊘ 4-8 Other Manual Transmissions In addition to the several transmissions already discussed, a variety of other designs have been used in automobiles, trucks, buses, and other heavy-duty equipment. Some of these have as many as ten forward speeds and two reverse speeds; they are combination units with a five-forward-speed-and-one-reverse-speed transmission and a two-speed auxiliary.

The Chevrolet Corvair has the engine mounted in the rear (Figs. 4-31 and 22-20), and this requires a somewhat different transmission arrangement. The transmission and differential (Chap. 23) are installed, as a unit, at the rear-axle location. Figure 4-32 is a sectional view of the transmission. It is a conventional three-speed transmission that is similar to

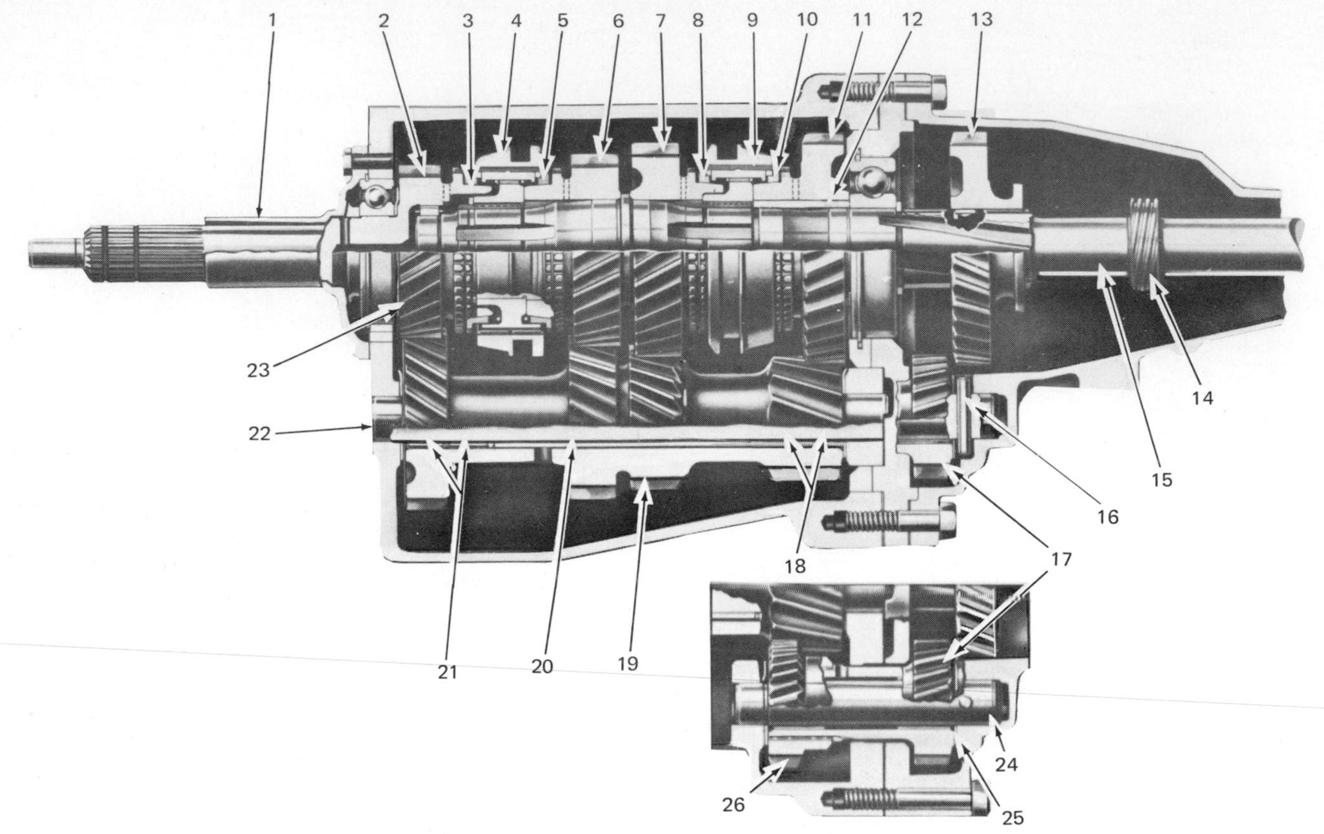

1. Bearing retainer
2. Main drive gear
3. Fourth-speed synchronizing ring
4. Third-and-fourth-speed-clutch assembly
5. Third-speed synchronizing ring
6. Third-speed gear
7. Second-speed gear
8. Second-speed synchronizing ring
9. First-and-second-speed-clutch assembly
10. First-speed synchronizing ring
11. First-speed gear
12. First-speed-gear sleeve
13. Reverse gear
14. Speedometer drive gear
15. Main shaft
16. Reverse-idler-shaft roll pin
17. Reverse idler gear (rear)
18. Countergear bearing roller
19. Countergear
20. Countershaft-bearing-roller spacer
21. Countershaft bearing roller
22. Countergear shaft
23. Oil slinger
24. Reverse idler shaft
25. Thrust washer
26. Reverse idler gear (front)

Fig. 4-23. Sectional view of a four-speed transmission. (*Chevrolet Motor Division of General Motors Corporation*)

others previously described except for some minor modifications necessary to adapt it to the rear mount. The Corvair was also supplied with a four-speed transmission of the fully synchronized type; it is similar to other four-speed transmissions already described. In addition, an automatic transmission was available for the Corvair as optional equipment.

The transmission shown in Fig. 4-33 is for use on a front-drive vehicle. The engine is mounted crosswise so that the transmission is also crosswise in the vehicle. Note that the clutch (at top) is of the hydraulically operated type. The transmission is a four-forward-speed unit.

Figure 4-34 is a sectional view of a transmission and differential for a car using a four-cylinder, rear-mounted, air-cooled engine.

Some cars are now being equipped with four- and five-speed transmissions, with the top speeds actually being overdrives. In other words, when the selector lever is moved to the fourth forward position (in the four-speed unit) or the fifth forward position (in the five-speed unit), the output shaft of the transmission turns faster than the input shaft. See ⊘ 4-9 and 4-10.

⊘ 4-9 Four-Speed Transmission with Overdrive This transmission is shown in cutaway view in Fig. 4-35. The gear ratios in the four forward speeds are as shown in Fig. 4-36. Note that in third, the ratio is 1:1. This means that the input and output shafts turn at the same speed. However, in fourth, the output shaft turns faster than the input shaft. The ratio

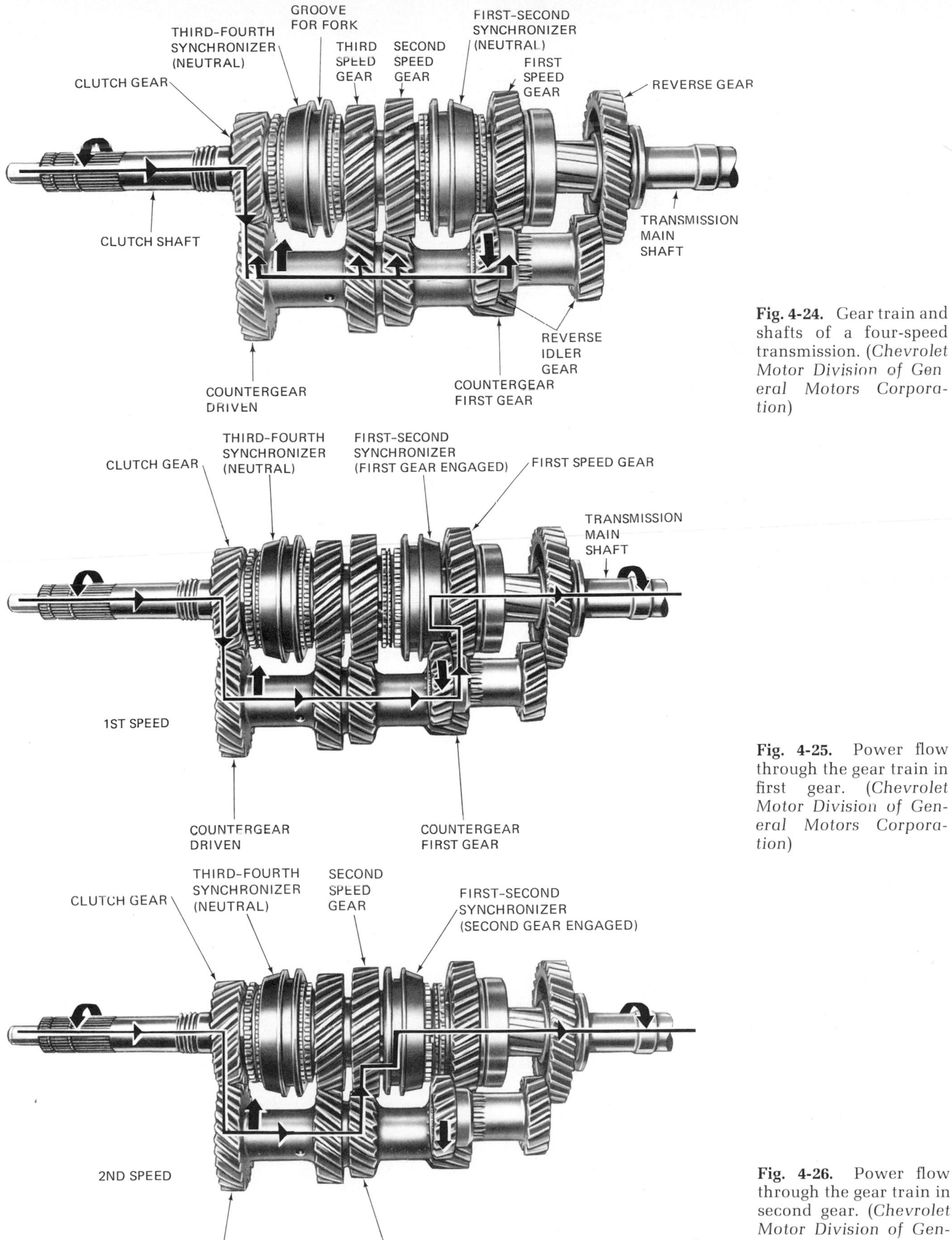

Fig. 4-24. Gear train and shafts of a four-speed transmission. (*Chevrolet Motor Division of General Motors Corporation*)

Fig. 4-25. Power flow through the gear train in first gear. (*Chevrolet Motor Division of General Motors Corporation*)

Fig. 4-26. Power flow through the gear train in second gear. (*Chevrolet Motor Division of General Motors Corporation*)

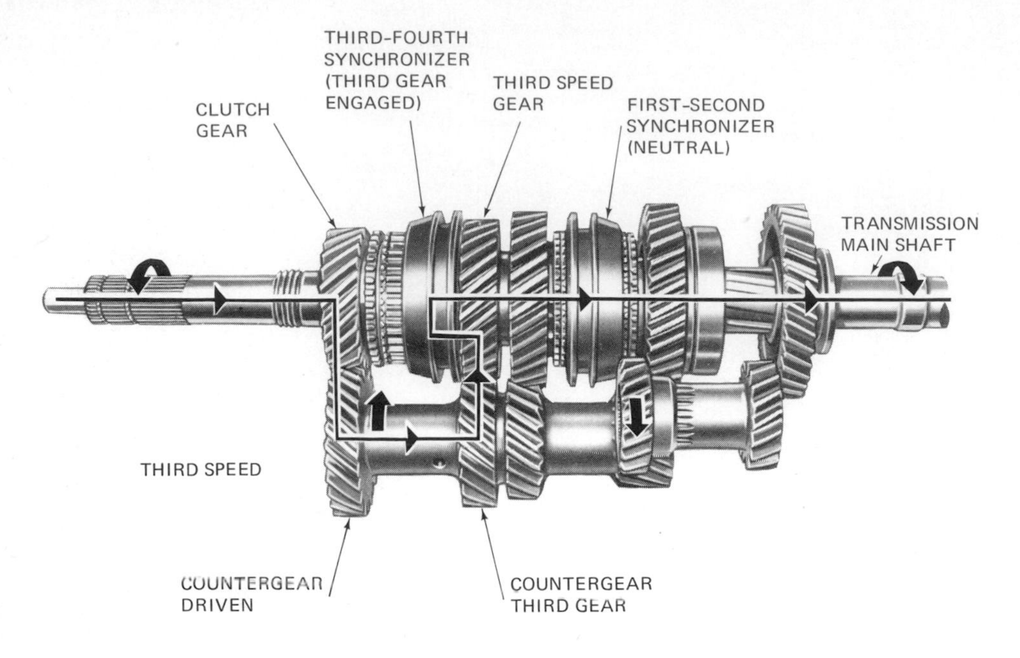

Fig. 4-27. Power flow through the gear train in third gear. (*Chevrolet Motor Division of General Motors Corporation*)

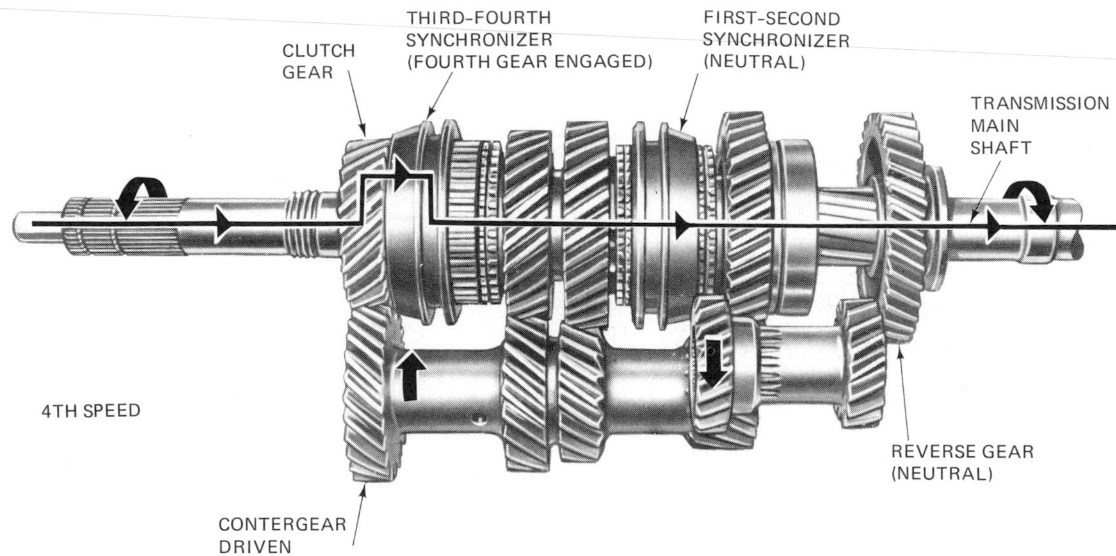

Fig. 4-28. Power flow through the gear train when in fourth gear. (*Chevrolet Motor Division of General Motors Corporation*)

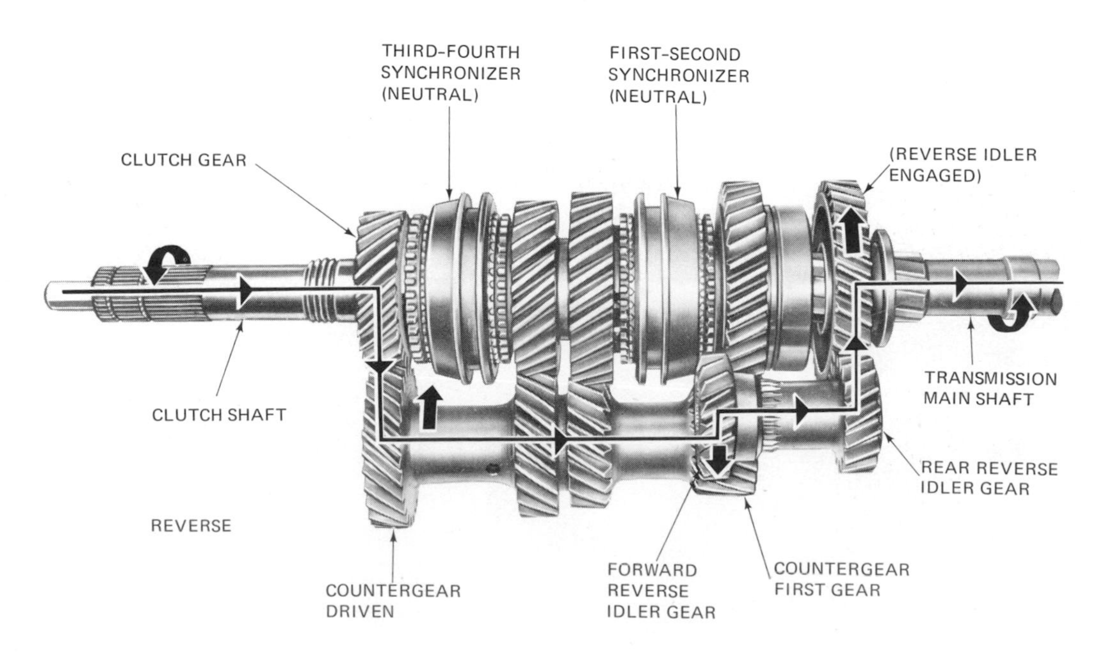

Fig. 4-29. Power flow through the gear train in reverse. (*Chevrolet Motor Division of General Motors Corporation*)

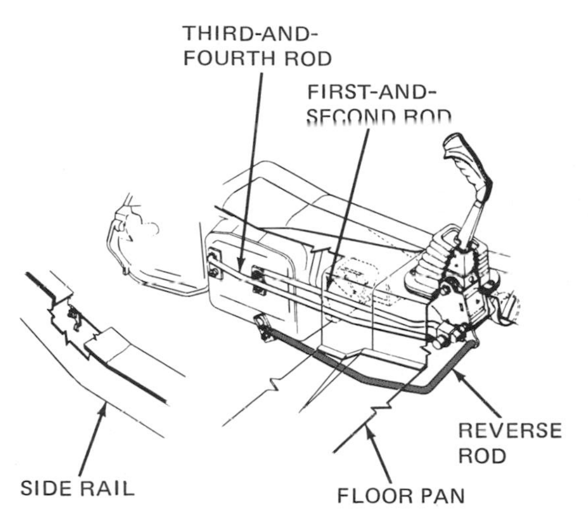

Fig. 4-30. Floorboard gearshift lever and linkage for a four-speed transmission. (*Chrysler Corporation*)

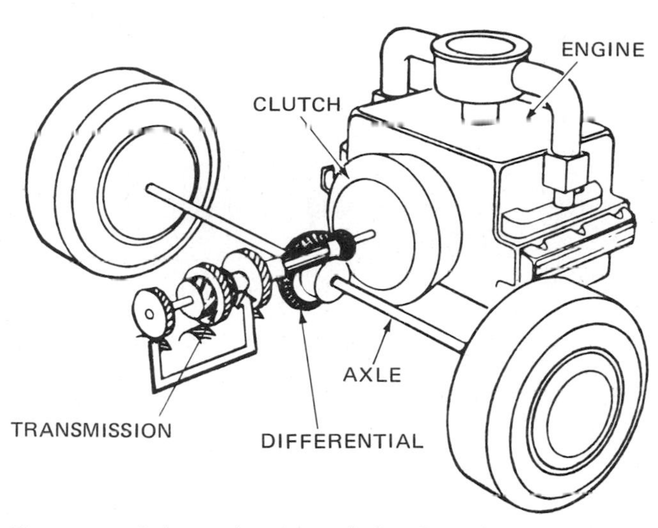

Fig. 4-31. Schematic view of the Corvair power train showing locations of the engine, clutch, and transmission. (*Chevrolet Motor Division of General Motors Corporation*)

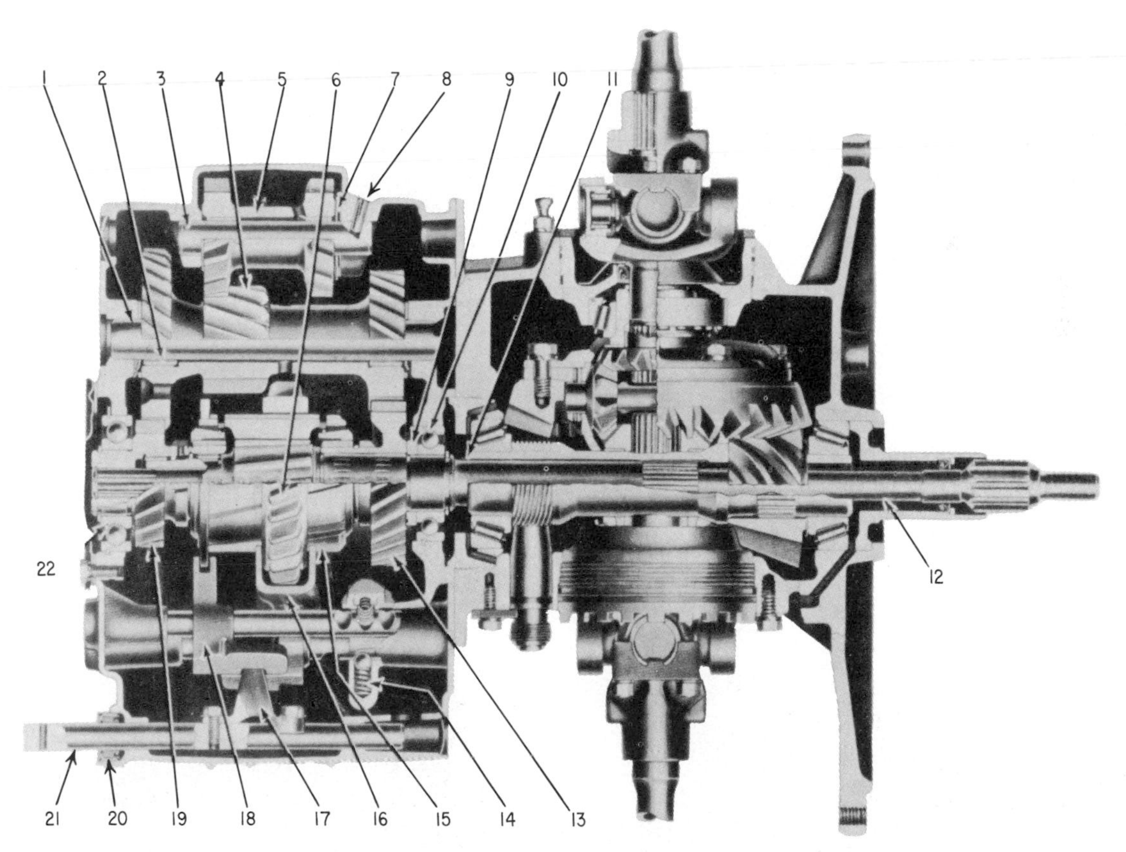

1. Countergear shaft
2. Countergear needle bearings
3. Reverse-idle-gear shaft
4. Countergear
5. Reverse idler gear
6. First-and-reverse sliding gear
7. Radial needle bearing
8. Reverse-idler-shaft retaining pin
9. Thrust washer
10. Main-shaft bearing
11. Main shaft
12. Clutch shaft
13. Second-speed gear
14. First-and-reverse detent spring and ball
15. Second-and-third-speed clutch
16. First-and-reverse shift fork
17. Manual-shift shaft finger
18. Second-and-third-speed shift fork
19. Clutch gear
20. Manual-shift shaft seal
21. Manual-shift shaft
22. Clutch-gear bearing

Fig. 4-32. Transmission for a rear-engine car. This assembly is called a *transaxle* by the manufacturer. (*Chevrolet Motor Division of General Motors Corporation*)

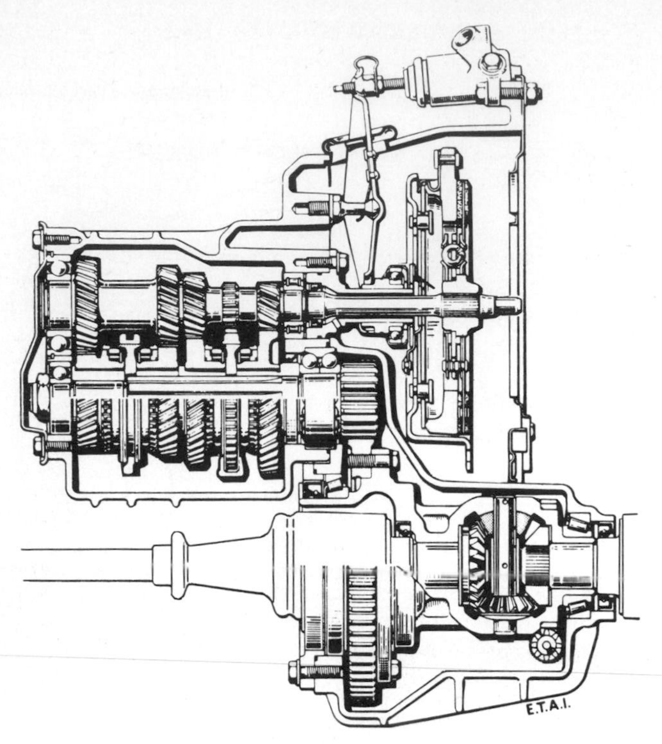

Fig. 4-33. Cutaway view of a four-forward-speed transmission for a cross-mounted engine used on a front-drive car. (*Simca*)

is 0.73:1.00. This means that the input shaft turns only 0.73 time to turn the output shaft one full turn. In Chap. 5, which describes the separate overdrive units, we explain the advantages of overdriving the output shaft. It reduces engine speed while maintaining highway speed. This reduces gasoline consumption and engine and accessory wear. For example, if the engine is turning 2,800 rpm to maintain highway speed and you shift to fourth, your engine speed would drop to about 2,100 rpm while still maintaining the same highway speed.

The overdrive is achieved by having a relatively large gear on the countershaft mesh with a relatively small gear on the main shaft. When the power flows through these two, the input shaft driving through the countershaft must turn only about $\frac{3}{4}$ turn (0.73 turn) to turn the output shaft once. The other three gear positions are conventional; that is, they are just like the three gear positions in the standard three-speed transmission described previously.

⊘ 4-10 Five-Speed Transmission with Overdrive This transmission offers the gear ratios shown in Fig. 4-37. Note that in fourth, the ratio is 1:1. This means that the input and output shafts turn at the same speed. However, in fifth, the output shaft turns faster. The ratio is 0.861:1.000. That is, the input shaft turns only 0.861 time to turn the output shaft one full turn. In Chap. 5, which describes separate overdrive units, we explain the advantages of overdriving the output shaft. It reduces engine speed at high car speeds, thus reducing gasoline consumption and engine and accessory wear.

Figure 4-38 shows the gears and shafts in the five-speed transmission. Note that the parts are shown for both four-speed and five-speed transmissions. The same parts are used for both except that there are extra parts for the five-speed unit. The two boxes show these additional parts. Note that part 29, the fifth-speed gear subassembly, and part 61, the countershaft fifth gear, are the two gears that come into play when the shift is made into the fifth speed. Note that gear 61 is larger than gear 29. That means that the input shaft must turn less than one complete revolution to turn the output shaft one revolution. The actual ratio is 0.861:1.000, as we have said. This means that if the engine is turning 3,000 rpm to maintain highway speed in fourth, it would have to turn only about 2,600 rpm to maintain the same speed in fifth.

Fig. 4-34. Top view of a transmission and differential for a rear-engine car. (*Volkswagen*)

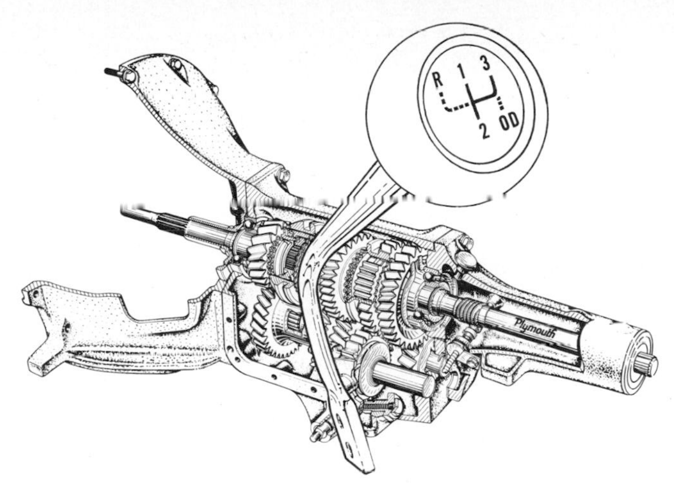

Fig. 4-35. Cutaway view of the four-speed transmission in which the fourth speed is an overdrive. In the fourth speed, the output shaft turns faster than the input shaft. In other words, the output shaft *overdrives* the input shaft. (*Chrysler Corporation*)

The five-speed transmission has synchronizers in all forward speeds. The first four forward speeds, as well as reverse, are very much like those in the four-speed transmission described in ⊘ 4-7. As noted previously, the main difference is that the five-speed transmission has extra gears to achieve overdrive in fifth.

⊘ 4-11 Transmission-controlled Spark Switch Several emission-control devices have now been added to cars. One is the transmission-controlled-spark (TCS) system. This system prevents vacuum advance in any gear but high. The switch that controls the system is screwed into a boss in the side cover of the transmission, as shown in Fig. 4-39. This switch is open in all gears but high, thereby preventing vacuum advance. In high, the switch is closed, allowing vacuum advance. The purpose of this system and how it works are described in *Automotive Emission Controls* and *Automotive Fuel, Lubricating, and Cooling Systems,* two other books in the McGraw-Hill Automotive Technology Series.

GEAR POSITION	GEAR RATIO	
	Input Shaft	Output Shaft
First	3.090	1.000
Second	1.670	1.000
Third	1.000	1.000 (Direct Drive)
Fourth	0.730	1.000 (Overdrive)

Fig. 4-36. Gear ratios of the four-speed transmission in which the fourth speed is an overdrive. (*Chrysler Corporation*)

⊘ 4-12 Backup Lights When the gearshift lever is moved to reverse, the linkage closes a switch that connects the backup lights to the battery. Thus, the lights come on automatically. This is a warning that the car is about to be backed. The backup lights also allow the driver to see behind the car.

⊘ 4-13 Speedometer Drive The speedometer is driven by a pair of gears found in the transmission-extension housing. One of these gears is mounted on the transmission main shaft (item 19 in Fig. 4-11). The other gear is mounted on the end of the flexible shaft connecting the speedometer to the transmission gear.

⊘ 4-14 Steering-Column Lock The steering-column lock requires that the transmission be in reverse gear before the ignition key can be removed from the ignition switch. In addition, the clutch safety switch will not permit the engine to be cranked unless the clutch pedal is pushed down far enough to completely disengage the clutch.

Check Your Progress

Progress Quiz 4-1 Here is your opportunity to check up on how well you remember the material you have just covered on various gearshift devices. If some of the questions stump you, go back and reread the pages that will give you the answers. This will help you remember the important facts.

Completing the Sentences The sentences that follow are incomplete. After each sentence there are several words or phrases, only one of which will correctly complete the sentence. Write each sentence in your notebook, selecting the proper word or phrase to complete it correctly.

1. The shifter mechanism (gearshift lever) on the steering column is normally connected to the transmission by: (*a*) a single link, (*b*) two or three linkages, (*c*) four linkages.

GEAR POSITION	GEAR RATIO	
	Input Shaft	Output Shaft
First	3.587	1.000
Second	2.022	1.000
Third	1.384	1.000
Fourth	1.000	1.000 (Direct Drive)
Fifth	0.861	1.000 (Overdrive)
Reverse	3.384	1.000

Fig. 4-37. Gear ratios of the five-speed transmission, showing the number of times the output shaft turns for each revolution of the input shaft. (*Toyota Motor Sales Company, Ltd.*)

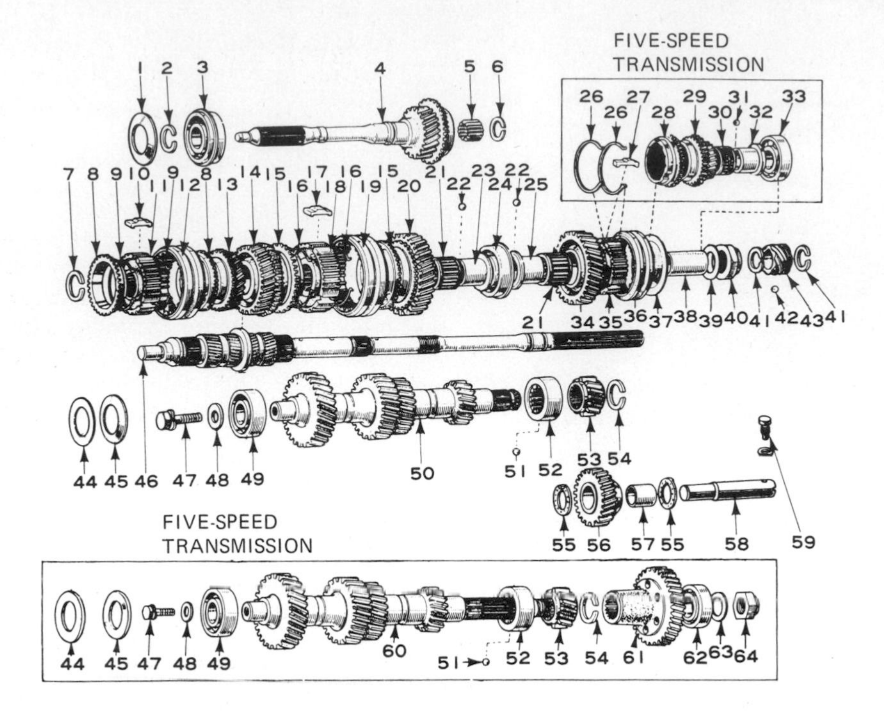

1. Gear-thrust-cone spring
2. Shaft snap ring
3. Radial ball bearing
4. Input shaft
5. Roller
6. Hole snap ring
7. Shaft snap ring
8. No. 1 synchronizer ring
9. No. 1 synchromesh-shifting-key spring
10. No. 2 synchromesh-shifting key
11. No. 2 transmission-clutch hub
12. No. 2 transmission-hub sleeve
13. Third-gear subassembly
14. Second-gear subassembly
15. No. 2 synchronizer ring
16. No. 1 synchromesh-shifting-key spring
17. No. 1 synchromesh shifting key
18. No. 1 transmission-clutch hub
19. No. 1 transmission-hub sleeve
20. First-gear subassembly
21. Needle roller bearing
22. Ball
23. First-gear bushing
24. Radial ball bearing
25. Reverse-gear bushing
26. No. 1 synchromesh-shifting-key spring
27. No. 3 synchromesh shifting key
28. No. 1 synchronizer ring
29. Fifth-gear subassembly
30. Needle roller bearing
31. Ball
32. Fifth-gear bushing
33. Radial ball bearing
34. Reverse gear
35. No. 3 transmission-clutch hub
36. No. 3 transmission-hub sleeve
37. Spacer
38. Spacer
39. Shim
40. Nut
41. Shaft snap ring
43. Speedometer drive gear
44. Shim
45. Gear-thrust-cone spring
46. Output shaft
47. Bolt with washer
48. Plate washer
49. Radial ball bearing
50. Countergear
51. Ball
52. Cylindrical roller bearing
53. Countershaft reverse gear
54. Shaft snap ring
55. Reverse-idler-gear thrust washer
56. Reverse idler gear
57. Bimetal-formed bushing
58. Reverse-idler-gear shaft
59. Shaft retaining bolt
60. Countergear
61. Fifth-gear countershaft
62. Radial ball bearing
63. Shim
64. Nut

Fig. 4-38. Gears and shafts in the four- and five-speed transmissions. The boxes enclose the additional parts in the five-speed unit. (*Toyota Motor Sales Company, Ltd.*)

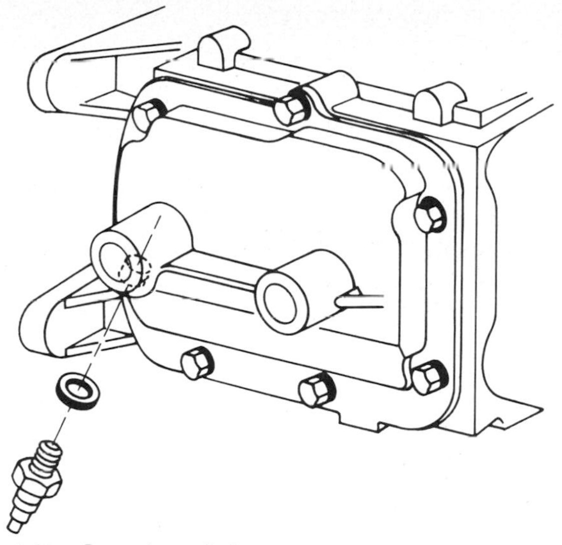

Fig. 4-39. Location of the transmission-controlled-spark switch on a transmission. (*Pontiac Motor Division of General Motors Corporation*)

2. Lifting the shift lever toward the steering wheel selects the: (*a*) first-reverse shift lever, (*b*) first-second shift lever, (*c*) second-third shift lever.
3. Moving the shift lever down away from the steering wheel selects the: (*a*) first-reverse shift lever, (*b*) first-second shift lever, (*c*) second-high shift lever.
4. To sum up the actions of the gearshift lever, it could be said that the first lever movement selects the gear and the second lever movement: (*a*) releases the gear, (*b*) shifts the gear, (*c*) engages the clutch.
5. Moving the gearshift lever parallel to the steering wheel with the lever lifted toward the wheel causes movement of the: (*a*) first-and-reverse gear, (*b*) first-and-second gear, (*c*) second-and-third gear.
6. The floor-shift type of lever mounts either on the floor or on: (*a*) the steering column, (*b*) a console, (*c*) the instrument panel.

CHAPTER 4 CHECKUP

NOTE: Since the following is a chapter review test, you should review the chapter before taking it.

You have been making good progress in your studies of automotive power trains and their components. Of these components, the transmission is one of the most important, and it is by far the most intricate. The automotive transmissions described in the chapters that follow may seem much more complicated than those covered in the chapter just completed. However, even though they may have more parts, they are no harder to understand. Thus, if you understand the construction and operation of the standard transmissions described in this chapter, you are well equipped to tackle the automatic transmissions described later in the book. Now find out how well you understand and remember the details of standard transmissions by answering the following questions. If any of the questions are hard to answer, just turn back into the chapter and reread the pages that will give you the information you need.

Completing the Sentences The sentences that follow are incomplete. After each sentence there are several words or phrases, only one of which will correctly complete the sentence. Write each sentence in your notebook, selecting the proper word or phrase to complete it correctly.

1. Synchronizing devices in the transmission synchronize gears about to be: (*a*) meshed, (*b*) demeshed, (*c*) stopped.
2. Synchronizing devices are normally used on the: (*a*) first-and-reverse, (*b*) reverse-and-second, (*c*) second-and-third gear positions.
3. Considering that the gearshift lever requires two separate motions to shift gears, the first movement: (*a*) moves the gear assembly, (*b*) selects the gear, (*c*) meshes the gears.
4. On the main shaft of the four-speed transmission described in the chapter, there are: (*a*) three, (*b*) four, (*c*) five external gears.
5. The three-speed transmission has: (*a*) one, (*b*) two, (*c*) three, (*d*) four shifter forks or yokes.
6. In the four-speed transmissions described in the chapter, there are: (*a*) four, (*b*) five, (*c*) six, (*d*) seven shift-lever positions, counting neutral and reverse.
7. The three-speed passenger-car transmission has: (*a*) three, (*b*) four, (*c*) five, (*d*) six shift-lever positions, counting neutral and reverse.
8. In the standard transmission, the countershaft gears turn: (*a*) faster than, (*b*) slower than, (*c*) at same speed as the clutch gear.

Listing Parts In the following, you are asked to list parts that go into various automotive transmissions. Write these lists in your notebook.

1. Make a list of the major parts used in a typical three-speed transmission.
2. Compare the illustrations of the three- and four-speed transmissions in the chapter and make a list of major parts that only the latter transmission contains.
3. List the parts that make up a typical linkage system between the steering-column gearshift lever and the transmission.

Purpose and Operation of Components In the following, you are asked to write the purpose and operation of the transmissions and components described in the chapter. If you have any difficulty in writing down your explanation, turn back into the chapter and reread the pages that will give you the answer. Then write your explanation. Don't copy; try to tell it in your own words, just as you might explain it to a friend. Write in your notebook.

1. What is the basic purpose of the transmission?

2. If a 30-tooth gear is meshed with a 12-tooth gear, what is the gear ratio? If the 30-tooth gear turns at 1,000 rpm, how fast will the 12-tooth gear turn?
3. What is torque?
4. If the 30-tooth gear in question 2 has 100 pound-feet [13.82 kg-m] of torque applied to it, what is the torque on the 12-tooth gear (ignoring friction)?
5. What manual transmission is most commonly used on cars?
6. Through what gears does the power flow in first in the transmission gearing shown in Figs. 4-10 to 4-12?
7. Through what gears does the power flow in second?
8. Through what gears does the power flow in reverse?
9. What is the purpose of the synchronizer? Explain how the synchronizer in Fig. 4-22 works.
10. What is meant by the expression "four on the floor"?
11. What are the two gearshift-lever locations for manual transmissions?
12. Describe the linkage actions when gears are shifted into second from first.

SUGGESTIONS FOR FURTHER STUDY

Examine disassembled transmissions and transmission parts. If you can find a transmission that has been removed from a car, take off the side or top cover so that you can see the gearing. Now, trace the power flow as you move the gears into the forward speeds and reverse. Make drawings in your notebook similar to those in this chapter, showing the gears through which the power flows. With the gears in each of the forward speeds, turn the input shaft (clutch gear). Count the number of times you must turn this shaft to turn the transmission main shaft (output shaft) once. This tells you the gear ratio in each of the gears.

Examine several manufacturers' shop manuals for manual transmissions. Make a list of car models and the types of manual transmissions they use. Late-model Chevrolets, for example, use three-speed Saginaw and four-speed Muncie and Saginaw transmissions. Write any important facts you learn in your notebook.

chapter 5

OVERDRIVES

This chapter discusses the purpose and function of the overdrive as well as its operation. The description of the planetary-gear system in the pages that follow is especially important because planetary-gear systems are used in many automatic transmissions. In order to understand automatic transmissions, you must understand the planetary-gear system. Thus, be sure you spend enough time studying planetary gears so that you understand how they operate.

⊘ **5-1 Purpose of Overdrive** In standard transmissions the high-gear position imposes a direct, or 1:1, ratio between the clutch shaft and the transmission main shaft. It is desirable at intermediate and high car speeds, however, to establish a still more favorable ratio between the two shafts so that the transmission main shaft will turn more rapidly than the clutch shaft. The result will reduce engine speed at high car speeds and provide more economical operation and less engine and accessory wear per car-mile.

For these reasons, several car manufacturers have supplied overdrive mechanisms as special equipment. Overdrives function to cause the main shaft to overdrive, or turn more rapidly than, the clutch shaft. Although the ratio varies somewhat on different cars, when brought into operation, the overdrive drops the engine speed about 30 percent without any change of car speed. Therefore, on a car where direct or high gear provides 40 mph [64.37 km/h] car speed at 2,000 engine rpm, the overdrive would drop the engine speed to 1,400 rpm and still maintain a car speed of 40 mph [64.37 km/h].

The overdrives used on modern cars are automatic, coming into operation when the car speed reaches a predetermined value, usually somewhere around 30 mph [48.28 km/h]. They contain a selective feature that permits the driver to remain in direct drive or, if preferable, to shift into overdrive by merely raising the foot from the accelerator momentarily. To come out of overdrive, the driver merely depresses the throttle past the wide-open position. This actuates a throttle switch that causes electric circuits to function and bring the car out of overdrive. Thus there are two separate controls: a centrifugal device, or governor, that places the car into overdrive when the cut-in speed is reached, and an electric control that brings the car out of overdrive.

⊘ **5-2 Overdrive Components** The overdrive is located just back of the transmission between the transmission and the drive shaft (Figs. 5-1 and 5-2). Essentially, it is made up of two parts: a planetary-gear system, and a freewheeling mechanism, together with the necessary controls and supports. An understanding of these devices is important because it is the basis for understanding the automatic transmissions described in following chapters. Such automatic transmissions commonly use planetary-gear systems. Let us examine the freewheeling mechanism and the planetary gear system in greater detail.

NOTE: There is one overdrive unit that is designed to be installed at the rear end of the drive shaft on the differential. With this arrangement, the overdrive can be used on both manually shifted and automatic transmissions. Although no automotive manufacturers have announced their intention to supply this rear-mounted overdrive, it is available for installation on present cars. Installation requires

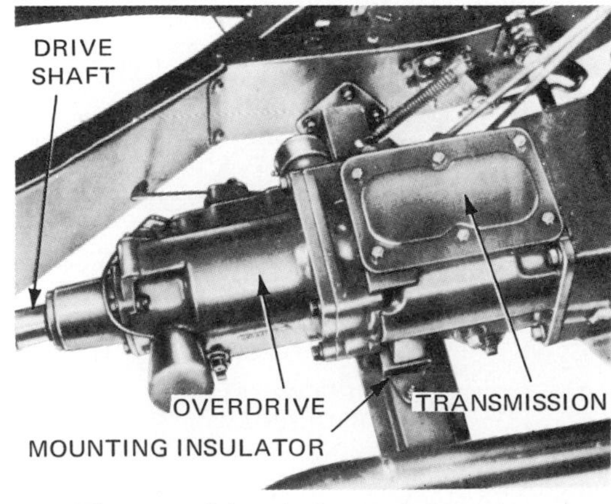

Fig. 5-1. The overdrive is located between the transmission and the drive shaft.

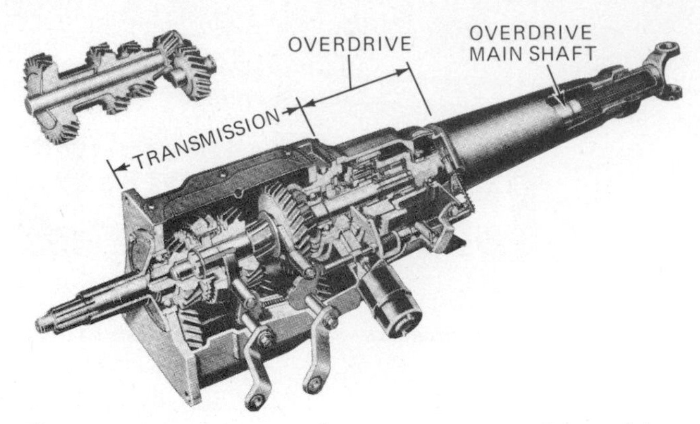

Fig. 5-2. Cutaway view of a transmission with overdrive. (*Ford Motor Company*)

shortening of the drive shaft and use of special adapters.

⊘ 5-3 Freewheeling Mechanism Essentially, the freewheeling mechanism is a coupling between two shafts that are in line with each other (Fig. 5-3). This coupling contains an inner and outer shell, or *race*, with rollers between them. The coupling is often called an *overrunning clutch*. The name comes from the action of the mechanism. When shaft A (Fig. 5-3) applies driving torque through the coupling, or overrunning clutch, the clutch acts as a solid drive and causes shaft B to turn at the same speed as shaft A. However, if shaft A should slow down or stop, shaft B could still turn faster than, or overrun, shaft A. In this case, the clutch "uncouples" the two shafts and thereby permits shaft B to overrun.

In the car, shaft A would be attached to the transmission output shaft and shaft B would be attached to the drive shaft. With the engine driving the car, the overrunning clutch would "clutch" and act as a solid drive so that shafts A and B would turn at the same speed. But with the accelerator pedal released so that the engine slows down, shaft B could then overrun shaft A and the car would therefore coast, or *freewheel*. This is the actual operation of freewheeling devices used on cars a number of years ago. In the overdrive, however, the action is somewhat different.

Figure 5-4 shows the inner parts of an overrunning clutch used in an overdrive. The only additional item needed to complete the overrunning clutch shown is an outer shell, or race, that encloses the rollers. Figure 5-5 shows the action of the overrunning clutch in end view. The clutch contains an inner shell, or race, which has a series of high spots, or cams, evenly spaced around its entire circumference. There is one cam for each roller. This inner race, with cams, is usually called the *clutch cam*. A number of hardened-steel rollers lie in the low areas between the high spots on the clutch cam. These rollers, in turn, are held in place by an outer shell, or race. There is also a roller retainer, or roller cage, as shown in Fig. 5-4, which simply retains the rollers in the proper relative positions. The inner race drives, and the outer race is driven. Also, the outer race can overrun, or turn faster than, the inner race.

When the inner race is driving and turning the outer race at the same speed, then the condition is as shown in Fig. 5-5*a*. The rollers have been turned

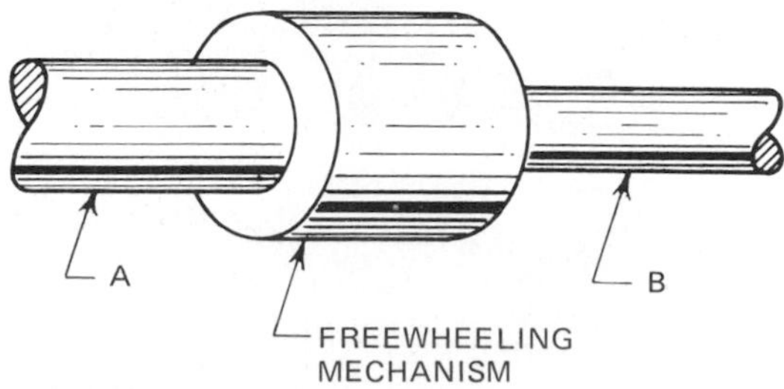

Fig. 5-3. The freewheeling mechanism provides solid drive when shaft A is turning shaft B. But if shaft A slows or stops, shaft B can still "freewheel," or turn faster than (overrun), shaft A.

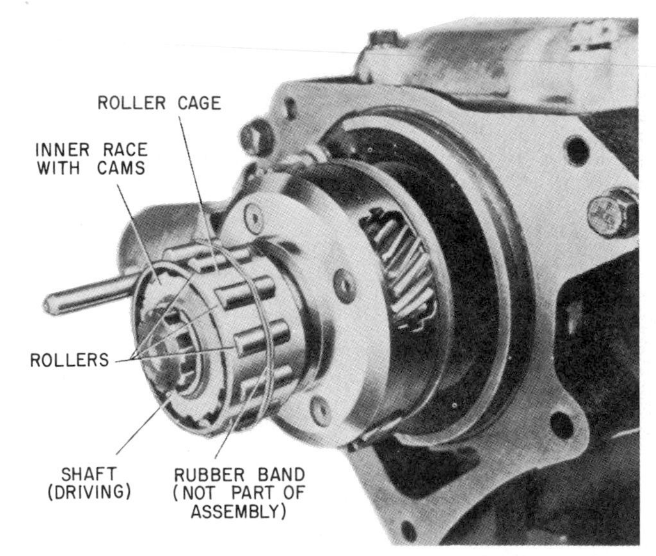

Fig. 5-4. Inner parts of an overrunning clutch, or freewheeling mechanism, used in an overdrive. The rubber band is not part of the assembly but is shown since it is temporarily holding the rollers in place. In the actual assembly, an outer race holds the rollers in place.

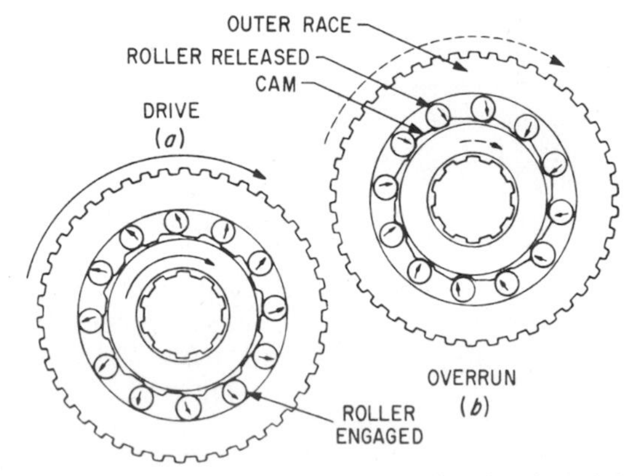

Fig. 5-5. Action in the overrunning clutch when (*a*) driving and (*b*) overrunning. (*American Motors Corporation*)

up onto the high spots, or cams, of the inner race, thus wedging between the two races. The outer race therefore is forced to turn with the inner race. The overrunning clutch acts as a solid drive.

However, if the inner race should slow down or stop, the outer race can continue to turn, or overrun, the inner race. As this happens, the rollers are rolled forward into the low spots on the clutch cam. There, they no longer wedge between the two races, and the outer race can overrun freely (Fig. 5-5*b*). But when the inner race speeds up again and catches up with the outer race, the rollers once more wedge so that the inner race drives the outer race; both races again turn at the same speed.

⊘ 5-4 Gear Combinations Before we discuss planetary gears, let us examine some gear combinations. In ⊘ 1-5 we learned that two meshing spur gears will turn in opposite directions. Also, we learned that when a big gear drives a small gear, the speed increases but the torque decreases. Conversely, when a small gear drives a big gear, the speed decreases but the torque increases. If you are not clear on these facts, review ⊘ 1-5. They are basic to what we are about to discuss: various gear combinations and planetary gears.

When two gears are in mesh, they turn in opposite directions, as we have said. But if another gear is put into the gear train, as shown in Fig. 5-6, the two outside gears turn in the same direction. The middle gear is called an *idler* gear. It doesn't do any work; it is idle.

If the space is too small for an idler gear like that shown in Fig. 5-6, you can still get rotation of both gears in the same direction by using an *internal gear* like that shown in Fig. 5-7. This gear is called an *internal gear* because the teeth are on the inside. However, it is usually called a *ring gear*. Most planetary gears have ring gears.

In Fig. 5-7 the small driving gear is shown meshed with the teeth on the inside of the large driven, or internal, gear. When the small gear rotates, its teeth push on the internal teeth of the ring gear, forcing the ring gear to rotate in the same direction as the driving gear.

The addition of one more gear to the ring-gear system illustrated in Fig. 5-7 turns it into a planetary

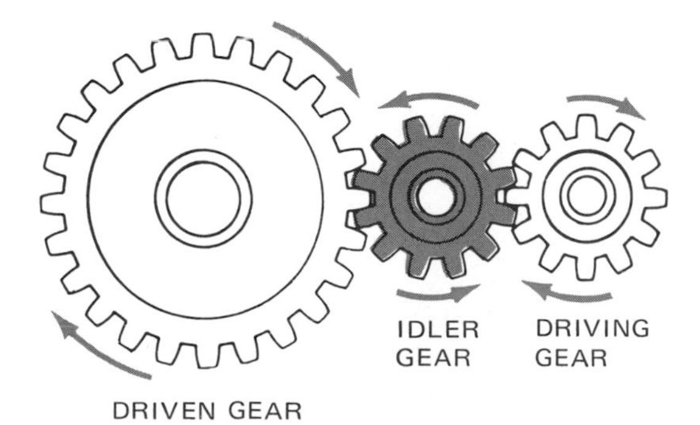

Fig. 5-6. The idler gear causes the driven gear to turn in the same direction as the driving gear.

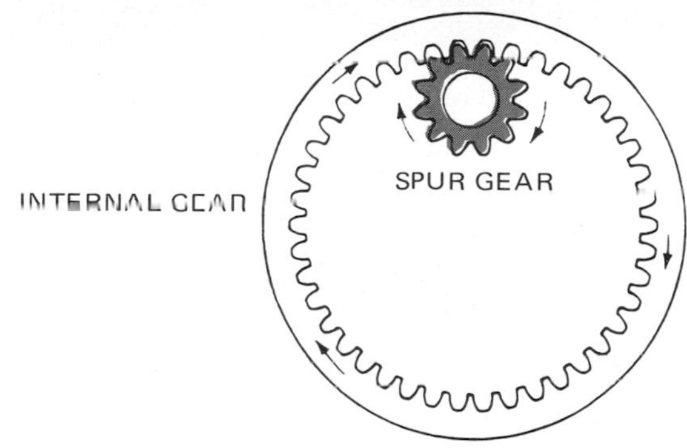

Fig. 5-7. If one internal gear is used with one external gear, both the driven and the driving gears turn in the same direction.

gearset. This gear is centered in the ring gear, as shown in Fig. 5-8. This gear is called the *sun gear* because it is in the center of the system just as the sun is in the center of our solar system. The sun gear meshes with the *spur gear,* and this spur gear, in turn, meshes with the internal, or ring, gear. The intermediate spur gear is called the *planet-pinion,* or *planet gear,* because it revolves around the sun gear like a planet revolves around the sun in our solar system. Notice one very important thing: The planet pinion rotates around the sun gear as it turns on its own axis. This is just what the planet Earth does as it revolves around the sun. Note the direction of rotation of each gear in the system as shown in Fig. 5-8.

⊘ 5-5 Planetary-Gear-System Operation Figure 5-8 shows a gear combination that can do several tricks. It can act as a speed reducer and a torque increaser. It can also act as a speed increaser and a torque reducer. Furthermore, it can act as a direction changer. All these variations are made possible by applying the input rotation to different gears and holding one of the other two gears stationary.

Before we study this gear combination, let's complete our planetary-gear system by adding another planet-pinion gear, as shown in Fig. 5-9. The two planet pinions are mounted on shafts that are

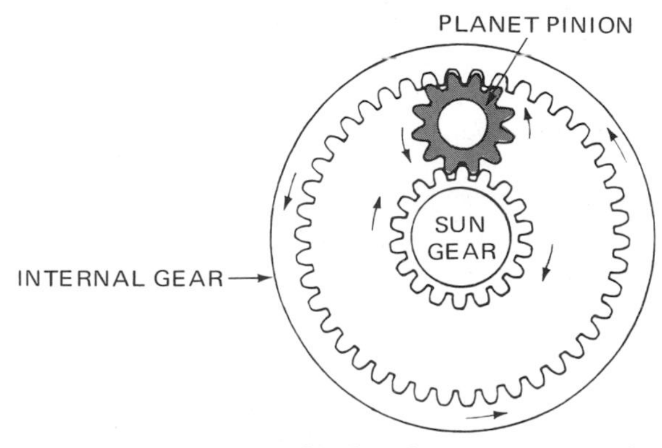

Fig. 5-8. If a sun gear is added to the arrangement shown in Fig. 5-7, the result is a simple planetary-gear system.

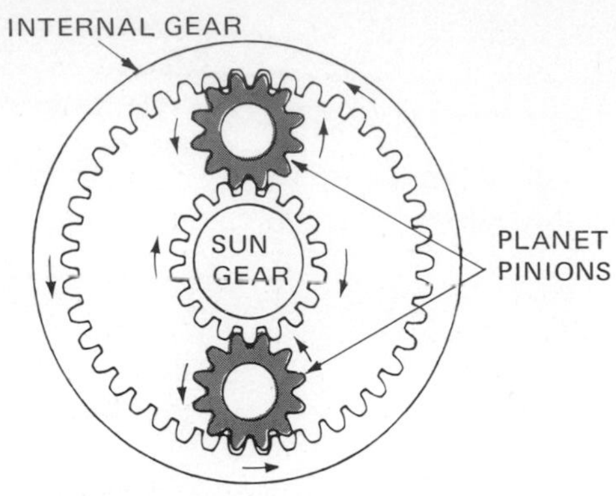

Fig. 5-9. To complete the planetary-gear system, we add a second planet pinion. This pinion balances the forces so that the system runs smoothly. Planetary-gear systems in overdrives and automatic transmissions have three or four planet pinions.

part of a plate called the *planet-pinion cage,* or *carrier,* which is illustrated in Fig. 5-10. The carrier is mounted on a shaft so that the shaft can rotate and carry the planet pinions around in a circle. When this happens, their teeth, meshed with both the sun gear and the ring gear, cause movement of all the various gears. Each of the gears and the planet-pinion carrier is called a *member.* There are three members in the planetary-gear system: the sun gear, the planet-pinion carrier with the planet pinions, and the ring gear.

When we hold one member stationary and turn another, the third member will produce either a speed increase, a speed reduction, or a direction reversal. Let's take a look at all of these possibilities.

1. SPEED INCREASE—SUN GEAR STATIONARY If we turn the planet-pinion carrier and hold the sun gear stationary, we get a speed increase at the ring gear. Here is how this happens. As the planet-pinion-carrier shaft is rotated, it carries the pinion shafts around with it. When this occurs, the planet pinions must rotate on their shafts because they are meshed with the stationary sun gear. That is, as the planet pinions move in a circle around the sun gear, they also rotate on their shafts. Before we find out how this movement affects the ring gear, let us see how many times the planet pinions rotate on their shafts while moving in one complete circle around the stationary sun gear.

The number of times a planet pinion rotates on its shaft when making one complete revolution around the sun gear depends on their gear ratio. Suppose the sun gear has 36 teeth and the planet pinion has 6 teeth. According to our previous discussions of gear ratios, we might assume that the ratio would be 36:6, or 6:1. We would then say that the planet pinion rotates six times. Is this correct? No. The planet pinion rotates *seven* times on its shaft as it completes one revolution around the stationary sun gear. Where did the extra rotation come from? It came from the revolution that the planet pinion makes as it circles the stationary sun gear. Look at Fig. 5-11 and follow the planet pinion as it moves from the top around the sun gear and back to its starting position. Notice that the planet pinion must rotate seven times on its shaft while making one revolution around the sun gear. *The seventh rotation comes from the change of position of the planet pinion as it moves around the sun gear.*

This situation holds true regardless of the gear ratio of the sun gear to the planet pinion. If they both have the same number of teeth, the planet pinion will rotate twice on its shaft while making one revolution around the stationary sun gear. (If you have two spur gears of the same size and with the same number of teeth, you can prove this to yourself.)

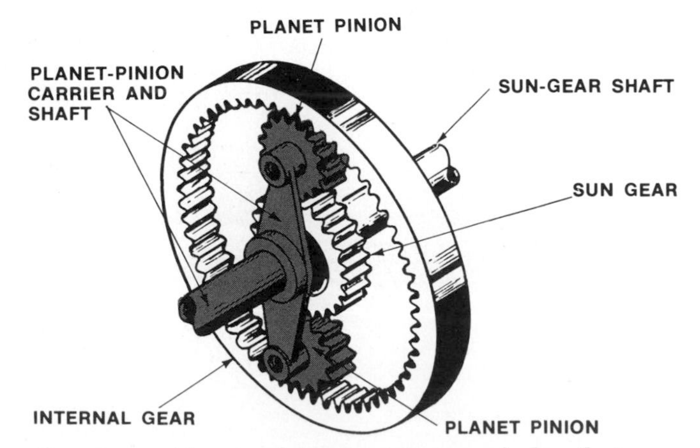

Fig. 5-10. Planet pinions rotate on shafts that are mounted on a planet-pinion carrier. The planet-pinion carrier is attached to a shaft that is exactly aligned with the sun-gear shaft. These shafts are exactly centered in the internal gear.

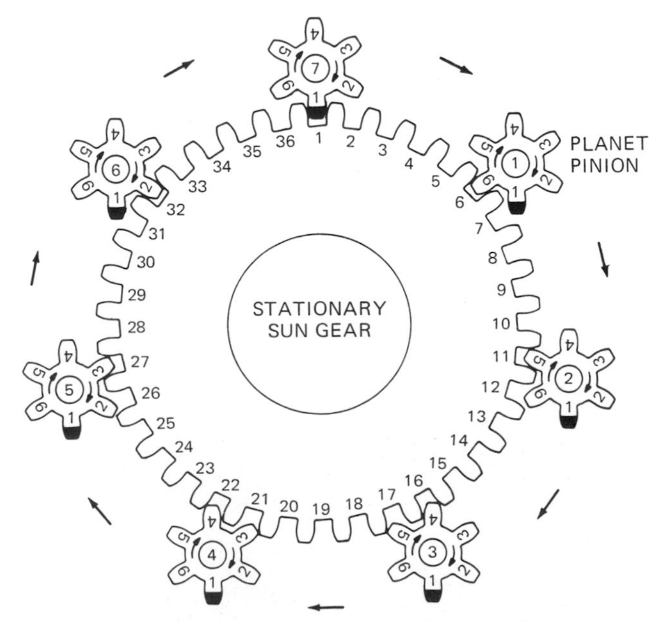

Fig. 5-11. The sun gear has 36 teeth; the planet pinion has 6 teeth. The positions marked 1 to 7 show the positions of the planet pinion as it rotates on its shaft and revolves around the sun gear. These positions prove that the planet pinion must rotate seven times as it makes one complete revolution around the sun gear.

Refer to Fig. 5-12. By the time that the planet pinion has revolved only one-quarter of the way around the stationary sun gear, it has already rotated $\frac{1}{2}$ turn on its shaft. By the time it has revolved halfway around the sun gear, it has completed one full turn on its shaft.

You can try this yourself with two coins (nickels, dimes, or quarters). Hold one coin, representing the sun gear, stationary. Put the other coin, representing the planet pinion, next to the stationary coin and roll it around the stationary coin. As you can see, the moving coin gets its extra rotation from the change in its position as it revolves around the stationary coin.

This extra rotation occurs regardless of the gear ratio of the sun gear and planet pinion. To take an extreme example, if the sun gear has 100 teeth, and the planet pinion has 10 teeth, the planet pinion rotates 11 times on its shaft as it makes one revolution around the stationary sun gear. Here's another example to think about: If the sun gear has 50 teeth and the planet pinion has 100 teeth, the planet pinion rotates, not $\frac{1}{2}$ turn, but $1\frac{1}{2}$ turns, as it revolves around the stationary sun gear. In this example, the sun gear is smaller than the planet pinion, but the extra revolution is still added despite the actual gear ratio.

Remember, we are talking about the first condition in the planetary-gear system: holding the sun gear stationary and rotating the planet-pinion carrier. This condition causes the planet pinions to rotate on their shafts as they revolve around the sun gear. This action causes the ring gear to rotate. *The ring gear will rotate in the same direction that the planet-pinion carrier is turning* and at a higher speed than the planet-pinion carrier.

Figure 5-13 shows how this speed increase is achieved. Remember, in studying this illustration, that the sun gear is stationary. The planet-pinion carrier rotates and forces the planet pinions to move in a circle around the sun gear. This action forces the planet pinions to rotate on their shafts. At any given instant, the pinion tooth meshed with the sun gear is stationary because the sun gear itself is stationary. Therefore, the planet pinion pivots around the stationary tooth. If the pinion shaft moves at a speed of 1 foot per second [0.305 m/s (meters per second)], then the outside tooth must move at twice this speed, or 2 feet per second [0.610 m/s]. This means that the ring-gear tooth meshed with this outside tooth must also move at 2 feet per second [0.610 m/s]. In Fig. 5-13, the internal gear rotates faster than the planet-pinion carrier.

To make planet-pinion-gear action a little clearer, let's take as an example the front wheel of a car rolling down the road at 30 mph [48.28 km/h]. The axle moves forward at 30 mph [48.28 km/h], but at any given instant, the point on the tire touching the road is stationary. At the same instant, though, the point on the tire exactly at the top of the wheel moves forward at 60 mph [96.56 km/h]; it moves twice as fast as the axle.

The ratio between the planet-pinion carrier and the internal gear can be altered by changing the sizes of the different gears. In Fig. 5-14, the ring gear makes one complete revolution while the planet-pinion carrier turns only 0.7 revolution. In other words, the ring gear turns faster than the planet-pinion carrier. The gear ratio between the two is 0.7:1. The system operates as a speed-increasing mechanism. The driven member, the ring gear, turns faster than the driving member, the planet-pinion carrier. *Notice that the ring gear rotates in the same direction as the planet-pinion carrier.* Notice also that the planetary-gear system in Fig. 5-14 has three, not two, planet pinions. This arrangement is common. Planet-pinion carriers in most overdrives and automatic transmissions have three or four planet pinions.

2. SPEED INCREASE—RING GEAR STATIONARY Another combination is to hold the ring gear stationary and turn the planet-pinion carrier. In this case, the sun gear is forced to rotate faster than the planet-pinion carrier and the system functions as a speed-increasing mechanism. The driven member, the sun gear, turns faster than the driving member, the planet-pinion carrier.

3. SPEED REDUCTION—SUN GEAR STATIONARY If we turn the ring gear while holding the sun gear stationary, the planet-pinion carrier will turn more slowly than the ring gear. This situation is just the opposite of the one described in item 1, in which

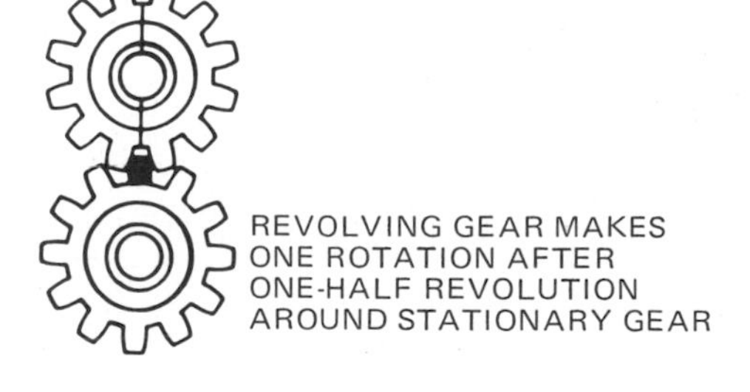

Fig. 5-12. Two spur gears having the same number of teeth. One spur gear serves as the sun gear, and the other serves as the planet pinion. After the planet pinion has made one-half revolution around the sun gear, it has made one complete rotation. By the time it has completed one revolution around the sun gear, it will have made two complete rotations.

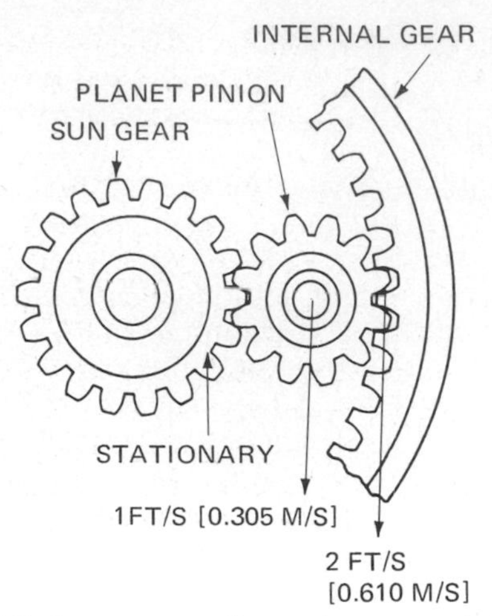

Fig. 5-13. If the sun gear is stationary and the planet-pinion carrier turns, the ring gear will turn faster than the carrier. The planet pinion pivots about the stationary teeth. If the center of the pinion shaft moves at 1 foot per second [0.305 m/s], the tooth opposite the stationary tooth will move at 2 feet per second [0.610 m/s] since it is twice as far away from the stationary tooth as the center of the shaft.

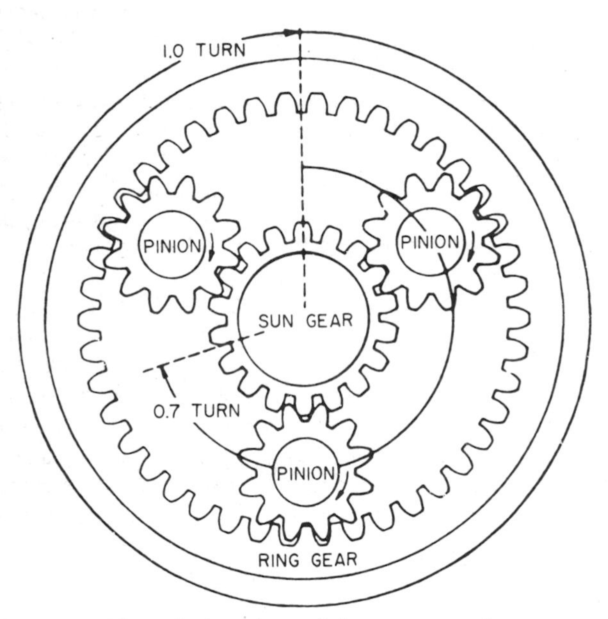

Fig. 5-14. The relative sizes of the gears, as shown, cause the ring gear to turn once while the planet-pinion carrier makes 0.7 turn with the sun gear held stationary.

the planet-pinion carrier turned the internal gear. Now, with the internal gear turning the planet-pinion carrier, the planetary-gear system functions as a speed-reducing mechanism. That is, the driven member, in this case the planet-pinion carrier, turns more slowly than the driving member, the ring gear.

4. *SPEED REDUCTION—RING GEAR STATIONARY* Let's see what happens if we hold the ring gear stationary and turn the sun gear. We find that the planet pinions turn on their shaft. They must also "walk around" the ring gear since they are in mesh with it. As they do this, the planet-pinion carrier is carried around. The carrier therefore rotates in the same direction as the sun gear but at a speed less than the sun-gear speed. In this case, the system functions as a speed-reducing mechanism. The driven member, the planet-pinion carrier, turns more slowly than the driving member, the sun gear.

5. *REVERSE 1* Another variation is to hold the planet-pinion carrier stationary and turn the ring gear. In this case, the planet pinions act as idlers and cause the sun gear to turn in the direction opposite to that of the ring-gear rotation. Here the system functions as a direction-changing system, with the sun gear turning faster than the ring gear.

6. *REVERSE 2* There is still one more combination. If the planet-pinion carrier is held stationary and the sun gear is turned, then the ring gear turns in a reverse direction, but slower than the sun gear.

7. *DIRECT DRIVE* If any two of the three members—the sun gear, carrier, or ring gear—are locked together, then the entire planetary-gear system is locked out and the input and output shafts turn at the same speed. That is, there is no change of speed through the system, and the gear ratio is 1:1. On the other hand, if no member is held stationary and no two members are locked together, then the system does not transmit power at all. The input shaft may turn, but the output shaft does not.

8. *CHART OF DRIVE CONDITIONS* All the planetary-gear-system operating conditions discussed previously are tabulated in the chart shown in Fig. 5-15. Consider condition 1: The letter T in this column indicates the driving gear which is the pinion carrier; H indicates that the sun gear is held stationary; D indicates that the ring gear is being driven; and I designates an increase of speed between the pinion cage, or carrier, and the ring gear.

The three conditions that you will find in automatic transmissions are listed in columns 3, 4, and 6. The planetary-gear system as used in automatic transmissions is designed to produce a speed change through the system. Also, the system provides direct drive by locking two of the members together.

9. *NAMES OF PLANETARY PARTS* Before we go further, we should mention that different manufacturers give different names for the parts of the planetary gears. For example, the ring gear is also called the *internal,* or an *annulus, gear.* The planet-

CONDITION	1	2	3	4	5	6
Ring gear	D	H	T	H	T	D
Carrier	T	T	D	D	H	H
Sun gear	H	D	H	T	D	T
Speed	I	I	L	L	IR	LR

D—Driven
H—Hold
I—Increase of speed
L—Reduction of speed
R—Reverse
T—Turn or drive

Fig. 5-15. Various conditions that are possible in the planetary-gear system if one member is held and another is turned.

pinion carrier is also called the *planet-pinion cage* and sometimes the *spider*. The planet pinions are also called *planet gears*. These different names are used alternatively in this lesson, in the pages that follow, and in manufacturers' shop manuals. You should know what they mean.

⊘ **5-6 Planetary-Gear System Applied to Overdrive** In the overdrive, the ring gear is attached to the output shaft and the three planet pinions are assembled into a cage that is splined to the transmission main shaft. The sun gear has an arrangement whereby it may be permitted to turn or it may be locked in a stationary position. When it is locked, the ring gear (and thus the output shaft) is forced to turn faster than the transmission main shaft. In other words, the output shaft *overdrives* the transmission main shaft. Various views of the overdrive parts show the planetary-gear members (Figs. 5-16 to 5-23).

⊘ **5-7 Overdrive Operation** Figure 5-16 shows the component parts of one type of overdrive in exploded view, with all major parts disassembled but lined up in their approximate relationship in the actual assembly. Figures 5-17 to 5-23 show various operating aspects of the overdrive assembly. Figures 5-17 and 5-18 illustrate the relationship of the overdrive parts and power path through the overdrive when it is in direct drive. The transmission main shaft and output shaft turn at the same speed, and the driving action is through the overrunning clutch, or freewheeling mechanism. In both illustrations, the power path is shown by a white line or by arrows. Note that the power path is directly to the clutch cam, which is splined to the transmission main shaft. From the clutch cam, it passes through the rollers to the outer shell, or race, which is attached to the output shaft. Figure 5-17 shows the overdrive in cutaway view, and Fig. 5-18 shows the overdrive partly cut away with the parts extended so that they can be seen more easily.

1. *GOING INTO OVERDRIVE* In Figs. 5-17 and 5-18, the overdrive is in direct drive but is ready to go into overdrive just as soon as the car speed is great enough and the driver momentarily releases the accelerator. Note that in the two illustrations the pawl in the solenoid is out of the way of the sun-gear control plate. It is held in this position by the blocker ring, as shown in Fig. 5-19*a*. The blocker ring is loosely assembled onto the sun-gear control plate so that it can turn a few degrees one way or the other.

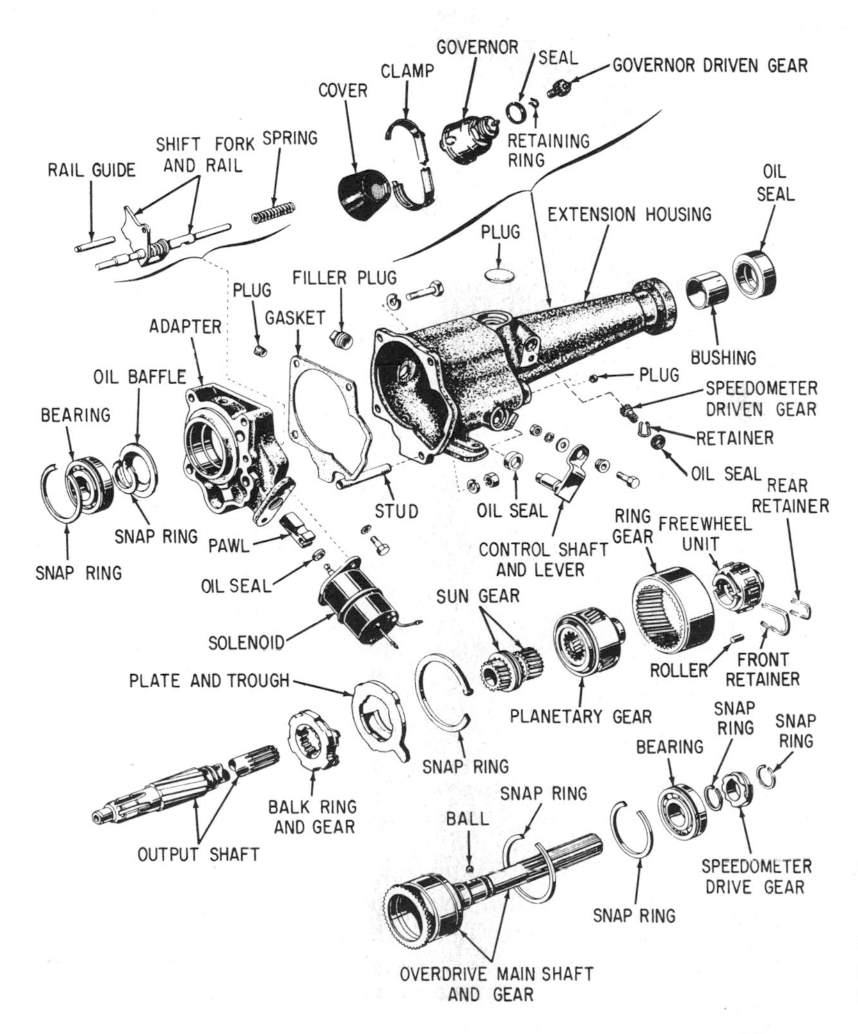

Fig. 5-16. Disassembled view of the overdrive unit. Note that the balk ring is also called the *blocker ring*. (*Ford Motor Company*)

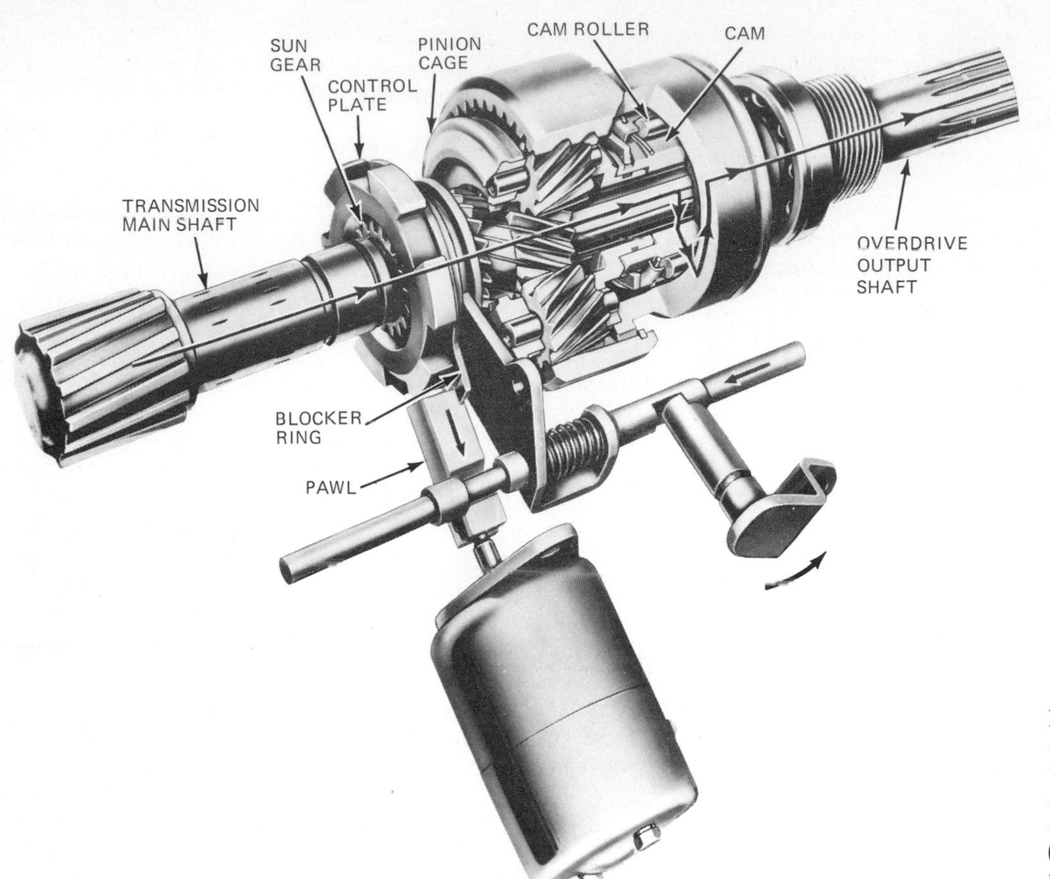

Fig. 5-17. Overdrive in direct drive with the transmission main shaft and output shaft turning at the same speed. (*American Motors Corporation*)

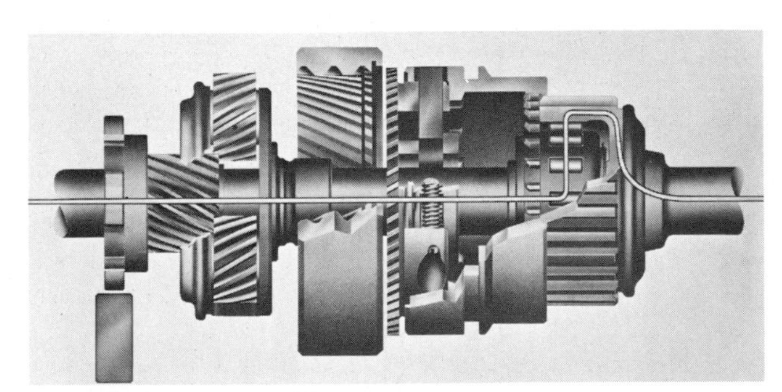

Fig. 5-18. Cutaway view of the overdrive unit with the car operating in direct drive. The assembly is shown in extended view with components separated so that all parts can be easily seen.

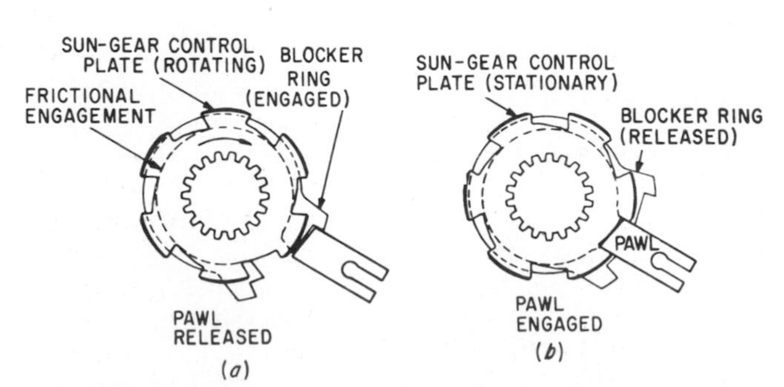

Fig. 5-19. Positions of the sun-gear control plate, blocker ring, and pawl with (*a*) the pawl released and (*b*) the pawl engaged. With the pawl engaged, the sun gear is held stationary.

When the car reaches overdrive cut-in speed (between 26 and 30 mph [41.84 and 48.28 km/h]), a governor driven from the overdrive output shaft closes electric contacts. This action connects the solenoid to the battery so that the solenoid is energized. This spring-loads the solenoid pawl so that it attempts to move upward and into a notch in the sun-gear control plate. However, the pawl is held away by the blocker ring, as shown in Fig. 5-19*a*.

When the driver momentarily releases the accelerator pedal, the engine speed drops. As it drops, the freewheeling mechanism goes into action, permitting the output shaft to overrun the transmission shaft. When this happens, the sun gear slows and then reverses directions. It does this because the ring gear (which rotates with the output shaft) begins to drive it through the planet pinions.

At the moment that the sun gear reverses directions, it moves the blocker ring around a few degrees, as shown in Fig. 5-19*b*. When this happens, the pawl can move inward and into the next notch that comes around on the sun-gear control plate. This movement locks the control plate in a stationary position. Figure 5-20 shows the three positions of the blocker ring and pawl.

Since the control plate is splined to the sun gear, this action also locks the sun gear in a stationary position. Now the car is in overdrive. With the sun gear locked, the power flow is as shown in Figs. 5-21 and 5-22. The transmission main shaft drives

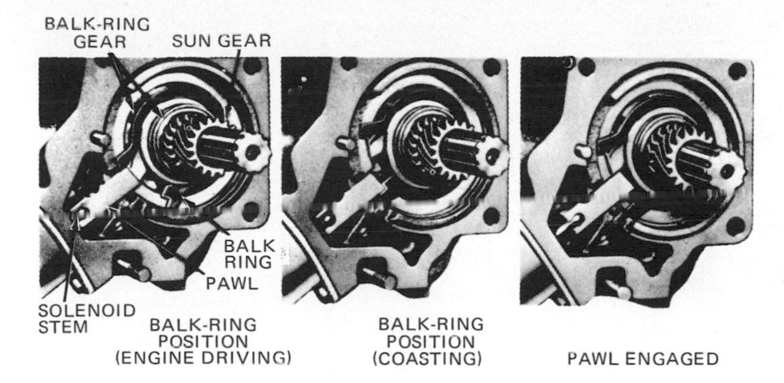

Fig. 5-20. The three positions of the blocker (balk) ring and solenoid pawl. (*Ford Motor Company*)

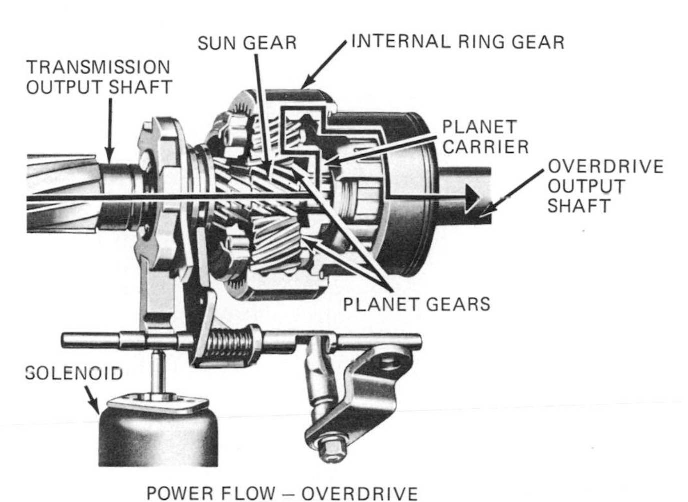

Fig. 5-21. With the sun gear held stationary and the planet-pinion cage driven by the transmission main shaft, the ring gear (and output shaft) is forced to turn faster than (or overdrive) the main shaft. (*Ford Motor Company*)

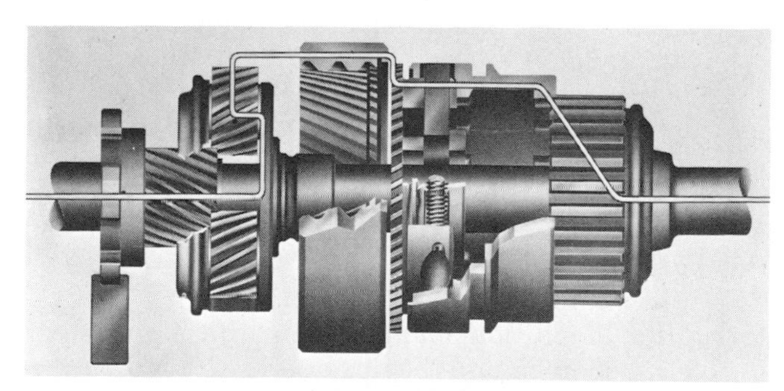

Fig. 5-22. Cutaway view of the planetary overdrive unit with the car operating in overdrive. The assembly is shown in extended view with components separated so that all parts may be seen easily.

through the planet-pinion cage (splined to the transmission main shaft) and causes the pinions to rotate around the sun gear, as shown in Fig. 5-13. The ring gear is attached to the output shaft through the outer race of the freewheeling mechanism. Thus the ring gear and output shaft overdrive (turn faster than) the transmission main shaft. Note that under this condition the freewheeling mechanism is freewheeling. The outer race is overrunning the clutch cam and thus the transmission main shaft.

2. COMING OUT OF OVERDRIVE To come out of overdrive, the driver merely pushes the accelerator all the way down. This would be the action a driver would take when an extra burst of power is required, for example, to pass another car. Pushing the accelerator all the way down causes the accelerator pedal to operate a kickdown switch. Operation of this switch produces two actions: First, it opens the solenoid circuit so that the solenoid attempts to withdraw the pawl from the sun-gear control plate. However, there is considerable pressure on the pawl since it is holding the sun gear stationary and the planet pinions are thrusting hard against the sun gear as they drive the ring gear. The second action of the kickdown switch momentarily relieves this drive, however, since the kickdown switch also opens the ignition circuit and thereby prevents the engine from delivering power. Since the engine will slow down if it is not delivering power, the driving thrust of the planet pinions on the sun gear is relieved almost instantly. This action frees the sun gear and the sun-gear control plate, and the drive pawl is pulled back by the spring in the solenoid. As the drive-pawl plunger in the solenoid "bottoms," it reestablishes the electric circuit to the ignition system so that the engine once again begins to deliver power. Now, with the sun gear unlocked, drive is again direct, as shown in Figs. 5-17 and 5-18.

The ignition system is disconnected for such a short time that the interruption of the flow of power is not noticeable. The entire sequence of events that takes place when the kickdown switch is closed may be completed in less than 1 second. The car goes from overdrive into direct drive very quickly.

To go into overdrive again, the driver merely releases the accelerator pedal momentarily, as already explained.

The electric controls involved in the overdrive action are described in ⊘ 5-8.

3. LOCKING OUT THE OVERDRIVE To lock out the overdrive, the car should be standing still or operating in direct drive. The driver pulls out a control knob on the car instrument panel. This action forces the control lever in the direction shown in Fig. 5-23. As the control lever moves in this direction, it forces the sun-gear cover plate and the sun gear to move toward the planet-pinion cage. The sun-gear teeth enter into mesh with internal teeth in the planet-pinion cage so that the two lock up.

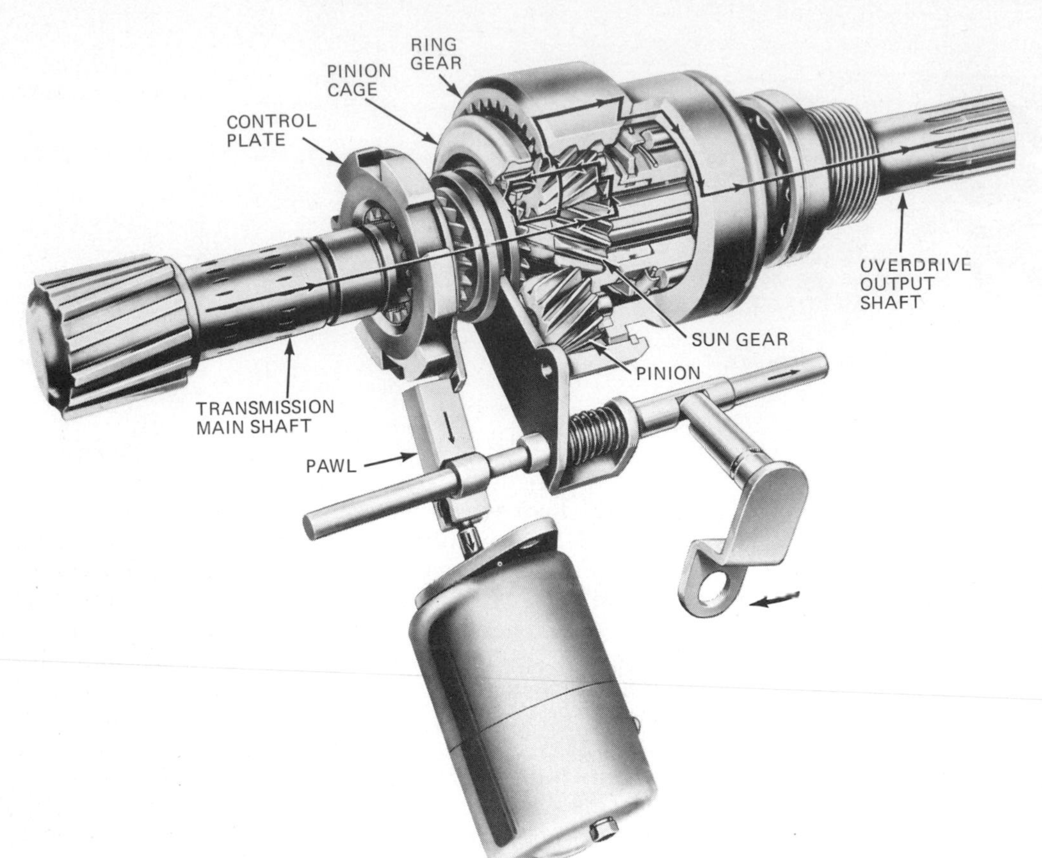

Fig. 5-23. Positions of the internal parts when the overdrive is locked out. Movement of an instrument-panel control shifts a control rod, which locks up the sun gear and planet-pinion cage so the two must turn together. (*American Motors Corporation*)

Under this condition, the sun gear and the pinion cage must turn together, and thus the entire assembly turns as a unit so that there can be no overrunning effect.

NOTE: The overdrive just explained is only one of several types that are used on automobiles. It is typical of them all, however, since all are essentially similar in construction and action.

⊘ 5-8 Overdrive Electric Controls Various types of electric controls of overdrives have been used, but essentially they have the same purpose: They must energize the solenoid as the car reaches cut-in speed. They must also disconnect the ignition circuit momentarily and at the same time open the solenoid circuit when the kickdown switch is closed as the driver wants to come out of overdrive.

Figure 5-24 shows a wiring circuit of the electric control system used with the overdrive described on previous pages. Let us trace this circuit in detail.

To go into overdrive, the driver pushes in the control knob on the instrument panel. This action places the system in the condition shown in Fig. 5-17 and connects the governor switch to the solenoid relay (through the lower contacts of the kickdown switch, as you will notice). When the car reaches cut-in speed, the governor closes its contacts, connecting the overdrive relay winding to the battery. The overdrive relay, in turn, closes its contacts, connecting the solenoid to the battery. Now the overdrive is ready to go into action. When the driver momentarily releases the pedal, the solenoid can send the pawl into a notch in the sun-gear control plate. This puts the transmission into overdrive.

To come out of overdrive, the driver pushes all the way down on the accelerator pedal, thus causing the lower contacts of the kickdown switch to open and the upper contacts to close. Opening the lower contacts opens the overdrive relay circuit. The overdrive relay therefore opens its contacts, opening the solenoid circuit. Also, closing the upper contacts on the kickdown switch directly grounds the ignition coil and thereby prevents ignition. With this interruption of ignition-system action, the engine stops delivering power and begins to slow down. As it does so, the thrust of the solenoid pawl is relieved and the spring pressure pulls the pawl out of the notch in the sun-gear control plate. When the solenoid pawl snaps into the "out" position, the contacts in the solenoid are opened, "ungrounding" the ignition coil and thereby permitting the ignition system to function again. The engine again begins to deliver power. This series of actions happens so fast that there is no noticeable lag in power delivery.

⊘ 5-9 Laycock Overdrive Figure 5-25 is a cutaway view of the Laycock J-type overdrive, made in England, that is used as an option on some American Motors Corporation six-cylinder cars. The overdrive can be engaged or disengaged automatically at about 35 mph [56 km/h] when the overdrive control is in

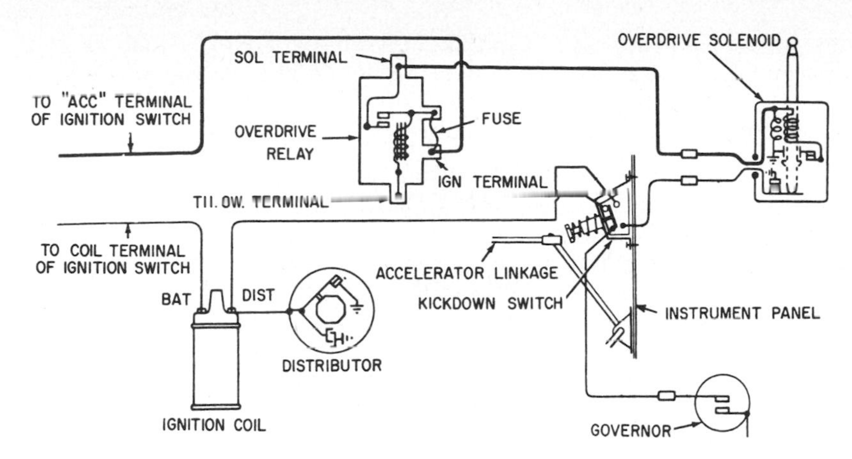

Fig. 5-24. Electric control circuit of the overdrive. (*Ford Motor Company*)

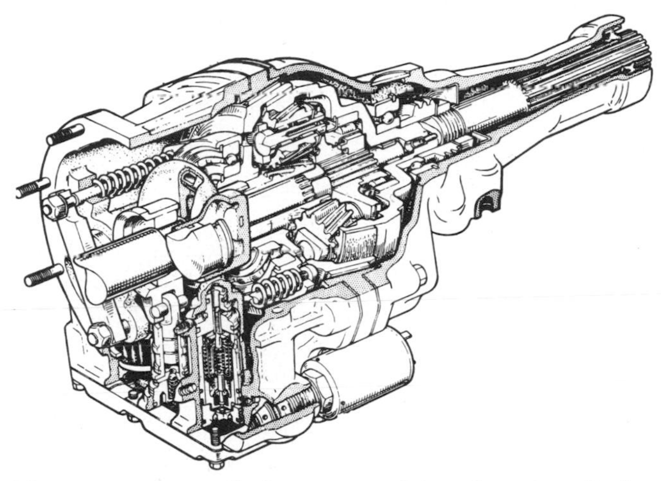

Fig. 5-25. Laycock J-type overdrive for six-cylinder engines. (*American Motors Corporation*)

the ON position. Also, the driver can engage or disengage it manually by pushing a button on the end of the turn-indicator lever.

Power flows from the transmission to the freewheeling, or overrunning, clutch through the planetary gears. When the overdrive goes into overdrive, hydraulic pressure causes the sliding-cone clutch to engage. This holds the sun gear stationary. Now, the input shaft, as it continues to rotate, carries the planet-pinion carrier around with it. The planet pinions therefore drive the ring gear, causing the ring gear to turn faster than the input shaft and planet-pinion carrier. In other words, the carrier overdrives the input shaft. Since the carrier is attached to the output shaft, the output shaft overdrives. Also, the system has a switch on the carburetor that permits the driver to push the accelerator all the way down and drop out of overdrive into third for passing. The saving in gasoline mileage is estimated to be at least 10 percent under the right operating conditions.

CHAPTER 5 CHECKUP

NOTE: Since the following is a chapter review test, you should review the chapter before taking it.

You have been making very good progress in your studies of automotive power trains and their components. In Chap. 4, you studied various transmissions used on cars, and in the chapter just finished you studied overdrives used with some of these transmissions. Actually, there is a double purpose in your studying the overdrive: First, of course, you should know how these devices are constructed and how they operate. Second, the component parts are used in one way or another in different automatic transmissions. Thus, a good understanding of the overdrive will lead to much easier understanding of the most intricate types of automatic transmissions. To give you a chance to check yourself on how well you understand and remember the material on the overdrive, the following questions have been included. If any of the questions cause you trouble, just turn back into the chapter and reread the pages that will clear up the difficulty.

Completing the Sentences The sentences that follow are incomplete. After each sentence there are several words or phrases, only one of which will correctly complete the sentence. Write each sentence in your notebook, selecting the proper word or phrase to complete it correctly.

1. The overdrive is located between the: (*a*) transmission and clutch, (*b*) transmission and drive shaft, (*c*) planetary gears and clutch.
2. The freewheeling mechanism contains: (*a*) a planetary gear, (*b*) a transmission, (*c*) an overrunning clutch, (*d*) a drive shaft.
3. The number of gears and pinions in the planetary-gear system described in the chapter is: (*a*) two, (*b*) three, (*c*) five, (*d*) eight.
4. The sun gear in the planetary-gear system meshes with the: (*a*) ring gear, (*b*) clutch gear, (*c*) planet pinions, (*d*) pinion cage.
5. Holding the sun gear stationary and turning the planet-pinion cage causes the ring gear to turn: (*a*) slower than, (*b*) faster than, (*c*) at the same speed as the planet-pinion cage.
6. In the overdrive, there is an arrangement whereby it is possible to lock stationary the: (*a*) ring gear, (*b*) sun gear, (*c*) planet-pinion cage.

7. In order to go into overdrive, the: (*a*) planet-pinion cage, (*b*) sun gear, (*c*) ring gear must be prevented from turning.
8. When coming out of overdrive, the overdrive electric control momentarily: (*a*) declutches the engine, (*b*) interrupts ignition-system action, (*c*) demeshes the sun gear.
9. If the ring gear is turned while the sun gear is held stationary, the planet-pinion cage will turn: (*a*) slower than, (*b*) at same speed as, (*c*) faster than the ring gear.
10. If the cage is held and the sun gear is turned, the ring gear will turn in a reverse direction but: (*a*) slower than, (*b*) at same speed as, (*c*) faster than the sun gear.

Listing Parts In the following, you are asked to list parts that go into various automotive transmissions discussed in the chapter. Write these lists in your notebook.

1. List the parts in the planetary-gear system.
2. List the parts in the overrunning clutch.
3. List the electrical components in the control circuit of the overdrive.
4. List the major parts in the overdrive.
5. List the parts through which the power passes when the car is in overdrive.

Purpose and Operation of Components In the following, you are asked to write the purpose and operation of the overdrives and overdrive components described in the chapter. If you have any difficulty in writing your explanation, turn back in the chapter and reread the pages that will give you the answer. Then write your explanation. Don't copy; try to tell it in your own words just as you might explain it to a friend. This is a good way to fix the explanation more firmly in your mind. Write in your notebook.

1. What is the purpose of the overdrive?
2. How does the planetary-gear system work when the sun gear is held stationary?
3. What is the purpose of the solenoid pawl? Describe the actions that permit the pawl to enter a notch in the sun-gear control plate.
4. Explain how the overdrive is locked out.
5. Explain what happens when the driver pushes the accelerator all the way down to come out of overdrive.
6. Explain the operation of the electric controls when the car speed increases to cut-in and the overdrive goes into operation.

SUGGESTIONS FOR FURTHER STUDY

If you are able to get hold of an overdrive or the parts of an overdrive, make a careful examination of the parts to determine how they fit together and operate. Note particularly the construction and operation of the overrunning clutch and planetary-gear system.

Study car manufacturers' manuals to learn more about overdrives. Although the unit described in the chapter is the most commonly used overdrive, other types have been used. Learn all you can about these other types by studying the manuals or by actually examining them, if possible. Be sure to write in your notebook any important facts you learn.

chapter 6

STANDARD-TRANSMISSION AND OVERDRIVE SERVICE

This chapter details the trouble-diagnosis, removal, overhaul, and reassembly procedures on various standard transmissions and overdrives. Following chapters discuss the operation and maintenance of the different semiautomatic and automatic transmissions in use on passenger cars.

⊘ 6-1 Standard-Transmission and Overdrive Trouble Diagnosis As a first step in any transmission or overdrive service, diagnosis of the trouble should be made in an attempt to pinpoint the trouble in the malfunctioning unit. Sometimes it is not possible to determine the exact location of a trouble, and the unit must be removed from the car so that it can be torn down and examined. At other times diagnosis will lead to the point of trouble so that it can be eliminated without major disassembly. It is also true that diagnosis may indicate that the transmission or overdrive must be removed. Nevertheless, it is a saving in time and effort in the long run always to check the operation of the assembly on the car to find the source of trouble. It may be that what was thought to be a case of transmission trouble is actually a trouble located in some other component of the car.

⊘ 6-2 Transmission and Overdrive Trouble-Diagnosis Chart The chart that follows lists the various troubles that might be blamed on the transmission or overdrive, together with their possible causes, checks to be made, and corrections needed. The chart is divided into two parts: Transmission Troubles and Overdrive Troubles. Most transmission troubles can be listed under a few headings, such as "hard shifting," "slips out of gear," "noises," and so on. In the chart, the section numbers are given where fuller explanations are found of how to find and eliminate the troubles.

NOTE: The troubles and possible causes are not listed in the chart in the order of frequency of occurrence; that is, item 1 (or item a) does not necessarily occur more often than item 2 (or item b).

TRANSMISSION AND OVERDRIVE TROUBLE-DIAGNOSIS CHART

(See ⊘ 6-3 to 6-10 for detailed explanations of the trouble causes and corrections listed below.)

COMPLAINT	POSSIBLE CAUSE	CHECK OR CORRECTION
TRANSMISSION TROUBLES		
1. Hard shifting into gear (⊘ 6-3)	a. Gearshift linkage out of adjustment	Adjust
	b. Gearshift linkage needs lubrication	Lubricate
	c. Clutch not releasing	Adjust (⊘ 3-4)
	d. Excessive clutch-pedal free play	Adjust (⊘ 3-4)
	e. Shifter fork bent	Replace or straighten
	f. Sliding gears or synchronizer tight on shaft splines	Replace defective parts
	g. Gear teeth battered	Replace defective gears
	h. Synchronizing unit damaged or springs improperly installed (after a service job)	Replace unit or defective parts; install spring properly
	i. Shifter tube binding in steering column	Correct tube alignment
	j. End of transmission input shaft binding in crankshaft pilot bushing	Lubricate; replace bushing

TRANSMISSION AND OVERDRIVE TROUBLE-DIAGNOSIS CHART (*Continued*)

COMPLAINT	POSSIBLE CAUSE	CHECK OR CORRECTION
2. Transmission sticks in gear (⊘ 6-4)	a. Gearshift linkage out of adjustment or disconnected	Adjust; reconnect
	b. Linkage needs lubrication	Lubricate
	c. Clutch not releasing	Adjust (⊘ 3-4)
	d. Detent balls (lockouts) stuck	Free
	e. Synchronizing unit stuck	Free; replace damaged parts
	f. Incorrect or insufficient lubricant in transmission	Replace with correct lubricant and correct amount
3. Transmission slips out of gear (⊘ 6-5)	a. Gearshift linkage out of adjustment	Adjust
	b. Insufficient lockout-spring pressure	Replace
	c. Bearings worn	Replace
	d. Excessive end play of shaft or gears	Replace worn or loose parts
	e. Synchronizer worn or defective	Repair; replace
	f. Transmission loose on clutch housing or misaligned	Tighten mounting bolts; correct alignment
	g. Clutch housing misaligned	Correct alignment
	h. Pilot bushing in crankshaft loose or broken	Replace
	i. Input-shaft retainer loose or broken	Replace
4. No power through transmission (⊘ 6-6)	a. Clutch slipping	Adjust (⊘ 3-4)
	b. Gear teeth stripped	Replace gears
	c. Shifter fork or other linkage part broken	Replace
	d. Gear or shaft broken	Replace
	e. Drive key or spline sheared off	Replace
5. Transmission noisy in neutral (⊘ 6-7)	a. Gears worn or tooth broken or chipped	Replace gears
	b. Bearings worn or dry	Replace; lubricate
	c. Defective input-shaft bearing	Replace
	d. Pilot bushing worn or loose in crankshaft	Replace
	e. Transmission misaligned with engine	Realign
	f. Countershaft worn or bent, or damaged thrust plate or washers	Replace worn or damaged parts
6. Transmission noisy in gear (⊘ 6-7)	a. Defective clutch friction disk	Replace
	b. Incorrect or insufficient lubricant	Replace with correct lubricant and correct amount
	c. Rear main bearing worn or dry	Replace or lubricate
	d. Gears loose on main shaft	Replace worn parts
	e. Worn or damaged synchronizers	Replace worn or damaged parts
	f. Speedometer gears worn	Replace
	g. Any condition noted in item 5	See item 5
7. Gears clash in shifting (⊘ 6-8)	a. Synchronizer defective	Repair or replace
	b. Clutch not releasing; incorrect pedal lash	Adjust
	c. Hydraulic system (hydraulic clutch) defective	Check cylinder; add fluid, etc.
	d. Excessive idle speed	Readjust
	e. Pilot bushing binding	Replace
	f. Incorrect gearshift-linkage adjustment	Adjust
	g. Incorrect lubricant	Replace with correct lubricant
8. Oil leaks (⊘ 6-9)	a. Foaming due to incorrect lubricant	Replace with correct lubricant
	b. Oil level too high	Use proper amount, no more
	c. Gaskets broken or missing	Replace
	d. Oil seals damaged or missing	Replace
	e. Oil slingers damaged, improperly installed, or missing	Replace correctly
	f. Drain plug loose	Tighten
	g. Transmission retainer bolts loose	Tighten
	h. Transmission or extension case cracked	Replace
	i. Speedometer-gear retainer loose	Tighten
	j. Side cover loose	Tighten
	k. Extension-housing seal worn or drive-line yoke worn	Replace

TRANSMISSION AND OVERDRIVE TROUBLE-DIAGNOSIS CHART (*Continued*)

COMPLAINT	POSSIBLE CAUSE	CHECK OR CORRECTION
OVERDRIVE TROUBLES		
	The overdrive may have any of the following troubles. In analyzing trouble on a car equipped with overdrive, be careful not to blame the overdrive for troubles in the transmission, or vice versa. For example, a certain overdrive trouble may prevent shifting the transmission into reverse. It would be easy to blame this on the transmission, whereas the fault would actually lie in the overdrive.	
1. Will not go into overdrive (⊘ 6-10)	a. Wiring defective b. Governor defective c. Kickdown switch defective d. Relay defective e. Solenoid defective f. Linkage to control knob on instrument panel out of adjustment g. Defect in overdrive, including gear jammed or broken, overrunning clutch defective, excessive shaft end play	Tighten connections; install new wiring Install new governor Install new switch Install new relay Install new solenoid Adjust Disassemble overdrive to eliminate defective part; tighten flange nut
2. Will not come out of overdrive (⊘ 6-10)	a. Wiring defective b. Kickdown switch defective c. Solenoid defective d. Pawl jammed e. Sun gear jammed	Tighten connections; install new wiring Install new switch Install new solenoid Free pawl Disassemble overdrive to eliminate jam and replace defective parts
3. Does not kick down from overdrive (⊘ 6-10)	a. Pawl jammed in sun-gear control plate b. Solenoid defective c. Relay defective d. Governor grounded e. Reverse-lockout switch grounded f. Kickdown switch defective g. Wiring defective h. Sun gear jammed i. Linkage to instrument-panel knob out of adjustment	Replace solenoid Replace solenoid Replace relay Replace governor Replace switch Replace switch Tighten connections; replace wiring Disassemble overdrive to eliminate jam Adjust
4. No power through overdrive (⊘ 6-10)	a. Planetary parts broken b. Overrunning clutch slipping	Replace defective parts Replace defective parts in overdrive
5. Noises in overdrive (⊘ 6-10)	a. Gears worn or chipped b. Main-shaft bearing worn or scored c. Overrunning-clutch parts worn or scored	Replace defective gears Replace Replace
6. Oil leaks (⊘ 6-10)	a. Excessive lubricant b. Loose mounting c. Defective or broken gaskets or oil seals	Put in only specified amount, no more Tighten mounting bolts Replace

⊘ 6-3 Hard Shifting into Gear Hard shifting into gear might be caused by improper linkage adjustment between the gearshift lever and the transmission. Improper adjustment might greatly increase the pressure necessary for gear shifting. The same trouble could result when the linkage is badly in need of lubrication and is rusted or jammed at any of the pivot points. Adjustment and lubrication of linkages is discussed in ⊘ 6-18 and 6-19.

Another cause of this trouble could be failure of the clutch to release completely. If the clutch linkage is out of adjustment or if other conditions, as outlined in ⊘ 3-4, prevent full clutch disengagement, then it will be difficult to shift gears into or out of mesh. Gear clashing will probably result since the engine will still be delivering at least some power through the clutch to the transmission. See ⊘ 3-4 for corrections of this sort of clutch trouble.

Inside the transmission, hard gear shifting could be caused by a bent shifter fork, sliding gear or synchronizer tight on the shaft splines, battered gear teeth, or a damaged synchronizing unit. A bent shifter fork, which might make it necessary to exert greater pressure in order to shift gears, should be replaced. The splines in the gears or on the shaft may become gummed up or battered from excessive wear so that the gear will not move easily along the shaft splines. If this happens, the shaft and gears should be cleaned or, if worn, replaced. If the gear teeth are battered, they will not slip into mesh easily.

Nothing can be done to repair gears with battered teeth; new gears will be required. The synchronizing unit could be tight on the shaft, or it could have loose parts or worn or scored cones; any of these conditions would increase the difficulty of meshing. To clear up troubles in the transmission, the transmission must be removed and disassembled. Sections in the latter part of the chapter describe these operations.

Another condition that can cause hard shifting is binding of the shifter tube in the steering column. The steering column must be partly disassembled so that the binding can be relieved.

⊘ 6-4 Transmission Sticks in Gear A number of the conditions that cause hard shifting into gear can also cause the transmission to stick in gear. For instance, improper linkage adjustment between the gearshift lever and the transmission, as well as lack of lubrication in the linkage, could make it hard to shift out of mesh. Adjustment and lubrication of linkages are discussed in ⊘ 6-18 and 6-19.

Another cause could be failure of the clutch to release completely. Improper clutch-linkage adjustment, as well as other conditions outlined in ⊘ 3-4 that prevent full release of the clutch, could make it hard to shift out of mesh. See ⊘ 3-4 for correction of this type of clutch trouble.

If the detent balls (or the lockout mechanism in the transmission) stick and do not unlock readily when shifting is attempted, it will be hard to shift out of gear. They should be freed and lubricated.

If the synchronizers do not slide freely on the shaft splines, then it will be hard to come out of mesh. The shaft and synchronizers should be cleaned or, if worn, replaced. Lack of lubricant in the transmission can cause gears to stick in mesh. See the latter part of this chapter for transmission removal, disassembly, reassembly, and replacement procedures.

⊘ 6-5 Transmission Slips out of Gear Improperly adjusted linkage between the gearshift lever and the transmission might produce pressure on the linkage in such a way that gears would work out of mesh. Linkage adjustment is outlined in ⊘ 6-18 and 6-19.

Worn gears or gear teeth may also increase the chances of gears coming out of mesh. Likewise, if the detent balls (or lockout mechanism in the transmission) lack sufficient spring pressure, there will be little to hold the gears in mesh and they may slip out. Worn bearings or synchronizers loose on the shaft tend to cause excessive end play or free motion that allows the gears to de-mesh.

In addition, if the transmission slips out of high gear, it could be due to misalignment between the transmission and the engine. This condition is serious and can soon damage the clutch as well as transmission parts. Misalignment can often be detected by the action of the clutch pedal; it causes clutch-pedal pulsations or a nervous pedal, as explained in ⊘ 3-6. ⊘ 3-15 describes the procedure of checking clutch-housing alignment; if the clutch housing is out of line, then the transmission will also be out of line.

⊘ 6-6 No Power through Transmission If the transmission is in mesh and the clutch is engaged and yet no power passes through the transmission, then the clutch could be slipping. ⊘ 3-2 describes various causes of clutch slippage. If the clutch is not slipping, then the trouble is in the transmission and the indication is that something serious has taken place which will require complete transmission overhaul. Conditions inside the transmission that would prevent power from passing through include gear teeth stripped from gear, a shifter fork or some other linkage part broken, a gear or shaft broken, and a drive key or spline sheared off. The transmission must be taken off and disassembled as explained in the latter part of the chapter so that the damaged or broken parts can be replaced.

⊘ 6-7 Transmission Noisy Several types of noise may be encountered in transmissions. Whining or growling, either steady or intermittent, may be due to worn, chipped, rough, or cracked gears. As the gears continue to wear, the noise may take on a grinding characteristic, particularly in the gear position that throws the greatest load on the worn gears. Bearing trouble often produces a hissing noise that will develop into a bumping or thudding sound as bearings wear badly. Metallic rattles could be due to worn or loose shifting parts in the linkage or to gears loose on shaft splines. Sometimes, if the clutch friction-disk-cushion springs or the engine torsional-vibration dampener are defective, the torsional vibration of the engine will carry back into the transmission. This noise would be apparent only at certain engine speeds.

In analyzing noise in the transmission, first note whether the noise is obtained in neutral with the car stationary or in certain gear positions. If the noise is evident with the transmission in neutral with the car stationary, disengage the clutch. If this does not stop the noise, then the chances are the trouble is not in the transmission at all (provided the clutch actually disengages and does not have troubles such as outlined in ⊘ 3-4). In this case, the noise is probably in the engine or clutch. But if the noise stops when the clutch is disengaged, then the trouble is probably in the transmission.

A squeal when the clutch is disengaged usually means that the clutch throw-out bearing needs lubrication or is defective. Also, a worn or dry pilot bushing in the crankshaft can become noisy. Noise can occur because the crankshaft continues to turn with the clutch disengaged but the clutch shaft itself (which pilots in the crankshaft bushing) stops turning.

Noise in neutral with the clutch engaged could

come from transmission misalignment with the engine, worn or dry bearings, worn gears, a worn or bent countershaft, or excessive end play of the countershaft. Notice that these are the parts which are in motion when the clutch is engaged and the transmission is in neutral.

Noise obtained in gear could result from any of the conditions noted in the previous paragraph. Also, it could be due to a defective friction disk in the clutch or a defective engine torsional-vibration dampener. In addition, the rear main bearing of the transmission could be worn or dry, gears could be loose on the main shaft, or gear teeth could be worn. Another cause of noise could be worn speedometer gears. Careful listening to notice the particular gear position in which the most noise is obtained is often helpful in pinpointing the worn parts that are producing the noise.

Worn transmission parts should be replaced after transmission removal and disassembly, as outlined in the latter part of the chapter.

⊘ 6-8 Gears Clash in Shifting Gear clashing that accompanies shifting may be due to failure of the synchronizing mechanism to operate properly. This condition might be caused by a broken synchronizer spring, incorrect synchronizer end play, or defective synchronizer cone surfaces. It could also be due to gears sticking on the main shaft or failure of the clutch to release fully. Gear clash can be obtained in low or reverse on many cars if a sudden shift is made to either of these gears while gears are still in motion. In some transmissions these two gear positions do not have synchromesh devices. In these cases, to prevent gear clash when shifting into either of these positions, it is necessary to pause long enough to allow the gears to come to rest. Of course, if the clutch is not releasing fully, then the gears will still be driven and may clash when the shift is made. Conditions that may prevent the clutch from releasing fully are discussed in ⊘ 3-4.

A worn or dry pilot bushing can keep the clutch shaft spinning even when the clutch is disengaged. This condition can cause gear clash when shifting. So can incorrect lubricant in the transmission.

The latter part of this chapter describes transmission removal and disassembly to replace defective synchromesh parts.

⊘ 6-9 Oil Leaks If the lubricant in the transmission case is not the correct type or if different brands of lubricant are put into the transmission, the lubricant may foam excessively. As it foams, it will completely fill the case and begin to leak out. The same thing might happen if the oil level is too high. In addition, if gaskets are broken or missing or if oil seals or oil slingers are damaged or missing, oil will work past the shafts at the two ends of the transmission. Also, if the drain plug is loose or if the transmission bearing retainer is not tightly bolted to the case, then oil will be lost. A cracked transmission or extension case will also leak oil. The right amount of the recommended oil should be used in the transmission to prevent excessive oil leakage due to foaming. The latter part of the chapter explains how to remove and disassemble the transmission so that defective gaskets, oil seals, and slingers can be replaced.

⊘ 6-10 Overdrive Troubles Certain conditions in the overdrive will cause such troubles as failure to go into overdrive, failure to come out of overdrive, inability to shift into reverse or to pull the instrument-panel control knob out from the overdrive position, power not passing through the overdrive, noises, and oil leaks. Troubleshooting these various conditions is detailed in following paragraphs. Refer to Fig. 6-1 to locate the various terminals and check spots mentioned.

1. *WILL NOT GO INTO OVERDRIVE*

a. With the ignition switch on, ground the KD terminal of the solenoid relay with a jumper lead. If the solenoid clicks, the relay and solenoid circuits are in operating condition. If no click is heard in the relay, check the fuse and replace if defective.

b. If the fuse is good, use a second jumper lead to connect the SOL and BAT terminals of the relay. If a click is now heard in the solenoid, the relay is probably at fault and should be repaired or replaced.

c. If the solenoid does not click in step *b*, check the wiring to terminal 4 of the solenoid and replace if necessary. If the wiring is not defective, the trouble is probably in the solenoid. Remove the solenoid cover, examine the solenoid contacts in series with the pull-in winding, and clean if necessary. Test again for clicks, as in step *b*, after replacing the solenoid cover and lead wires. Replace the solenoid if the trouble has not been corrected.

d. If the relay and solenoid circuits are in good condition, as determined in step *a*, leave the ignition switch on and make sure the manual control knob is in the overdrive position. Ground one and then the other of the two terminals next to the stem of the kickdown switch (identified as SW and REL). If the solenoid clicks when one terminal is grounded but not the other, replace the switch. If the solenoid does not click when either of the terminals is grounded, check the wiring between the relay and the kickdown switch and replace if defective.

e. If the solenoid clicks as each terminal is grounded in step *d*, ground the governor-switch terminal. If the solenoid clicks, the governor switch may be defective. If the solenoid does not click, check the wiring between the kickdown and governor switches and replace if necessary.

2. *WILL NOT COME OUT OF OVERDRIVE*

a. Remove the connection to the KD terminal of the relay. If this releases the overdrive, look for a grounded control circuit between the relay and governor switch.

b. If the overdrive is not released in step *a*, disconnect the lead to the SOL terminal of the relay. If this releases the overdrive, replace the relay.

3. DOES NOT KICK DOWN FROM OVERDRIVE

a. With the engine running, connect a jumper lead between terminal 6 of the solenoid and the ground. Operate the kickdown switch by hand. This should stop the engine. If it does, the solenoid is probably defective, and it should be checked for dirty ground-out contacts or other defects within the ground-out circuit of the solenoid (Fig. 6-1). Clean the contacts or replace the contact plate, as required.

b. If the engine does not stop in step *a*, ground one and then the other of the two terminals (identified as IGN and SOL) farthest from the stem of the kickdown switch. The engine should stop when one of the two terminals (IGN) is grounded. If the engine does not stop when either of the terminals is grounded, the wiring or connections to the switch between the switch and coil are defective. When the other terminal (SOL) is grounded, the engine should stop when the kickdown switch is operated. If the engine does not stop when the kickdown switch is operated with the second terminal grounded, the kickdown switch is defective. If the trouble is in the kickdown switch, adjust the linkage to give more travel of the switch rod. If this does not correct the trouble, replace the kickdown switch.

If the kickdown switch operates as it should, check for an open circuit in the wiring between the kickdown switch and terminal 6 of the solenoid.

c. If the trouble is not located by the above checks, the upper contacts of the kickdown switch may not be opening. To check for this condition, ground the overdrive control circuit at the governor switch. This should cause the solenoid to click. Operate the kickdown switch by hand. This should cause a second click as the solenoid releases. If there is no second click, adjust the linkage to give more travel of the switch rod. If this does not correct the trouble, replace the kickdown switch.

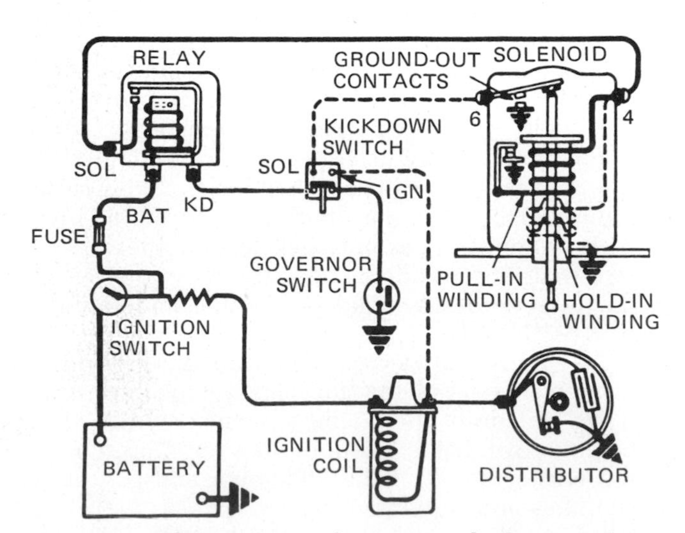

Fig. 6-1. Electric control circuit of the overdrive. *(Chevrolet Motor Division of General Motors Corporation)*

4. NO POWER THROUGH OVERDRIVE This condition could result from broken planetary parts or worn or broken overrunning-clutch components which permit the clutch to slip in the driving position. Either of these conditions requires disassembly of the overdrive so that defective parts can be replaced.

5. NOISES IN OVERDRIVE Noises in the overdrive can arise from conditions similar to those in the transmission which produce noise. Thus, worn or chipped gears, worn bearings, or worn or scored overrunning-clutch parts will cause noise. Damaged parts must be replaced after the unit is disassembled.

6. OIL LEAKS Oil will leak from the overdrive if there is excessive lubricant, if the mounting is loose, or if there are defective gaskets or oil seals. Defective gaskets or oil seals must, of course, be replaced.

Check Your Progress

Progress Quiz 6-1 Knowledge of what to check for and what to expect when certain troubles occur helps greatly in diagnosing troubles. If you are acquainted with the various possible conditions that could cause a certain trouble, you will find it much easier to locate the cause quickly. Thus, the material just covered will prove of great value to you as you go into the service shop and begin to service cars with various types of transmission troubles. The following quiz will help you find out just how well you are remembering the material. It will also help you to review the material so that the important points will be fixed more firmly in your mind. If any of the questions stump you, reread the pages that will give you the answer.

Correcting Troubles Lists The purpose of this exercise is to help you spot related and unrelated troubles in a list. For example, in the list "hard shifting into gear: clutch not releasing, linkage out of adjustment, differential defective, shifter fork bent, sliding gear tight on shaft, gear teeth battered," you can see that "differential defective" does not belong because it is the only condition that would not cause hard shifting into gear. Any of the other conditions in the list could cause this trouble.

In each of the following lists, you will find one item that does not belong. Write each list in your notebook, but *do not write* the item that does not belong.

1. Hard shifting into gear: clutch not releasing, linkage out of adjustment, shifter fork bent, sliding-gear

teeth battered, gear tight on shaft, crankshaft worn, synchronizing unit damaged.
2. Transmission sticks in gear: clutch not releasing, linkage out of adjustment, transmission shifter lock-out stuck, distributor jammed, gears tight on shaft splines.
3. Transmission slips out of first or reverse: clutch not releasing, linkage out of adjustment, gear loose on shaft, gear teeth worn, excessive end play of gears, bearings worn.
4. Transmission slips out of second: linkage out of adjustment, idler gear loose, gear teeth worn, excessive end play of main shaft.
5. Transmission slips out of high: clutch slipping, transmission misaligned, gears or bearings worn, excessive main-shaft end play.
6. No power through transmission: clutch slipping, gear teeth stripped, shifter-linkage part broken, gear or shaft broken, drive shaft bent.
7. Transmission noisy in neutral: transmission not aligned with engine, synchromesh cones worn, bearings or gears worn, countershaft bent or has excessive end play.
8. Transmission noisy in gear: bearings or gears worn, clutch friction disk defective, engine torsional-vibration dampener defective and allowing engine vibration to appear in transmission, gears loose on main shaft, steering gear worn.
9. Gears clash in shifting: clutch not releasing, synchronizer defective, gears sticky on main shaft, overrunning clutch slipping.
10. Oil leaks from transmission: foaming, excessive lubricant, gaskets defective, oil seals damaged, clutch slipping.

⊘ 6-11 Transmission Removal and Installation Because of the variations in construction of transmissions on different types of automobile, different procedures must be followed in the removal, disassembly, repair, assembly, and installation of their transmissions. These operations require about 5 to 7 hours, the difference in time being due to variations in the procedures. Basically, however, the procedures are similar, although it may be helpful to refer to the manufacturer's shop manual before attempting such work. In general, the following steps are required:

1. Drain lubricant from the transmission.
2. Disconnect the rear axle, the front end of the drive shaft, or the universal joint, according to type. Where needle bearings are used, tape the bearing retainers to the shaft to avoid losing the needles.
3. Disconnect the shifting linkages from the transmission, handbrake linkage or spring, and speedometer cable. On some floor models, you must remove the shift controls before you can remove the transmission.
4. Install an engine support where specified (see Fig. 6-2). On some cars, it is necessary to loosen the engine mounts and raise the rear end of the engine enough to permit removal of a supporting frame member.

Fig. 6-2. Installing engine support prior to removal of the transmission.

5. Remove the attaching bolts or stud nuts. Where recommended, two pilot or guide pins should be used (Fig. 6-3). These pins are substituted for transmission bolts and prevent damage to the clutch friction disk as the transmission is moved back. Move the transmission toward the rear until the main gear shaft clears the clutch disk. Then lower the transmission to the floor.

CAUTION: The transmission is heavy. Always use a transmission jack if one is available. If not, get another person to help you lower the transmission from the car. Do not support the transmission on the hub of the clutch friction disk. Doing so will damage the friction disk. That is the purpose of the guide pins—to carry the weight of the transmission until the shaft splines have slipped out of the disk hub.

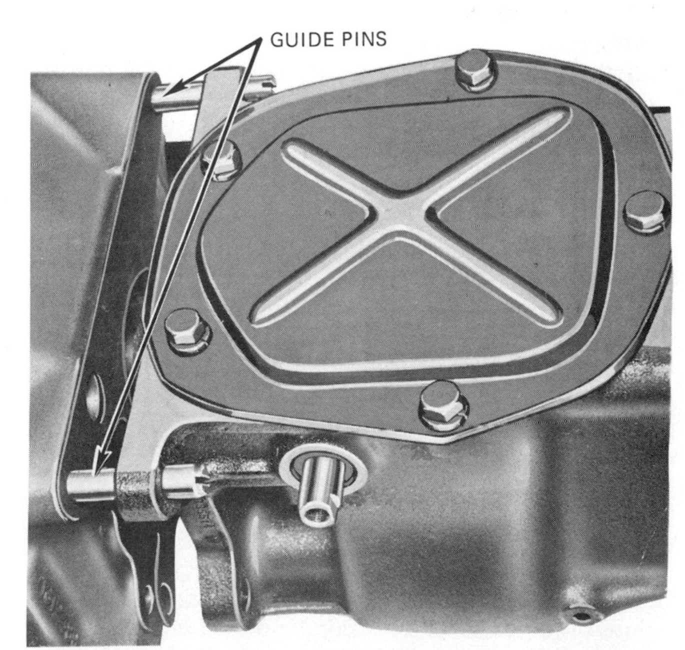

Fig. 6-3. Using guide, or pilot, pins in transmission removal or replacement. The pins maintain transmission alignment with the clutch as the transmission is moved backward or forward so that the clutch will not be damaged. (*Buick Motor Division of General Motors Corporation*)

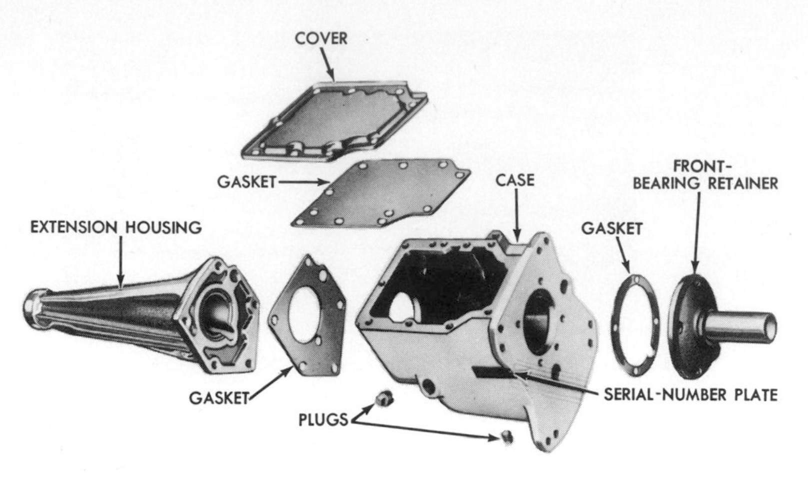

Fig. 6-4. Transmission case and related parts. (*Ford Motor Company*)

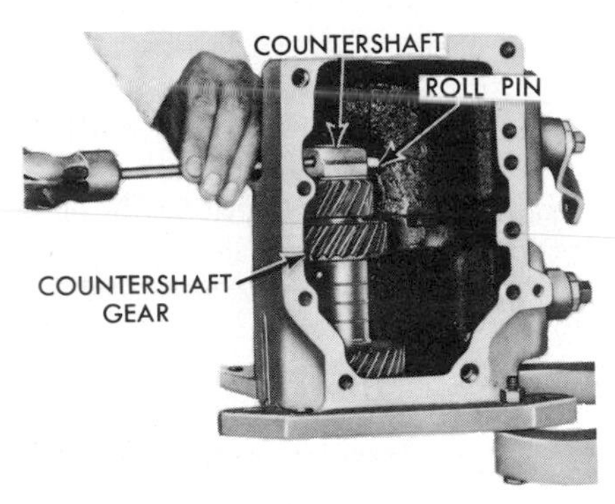

Fig. 6-5. Driving out the countershaft roll pin. (*Ford Motor Company*)

6. With the transmission out, inspect the clutch, flywheel, and flywheel bolts for tightness. Check the transmission input-shaft pilot bushing in the crankshaft.
7. In general, installation is the reverse of removal. Just before installation, shift the transmission into each gear and turn the input shaft to see that the transmission works as it should. Be sure the matching faces of the transmission and flywheel housing are clean. Put a small amount of lubricant on the splines of the input shaft. Prealign the splines on the input shaft and the friction-disk hub by turning the input shaft so that the splines line up. Install guide pins, and lift the transmission. Slide the transmission forward into position. Turn the shaft, if necessary, to secure alignment of the shaft and friction-disk-hub splines. Put the bolts into place, and tighten them to the correct tension. Replace the guide pins with bolts, and tighten them.

CAUTION: If the transmission does not fit snugly against the flywheel housing, or if you cannot move it easily into place, do not force it. It may be that the splines on the shaft and hub are not aligned. Or perhaps roughness, dirt, or a loose retainer ring in the transmission may be blocking the transmission. If the bolts are tightened under such conditions, the transmission case may break. And there will not be alignment. Instead, move the transmission back and try to determine the cause of the trouble.

8. As a final step in the installation procedure, fill the transmission with the proper kind and amount of lubricant. Then attach and adjust the gear-shift linkage.

⊘ 6-12 Transmission Disassembly Transmission service requires a number of special tools, and transmission disassembly should not be attempted without them. The following pages describe and illustrate these tools in use. Because transmission construction varies with different cars, it is suggested that the manufacturer's shop manual be referred to before disassembly is attempted. Typical disassembly and overhaul procedures follow.

⊘ 6-13 Disassembly of a Three-Speed Fully Synchronized Transmission A three-speed fully synchronized transmission is described in ⊘ 4-5 and shown in Figs. 4-10 and 4-11. In this section, we describe the disassembly of a similar unit.

With the transmission in a holding fixture, remove the capscrews, cover gasket, extension housing, and gasket from the transmission case (Fig. 6-4).

From the front of the case, remove the front-bearing retainer and gasket after removing the attaching screws. Remove the lubricant filler plug from the side of the case and, working through the plug opening, drive the roll pin out of the countershaft and case with a $\frac{1}{4}$-inch [6.35 mm] punch (Fig. 6-5). Hold the countershaft gear with a hook and use a dummy shaft, as shown in Fig. 6-6, to push the countershaft out the rear of the case. Lower the countershaft gear with thrust washers on the dummy shaft to the bottom of the case.

Pull the input (clutch) gear forward until it contacts the case and then remove the large snap ring. On some models, the gear can now be removed from the front of the case. On others, it must be removed from the top of the case after the output-shaft assembly has been removed. Remove the snap ring, speedometer drive gear, and drive-gear lock ball from the output shaft.

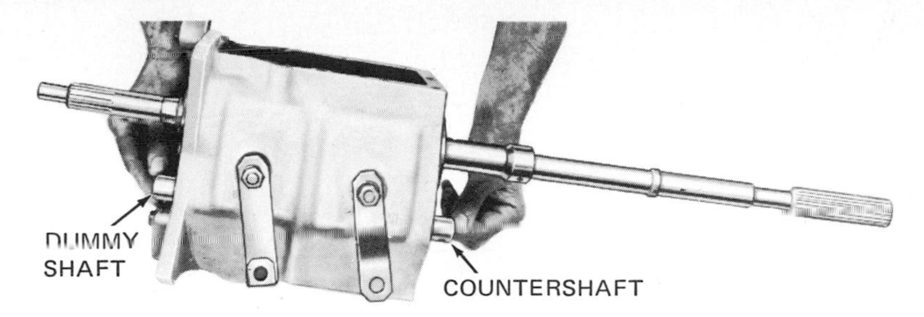

Fig. 6-6. Pushing out the countershaft with a dummy shaft. (*Ford Motor Company*)

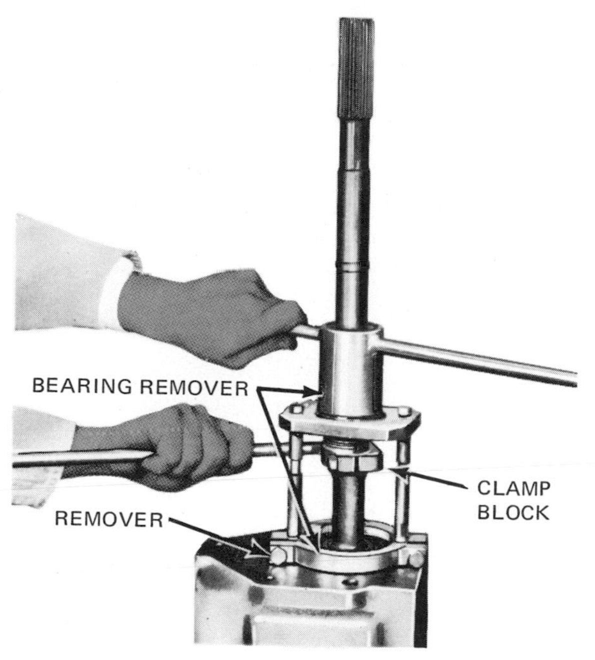

Fig. 6-7. Removing the output-shaft bearing. (*Ford Motor Company*)

Remove the output-shaft rear-bearing snap ring and then use the special tool to remove the bearing (Fig. 6-7). Put both shift levers in neutral (center) and remove the setscrew, detent spring, and plug (Fig. 6-8).

Remove the first-and-reverse shift-fork setscrew and slide the first-and-reverse shift rail out the rear of the case (Fig. 6-8). Slide the first-and-reverse synchronizer forward as far as possible and then rotate the shift fork up and take it out of the case.

Move the second-and-third shift fork to the second-speed position and remove the setscrew from the fork. Rotate the shift rail 90 degrees, as shown in Fig. 6-9, and lift the interlock plug (Fig. 6-8) from the case with a magnet. Tap on the end of the second-and-third shift rail to knock out the expansion plug and remove the shift rail. Rotate the shift fork upward and lift it from the case.

Now the output-shaft assembly is free, and it can be removed through the top of the case. If the clutch gear has not been removed, take it out from the top of the case.

Lift the reverse idler gear and thrust washers from the case. Lift the countershaft gear, thrust washers, and dummy shaft from the case.

Disassemble the output shaft by removing the snap rings. Follow Fig. 6-10.

Figures 6-11 and 6-12 will serve as guides to the disassembly and reassembly of the synchronizers. Do not mix the inserts and insert springs between the two synchronizers. If the tip of the rear insert spring (Fig. 6-11) of the first-and-reverse synchronizer is less than 0.120 inch [3.048 mm], replace the spring. When assembling this spring, make sure the inserts are properly located (Fig. 6-13).

If the input- (clutch-) shaft bearing requires replacement, it can be pressed off the shaft and a new one pressed on. The countershaft roller bearings can be examined by removing the dummy shaft. Then, to replace the rollers, coat the bore in each end of the countergear with grease, insert the dummy shaft, and install the rollers (25) and retainers in each end of the countergear.

⊘ 6-14 Inspection of Transmission Parts As a first step in inspecting transmission parts, wash them, except for the ball or roller bearings and seals, in a suitable solvent. Brush or scrape all dirt from the parts. Do not damage any part with the scraper. Dry parts with compressed air.

CAUTION: Do not clean, wash, or soak transmission seals in cleaning solvents.

To clean the bearings, rotate them slowly in cleaning solvent to remove all lubricant. Hold the bearing assembly stationary so it will not rotate and dry it with compressed air. Immediately lubricate bearings with approved transmission lubricant and wrap them in clean, lintfree cloth or paper.

Clean the magnet at the bottom of the case with kerosene or mineral spirits.

1. *PARTS INSPECTION* Check the transmission case for cracks and worn or damaged bearings, bores, or threads. Check the front of the case for nicks or burrs that could cause misalignment of the case with the flywheel housing. Remove all nicks with a fine stone.

If any cover is bent, replace it. Make sure the vent hole is open. Check the condition of the shift levers, forks, shift rails, and the lever and shafts.

Check the ball bearings as explained in item 2. Replace roller bearings that are broken, worn, or rough.

Replace the countershaft-gear assembly if gear

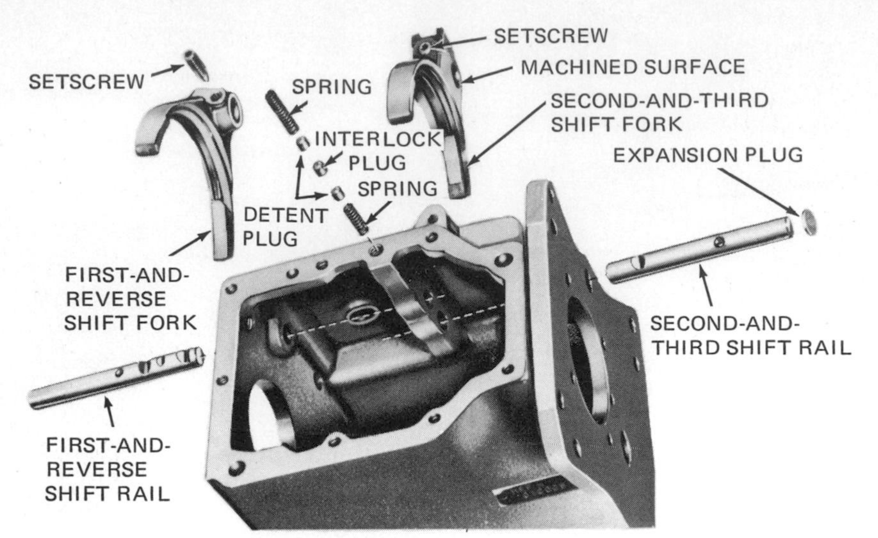

Fig. 6-8. Shift rails, forks, and interlock arrangement. (*Ford Motor Company*)

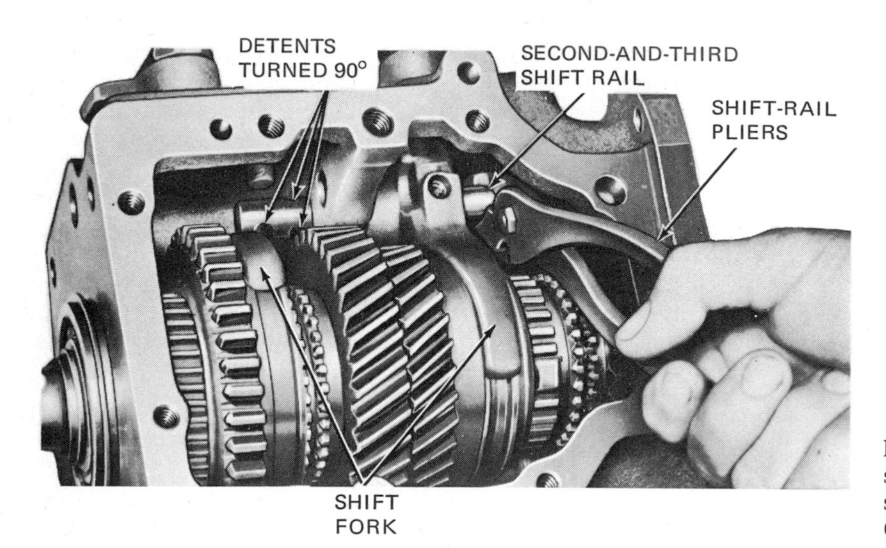

Fig. 6-9. Rotating the second-and-third-speed shift rail. (*Ford Motor Company*)

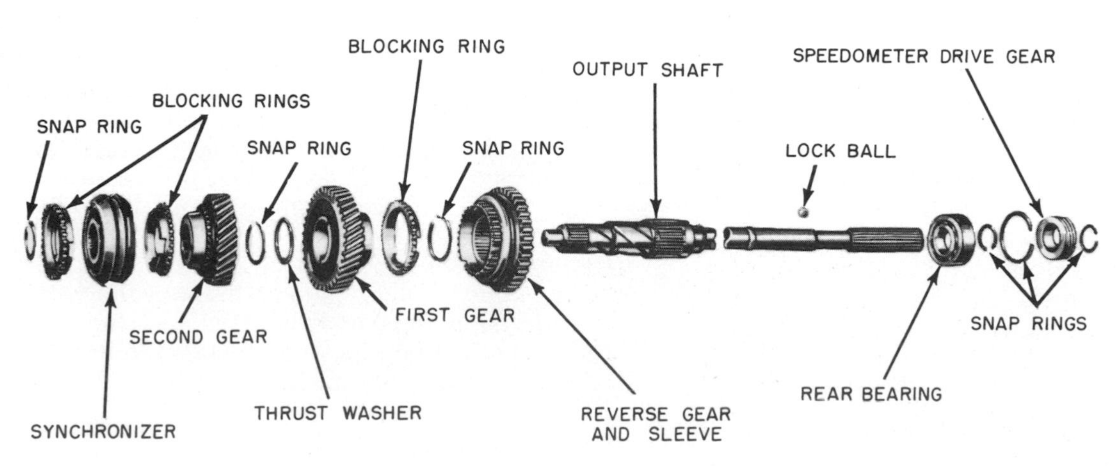

Fig. 6-10. Relationship of parts on the output shaft. (*Ford Motor Company*)

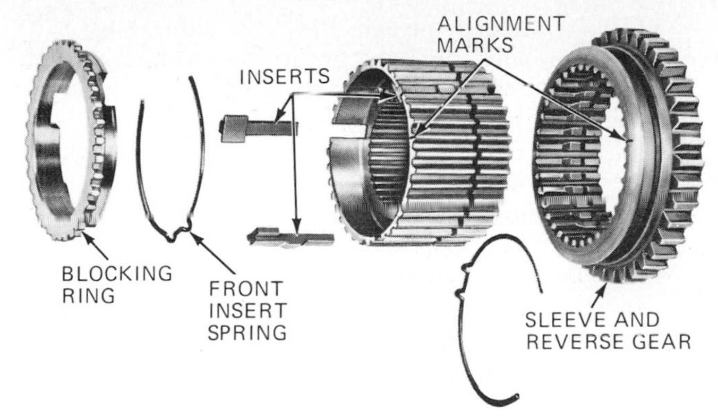

Fig. 6-11. Disassembled first-and-reverse synchronizer. (*Ford Motor Company*)

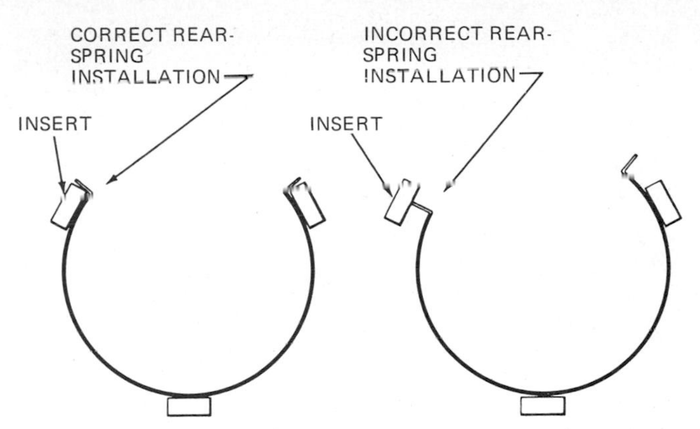

Fig. 6-13. Correct and incorrect installation of the first-and-reverse synchronizer insert spring. (*Ford Motor Company*)

teeth are broken, chipped, or worn. Replace the shaft if it is bent, worn, or scored. Replace the reverse-idler gear or sliding gear if teeth are damaged. Replace the reverse-idler shaft if it is bent, worn, or scored.

Replace the input shaft and gear if the splines are damaged or if the teeth are chipped, worn, or broken. If the roller-bearing surface in the bore of the gear is worn or rough, or if the cone surface is damaged, replace the gear and the gear rollers.

Check the synchronizer sleeves for free movement on their hubs. Make sure that the alignment marks (if present) are properly indexed. Check the synchronizer blocking rings for widened index slots, rounded clutch teeth, and internal surfaces that are too smooth (they must have machined grooves). Put the blocker ring on the cone and measure the distance between the face of the blocker ring and the clutch teeth on the gear. The distance must be not less than 0.020 inch [0.508 mm].

Replace the speedometer-drive gear if the teeth are stripped or damaged. Be sure you install a replacement gear of the correct size. Otherwise, the speedometer and odometer will not register correctly.

Replace the output shaft if there is any evidence of wear or spline damage. Inspect the bushing and seal in the extension housing, and replace it if it is worn or damaged.

NOTE: The bushing and seal must be replaced *after* the extension housing has been installed on the transmission.

Replace the seals in the input-shaft-bearing retainer and on the cam and shafts.

2. *BALL-BEARING INSPECTION* There are four checks:

a. Inner-Ring Raceway Hold the outer ring stationary and rotate the inner ring three times. Examine the raceway of the inner ring for pits or spalling (Fig. 6-14). Note the types of damage and those that require a bearing replacement.

b. Outer-Ring Raceway Hold the inner ring stationary and rotate the outer ring three times. Examine the outer ring raceway for damage (Fig. 6-14). Note the types of damage and those that require bearing replacement.

c. External Surfaces Replace the bearing if there are radial cracks on the front or rear faces of the outer or inner rings. Replace the bearing if there are cracks on the outside diameter or outer ring (check carefully around the snap-ring groove). Also replace the bearing if the ball cage is cracked or deformed.

d. Spin Test Lubricate the bearing raceways with a little clean oil. Turn the bearing back and forth slowly to coat the raceways and balls. Hold the bearing by the inner ring in a vertical position. Some vertical movement between the inner and outer rings is okay. Spin the outer ring several times by hand. *Do not use an air hose!* If you

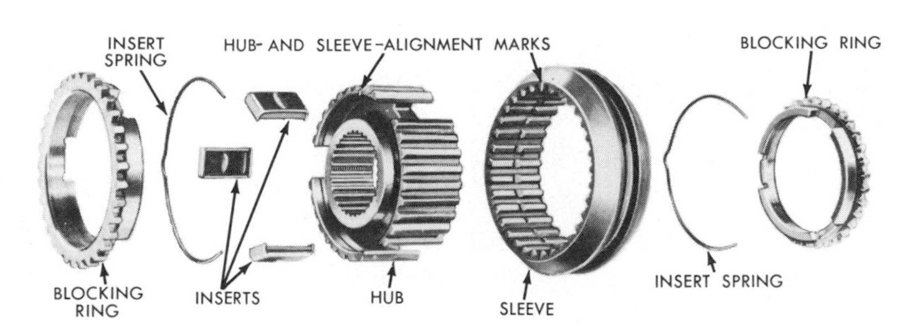

Fig. 6-12. Disassembled second-and-third synchronizer. (*Ford Motor Company*)

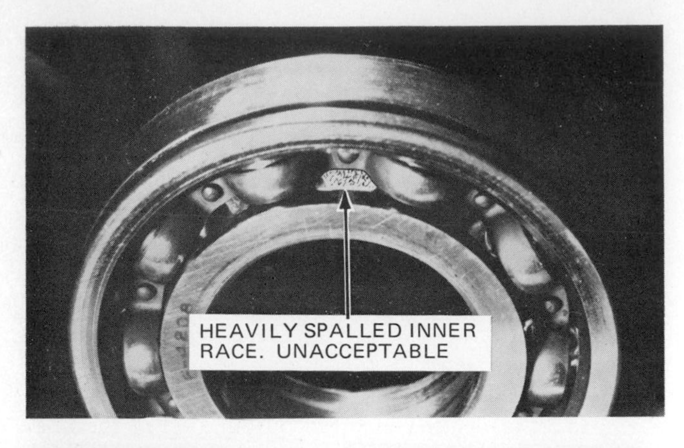

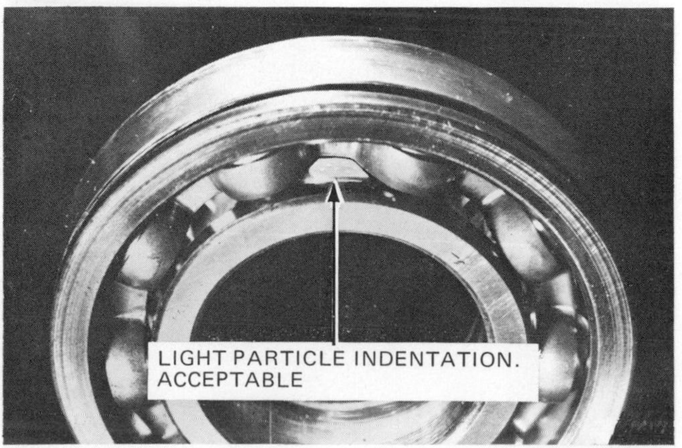

Fig. 6-14. Inspection of ball bearings. (*Ford Motor Company*)

notice roughness or vibration, or if the outer ring stops abruptly, reclean the bearing, relubricate it, and spin it again. Roughness is usually caused by particles of dirt in the bearing. If the bearing is still rough after cleaning and lubricating it three times, discard it.

Now hold the bearing by the inner ring in a horizontal position with the snap-ring groove up. Spin the outer bearing several times by hand, as described previously. If the bearing is still rough after cleaning and relubricating it three times, discard it.

⊘ 6-15 Assembly of a Three-Speed Fully Synchronized Transmission After all parts have been inspected as noted (⊘ 6-14), the transmission is reassembled as follows.

Lay the countershaft assembly with the dummy shaft in place in the bottom of the case. Reassemble the output shaft (except for the rear bearing and the speedometer-drive gear) as shown in Fig. 6-10. Make sure that all parts are restored in their original relationship and as shown in Fig. 6-10. All machined surfaces and splines should be lubricated with transmission lubricant on assembly.

Coat the bore of the clutch-gear shaft with a *thin* film of grease (a thick film will plug lubricant holes) and install 15 bearing rollers. On some models the clutch-gear assembly may now be installed through the top of the case. On others, it must be installed through the front of the case after the output shaft is installed. When installed, secure the clutch-gear assembly with the snap ring in the bearing groove.

Position the output-shaft assembly in the case, and position the second-and-third shifter fork on its synchronizer. Put a detent spring and plug in the case (Fig. 6-8). Move the second-and-third-speed synchronizer to the second-speed position, align the fork, and install the second-and-third-speed shift rail. The detent plug must be pushed down to allow the rail to move into place. Move the rail forward until the plug engages the notch in the rail. Attach the fork to the shaft with the setscrew. Move the synchronizer to the neutral position.

Install the interlock plug in the case. With the second-and-third shift rail in neutral position, the interlock plug should be slightly below the surface of the first-and-reverse shift-rail bore. Move the first-and-reverse synchronizer forward to the first-speed position. Put the first-and-reverse shift fork into position and install the shift rail. Move the rail in until the center (neutral) notch is aligned with the detent bore. Attach the fork with a setscrew. Install the other detent plug and spring and secure with a setscrew turned down flush with the case.

Put a new expansion plug in the front of the case.

Hold the clutch (input) shaft and blocking ring in position and move the output shaft forward to seat the pilot in the clutch roller bearings. Tap the clutch-gear bearing into place while holding the output shaft to prevent the rollers from dropping.

Install the front-bearing retainer with a new gasket, making sure the oil-return slot is at the bottom.

Install the large snap ring on the rear bearing and press the bearing into place with the tool shown in Fig. 6-7 (except that the replacer, not the remover, is used at the bearing). Secure the bearing on the shaft with a snap ring. Hold the speedometer-drive-gear lock ball in the shaft detent and slide the gear into place. Secure with a snap ring.

Turn the transmission to a vertical position and, working through the drain hole in the bottom of the case, align the bore of the countershaft gear and the thrust washers with the bore of the case, using a screwdriver. Working from the rear of the case, push the countershaft into position, pushing the dummy shaft out ahead of it. Before the countershaft is completely in position, make sure the roll-pin holes in the case and shaft will align. Then push the countershaft into place and install the roll pin.

Coat the new extension-housing gasket with sealer, put it into place, and attach the extension. Coat the screw threads with sealer before installing screws.

Install the filler and drain plugs in the case. Put the transmission in gear and turn it to a horizontal position. Pour transmission lubricant over all moving parts while turning one of the shafts. Coat the cover gasket with sealer and install the cover and gasket, coating screw threads with sealer before installing screws.

Check the transmission in all gear positions before installing it on the car.

⊘ 6-16 Disassembly of a Four-Speed Fully Synchronized Transmission A four-speed fully synchronized transmission is described in ⊘ 4-7 and illustrated in Figs. 4-23 to 4-30. Figure 6-15 is a disassembled view of one model, the disassembly procedures of which are outlined below.

First, remove the transmission side cover, as follows: Shift the transmission into second speed by moving the 1-2 shifter lever into forward position. Then remove the cover assembly and allow oil to drain.

Remove four bolts and two bolt lock strips from the front-bearing retainer (1 in Fig. 6-15) and gasket. Use a special tool to remove the main-drive-gear retaining nut (Fig. 6-16). To do this you must lock up the transmission by shifting into two gears.

Put the transmission gears in neutral and drive the lockpin from the reverse-shifter-lever boss (Fig. 6-17). Pull the shifter shaft out about $\frac{1}{8}$ inch [3.18 mm] to disengage the shift fork from reverse gear.

Remove the six bolts attaching the case extension to the case. Tap the extension with a soft hammer away from the transmission case. When the reverse-idler shaft is out as far as it will go, move the extension to the left so that the reverse fork clears the reverse gear. Now, the extension and gasket can be removed.

Remove the reverse-idler gear, flat thrust washer, shaft, roll spring pin, speedometer gear, and reverse gear.

Slide the 3-4 synchronizer clutch sleeve to fourth-speed position (forward), as shown in Fig. 6-18. Now, carefully remove the rear-bearing retainer and entire main-shaft assembly from the base by tapping the bearing retainer with a soft hammer.

Unload the 17 bearing rollers from the main drive gear and then remove the fourth-speed-synchronizer blocker ring. Lift the front half of the reverse-idler gear, and remove the tanged thrust washer from the case.

Use an arbor press to press the main drive gear down from the front bearing (Fig. 6-19). From inside the case, tap out the front bearing and snap ring. Also from the front of the case, press out the countershaft with a special tool (Fig. 6-20), and then remove the countergear and both tanged washers.

Remove 112 rollers, six 0.070-inch [1.778 mm] spacers, and the roller spacer from the countergear. Remove the main-shaft front snap ring as shown in Fig. 6-21 and slide the third-and-fourth-speed clutch assembly, third-speed gear, and synchronizing ring from the front of the main shaft.

Spread the rear-bearing-retainer snap ring and press the main shaft out of the retainer. Remove the main-shaft rear snap ring. Support the second-speed gear and press on the rear of the main shaft to remove the rear bearing, first-speed gear and sleeve, first-speed synchronizing ring, 1-2 speed synchronizing ring, and second-speed gear.

Inspect the transmission parts (see ⊘ 6-14).

⊘ 6-17 Reassembly of a Four-Speed Fully Synchronized Transmission

1. ASSEMBLE THE MAIN SHAFT Install the second-speed gear with the hub toward the rear of the shaft. Install the 1-2 synchronizer clutch assembly (taper to rear, hub to front), with a synchronizer ring on either side of the clutch assembly so that their keyways line up with the clutch keys (Fig. 6-22). Use a pipe of the correct diameter (Fig. 6-23) to press the first-gear sleeve onto the main shaft.

Install the first-speed gear (hub to front) and use a pipe of the correct diameter to press the rear bearing on. Install the snap ring of the correct thickness to get maximum distance of 0 to 0.005 inch [0.127 mm] between the ring and rear face of the bearing.

Install the third-speed gear (hub to front) and synchronizer ring (notches to front). Install the third-and-fourth-speed-gear clutch assembly (hub and sliding sleeve) with both sleeve taper and hub toward the front. Make sure the keys in the hub correspond to the notches in the synchronizer ring.

Install the snap ring in the groove in the main shaft in front of the third-and-fourth-speed clutch assembly, with ends of the snap ring seated behind the spline teeth. Install the rear-bearing retainer.

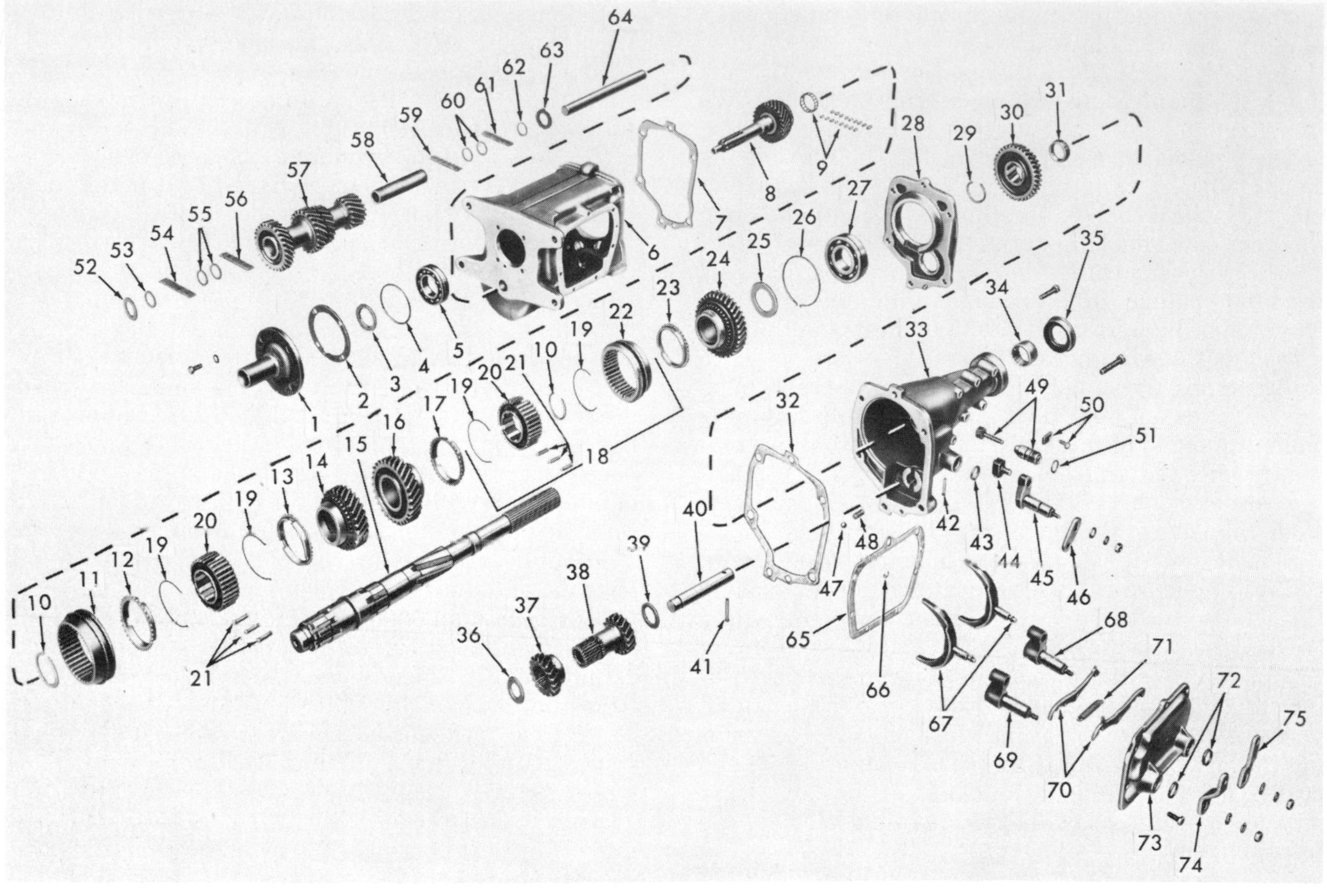

1. Bearing retainer
2. Gasket
3. Bearing retaining nut
4. Bearing snap ring
5. Main-drive-gear bearing
6. Transmission case
7. Rear-bearing-retainer gasket
8. Main drive gear
9. Bearing rollers (17), cage
10. Snap ring
11. Third-and-fourth-speed-clutch sliding sleeve
12. Fourth-speed-gear synchronizing ring
13. Third-speed synchronizing ring
14. Third-speed gear
15. Main shaft
16. Second-speed gear
17. Second-speed-gear synchronizing ring
18. First-and-second-speed-clutch assembly
19. Clutch key spring
20. Clutch hub
21. Clutch keys
22. First-and-second-speed-clutch sliding sleeve
23. First-speed-gear synchronizing ring
24. First-speed gear
25. First-speed-gear sleeve
26. Rear-bearing snap ring
27. Rear bearing
28. Rear-bearing retainer
29. Selective-fit snap ring
30. Reverse gear
31. Speedometer drive, clip
32. Rear-bearing-retainer-to-case extension gasket
33. Case extension
34. Extension bushing
35. Rear oil seal
36. Reverse idler front thrust washer (tanged)
37. Reverse idler gear (front)
38. Reverse idler gear (rear)
39. Flat thrust washer
40. Reverse idler shaft
41. Reverse-idler-shaft roll pin
42. Reverse-shifter-shaft lockpin
43. Reverse-shifter-shaft lip seal
44. Reverse shift fork
45. Reverse shifter shaft and detent plate
46. Reverse shifter lever
47. Reverse-shifter-shaft detent ball
48. Reverse-shifter-shaft ball detent spring
49. Speedometer driven gear and fitting
50. Retainer and bolt
51. O-ring seal
52. Tanged washer
53. Spacer
54. Bearing rollers (28)
55. Spacer
56. Bearing rollers (28)
57. Countergear
58. Countergear roller spacer (seam type)
59. Bearing rollers (28)
60. Spacer
61. Bearing rollers (28)
62. Spacer
63. Tanged washer
64. Countershaft
65. Gasket
66. Detent-cam retainer ring
67. Forward-speed shift forks
68. First-and-second-speed-gear shifter shaft and detent plate
69. Third-and-fourth-speed-gear shifter shaft and detent plate
70. Detent cams
71. Detent-cam spring
72. Lip seals
73. Transmission side cover
74. Third-and-fourth-speed shifter lever
75. First-and-second-speed shifter lever

Fig. 6-15. Disassembled four-speed transmission. (*Chevrolet Motor Division of General motors Corporation*)

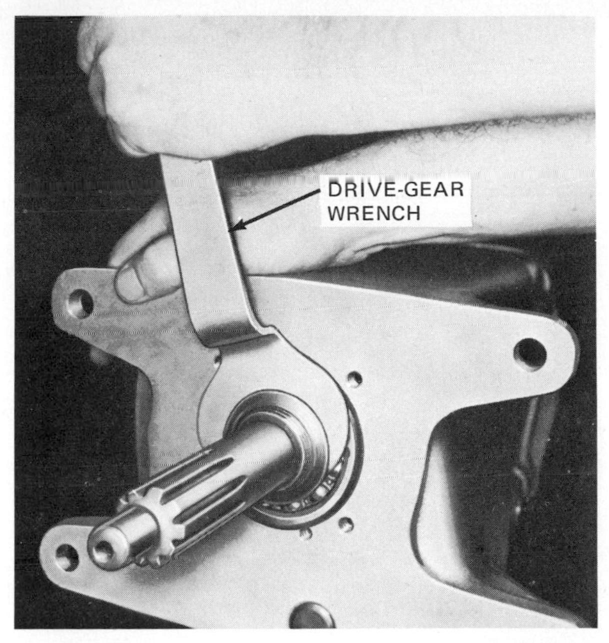

Fig. 6-16. Removing the main-drive-gear retaining nut. (*Chevrolet Motor Division of General Motors Corporation*)

Spread the snap ring in the plate to allow the snap ring to drop around the rear bearing. Press on the end of the main shaft until the snap ring engages the groove in the rear bearing.

Install the reverse gear, shift collar to rear, and the two antirattle springs. Install the retaining clip and speedometer gear.

2. *ASSEMBLE THE COUNTERGEAR* Install the roller spacer in the countergear, then install the rollers, as follows: Use heavy grease to retain the rollers and install a spacer, 28 rollers, a spacer, 28 more rollers, and then another spacer. Then at the other end, install a spacer, 28 rollers, a spacer, 28 more rollers, and another spacer. Insert a special tool into the countergear.

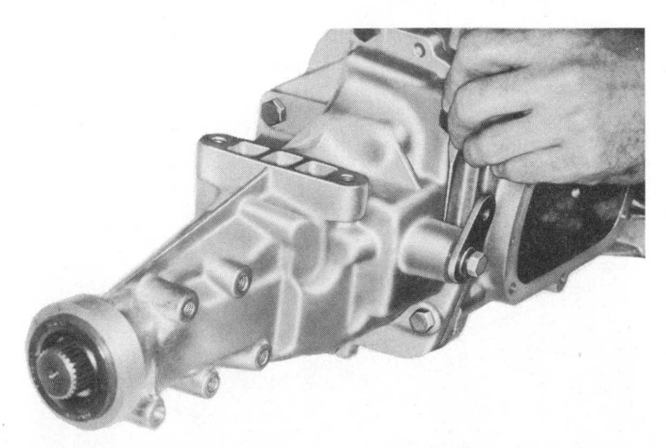

Fig. 6-17. Removing the reverse-shifter-shaft lockpin. (*Chevrolet Motor Division of General Motors Corporation*)

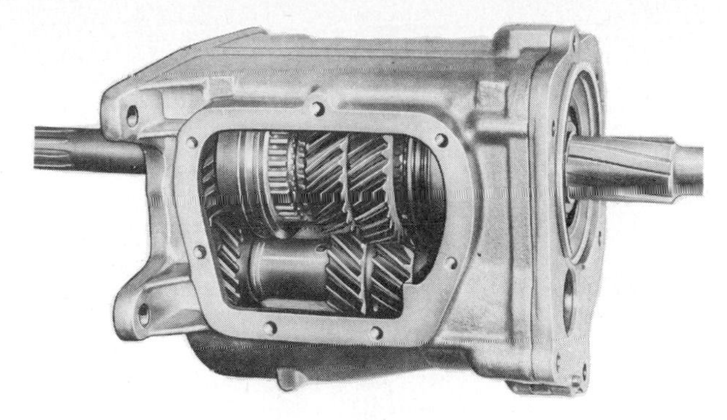

Fig. 6-18. Third-and-fourth-speed synchronizer clutch sleeve in fourth-gear position. (*Chevrolet Motor Division of General Motors Corporation*)

3. *ASSEMBLE THE TRANSMISSION* Lay the transmission case on its side, with the side cover opening toward you. Put the countergear tanged washers in place, retaining them with grease. Make sure the tangs are in the case notches.

Put the countergear in the bottom of the case. Turn the case on end, front end down. Lubricate the countershaft and start it into the case. The flat on the end of the shaft should face the bottom of the case. Align the countergear and press the countershaft into the case until the flat on the shaft is flush with the rear of the case. Make sure the thrust washers remain in place.

Check the end play of the countergear by attaching a dial indicator to the case (Fig. 6-24). If the end play is greater than 0.025 inch [0.63 mm], install new thrust washers.

Into the main drive gear, install the pinion carrier and 17 roller bearings, using grease to hold them in place. Install the main drive gear and pilot the bearings through the side cover opening and into position in front of the case.

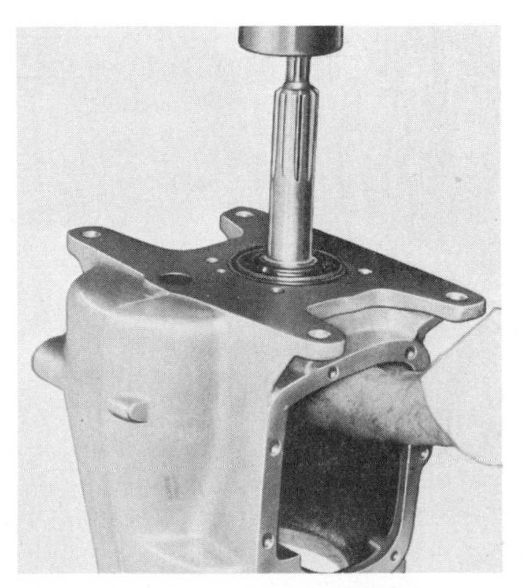

Fig. 6-19. Removing the main drive gear. (*Chevrolet Motor Division of General Motors Corporation*)

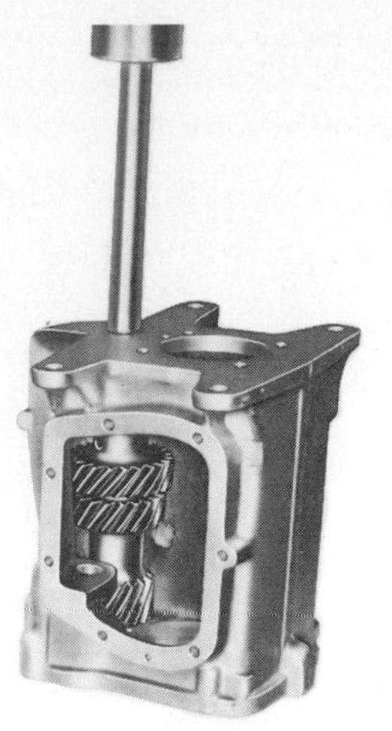

Fig. 6-20. Removing the countershaft with a special tool. (*Chevrolet Motor Division of General Motors Corporation*)

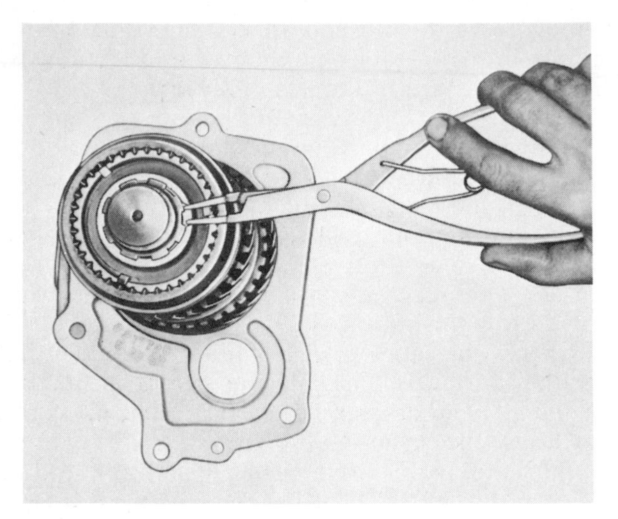

Fig. 6-21. Removing the main-shaft front snap ring. (*Chevrolet Motor Division of General Motors Corporation*)

Fig. 6-22. Installing the synchronizing ring. (*Chevrolet Motor Division of General Motors Corporation*)

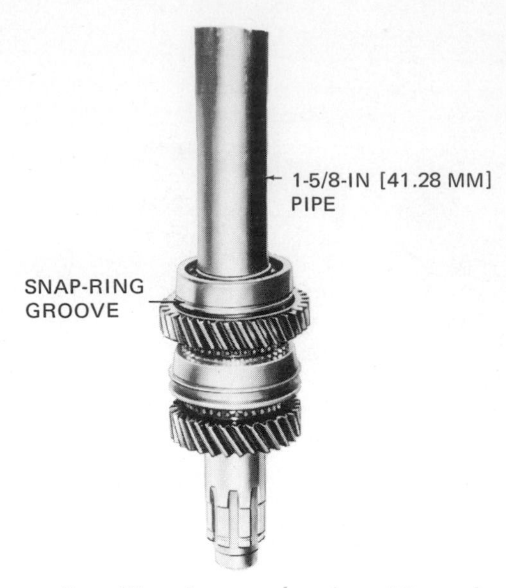

Fig. 6-23. Installing the rear bearing. (*Chevrolet Motor Division of General Motors Corporation*)

Put the gasket in position on the front face of the rear-bearing retainer. Install the fourth-speed synchronizing ring on the main drive gear with the notches toward the rear of the case.

Slide the 3-4 synchronizing clutch sleeve forward into the fourth-speed detent position (Fig. 6-18). Lower the main-shaft assembly into the case. Make sure the notches on the fourth-speed synchronizing ring correspond to the keys in the clutch assembly (Fig. 6-25). Also, make sure the main drive gear engages both the countergear and the antilash plate (on standard-ratio models).

With the guide pin in the rear-bearing retainer aligned with the hole in the rear of the case, tap the

Fig. 6-24. Checking countergear end play. (*Chevrolet Motor Division of General Motors Corporation*)

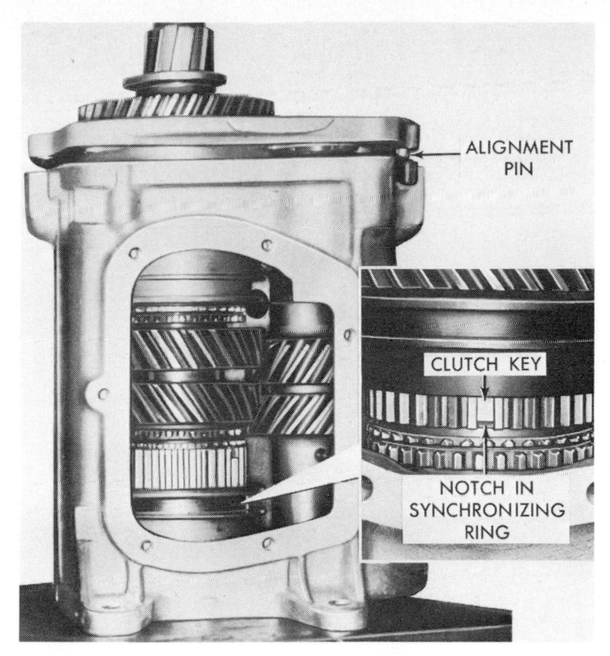

Fig. 6-25. Installing the main-shaft assembly. (*Chevrolet Motor Division of General Motors Corporation*)

rear-bearing retainer into position with a soft hammer.

From the rear of the case, insert the rear reverse-idler gear, engaging the splines with the front gear inside the case. Stick a gasket on the rear face of the rear-bearing retainer with grease.

Install the remaining flat thrust washer on the reverse-idler shaft, and install the shaft, roll pin, and thrust washer into the gears and front boss of the case. Make sure the front tanged washer stays in place. The roll pin should be in a vertical position.

Pull the reverse-shifter shaft to the left side of the extension and rotate the shaft to bring the reverse-shift fork forward (to reverse detent position). Start the extension onto the transmission shaft (Fig. 6-26) while pushing in on the shifter shaft to engage the shift fork with the reverse gearshift collar. Then pilot the reverse idler shaft into the extension housing to permit the extension to slide onto the case. Install the six extension-and-retainer-to-case bolts and torque to specifications.

Push or pull the reverse-shifter shaft to line up

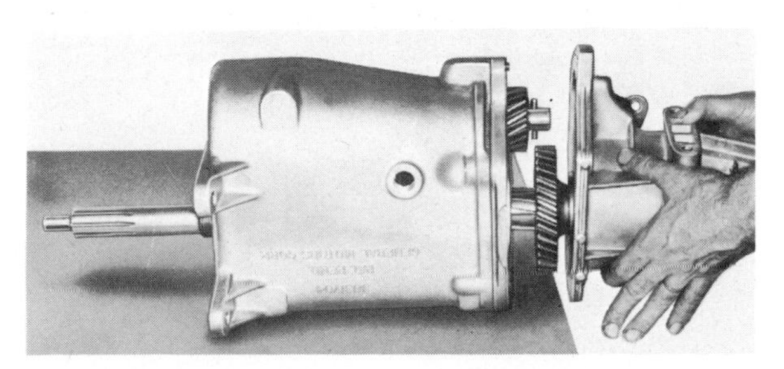
Fig. 6-26. Installing the case extension. (*Chevrolet Motor Division of General Motors Corporation*)

the groove in the shaft and holes in the boss, and drive in the lockpin. Install the shifter lever.

Press the bearing onto the main drive gear (snap-ring groove to front) and into the case until several main-drive-gear threads are exposed. Lock the transmission by shifting into two gears. Install the main-drive-gear retaining nut and torque to specifications. Stake the nut into place at the gear-shaft hole. Do not damage threads on the shaft.

Install the main-drive-gear bearing retainer, gasket, four retaining bolts, and two strip-bolt-lock retainers. Use a sealer on the bolts and torque to specifications.

Shift the main-shaft 3-4 sliding clutch sleeve into neutral and the 1-2 sliding clutch sleeve into second. Install the side-cover gasket and put the side cover in place. Be sure the dowel pin aligns. Install attaching bolts and torque to specifications.

⊘ 6-18 Floor-Shift-Lever Adjustments Figure 6-27 shows a typical floor-shift lever for a four-speed transmission. Figure 4-9 shows a similar linkage in disassembled view. These illustrations can serve as a guide in case the transmission linkage is disassembled. Whenever linkage is disconnected, the linkage joints should be lubricated with a little chassis grease. Linkage adjustment for the model shown in Fig. 6-27 follows.

Position the selector lever in neutral and loosen the three swivel-nut assemblies. Insert a $\frac{1}{4}$-inch-[6.35 mm] diameter gauge pin into the bracket and holes in the control levers, as shown at the bottom of Fig. 6-27, to align them in the neutral position. Tighten the swivel-nut assemblies on the linkage rods and remove the gauge pin. Check the complete shift pattern with the engine off. Try it again with the engine on.

⊘ 6-19 Steering-Column Shift-Lever Adjustments Figure 6-28 shows a typical steering-column shift lever with its linkage to the transmission. Linkage adjustments are made at the swivels on the ends of the two shifter-tube levers at the bottom of the shifter tube.

Whenever gearshift linkage has been disconnected, adjust it by moving the two transmission shift levers until the transmission is in neutral. Both neutral detents must be engaged. To check, start the engine with the clutch disengaged and engage the clutch slowly. If the transmission is in neutral, stop the engine and proceed as follows.

Move the selector lever at the wheel to neutral. Align the first-and-reverse shifter-tube lever with the second-and-third shifter-tube lever on the mast jacket (view C in Fig. 6-28). Install both control rods on the transmission shifter levers and secure them with retaining clips (view A).

Put the swivel on the end of the first-and-reverse shifter control rod and adjust until the swivel is in position to enter the mast-jacket shifter-lever hole.

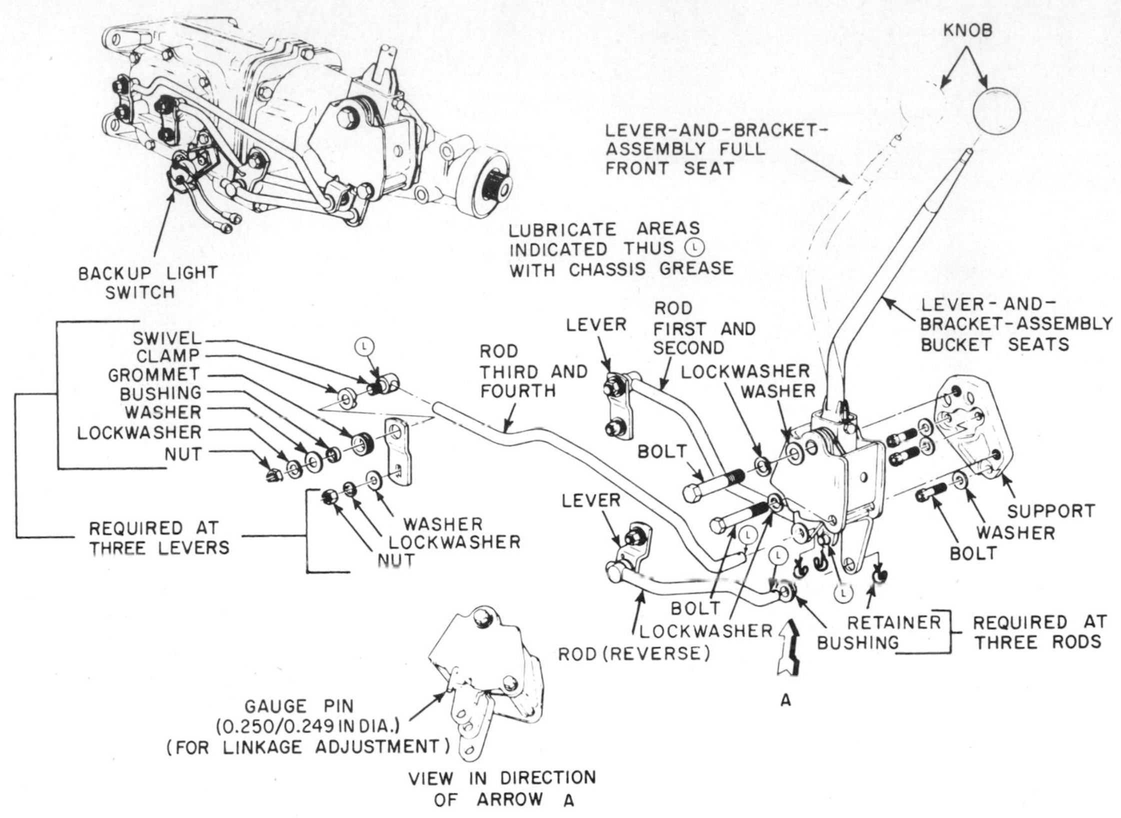

Fig. 6-27. Gearshift linkage for a four-speed transmission. (*Pontiac Motor Division of General Motors Corporation*)

Put the retaining clamp on the swivel, install the swivel in the lever hole, and secure with a swivel nut (view C).

Install the lower second-and-third shifter control rod between the idler lever (view B) and the transmission lever. Then install the upper second-and-third shifter control rod between the idler lever and the swivel in the second-and-third shifter-tube lever. Be sure the shifter-tube levers remain aligned in the neutral position. Check the adjustments by shifting into all gear positions.

⊘ 6-20 Overdrive Service Overdrive disassembly and reassembly varies from model to model. When you are working on a specific model, the applicable shop manual should be followed carefully. Disassembly of the overdrive used (with slight variations) on many cars such as Chevrolet and Ford is detailed below. (Refer to Fig. 5-16 for an exploded view of this unit.)

1. *DISASSEMBLY* With transmission and overdrive assembly off the car and mounted on transmission stand, disassemble as follows:

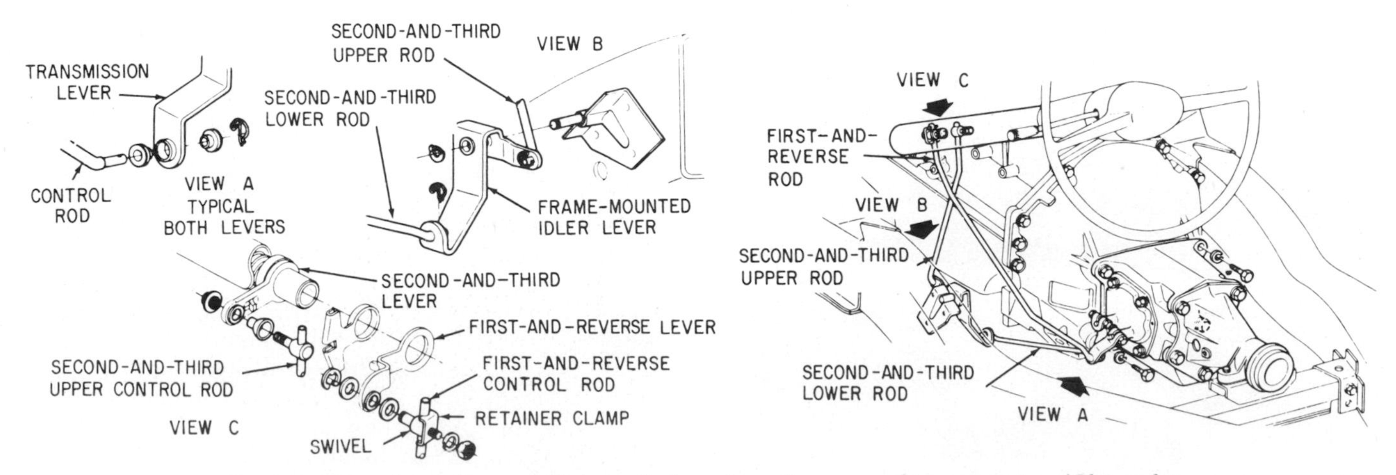

Fig. 6-28. Steering-column shift-lever linkage for a three-speed transmission. (*Chevrolet Motor Division of General Motors Corporation*)

a. Remove the companion-flange nut and flange from the end of the shaft with a special holding tool and puller.
b. Disconnect the wires and take off the governor (Fig. 6-29), and overdrive lockout switch, if so equipped.
c. Drive out the tapered locating pin (1 in Fig. 6-30) and pull the shift shaft out as far as possible to disengage the operating cam from the shift rail. Then remove the four cap screws holding the overdrive housing to the transmission-case adapter. Tap the end of the overdrive shaft lightly with a lead hammer while removing the housing so that the shaft will not come off with the housing and spill the freewheeling rollers.
d. Hold the adapter plate to the case by replacing the cap screw in the upper right-hand hole. Then remove the shift-rail reverse-lockup spring, shift lever, shift shaft, and shaft oil seal. Now, removal of the rear oil seal permits removal of the two snap rings and the overdrive-shaft rear bearing.
e. Remove the speedometer and governor drive gears from the shaft. The overdrive main shaft can now be pulled out. Hold one hand under the assembly to catch the rollers as they drop out (Fig. 6-31).

CAUTION: If the adapter plate is allowed to move away from the transmission case, the transmission main-shaft pilot needle bearings will drop out of the clutch gear. If this happens, the transmission will have to be disassembled in order to replace the bearings.

f. The ring gear can now be taken off the overdrive shaft by removing the large snap ring.
g. With the freewheeling rollers out of the cage, remove the retaining clip (Fig. 6-32) and take off the freewheeling unit and pinion-cage assembly from the transmission main shaft.
h. To separate the freewheeling unit and the pinion-cage assembly, take out the retaining clip (Fig. 6-33).

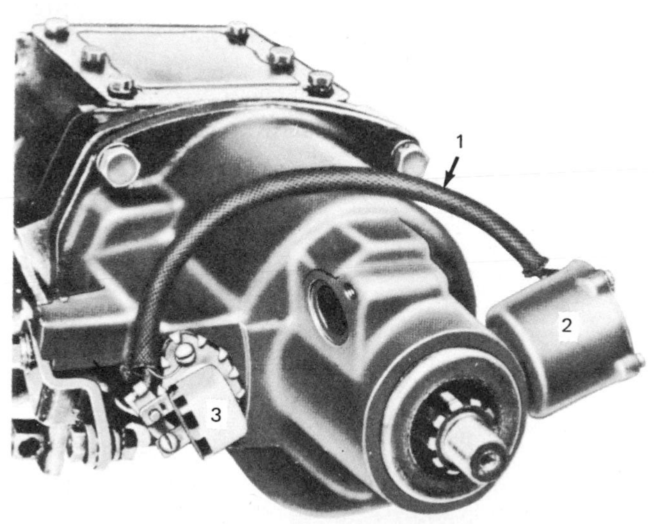

Fig. 6-29. Overdrive on transmission showing wire (1) between governor (2) and overdrive lockout switch (3) the lockout switch was not used on all overdrive installations.

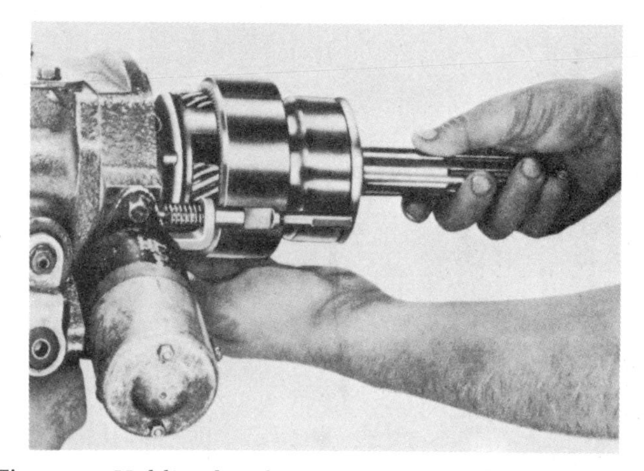

Fig. 6-31. Holding hand under the assembly to catch the rollers as the main shaft is removed.

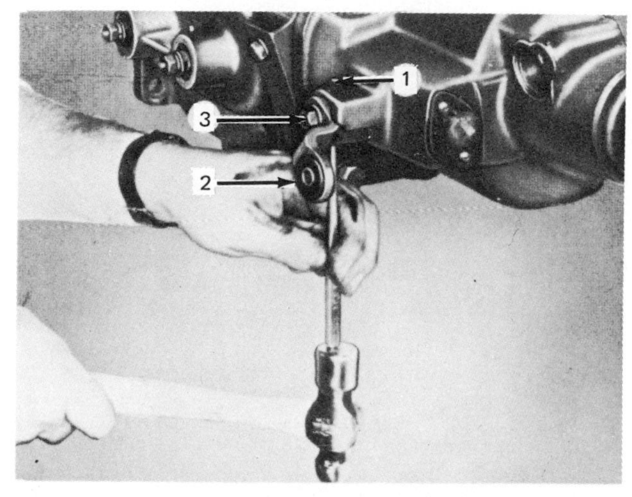

Fig. 6-30. Removing locating pin (1) so that shaft lever (2) with shift shaft (3) can be pulled out far enough to disengage operating cam from shift rail.

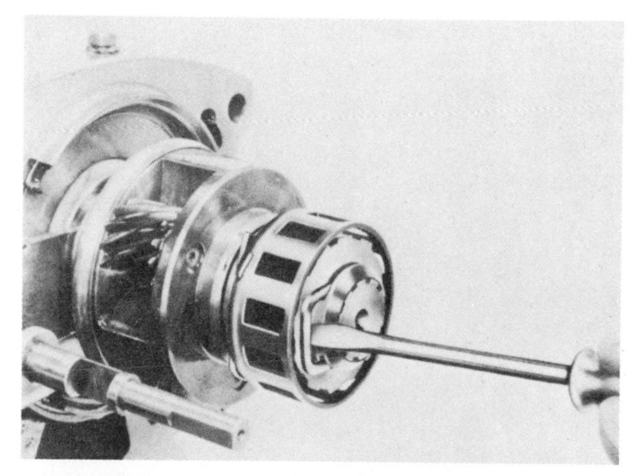

Fig. 6-32. Removing the retaining clip holding the freewheeling unit in place.

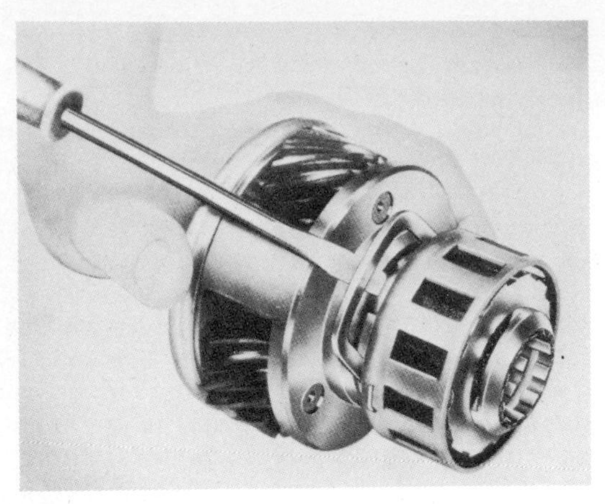

Fig. 6-33. Separating the freewheeling unit from the pinion-cage assembly by removing the retaining clip.

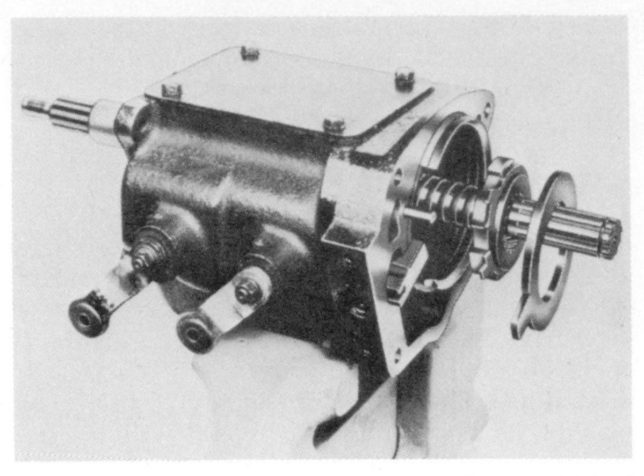

Fig. 6-35. Removing the sun-gear-cover-plate-and-blocker assembly.

i. The overdrive sun gear (Fig. 6-34) and shift-rail assembly can now be removed. The shift collar can be taken off the sun gear by removing the snap rings.

j. The overdrive solenoid can be taken off by removing the cap screws and turning the solenoid one-fourth turn.

k. Take out the large snap ring and then pull the sun-gear cover plate and blocker assembly out (see Fig. 6-35).

l. If the transmission also requires disassembly, the main shaft, with adapter plate and gears, can be removed as a unit from the transmission case.

2. *INSPECTION OF PARTS* Inspection of overdrive parts is very similar to inspection of transmission parts, as outlined in ⊘ 6-14. Note particularly the condition of the oil seals, gears, and bearings. In addition, the electrical condition of the electrical components should be noted. That is, the operation of the governor, solenoid, relay, and switches should be checked. These checks require such electrical testing instruments as those described in *Automotive Electrical Equipment,* another book in the McGraw-Hill Automotive Technology Series.

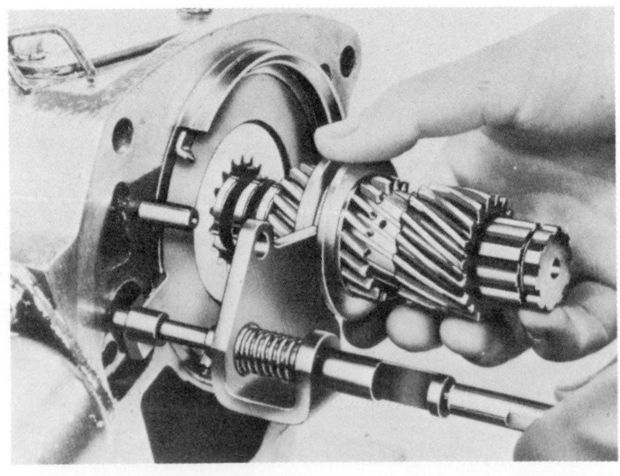

Fig. 6-34. Removing the sun gear and shift-rail assembly.

3. *REASSEMBLY OF OVERDRIVE* Reassembly is essentially the reverse of disassembly. When replacing the rollers, use a little Lubriplate or wheel-bearing grease to hold them in position during the reassembly process. Also, be sure to use new gaskets. When installing the sun-gear blocker assembly, make sure the blocker ring and pawl are properly positioned, as shown in Fig. 6-36. Also, make sure that the freewheeling retainer springs are installed correctly. They should be installed so that when viewed from the rear, the ends of the springs that are outside point in a clockwise direction (Fig. 6-37).

Fig. 6-36. Proper position of the blocker ring and pawl when the overdrive is assembled.

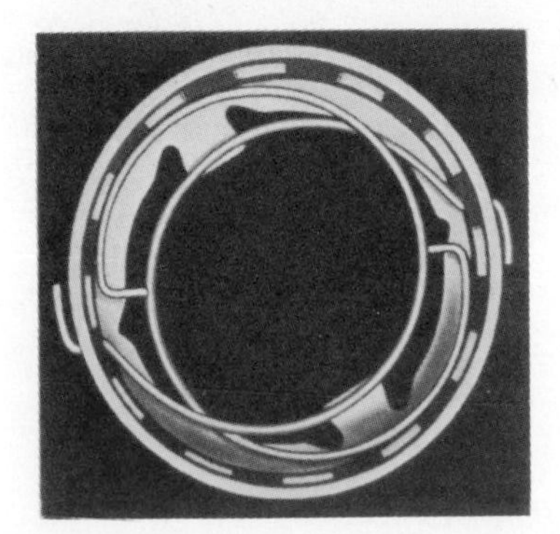

Fig. 6-37. Proper position of the freewheeling retainer springs.

CHAPTER 6 CHECKUP

NOTE: Since the following is a chapter review test, you should review the chapter before taking it.

Some automotive technicians are transmission specialists and spend a great part of their time working on transmissions. However, any good, all-round automotive technician should know how to diagnose transmission troubles and correct them, if necessary, by removing and disassembling the transmission. The chapter you have just finished describes different transmission troubles, methods of tracking down their causes, and repair procedures to eliminate these causes. Now, in the checkup that follows, these different transmission troubles and services are highlighted. Thus, the checkup will help you fix these troubles and services in your mind. If you are not sure about the correct answers to any of the questions, turn back into the chapter and review the pages that will answer them for you. Write the answers in your notebook. The act of writing the answers will help you to remember them, and your notebook will become a valuable source of information you can refer to quickly in the future.

Completing the Sentences The sentences that follow are incomplete. After each sentence there are several words or phrases, only one of which will correctly complete the sentence. Write each sentence in your notebook, selecting the proper word or phrase to complete it correctly.

1. Improperly adjusted gearshift linkage could cause: (*a*) hard shifting into gear, (*b*) clutch slippage, (*c*) noise in neutral.
2. Misalignment between the transmission and the engine is most liable to cause: (*a*) no power through transmission, (*b*) transmission slipping out of high gear, (*c*) transmission slipping out of first.
3. Transmission noise in neutral could result from: (*a*) misadjusted gear linkage, (*b*) worn gears, (*c*) worn speedometer gears, (*d*) loose drive shaft.
4. Clashing of gears when shifting into high could result from: (*a*) engine misalignment, (*b*) a sheared drive key, (*c*) a defective synchronizer.
5. Failure of the overdrive to go into overdrive could be due to: (*a*) incorrect clutch-pedal adjustment, (*b*) a defect in the electric circuit, (*c*) incorrect gearshift-linkage adjustment.
6. Pilot or guide pins are used in transmission removal to prevent damage to the: (*a*) clutch shaft, (*b*) clutch friction disk, (*c*) countershaft, (*d*) gear-shift linkage.
7. Transmission noise in gear could result from a: (*a*) bent shifter fork, (*b*) worn rear transmission bearing, (*c*) clutch that is not releasing.
8. Ball bearings with discolored or cracked balls or races should be: (*a*) relubricated, (*b*) replaced, (*c*) reoiled, (*d*) recleaned.

Troubles and Service Procedures In the following, you are asked to list the troubles causes and service procedures. Do not copy from the book; try to write them in your own words. Writing the trouble causes and service procedures in your own words will help you to remember the causes and procedures better. You will thereby be greatly benefited when you go into the automotive shop. Write in your notebook.

1. List conditions that could result from the gearshift linkage being out of adjustment.
2. List conditions that could result from worn gears in the transmission.
3. List conditions that could result from lack of lubrication in the transmission and gearshift linkage.
4. List conditions that could cause hard shifting.
5. List conditions that could cause the transmission to slip out of gear.
6. List possible causes of noise in the transmission.
7. Describe the electrical checks to be made if the overdrive does not go into overdrive.
8. Describe the electrical checks to be made if the overdrive does not come out of overdrive.
9. Describe the electrical checks to be made if shift into reverse cannot be made and overdrive instrument-panel knob is jammed in OD position.
10. Refer to the shop manual of a specific model of car and list the steps to be taken to remove, disassemble, reassemble, and replace the transmission.
11. List the checks to be made on the parts of a disassembled transmission.
12. Refer to the shop manual of a specific model of car and list the steps to be taken to adjust the gearshift linkage.

SUGGESTIONS FOR FURTHER STUDY

When you are in the automotive service shop, pay special attention to the people working on transmissions so that you can learn more about how various transmission-repair and readjustment jobs are done. Notice the special tools required and how these tools are used.

Study different car manufacturers' manuals in order to become more familiar with different transmission constructions and service procedures. Be sure to write in your notebook any important facts you learn so that you will have a permanent record of them.

chapter 7

TORQUE CONVERTERS

This chapter describes the operation of torque converters used in automatic transmissions. Later chapters describe specific makes and models of automatic transmissions.

For many years, engineers and inventors have searched for means of making gear shifting easier. The introduction of synchronizing devices such as the synchromesh was one result of their efforts. This mechanism makes it virtually impossible to clash gears when shifting. However, it was still necessary to operate a clutch to interrupt the flow of power and move a lever to shift the gears from one meshing position to another.

In recent years, automatic devices have been developed which eliminate this job. That is, the gear ratio through the transmission is changed semiautomatically or automatically in accordance with car and engine speed and the driver's wishes.

With this type of gear-ratio-changing device, the power flow through the transmission is not interrupted, even momentarily, during a gear-ratio change. Continuous power flow is achieved by hydraulically actuated clutches or brake bands, as we shall explain later. If the gear-ratio change is made too abruptly, the occupants of the car will be jarred. If it is done too softly, the friction elements in the clutches and bands will be destroyed by heat. To eliminate these problems, the control of the clutches and brake bands must be precise.

A torque converter (a special type of fluid coupling) couples the engine with the power train. Since it transmits engine-power output through a fluid, it allows the transmission to be in gear at engine idle speed. This is because the velocity of the fluid in the torque converter is too low during engine idle to drive the car.

⊘ 7-1 Hydraulics Before we discuss the torque converter and the various transmissions with which it is used, we should first understand something about *hydraulics*. Hydraulics is the science of liquids, such as water or oil. Our special interest, insofar as automatic transmissions are concerned, is in the pressures that can be exerted by liquids.

⊘ 7-2 Incompressibility of Liquids If a gas, such as air, is put under pressure, it can be compressed into a smaller volume (Fig. 7-1). However, applying pressure to a liquid cannot change its volume. Liquids are incompressible.

⊘ 7-3 Transmission of Motion by Liquid Since liquid is not compressible, motion can be transmitted by liquid. For example, Fig. 7-2 shows two pistons in a cylinder with a liquid between them. When the applying piston is moved 8 inches [203.2 mm] into the cylinder, as shown, then the output piston will be pushed the same distance along the cylinder. In Fig. 7-2, you could substitute a solid connecting rod between pistons A and B and get the same result. But the advantage of such a system is that motion can be transmitted between cylinders by a tube (Fig. 7-3). In Fig. 7-3, as the applying piston is moved, liquid is forced out of cylinder A, through the tube, and into cylinder B. This causes the output piston to move in its cylinder.

⊘ 7-4 Transmission of Pressure by Liquid The pressure applied to a liquid is transmitted by the liquid in all directions and to every part of the liquid. For example (Fig. 7-4), when a piston with 1 square inch [6.452 cm^2 (square centimeters)] of area applies a force of 100 pounds [45.359 kg] on a liquid, the pressure on the liquid is 100 psi (pounds per square inch) [7.031 kg/cm^2 (kilograms per square

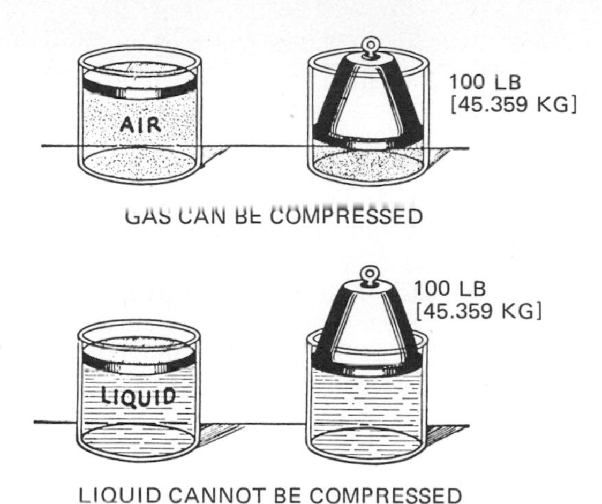

Fig. 7-1. Gas can be compressed when pressure is applied. Liquid, however, cannot be compressed when pressure is applied. (*Pontiac Motor Division of General Motors Corporation*)

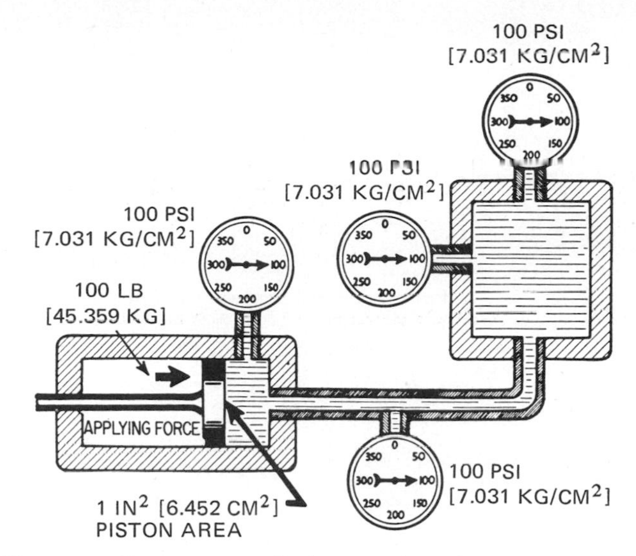

Fig. 7-4. Pressure applied to a liquid is transmitted equally in all directions. (*Pontiac Motor Division of General Motors Corporation*)

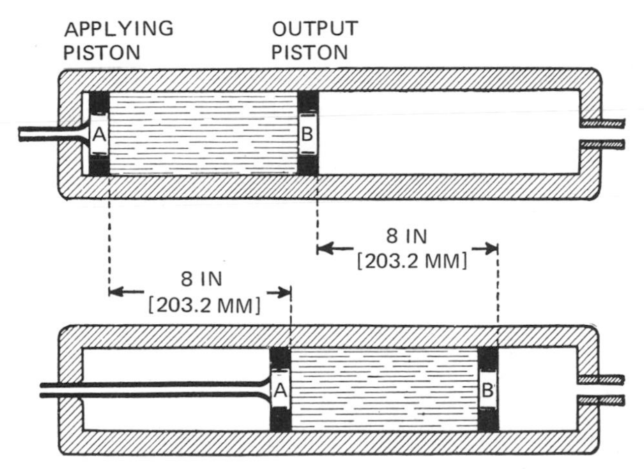

Fig. 7-2. Motion can be transmitted by liquids. When the applying piston A is moved 8 inches [203.2 mm], then the output piston B is also moved 8 inches [203.2 mm]. (*Pontiac Motor Division of General Motors Corporation*)

centimeter)]. This pressure will be registered throughout the entire hydraulic system. If the area of the piston is 2 square inches [12.903 cm^2] and the piston applies a force of 100 pounds [45.359 kg], then the pressure is only 50 psi [3.515 kg/cm^2] (Fig. 7-5).

With an input-output system (Fig. 7-6), we can determine the force applied to any output piston by multiplying the pressure in pounds per square inch by the area of the output piston in square inches. For example, the pressure shown in Fig. 7-6 is 100 psi [7.031 kg/cm^2]. The output piston to the left has an area of 0.5 square inch [3.226 cm^2]. Thus, the output force on this piston is 100 times 0.5, or 50 pounds [22.680 kg]. The center piston has an area of 1 square inch [6.452 cm^2], and its output force is therefore 100 pounds [45.359 kg]. The right-hand piston has an area of 2 square inches [12.903 cm^2], and its output

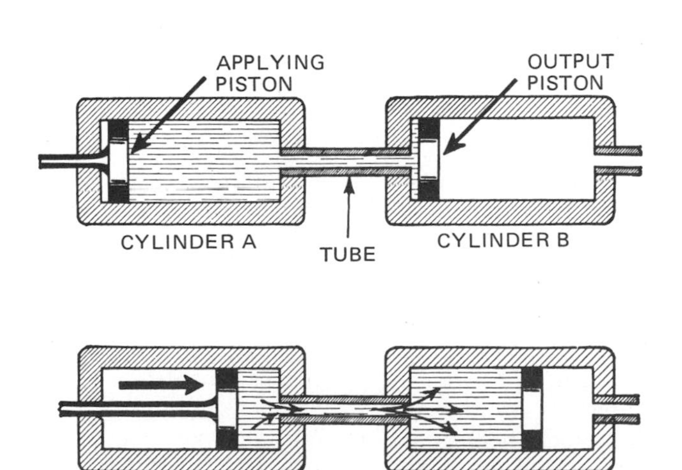

Fig. 7-3. Motion may be transmitted through a tube from one cylinder to another by hydraulic pressure. (*Pontiac Motor Division of General Motors Corporation*)

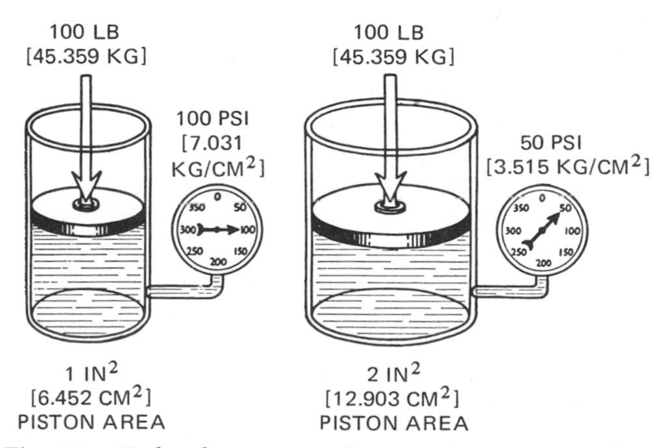

Fig. 7-5. Hydraulic pressure (in pounds per square inch [kilograms per square centimeter]) in the system is determined by dividing the applying force (in pounds [kilograms]) by the area (in square inches [square centimeters]) of the applying piston. (*Pontiac Motor Division of General Motors Corporation*)

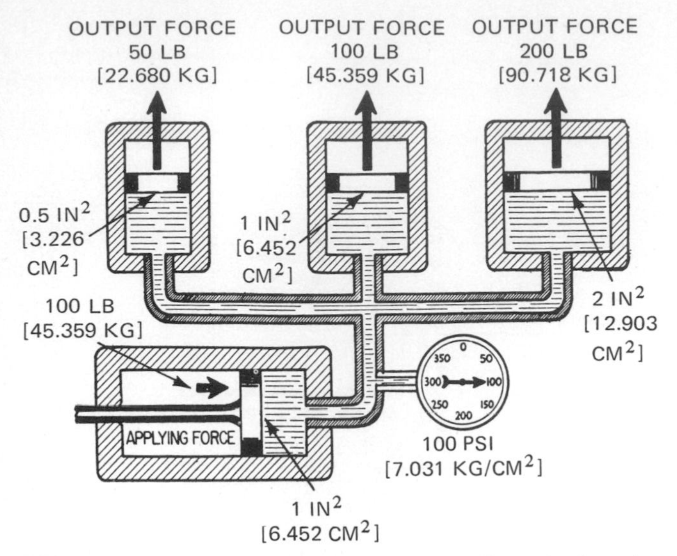

Fig. 7-6. The force (in pounds [kilograms]) applied to the output piston is the pressure in the system in pounds per square inch times the area (in square inches [square centimeters]) of the output piston. (*Pontiac Motor Division of General Motors Corporation*)

force is therefore 200 pounds (100 × 2) [90.718 kg]. The bigger the output piston, the greater the output force. If the area of the piston were 100 square inches [645.16 cm²], for example, then the output force would be 10,000 pounds [4,535.9 kg]. Likewise, the higher the hydraulic pressure, the greater the output force. If the hydraulic pressure on the 2 square inch [12.903 cm²] piston went up to 1,000 psi [70.3 kg/cm²], then the output force on the piston would be 2,000 pounds [907.18 kg].

In all the preceding illustrations, a piston-cylinder arrangement has been shown as the means of producing the pressure. However, any sort of pump can be used. In studying the various automatic transmissions, we shall learn that two types of pumps are used (gear and rotor).

⊘ 7-5 Hydraulic Valves A simple application of the preceding principles is found in a pressure-regulator valve (Fig. 7-7). The valve is spring-loaded and is essentially a small piston that can move back and forth in a cylinder bore. The valve action produces a constant pressure from a variable pressure source. For example, suppose the pressure source is an oil pump driven by an automobile engine. When the engine is operated at high speed, the oil pump is also operated at high speed and puts out a large volume of oil. This action would produce a high pressure if the pressure-regulator valve did not dump part of this oil in the following manner.

As the pressure goes up, there is an increasing force on the valve. Finally, when the preset value is reached, the oil pressure is great enough (in pounds per square inch or kilograms per square centimeter) to overcome the spring force. The valve is moved back in its cylinder bore. As it moves back, the valve uncovers an opening, or port, which is connected to a low-pressure return line to the oil reservoir. Now, part of the oil from the pump can flow through this return line. This reduces the pressure so that the valve starts forward again (moved by the spring force). However, as it moves forward, the valve partly shuts off the port to the return line. Since less oil can now escape, the oil pressure goes up and the valve is again moved back. Actually, the valve does not normally move back and forth as just described. Instead, it seeks and finds the position at which the oil pressure just balances the spring force. Then, if the incoming oil volume changes (because of a change in pump speed), the valve position will change. In action, the valve maintains a constant output pressure by dumping a smaller or greater part of the oil from the pump. As the pump speed goes up, for example (which means the pump delivers more oil), the valve moves away to open the port wider and permit more oil to flow into the return line.

The output-pressure and pump-pressure lines do not actually have to come to the piston as shown in Fig. 7-7*a*. The arrangement can be as shown in Fig. 7-7*c*. Here, the pump pressure entering the output-pressure line is reduced because the piston has positioned itself so as to dump the excess oil

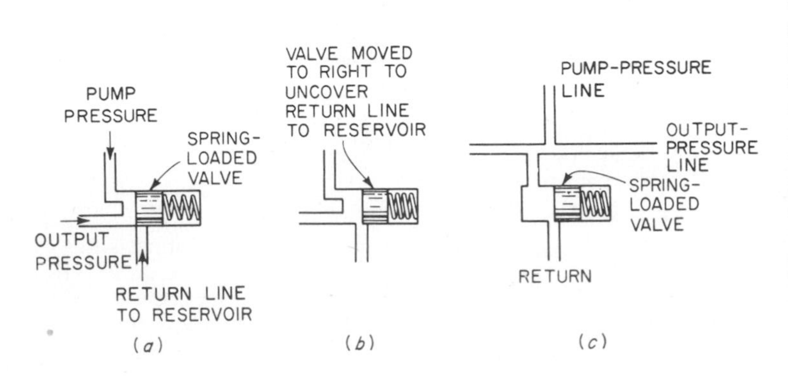

Fig. 7-7. (*a*) A pressure-regulator valve. As pump pressure increases, the spring-loaded valve moves back against spring pressure [as at (*b*)], dumping more of the oil from the pump into the return line. This action maintains a constant output pressure. (*c*) Arrangement in which neither the output-pressure line nor the pump-pressure line is directly connected to the pressure-regulator valve.

from the output-pressure line. Just enough oil is dumped to maintain the pressure in the output line to the value represented by the piston spring. In the actual transmission, the arrangement shown in Fig. 7-7c is used, but there is one or more boost valve lined up with the pressure-regulator valve. Under certain conditions, a boost valve will have oil pressure applied to it. This causes the boost valve to add to the spring pressure. The result is that the output pressure is boosted, or goes up. More on this later, when we describe actual transmissions.

⊘ 7-6 Balanced Valves

In automatic transmissions, balanced valves are used to produce pressure changes that are proportional to the movement of mechanical linkage or variations in spring force. A balanced valve is shown in Fig. 7-8. It contains a valve spool, which is essentially a solid cylinder with an undercut section. The valve spool moves back and forth in a cylinder. Oil pressure works against it on one end, and spring force against it on the other end. In the centered position shown in Fig. 7-8, input oil, under constant pressure, passes into the cylinder, around the undercut section of the valve spool, through the bypass line, and then out through the return line and the output-pressure line. Input pressure is held constant by a pressure-regulator valve, as explained in ⊘ 7-5. Let us see how variations in spring force can cause variations in output pressure.

In Fig. 7-9a, the mechanical linkage has been moved so as to increase the spring force against the end of the valve spool. The valve spool therefore moves (to the left in Fig. 7-9a). This movement tends to close the return line. As the return line is closed, pressure begins to build up on the output end of the valve (just as in the pressure-regulator valve discussed previously). The increasing pressure acts against the output end of the valve, and this pressure works against the spring force. When a balanced condition is reached (output pressure balances spring pressure), the valve stops moving.

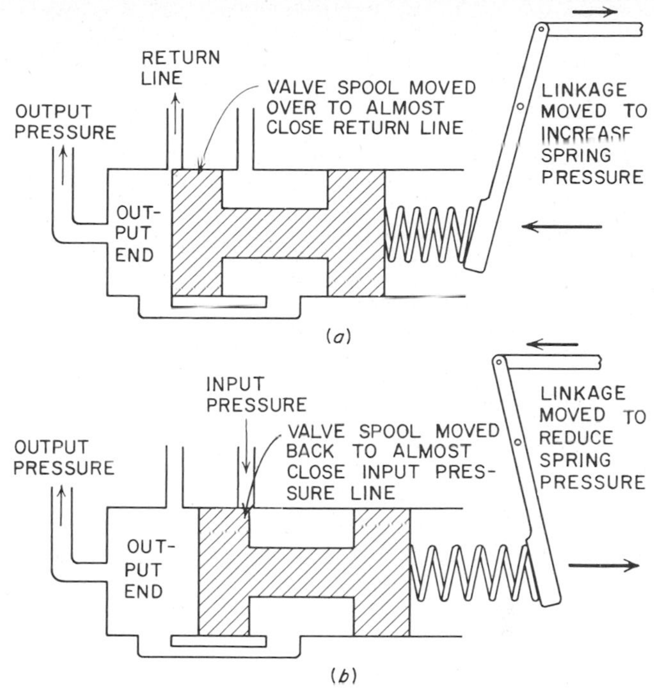

Fig. 7-9. If the linkage is moved to increase the spring force (a), then the valve spool will be moved toward the output end to partly close off the return line and permit a hydraulic-pressure increase to balance the spring-pressure increase. However, if the linkage is moved to reduce the spring force (b), then the hydraulic pressure forces the valve spool away from the output end to partly close off the input-pressure lines; so hydraulic pressure is reduced to balance the spring-force reduction.

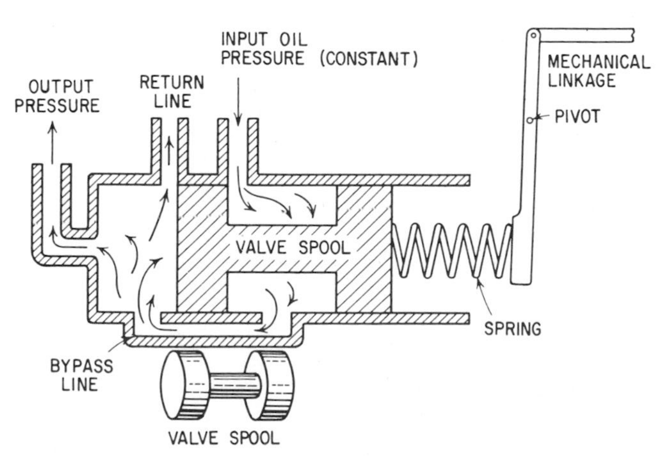

Fig. 7-8. Schematic layout of a balanced valve. The small drawing at the bottom is a perspective view of the valve spool. It is included to give you a better idea of what the valve spool looks like.

If the spring pressure is reduced by movement of the linkage (Fig. 7-9b), then the pressure on the output end of the valve can move the valve (to the right in the illustration). As the valve moves, it tends to close off the input-pressure line. Consequently, no oil can enter and a lower pressure results. The pressure falls off until a balanced condition results (output pressure balances spring force). Then the valve stops moving.

To show exactly how the valve might work, let us look at two examples. Suppose the output end of the valve has an area of 1 square inch [6.452 cm^2] and that the input pressure is 100 psi [7.03 kg/cm^2]. Now, let us apply a spring force of 10 pounds [4.536 kg] on the valve spool (see Fig. 7-10). In order for the valve to balance, there must be only 10 pounds [4.536 kg] of force on the output end. The oil pressure, as it enters and goes through the bypass to the output end, forces the valve to the right (in Fig. 7-10). This movement tends to shut off the input pressure. The pressure increases until a balance is attained. If the force on the spring reaches 100 pounds [45.359 kg], the full input pressure (of 100 psi [7.03 kg/cm^2]) can pass unhampered through the valve. This pressure then becomes the output pres-

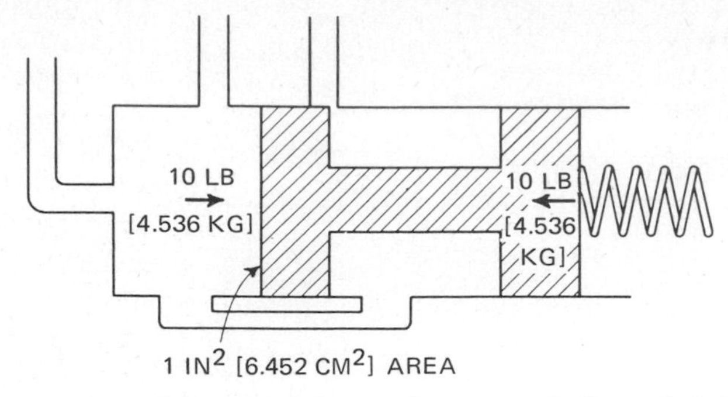

Fig. 7-10. If a spring force of 10 pounds [4.536 kg] is applied, the valve will be removed to the right just enough to admit an input pressure of 10 pounds [4.536 kg] (output end having an area of 1 square inch [6.452 m²]).

sure. However, a spring force above 100 pounds [45.359 kg] cannot raise the pressure above 100 psi [7.031 kg/cm²], since that is all the pressure (input pressure) available.

⊘ 7-7 Control by Varying Hydraulic Pressure In the typical hydraulic system of an automatic transmission, there are valves that produce a varying output pressure which depends not only on spring force, but also on a varying hydraulic control pressure. A typical arrangement is shown in Fig. 7-11. The control pressure enters the valve in a space between two lands of different sizes. In the example shown, one face has an area of $\frac{1}{2}$ square inch [3.226 cm²] while the other has an area of only $\frac{1}{4}$ square inch [1.613 cm²]. The effect is exactly as though the control pressure were working on the end of a valve with an area of $\frac{1}{4}$ square inch [1.613 cm²].

Now if the control pressure were 20 psi [1.406 kg/cm²], there would be an effective hydraulic pressure on the valve (to the left in Fig. 7-11) of 5 pounds ($20 \times \frac{1}{4}$) pounds [2.268 kg]. Suppose the spring produced a force of 10 pounds [4.536 kg]. The result would be that the valve would be pushed to the left with a total pressure of 15 pounds [6.804 kg]. The input pressure would therefore be reduced to 15 psi [1.054 kg/cm²], by the valve action, which would be the output pressure. If the control pressure went up to 60 psi [4.218 kg/cm²], then its effect on the valve would amount to 15 ($60 \times \frac{1}{4}$) pounds [6.804 kg] of pressure. With the added force of 10

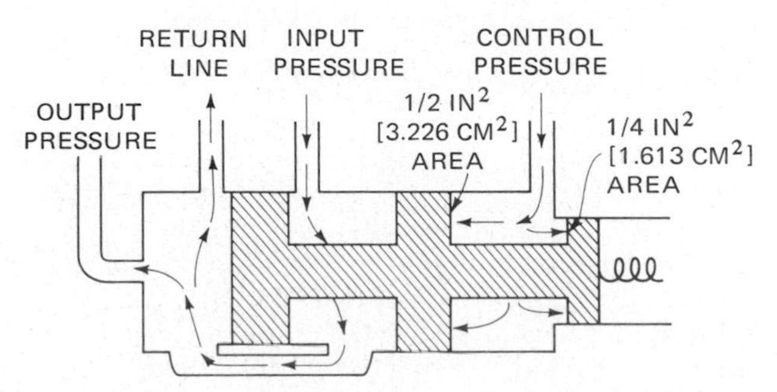

Fig. 7-11. Balanced valve using a varying hydraulic pressure to control output pressure.

pounds [4.536 kg] from the spring, the output pressure would be 25 psi [1.757 kg/cm²].

You can see that by varying the control pressure, you get the same effect as though you were varying the spring force. The output pressure varies in proportion to the control pressure. When you study the hydraulic systems of automatic transmissions, you will find several valves that use this principle. In other words, there are valves that have a difference in face areas between lands, and a varying oil pressure is introduced into this space.

⊘ 7-8 Brake Drums and Bands In automatic transmissions, brake drums and bands are used to control planetary gears and cause a change of gear ratio through the transmission. You will recall from our discussions of planetary gears (⊘ 5-5) that if the ring gear is held stationary, a speed increase or reduction can be achieved. If the sun gear is driving when the internal or ring gear is held stationary, the planet-pinion carrier will turn slower than the sun gear and in the same direction.

A typical brake band is shown in Fig. 7-12. It is made of steel and has a lining of tough asbestos material bonded to it. In one transmission arrangement, the brake band fits loosely around a drum that is part of the ring gear somewhat as shown in Fig. 7-13. The brake band is applied by hydraulic pressure under certain conditions to bring the ring gear to a halt. This produces a speed change, as noted above.

⊘ 7-9 Servos and Brakes The device that hydraulically applies the brake band is called a *servo*. There are several servos in automatic transmissions, and they are used in other places in automobiles, for

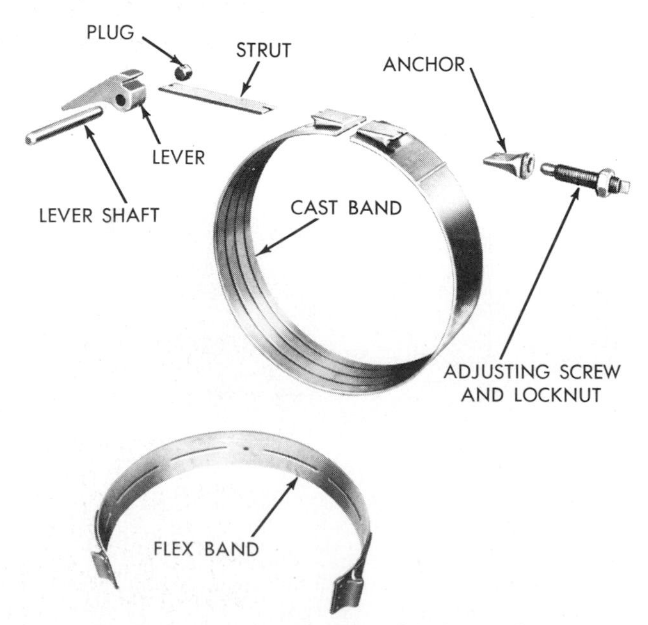

Fig. 7-12. Brake band with related linkages for an automatic transmission. (*Chrysler Corporation*)

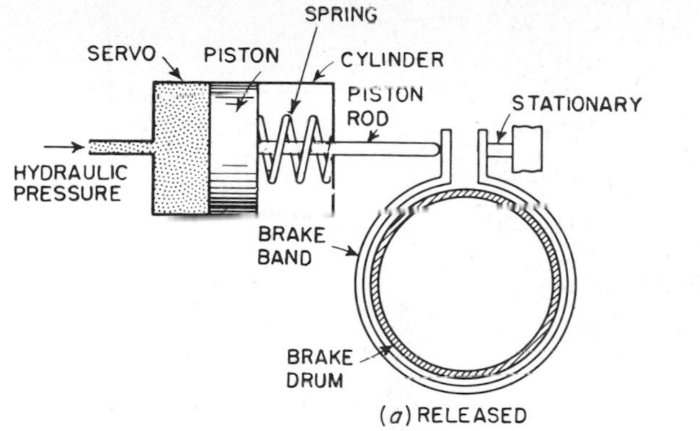

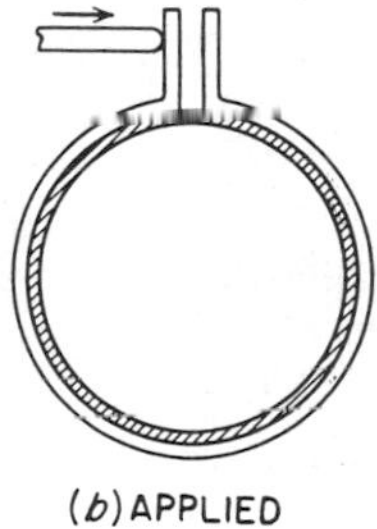

Fig 7-13. The servo consists of a piston that moves in a cylinder in accordance with changing hydraulic pressure. This movement may be used to perform some job, such as applying a brake.

instance, in the brakes. A transmission servo is shown in Fig. 7-13, which illustrates the operating servo, brake band, and brake drum mentioned in ⊘ 7-8.

Essentially, the servo is a cylinder and piston. Hydraulic pressure can be admitted to one or the other end of the cylinder. This pressure causes the piston to move; mechanical linkage from the piston then causes movement of some mechanism. Thus, hydraulic pressure is used to produce a mechanical action. In Fig. 7-13, if hydraulic pressure is applied to the piston, the piston will move (to the right in the illustration). This action causes the piston rod to apply the brake band. Thus, the brake drum can be made to stop turning by increasing the hydraulic pressure. In fact, the hydraulic brakes used in automobiles work on exactly this principle. Operation of a foot pedal increases the hydraulic pressure in the master brake cylinder, and this increasing hydraulic pressure causes pistons in the wheel cylinders to move. As the wheel-cylinder pistons move, they force the brake shoes against the brake drums or disks so that the car is braked.

Figure 7-14 is a sectional view of an actual servo for controlling the brake band in an automatic transmission. The *apply pressure* for applying the brake band comes from the control valves in the hydraulic system. Hydraulic fluid flows from the hydraulic system through a hole in the piston stem and fills the chamber on the apply side of the piston. This pressure forces the piston to move (to the left in Fig. 7-14). As the piston moves, it tightens the brake band around the brake drum, forcing it to stop rotating. Then, when engine speed, car speed, and amount of throttle opening call for an upshift, the hydraulic-system valves cut off the hydraulic pressure to the apply side of the servo piston. At the same time, hydraulic pressure is applied to the release side of the piston. This release-side pressure plus the tension of the spring forces the piston back (to the right in Fig. 7-14) so that the brake band loosens on the brake drum. The drum and ring gear are then free to revolve. This prepares the transmission for an upshift. However, something else must happen to produce this upshift in addition to releasing the ring gear. We shall come to that soon.

⊘ 7-10 Accumulators To prevent sudden application of the brake band, which would produce a harsh shift, an accumulator is used. Figure 7-15 shows one early version of an accumulator. This device produces a rapid but smooth band application. To start with, when oil is directed to the servo by the hydraulic control system, the oil enters

Fig. 7-14. Brake band with servo that operates to apply the brake band. (*Ford Motor Company*)

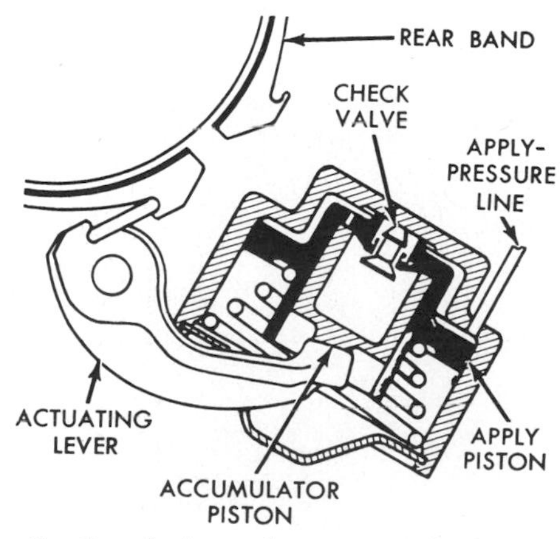

Fig. 7-15. Sectional view of an accumulator. (*Ford Motor Company*)

back of the apply piston. At first, the apply piston is held back by the heavy spring. But the oil can pass through the check valve and apply pressure against the accumulator piston. This action causes the accumulator piston to move down and operate the actuating lever so that the brake band is engaged. Then, the hydraulic pressure moves the apply piston so that it adds its force against the brake band, engaging it hard for the full braking effect. The combination gives a smooth band application. In many automatic transmissions, the accumulator has become a separate valve interposed in the hydraulic circuit between the pressure source and the servo. We shall discuss accumulators again when we discuss specific models of automatic transmissions.

NOTE: In many automatic transmissions, the accumulators are used to cushion the shock of clutch engagement instead of being used in connection with the brake bands. This is true of the General Motors Model 350 Turbo Hydra-Matic automatic transmission, for example, which uses two accumulators in connection with the two clutches that produce upshifts from first to second and from second to third. See ⊘ 7-11.

⊘ 7-11 Clutches Hydraulically operated clutches provide another means for locking up, or releasing, a rotating member of the planetary-gear system. The clutches used in automatic transmissions are the multiple-disk or -plate type. Figure 7-16 shows a clutch of this type in sectional view, and Fig. 7-17 shows a disassembled clutch. The plates are alternately attached to an outer housing, or drum, and an inner hub. When the clutch is not applied, the two members can rotate independently of each other. However, when the clutch is applied, the plates are forced together and the friction between them locks up the two members so that they must rotate together.

Application of the clutch is produced by oil pressure, which forces an annular piston to push the plates together. The annular piston is a ring that fits snugly into the *clutch-drum assembly,* also called

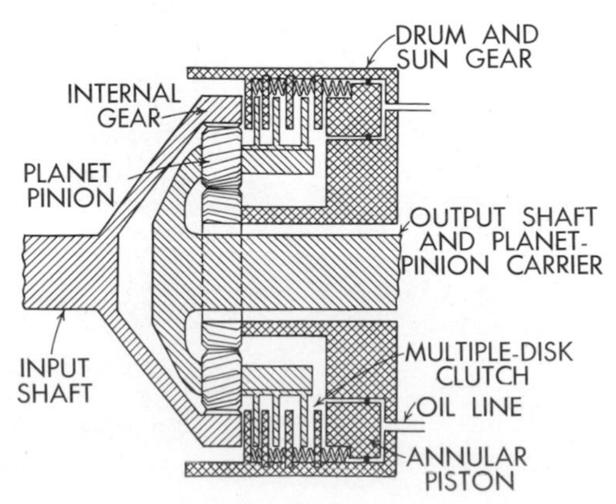

Fig. 7-16. Sectional view of a multiple-disk clutch used in automatic transmissions.

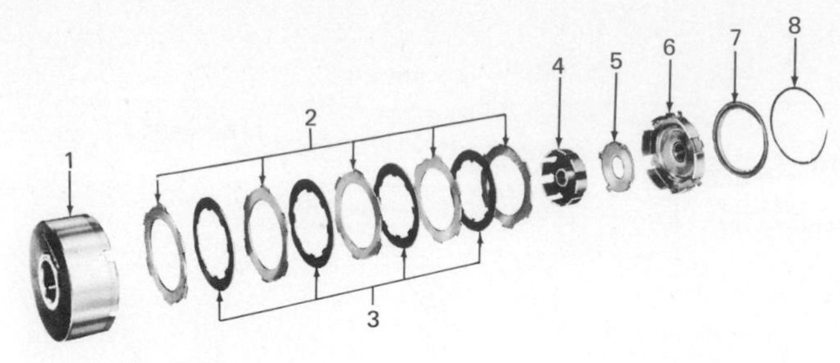

1. Clutch-drum assembly
2. Clutch driven plates
3. Clutch drive plates
4. Clutch hub
5. Clutch-hub thrust washer
6. Low-sun-gear-and-thrust-flange assembly
7. Clutch-flange retainer
8. Retainer snap ring

Fig. 7-17. Clutch parts. (*Chevrolet Motor Division of General Motors Corporation*)

the *piston-retainer assembly.* This piston, which is ring-shaped, is shown in Fig. 7-16 in sectional view. An actual piston can be seen in the disassembled view of a clutch drum shown in Fig. 7-18. The piston fits snugly into the drum, sealed by a sliding seal to the drum's inner surface. It is spring-loaded away from the plates by a series of springs that rest against the piston return seat. Thus, the oil pressure must overcome the spring force to move the piston and cause it to force the plates together so that the clutch engages.

Now, as you can see, there are two separate lockup mechanisms used in automatic transmissions: bands and clutches. Later we shall see how these are used to produce the shifting action in transmissions. Actually, the clutches and brake bands are used in a variety of ways in different models of automatic transmissions.

NOTE: The arrangement described in this chapter is only one of several arrangements used in auto-

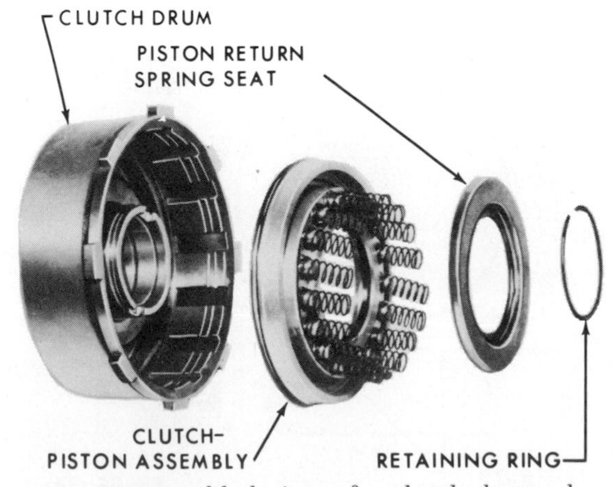

Fig. 7-18. Disassembled view of a clutch drum, showing the piston, springs, and return seat. (*Oldsmobile Division of General Motors Corporation*)

matic transmissions. In some transmissions, when the band is applied, it holds the ring gear, or planet-pinion carrier, stationary. Different transmissions may lock different members together when the clutch is applied. The principle is the same in all transmissions, however. There is gear reduction when the band is applied, and there is direct drive when the clutch is applied.

Check Your Progress

Progress Quiz 7-1 The hydraulic fundamentals we have covered in the past few pages apply to all types of automatic transmissions. Thus, you should have a good understanding of those fundamentals before you proceed with your studies of transmission. The following quiz will help you review the fundamentals and thus fix them more firmly in your mind. If you have any trouble answering the questions, review the past few pages. Remember, good students may reread their lessons several times in order to make sure they have learned all the essential details.

Completing the Sentences The sentences that follow are incomplete. After each sentence there are several words or phrases, only one of which will correctly complete the sentence. Write each sentence in your notebook, selecting the proper word or phrase to complete it correctly.

1. If a liquid is put under pressure, it will: (*a*) compress, (*b*) not compress, (*c*) increase in volume.
2. Since a liquid is incompressible, it can be used to transmit: (*a*) motion and pressure, (*b*) rotation and speed, (*c*) vacuum and heat.
3. A piston with an area of 5 in^2 is pushed into a cylinder filled with a liquid with a force of 200 lb; the pressure developed is: (*a*) 40 psi, (*b*) 200 psi, (*c*) 1,000 psi.
4. A piston with an area of 0.25 in^2 is pushed into a cylinder filled with liquid with a force of 200 lb; the pressure developed is: (*a*) 50 psi, (*b*) 200 psi, (*c*) 800 psi.
5. An output piston against which hydraulic pressure is working has an area of 12 in^2, and the hydraulic pressure is 240 psi. The force on the piston is: (*a*) 20 lb, (*b*) 240 lb, (*c*) 2,400 lb, (*d*) 2,880 lb.
6. If the pressure-regulating valve described in the chapter has an area of 2 in^2 and the spring imposes a force of 25 lb on it, then the regulating pressure will be: (*a*) 12.5 psi, (*b*) 25 psi, (*c*) 50 psi.
7. If the spring force on the balanced valve is 25 lb, the input pressure is 100 psi, and the area of the output end of the valve is 2 in^2, then the output pressure will be: (*a*) 12.5 psi, (*b*) 25 psi, (*c*) 50 psi.
8. If the control pressure on the valve shown in Fig. 7-11 is 100 psi, and the spring force is 20 lb, then the output pressure will be: (*a*) 120 psi, (*b*) 60 psi, (*c*) 45 psi.
9. The brake band is operated by: (*a*) the clutch piston, (*b*) the output valve, (*c*) a servo.
10. The clutch disks are forced together by movement of the: (*a*) clutch springs, (*b*) clutch piston, (*c*) servo piston.

⊘ 7-12 Operation of Fluid Coupling The clutch used in automotive power trains with manually shifted transmissions (not automatic) is a mechanical coupling. When engaged, it couples the engine to the transmission. The fluid coupling is a hydraulic coupling. It is always "engaged"; but since the coupling is produced by a fluid, the driven member can slip, or turn slower than, the driving member.

1. *FLUID COUPLING* A simple fluid coupling can be illustrated with two electric fans. If the fans are placed a few inches apart and facing each other and if one fan is plugged in so that it runs, the current of air from it will cause the blades of the other fan to turn (Fig. 7-19). In this case the air acts as the fluid; but since the two fans are not enclosed or closely coupled, this sort of fluid coupling is not efficient.

To make a more efficient fluid coupling, oil is used as the fluid and the two halves, or members, of the coupling are mounted very close together and enclosed in a housing. Figure 7-20 shows the two members of a fluid coupling. Note that they resemble a hollowed-out doughnut sliced in half, with blades or vanes set radially into the hollow halves. Figure 7-21 is a cross section of a fluid coupling. The driving member is attached to the engine crankshaft, and the driven member is attached to the transmission shaft. This shaft is, in turn, connected through gearing and the drive shaft to the differential and rear wheels.

The hollow space in the two members is filled with oil. When the driving member begins to rotate (as the engine is started and runs), the oil is set into motion. The vanes in the driving member start to carry the oil around with them. As the oil is thus spun, it is thrown outward, or away from the shaft, by centrifugal force. Thus, the oil moves outward in the driving member in a circular path, as shown by the dotted arrows in Fig. 7-21. Also, since the oil is being carried around with the rotating driving

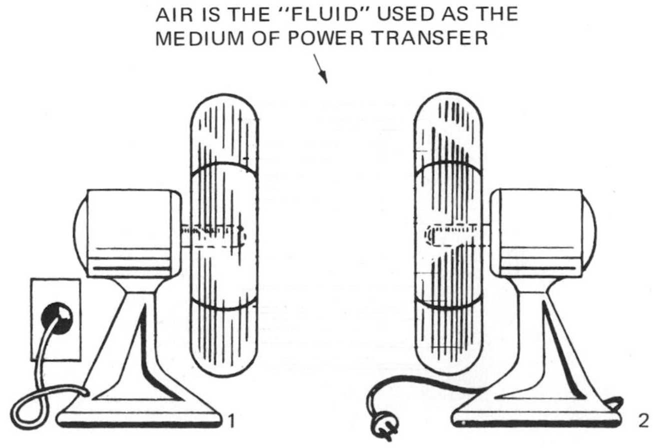

Fig. 7-19. Rotation of fan 1 causes fan 2 to rotate. This is a simple *fluid coupling*, with air serving as the fluid.

Fig. 7-20. Simplified version of two members of a fluid coupling. (*Chevrolet Motor Division of General Motors Corporation*)

member, it is thrown into the driven member at an angle, as shown in Fig. 7-22. Thus, the oil strikes the vanes of the driven member at an angle, as shown, thereby imparting torque, or turning effort, to the driven member. The faster the driving member turns, with the driven member stationary, the harder the oil strikes the vanes of the driven member. The harder the oil strikes the vanes, the greater the turning effort imparted to the driven member (within limits).

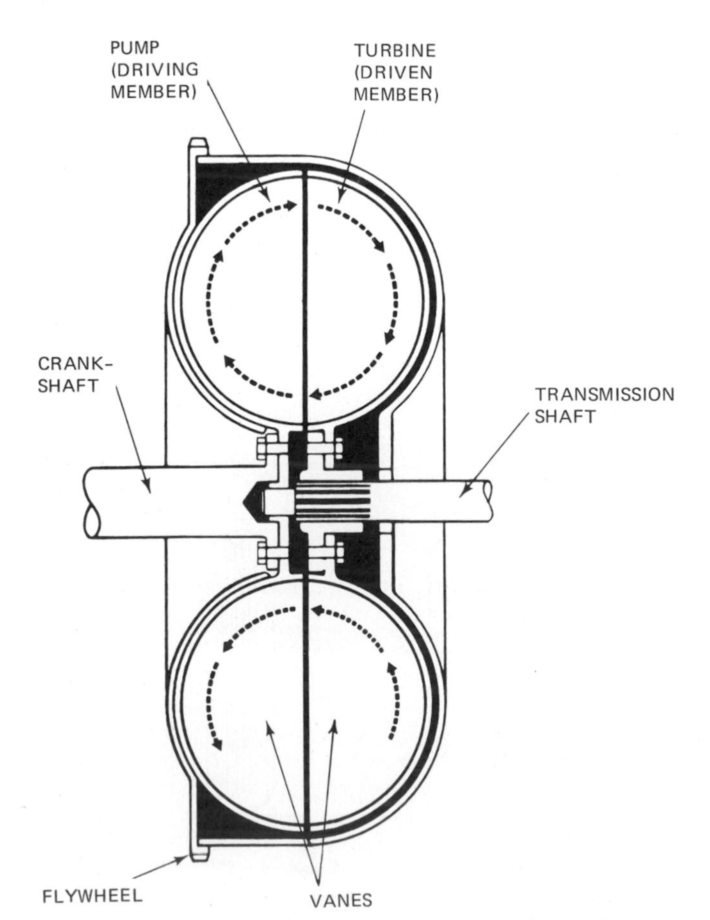

Fig. 7-21. Cross-sectional view of a fluid coupling.

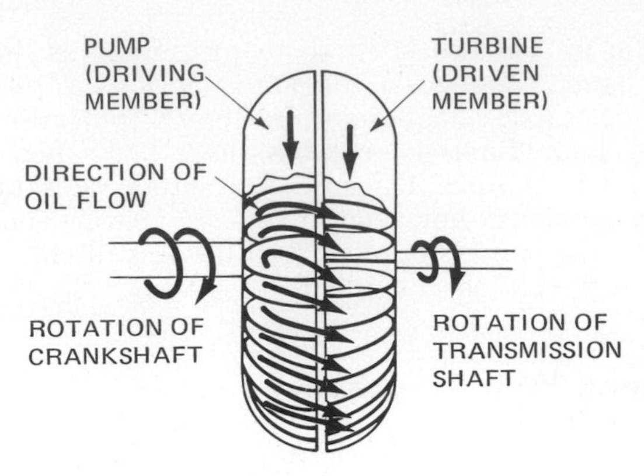

Fig. 7-22. Fluid coupling in action. Oil is thrown from the driving member into the driven member. The outer casings have been cut away so that the vanes can be seen.

This turning effort, carried to the wheels, sets the car into motion. As the driven member approaches the speed of the driving member, the effective force of the oil on the driven-member vanes is reduced. If the two members turn at the same speed, the oil does not pass from one to the other member. With no oil passing from one member to the other, no power is being transmitted through the coupling. Thus, this *same-speed* condition will not exist when the engine is driving the car. The driving member always must be turning a little faster than the driven member in order for engine power to flow through the fluid coupling to the car wheels.

However, if the engine speed is reduced so that the car begins to drive the engine, there will be a point at which both members turn at the same speed. Then, as the engine slows further, the driven member will temporarily become the driving member (since the car will be driving it). As this happens, the normally driven member will begin to pass oil into the normally driving member. Now, the engine will exert a braking effect on the car (the same condition that results in a clutch-equipped car when you release the accelerator and coast).

2. GUIDE RING The simple fluid coupling just described would not be very efficient under many conditions because of the *turbulence* that would be set up in the oil. Turbulence is a state of violent random motion or agitation. Thus, under certain conditions (when there is considerable difference in speed between the driving and driven members), the oil would be striking the vanes of the driven member with great force. This action would cause the oil to swirl about in all directions, particularly in the center sections of the members (Fig. 7-23). To reduce this turbulence and thereby make the coupling more efficient, fluid couplings use a split guide ring centered in the members (Fig. 7-24). The guide ring looks much like a hollowed-out doughtnut, sliced in half. Each half is attached to the vanes of one of the coupling members. With this arrangement, the oil does not have a chance to set up the turbulences, as shown in Fig. 7-23.

3. OPERATING CHARACTERISTICS OF THE

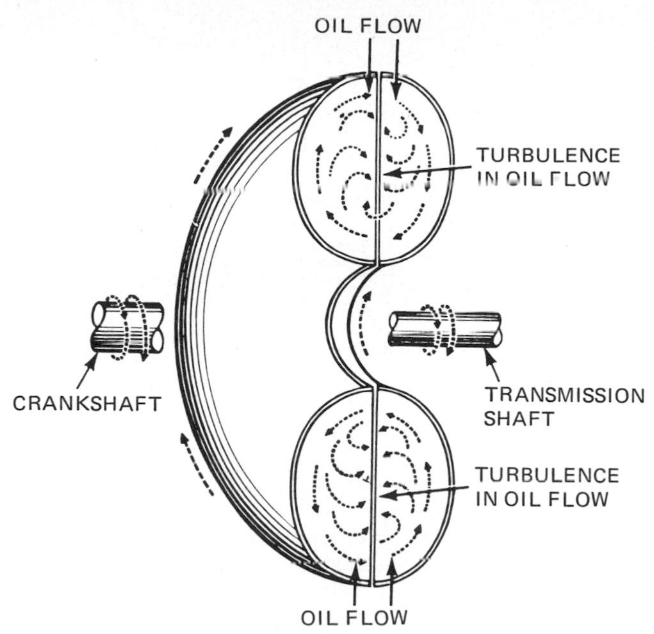

Fig. 7-23. Turbulence in oil flow in center sections of fluid-coupling members.

FLUID COUPLING Essentially, the fluid coupling is a special form of clutch which provides a smooth, vibrationless coupling between the engine and transmission. It operates at maximum efficiency when the driven member approaches the speed of the driving member. If there is a big difference in the speeds of the two members, power is lost and efficiency is low. In the following sections on torque converters, we shall examine the reason for this loss of efficiency. The fluid coupling was used with a gear-shifting mechanism and a form of transmission which provides the varying gear ratios between the engine and the rear wheels. The fluid coupling described above is no longer used in automobiles. Instead, a special type of fluid coupling, with extra members, is used. It is called a *torque converter.*

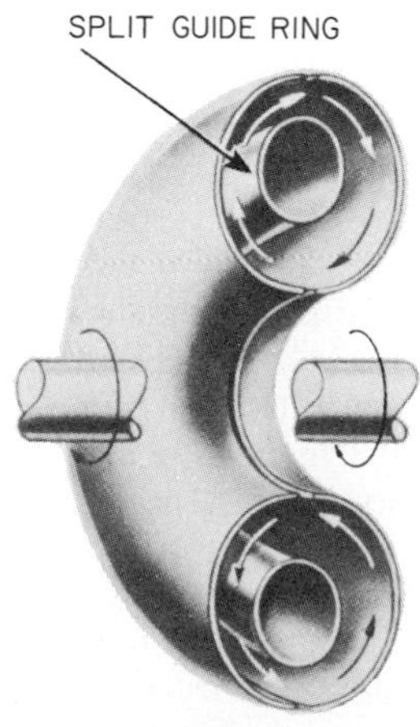

Fig. 7-24. Split guide ring designed to reduce oil turbulence. (*Chevrolet Motor Division of General Motors Corporation*)

⊘ 7-13 Torque Converter—A Special Type of Fluid Coupling The torque converter is a special form of fluid coupling. It has both a driving member and a driven member, each equipped with vanes between which oil passes. However, the fluid coupling described previously has flat vanes. It is essentially a special form of clutch that transmits torque at maximum efficiency when both members are turning at approximately the same speed. When the driving member turns appreciably faster than the driven member, the efficiency with which torque is delivered to the driven member is lowered. Here's the reason:

When the driving member is turning considerably faster than the driven member, oil is thrown onto the vanes of the driven member with considerable force. Oil strikes the driven-member vanes and splashes, or bounces back, into the driving member. In other words, this effect actually causes the oil to work against the driving member. Thus, when there is a big difference in driving and driven speeds, a good part of the driving torque is used in overcoming this "bounce-back" effect. Torque is lost; there is a torque reduction through the fluid coupling.

The situation is greatly different in the torque converter. This device is designed so as to prevent or reduce to a minimum the bounce-back effect. The result is that torque is not reduced when there is a large difference between driving and driven speeds. Quite the reverse: The torque is increased, or multiplied, in the torque converter.

⊘ 7-14 Torque Converter The torque converter acts, in a sense, like a gear transmission with a large number of gearshift positions. That is, it can transmit torque at a 1:1 ratio. Or, under certain conditions, it can increase, or multiply, this torque so that more torque is delivered than is applied. This ability can be compared with a transmission in low gear. In low gear, the speed through the transmission is reduced, which increases the torque. (If the relationship between speed reduction and torque increase is not clear, reread ⊘ 1-5.) In a like manner, speed reduction through the torque converter means a torque increase. Torque output is greater than the torque input; there is torque multiplication.

⊘ 7-15 Torque-Converter Action The torque converter provides varying drive ratios between the driving and the driven members, thus providing varying amounts of torque increase. It accomplishes this action by means of curved vanes in the driving and driven members and by the use of one or more extra members (between the driving and driven members). These extra members act to reduce the splashing, or bounce-back, effect mentioned in ⊘ 7-13. You will recall that this effect causes torque loss in a fluid coupling since the bouncing oil strikes the forward faces of the driving-member vanes and thereby tends to slow down the driving member.

NOTE: In the torque converter, the driving member is usually referred to as the *pump* (also called the *impeller*) and the driven member is called the *turbine*. In operation, the pump drives the turbine.

1. *CURVED VANES* As mentioned, the vanes in the driving and driven members of the torque converter are curved. This curvature is shown in Fig. 7-25. Note that the curving vanes allow the oil to change directions gradually as it passes from the driving to the driven member. The heavy arrows show the oil paths. The small arrows indicate the driving force with which the oil strikes the vanes of the driven member. It can be seen that the oil, which is moving around with the driving member, is thrown with a forward motion, or velocity, into the driven member. As the oil is passed into the driven member, it "presses" forward all along the vanes, as shown by the small arrows. This action produces the push which causes the driven member to rotate.

2. *BOUNCE-BACK EFFECT WITHOUT ADDITIONAL MEMBERS* When the two members are revolving at about the same speed, there is relatively little movement of oil between them. This is similar to the action in the fluid coupling. However, when the driving member is revolving considerably faster than the driven member, the oil is thrown forward with considerable velocity into the driven member. You will recall the difficulty this action produces in the fluid coupling because of the bounce-back effect. You can see how this effect occurs in Fig. 7-26, where the front parts of the vanes and guide ring have been cut away so that the inner ends of the vanes can be seen. Compare this illustration with Fig. 7-22.

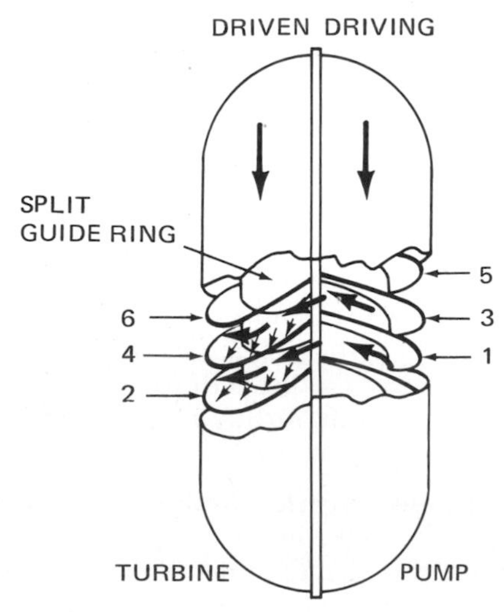

Fig. 7-25. Simplified cutaway view of two members of a torque converter. The heavy arrows show how oil circulates between driving- and driven-member vanes. In operation, the oil is forced by vane 1, downward toward vane 2; thus, it pushes downward against vane 2, as shown by small arrows. Oil then passes around behind the split guide ring and into the driving member again, or between vanes 1 and 3. Then it is thrown against vane 4 and continues this circulatory pattern, passing continuously from one member to the other.

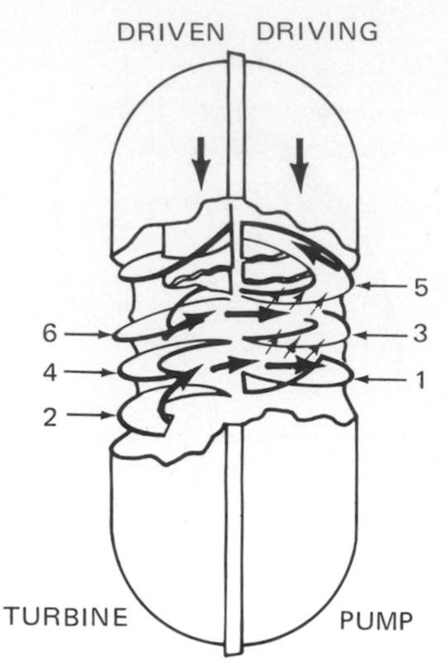

Fig. 7-26. This illustration shows what would happen if the vanes in Fig. 7-25 were continuous. Actually, the inner ends of the vanes are not as shown here but are as pictured in following illustrations. Here, the split guide and outer ends of the vanes have been cut away. If the vanes were as shown here, the oil leaving the trailing edges of the driven member would be thrown upward against the forward faces of the driving-member vanes. It would thus oppose the driving force. This effect, shown by the small arrows, would cause wasted effort and loss of torque.

NOTE: In actuality, the inner ends of the vanes are not shown in Fig. 7-26. This illustration is included merely to show what would happen if the vanes were continuous and if there were no additional members in the assembly.

As the oil passes through the driven member, as shown by the heavy arrows (Fig. 7-26), it moves along the curved vanes of the driven member. It is still moving rapidly as it leaves the trailing edges of the driven-member vanes. (The trailing edges are the back edges, or the edges that the oil passes last as it leaves the member.) However, note that the oil has changed directions. The curved vanes of the driven member have caused the oil to leave the trailing edges of the vanes so that it is thrown against the forward face of the driving-member vanes. This is shown by the small arrows in Fig. 7-26. Thus, the oil opposes the driving force of the driving member. With a big difference in speeds between the driving and driven members, this opposing force would use up a good part of the power applied by the driving member. Thus, some means of reducing this effect must be used if the torque converter is to function efficiently when the driving member turns considerably faster than the driven member.

A simple method of overcoming this effect and at the same time increasing the push (torque) on the driven member is given in Fig. 7-27. In Fig. 7-27, a jet of oil is shown striking a hemispherical bucket

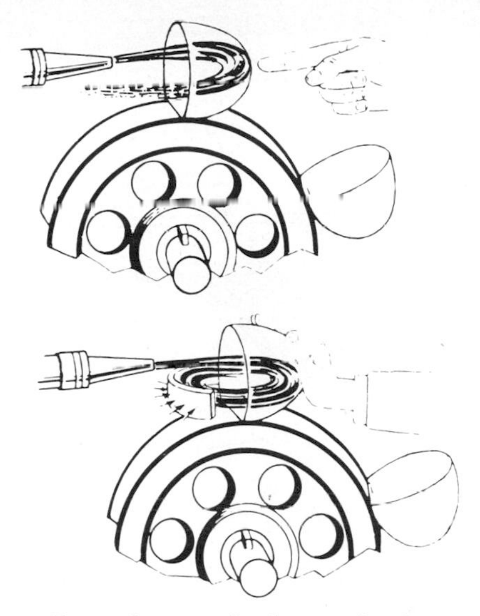

Fig. 7-27. Effect of a jet of oil on a bucket attached to a wheel. If the oil enters and leaves as at (*a*), the push imparted to the bucket and wheel is small. But if the oil jet is redirected into the bucket by a curved vane as at (*b*), the push is increased. (*Chrysler Corporation*)

attached to the rim of a wheel. This action can be compared with the oil being thrown from the driving member into the curved vanes of the driven member in the torque converter. The jet of oil swirls around the curved surface of the bucket and leaves the bucket with almost the same velocity as it had when entering the bucket. In exactly the same way, the oil is thrown into the driven member (Fig. 7-26); the curved vanes of the driven member react against it, change its direction, and throw it back to the driving member.

In Fig. 7-27*a*, the oil gives up very little of its energy to the bucket, and there will be only a small push on the bucket. That is, there will be little torque applied to the driven member.

Furthermore, it is this returning oil which bounces back into the driving member and causes the loss of torque, as already explained. However, we can prevent this effect if we install a stationary curved vane, as shown in Fig. 7-27*b*. Now the oil, as it comes out of the bucket, enters the stationary curved vane and is again reversed in direction. It is once more directed into the bucket. Theoretically, the oil can make many complete circuits between the bucket and the vane. Each time it reenters the bucket, it imparts a further push (or added torque) to the bucket. This effect is known as *torque multiplication.*

In an actual torque converter, there are stationary curved vanes which reverse the direction of the oil as it leaves the driven member. Thus, the oil enters the driving member in a "helping" direction, passes through, and then reenters the driven member, where it gives the vanes of the driven member another "push." In other words, the oil is repeatedly redirected into the driven member, each time adding torque to the driven member. Thus, there will be an increase in torque; torque has been multiplied. Following paragraphs explain how this effect is achieved in actual torque converters.

3. *CHANGE IN DIRECTION OF OIL FLOW* In an actual torque converter, one or more additional members are placed between the trailing edges of the driven-member vanes and the leading edges of the driving-member vanes. The trailing edges are the edges of the vanes that the oil passes last as it leaves a member. The leading edges are the edges onto which the oil first flows. The additional member or members in the torque converter have curved vanes that change the direction of the oil into a "helping" direction instead of a "hindering" direction. The effect illustrated in Fig. 7-27 is thus achieved.

4. *NUMBER OF MEMBERS* Today torque converters generally have three members: the driving member, or pump; the driven member, or turbine; and the stator. It is the stator that changes the direction of oil flow into a helping direction, as shown in Fig. 7-27. This action is explained further in ⊘ 7-16.

⊘ 7-16 Three-Member Torque-Converter Action Figure 7-28 is a partial cutaway view of an automatic transmission using a torque converter with three members. Figure 7-29 shows the three members detached from the assembly and separated. Figure 7-30 is a simplified drawing of a torque converter. The stator is mounted on a freewheeling mechanism (overrunning clutch) which permits it to run free when both torque members are turning at about the same speed. However, when torque increase (and drive reduction) takes place, the stator stops and acts as a reactor. It turns the oil from the trailing edges of the turbine into a "helping" direction before it enters the pump. In this torque converter, the maximum torque increase that is produced is slightly over 2:1.

As turbine speed approaches pump speed, the torque increase gradually drops off until it becomes 1:1 as the turbine and pump speeds reach a ratio of approximately 9:10. At this point, the oil begins to strike the back faces of the vanes in the stator so that the stator begins to turn. Thus, in effect, the stator "gets out of the way" of the oil and thereby no longer enters into torque-converter action. The converter therefore acts simply as a fluid coupling under these conditions.

A typical overrunning-clutch construction is shown in cutaway view in Fig. 7-31. It is similar in principle to the freewheeling mechanism used in overdrives (⊘ 5-3). The overrunning clutch contains a series of rollers placed between the stator and the shaft. The rollers permit the stator to overrun, or freewheel, when the oil strikes the back of the stator blades. However, when the oil strikes the front of the blades, it attempts to turn the stator in the opposite direction. This action locks up the overrunning clutch, and so the stator is held stationary. Thus the

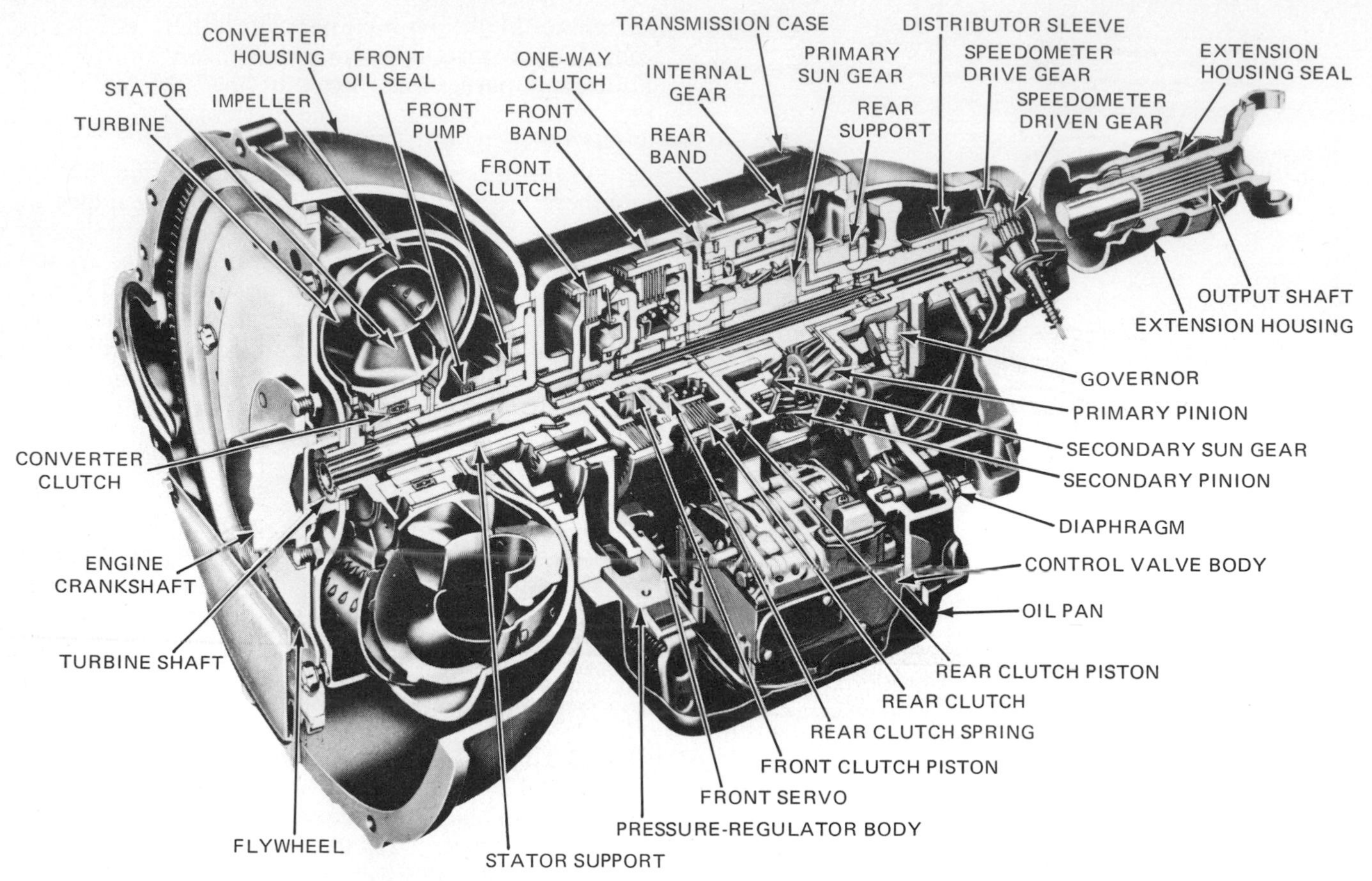

Fig. 7-28. Partial cutaway view of an automatic transmission which has a three-member torque converter. (*Ford Motor Company*)

stator acts as a reaction member, as already described.

Some overrunning clutches use sprags instead of rollers. The sprags are shaped somewhat like flattened rollers (Fig. 7-32). A series of sprags are placed between the inner and outer races (Fig. 7-33). They are held in place by two springs positioned in the sprag notches. During overrunning, when the stator action is not needed, the inner race is not locked up. The inner race is part of the stator, and thus the stator can spin freely. However, when the stator action is needed, the oil is directed into the stator vanes. This attempts to spin the stator backward. When this happens, the sprags jam between the outer and inner races, as shown in Fig. 7-32, to lock the stator in a stationary position.

Modern automatic transmissions use three-member torque converters. They are described in following chapters.

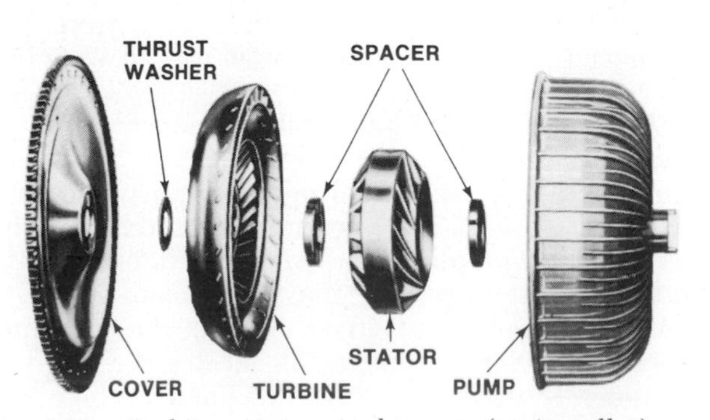

Fig. 7-29. Turbine, stator, and pump (or impeller) used in a torque converter. (*Ford Motor Company*)

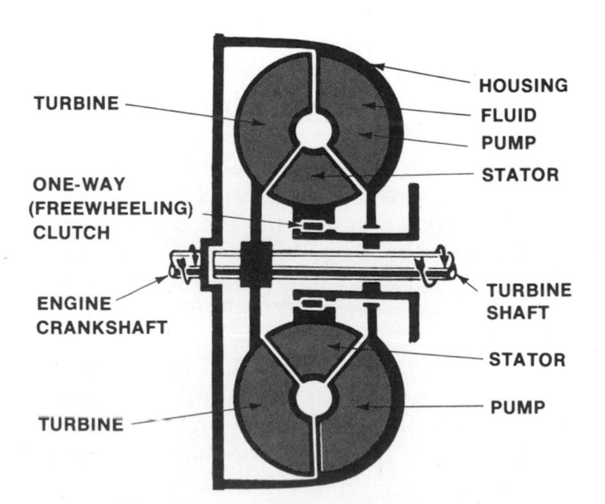

Fig. 7-30. Simplified drawing of a torque converter, showing locations of the pump, stator, and turbine. (*Ford Motor Company*)

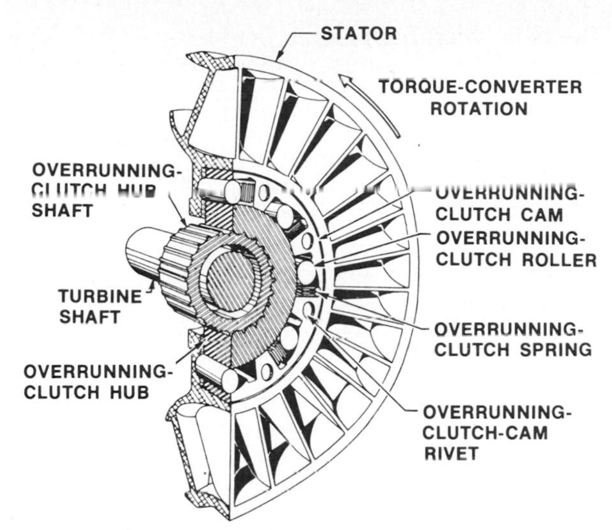

Fig. 7-31. Details of an overrunning clutch used to support a stator in a torque converter. (*Chrysler Corporation*)

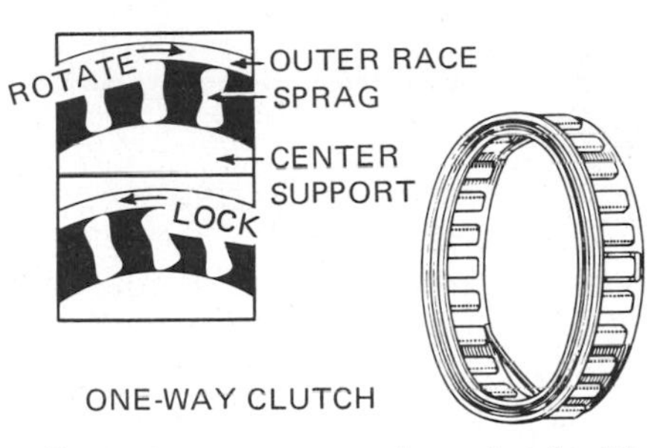

Fig. 7-33. Sprag-type overrunning clutch. The lockup action is shown to the left, and the sprag assembly is shown to the right. (*Toyota Motor Sales Company, Ltd.*)

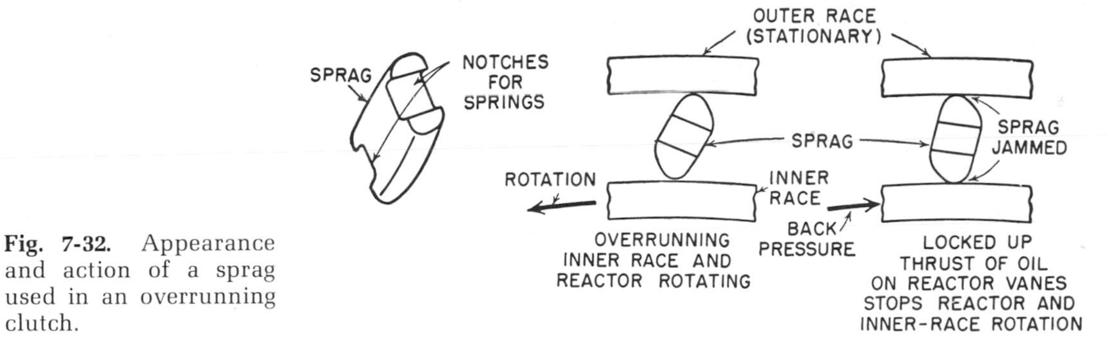

Fig. 7-32. Appearance and action of a sprag used in an overrunning clutch.

CHAPTER 7 CHECKUP

NOTE: Since the following is a chapter review test, you should review the chapter before taking it.

You have been making good progress in your studies of automatic transmissions and, with the completion of this chapter on fluid couplings and torque converters, are ready to start your studies of the automatic transmissions using these two devices. The following checkup will not only help you review the important details of these two devices, but will also assist you in remembering the important points about them. Reviewing the important details and writing the essential data on fluid coupling and torque converters will fix the information in your mind. Also, since you will be writing in your notebook, you will be making your notebook a valuable source of information you can refer to quickly in the days to come.

Completing the Sentences The sentences that follow are incomplete. After each sentence there are several words or phrases, only one of which will correctly complete the sentence. Write each sentence in your notebook, selecting the proper word or phrase to complete it correctly.

1. The purpose of the fluid coupling is to act as a: (*a*) synchronizing device, (*b*) automatic gear changer, (*c*) flexible power-transmitting coupling.
2. The fluid coupling consists essentially of two: (*a*) doughnuts, (*b*) vaned members, (*c*) guide rings, (*d*) driving shafts.
3. In the fluid coupling, oil passes from the driving member to the: (*a*) coupling, (*b*) vanes, (*c*) driven member, (*d*) gear.
4. The purpose of the guide ring in the fluid coupling is to reduce oil: (*a*) movement between members, (*b*) turbulence, (*c*) level in coupling.
5. The fluid coupling has maximum efficiency when the driving and driven members are turning at: (*a*) high speed, (*b*) low speed, (*c*) different speeds, (*d*) about the same speed.
6. The vanes in the torque converter are: (*a*) flat, (*b*) vertical, (*c*) curved.
7. In the torque converter, the additional member is placed between the pump and the: (*a*) stator, (*b*) rotor, (*c*) turbine.
8. The typical torque converter has: (*a*) three members, (*b*) four members, (*c*) five members.
9. Two types of overrunning clutches used in torque converters are roller and: (*a*) ball, (*b*) sprag, (*c*) needle bearing.
10. The two controlling devices in the transmission operated by hydraulic pressure are the brake bands and: (*a*) pistons, (*b*) clutches, (*c*) gears.

Purpose and Operation of Components In the following you are asked to write the purpose and operation of the various components in the fluid couplings and torque converters described in the chapter. If you have any difficulty in writing your explanations, turn back into the chapter and reread the pages that will give you the answer. Don't copy; try to tell it in your own words. That is a very good way to fix the explanation in your mind. Write in your notebook.

1. Explain the purpose of the fluid coupling.
2. Describe the fluid coupling.
3. What is the purpose of the guide ring used in some fluid couplings, and how does the guide ring work?
4. Explain why the fluid coupling is most efficient when both members are turning at about the same speed.
5. Explain why the fluid coupling is inefficient when the two members are turning at greatly different speeds.
6. Explain the bounce-back effect.
7. Explain how an additional member in the torque converter overcomes the bounce-back effect.
8. Describe what happens to the stator in the torque converter when the pump and turbine are turning at greatly different speeds; when they are turning at about the same speed.
9. Describe the construction and operation of a roller-type overrunning clutch; of a sprag-type overrunning clutch.

SUGGESTIONS FOR FURTHER STUDY

If possible, try to examine complete transmissions as well as the fluid couplings and torque converters used on them and described in this chapter. Many fluid couplings and torque converters are now welded units and cannot be opened for examination. However, many school automotive shops have cutaway or disassembled units which will be of great help to you in understanding these units. Examine them carefully. Also, study carefully any shop manuals you can find which cover the operation and service of these various transmissions. Be sure to write the important facts in your notebook.

chapter 8

AUTOMATIC-TRANSMISSION FUNDAMENTALS

This chapter discusses the fundamentals of automatic-transmission action. We have already looked into torque-converter action and the two actuating devices that produce shifting in the transmission: brake bands and multiple-disk clutches. We have also reviewed some basic hydraulics as it applies to the valves in automatic transmissions. All this was covered in Chap. 7. Now, we shall put all this together to see how automatic transmissions work. Later chapters will describe specific models and makes of automatic transmissions.

⊘ 8-1 Automatic Transmissions Automatic transmissions do the job of shifting gears without assistance from the driver. They start out in low as the car begins to move forward. Then they shift from low gear into intermediate (where present) and then high gear as the car picks up speed. The actions are produced hydraulically, that is, by oil pressure.

There are two basic parts to the automatic transmission: the torque converter, covered in Chap. 7, and the gear system. The torque converter passes the engine power to the gear system, that is, to the planetary-gear system (Figs. 8-1 and 8-2).

Although there are several variations of automatic transmissions, all work in the same general manner. All have selector levers on the steering column or console, or buttons on the steering column or instrument panel. In some automatic transmissions, there are five selector-lever positions: P (park), R (reverse), N (neutral), D (drive), and L (low). In addition, most transmissions have a second D position so that there are D1 and D2 positions (D2 is called S, or *super* range). In P, the transmission is locked up so that the car cannot move. In N, no power flows through the transmission but the locking effect is off. In L, there is a gear reduction through the transmission, which provides extra torque at the wheels for a hard pull or for braking when going down a long hill. In the D position (or positions), the transmission automatically shifts up or down according to car speed and throttle position. All these positions are covered on later pages.

Some automatic transmissions have a select-shift feature which enables the driver to select the desired gear. The transmission will then stay in this gear until the driver manually shifts out of it.

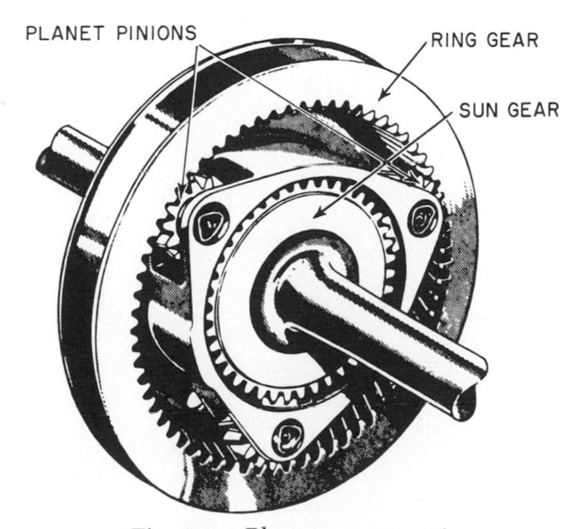

Fig. 8-1. Planetary gearset.

⊘ 8-2 Planetary Gears in Automatic Transmissions We have already described planetary gears in some detail in ⊘ 5-5, and you should review that section if you are not sure you understand planetary-gear construction and operation. So far, the planetary gears we have talked about are *simple* planetary gears (Fig. 8-1). They consist of a ring gear, a sun gear, and planet pinions held in a planet-pinion carrier. As we shall learn later, some automatic-transmission planetary gears are *compound* planetary gears. They consist of a sun gear, two sets of planetary gears with carriers, and two ring gears. Before we discuss their operation, however, let us review what we know about planetary-gear action.

You will remember that holding one of the three members of the simple planetary gearset (sun gear, planet-pinion cage, or ring gear) stationary turns the assembly into a speed-reducing or speed-increasing unit (according to which member is held stationary).

Also, locking up two of the members together produces direct drive through the gearset. In addition, if the planet-pinion cage is held stationary and the ring gear or sun gear is turned, the planet pinions act as reverse-idler gears. The direction of rotation is reversed through the gearset. These, briefly, are the three actions of the planetary gears in the automatic transmission. They produce either gear reduction, direct drive, or reverse.

Figure 8-2 shows one arrangement for carrying the power flow from the torque converter into the planetary gears. This is only one of three possible arrangements. The power can flow to the ring gear, the carrier, or the sun gear.

⊘ 8-3 Planetary-Gear-System Control We have seen how holding one member of the planetary-gear system stationary can produce gear reduction or reverse and also how locking up two members can produce direct drive. Now let us examine the means used to hold one or another of the members stationary and lock two members together to establish a 1:1 gear ratio (both input and output shafts turning at same speed) through the system.

Figure 8-3 illustrates the two mechanisms used in a planetary-gear system to achieve the various conditions: a multiple-plate clutch and a brake band and brake drum (these are described in Chap. 7). Note that the input shaft has the internal gear (ring gear) on it and the planet-pinion cage (or carrier) is attached to the output shaft. The sun gear is separately mounted, and it has a brake drum as part of the sun-gear assembly. The sun gear and the brake drum turn together as a unit.

The brake band is operated by a servo, as shown in Fig. 8-4. The servo consists of two pistons on a single stem (or rod) all mounted in a cylinder and linked to the brake band positioned around the brake drum. Without pressure on either side of the servo pistons, the spring force holds the brake band away from the brake drum. The sun gear is therefore not held stationary. However, under the proper operating conditions, oil under pressure is admitted to the left part of the cylinder, as shown in Fig. 8-5, forcing the piston to the right. This action compresses the spring and tightens the brake band on the drum, thus bringing the drum and the sun gear to a halt. The sun gear is therefore held stationary. With the ring gear turned and the sun gear held

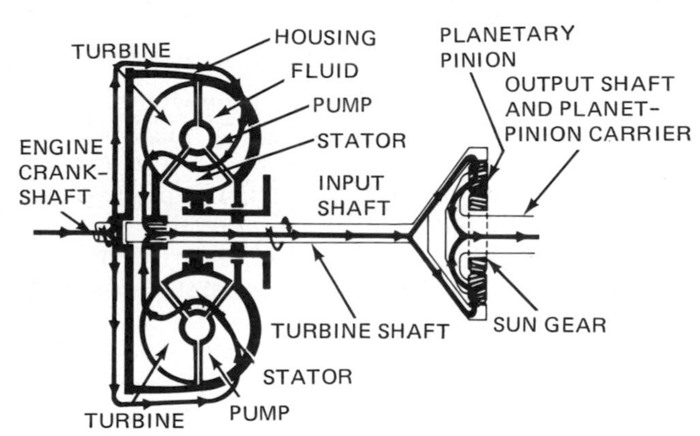

Fig. 8-2. Power flow through the torque converter to the planetary gears.

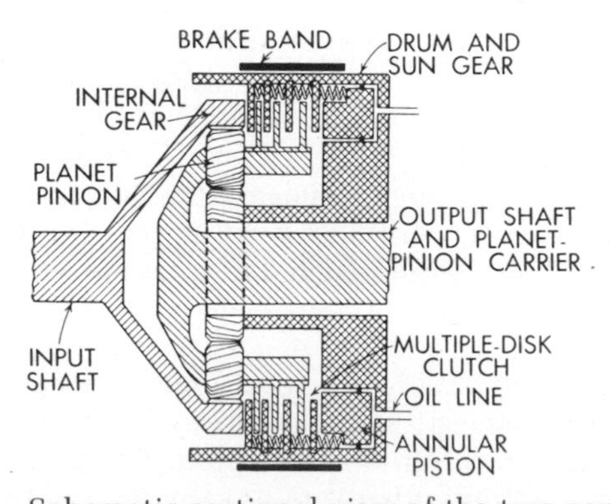

Fig. 8-3. Schematic sectional view of the two controlling mechanisms used in the front planetary gearset in one model of automatic transmission. One controlling mechanism consists of a brake drum and brake band; the other is a multiple-disk clutch.

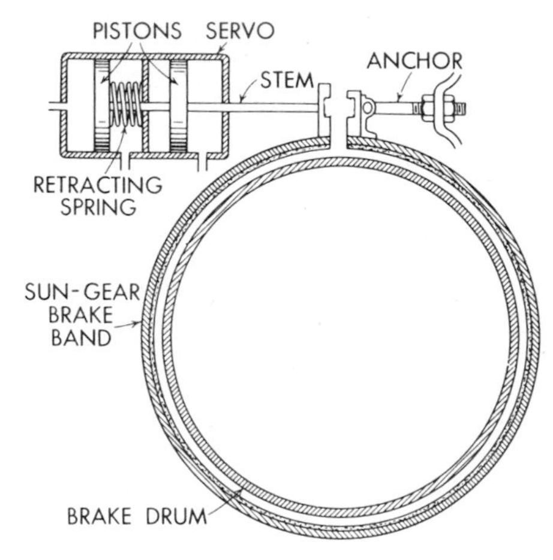

Fig. 8-4. Arrangement of the servo, brake band, and brake drum.

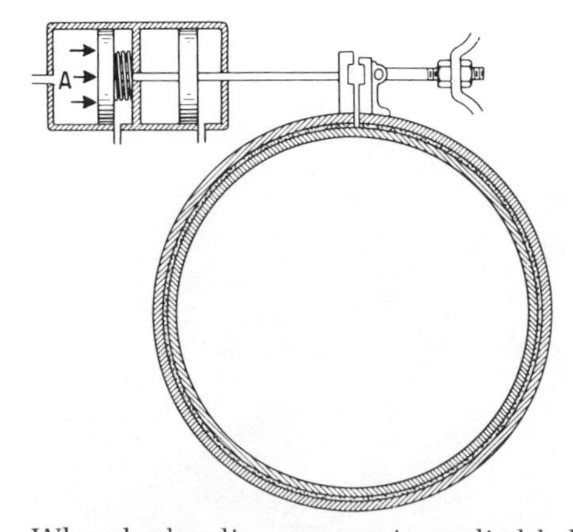

Fig. 8-5. When hydraulic pressure is applied behind the piston at A, the piston and rod move to the right (in the illustration) to cause the brake band to tighten on the brake drum. The brake drum (and sun gear) is thus held stationary.

stationary, condition 3 (Fig. 5-15) is obtained and there is speed reduction, or gear reduction, through the system.

NOTE: The arrangement described here (the brake band holding the sun gear stationary) is only one of several arrangements used in automatic transmissions. In some transmissions, when the brake band is applied, it holds the ring gear or the planet-pinion carrier stationary. Different transmissions may lock different members together when the clutch is applied. The principle is the same in all transmissions, however. There is gear reduction when the band is applied, and there is direct drive when the clutch is applied. Note also that in some transmissions the brake-band function is taken over by another clutch. We describe these variations in later chapters.

Under certain other conditions, the brake band must be released. This action takes place when the hydraulic control system directs oil into the right-hand chambers of the servo cylinder. This forces the pistons to move to the left (Fig. 8-6). At the same time, the clutch is actuated so as to lock the sun gear and planet-pinion cage together. The clutch consists of a series of clutch plates alternately splined to the cage and to the inner face of the brake drum (see Figs. 7-17, 8-3 and 8-7). When the clutch is disengaged, these plates are held apart. However, simultaneously with the releasing of the brake band, as explained above, oil is directed into the chamber back of the annular (ringlike) piston in the brake-drum–sun-gear assembly. This action forces the piston to the left (Fig. 8-7) so that the clutch plates are forced together. Then, friction between the plates locks the sun gear and planet-pinion cage together. When this happens, the planetary-gear system acts like a direct-drive coupling and both the input and output shafts turn at the same speed (1:1 gear ratio).

To achieve reverse in the planetary-gear system

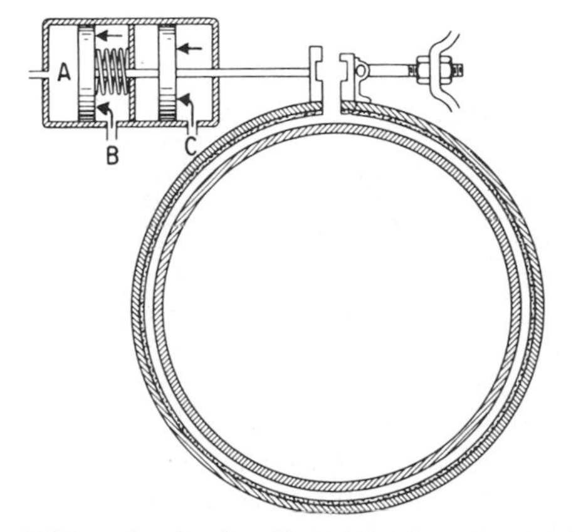

Fig. 8-6. When the hydraulic control system admits oil under pressure at B and C, the pressure forces the pistons to the left (in the illustration). This action causes the brake to release the brake drum and sun gear so that the sun gear can rotate.

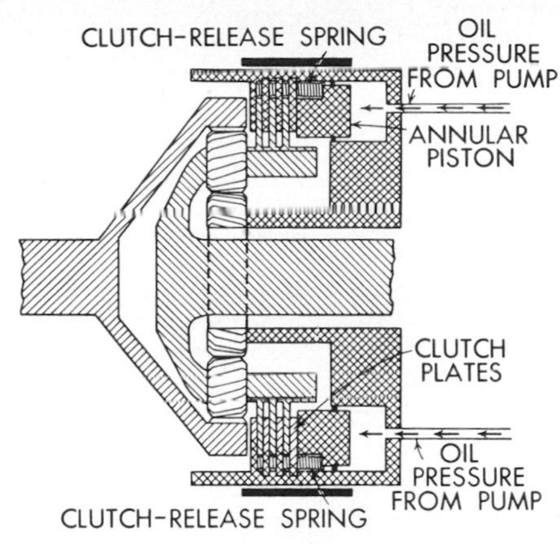

Fig. 8-7. One set of clutch plates is splined to the sun-gear drums. The other set is splined to the planet-pinion carrier. When the hydraulic control system directs oil behind the annular piston, as shown by the arrows, the clutch plates are forced together so that the sun gear and planet-pinion carrier are locked together.

shown in Figs. 8-1 to 8-7, it would be necessary to hold the planet-pinion cage stationary and allow both the ring gear (which is on the input shaft) and the sun gear and drum to rotate. With this arrangement, the sun gear would rotate in the reverse direction. Actually, more complex arrangements are used in automatic transmissions, as we shall see.

⊘ 8-4 Shift Control Figure 8-8 is a simplified diagram of a hydraulic control circuit for a single planetary gearset in an automatic transmission. Later we shall look at the circuits for automatic transmissions that use two or more planetary gearsets. As we have noted, automatic transmissions use more than one planetary gearset.

The major purpose of the hydraulic circuit is to control the shift from gear reduction to direct drive. The shift must take place at the right time, and this depends on car speed and throttle opening. These two factors produce two varying oil pressures that work against the two ends of the shift valve.

The shift valve is a *spool valve* inside a bore, or hole, in the valve body. Figure 8-9 shows the spool valve. Pressure at one end of the spool valve comes from the *governor*.

NOTE: A *governor* is a device that controls, or governs, another device. In the hydraulic circuit, the governor controls pressure on one end of the shift valve.

Pressure at the other end of the spool valve changes as vacuum in the intake manifold changes. Let us explain. First, the governor pressure changes with car speed. The governor is driven from the transmission by the output shaft. As output-shaft speed and car speed increase, the governor pressure increases proportionally. This pressure works against the right end of the shift valve, as shown in Fig. 8-8.

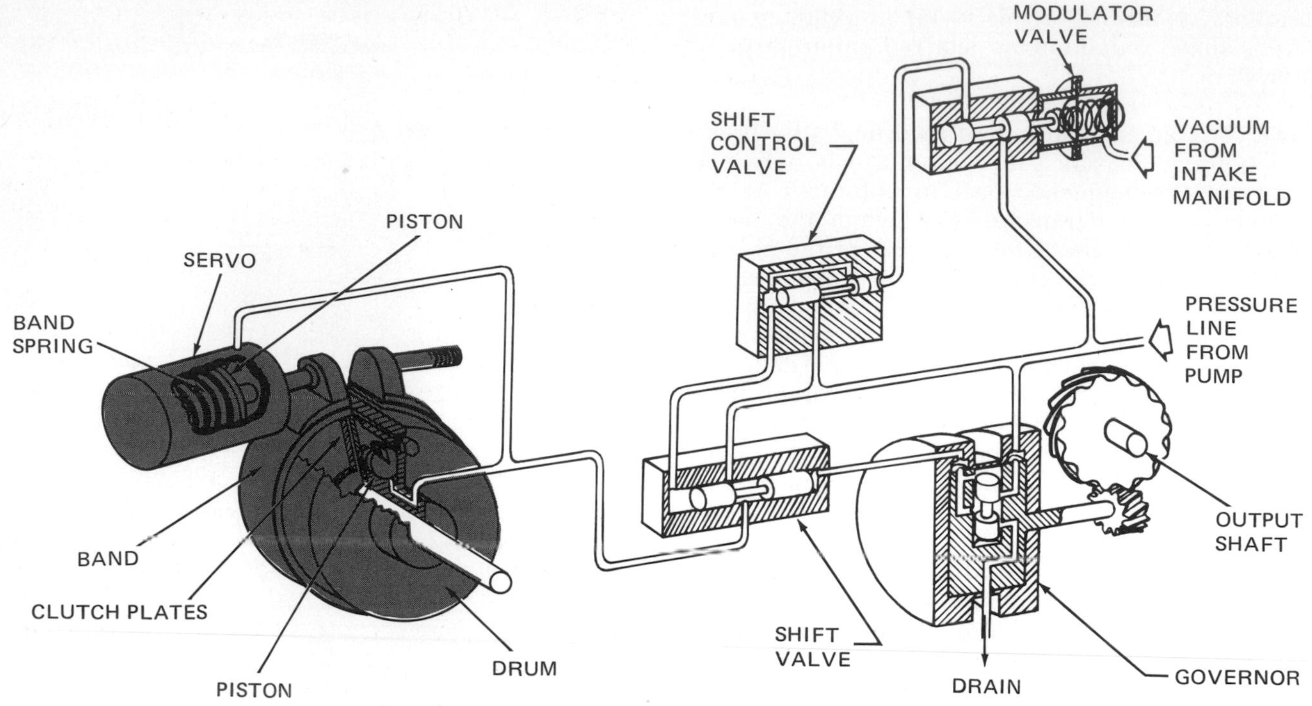

Fig. 8-8. Schematic diagram of the hydraulic control system for the brake-band servo and clutch. In the system shown, the band is normally on, and the clutch off; this arrangement produces gear reduction. But when the shift valve is moved, pressure from the oil pump is admitted to the front of the brake-band piston and to the clutch piston. This movement causes the band to release and the clutch to apply. Now, with the clutch locking two planetary members together, the planetary system goes into direct drive.

The governor pressure is actually a modified line pressure. That is, a pump in the transmission produces the line pressure. This line pressure passes into the governor. The governor then releases part of the pressure to the right end of the shift valve. The higher the car speed, the more pressure released by the governor. It is this modified pressure, the *governor pressure,* that works on one end of the shift valve.

Working on the left end of the shift valve is a pressure that changes as intake-manifold vacuum changes. Line pressure enters the modulator valve at the upper right in Fig. 8-8. The modulator valve contains a spool valve attached to a spring-loaded diaphragm. Vacuum increases in the intake manifold when the throttle is partly closed. This vacuum pulls the diaphragm in and moves the modulator spool valve to the right. The motion cuts off the line pressure going to the shift-control valve. When this happens, the shift-control valve moves to the right, cutting off pressure from the left end of the shift valve. This means that the shift valve is pushed to the left by governor pressure. As a result, line pressure can pass through the shift valve. Therefore, line pressure is applied to the clutch and servo at the planetary gearset. With this condition, the band is released and the clutch is applied. This puts the planetary gearset into direct drive.

Fig. 8-9. Spool valve for a shift valve.

1. *SHIFT ACTION* Now let's see how the hydraulic control circuit works. To start with, there is no pressure going to the planetary-gear controls. The clutch is released, and the band is applied by the heavy spring in the servo.

NOTE: Remember that Fig. 8-8 is a *simplified* version of the actual system found in modern automatic transmissions. In modern transmissions oil pressure is used to help the spring hold the band tight. The basic principles, however, are as shown in Fig. 8-8.

With the clutch released and the band applied, the planetary gearset is in gear reduction or low. As car speed increases, the governor releases more and more pressure. This pressure is applied to one end of the shift valve, as mentioned. The pressure on the other end of the shift valve depends on intake-manifold vacuum, that is, on engine speed plus throttle opening. As long as the throttle is held open, there is little manifold vacuum. The pressure on the left end of the shift valve is high. This high pressure holds the planetary gearset in low for good acceleration. However, as car speed continues to increase, the governor pressure becomes great enough to push the shift valve to the left. This lets line pressure through to the planetary gearset. Now the band is released and the clutch is applied. This shifts the planetary gearset into direct drive.

The upshift will also take place if the throttle is partly closed after the car reaches intermediate speed. Closing the throttle increases the intake-manifold vacuum. The vacuum cuts off line pressure to the shift-control valve. When this happens, the shift-control valve moves to the right, cutting off pressure to the left end of the shift valve. The shift valve then moves to the left, pushed by governor pressure. This applies line pressure to the planetary gearset. The clutch applies and the band releases Now the planetary gearset shifts into direct drive.

There is a reason for this roundabout way of getting pressure to the left end of the shift valve. It is to vary the point of upshift according to driving conditions. When the vehicle is accelerating, the driver wants high engine torque. The gears must stay in low. Then, when the vehicle reaches the desired cruising speed, less torque is needed and so the driver eases up on the throttle. This action increases the intake-manifold vacuum so that the upshift takes place. As you can see, the upshift can take place at any speed from medium to high. Also, for fast acceleration, the driver opens the throttle. This action reduces the intake-manifold vacuum. As a result, the planetary gearset drops into gear reduction to increase torque.

2. *OTHER VALVES* The preceding discussion is a very simplified description of how the upshifting is accomplished. The actual valves are more complicated, having springs for initial loading of the valves. There are other valves besides the ones mentioned to ease the shifts, regulate pressures, time downshifts, and so on. There is also a manual-shift valve.

3. *MANUAL-SHIFT VALVE* The manual-shift valve is controlled by the driver by movement of a shift lever on the steering column or console. As the manual valve is moved, it opens or closes various lines that direct oil pressure to the valves in the transmission. For instance, when the valve is moved to D, or drive, oil pressure is directed to the transmission valves so that they are ready to shift into D whenever the speed and throttle conditions are right. (This action has already been discussed in previous paragraphs.) If the manual-shift valve is placed in L, or low, the oil pressure is directed to the valves so that the modulator valve is blocked from producing an upshift.

⊘ 8-5 Transmission Fluid We have used the term "oil" throughout our discussion of transmission op-

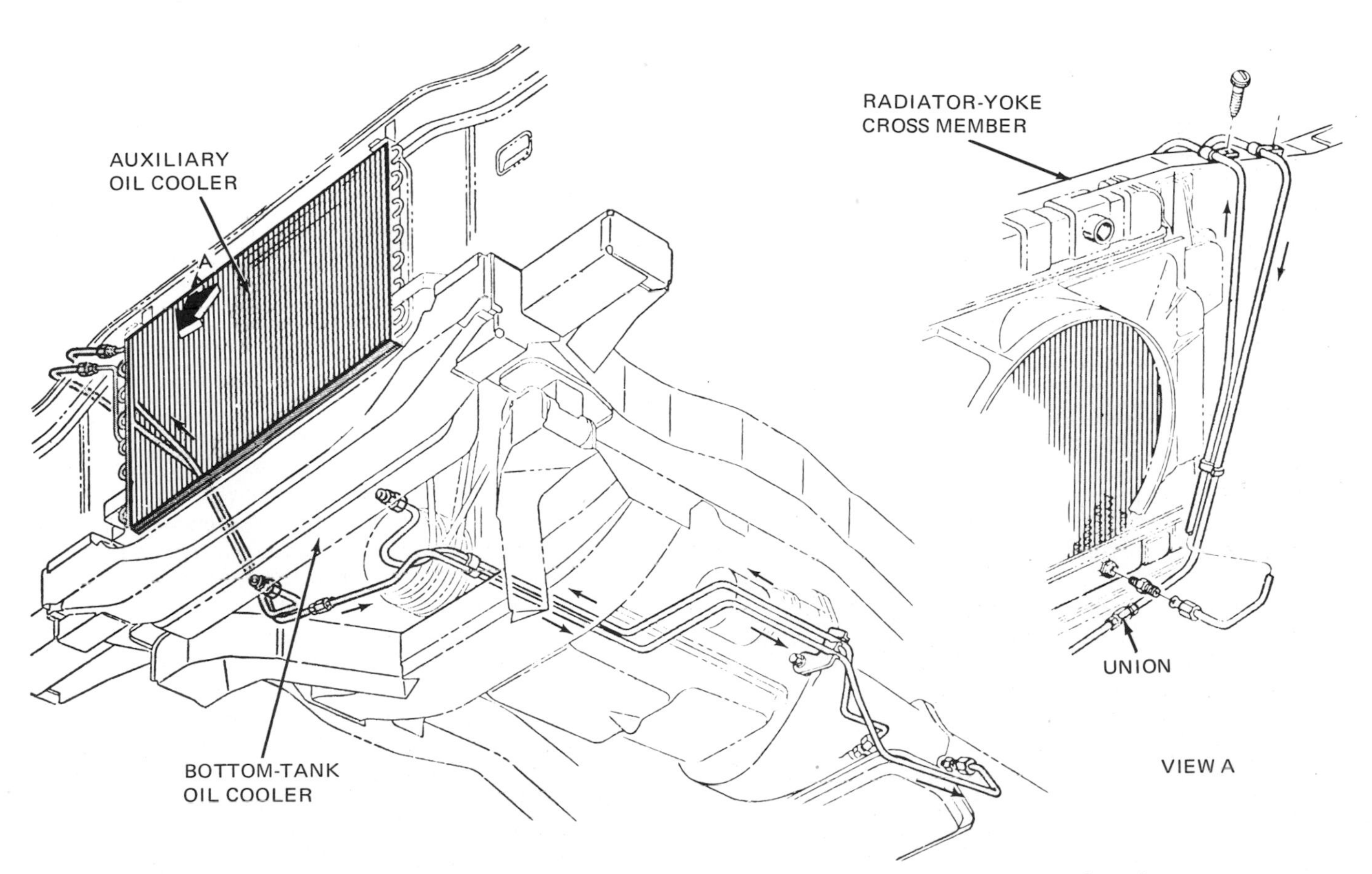

Fig. 8-10. Two views of the connections of the tubes from the transmission to the oil cooler in the bottom tank of the cooling-system radiator. (*Chrysler Corporation*)

eration. *Transmission fluid* could be considered a form of oil, but it is a very special sort of oil. Transmission fluid has several additives, such as viscosity-index improvers, oxidation and corrosion inhibitors, extreme-pressure and antifoam agents, detergents, dispersants, friction modifiers, pour-point depressants, and fluidity modifiers. The oil is dyed red so that it will not be confused with other automotive lubricants. Also, if leakage occurs, it is easy to tell whether it is engine oil or transmission fluid that is leaking.

CAUTION: It is extremely important to use the proper transmission fluid recommended by the automobile manufacturer. Use of a transmission fluid that is not on the recommended list can cause serious transmission trouble.

⊘ 8-6 Transmission-Fluid Coolers Transmission fluid may become very hot, especially under severe operating conditions. Thus, cars are equipped with transmission-fluid coolers, as shown in Figs. 8-10 (p. 111) and 8-11. The transmission fluid circulates through a tube located in the engine-cooling-system radiator. Heat from the transmission fluid passes to the coolant circulating through the engine cooling system. This process cools the transmission fluid. In addition, a vehicle used for towing or trailering is frequently equipped with an auxiliary transmission-fluid cooler.

⊘ 8-7 Automatic-Transmission Variations A basic automatic-transmission model may be used, with small variations, with a variety of engines and automobiles. As we have seen, the operation of the clutches and brake bands in a transmission is due to hydraulic pressures. The timing and amount of pressure can be changed by changing valve springs, hydraulic connections, and servos. In this way a basic model can be adapted to suit a particular engine and automobile. In other words, the planetary-gear system, clutches, and brake bands can remain essentially the same. But the valves and other components of the hydraulic system can be altered to suit engine requirements. For example, when the transmission is used with one engine, the shift points might be designed to be on the high side. But with another engine having a different torque output curve, best car operation might require that the transmission shift at a somewhat lower speed.

Also, the clutches may differ with different engines. For example, with a high-output engine, the transmission clutches may have more plates for greater holding power to handle the higher engine outputs.

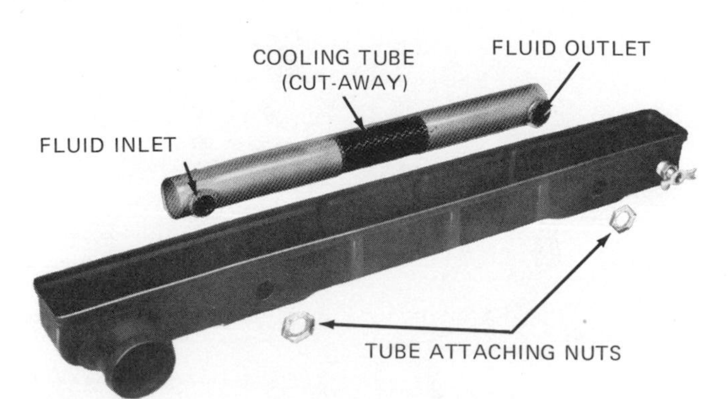

Fig. 8-11. Lower tank of the radiator and oil-cooler tube, showing how the tube fits into the lower tank. (*Chrysler Corporation*)

CHAPTER 8 CHECKUP

NOTE: Since the following is a chapter review test, you should review the chapter before taking it.

You have been making steady progress in your studies of automatic transmissions and are now ready to start studying specific transmission models used on Chrysler, Ford, and General Motors cars. The chapter you have just finished covers some of the fundamentals that apply to all automatic transmissions. Now, take the following test to check up on yourself and find out how well you remember the material you studied in the chapter. Reviewing the important details and writing the essential data on the transmissions will fix the information in your mind. Also, since you are writing in your notebook, you will be making your notebook a valuable source of information you can refer to quickly in the future.

Completing the Sentences The sentences that follow are incomplete. After each sentence there are several words or phrases, only one of which will correctly complete the sentence. Write each sentence in your notebook, selecting proper word or phrase to complete it correctly.

1. The selector levers for most automatic transmissions have: (*a*) four, (*b*) five, (*c*) six, (*d*) seven positions.
2. The selector-lever position in which there is no power flow through the transmission but the locking effect is off is: (*a*) P, (*b*) D, (*c*) N, (*d*) R.
3. The typical planetary gearset consists of planet pinions, a ring or internal gear, and: (*a*) a sun gear, (*b*) pinions, (*c*) a clutch, (*d*) a brake band.
4. The three actions of the planetary gears in an automatic transmission are to provide reverse, direct drive, and: (*a*) high gear, (*b*) gear reduction, (*c*) gear increase.
5. The two control mechanisms used with a planetary-gear system are: (*a*) valve and pump, (*b*) torque converter and sun gear, (*c*) bands and clutches.
6. The brake band is actuated by a: (*a*) servo, (*b*) clutch piston, (*c*) manual valve.
7. The ring-shaped part in the clutch which puts the pressure on the clutch disks is called the: (*a*) spring, (*b*) piston, (*c*) ring gear.

8. The two controlling factors that cooperate to produce gear shifting are: (*a*) hydraulic pressure and governor pressure, (*b*) car speed and governor pressure, (*c*) throttle opening and car speed.
9. The bellows in the modulator valve is actuated by: (*a*) intake-manifold vacuum, (*b*) car speed, (*c*) engine speed.
10. The valve which is controlled by linkage to the selector lever is called the: (*a*) shift valve, (*b*) manual valve, (*c*) shift-control valve.

Purpose and Operation of Components In the following, you are asked to write the purpose and operation of the automatic-transmission controls described in the chapter. If you have any difficulty in writing your explanation, turn back into the chapter and reread the pages that will give you the answers. Then write your explanation. Don't copy; try to tell it in your own words. This is a good way to fix the explanation firmly in your mind. Write in your notebook.

1. Explain the purpose of the different selector-lever positions.
2. Explain how the planetary-gear system produces gear reduction when the brake band is applied.
3. Explain how the planetary-gear system produces direct drive when the clutch is applied.
4. How can reverse be achieved in the planetary-gear system?
5. Explain how the two controlling factors, car speed and throttle opening, control gear shifting.
6. Explain how the modulator valve operates.
7. What is the purpose of dyeing the transmission fluid red?

SUGGESTIONS FOR FURTHER STUDY

If possible, examine the complete transmissions and also the transmission parts and controls described in this chapter. Many school automotive shops have these transmissions and parts available, and often automotive service shops have defective parts or units that you can examine. Study carefully any shop manuals from the manufacturers of automatic transmissions to gain a more complete idea of the construction and operation of these units. Be sure to write in your notebook any important facts you learn.

chapter 9

GENERAL MOTORS TYPE 300 AUTOMATIC TRANSMISSION

This chapter describes one of the simpler automatic transmissions using a torque converter. It is a two-speed unit with one upshift from low to direct and also reverse. This transmission has been used, in various modifications, in many General Motors cars and is known by several names, such as the Buick Super Turbine 300, Chevrolet Aluminum Powerglide, Oldsmobile Jetaway, and Pontiac M-35. It is most widely known by the name Powerglide, *and we shall use this term in this chapter.*

⊘ 9-1 Powerglide Operating Characteristics A selector lever and quadrant on the steering column, as shown in Fig. 9-1, or on the floorboard console has five positions: P (park), R (reverse), N (neutral), D (drive), and L (low). In P, a ratchet in the transmission locks the transmission output shaft and thus the drive shaft and rear wheels. In R, the transmission reverses the direction of drive-shaft and rear-wheel rotation so that the car is backed up. In N, there is no drive through the transmission. In D, the car will start forward in low gear and then upshift to direct, or high, gear when car speed and engine output are properly related. In L, the car will start forward in low gear and stay in low gear; the transmission will not upshift.

The selector lever is connected by linkage to a manual valve in the transmission. When the selector lever is moved, the manual valve is shifted so that oil pressure from the transmission hydraulic system operates the clutches and brake band in the transmission to produce the gear selected by the driver.

The transmission has a forced downshift feature. If the driver depresses the throttle all the way and the car speed and engine output have the proper relationship, the transmission will downshift. This action provides a shift into low gear, often called *passing gear,* for quick acceleration.

⊘ 9-2 Torque-Drive Transmission The Torque-Drive transmission is the same as the Powerglide except that the automatic shifting is removed. This transmission is for drivers who wish to "shift for themselves." When the driver shifts into L (low), the transmission stays in that gear. When the driver shifts into D (drive), the transmission stays in that gear. The modifications made to the Powerglide to produce the Torque Drive are relatively minor. The main changes are in the hydraulic system, where certain valves producing upshifts and downshifts are eliminated.

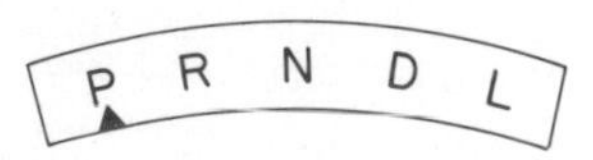

Fig. 9-1. Selector-lever quadrant on the steering column. The transmission is in P (park).

⊘ 9-3 Powerglide Construction and Operation Figure 9-2 is a cutaway view of the Powerglide transmission. Figure 9-3 is a sectional view with all the parts named. The transmission contains two multiple-plate clutches and a brake band to control the planetary-gear system. The text has already described the construction and operation of these clutches and brake bands in ⊘ 7-8 to 7-11. If you are not clear on these details, reread these sections. The torque converter used in the Powerglide is of the three-member type described in ⊘ 7-12 to 7-16. Engine power passes through the torque converter and enters the planetary-gear system of the transmission.

The planetary-gear system used in the Powerglide transmission is more complicated than the one described in Chap. 5. It is called a *compound planetary-gear system* because it has extra gears. You should have a clear understanding of the simple planetary-gear system used in the overdrive and described in detail in ⊘ 5-5. Chapter 8 also described how clutches and brake bands can control planetary-gear systems so as to cause them to produce gear reduction, reverse, and direct drive.

Figures 9-4, 9-6, and 9-7 show the three drive

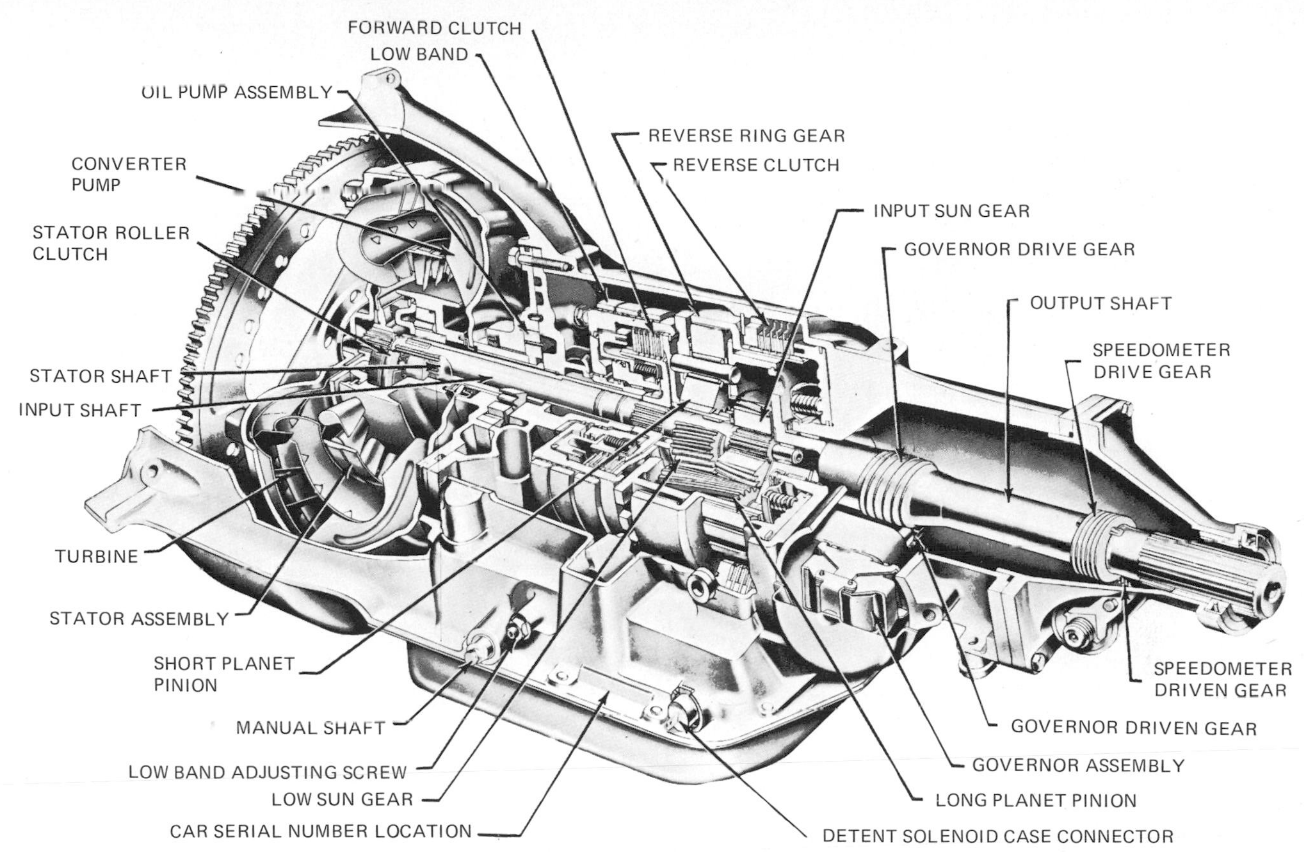

Fig. 9-2. Powerglide automatic transmission cut away so that internal parts can be seen. (*General Motors Corporation*)

conditions in the transmission. They also illustrate the more complicated (compound) planetary-gear system used in the Powerglide. Notice that there are two sun gears: the input sun gear and the low sun gear. The input sun gear is splined to the input shaft so that it must turn with the input shaft. The low sun gear is splined to the forward-clutch drum so that they must turn, or remain stationary, together. The forward-clutch hub is splined to the input shaft so that it must turn with the input shaft.

The Powerglide has two types of planet pinions: long and short. Both are carried on the same planet carrier, which is splined to the transmission output shaft. The input sun gear is meshed with the planet long pinions. The planet long pinions are meshed with the planet short pinions. The planet short pinions are meshed with the low sun gear.

It is important to understand how these gears are meshed and how they relate to each other. Figure 9-5 will help you understand. Note that in this illustration the two sun gears are partly cut away so that both can be seen. In the actual assembly the input sun gear is behind the low sun gear. Remember: The input sun gear is meshed with the planet long pinions. The low sun gear is meshed with the planet short pinions. The planet short pinions and planet long pinions are meshed with each other.

Now let's look at the operation of the Powerglide transmission in the three gears: low, drive, and reverse.

1. *POWER FLOW IN L (LOW) RANGE* This is shown in Figs. 9-4 and 9-5. With the transmission in low, the forward clutch is released and the brake band is applied. Thus, the clutch drum and low sun gear are held stationary. The power flow is from the input sun gear, through the planet long pinions, and the planet short pinions. Since the planet short pinions are meshed with the stationary low sun gear, these pinions must walk around the low sun gear as they rotate. Thus, they carry the planet-pinion carrier around with them in the direction shown. Figure 9-5 shows the directions in which the gears and pinions rotate. Gear reduction is achieved through the combination of short and long pinions and the fact that the planet-pinion carrier is being driven through the planet pinions.

2. *POWER FLOW IN D (DRIVE) RANGE* The forward clutch is applied and the low band is released in drive, as shown in Fig. 9-6. The low sun gear is locked to the input shaft so that the entire planetary-gear system must turn as a unit. There is no gear reduction through the transmission.

3. *POWER FLOW IN R (REVERSE)* This is shown in Fig. 9-7. The low band and forward clutch are released and the reverse clutch is applied. The reverse ring gear is held stationary by the clutch. In

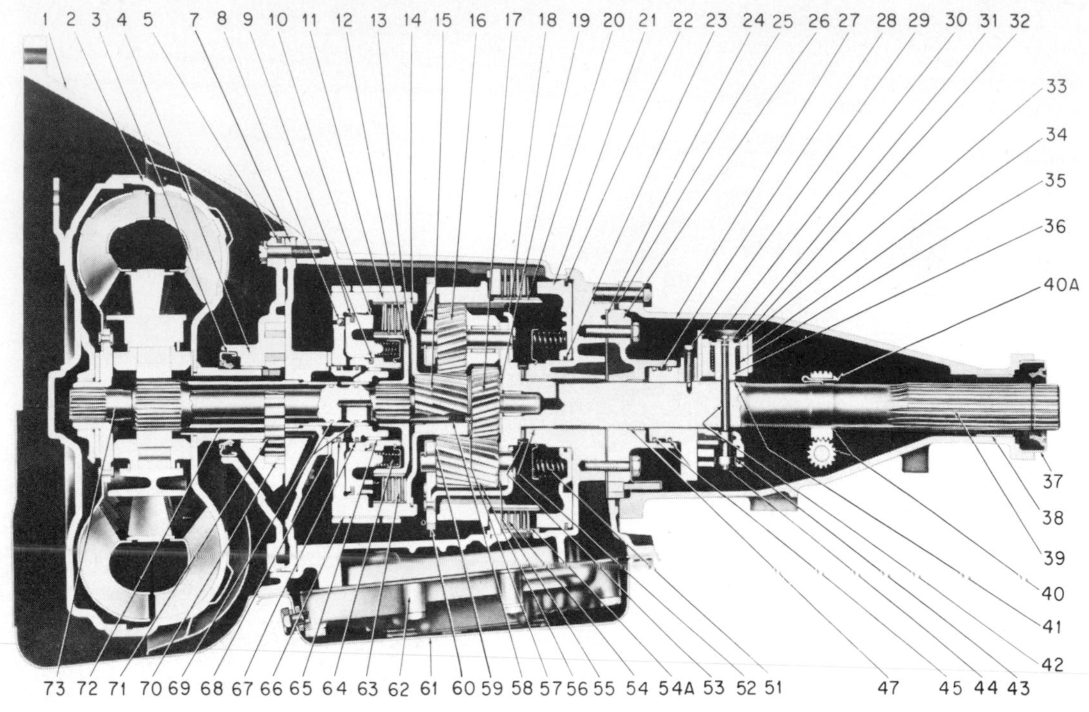

1. Transmission case
2. Welded converter
3. Front-oil-pump-seal assembly
4. Front-oil-pump body
5. Front-oil-pump body square-ring seal
7. Front-oil-pump cover
8. Clutch-relief-valve ball
9. Clutch-piston inner and outer seal
10. Clutch piston
11. Clutch drum
12. Clutch hub
13. Clutch-hub thrust washer
14. Clutch-flange retainer ring
15. Low-sun-gear-and-clutch-flange assembly
16. Planet short pinion
17. Planet input sun gear
18. Planet carrier
19. Planet-input-sun-gear thrust washer
20. Ring gear
21. Reverse piston
22. Reverse-piston outer seal
23. Reverse-piston inner seal
24. Extension seal ring
25. Rear-pump wear plate
26. Rear pump
27. Extension
28. Governor hub
29. Governor-hub drive screw
30. Governor body
31. Governor-shaft retainer clip
32. Governor-outer-weight retainer ring
33. Governor-inner-weight retainer ring
34. Governor outer weight
35. Governor spring
36. Governor inner weight
37. Extension rear oil seal
38. Extension rear bushing
39. Output shaft
40. Speedometer drive and driven gear
40A. Speedometer-drive-gear retaining clip
41. Governor-shaft belleville springs
42. Governor shaft
43. Governor valve
44. Governor-valve retaining clip
45. Governor-hub seal rings
47. Rear-pump bushing
51. Reverse-piston return springs, retainer and retainer ring
52. Transmission-rear-case bushing
53. Output-shaft thrust bearing
54. Reverse-clutch pack
54A. Reverse-clutch cushion spring (waved)
55. Pinion thrust washer
56. Planet long pinion
57. Low-sun-gear needle thrust bearing
58. Low-sun-gear bushing (splined)
59. Pinion thrust washer
60. Parking lock gear
61. Transmission oil pan
62. Valve body
63. High-clutch pack
64. Clutch-piston-return spring, retainer, and retainer ring
65. Clutch-drum bushing
66. Low brake band
67. High-clutch seal rings
68. Clutch-drum thrust washer (selective)
69. Turbine-shaft seal rings
70. Front-pump driven gear
71. Front-pump drive gear
72. Stator shaft
73. Input shaft

Fig. 9-3. Sectional view of a Powerglide automatic transmission. (*Chevrolet Motor Division of General Motors Corporation*)

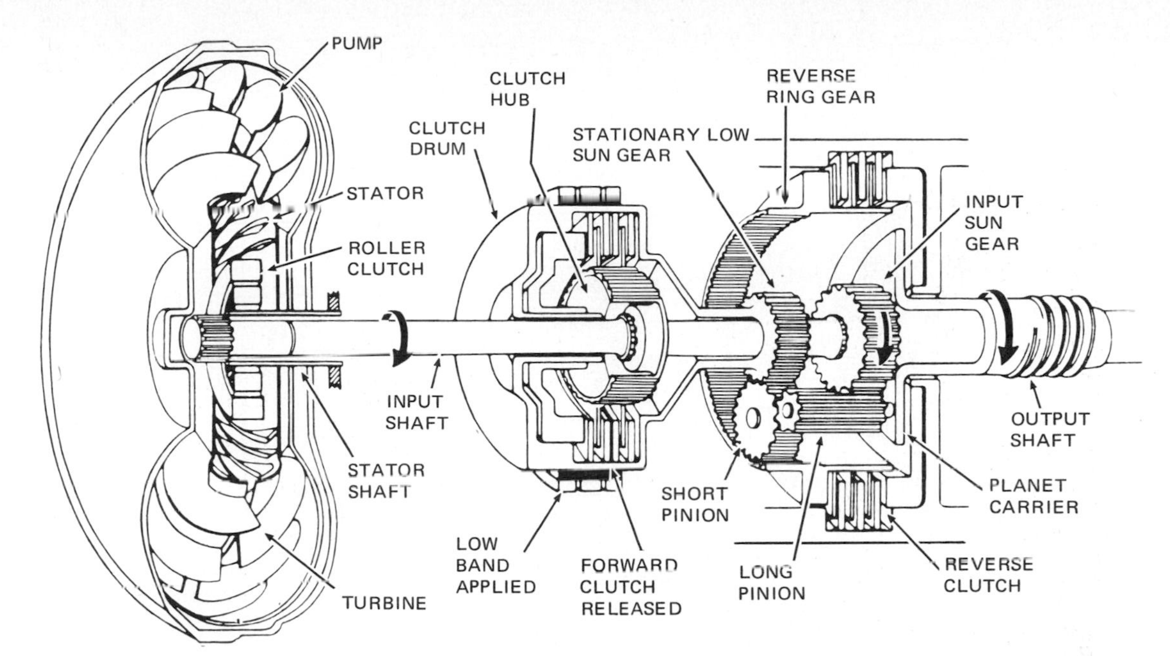

Fig. 9-4. Power flow in L (low) range. The forward clutch is released and the low band is applied so that the low sun gear is held stationary. (*Pontiac Motor Division of General Motors Corporation*)

early models, a brake band and drum were used instead of the clutch. The clutch, having greater holding power, was used on later models so that the transmission could be used with higher-output engines.

With either the clutch or band holding the ring gear stationary, as the planet short pinions rotate they are forced to circle around inside the stationary ring gear. You can see how this happens in Fig. 9-8. As the short pinions walk around the stationary ring gear, they carry the planet carrier around in the reverse direction. The planet carrier therefore turns the output shaft in the reverse direction so that the car is backed up. There is gear reduction through the transmission in reverse.

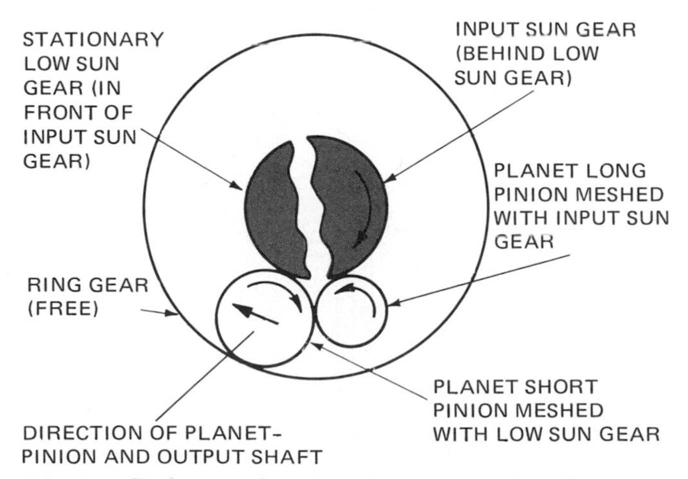

Fig. 9-5. End view of gears in the transmission planetary-gear system with the system in low gear. The two sun gears are shown partly cut away so that their locations can be seen. The input sun gear is behind the low sun gear.

⊘ 9-4 Hydraulic Control System We shall now examine the hydraulic control system and find out how the clutches and brake band are controlled to produce the drive ratio shown in Figs. 9-4 to 9-8. We have already learned a good deal about hydraulics and valves. We learned that motion and pressure can be transmitted by hydraulic means. Hydraulic pressure is used to apply the brake band and clutches. In our studies of the Powerglide hydraulic system, we shall look first at the pressure producer—the pump—and find out how its pressure is controlled. We shall then learn how the hydraulic system uses this pressure to produce the effects needed to make the shifts. Refer to Plates 1 to 3 in the color insert for the complete hydraulic circuit.

⊘ 9-5 Oil Pump The transmission fluid is a *special* type of oil, as mentioned previously. The terms *oil* and *fluid* are used interchangeably in the trade, and we shall use them both in this book. The oil pump used in the transmission is the gear type. Figure 9-9 shows a pump of this type with the cover removed. The illustration shows clearance being checked between the outer gear and the pump case. Figure 9-10 shows the gears alone, together with the crescent that forms part of the pump.

In operation, the inner gear is driven, which causes the outer gear to turn also. As the gears rotate, the teeth move away from mesh. Oil is "sucked" into the spaces where the teeth move apart through an oil port positioned at that place. (Actually, the oil is not "sucked" in. Atmospheric pressure pushes the oil in to fill the spaces produced by the separating teeth.) Then, the oil-filled spaces between the teeth are carried around to the other side of the crescent.

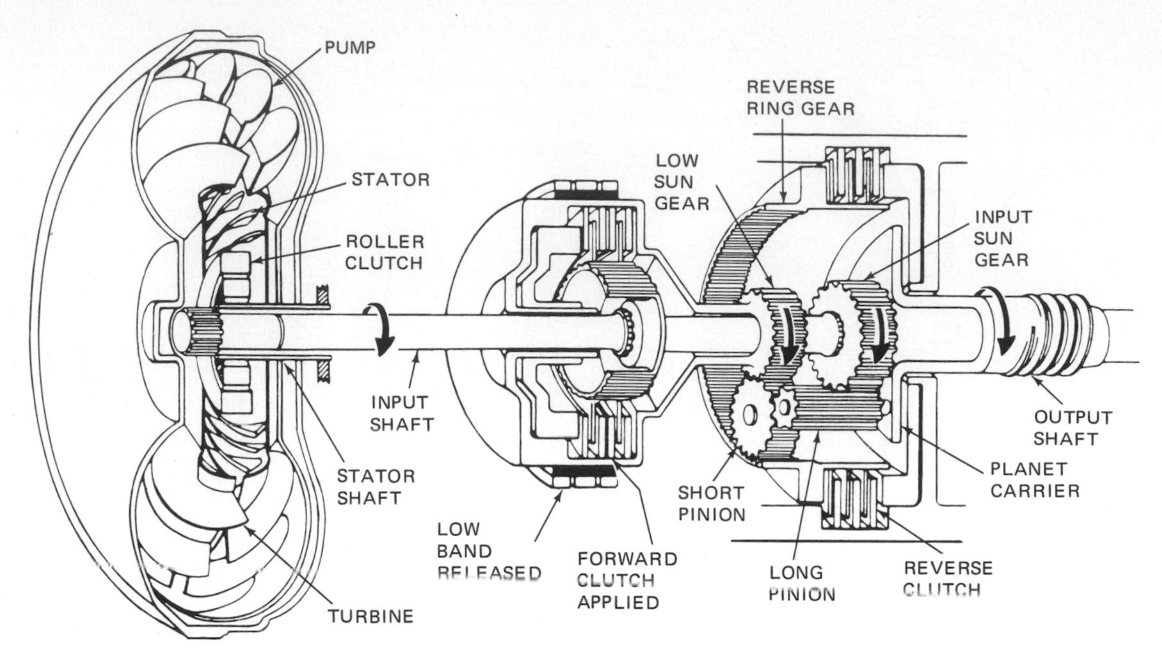

Fig. 9-6. Power flow in D (drive) range. The low band is released and the forward clutch is applied, locking the low sun gear to the input shaft. (*Pontiac Motor Division of General Motors Corporation*)

Here, as the teeth begin to come together, pressure is created on the oil and it is forced out of the pump through the outlet oil port.

⊘ 9-6 Pressure Regulation The oil pump, left to itself, would produce a fairly low pressure at low engine speeds and an extremely high pressure at high engine speeds. The pump is driven through the torque-converter case and thus turns at engine speed. Thus, the pump turns at varying speeds. But the varying pump pressures that can result are not satisfactory for the hydraulic control system. The pressure must be regulated.

A simple pressure-regulator valve is shown in Fig. 9-11. The oil pressure from the oil pump pushes down on the valve. Opposing this motion is the calibrated spring. As the oil pressure increases, the valve is pushed down. Finally, when the specified pressure is reached, the valve is pushed down far enough for it to clear the return oil line to the oil sump. When this happens, part of the oil from the

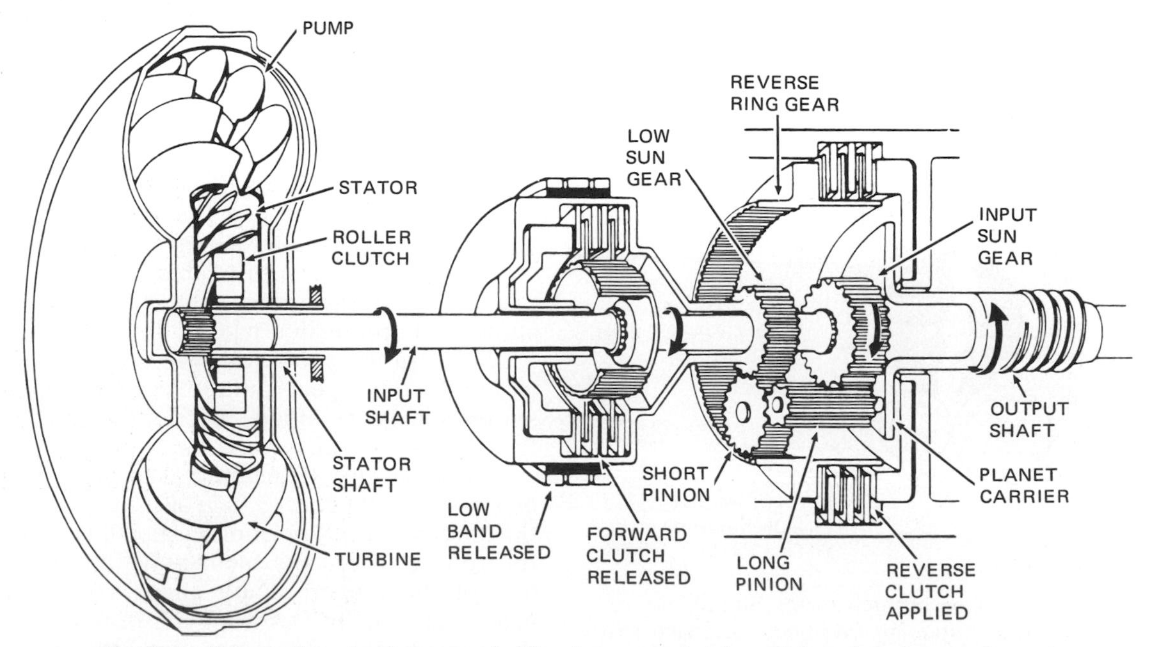

Fig. 9-7. Power flow in R (reverse). The forward clutch and low band are released and the reverse clutch is applied to hold the reverse ring gear stationary. (*Pontiac Motor Division of General Motors Corporation*)

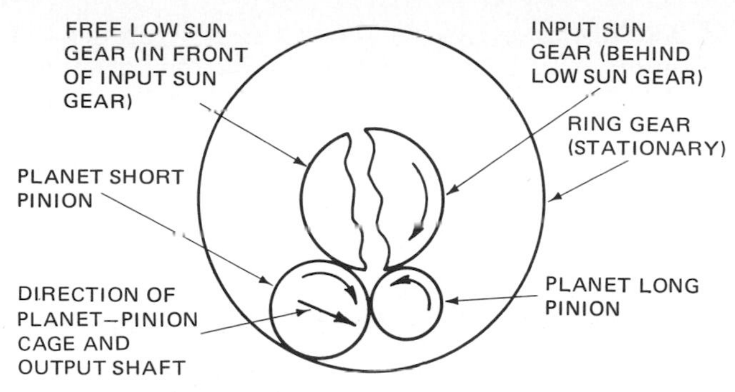

Fig. 9-8. End view of gears in the transmission planetary-gear system with the system in reverse gear.

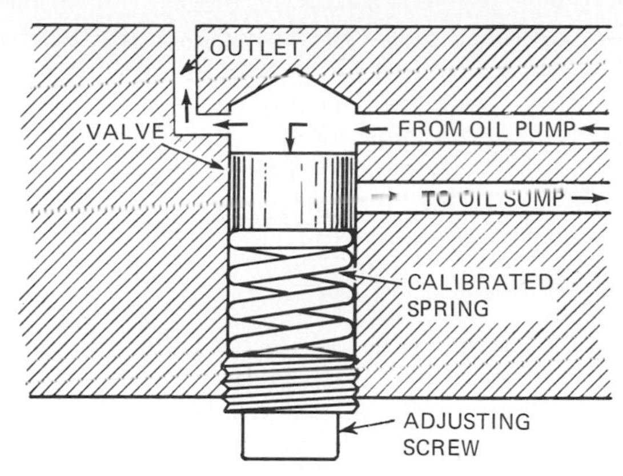

Fig. 9-11. Simple pressure regulator. (*Chevrolet Motor Division of General Motors Corporation*)

Fig. 9-9. Pump body with cover removed to show the two gears set into the cavity in the body. The technician is checking the clearance between the driven gear and the pump body. (*Chevrolet Motor Division of General Motors Corporation*)

oil pump is dumped through the return line, thus preventing any further increase of pressure. Thus, the outlet pressure is kept from increasing beyond the specified value. This value can be changed by turning the adjusting screw in or out. If the screw is turned in, thus increasing the pressure of the calibrated spring on the valve, the oil pressure will increase. If the adjusting screw is backed out, the oil pressure will decrease.

Actual pressure-regulator valves are more complicated than the simple one shown in Fig. 9-11. They are the spool type, such as the one shown in Fig. 9-12. In Fig. 9-12, there is no oil pressure and the spring has pushed the valve to the bottom of its bore. When the engine is started, the pump directs oil to the pressure-regulator valve. The oil flows through the space between the spool lands. From there, it flows through the exit, or regulated-pressure, line. The oil also flows through an orifice to a chamber under the valve. As pressure builds up, the valve is raised against the valve-spring pressure. When regulated pressure is reached, the valve is raised enough so that the space between the two lands starts to clear the return port to the exhaust

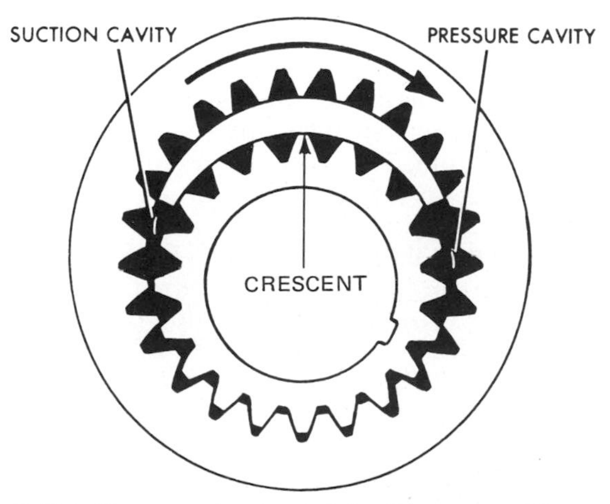

Fig. 9-10. Gears and crescent in the pump. The gears rotate in the direction shown by the arrow, but the crescent is stationary because it is part of the pump body. (*Chevrolet Motor Division of General Motors Corporation*)

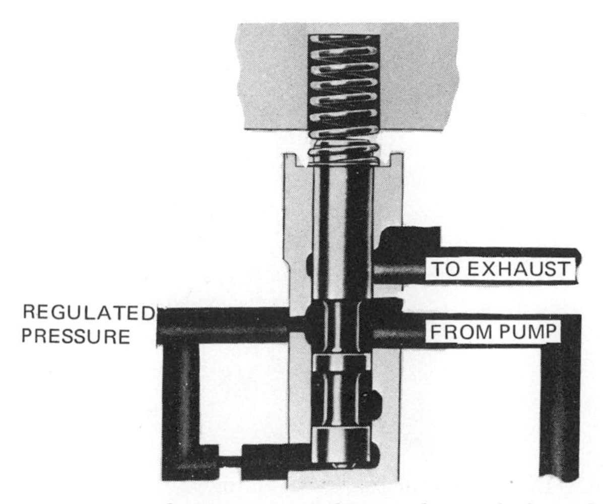

Fig. 9-12. Typical pressure-regulator valve with the valve at rest. (*Chevrolet Motor Division of General Motors Corporation*)

or the sump (Fig. 9-13). Now, part of the oil is dumped through this line to prevent any further rise in pressure. Note that as the valve moves up, the lower land partly cuts off the port from the pump and also the outlet, or regulated-pressure, line. In operation, the valve positions itself so that the regulated pressure is maintained at the value for which the valve spring is calibrated.

⊘ 9-7 Oil Cooler The oil used in the transmission gets hot. Part of the heat is from the engine, but much of it comes from the workings of the transmission itself. In the transmission, the oil is tossed back and forth between the pump and turbine of the torque converter, forced through the oil pump and various lines, and also serves as the lubricant for all the moving parts in the transmission. The result of all this activity is that the oil gets hot. Rapid stirring and movement of any fluid will heat it up. This is because of the frictional effect of the layers of fluid passing over each other. To prevent overheating of the oil and of the transmission itself, the transmission usually has an oil-cooling system such as the one shown in Fig. 9-14. The oil circulates through a small radiator that is part of the engine cooling system. In Fig. 9-14, the transmission oil-cooler radiator is located inside the right-hand engine-cooling-system radiator tank. It is immersed in the coolant of the cooling system. This arrangement permits the engine cooling system to keep the transmission oil and thus the transmission at a sufficiently cool operating temperature.

⊘ 9-8 Valves With minor exceptions, all the valves in the hydraulic system are located in the valve body, which is at the bottom of the transmission, as shown in Figs. 9-2 and 9-3. These valves move in their bores under the influence of various pressures to produce the shifts in the transmission. More on that later.

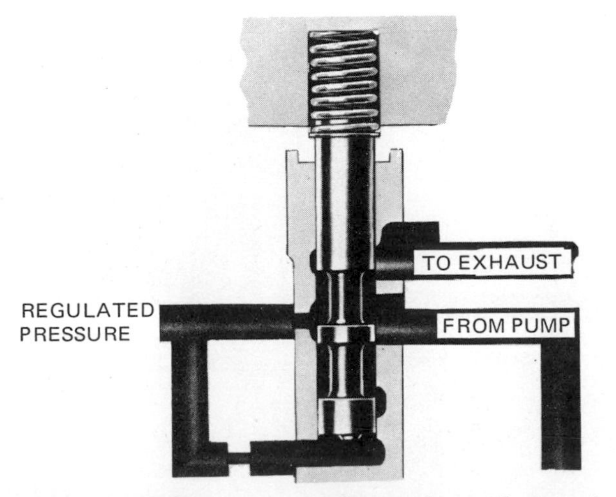

Fig. 9-13. Typical pressure-regulator valve with the valve in action. Note that the valve has been pushed up, compressing the spring. (*Chevrolet Motor Division of General Motors Corporation*)

⊘ 9-9 Oil-Pump Circuit Let us now examine the oil-pump circuit in the transmission. This is shown in Fig. 9-15. The pump draws oil from the sump, through the screen, and sends it into the mainline. The pressure regulator senses this pressure, as noted in ⊘ 9-6. The pressure, entering the cavity back of land 1, moves the valve so that part of the oil is dumped into the return line to the sump. This action keeps the pressure from rising above specified values. The oil-pump priming valve clears the oil of any air that might have become trapped in it.

As the valve moves, land 2 clears the port leading to the converter, as shown in Fig. 9-16. Now, oil flows to the converter, filling it to replace any oil that may have leaked down during a period of idleness.

⊘ 9-10 Lubrication Circuit Oil under pressure is supplied to the bearings and bushings in the transmission. The oil is supplied to the input shaft between two seal rings, where it enters a main oil gallery or passage drilled from the rear of the shaft. This gallery supplies the lubrication needed by the high-and-reverse clutch plates, planetary-gear system, and various bushings and thrust-washer surfaces.

⊘ 9-11 Mainline-Pressure Circuit As the converter and lubrication circuits are filled, the mainline pressure increases further. This causes the pressure-regulator valve to move farther to the right (in Fig. 9-15). This movement allows mainline oil to be dumped into the return circuit to the sump, as already explained, to prevent any further rise in mainline pressure.

The mainline pressure is delivered to the throttle valve, detent valve, vacuum-modulator valve, and manual valve. We shall look at each of these valves in detail in the following sections.

⊘ 9-12 Vacuum Modulator and Booster-Valve Circuit The vacuum modulator changes, or *modulates*, the mainline pressure to meet the needs of changing engine loads. This modulation is controlled by hydraulic pressure, spring pressure, and intake-manifold vacuum. The vacuum modulator includes a spool valve, an airtight diaphragm, and a spring, as shown in Fig. 9-17. The vacuum chamber is connected to the intake manifold by a tube. The spring tends to keep the spool valve bottomed. In this position, oil is delivered through a drilled passage in the valve to the blind cavity at the left end of the valve. The mainline oil pressure in this cavity tends to oppose spring pressure and tries to move the spool valve to the right. At the same time, mainline pressure flows between the valve lands, as shown in Fig. 9-17, to the end of the booster valve. The booster valve is at the end of the pressure-regulator valve.

The mainline pressure that flows through the modulator valve is called *modulator pressure*. It forces the booster valve to the left in Fig. 9-17, and

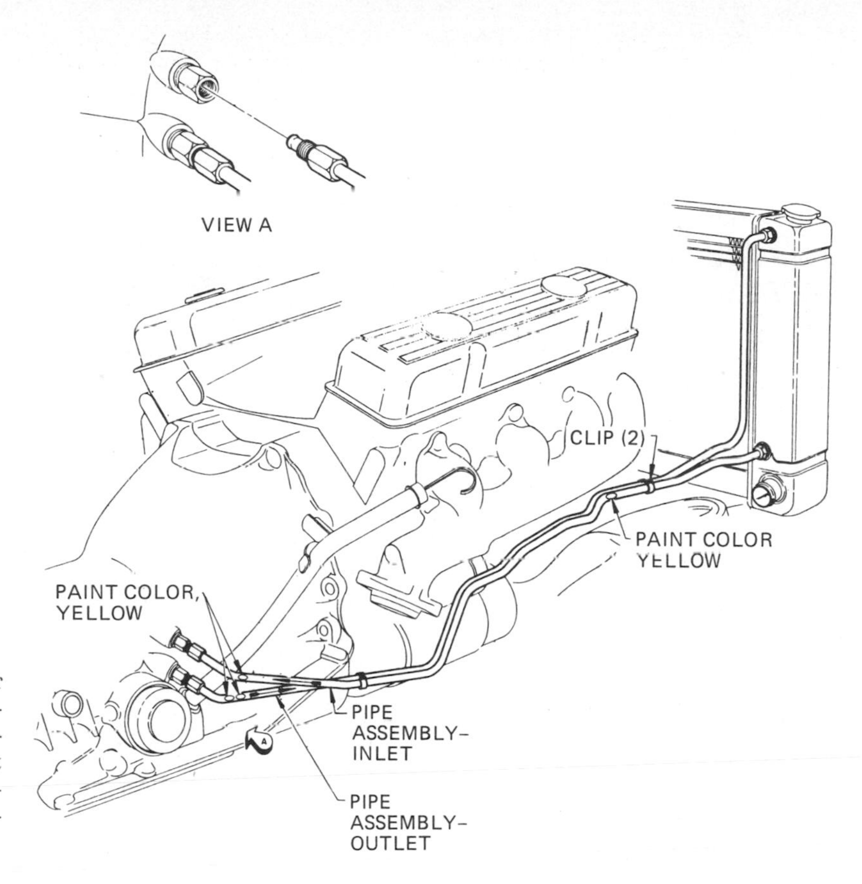

Fig. 9-14. Oil-cooler lines between the transmission and the engine-cooling radiator. (*Buick Motor Division of General Motors Corporation*)

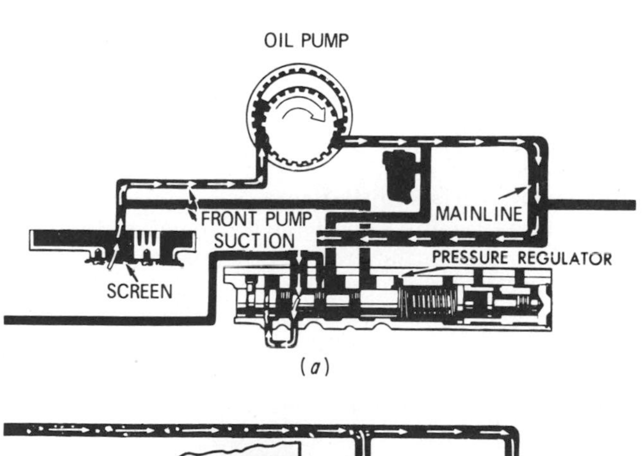

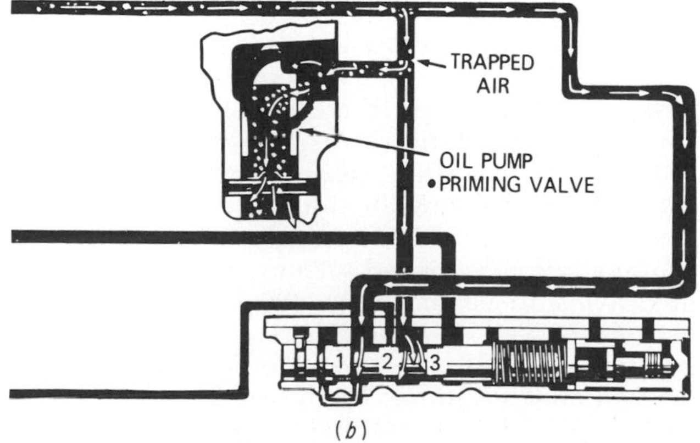

Fig. 9-15. Oil-pump circuits: (*a*) complete circuit; (*b*) priming valve circuit. (*Chevrolet Motor Division of General Motors Corporation*)

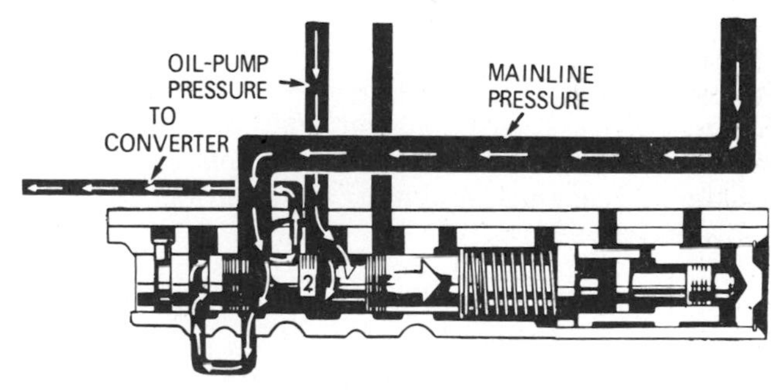

Fig. 9-16. Converter-charging circuit. (*Chevrolet Motor Division of General Motors Corporation*)

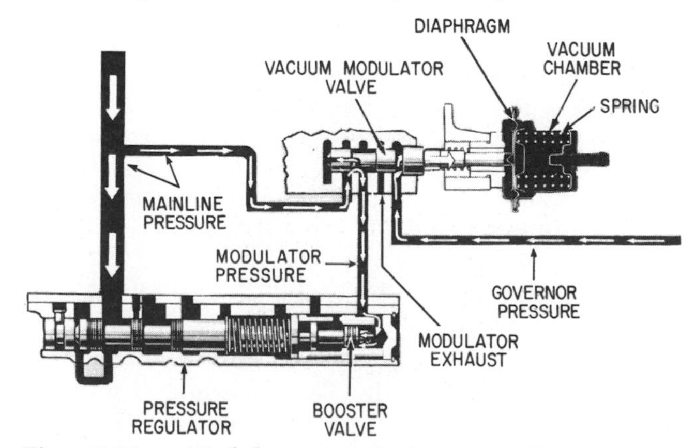

Fig. 9-17. Modulator and booster-valve circuits. (*Chevrolet Motor Division of General Motors Corporation*)

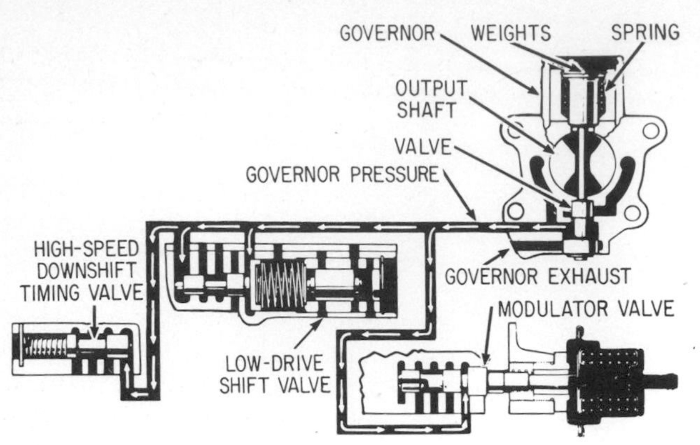

Fig. 9-18. Governor circuits. (*Chevrolet Motor Division of General Motors Corporation*)

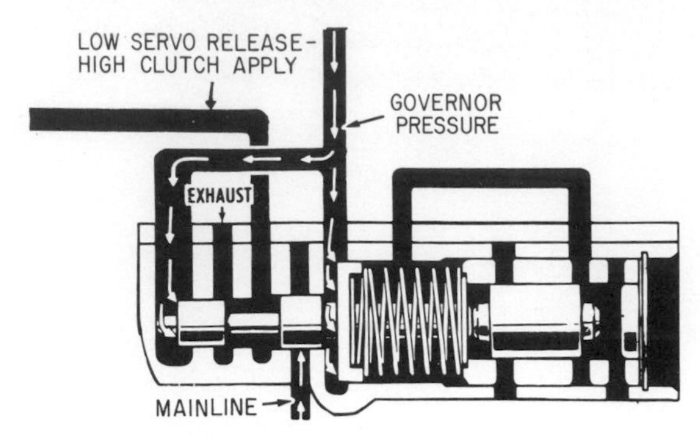

Fig. 9-19. Low-drive shift-valve circuits. (*Chevrolet Motor Division of General Motors Corporation*)

this movement adds to the spring pressure on the pressure-regulator valve. The result is that the regulated, or mainline, pressure increases. This is precisely the effect desired because higher mainline pressure and greater band and clutch holding power are needed when high engine power is developed. And high engine power is developed with low intake-manifold vacuum.

To see what we mean, consider the effect if the throttle is closed so that a high intake-manifold vacuum develops. This vacuum is connected to the vacuum-modulator vacuum chamber. This causes the diaphragm to move (to the right in Fig. 9-17) against the spring pressure. As the diaphragm moves to the right, it moves the spool valve with it. The land on the valve therefore partly cuts off the mainline pressure the valve has been passing to the booster valve. That is, the modulator pressure decreases. With less pressure on the end of the booster valve, the regulated, or mainline, pressure also decreases.

There you have it: There is increased mainline pressure with open-throttle, high-engine-output conditions. And there is less mainline pressure with closed-throttle, low-engine-output conditions.

Governor pressure also has some effect on modulator pressure. Governor pressure, as we shall see in ⊘ 9-13, varies directly with car speed. Governor pressure flows to a cavity between two lands on the spool valve, as shown in Fig. 9-17. The face of the land to the right is larger than the face of the land to the left. Thus, governor pressure, acting against the two faces, tends to move the valve to the right. This tends to reduce modulator pressure and thus mainline pressure. This action compensates for the effects of very low intake-manifold vacuum, which could result in high mainline pressures that are not needed at high car speeds.

⊘ 9-13 Governor Circuit The governor provides a pressure proportional to car speed. As car speed increases, governor pressure increases. Figure 9-18 shows the governor circuit. The governor is mounted on the transmission output shaft and thus rotates at varying speeds directly related to car speed. The governor contains spring-loaded weights connected by a through-shaft to a valve, as shown.

At rest, the governor weights are in toward the output shaft and the mainline pressure is shut off through the governor. However, when the car starts to move, the centrifugal force acting on the weights makes them move outward against the spring pressure. This action moves the governor valve inward so that it begins to pass mainline pressure. This pressure exits as governor pressure. The faster the car moves, the farther out the weights move against the spring pressure. And the farther in the valve moves to pass more mainline pressure. In other words, governor pressure increases with car speed.

This governor pressure flows to the low-drive shift valve and to the vacuum-modulator valve, as already explained.

⊘ 9-14 Low-Drive Shift Valve The governor pressure, acting on the low-drive shift valve, tends to move it so that an upshift will occur (see Fig. 9-19). Opposing this are spring pressure and throttle-valve pressure. Throttle-valve pressure is simply mainline pressure that has passed through the throttle valve. The throttle valve, being linked to the accelerator, moves as the carburetor throttle is opened or closed. As the carburetor throttle is opened, the throttle valve in the hydraulic system passes more mainline pressure, which becomes the throttle-valve, or TV, pressure, varying with throttle opening.

Thus, the governor pressure works on one end of the low-drive shift valve, and spring and throttle-valve pressures work on the other end. The low-drive shift valve remains stationary until the right combination is achieved. When this happens, the low-drive shift valve moves and pressure is passed through the valve to the release side of the servo and to the high clutch. Now, the band releases and the high clutch applies, producing an upshift from low to drive. Figure 9-20 shows a typical linkage from the throttle lever on the carburetor to the lever on the transmission that is connected to the transmission throttle valve.

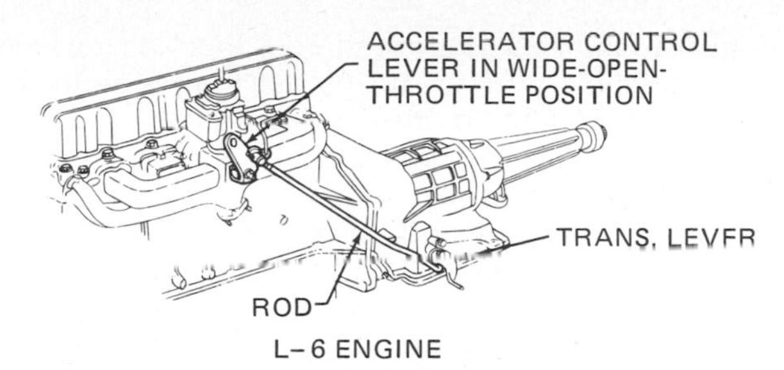

Fig. 9-20. Linkage between the carburetor control lever and the transmission lever. (*Chevrolet Motor Division of General Motors Corporation*)

⊘ 9-15 Manual Valve The manual valve is connected by linkage to the selector lever on the steering column or floorboard console. Figure 9-21 shows the linkage between the selector lever on the steering column and the lever on the transmission which moves the manual valve. Figure 9-22 shows the manual valve in the Powerglide transmission. The valve, which is a spool valve, slides back and forth in its cylinder bore as the selector lever is moved. In each position that the selector is placed, the manual valve directs oil pressure to the other valves and to the clutches and band servo to produce the drive conditions selected. Following sections and illustrations explain these conditions.

⊘ 9-16 Hydraulic Circuit in Low Gear, Drive Range Plate 1 shows the complete circuit, with the different pressures and oil flows indicated in different patterns and colors. The conditions shown in Plate 1 (in the insert) are the actions with the transmission in low gear in drive range. This is what happens when the manual valve is moved to D (drive) and the car is first starting out in low gear. With this condition, the manual valve is directing

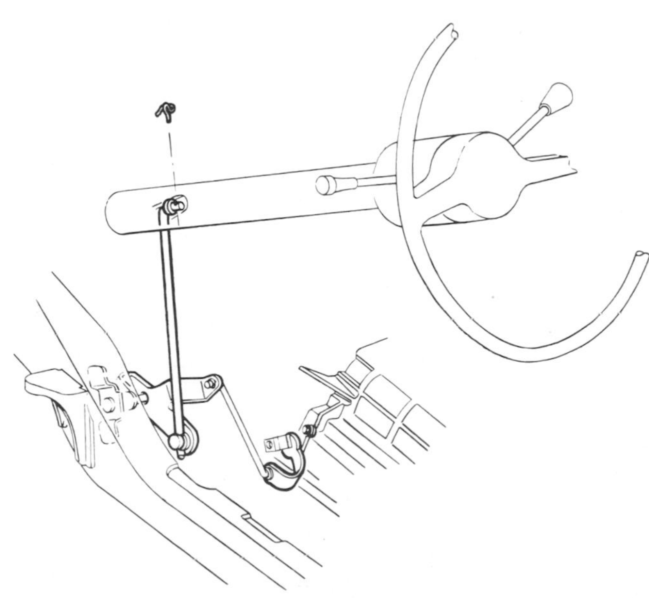

Fig. 9-21. Linkage between the selector lever and the manual-valve lever on the transmission. (*Chevrolet Motor Division of General Motors Corporation*)

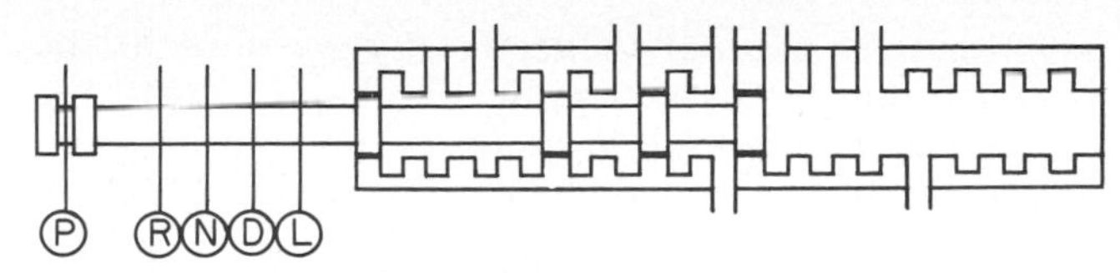

Fig. 9-22. Manual valve. (*Chevrolet Motor Division of General Motors Corporation*)

mainline pressure to the shift valve and low servo. Since the oil pump is driven by the converter hub, it develops oil pressure whenever the engine is running. Thus, mainline pressure is available to operate the low servo. The servo applies the band to hold the forward-clutch drum and low sun gear stationary so that the transmission is in low gear, as shown in Fig. 9-4.

There is no governor pressure until the car starts to move, and so governor pressure does not enter into the hydraulic-system action shown in Plate 1.

⊘ 9-17 Hydraulic Circuit in Direct Drive As car speed increases, the transmission will upshift to produce the conditions in the hydraulic system shown in Plate 2. We have already noted that the upshift is produced by a combination of governor pressure and throttle-valve pressure acting on the low-drive shift valve. When governor pressure is great enough to overcome the low-drive shift-valve spring, plus the throttle-valve pressure, the shift valve moves. This opens the circuits to the forward, or high, clutch apply and the low servo release, as shown. Now, the low-servo-release pressure forces the servo piston to move so as to release the band. At the same time, the high or forward clutch applies so that the transmission shifts into direct.

Note that there is a downshift timing valve between the clutch and the servo. The downshift timing valve times the pressure flow to the servo so that the band will not release too quickly. It releases just as the clutch applies so that a smooth upshift occurs.

⊘ 9-18 Detent Downshift The driver can force a downshift from direct drive to low, below about 60 mph [96.6 km/h], by pushing the accelerator pedal all the way down. This forces the throttle valve all the way in so that throttle-valve pressure is increased to maximum. This TV pressure, acting on the low-drive shift valve, causes it to shift into the low position. Now, the forward clutch is released and the band is applied so that the transmission shifts into low.

If the speed is above the shifting point—above about 60 mph [96.6 km/h]—the governor pressure is great enough to prevent the low-drive shift valve from moving. However, if the car speed (governor pressure) is below this point, the shift can occur.

⊘ 9-19 Manual Low In manual low, the selector lever and manual valve are moved to L (low). In this position, the manual valve directs pressure to the servo so that the servo applies the band to shift the transmission into low. Pressure is thus cut off from the forward clutch so that it cannot operate. As a result, the transmission stays in low.

⊘ 9-20 Reverse When the selector lever and manual valve are moved to R (reverse), the manual valve directs mainline pressure to the reverse clutch and cuts it off from the servo and forward clutch. This action causes the reverse clutch to apply and results in the condition shown in Fig. 9-7. In other words, the reverse ring gear is held stationary and the direction of rotation is reversed through the transmission. In addition, mainline pressure is directed to the end of the booster valve, causing the booster valve to move the pressure-regulator valve so that mainline pressure is increased. This increased pressure is required to ensure sufficient holding power in the reverse clutch. The situation in the hydraulic system is shown in Plate 3.

CHAPTER 9 CHECKUP

NOTE: Since the following is a chapter review test, you should review the chapter before taking it.

The chapter you have just finished describes in detail the Powerglide, one of the most widely used two-speed automatic transmissions. A good understanding of this transmission will help you understand the more complex transmissions discussed in following chapters. Thus, you should consider this chapter the foundation on which to build your full understanding of all automatic transmissions. It is therefore important for you to remember all the essential details discussed in this chapter. The tests that follow will help you retain these details.

Completing the Sentences The sentences that follow are incomplete. After each sentence there are several words or phrases, only one of which will correctly complete the sentence. Write each sentence in your notebook, selecting the proper word or phrase to complete it correctly.

1. The selector lever for the Powerglide has the following positions: (*a*) P N R D S L, (*b*) P R N D L, (*c*) P R N D L1 L2.
2. The main difference between the Powerglide and the Torque Drive is that in the Torque Drive: (*a*) there are more valves, (*b*) low range has been eliminated, (*c*) automatic shifting has been eliminated.
3. To control the planetary-gear system in the Powerglide, there are two clutches and: (*a*) one brake band, (*b*) two brake bands, (*c*) a reverse clutch.
4. The Powerglide has two sun gears; the input sun gear is splined to the input shaft, and the low sun gear is splined to the: (*a*) ring gear, (*b*) output shaft, (*c*) forward-clutch drum, (*d*) reverse-clutch drum.
5. The input sun gear is meshed with the: (*a*) ring gear, (*b*) planet short pinions, (*c*) planet long pinions.
6. The low sun gear is meshed with the: (*a*) ring gear, (*b*) planet short pinions, (*c*) planet long pinions.
7. When the forward clutch is released and the brake band is applied, the transmission is in: (*a*) low, (*b*) direct drive, (*c*) reverse.
8. When the forward clutch is applied and the brake band is released, the transmission is in: (*a*) low, (*b*) direct drive, (*c*) reverse.
9. The upshift is produced by a combination of governor pressure and throttle-valve pressure acting on the: (*a*) pressure valve, (*b*) low-drive shift valve, (*c*) detent shift valve.
10. The vacuum modulator is connected: (*a*) by linkage to the throttle, (*b*) to the pump, (*c*) to the intake manifold.

Purpose and Operation of Components In the following, you are asked to write the purpose and operation of various components in the Powerglide. If you have any difficulty in writing your explanations, turn back into the chapter and reread the pages that will give you the information you need. But don't copy; try to tell it in your own words. That will help you fix the explanation in your mind.

1. Describe the power flow through the transmission in L; in D; in R.
2. Explain how the sun gears, ring gear, and planet pinions are meshed.
3. To what is the input sun gear splined?
4. To what is the low sun gear splined?
5. Explain how the oil pump works.
6. Explain how pressure regulation is achieved.
7. Describe the oil-cooler system.
8. What is the purpose of the vacuum modulator? How does it work?
9. Explain how the governor works.
10. Explain how the low-drive shift valve works.
11. Explain how a detent downshift takes place.
12. Study the hydraulic circuit showing conditions in low gear, drive range, and make a sketch showing oil flow to the various valves and components of the hydraulic circuit.
13. Follow the instructions in question *12* for conditions in direct drive.
14. Follow the instructions in question *12* for conditions in reverse.

SUGGESTIONS FOR FURTHER STUDY

If possible, try to examine the complete transmission as well as the various parts used in it which were described in the chapter you have just finished. Torque converters are welded units and cannot be opened for examination. However, many school shops have cutaways or disassembled converters which you can examine to help you understand these units. Also, study carefully any shop manuals you can find which cover the operation and service of the transmission. Be sure to write in your notebook any important facts you learn.

chapter 10

ALUMINUM POWERGLIDE TRANSMISSION SERVICE

This chapter covers the trouble-diagnosis and service procedures for the Aluminum Powerglide two-speed automatic transmission. The chapter is divided into three parts: "Normal Maintenance and Adjustments," "Powerglide Trouble Diagnosis," and "Powerglide Transmission Overhaul." The service procedures outlined cover, in general, all the variations of the basic Powerglide, including the Buick Super Turbine 300, Oldsmobile Jetaway, and Pontiac M-35. Also, the instructions cover the no-shift Torque Drive.

Normal Maintenance and Adjustments

The following includes checking oil level, draining and refilling the transmission, low-band adjustment, and linkage adjustments.

⊘ 10-1 Oil-Level Check The transmission oil level should be checked every 6,000 miles [9,656 km] with the engine idling, selector lever in N (neutral), parking brake set, and transmission at operating temperature. If the fluid level is low, add enough fluid to bring the level to the full mark on the dipstick.

CAUTION: Do not overfill. Overfilling causes aeration and foaming of the fluid, which will produce loss of drive, improper band and clutch application, and overheating of the transmission.

CAUTION: Use only Dexron automatic transmission fluid. Use of any other kind of fluid can cause complete transmission failure.

If the transmission requires frequent addition of fluid, check for leakage sources—seals, gaskets, cracks in case, and so on.

⊘ 10-2 Oil Change Every 24,000 miles [38,624 km], or 12,000 miles [19,312 km] in severe service,[1] drain the oil and refill with new Dexron fluid. Proceed as follows:

1. Drain the oil after a period of operation by lifting the car on a hoist, placing the drain collector underneath, and removing the oil-pan plug.
2. Install the plug, lower the car, and add 2 quarts [1.892 l (liters)] Dexron fluid.

[1] Heavy city traffic, taxicab, patrol car, where car sits and idles for long periods, or where car pulls a trailer.

3. With the engine idling, move the selector lever to all positions and then back to N. Recheck the fluid level and add fluid as necessary to bring the level up to full.

⊘ 10-3 Low-Band Adjustment The low band should be adjusted at the first transmission oil change or when the low band has been slipping.

1. With the selector in neutral, raise the car on a hoist.
2. Remove the protective cap from the adjusting screw.
3. Use a wrench to loosen the locknut $\frac{1}{4}$ turn, as shown in Fig. 10-1.
4. Hold the nut and use the band adjusting tool as shown, to adjust band screw to 70 pound-inches [1.26 kg-mm (kilogram-millimeters)]. Then back off

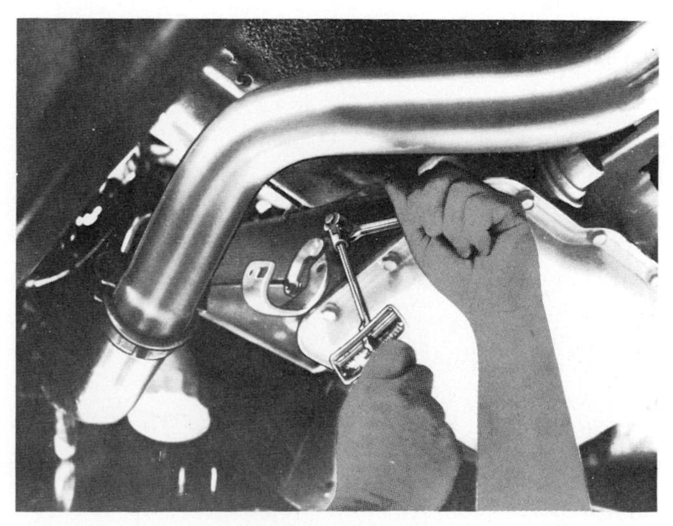

Fig. 10-1. Adjusting the low band. (*Chevrolet Motor Division of General Motors Corporation*)

the screw 4 turns if the transmission has been used more than 6,000 miles [9,656 km] or 3 turns if it has been used less than 6,000 miles [9,656 km].
5. Tighten the locknut and install the protective cap.

⊘ 10-4 Column-Type Shift Control Linkage Figures 10-2 and 10-3 show two varieties of column-type shift-control linkages. To adjust, proceed as follows:

1. Loosen the adjustment clamp at the cross shaft. See Fig. 10-2.
2. Rotate the transmission lever clockwise to drive position.
3. Set the selector lever to D (drive) and remove any free play by rotating and holding the cross shaft upward and pulling the shift rod down.
4. Tighten the clamp and recheck the adjustment.
5. Readjust the indicator needle if necessary to agree with the transmission detent positions.
6. Readjust the neutral safety switch if necessary.

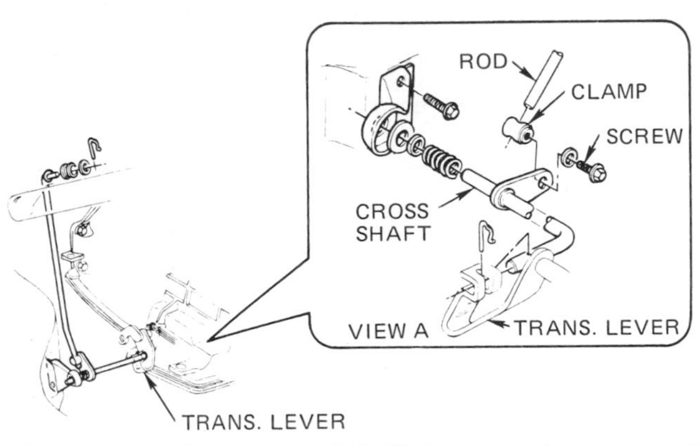

Fig. 10-2. Column-type shift linkage used on several Chevrolet models. (*Chevrolet Motor Division of General Motors Corporation*)

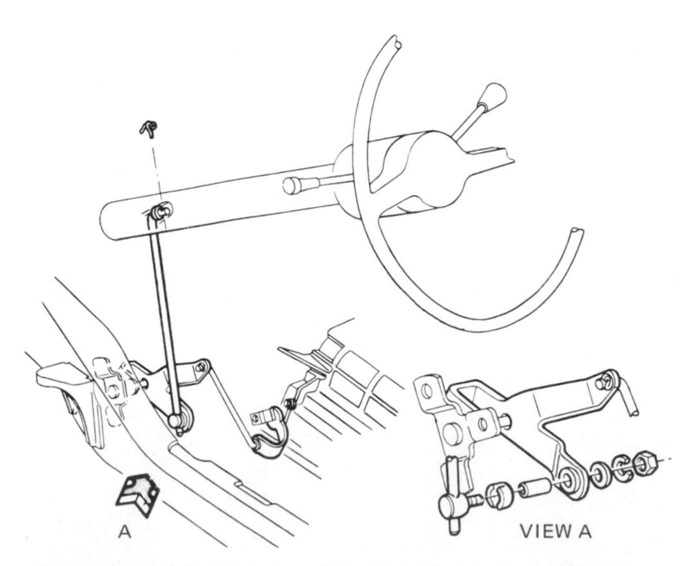

Fig. 10-3. Column-type shift linkage used on the Nova. (*Chevrolet Motor Division of General Motors Corporation*)

CAUTION: Incorrect linkage adjustment can cause transmission failure. With incorrect adjustment, the transmission can operate with the manual valve not quite in its proper position. Such operation can reduce pressure to the band and clutches so they will only partly engage, slip, and soon burn out.

⊘ 10-5 Console-Type Shift-Control Linkage Figure 10-4 is the linkage for a console-type shift control. Adjust by placing the selector lever in D and raising the car on a hoist. Disconnect the cable from the transmission lever. Move the lever to drive by rotating it counterclockwise to L and then back one detent. Measure the distance from the rearward face of the attachment bracket to the center of the cable attachment pin. This dimension should be $5\frac{1}{2}$ inches [135.2 mm]. Adjust the pin if necessary. Install the cable to the transmission lever, lower the car, and check the adjustment.

⊘ 10-6 Throttle-Valve Linkage Figure 10-5 shows the throttle-valve linkage for six- and eight-cylinder engines. On the six-cylinder engine, the swivel is adjusted to allow the transmission lever to be against the internal stop when the accelerator control lever is in the wide-open-throttle position. On the V-8, proceed as follows:

1. Remove the air cleaner and disconnect the accelerator linkage at the carburetor.
2. Disconnect the accelerator-return and TV-rod springs.
3. With the right hand, pull the TV upper rod forward until the transmission is through the detent. With the left hand, open the carburetor to wide-open-throttle position. The carburetor must reach wide-open throttle at the same time the ball stud makes contact with the end of the slot in the upper TV rod. Adjust the swivel on the end of the upper TV rod as necessary.
4. Connect and adjust the accelerator linkage as necessary, and check for freedom of movement in linkage.

⊘ 10-7 Towing Instructions If the transmission, drive line, and axle are in good condition, the car may be towed in neutral at speeds of up to 35 mph [56.33 km/h] for not more than 50 miles [80.47 km]. For higher speeds and longer distances or if transmission trouble is suspected, the drive shaft should be disconnected or the car should be towed with the rear wheels off the ground. On a rear-end tow, tie the steering wheel in the straight-ahead position so that the car will not wander.

Powerglide Trouble Diagnosis

Operation of the transmission can be affected by a number of factors. When trouble is experienced, the transmission operation should be checked either on

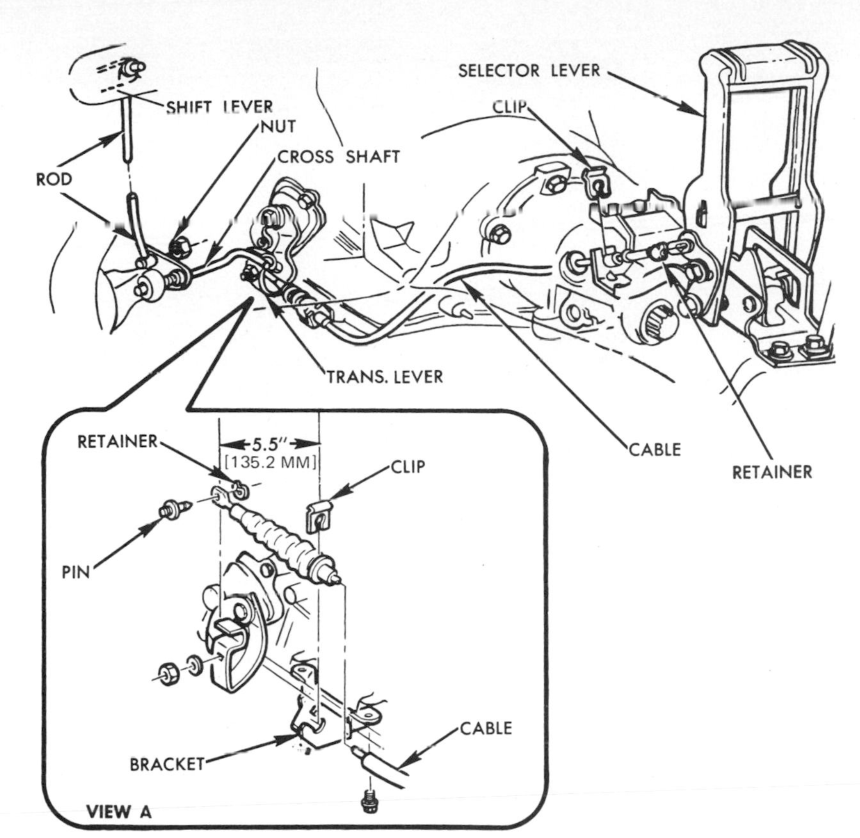

Fig. 10-4. Console-type shift-control linkage. (*Chevrolet Motor Division of General Motors Corporation*)

a chassis dynamometer or on the road. The miles per hour at which shift points occur should be noted and compared with the specifications in the car shop manual. A typical shift-point table is shown in Fig. 10-6. In addition, pressure checks can be made to assist in diagnosis.

NOTE: An engine that is not performing satisfactorily might make the transmission look bad. Therefore, tune up the engine if it needs attention before testing the transmission.

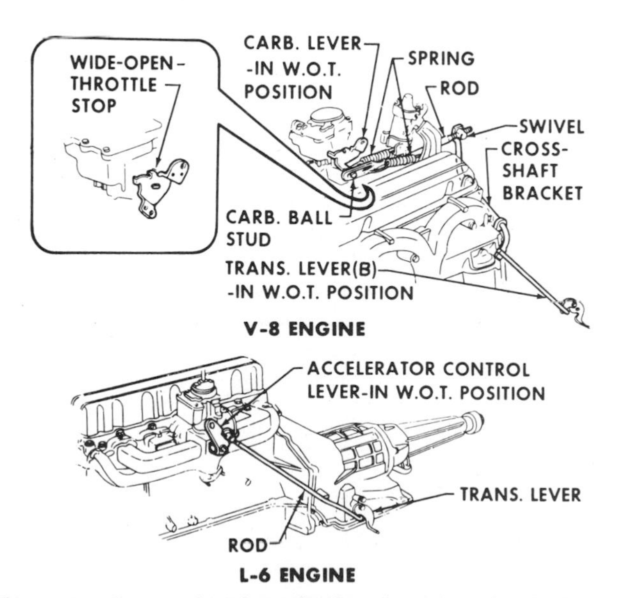

Fig. 10-5. Six- and eight-cylinder throttle-valve linkages. WOT means *wide-open throttle*. (*Chevrolet Motor Division of General Motors Corporation*)

CAUTION: If you road-test a car, obey all traffic laws. Drive safely. Use a chassis dynamometer (⊘ 1-9), if available, so that you can make all the tests in the shop.

⊘ 10-8 Pressure Checks The following four pressure checks are to be made with the gauge connected to the low-servo-apply-pressure test point shown in Fig. 10-7. Transmission must be at operating temperature, and linkages must be properly adjusted. See ⊘ 10-4 to 10-6.

1. Wide-open-throttle upshift pressure
2. D overrun (coasting) pressure
3. Idle pressure in D
4. Manual L pressure

To get the pressure gauge into the driver's compartment where it can be watched, push the hose up past the steering-mast seal to a convenient place and then connect the gauge. You will also need an intake-manifold-vacuum gauge for the second pressure check. Specified pressures for one line of cars are given in the table in Fig. 10-8. Proceed with pressure checks as follows:

1. Check wide-open-throttle upshift pressure and compare it with the pressures given in Fig. 10-8.
2. Check the D overrun, or coasting, pressure with the vehicle coasting in D at 25 to 20 mph [40.234 to

CHEVROLET

ENGINE	THROTTLE POSITION	mph IN DRIVE: UPSHIFT MIN.	UPSHIFT MAX.	DOWNSHIFT MAX.	DOWNSHIFT MIN.
L-6 (250) Base 3.08 Axle F78-15 Tire	Closed throttle	16	21	17	13
	Detent touch	45	56	41	26
	Wide-open throttle	54	63	56	63
L-6 (250) Base 3.08 Axle G78-15 Tire	Closed throttle	17	23	18	14
	Detent touch	45	58	42	27
	Wide-open throttle	55	64	57	48
V-8 (350) Base 2.78 Axle F78-15 Tire	Closed throttle	19	26	20	15
	Detent touch	60	75	49	31
	Wide-open throttle	74	88	70	60
V-8 (350) Base 2.73 Axle G78-15 Tire	Closed throttle	19	27	20	16
	Detent touch	61	77	50	32
	Wide-open throttle	76	82	71	61
V-8 (350) L48 3.08 Axle G78-15 Tire	Closed throttle	17	24	18	14
	Detent touch	54	68	44	28
	Wide-open throttle	67	76	63	54

Fig. 10-6. Powerglide shift points for different car models. (*Chevrolet Motor Division of General Motors Corporation*)

32.187 km/h] and with intake-manifold vacuum about 20 inches [508 mm]. Pressure should be as shown in Fig. 10-8.

3. Check the idle pressure in D with the car station-

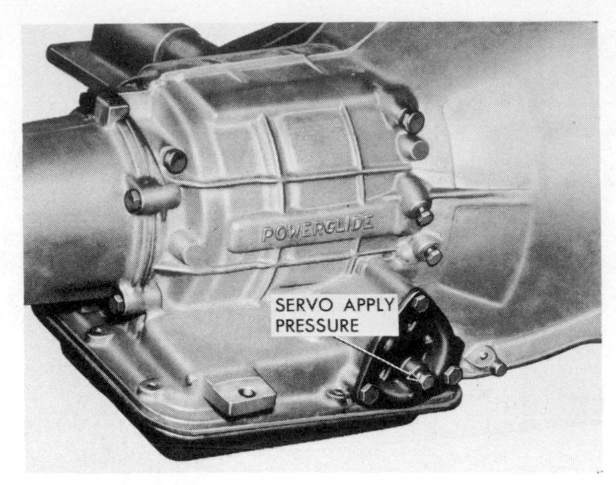

Fig. 10-7. Point at which to connect the pressure gauge for a transmission pressure check. (*Chevrolet Motor Division of General Motors Corporation*)

ary, brakes applied, and selector lever in D. If the pressure is not within the range shown, the following could be the cause:

a. Pressure-regulator valve stuck
b. Vacuum-modulator valve stuck
c. Boost valve stuck
d. Leak at the low-servo piston rod—between the ring and bore or between rod and bore
e. Leak at valve-body gaskets
f. Excessive oil-pump clearances
g. Leak in transmission-case passages due to porosity

4. To check the manual L pressure, use a tachometer and adjust engine speed to 1,000 rpm with the car stationary. If pressures are not within the range shown in Fig. 10-8, the following could be the cause:

a. Partly plugged oil screen
b. Broken or damaged low-servo ring
c. Pressure-regulator valve stuck
d. Leak at valve-body gaskets
e. Leak at servo center
f. Excessive oil-pump clearances

⊘ 10-9 Powerglide Trouble-Diagnosis Chart The chart on pp. 129–130 lists various operating troubles and their possible causes.

ENGINE	WIDE-OPEN THROTTLE UPSHIFT	IDLE PRESSURE IN DRIVE RANGE: 16 inHg	10 inHg	MANUAL-LOW RANGE @ 1,000 rpm	DRIVE-RANGE OVERRUN (COAST) @ 20–25 mph (approx. 20 inHg)
L-4	86–101	48–53	55–73	105–117	48–53
L-6 230	88–106	52–72	79–97	106–117	48–53
L-6 250	88–106	52–72	79–97	106–117	48–53
V-8 307	90–109	55–75	83–102	125–138	48–53
V-8 350	90–109	55–75	83–102	125–138	48–53

Fig. 10-8. Table of pressures (in pounds per square inch) for one line of cars. (*Chevrolet Motor Division of General Motors Corporation*)

POWERGLIDE TROUBLE-DIAGNOSIS CHART

TROUBLE	POSSIBLE CAUSE
1. No drive in any selector position	a. Low oil level b. Clogged oil suction screen c. Defective pressure regulator valve d. Defective oil pump e. Broken input shaft f. Sticking oil-pump priming valve
2. Engine flares on standstill starts and acceleration lags	a. Low oil level b. Clogged oil suction screen c. Improper band adjustment d. Disengaged or broken band-apply linkage e. Worn band facing f. Blocked servo-apply passage g. Broken or leaking servo piston ring h. Stator in converter not holding (rare)
3. Engine flares on upshift	a. Low oil level b. Improper band adjustment c. Plugged vacuum-modulator line d. Clogged oil suction screen e. Blocked high-clutch feed orifice f. Worn high-clutch plates g. Sticking or hung-up high-clutch piston h. Relief ball not sealing in high-clutch drum
4. No upshift	Low band not releasing, probably due to: 1. Stuck low-and-drive valve 2. Defective governor 3. Stuck or maladjusted throttle valve 4. Maladjusted manual-valve lever
5. Harsh upshift	a. Improper carburetor-to-transmission TV-rod adjustment b. Improper band adjustment
6. No drive in reverse	a. Improper manual-valve-lever adjustment b. Stuck reverse-clutch piston c. Worn-out reverse-clutch plates d. Reverse clutch leaking excessively e. Blocked reverse-clutch-apply orifice
7. Improper shift points	a. Improper carburetor-to-transmission linkage adjustment b. Improper throttle-valve (TV) adjustment c. Defective governor
8. Oil leaks	a. Transmission case and extension housing 1. Extension-housing oil seal 2. Shifter-shaft oil seal 3. Speedometer driven gear fitting 4. Pressure taps 5. Oil-cooler pipe connections 6. Porosity in case and/or housing 7. Vacuum modulator assembly (very smoky exhaust indicates a ruptured vacuum-modulator diaphragm) b. Transmission oil-pan gasket c. Converter cover pan 1. Oil-pump attaching bolts 2. Oil-pump seal ring 3. Oil-pump oil seal 4. Plugged oil drain in oil pump 5. Porosity in transmission case
9. Oil forced out filler tube	a. Oil level too high with aeration and foaming caused by planet carrier running in oil b. Water in oil c. Leak in oil-pump suction circuits

POWERGLIDE TROUBLE-DIAGNOSIS CHART (*Continued*)

TROUBLE	POSSIBLE CAUSE
	d. Broken or disconnected vacuum-modulator line e. Leaking vacuum-modulator diaphragm f. Sticking vacuum-modulator valve g. Sticking booster valve in valve body
10. Harsh closed-throttle (coast) downshift	a. High engine idle speed b. Improper band adjustment c. Malfunction of downshift timing valve d. High mainline pressure, probably due to: 1. Vacuum-modulator line broken or disconnected 2. Modulator diaphragm ruptured 3. Sticking vacuum-modulator valve, regulator valve, or booster valve
11. No downshift	a. Low TV pressure b. High governor pressure c. Sticking low-and-drive shift plug d. Sticking low-and-drive shift valve
12. Clutch failure—burned plates	a. Band adjusting screw backed off more than specified b. Improper order of clutch-plate assembly c. Extended operation with low oil level d. Stuck relief ball in clutch drum e. Abnormally high speed upshift, probably due to: 1. Improper governor action 2. Transmission operated at high speed in manual "low"
13. Car creeps excessively in drive range	Idle speed set too high
14. Car creeps in neutral	a. Improper manual-valve-lever adjustment b. High clutch or low band not released

Check Your Progress

Progress Quiz 10-1 The following questions will help you find out how well you are understanding what you are reading. If you have any trouble answering the questions, just reread the past few pages and try the questions again. The combination of reviewing the material and taking the quiz will help you to remember the essential facts.

Completing the Sentences The sentences that follow are incomplete. After each sentence there are several words or phrases, only one of which will correctly complete the sentence. Write each sentence in your notebook, selecting the proper word or phrase to complete it correctly.

1. In normal service, the oil should be changed every: (*a*) 12,000 miles [19,312 km], (*b*) 24,000 miles [38,624 km], (*c*) 36,000 miles [57,936 km].
2. If the transmission has been run more than 6,000 miles [9,656 km], then in adjusting the low band the screw should be backed off: (*a*) one turn, (*b*) two turns, (*c*) three turns, (*d*) four turns.
3. The four pressure checks are wide-open-throttle upshift, idle pressure in D, D overrun pressure, and: (*a*) reverse pressure, (*b*) manual L pressure, (*c*) manual L2 pressure.
4. In Powerglide trouble diagnosis, two recommended checks are pressure checks and: (*a*) clutch pressure, (*b*) band application, (*c*) shift points.

Correcting Troubles Lists The purpose of this exercise is to help you spot the unrelated trouble in a list. For example, in the list "No drive in any selector position: clogged oil-suction screen, defective oil pump, broken input shaft, clutch not releasing," you can see that "clutch not releasing" does not belong because it is the only condition in the list that would not be blamed for the trouble. Any of the other items in the list could cause "no drive in any selector position."

In each of the following lists, you will find one item that does not belong. Write each list in your notebook, but *do not write* the item that does not belong.

1. Engine flares on standstill and acceleration lags: low oil level, improper band adjustment, band not holding, blocked servo-apply passage, servo not working, forward clutch locked.
2. No upshift: low band not releasing, defective governor, stuck low-drive valve, stuck throttle valve, clutch not releasing, maladjusted manual-valve lever.
3. No drive in reverse: improper manual-valve-lever adjustment, stuck reverse-clutch piston, band not

holding, reverse-clutch plates worn, reverse clutch leaking excessively.
4. Car creeps in neutral: improper manual-valve-lever adjustment, brake band not applying, high clutch not released, low band not released.
5. Harsh upshift: improper carburetor-to-transmission TV-rod adjustment, forward clutch not applying, improper band adjustment.

Powerglide Transmission Overhaul

If the transmission requires major service, it must be removed from the car and installed on a repair fixture as shown in Fig. 10-9. Removal, overhaul, and reinstallation procedures follow.

⊘ 10-10 Transmission-Service Precautions Cleanliness is of the utmost importance in transmission work. The smallest, almost invisible, piece of dirt or lint from a cleaning rag can cause a valve to hang up and prevent normal transmission action. Under some conditions, this situation could cause fatal damage to the transmission. Other things to watch include:

1. Never mix parts between transmissions. Keep all parts belonging to the transmission you are working on in one place so that they will not get mixed with parts from another transmission.
2. Clean the outside of the transmission thoroughly before starting to work on it. One recommendation is to plug all openings and use steam to clean the transmission. The workbench, tools, your hands, and all parts must be kept clean at all times. Do not allow dust to blow in from outside or from other shop areas and settle on the transmission parts.

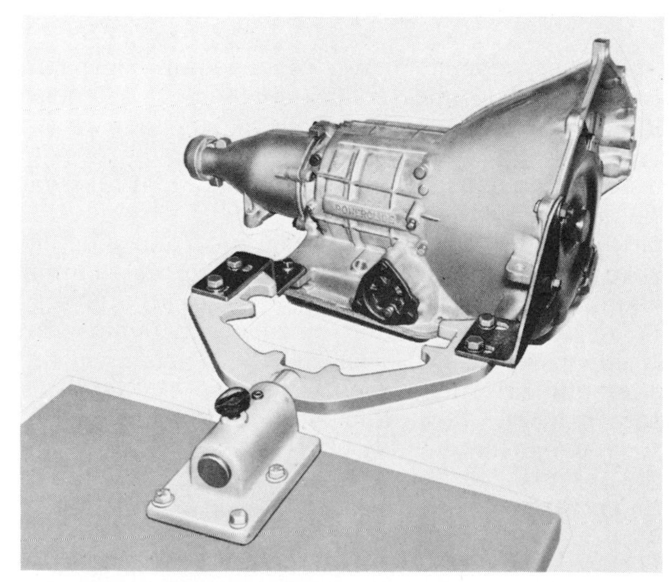

Fig. 10-9. Transmission mounted in a repair fixture. (*Chevrolet Motor Division of General Motors Corporation*)

3. Before installing screws into aluminum parts such as the case, dip the screws into transmission fluid to lubricate them. Lubrication will prevent their galling the threads and seizing.
4. If the threads in an aluminum part are stripped, repair can be made with a Heli-Coil, as shown in Fig. 10-10. First, the hole is drilled and then tapped with a special Heli-Coil tap. Finally, a Heli-Coil is installed to bring the hole back to its original thread size.
5. Special care and special tools must be used to protect the seals during the assembly procedure to prevent damage to the seals. The slightest flaw in a seal or a sealing surface can result in an oil leak and transmission trouble.
6. The aluminum case and other parts are relatively soft and can be easily scratched, nicked, or burred. Use great care in handling them.
7. Discard all old O sealing rings, gaskets, and oil seals that are removed, and use new ones for reassembly.
8. During disassembly, clean and inspect all parts as explained in following paragraphs.
9. During reassembly, lubricate all internal parts with transmission fluid.

⊘ 10-11 Cleaning and Inspecting Transmission Parts As parts are removed from the transmission, clean and inspect them as explained below. Wash all metal parts in solvent and blow them dry with compressed air. Do not use solvents that could damage clutch facings. Never wipe parts with shop cloths or paper because these could leave lint that could cause transmission trouble. Check small passages with small wire such as tag wire. Inspect parts as follows:

1. Check linkage and pivot points for excessive wear.
2. Check bearing and thrust faces for wear and scoring.
3. Check mating surfaces of castings and end plates for burrs or irregularities that could cause poor seating and oil leaks. To remove burrs, lay a piece of crocus cloth on a very flat surface such as a piece of glass. Then lay the part on the crocus cloth, flat surface down. Move the part back and forth in a figure-8 pattern. This procedure is called *lapping* and will bring the metal surface to a smooth finish.
4. Check for damaged grooves or lands where O rings seat. Irregularities here can cause serious oil leaks.

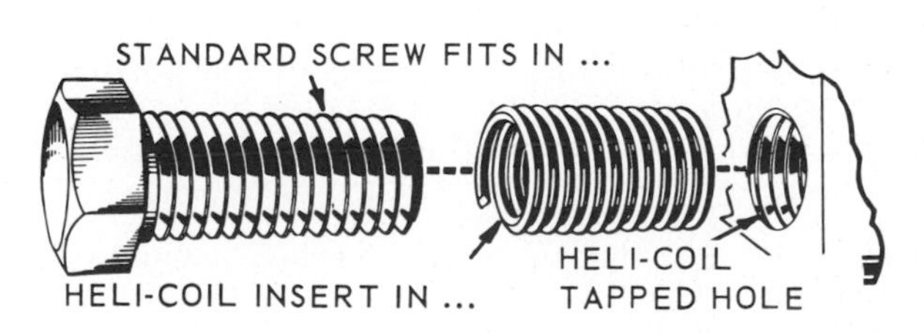

Fig. 10-10. Heli-Coil installation. (*Chrysler Corporation*)

5. Check castings for cracks and sand holes and for damaged threads. Thread repairs can often be made with Heli-Coils, as already noted and shown in Fig. 10-10.
6. Check gear teeth for chipping, scores, or wear.
7. Make sure valves are free of burrs and that the shoulders are square and sharp. Burrs can be removed from bores by honing. Valves must slide freely in the bores.
8. Check the facings on composition clutch plates for damaged surfaces and loose facings. If you can remove flakes of facing with your thumbnail or if the plates look scored, worn, or burned, replace the plates. Some discoloration is normal, however, and is not a sign of failure.
9. Inspect springs for distortion or collapsed coils.
10. Check bushings and test for wear by inserting the mating parts into the bushings and noting the amount of looseness. Worn, scored, or galled bushings should be replaced. You will need special tools to remove and replace bushings.
11. If the transmission shows evidence that foreign material has been circulating in the oil, all the old oil must be removed from the transmission, torque converter, and oil cooler. All parts must be cleaned and the oil cooler, oil-cooler lines, and torque converter must be flushed out.[2] Install a new oil filter on reassembly.

⊘ 10-12 Transmission Removal A typical procedure for removing a Powerglide transmission from a car follows. The procedure may vary somewhat from car to car because of individual differences in engine and accessory mounting arrangements.

1. Raise vehicle on hoist and remove drain plug to drain oil. Draining the oil can also be done after the transmission is removed.
2. Disconnect oil-cooler lines (where present), the vacuum-modulator line, and the speedometer drive cable, and tie the lines up out of the way.
3. Disconnect the manual and TV control-lever rods from the transmission.
4. Disconnect the drive shaft from the transmission.
5. Put the transmission jack in position under the transmission.
6. Disconnect the engine rear mount from the extension housing. Disconnect the transmission-support cross member and slide it rearward.
7. Remove the converter underpan. Mark the flywheel and converter so that they can be reattached in the same relationship. Remove the flywheel-to-converter attaching bolts.
8. Support the engine at the oil-pan rail with a jack or other suitable brace capable of supporting the engine when the transmission is removed.
9. Lower the rear of the transmission slightly so that the upper transmission-housing-to-engine attaching bolts can be reached. Remove upper bolts.

CAUTION: On V-8 engines, do not lower the engine too far or the distributor may be forced against the fire wall and damaged. Have an assistant topside to watch.

10. Remove the rest of the transmission-housing-to-engine attaching bolts.
11. Move the transmission to the rear and downward to remove it from the car. Mount the transmission on a repair fixture, as shown in Fig. 10-9.

NOTE: Watch the converter when moving the transmission rearward. If the converter does not move, gently pry it free of the flywheel.

CAUTION: Keep the front of the transmission up to prevent the converter from falling out. Install a C clamp or similar tool to prevent the converter from falling off the transmission while moving it from the car to the repair fixture.

⊘ 10-13 Transmission Disassembly Figure 10-11 is a disassembled view of the transmission case. Figure 10-12 shows the internal parts of the transmission. Follow these two illustrations when disassembling or reassembling the transmission.

1. With the transmission in a repair fixture, as shown in Fig. 10-9, drain the oil if it has not been drained. Remove the C clamp or tool and slide the converter off the shaft.

NOTE: If transmission overhaul is necessary because of clutch failure, then a new torque converter is also required. The old converter is probably loaded with debris from the failure that cannot be flushed out.

2. If the speedometer driven gear needs replacement, loosen the bolt and retainer and remove the gear.
3. Remove the extension housing and square-cut seal ring by removing five bolts.
4. If the rear oil seal needs replacement, pry it from the housing.
5. Depress the clip shown in Fig. 10-13 and slide the speedometer drive gear from the clip and output shaft.
6. On the weight side of the governor, remove the E clip from the governor shaft. Then remove the shaft and governor valve from the opposite side, as shown in Fig. 10-14. Remove the urethane washer from the governor body.
7. Loosen the drive screw from the governor hub and remove the governor assembly, as shown in Fig. 10-15.
8. Remove the four bolts that fasten the governor support to the transmission case, and then remove the support body, gasket, and extension seal ring.
9. Rotate the transmission so that the pump end is

[2] If clutch failure has occurred, the shop manual recommends replacement of the torque converter since the old converter probably contains so much debris that it cannot be flushed out.

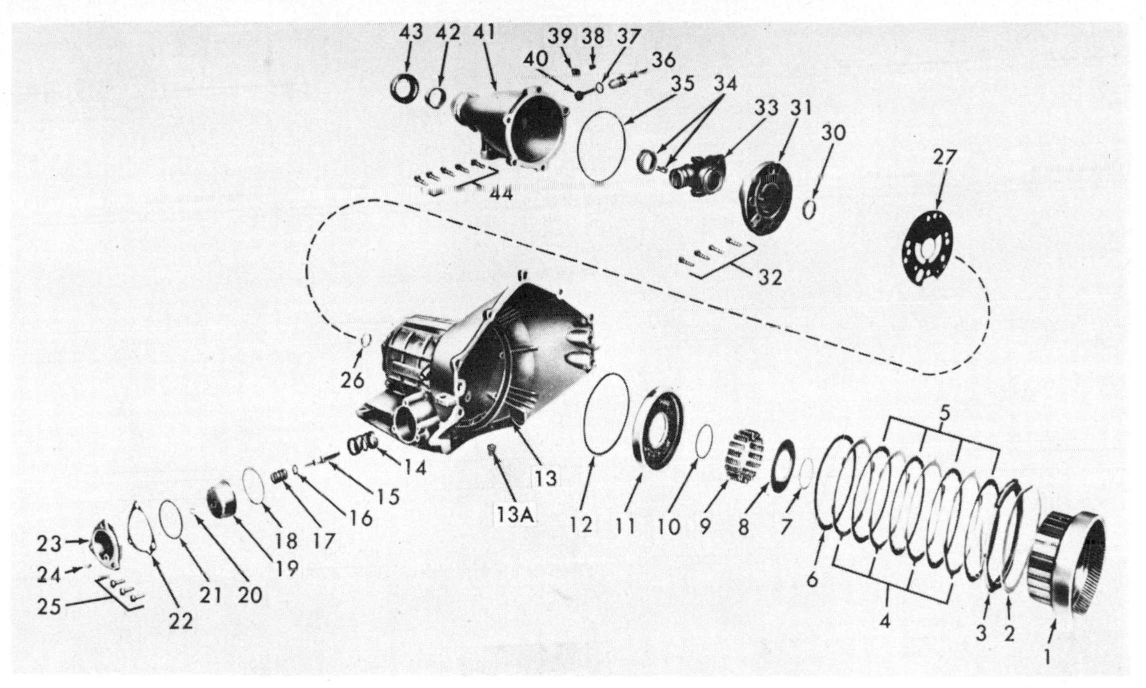

1. Reverse ring gear
2. Reverse-clutch-pack snap ring
3. Reverse-clutch pressure plate
4. Reverse-clutch reaction plates
5. Reverse-clutch drive plates
6. Reverse-clutch cushion spring
7. Reverse-clutch-piston-return-spring-retainer snap ring
8. Reverse-clutch-piston-return-spring retainer
9. Reverse-clutch-piston return springs
10. Reverse-clutch-piston inner seal
11. Reverse-clutch piston
12. Reverse-clutch-piston outer seal
13. Transmission case

13A. Transmission-case screen

14. Servo-piston return spring
15. Servo-piston rod
16. Servo-piston-apply-spring seat
17. Servo-piston apply spring
18. Servo-piston seal ring
19. Servo piston
20. Servo-piston-rod-spring retainer
21. Servo-cover seal
22. Servo-cover gasket
23. Servo cover
24. Servo-cover plug
25. Servo-cover bolts
26. Transmission-case bushing
27. Gasket
30. Governor-support bushing
31. Governor support
32. Governor-support-to-case attaching bolts
33. Governor assembly
34. Speedometer drive gear and clip
35. Seal
36. Speedometer-shaft fitting
37. Speedometer-shaft-fitting oil seal
38. Lock-plate attaching screw
39. Lock plate
40. Speedometer driven gear
41. Transmission extension
42. Extension bushing
43. Extension oil seal
44. Extension-to-case attaching screws

Fig. 10-11. Transmission case and related parts. (*Chevrolet Motor Division of General Motors Corporation*)

up. Remove the seven pump attaching bolts. Note that the bolt holes are offset so that the pump will mount only one way on reassembly.

10. Install slide hammers in the two threaded holes in the pump, as shown in Fig. 10-16, and remove the pump and stator-shaft assembly.

NOTE: The pump bolts have special sealing washers which must be in place when the pump is installed.

11. Remove the oil-pump gasket from the case (or from pump) and remove the tanged selective-fit thrust washer from the pump cover.

12. Back off the band-anchor adjusting screw to release tension on the low band. Then, with transmission horizontal, grasp the transmission input shaft and carefully work it and the clutch assembly out of the case, as shown in Fig. 10-17 (p. 136). Do not lose the low-sun-gear splined bushing from the end of the input shaft. The sun-gear thrust washer will probably remain in the planet carrier.

CAUTION: Do not damage the machined face of the clutch drum!

13. Remove the low brake band, apply and anchor struts, and adjusting screw.

14. Remove the planet carrier and output-shaft thrust caged bearing from the front of the transmission.

15. Remove the reverse ring gear if it did not come out with the planet carrier.

16. Use a large screwdriver to remove the reverse-clutch-pack snap ring. Then lift out the clutch plates and waved cushion spring.

17. Install the spring-compressor tool, as shown in Fig. 10-18 (p. 136), to compress the reverse-piston return springs and retainer. Remove the snap ring, take

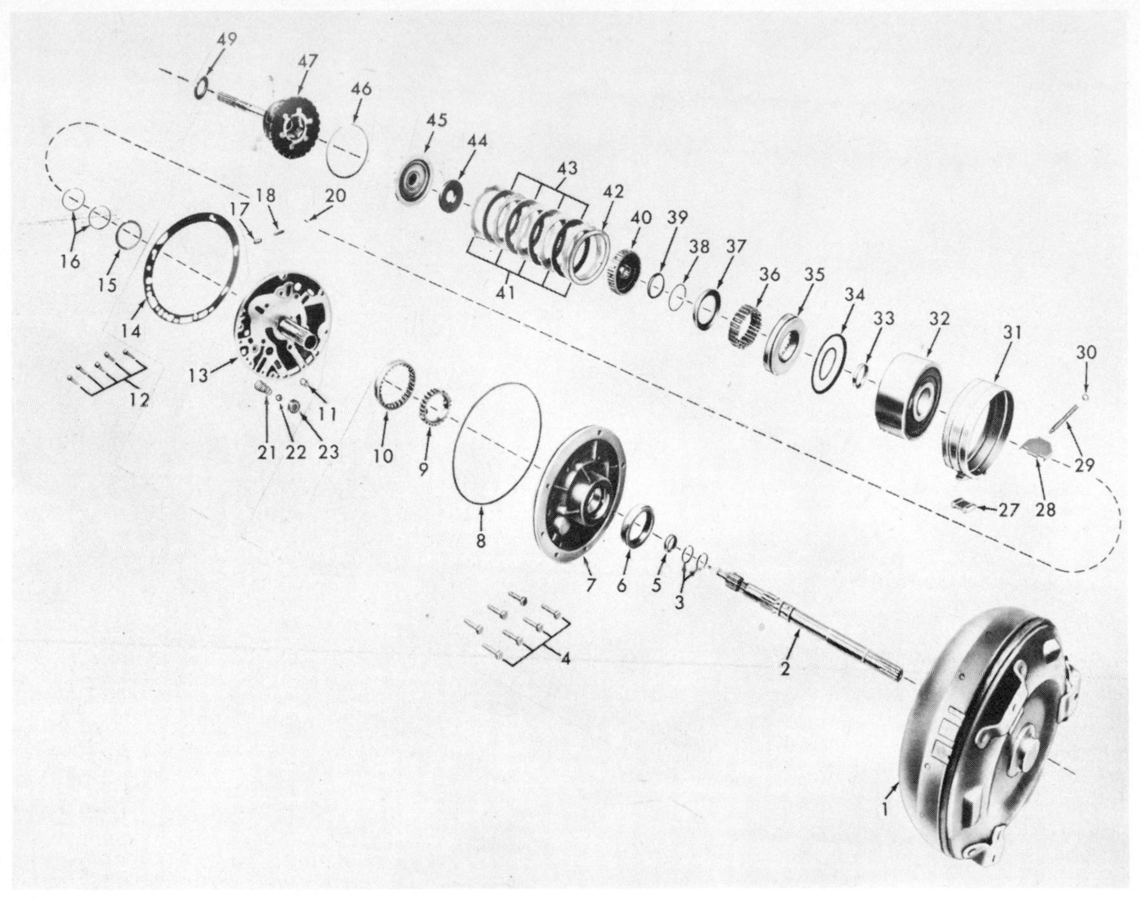

1. Converter assembly
2. Input shaft
3. Input-shaft oil seals
4. Oil-pump-to-case attaching bolts and sealing washers
5. Low-sun-gear bushing
6. Pump oil seal
7. Oil-pump body
8. Pump-to-case oil seal
9. Oil-pump drive gear
10. Oil-pump driven gear
11. Downshift timing valve
12. Oil-pump-cover-to-pump-body attaching screws
13. Oil-pump-cover-and-converter stator shaft
14. Oil-pump gasket
15. Clutch-drum thrust washer (selective fit)
16. High-clutch seal rings
17. Pump priming valve
18. Pump-priming-valve spring
20. Pump-priming-valve-spring retaining pin
21. Oil-cooler-bypass-valve spring
22. Oil-cooler bypass valve
23. Oil-cooler-bypass-valve seal
27. Band apply strut
28. Band anchor strut
29. Band anchor adjusting screw
30. Band-anchor-adjusting-screw nut
31. Low brake band
32. Clutch drum
33. Clutch-drum bushing
34. Clutch-piston outer and inner seals
35. Clutch piston
36. Clutch-return springs
37. Clutch-spring retainer
38. Clutch-spring retainer snap ring
39. Clutch-hub front thrust washer
40. Clutch hub
41. Clutch driven plates (flat)
42. Clutch cushion spring (waved)
43. Clutch drive plates (waved)
44. Clutch-hub rear thrust washer
45. Low-sun-gear-and-clutch-flange assembly
46. Clutch-flange retainer ring
47. Planet-carrier-and-output-shaft assembly
49. Output-shaft thrust bearing

Fig. 10-12. Internal mechanism of the transmission. (*Chevrolet Motor Division of General Motors Corporation*)

off the tool, and remove the piston-spring retainer and 17 springs.

18. To remove the reverse piston, apply compressed air to the reverse-piston port at the rear of the transmission, as shown in Fig. 10-19.

19. Remove inner and outer seals from the piston.

20. From the case, remove three servo cover-to-case attaching bolts, cover, seal, gasket, servo-piston assembly, and return spring.

21. Turn the transmission so that the oil pan is up and remove the oil-pan attaching screws, oil pan, and gasket. Remove and discard the old oil screen.

22. Remove the vacuum modulator and gasket and the vacuum-modulator plunger, dampening spring, and valve, as shown in Fig. 10-20.

23. Remove the detent guide plate from the valve body by removing two attaching bolts. Remove the range-selector detent-roller spring.

24. Note the locations of the remaining valve-body attaching bolts, as shown in Fig. 10-21. Remove these

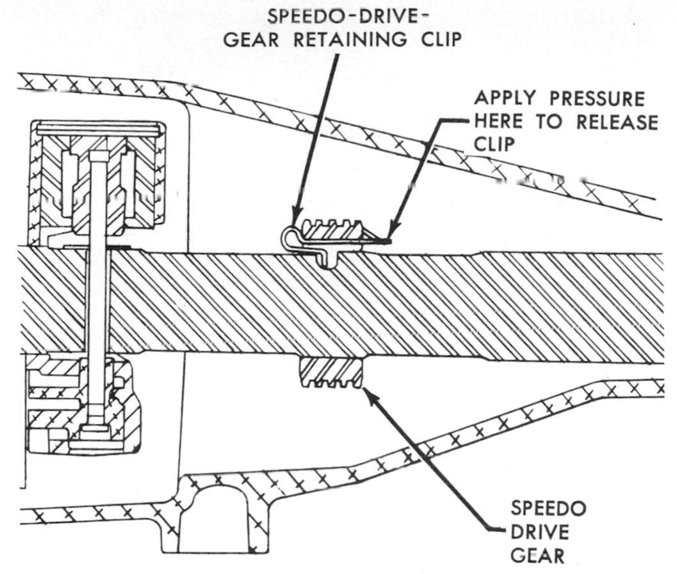

Fig. 10-13. How to remove the speedometer drive gear. (*Chevrolet Motor Division of General Motors Corporation*)

bolts and lift off the valve body and gasket, disengaging the servo-apply tube from the transmission case as the valve body is lifted off.

25. If necessary, the manual levers may be removed. See Fig. 10-22 for location of parts.

26. Inspect and overhaul subassemblies, as explained in following sections.

⊘ 10-14 Converter Check The converter is a welded assembly and must be replaced as a unit if

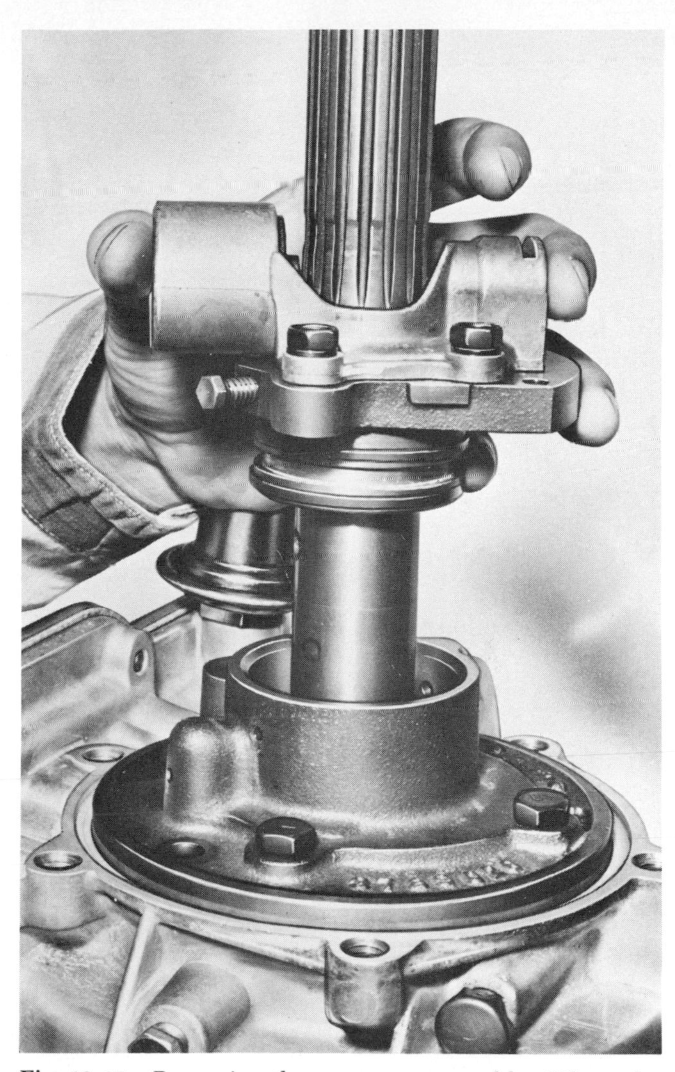

Fig. 10-15. Removing the governor assembly. (*Chevrolet Motor Division of General Motors Corporation*)

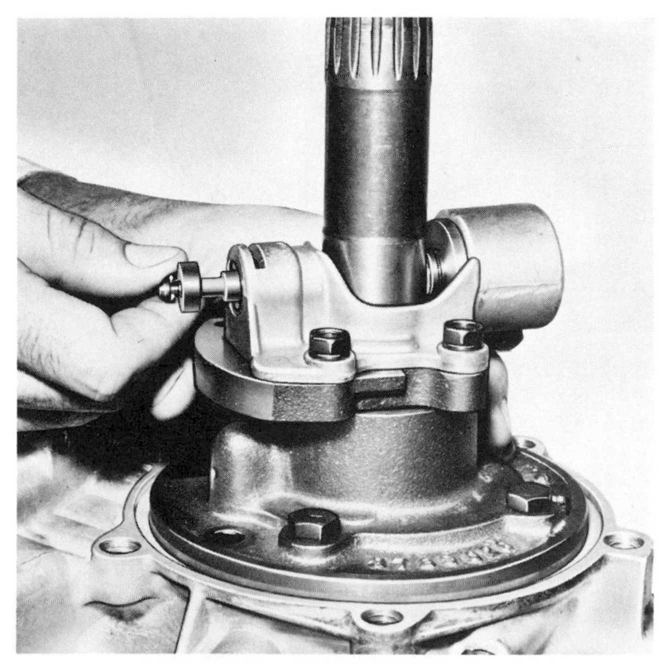

Fig. 10-14. Removing the governor valve and shaft. (*Chevrolet Motor Division of General Motors Corporation*)

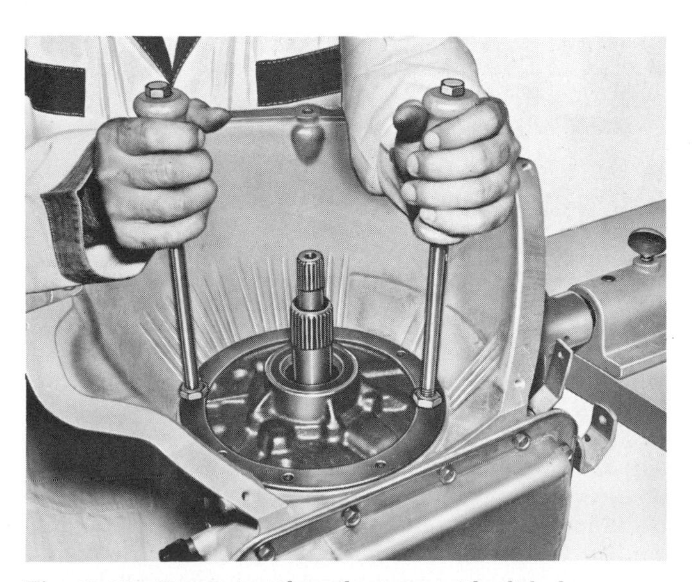

Fig. 10-16. Removing the oil pump with slide hammers. (*Chevrolet Motor Division of General Motors Corporation*)

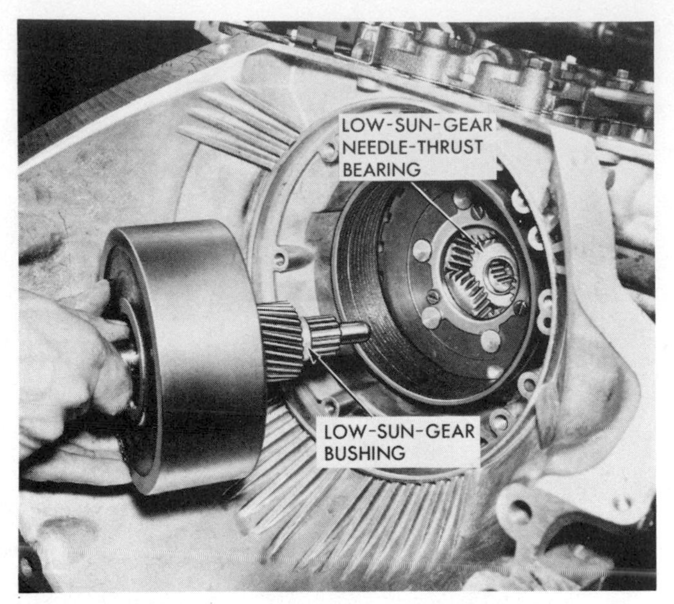

Fig. 10-17. Removing the clutch assembly and input shaft. (*Chevrolet Motor Division of General Motors Corporation*)

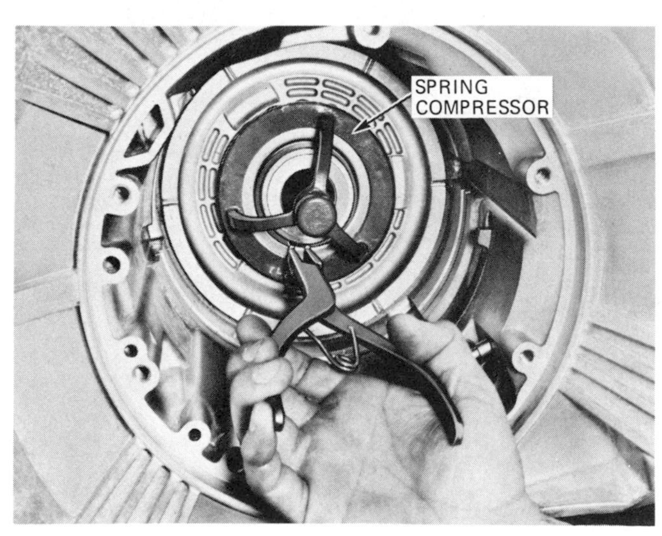

Fig. 10-18. Removing the reverse-piston-spring-retainer snap ring. (*Chevrolet Motor Division of General Motors Corporation*)

it has any defects. The converter can be tested for leaks by sealing it with a special test fixture, filling it with air at 30 psi [2.11 kg/cm^2], and submerging it in water. If it leaks, discard it.

⊘ 10-15 Vacuum-Modulator Assembly The vacuum modulator cannot be repaired. If it has any defects, it must be discarded and a new assembly installed. The vacuum-modulator diaphragm can be checked with a vacuum source to see if it leaks and if the assembly works. If the diaphragm has been leaking, transmission oil will be pulled through it and into the engine intake manifold. Such a condition will cause smoky engine exhaust and loss of transmission oil.

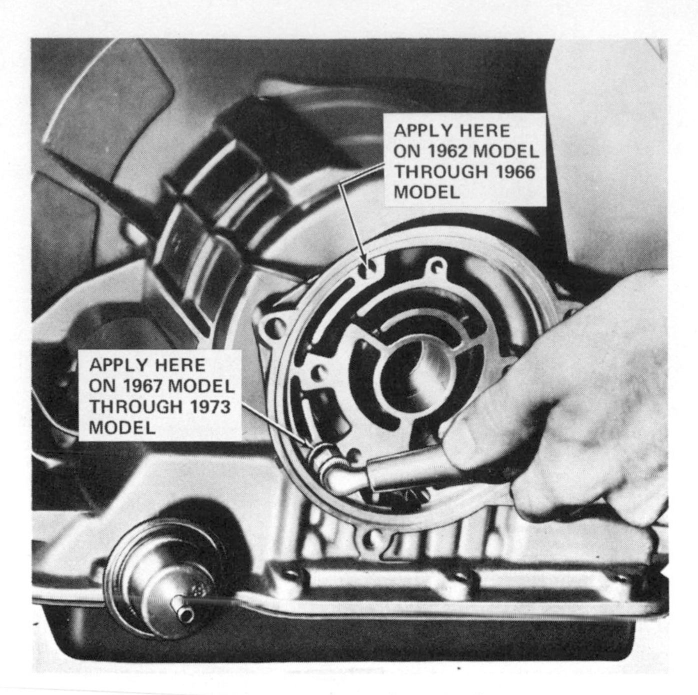

Fig. 10-19. Applying compressed air to the reverse-piston port to remove the piston. (*Chevrolet Motor Division of General Motors Corporation*)

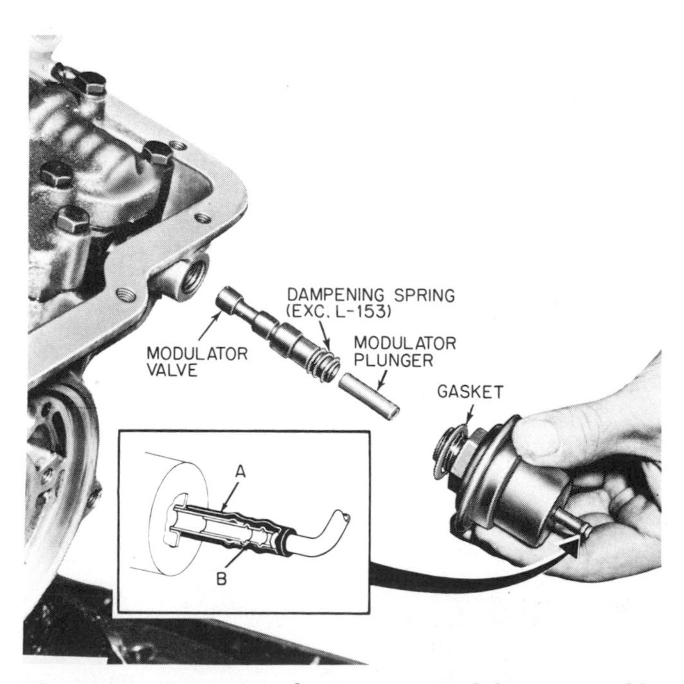

Fig. 10-20. Removing the vacuum-modulator assembly and related parts. (*Chevrolet Motor Division of General Motors Corporation*)

⊘ 10-16 Extension-Housing Assembly The rear oil seal should be removed and a new oil seal installed if it is damaged in any way. If the rear bushing is worn or otherwise defective, use special tools to remove it and install a new bushing.

⊘ 10-17 Governor Assembly Figure 10-23 is a disassembled view of the governor and its related parts.

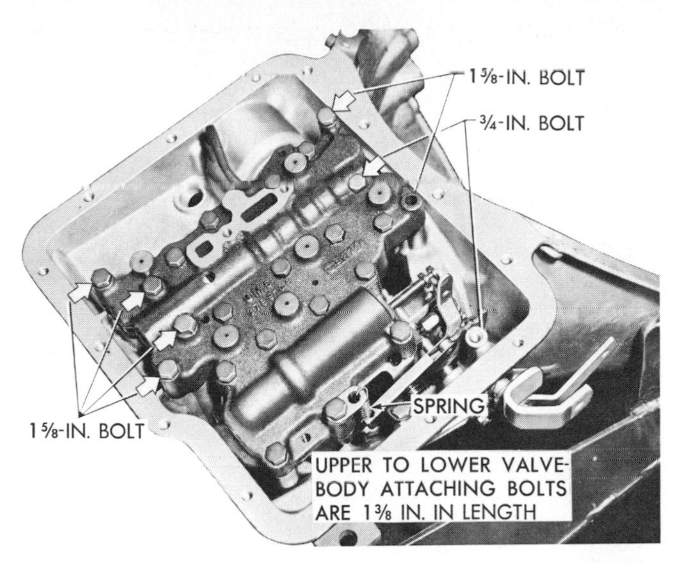

Fig. 10-21. Locations of valve-body attaching bolts. (*Chevrolet Motor Division of General Motors Corporation*)

The valve and shaft were removed during transmission disassembly. To complete the disassembly, remove the snap rings. If it is necessary to separate the body from the hub, remove the four bolts.

NOTE: The governor is a factory-balanced unit. If body replacement is necessary, the body-and-hub assembly must be replaced as a single unit.

Clean parts and check for wear, nicks, or other damage. To reassemble the governor, install the inner weight and spring in outer weight, retain it with a snap ring, and install the assembly in the body, retaining it with the large snap ring.

⊘ 10-18 Valve Body Figure 10-24 is a disassembled view of the valve body. The upper and lower valve bodies are attached by bolts which are removed to separate the bodies, gaskets, and transfer plate. Most of the valves and springs are held in place by snap rings which can be removed to remove the valves. To remove the throttle valve and detent valve, first take out six bolts and remove the upper-valve-body plate. Then remove the retaining pin by wedging a thin screwdriver between its head and the valve body. Now, the throttle valve will fall out when the valve body is tilted. If necessary, disassemble the detent-valve assembly by removing the E clip.

CAUTION: Do not disturb the setting of the adjustment jam nut at the end of the throttle-valve train. This nut is factory preset and normally does not require adjustment. However, if adjustment is needed, refer to ⊘ 10-25.

Clean and check all parts. Look for burrs, cracks, dents, or other damage that could cause valve malfunction. Valves should slide easily in their bores. Check springs for distortion or collapsed coils. To reassemble the valve body, refer to Fig. 10-24 for relationship of parts.

CAUTION: Do not nick, scratch, or otherwise damage the valves or valve body. Even slight nicks can cause a valve to hang up.

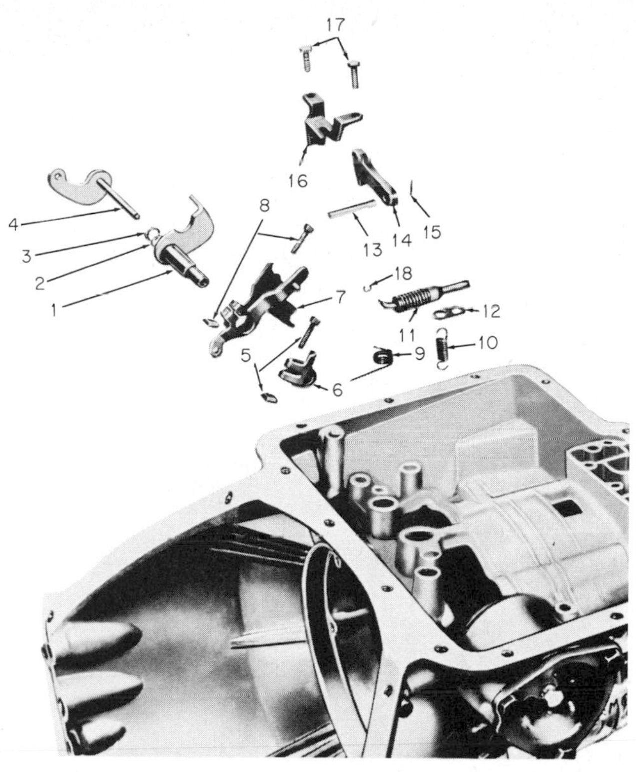

1. Park-lock-and-range-selector outer lever and shaft
2. Throttle-valve-control-shaft oil seal
3. Throttle-valve-control-shaft washer
4. Throttle-valve-control lever and shaft
5. Throttle-valve-control-inner-lever-to-control-shaft attaching screw and nut
6. Throttle-valve control inner lever
7. Park-lock-and-range-selector inner lever
8. Park-lock-and-range-selector-inner-lever attaching screw and nut
9. Park-lock-pawl disengaging spring
10. Range-selector detent roller spring
11. Park-lock actuator assembly
12. Range-selector-detent-roller-spring retainer
13. Park-lock-pawl shaft
14. Park-lock pawl
15. Park-lock-pawl-shaft retaining ring
16. Park-lock-pawl reaction bracket
17. Park-lock-pawl-reaction-bracket attaching bolts
18. Park-lock-actuator-to-park-lock-and-range-selector-inner-lever retaining clip

Fig. 10-22. Transmission levers and related parts. (*Chevrolet Motor Division of General Motors Corporation*)

⊘ 10-19 Oil-Pump Assembly Numbers 6 to 23 in Fig. 10-12 show the various pump parts. To disassemble the pump, remove the five pump-cover-

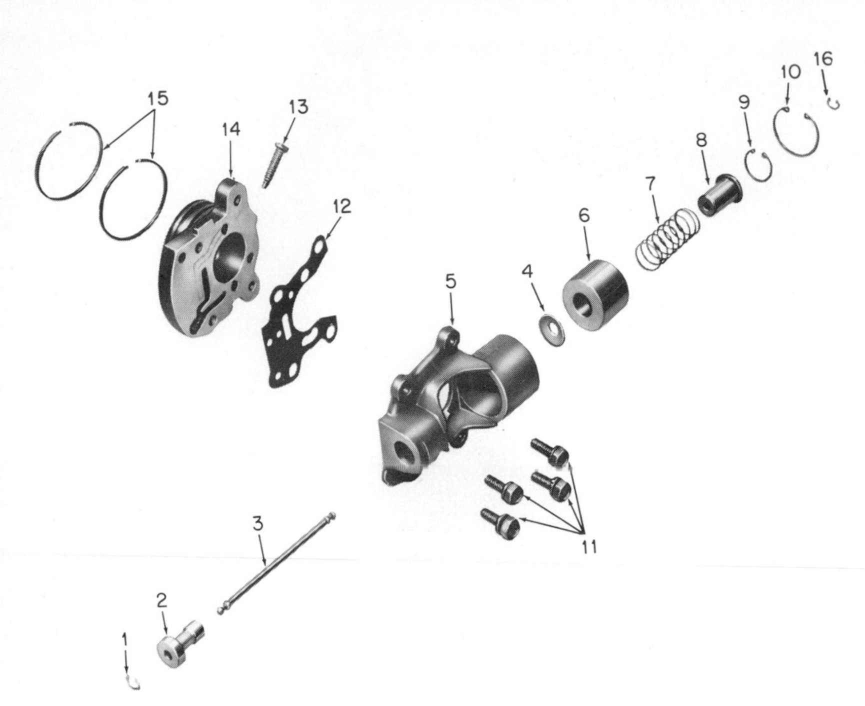

1. Valve-to-shaft retaining snap ring
2. Valve
3. Shaft
4. Urethane washer
5. Body
6. Outer weight
7. Spring
8. Inner weight
9. Inner-weight-to-outer-weight retaining snap ring
10. Outer-weight-to-body retaining snap ring
11. Body-to-hub screws and lock washers
12. Gasket
13. Hub-drive screw
14. Hub
15. Hub oil-seal ring
16. Inner-weight-to-shaft retaining snap ring

Fig. 10-23. Disassembled view of the governor and related parts. (*Chevrolet Motor Division of General Motors Corporation*)

to-body bolts. Separate the cover from the body and take out the gears.

CAUTION: Handle gears carefully; they are not heat-treated.

If the pump cover contains valves, they may be removed as required. Inspect all parts after cleaning them in solvent. Check the body bushing for galling or wear. Clearance between the body bushing and converter-pump hub should be no greater than 0.005 inch [0.13 mm]. Check as shown in Fig. 10-25. If the bushing is worn, replace the pump body.

With the parts cleaned, put the gears into the pump body and make these three checks:

1. Check the clearance between the outside diameter of the driven gear and the pump body, as shown in Fig. 10-26. The clearance should be between 0.0035 and 0.0065 inch [0.093 and 0.163 mm].
2. Check the clearance between the inside diameter of the driven gear and the pump crescent, as shown in Fig. 10-27. The clearance should be between 0.003 and 0.009 inch [0.08 and 0.23 mm].
3. Check the gear end clearance between a straightedge held across the face of the pump body and the gears, as shown in Fig. 10-28. The clearance should be between 0.0005 and 0.0015 inch [0.013 and 0.043 mm]. Excessive clearance requires replacement of the gears and pump body.

To reassemble the pump, lubricate parts generously with transmission fluid and install the gears, with recessed side of the drive-gear lugs down or facing the converter. Set the pump cover in place and loosely install two attaching bolts. Put the pump assembly minus the rubber oil-seal ring into the transmission case upside down. Install the other three bolts and tighten all five bolts to 20 pound-feet [2.764 kg-m].

Remove the pump assembly from the case, install the rubber oil-seal ring in its groove in the pump body, and install two high-clutch oil-seal rings onto the hub of the pump cover.

⊘ 10-20 Clutch-Drum Assembly Numbers 32 to 46 in Fig. 10-12 show the parts in the clutch-drum assembly. Start the disassembly by removing the clutch-flange retainer ring from the clutch drum. Then proceed as follows:

CAUTION: Do not damage in any way the machined surfaces on the clutch drum. Do not allow tools or other parts to scratch or nick these surfaces.

1. Remove the low-sun-gear-and-clutch-flange assembly from the clutch drum. Remove the hub rear thrust washer.
2. Lift out the clutch hub, clutch pack, and hub front thrust washer.
3. Use a spring compressor, as shown in Fig. 10-29, to compress the springs so that the snap ring can be removed. Then remove the retainer and springs.

NOTE: Use a piece of clean cardboard, as shown in Fig. 10-29, to protect the machined face of the clutch drum.

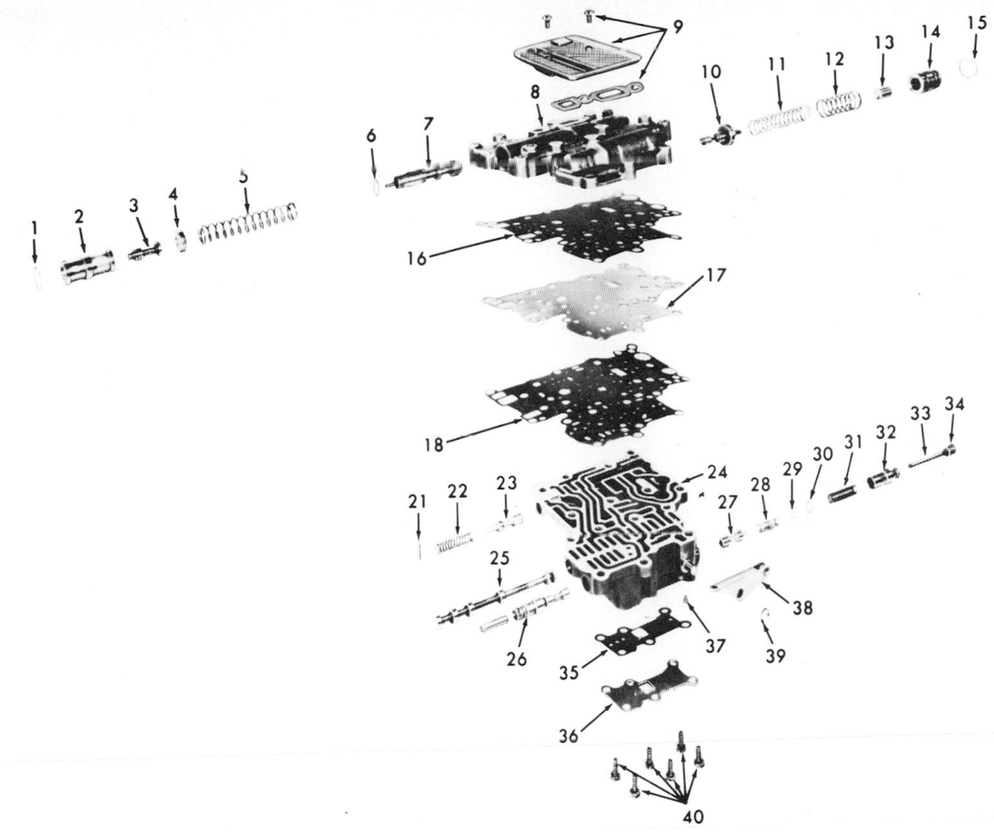

1. Snap ring
2. Hydraulic-modulator-valve sleeve
3. Hydraulic modulator valve
4. Pressure-regulator-spring retainer
5. Pressure-regulator spring
6. Pressure-regulator-spring seat
7. Pressure-regulator valve
8. Lower valve body
9. Suction screen, gasket and attaching screws
10. Low-and-drive valve
11. Low-and-drive-valve inner spring
12. Low-and-drive-valve outer spring
13. Low-and-drive regulator valve
14. Low-and-drive-regulator-valve sleeve and cap
15. Snap ring
16. Transfer-plate-to-lower-valve-body gasket
17. Transfer plate
18. Transfer-plate to-upper-valve-body gasket
21. High-speed-downshift-timing-valve stop pin
22. High-speed-downshift-timing-valve spring
23. High-speed-downshift timing valve
24. Upper valve body
25. Manual control valve
26. Vacuum-modulator valve. Plunger and spring (exc. L4)
27. Throttle valve
28. Throttle-valve spring
29. Throttle-valve-spring seal
30. Throttle-valve-spring-regulator guide washer
31. Detent-valve spring
32. Detent valve
33. Throttle-valve spring regulator
34. Throttle-valve-spring-regulator nut
35. Upper-valve-body plate gasket
36. Upper-valve-body plate
37. Detent-valve-and-spring retaining stud
38. Range-selector detent lever
39. Snap ring
40. Upper-valve-body-plate-to-upper-valve-body attaching bolts and washer

Fig. 10-24. Exploded view of the valve body. (*Chevrolet Motor Division of General Motors Corporation*)

4. Lift up on the piston with a twisting motion to remove it from the drum. Remove the inner and outer seals from the piston.

5. Inspect as follows after washing all parts in cleaning solvent and air-drying them.

a. Check the steel ball in the clutch drum. It must be free to move and the orifice must be open. If the ball is loose enough to fall out or does not rattle, replace the clutch drum. Do not attempt to replace or stake the ball.

b. If the drum bushing is worn, use a special tool to press it out and press in a new bushing. Press in only until the face of the tool just touches the front face of the drum.

c. Check the clutch plates for burning, wear, pitting, or metal pickup. Plates should be a free fit in the splines or slots.

6. As a first step in reassembly, lubricate new inner and outer piston seals with transmission fluid and install them both with the seal lips down or toward the front of the transmission.

7. Install the clutch piston into the drum with a twisting motion and push it all the way in. Use a wire loop, if you wish, to help get the seal started in the drum.

8. Put springs into place; lay the spring retainer and snap ring on top of them. Use the spring compressor, as shown in Fig. 10-29, to install the snap ring.

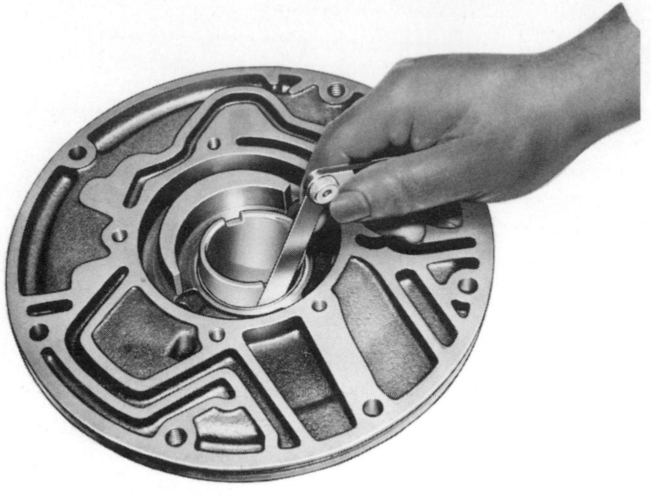

Fig. 10-25. Checking clearance between the body bushing and the converter-pump hub. (*Chevrolet Motor Division of General Motors Corporation*)

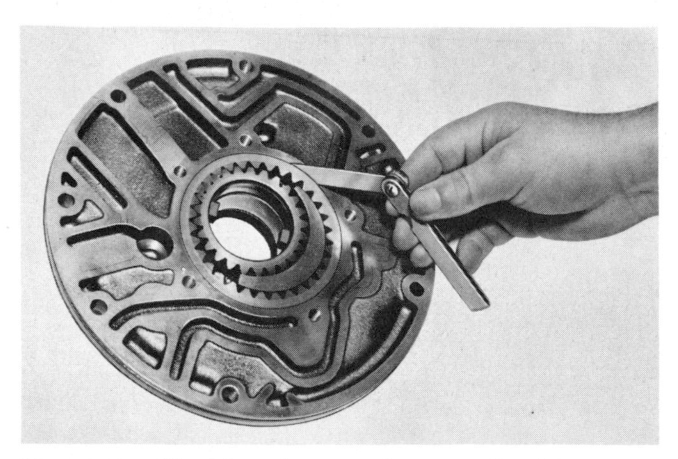

Fig. 10-26. Checking clearance between the driven gear and the pump body. (*Chevrolet Motor Division of General Motors Corporation*)

Fig. 10-27. Checking clearance between the driven gear and the crescent. (*Chevrolet Motor Division of General Motors Corporation*)

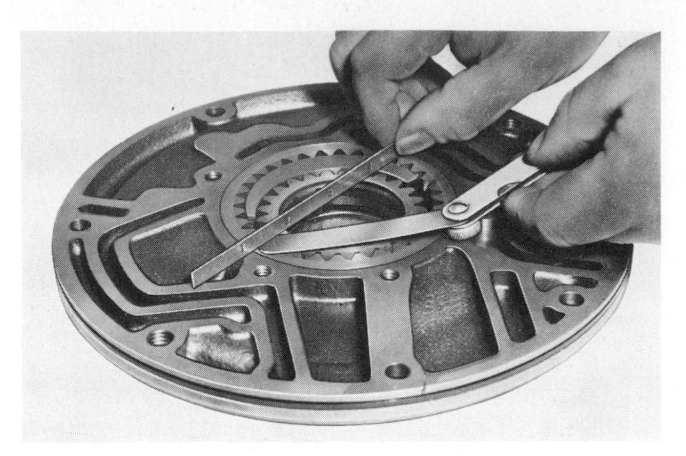

Fig. 10-28. Checking gear end clearance. (*Chevrolet Motor Division of General Motors Corporation*)

9. Install the clutch hub front thrust washer with its lip toward the clutch drum, and put the clutch hub into the clutch drum.

10. If the waved clutch-cushion spring is used, install it.

11. On some V-8 engines, the first steel driven plate to the rear of the cushion spring is a selective-fit driven plate. Refer to the applicable shop manual for specifications. A typical procedure to determine the proper thickness of the plate follows:

a. Stack the clutch pack, omitting the selective-fit plate, and measure its height.

b. If the stack height is 0.872 to 0.903 inch [22.15 to 22.93 mm], install a selective-fit driven plate that is 0.060 inch [1.52 mm] thick.

c. If the stack height is 0.798 to 0.872 inch [20.27 to 22.15 mm], install a selective-fit driven plate that is 0.090 inch [2.29 mm] thick.

12. Then install the plates, alternating drive and driven plates.

13. Install the clutch-hub rear thrust washer into the

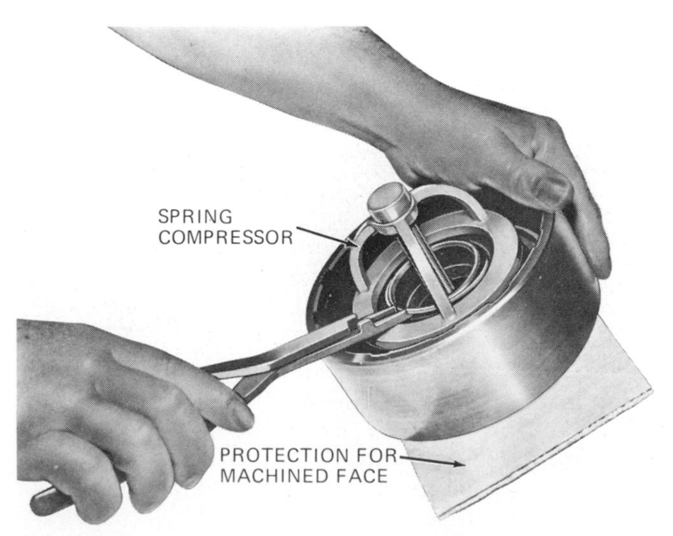

Fig. 10-29. Removing the clutch-spring-retainer snap ring. (*Chevrolet Motor Division of General Motors Corporation*)

low sun gear (secure it with petroleum jelly) and put the low-sun-gear-and-clutch-flange assembly into the clutch drum. Install the snap ring.
14. Check to see if the clutch hub will turn. It should turn freely.

⊘ 10-21 Planet-Carrier Assembly Two different planet-carrier assemblies are used in Powerglides: large and small. In the large planet-carrier assembly, the ends of the pinion shafts are flared at each end for retention in the carrier. This unit cannot be disassembled. If any defect is apparent, the complete assembly must be replaced. Checks should be made of the gear teeth, bearings, and bushings. Also, check the end clearance of the planet gears, as shown in Fig. 10-30. Replace the complete assembly if there is excessive wear or damage to any part.

To disassemble and repair the small planet-carrier assembly, refer to the appropriate shop manual.

⊘ 10-22 Transmission Case Wash the transmission case thoroughly and air-dry it. Make sure all oil passages are clear. Blow them out with compressed air. Check the case for cracks and stripped threads. If the shifter-shaft oil seal is leaking or damaged, replace it. If the rear bushing is worn, it should be replaced with a special driver. Be very careful in driving the new bushing into place because excessive force could crack the aluminum casting.

Powerglide Transmission Reassembly

Use transmission fluid or petroleum jelly to retain washers and other parts during transmission reassembly. Lubricate all bearings, ring seals, and clutch plates prior to reassembly.

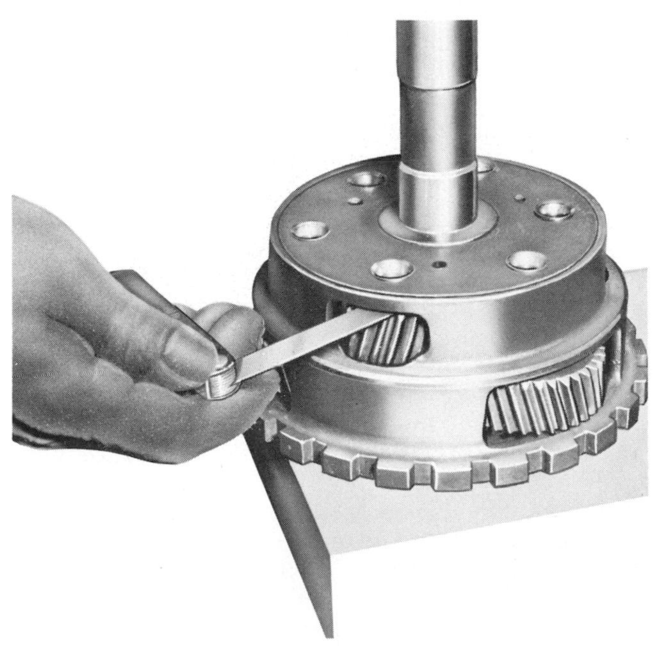

Fig. 10-30. Checking the planet-gear end clearance. *(Chevrolet Motor Division of General Motors Corporation)*

⊘ 10-23 Transmission Reassembly Procedure To reassemble the transmission, proceed as follows:

1. If the manual linkage was removed, refer to Fig. 10-22 for the relationship of parts when reinstalling it.
2. Thread the low-band adjusting screw partly into the case.
3. Lubricate and install the inner and outer seals on the reverse piston. The lips should face to the rear of the case. Install the piston with a twisting motion. Use a wire loop or feeler gauge to start the seal into the case.
4. Install 17 springs, retainer, and snap ring. Use a compressor tool, as shown in Fig. 10-18, so that the snap ring can be installed. Remove the tools.

CAUTION: Be careful that the spring retainer does not catch on the edge of the internal hub or in the snap-ring groove in the case.

5. Install the large waved cushion spring and clutch pack, starting with a reaction plate and alternating with the drive (faced) plates. The notched lug of each reaction plate must be installed in the 7 o'clock groove in the case, as shown in Fig. 10-31. Then install the pressure plate, with the lug having the dimple in the same groove, as shown.
6. Install the clutch-plate retainer ring.
7. With the rear of the transmission down, align the internal lands and grooves of the reverse-clutch faced plates. Then engage the ring gear with these plates. This engagement must be made by feel while you jiggle and turn the ring gear.
8. Put the output-shaft thrust bearing over the out-

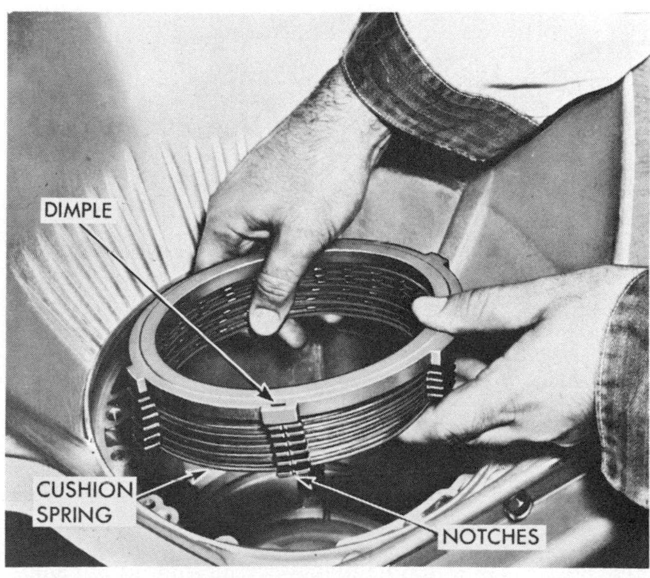

Fig. 10-31. Installing clutch plates. *(Chevrolet Motor Division of General Motors Corporation)*

put shaft and install the planetary carrier and output shaft in the transmission case, as shown in Fig. 10-32.

9. Turn the transmission horizontal. Make sure the two input-shaft seal rings are in place on the shaft. Install the clutch drum, machined face first, onto the input shaft, and install the low-sun-gear splined bushing against the shoulder of the input shaft.

10. Put the clutch-drum-and-input-shaft assembly into the case, aligning the thrust-needle bearing on the input shaft and indexing the low sun gear with the short pinions on the planet carrier.

11. To check end play, temporarily remove the rubber oil-seal ring from the pump body and install the clutch-drum selective thrust washer, gasket, and pump in the case. Install and tighten two pump-to-case bolts. Install a dial indicator, as shown in Fig. 10-33. With the plunger of the dial indicator resting on the end of the input shaft, push up on the output shaft and note the indicator-needle movement. The end play, as shown on the indicator, should be between 0.028 and 0.059 inch [0.71 and 1.50 mm]. If the end play is not correct, select a thinner or thicker selective thrust washer and repeat the check. Washers of these thicknesses are available: 0.061, 0.078, 0.092, and 0.106 inch [1.55, 1.98, 2.34, and 2.69 mm]. Remove the pump and washer from the case.

12. Install the servo piston, piston ring, and spring in the servo bore. Use a new gasket and O ring, and install the servo cover.

13. With the transmission horizontal, install the low brake band, anchor, and apply strut between the servo piston rod and band. Seat the servo piston in the case. If necessary, turn the low-band adjusting screw in far enough to keep struts from falling out of position.

Fig. 10-32. Installing the planet carrier and output shaft. (*Chevrolet Motor Division of General Motors Corporation*)

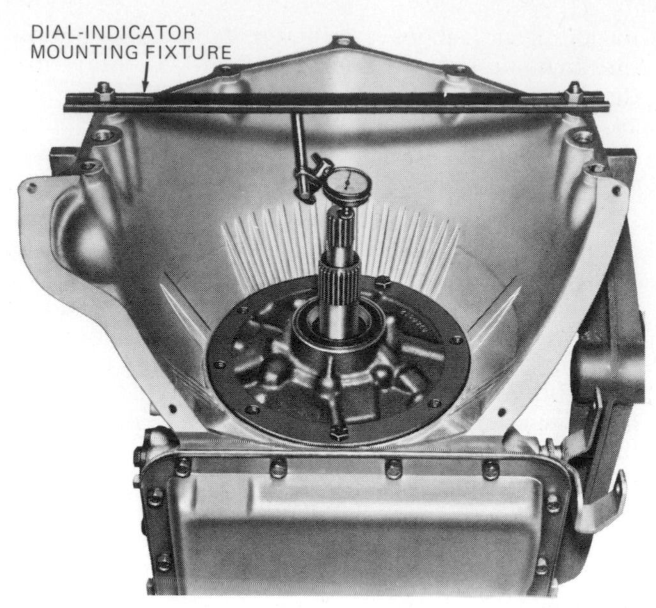

Fig. 10-33. Using a dial indicator to check transmission end play. (*Chevrolet Motor Division of General Motors Corporation*)

14. Install the servo cover and gasket with three bolts torque to 20 pound-feet [2.764 kg-m]. Make sure the gasket is properly aligned with the three bolt holes and one drain-back passage in the case.

15. Install the oil-seal ring in the groove around the pump body, put the selective thrust washer in place, and install the oil-pump gasket and oil pump. Secure with seven attaching bolts torqued to 15 pound-feet [2.073 kg-m]. Make sure the bolt sealing washers are in good condition. Replace any that are not.

16. Install the governor-and-hub assembly, as shown in Fig. 10-15, after the governor support and gasket have been attached, as shown. Then, align the hole in the governor hub with the hole in the output shaft and install the drive screw, tightening it to 8 pound-feet [1.106 kg-m]. Put the urethane washer in the governor between the outer weight and output shaft, and install the governor shaft and valve, as shown in Fig. 10-15. Secure with an E clip.

17. Install the speedometer drive gear, as shown in Fig. 10-13. Put a retaining clip on the shaft, with tang in hole, depress the clip, and slide the speedometer gear into place.

18. Install the valve body, using a new gasket and guiding the servo-apply line into its hole in the case. Make sure the range-selector inner lever properly picks up the manual control lever. Secure with valve-body-to-case bolts.

19. Install the new screen and gasket. Install the guide plate, making sure the inner lever properly picks up the manual valve. Install attaching bolts.

20. Install the vacuum-modulator valve and the vacuum modulator and gasket.

21. Install the oil pan, using a new gasket.

22. Install the converter, rotating the converter so that it engages in the drive lugs of the oil-pump drive

gear. Install a C clamp or similar tool to hold the converter in place.
23. Adjust the low band and, if necessary, the throttle valve, as explained in following sections. Then install the transmission on the car.

⊘ 10-24 Low-Band Adjustment Tighten the low-servo adjusting screw to 70 pound-inches [1.26 kg-mm], as shown in Fig. 10-34. Rotate the input and output shafts while the screw is being tightened to make sure the band is centered. Back off three complete turns for a band that has been used less than 6,000 miles [9,656 km] or four turns for a band that has been used more than 6,000 miles [9,656 km].

CAUTION: The amount of backoff is not approximate; it must be exactly the number of turns stated!

⊘ 10-25 Throttle-Valve Adjustment There is no way to check throttle-valve pressures. If the operation of the transmission indicates that the pressures are not right, adjustment can be made by adjusting the position of the jam nut on the throttle-valve assembly, as shown in Fig. 10-35. To raise the pressure 3 psi [0.210 kg/cm^2], back off the jam nut one full turn. Tightening the jam nut one full turn will lower the throttle-valve pressure 3 psi [0.210 kg/cm^2]. A difference of 3 psi [0.210 kg/cm^2] will cause a change of about 2 to 3 mph [3.219 to 4.828 km/h] in the wide-open-throttle upshift point.

NOTE: Use care when making this adjustment because there is no way to check actual throttle-valve pressures.

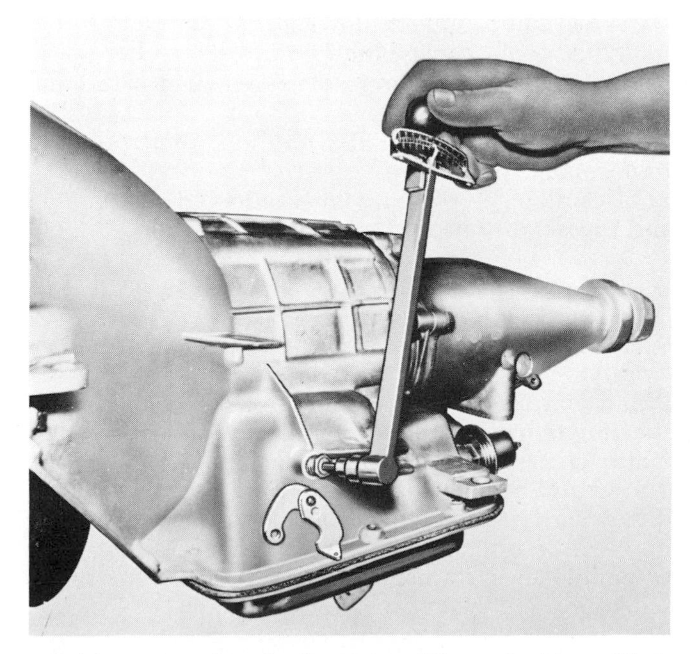
Fig. 10-34. Low-band adjustment. (*Chevrolet Motor Division of General Motors Corporation*)

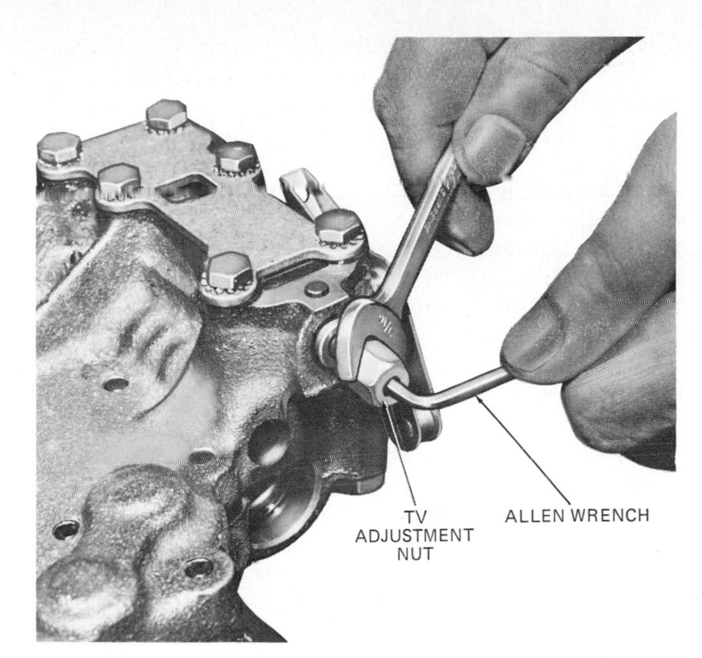

Fig. 10-35. Adjusting throttle-valve pressure. (*Chrysler Corporation*)

⊘ 10-26 Installing Transmission in Car Mount the transmission on the transmission jack and remove the converter holding tool.

CAUTION: Do not allow the converter to move forward after the tool is removed.

1. Raise the transmission into place at the rear of the engine and install the transmission-case-to-engine upper mounting bolts. Then install the rest of the mounting bolts. Torque to 35 pound-feet [4.837 kg-m].
2. Remove the support from under the engine and raise the rear of the transmission up to the final position.
3. Align the scribe marks on the flywheel and converter cover. Install attaching bolts and torque to 30 to 35 pound-feet [4.146 to 4.837 kg-m].
4. Install the converter underpan.
5. Reinstall the transmission-support cross member to the transmission and frame.
6. Remove the transmission jack.
7. Connect the drive shaft.
8. Connect the manual and throttle-valve control-lever rods to the transmission.
9. Connect oil-cooler lines (if used), vacuum-modulator line, and speedometer cable to the transmission.
10. Refill the transmission through the filler tube, using the specified transmission fluid and following the procedure outlined earlier in the chapter.
11. Check the transmission for proper operation and leakage. If necessary, adjust linkages.
12. Lower the vehicle and remove it from the hoist.

CHAPTER 10 CHECKUP

NOTE: Since the following is a chapter review test, you should review the chapter before taking it.

The service procedures on the Powerglide, as outlined in this chapter, also apply, in a general way, to other automatic transmissions. Thus, a careful study of this chapter will help you lay the foundation for the service work on automatic transmissions you will do in the shop. It must be remembered, however, that each transmission model is unique and requires a specialized service procedure all its own. However, the general comments about cleanliness, thread repair, and cleaning and inspecting transmission parts will apply to all transmissions. In the checkup that follows, Powerglide trouble diagnosis and service are highlighted. If you are not sure about the correct answers, reread the pages in the chapter that will give you the information you need. Write the answers in your notebook. This will help you remember them. Also your notebook will become a valuable source of information you can refer to quickly in the future.

Completing the Sentences The sentences that follow are incomplete. After each sentence there are several words or phrases, only one of which will correctly complete the sentence. Write each sentence in your notebook, selecting the proper word or phrase to complete it correctly.

1. After the transmission is removed from the car, the converter may be removed by: (*a*) taking out four bolts, (*b*) slipping it off, (*c*) prying it loose.
2. In diagnosing troubles, two checks to be made are: (*a*) shift points and pressure, (*b*) throttle valve and pressure, (*c*) vacuum and shift points.
3. The most important precaution to follow in automatic-transmission work is: (*a*) replace all springs and seals, (*b*) install all new clutch plates, (*c*) keep everything absolutely clean.
4. To remove the reverse piston from the case, use: (*a*) a spring compressor, (*b*) a puller, (*c*) compressed air.
5. Transmission oil will be burned in the engine if the: (*a*) transmission is overfilled, (*b*) extension-housing seals leak, (*c*) vacuum-modulator diaphragm leaks.
6. In reassembling the pump, the following clearances should be checked: between the outside diameter of the driven gear and the pump body, between the inside diameter of the driven gear and the pump crescent, and: (*a*) between the inside and outside gears, (*b*) gear end clearance, (*c*) between the pump cover and the pump body.
7. If the steel ball in the clutch drum is loose enough to fall out: (*a*) install a new ball, (*b*) stake the ball in place, (*c*) get a new clutch drum.
8. During reassembly, one of the first things to be reinstalled in the case is the: (*a*) clutch drum, (*b*) brake band, (*c*) reverse-clutch piston.

Troubles and Service Procedures In the following, write the required trouble causes and service procedures in your notebook. Do not copy from the book; try to write them in your own words. Writing the trouble causes and service procedures in your own words will help you remember them. You will thereby be greatly benefited when you go into the automotive shop.

1. Describe the procedure of adding oil; of changing oil.
2. Explain how to adjust the low band.
3. Explain a typical shift-control-linkage adjustment.
4. Describe a typical throttle-valve-linkage adjustment.
5. What towing precautions should be observed?
6. Explain how to make the pressure checks of the Powerglide.
7. What are the possible causes of no drive in any selector position?
8. What are the possible causes of engine flare on standstill starts combined with lagging acceleration?
9. What could cause the engine to flare on upshifts?
10. What could prevent upshifting?
11. What are the causes of a harsh upshift?
12. With no drive in reverse, what should be checked?
13. If the shift points are incorrect, what could be the cause of the trouble?
14. Make a list of points where oil could leak.
15. What could cause oil to be forced out of the filler tube?
16. What could cause harsh closed-throttle downshifts?
17. What could prevent downshifts?
18. What could cause clutch failures?
19. List the precautions to observe when servicing the Powerglide.
20. Describe a typical Powerglide removal-and-replacement procedure.
21. Describe a typical Powerglide disassembly-and-reassembly procedure.

SUGGESTIONS FOR FURTHER STUDY

When you are in the automotive service shop, pay special attention to the mechanics working on automatic transmissions. Notice the special tools required and how they are used. Study shop manuals that describe the service procedures for the various Powerglide transmissions to learn more about how to handle these transmissions. Write in your notebook any important facts you learn.

chapter 11

BORG-WARNER AUTOMATIC TRANSMISSION

This chapter describes the construction and operation of the Borg-Warner automatic transmission, which uses a compound planetary gearset. This planetary gearset is similar to that used in the Powerglide transmission, discussed in Chap. 9. That is, it is a compound planetary-gear system, with two sun gears and two sets of planet pinions, one long and one short. The Borg-Warner automatic transmission provides three forward speeds and reverse (the Powerglide, remember, has only two forward speeds). Chapter 12 discusses the service procedures for the Borg-Warner transmission.

⊘ 11-1 Borg-Warner Transmission Construction Figure 11-1 is a sectional view of the Borg-Warner automatic transmission. Use this illustration as a reference as you read the following explanations of how the transmission operates. In Fig. 11-1, locate the clutches, brake bands, and planetary-gear parts.

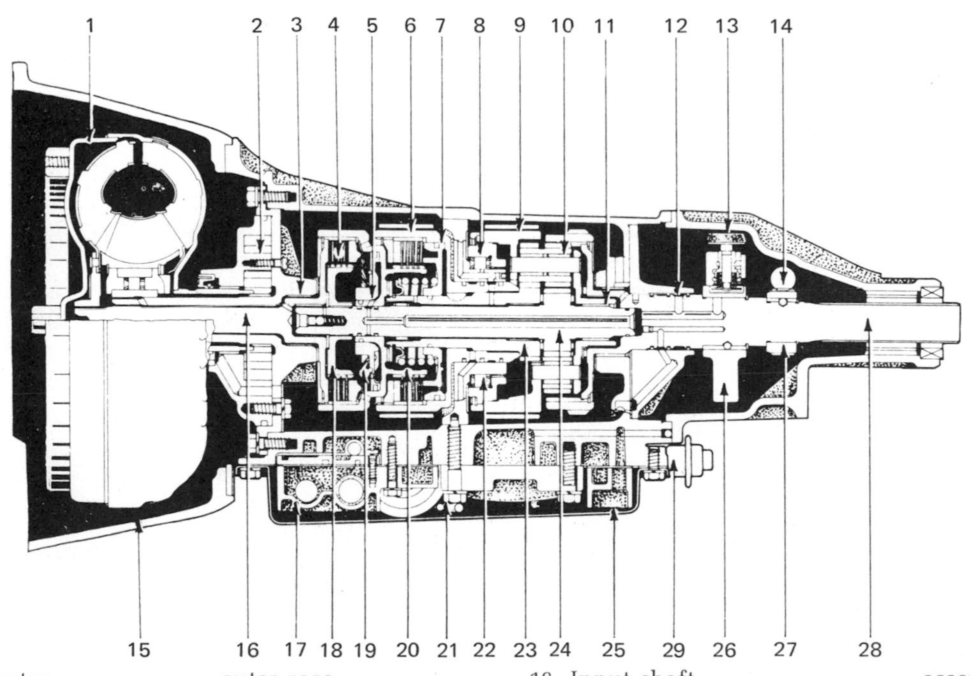

1. Torque converter
2. Oil-pump gear
3. Pump adapter and converter support
4. Front-clutch plate
5. Front-clutch piston
6. Front brake band
7. Front drum
8. One-way clutch outer race
9. Rear brake band
10. Long pinion
11. Planet cover
12. Rear adapter
13. Extension housing
14. Speedometer driven gear
15. Converter housing
16. Input shaft
17. Valve-body assembly
18. Front-clutch hub
19. Front-clutch spring
20. Front-clutch cylinder
21. Oil tube
22. One-way clutch assembly
23. Reverse sun gear
24. Forward sun gear
25. Oil pan
26. Governor assembly
27. Speedometer drive gear
28. Output shaft
29. Vacuum diaphragm

Fig. 11-1. Sectional view of the Borg-Warner automatic transmission which uses a compound planetary gearset. (*Chrysler Corporation*)

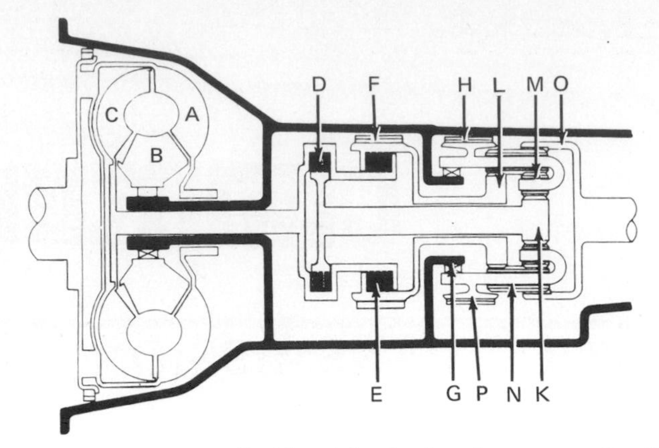

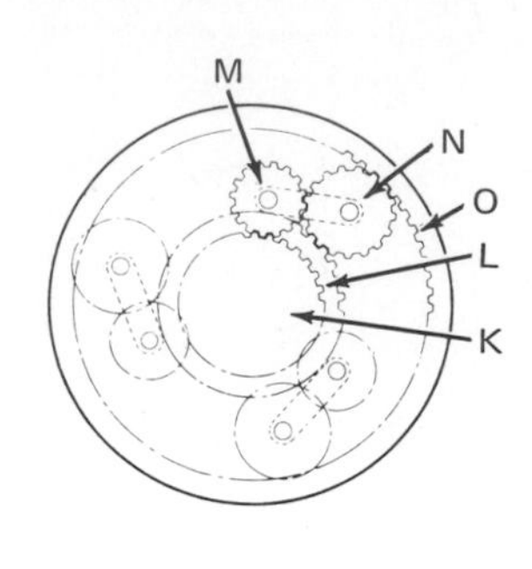

A. Pump
B. Stator
C. Turbine
D. Front clutch
E. Rear clutch
F. Front brake band
G. One-way clutch
H. Rear brake band
K. Forward sun gear (No. of teeth: 28)
L. Reverse sun gear (No. of teeth: 32)
M. Short pinion (No. of teeth: 16)
N. Long pinion (No. of teeth: 17)
O. Ring gear (No. of teeth: 67)
P. Planet-pinion carrier

Fig. 11-2. Simplified drawing showing locations of transmission components. (*Chrysler Corporation*)

To make location of parts easier, refer to Fig. 11-2, which is a simplified drawing showing the relationship of the components. As you can see, the transmission includes a three-member torque converter; front clutch; rear clutch, one-way, or overrunning, clutch; front brake band; rear brake band; and compound planetary gearset.

The hydraulic system, which we shall discuss later, includes several valves that control shifting. The driver selects the lever position to provide the desired driving condition. In some models, the selector lever is on the steering column, as, for example, in the Powerglide described previously. These models have the selector quadrant shown in Fig. 11-3. In other models, the selector lever is mounted in a console on the floor of the front seat (Fig. 11-4). In L (low), the transmission will start in low gear to provide a gear ratio of 2.393:1. It can shift to second (in L) to provide a gear ratio of 1.450:1. When the selector lever is moved to D (drive), the transmission will be in low gear when first starting out. The transmission will then shift to second and to high as car speed increases. The shift points vary with car speed and the amount of throttle opening (Fig. 11-5). The hydraulic system, described later, controls these shift points.

⊘ 11-2 Power Flow in Low When the driver moves the selector lever to L (low), the hydraulic system applies the front clutch and rear band (Fig. 11-6). See also Plate 6. With the front clutch applied, power flows from the torque converter to the forward sun gear. The rear band, when applied, holds the planetary-pinion carrier. Therefore power flows through the short pinions to the long pinions and then to the ring gear. The ring gear is attached to the output shaft. Since the planet pinions act only as idlers, the gear ratio is the ratio between the ring gear (67 teeth) and the forward sun gear (28 teeth), or 67:28. This produces a gear reduction of 2.393:1.

For another look at the power flow through the transmission, refer to Fig. 11-2. Power flows from the turbine of the torque converter (C) to the front clutch (D), forward sun gear (K), short pinions (M), long pinions (N), and ring gear (O). Note that the long pinions are also meshed with the reverse sun gear (L) but that the reverse sun gear does not enter into the operation in L. Note that, in L, the engine can brake the car, as when going down a long hill. With the selector lever in L, the transmission can shift from low to second or from second to low (1-2 upshift or 2-1 downshift), but it cannot shift up to direct drive.

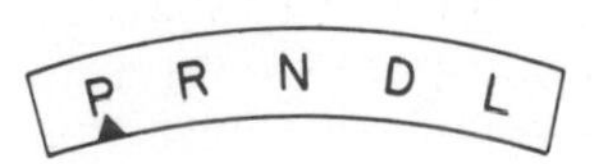

Fig. 11-3. Selector-lever quadrant on the steering column. Transmission is in P (park).

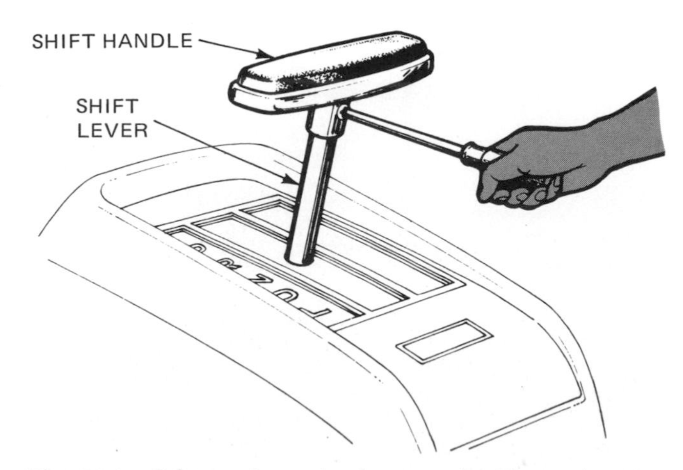

Fig. 11-4. Selector lever in the console. The technician is using a screwdriver to loosen the handle. (*Chrysler Corporation*)

CAR SPEED, mph

	1-2 UPSHIFT	2-3 UPSHIFT	3-2 DOWNSHIFT	3-1 DOWNSHIFT (2-1 DOWNSHIFT)
When accelerator pedal is slightly depressed (zero throttle)	8	10	...	6 (3-1)
When accelerator pedal is fully depressed (full throttle)	37	40	35	6 (2-1)
When accelerator pedal is depressed past full throttle (kickdown throttle)	61	96	81	25 (2-1)

Fig. 11-5. Upshifts and downshifts as related to car speed. (*Chrysler Corporation*)

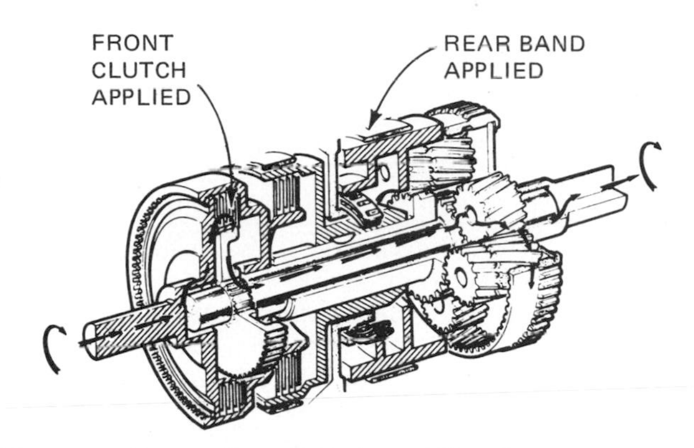

Fig. 11-6. Power flow in L, low. (*Chrysler Corporation*)

⊘ 11-3 Power Flow in Low Drive When the driver moves the selector lever to D (drive) and the car first moves out, the transmission is in D low (low drive) (Fig. 11-7). In this position, the front clutch and the one-way, or overrunning, clutch are holding. The effect is the same as when the selector lever is in L, with one exception: The planet-pinion carrier is kept from rotating backward by the one-way clutch (instead of being held stationary by the rear band, as in L). The gear ratio is the same (2.393:1). The purpose of having the planet-pinion carrier held by the one-way clutch is that when the car has gained speed and, in effect, is coasting, the pinion carrier can freewheel. That is, the one-way clutch allows the planet-pinion carrier to run free of the engine. There is no engine braking effect in D low. The freewheeling provides a smooth upshift from first to second and a smooth downshift from second to first.

Refer again to Fig. 11-2 to follow the power flow through the transmission. Power flows from the turbine (C) to the front clutch (D), forward sun gear (K), short pinions (M), long pinions (N), and ring gear (O). The planet-pinion carrier (P) is kept from turning backward by the one-way clutch (G).

⊘ 11-4 Power Flow in Second Drive When the upshift in D to second (second drive) takes place (produced by the hydraulic system), the front clutch remains applied but the overrunning clutch releases the rear carrier when the front band applies (Fig. 11-8). This locks the reverse sun gear. Now (refer to Fig. 11-2) power flows from the turbine (C) to the front clutch (D), forward sun gear (K), short pinions (M), long pinions (N), and ring gear (O). At first glance, this power flow looks exactly like the power flow in D low.

However, the difference here is that the planet-pinion carrier has been released by the release of the rear band. As a result, it is free to rotate forward around the reverse sun gear. Now, the planet pinions are no longer idlers as in D low. The gear ratio is the sum of the ratios among the forward sun gear, reverse sun gear, pinions, and ring gear. The combined gear ratio provides a gear reduction of 1.450:1 in D second.

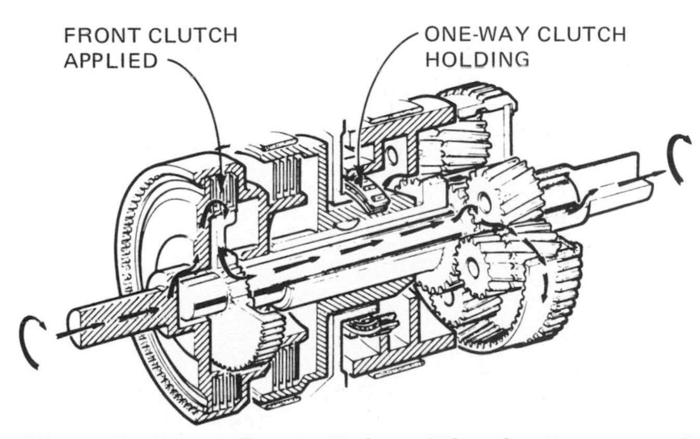

Fig. 11-7. Power flow in D, low. (*Chrysler Corporation*)

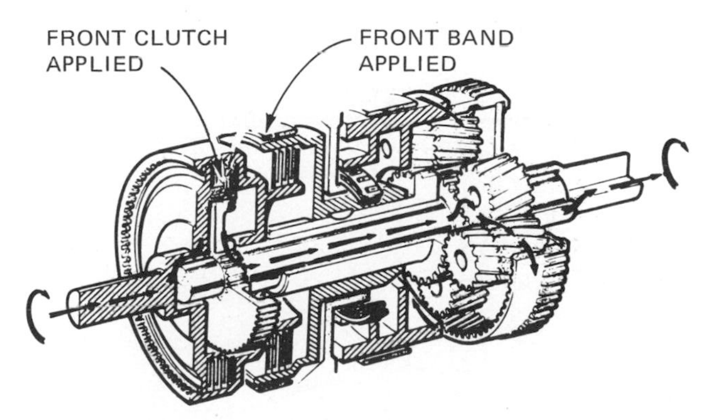

Fig. 11-8. Power flow in D, second. (*Chrysler Corporation*)

NOTE: The transmission can also shift into second with the selector lever in L. This can happen when engine and car speeds increase sufficiently. The essential difference between being in second in D and being in second in L is that, in second L, the transmission cannot shift into third. More on this later when we describe the hydraulic system.

⊘ 11-5 Power Flow in Direct Drive When the upshift to direct drive takes place, both clutches are applied (Fig. 11-9). This locks the two sun gears together so that the whole assembly must turn as a unit. This action transmits the engine power directly to the output shaft with a gear ratio of 1.000:1.

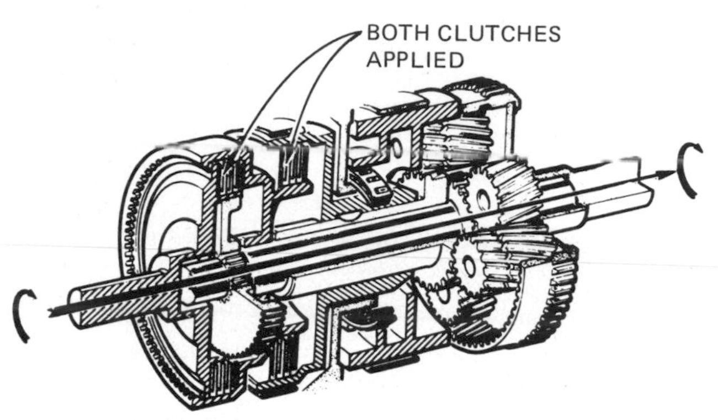

Fig. 11-9. Power flow in D, direct. (*Chrysler Corporation*)

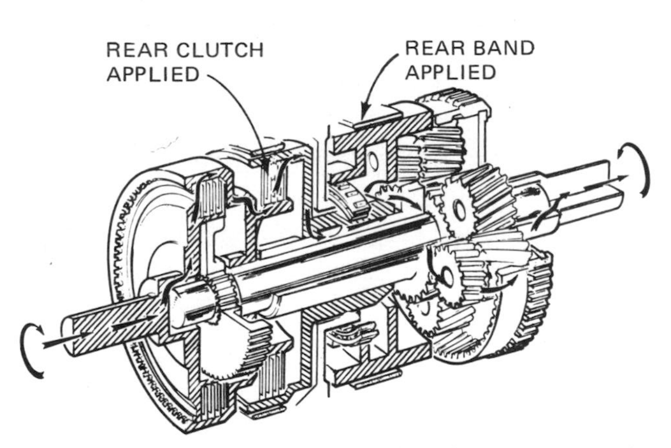

Fig. 11-10. Power flow in R. (*Chrysler Corporation*)

⊘ 11-6 Power Flow in Reverse When the driver moves the selector lever to R (reverse), the hydraulic system applies the rear clutch and rear band (Fig. 11-10). The rear band locks the planet-pinion carrier so that the long pinions work as idler gears. The rear clutch locks the reverse sun gear to the input shaft. Therefore (refer to Fig. 11-2) power flows from the turbine (C) to the rear clutch (E), reverse sun gear (L), long pinions (N), and ring gear (O). Note that with the single set of long pinions acting as idlers, the ring gear turns in the reverse direction. The gear ratio in reverse is 2.094:1.

⊘ 11-7 Hydraulic System The hydraulic system is shown in Plates 4 to 8. Keep them in front of you as you read the following explanations. Figures 11-11 to 11-25 show all the valves in the system, and we shall explain how they work. First, let us look at some of the features of the hydraulic system. The transmission has a pump (2 and 3 in Fig. 11-1), which is a gear-type pump similar to the one used in the Powerglide (Fig. 9-9). It supplies full pump pressure to the primary and secondary regulator valves. These two valves modify the pump pressure according to various operating conditions. Refer to Plate 4 for the location of these valves and other components as we describe the operation of the hydraulic system in detail.

⊘ 11-8 Governor The governor is located on the output shaft (see 26 and 28 in Fig. 11-1). The governor releases increasing pressure as output-shaft, and car speed, increases. Figure 11-11 shows the governor. At low speed, the weight is held in by the spring. As speed increases, centrifugal force causes the weight to move out. This action moves the valve so that more of the line pressure can pass through the valve. This governor pressure is applied to various valves in the hydraulic system, as we shall ex-

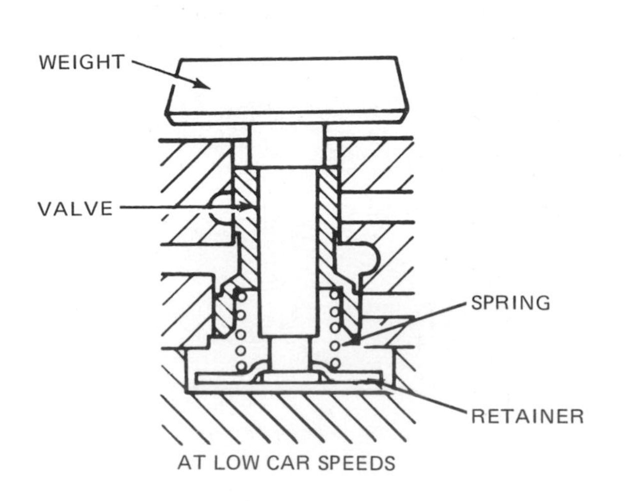

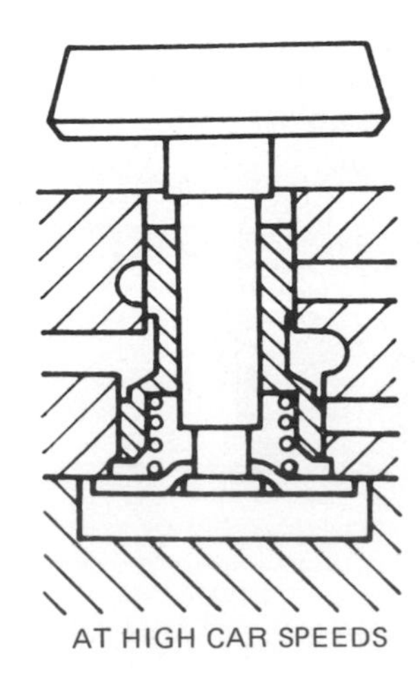

Fig. 11-11. Governor operation. (*Chrysler Corporation*)

plain later. The governor pressure influences these valves and helps determine when shifts take place.

⊘ 11-9 Hydraulic System in Low Drive Plate 4 shows conditions with the selector lever in D (drive) and the transmission in low gear. This is the same condition as the one shown in Fig. 11-7. We start our examination of the hydraulic system by locating and describing the purpose of each of the valves. We discuss each of these valves in detail later.

1. *PRIMARY AND SECONDARY REGULATOR VALVES* These valves take pump pressure and reduce it to line pressure. Line pressure actuates the clutches and brake bands and is increased by the regulator valves to provide more holding power at high loads.
2. *MANUAL VALVE* The manual valve (Plate 4, center) is moved by the movement of the selector lever. In Plate 4, it is shown in the D (drive) position. Note that the D is just above the small diameter on the left end of the spool valve. In this position, the valve admits line pressure to the front clutch (see Fig. 11-7).
3. *GOVERNOR* The governor (not shown in Plate 4) is attached to the output shaft (26 and 28 in Fig. 11-1). As car speed increases, the governor releases an increasingly larger percentage of the line pressure. This is called *governor pressure.* It helps determine when upshifts from low to second and from second to direct occur. It also helps determine when downshifts occur.
4. *THROTTLE VALVE* The throttle valve (Plate 4, upper right) releases varying amounts of line pressure, according to the amount the throttle is open. This pressure, called *throttle pressure,* helps determine when upshifts or downshifts occur.
5. *MODULATOR VALVE* The modulator valve (Plate 4, lower right) serves as a go-between, in effect, between the throttle pressure and governor pressure. It is subject to both pressures and produces a resulting *modulator-valve pressure* that works on the primary regulator valve. This pressure causes the primary regulator valve to vary line pressure to suit different operating conditions.
6. *1-2 AND 2-3 SHIFT VALVES* The 1-2 and 2-3 shift valves, controlled by other valves and by throttle and governor pressures, produce the upshifts and downshifts in D (drive). They direct line pressure to the clutches and brake-band servos to apply the clutches or bands (see Figs. 11-7 to 11-9).
7. *SERVO-ORIFICE CONTROL VALVE* The servo-orifice control valve paces the actions of the shift valves in accordance with operating conditions. It produces fast 2-3 upshifts or 3-2 downshifts under some conditions, and it produces slow upshifts or downshifts under other conditions.
8. *DOWNSHIFT VALVE* The downshift valve works when the throttle is opened wide. When the throttle is "floored," a switch connects the kickdown solenoid to the battery. This connection operates the downshift valve so that line pressure is applied to the shift valves, causing a downshift. The action puts the transmission into a lower gear for quick passing on the highway, for example.

NOTE: The preceding discussion is only an introduction to the valves in the hydraulic system. In the sections that follow, we shall discuss each valve in detail and determine its function. Remember that when you read this material for the first time, it may be confusing. Therefore, read these sections more than once to understand how everything works.

⊘ 11-10 Primary and Secondary Regulator Valves We start our detailed examination of the hydraulic system and its valves by looking at the primary and secondary regulator valves (Plate 4, left). These valves, influenced by throttle pressure (amount of throttle opening) and governor pressure (car speed), modify the pump pressure. They receive pump pressure and reduce it to line pressure. Line pressure must be changed to produce correct operation of the other valves in the system. Line pressure is changed by the two regulator valves.

⊘ 11-11 Primary Regulator Valve The primary regulator valve (Fig. 11-12) includes a spool valve that is considerably more complex than the one shown in Fig. 7-8. The valve is subjected to four forces: spring pressure, throttle pressure (which varies with throttle opening), line pressure, and modulator-valve pressure (which varies with car speed).

Figure 11-12 shows two positions of the primary regulator valve. To the left, we see its position at low speed. The pressure is controlled by the governor (see Plate 4). Line pressure, entering at port 1, is also low. Therefore, there is not much of an upward push on the valve. It is held down by the spring and also throttle pressure. Note that line pres-

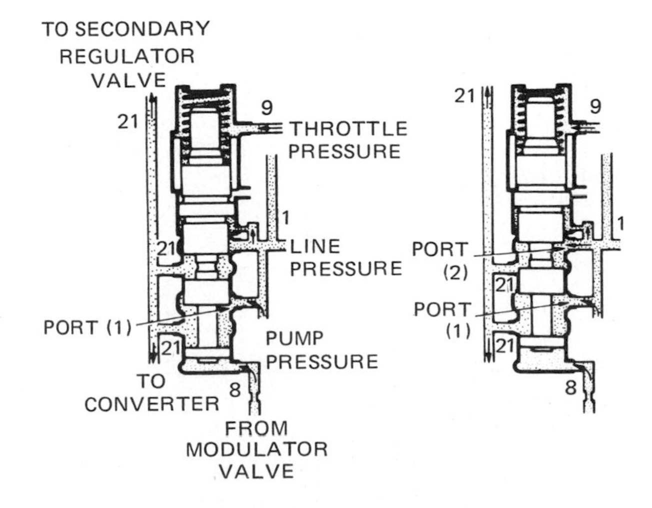

Fig. 11-12. Primary-regulator-valve operation. (*Chrysler Corporation*)

sure, passing through port 1, flows out exit 21. This pressure passes to the secondary regulator valve. The secondary regulator valve prevents excessive pressure from passing into the converter and lubricating system, as explained in ⊘ 11-12.

To the right in Fig. 11-12, we see the valve position at high speed. The governor has released more pressure through the modulator valve, and so this pressure pushes up on the bottom of the primary regulator valve. At the same time, the throttle pressure, which increases as the throttle is opened wider, is pushing down on the top of the regulator valve. But the modulator-valve pressure overcomes the spring and throttle pressure and pushes the regulator valve up, as shown to the right. Note that both ports 1 and 21 are now open to the secondary valve. The line pressure has increased because the pump is being driven faster and is putting out a great deal more oil. However, the two open ports take care of this additional oil, passing it up to the secondary valve, where part of it is dumped back to the pump, as explained later. This arrangement prevents excessive line pressure.

⊘ 11-12 Secondary Regulator Valve Plate 4 shows the position of the secondary regulator valve in the hydraulic system (to the left). Figure 11-13 shows the secondary valve. To the left, we see its position when both engine speed and pump discharge are low. The primary valve is doing all the regulating. However, when engine speed increases so that the pump discharge is high, the valve position is as shown to the right. Much more oil flows from the pump, passing through both ports 1 and 21 in the primary regulator valve (Fig. 11-12, right). With this condition, the increased pressure and flow force the secondary regulator valve up, as shown to the right in Fig. 11-13. Now, the additional oil flows out port 24 and back to the pump. This dumping of the oil prevents an excessive increase in oil pressure.

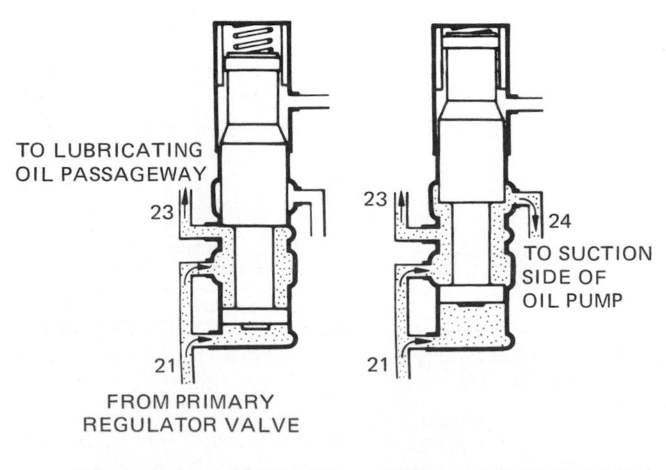

Fig. 11-13. Secondary-regulator-valve operation. (*Chrysler Corporation*)

⊘ 11-13 Throttle Valve and Vacuum Diaphragm Locate the throttle valve and vacuum diaphragm toward the upper right in Plate 4. Figure 11-14 shows the assembly. The throttle valve takes in line pressure at port 1 and reduces it in accordance with throttle position (intake-manifold vacuum). The throttle-valve position is controlled by a spring at one end and the vacuum diaphragm at the other. When intake-manifold vacuum is high, indicating a closed throttle, the diaphragm is pulled to the right (Fig. 11-14). Such movement allows the spring to push the valve to the right. This action cuts off most of the line pressure coming in at port 1. Thus, the pressure leaving the throttle valve at port 9 is comparatively low. If you trace the circuit in Plate 4, you will see that this pressure goes to the 1-2 and 2-3 shift valves. This pressure helps determine when shifts take place, as we explain in ⊘ 11-17 and 11-18.

With fairly light throttle-valve pressure at the shift valves, the shifts can take place at relatively low speeds. On the other hand, if the throttle is opened wide, the shifts take place at high speeds. The reason for this is that, with an opened throttle, intake-manifold vacuum is low. The diaphragm spring is able to push the diaphragm to the left (Fig. 11-14). This pushes the throttle valve to the left. In that position, it uncovers more of the line-pressure port (1). Thus, more pressure flows through the throttle valve (via port 9). This increased pressure, working on the shift valves, prevents shifting until higher speeds are reached.

NOTE: Since all the valves are interconnected and work together, you may not understand completely just how everything works until you have read the explanations of all the valves (⊘ 11-10 to 11-18). Then you will probably need to study these sections one or two more times in order to gain a full understanding. Once you understand how everything works, it will seem so simple that you will probably be asking yourself why it seemed so complicated the first time around!

⊘ 11-14 Modulator Plunger and Valve Locate the modulator-plunger-and-valve assembly in Plate 4. Figure 11-15 also shows this assembly. Note that there are two separate valves: the modulator valve

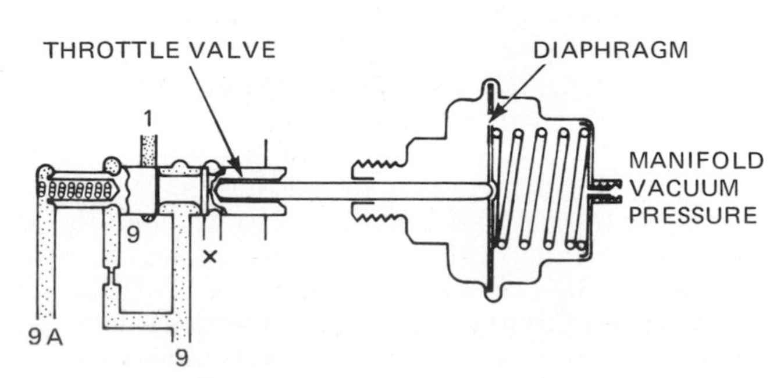

Fig. 11-14. Throttle valve and vacuum diaphragm. (*Chrysler Corporation*)

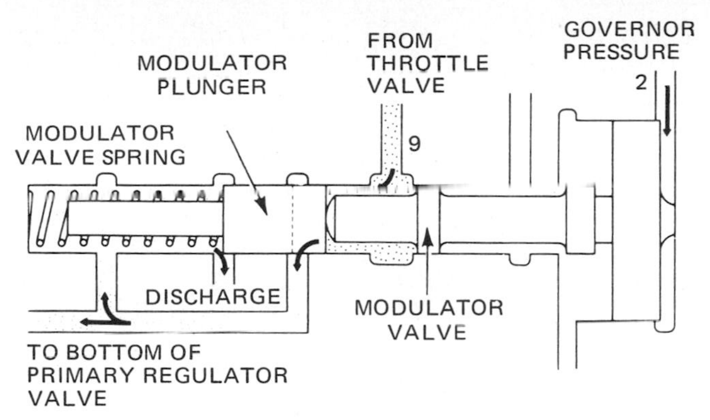

Fig. 11-15. Modulator plunger and valve positions at low speed. (*Chrysler Corporation*)

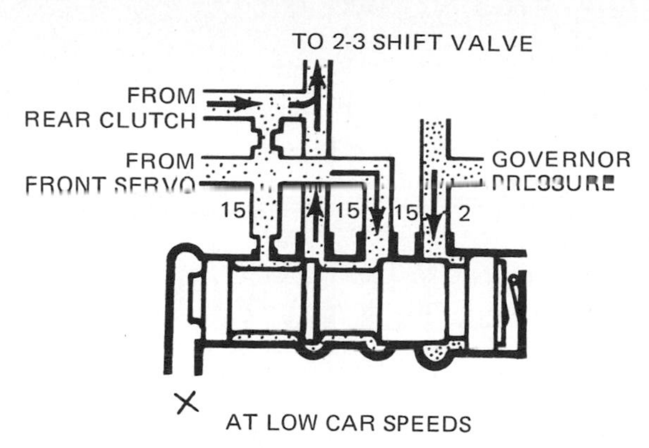

Fig. 11-17. Servo-orifice-control-valve position when a 2-3 upshift occurs at low car speeds. (*Chrysler Corporation*)

and the modulator plunger. The major actuating force on the modulator valve is governor pressure. It enters at port 2. In low gear, with only a little throttle opening, the output shaft and governor cannot turn fast enough to produce a high governor pressure. Thus only a small pressure works on the modulator valve.

The modulator valve and plunger take the positions shown in Fig. 11-15. In these positions, they allow only a small part of the pressure from the throttle valve (9) to pass through and proceed to the bottom of the primary regulator valve. However, if the throttle is opened, then the throttle valve will release more pressure, forcing the modulator plunger to the left. This action releases more pressure to the bottom of the primary regulator valve (through port 8 in Fig. 11-16). Also, as governor speed and pressure increase, they can force the modulator valve to the left, as shown in Fig. 11-16. This action is tied in with the shift from first to second, as described in ⊘ 11-17.

⊘ 11-15 Servo-Orifice Control Valve The servo-orifice control valve (Plate 4, upper left) is illustrated in Fig. 11-17. This valve times the release of the front band for a 2-3 upshift or the application of the front band for a 3-2 downshift. The servo for

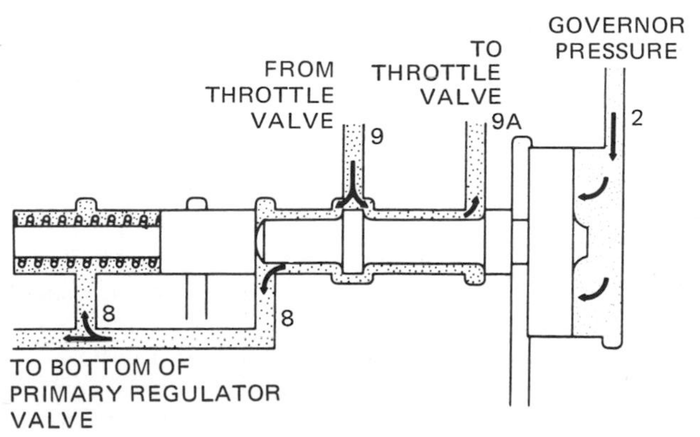

Fig. 11-16. Modulator plunger and valve positions at high speed and high governor pressure. (*Chrysler Corporation*)

the front band is similar to the one shown in simplified form in Fig. 8-4. It has two chambers and two pistons. One of these chambers is used to apply the band, and it is fed line pressure through the 1-2 valve (via port 19—see Plate 4 or 5, upper right). The other chamber is the release chamber, and it is fed line pressure through the servo-orifice control valve. Refer to Plate 7 to see how this control valve feeds line pressure to the front-servo release chamber when the transmission is in direct drive.

1. 2-3 UPSHIFT Figure 11-17 shows the conditions when a 2-3 upshift occurs at low car speeds. Governor pressure enters to the right at port 2. This pressure is low, and so the valve takes the position shown. Meantime, line pressure is sent from the 2-3 shift valve to the servo-orifice control valve. It enters the valve as shown in Fig. 11-17. Note that this line pressure also goes to the rear clutch. The pressure passes through the control valve and emerges at the right-hand port 15. It then goes to the front-servo release chamber. Thus, the band is released at the same time that the clutch is applied.

At high car speed, there must be a slight delay in the release of the band. The reason for this is that the engine is also turning at high speed and everything happens very rapidly. If the band releases before the rear clutch applies, the engine would be allowed to "run up." That is, the engine would run free without a load for a moment. Therefore, the engine would increase its speed very rapidly. Then, when the rear clutch applies, it will tend to grab and suddenly slow the engine. The effect would be a rough shift.

To prevent this condition, the servo-orifice control valve acts to restrict the pressure flow to the front-servo release chamber. See Fig. 11-18. At high car speed, governor pressure is great enough to push the control valve to the right, as shown. This action shuts off the unrestricted passage (the right port 15) to the front-servo release chamber. Now, the pressure must pass through the restricting orifice (the left port 15). This action slows the front-brake-band release enough to make sure that the rear clutch applies before the front band completely releases. This prevents engine runup.

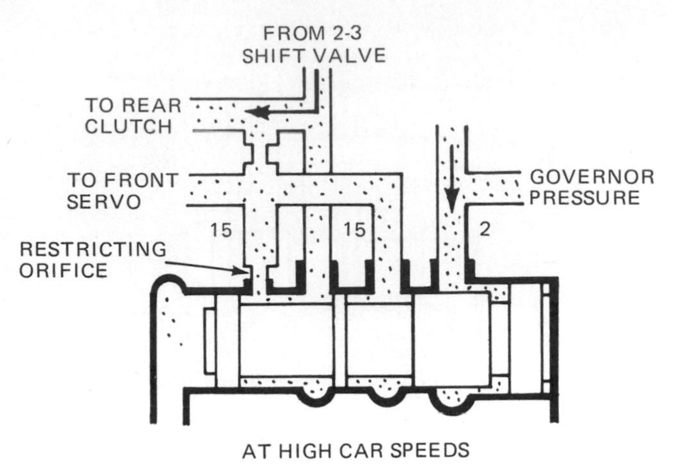

Fig. 11-18. Servo-orifice-control-valve position when a 2-3 upshift occurs at high car speeds. (*Chrysler Corporation*)

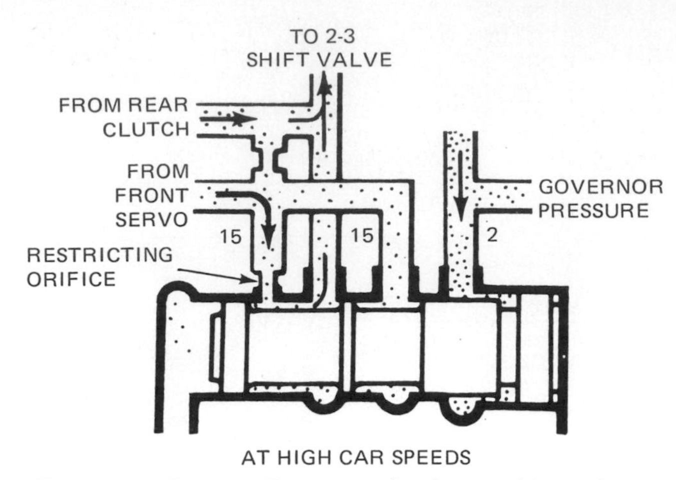

Fig. 11-20. Servo-orifice-control-valve position when a 3-2 downshift occurs at high car speeds. (*Chrysler Corporation*)

2. *3-2 DOWNSHIFT* When downshifting from 3 to 2 at low car speed, the condition is as shown in Fig. 11-19. This condition is the same as for a 2-3 upshift at low car speed insofar as the control-valve position is concerned. However, in this case, the control valve releases pressure from the rear clutch and the front-servo release chamber.

Refer to Plate 5 for a moment and trace this pressure release. Note that the circuit from the rear clutch and front-servo release chamber to the 2-3 shift valve is shown without color. This means there is no pressure. Trace this circuit through the 2-3 valve and down to the manual valve. Note that pressure exits from the manual valve at X. X means the oil in the line is released to the sump from which the pump picks up oil.

With unrestricted flow, the release chamber of the front servo releases the front band so the front band applies at the same time that the rear clutch releases. This prevents engine runup.

At high car speed (Fig. 11-20), the 3-2 downshift requires a slower release of the pressure in the front-servo release chamber. This action prevents the application of the front band before the rear clutch releases. If both the front band and rear clutch apply at the same time, engine tie up would occur. That is, engine braking would occur that would slow the engine. Then, when the front band applies, there would be a sudden jerking action that would result in a rough shift.

Figure 11-20 shows how this slow brake-band application is accomplished. Governor pressure pushes the control valve to the right, as shown. This action puts the restricting orifice into the circuit from the release chamber of the front servo. As a result, the pressure is released slowly and the front band therefore applies slowly. This ensures full rear-clutch release before the front band is fully applied. Thus, a smooth 3-2 shift results.

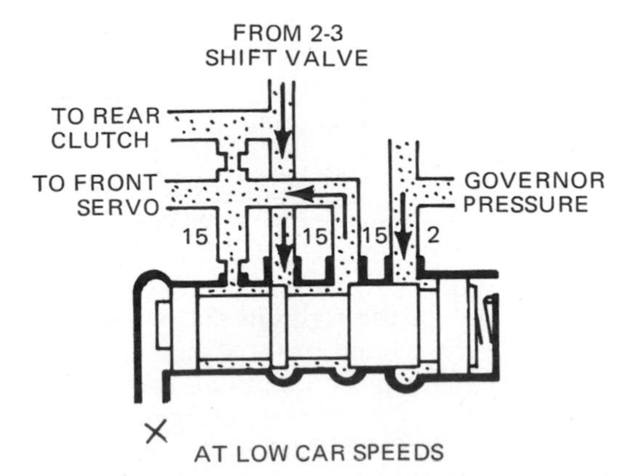

Fig. 11-19. Servo-orifice-control-valve position when a 3-2 downshift occurs at low car speeds. (*Chrysler Corporation*)

⊘ 11-16 Downshift Valve and Kickdown Solenoid When the throttle is pushed to wide open, a kickdown switch connecting the kickdown solenoid to the battery closes. The kickdown solenoid then draws in its plunger, allowing the downshift valve to move to the left. See Fig. 11-21. To see what this action causes, refer also to Plate 4. Note that port 5 opens as the downshift valve moves to the left. This action dumps part of the line pressure, and so line pressure decreases. As the same time, port 11 opens to the plunger side of each shift valve. This action moves the shift valves to the left. Thus, the transmission downshifts to second or low, according to car speed.

⊘ 11-17 1-2 Shift Valve and Plunger The 1-2 shift valve and plunger control the 1-2 or 2-1 shifts according to governor pressure and throttle pressure. In first speed (Figs. 11-7 and 11-22), the 1-2 shift valve and 1-2 shift plunger (two separate spool valves) are positioned as shown. They are held in this position by line pressure (5) and throttle pressure from the 2-3 shift valve (10 and 10A) until car speed increases enough. As car speed increases, governor pressure

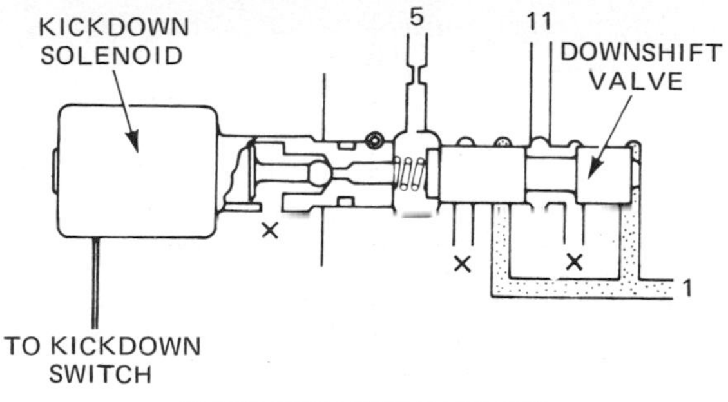

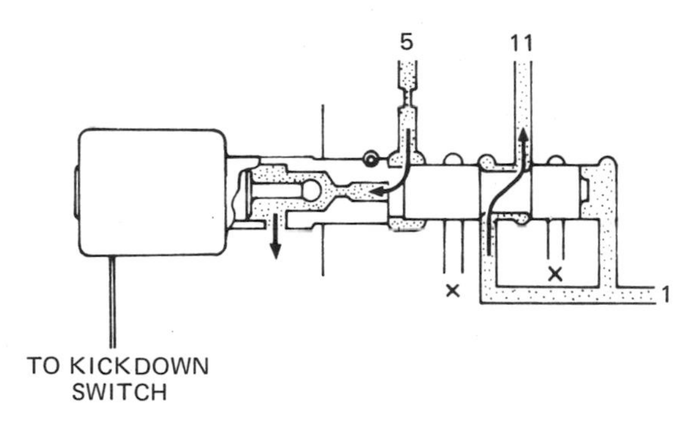

Fig. 11-21. Downshift valve and kickdown solenoid, showing the OFF and ON positions. (*Chrysler Corporation*)

increases. When governor pressure is great enough to overcome throttle pressure, the 1-2 shift valve and plunger move to the right (Fig. 11-23). See also Fig. 11-8 and Plate 5. Note that this movement admits line pressure through port 5 to port 19, which then feeds line pressure to the front servo. The *apply* piston of the servo is operated so that the front band applies. Thus, the transmission upshifts from 1 to 2.

Note that the greater the throttle pressure, the greater the governor pressure must be to overcome it. In other words, as the throttle is opened wider (to increase throttle pressure), car speed (governor pressure) must increase to overcome the throttle pressure.

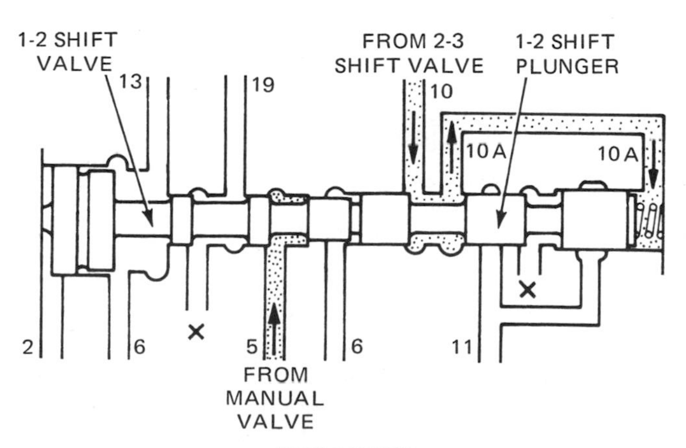

Fig. 11-22. 1-2 shift valve and plunger positions in first speed (low). (*Chrysler Corporation*)

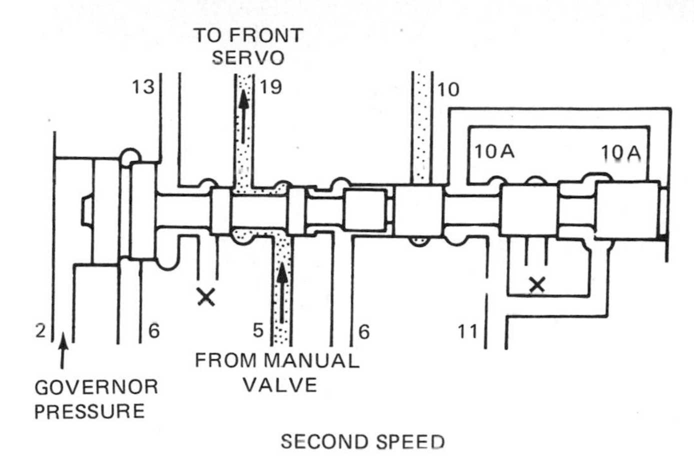

Fig. 11-23. 1-2 shift valve and plunger positions in upshift to second. (*Chrysler Corporation*)

In addition, when the throttle is pushed down, with the transmission in second, throttle pressure increases. If throttle pressure increases enough to overcome governor pressure (car speed), the transmission will downshift from 2 to 1. Even if the throttle is held steady, if car speed and governor pressure decrease enough (as they might if the car is going up a hill), downshift will occur. Note that downshift can also occur if the throttle is opened wide so that the kickdown solenoid is actuated (Fig. 11-21).

⊘ 11-18 2-3 Shift Valve and Plunger The 2-3 shift valve and plunger (Fig. 11-24) control the 2-3 upshift or 3-2 downshift. Plate 7 shows conditions in the hydraulic system in direct drive. See also Fig. 11-9. Both clutches apply, locking the two sun gears together. The 2-3 shift valve and plunger are controlled by three pressures: throttle pressure, governor pressure, and line pressure (Fig. 11-24). Throttle pressure works on the plunger (to the right) and through the spring on the 2-3 shift valve. Governor pressure works on the 2-3 shift valve (to the left). Line pressure (from the manual valve) works on the center part of the 2-3 shift valve.

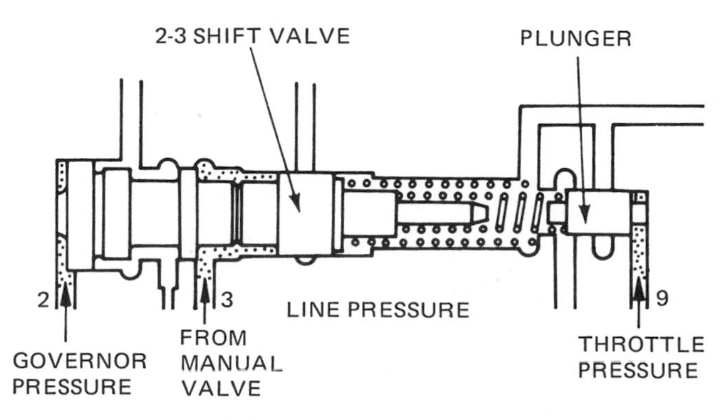

Fig. 11-24. 2-3 shift valve and plunger positions in upshift in direct drive at low car speed and small throttle opening. (*Chrysler Corporation*)

Note that the stepup diameter of the shift valve to the left of port 3 (where line pressure enters) is larger than the stepup diameter to the right of port 3. This condition means that the line pressure has more area to work on the left-hand part of the 2-3 shift valve. Thus, line pressure tends to push the 2-3 shift valve to the left. This is an important factor in the shifting action.

As governor pressure (car speed) increases, it becomes great enough to overcome throttle pressure. Thus, the 2-3 shift valve moves to the right. As a result, line pressure from the manual valve (via port 3) goes to the servo-orifice control valve. You will remember (from ⊘ 11-15) that the servo-orifice control valve times the release of the front band for a 2-3 upshift. Line pressure from the 2-3 shift valve also goes to the rear clutch (via port 15). See Plate 7. Therefore, the front band releases and the rear clutch applies. Thus, the transmission upshifts from 2 to 3, or into direct drive.

Note that the greater the throttle pressure, the greater the governor pressure must be to overcome it. In other words, as the throttle is opened wider (to increase throttle pressure), car speed (governor pressure) must increase also before an upshift takes place.

Also, when the throttle is pushed down, with the transmission in direct drive, throttle pressure increases. If throttle pressure increases enough to overcome governor pressure (car speed), the 2-3 shift valve will move to the left (Fig. 11-25) so that the transmission will downshift from 3 to 2. Even if the throttle is held steady, if car speed and governor pressure decrease enough (as they might if the car is going up a hill), downshift will occur. Note that downshift can also occur if the throttle is opened wide so that the kickdown solenoid is actuated (see Fig. 11-21).

⊘ 11-19 Range Control Valve The range control valve, which is shown near the top center of Plates 4 and 7, does not enter into the action of the transmission we are describing in this chapter. This transmission is a five-position unit (P, R, N, D, and L), and the range control valve is not needed. The control-valve body used with this transmission, however, is also used with a six-position transmission (P, R, N, D, 2, and 1). The range control valve works with the six-position transmission to control the additional selector-lever and shift positions.

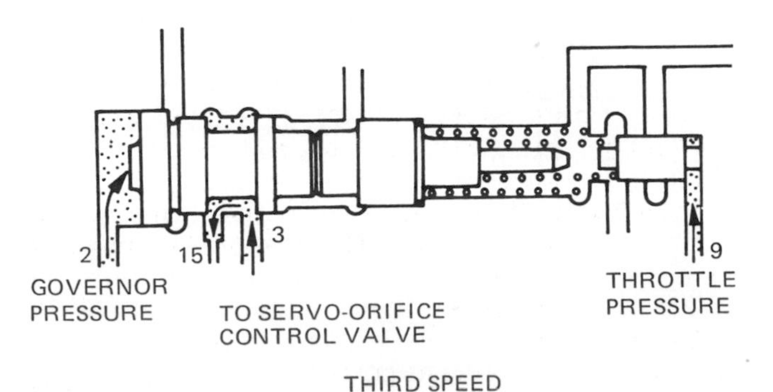

Fig. 11-25. Shift valve and plunger positions in upshift to direct drive at third or high speed. (*Chrysler Corporation*)

⊘ 11-20 Hydraulic Control Circuit in Low Drive The hydraulic control circuit is shown in low drive in Plate 4. We have already discussed this plate in some detail. The manual valve is in D, feeding line pressure to both shift valves. Throttle pressure from the throttle valve is fed to both shift valves. Governor pressure is also fed to both shift valves. As we have already explained, throttle pressure and governor pressure "fight" to control the upshift. When governor pressure overcomes throttle pressure, the transmission upshifts from low to second.

⊘ 11-21 Hydraulic Control Circuit in Second Drive Plate 5 shows what happens in the hydraulic control circuit at the 1-2 shift valve when the transmission upshifts from low to second. The rest of the circuit is approximately as shown in Plate 4.

⊘ 11-22 Hydraulic Control Circuit in Direct Drive Plate 7 shows the hydraulic circuit after the transmission has upshifted from second to direct. Both clutches apply and both brake bands release (see Fig. 11-9). We have already described the operation of the valves which produce the upshift (⊘ 11-18).

⊘ 11-23 Hydraulic Control Circuit in Low When the selector lever is moved to L (low), the conditions are as shown in Plate 6. Note that line pressure is cut off from the 2-3 shift valve, and so the transmission cannot shift into direct drive. Also, line pressure is fed to the rear servo so that the rear brake band applies. See Fig. 11-6, which shows that in low L, the front clutch and rear band apply. Thus, there is firm drive through the transmission, and so the engine can brake the car.

Note that the transmission can upshift into second with the selector lever in L. This upshift can occur if governor pressure (car speed) increases enough to overcome throttle pressure. When this happens, the 1-2 shift valve moves to the right (see Fig. 11-23), causing the front clutch to remain applied while the front band also applies (Fig. 11-8).

In both low L and second L, the drive is firm through the transmission so that the engine can brake the car, say, when coming down a hill.

⊘ 11-24 Hydraulic Control Circuit in Reverse When the selector lever is moved to R (reverse), the conditions shown in Plate 8 result. Note how line pressure is fed to the two shift valves, causing them to apply the rear clutch, release the front brake band and front clutch, and apply the rear brake band.

These actions produce the conditions shown in Fig. 11-10. The output shaft turns in the reverse direction.

⊘ 11-25 Parking When the selector lever is moved to P (park), a mechanical linkage is actuated. This linkage is shown in Fig. 12-31. Movement of the linkage causes a parking pawl to drop into the teeth on the outside of the ring gear attached to the output shaft. Figure 12-29 shows these outside teeth. Figure 12-34 shows the parking pawl in the parking position. Note that the pawl has dropped down between teeth on the ring gear. In this position, the output shaft, drive train, and car wheels are locked in a stationary position.

CHAPTER 11 CHECKUP

The Borg-Warner transmission covered operates on the same general principles as other automatic transmissions. Also, each automatic transmission has similar valves in their hydraulic systems. These valves provide smooth shifts, up or down, as car speed and throttle opening demand. However, there are variations among the different automatic transmissions. Knowing what these special variations are is the mark of good automatic-transmission technicians. Thus, we have discussed the valves in detail and how they do their jobs. The tests that follow will help you check up on yourself so that you will know whether or not you fully understand what the Borg-Warner transmission is all about. If you are not sure about an answer, reread the pages in the chapter that will give you the information you need. Write your answers in your notebook. This will help you to remember them. Also, your notebook will become a valuable source of information you can refer to quickly in the future.

Completing the Sentences The sentences that follow are incomplete. After each sentence there are several words or phrases, only one of which will correctly complete the sentence. Write each sentence in your notebook, selecting the proper word or phrase to complete it correctly.

1. The transmission has two multiple-plate clutches, a one-way clutch, and: (*a*) a brake band, (*b*) two brake bands, (*c*) three brake bands.
2. When the selector lever is at L, the transmission can shift between: (*a*) 1 and 2, (*b*) 1 and 3, (*c*) 2 and 3.
3. In L low, the rear band applies and the: (*a*) front clutch releases, (*b*) rear clutch applies, (*c*) front clutch applies.
4. The main difference between L low and D low is that in D low, the planet-pinion carrier is kept from rotating by the: (*a*) one-way clutch, (*b*) rear brake band, (*c*) front brake band.
5. In D low, the planet-pinion carrier: (*a*) turns in the reverse direction, (*b*) can freewheel, (*c*) is locked to the output shaft.
6. In D second, the front clutch remains applied and the: (*a*) rear band applies, (*b*) front band applies, (*c*) front band releases.
7. In D direct, or third: (*a*) both clutches apply, (*b*) both bands apply, (*c*) both clutches release.
8. In the hydraulic system, there: (*a*) is one regulator valve, (*b*) are two regulator valves, (*c*) are three regulator valves.
9. The two major forces that influence the operation of the regulator valves are: (*a*) throttle pressure and governor pressure, (*b*) pump pressure and line pressure, (*c*) input pressure and car speed.
10. The throttle valve is controlled by throttle position (or intake-manifold vacuum) and: (*a*) line pressure, (*b*) a spring, (*c*) governor pressure.
11. The major actuating force on the modulator plunger and valve is: (*a*) throttle pressure, (*b*) line pressure, (*c*) governor pressure.
12. The valve that times the 2-3 upshift or the 3-2 downshift is called the: (*a*) timing valve, (*b*) manual valve, (*c*) servo-orifice control valve.
13. The servo-orifice control valve times the release or application of the: (*a*) front band, (*b*) rear band, (*c*) front clutch.
14. The three pressures that control the action of the 1-2 shift valve are line pressure: (*a*) governor pressure, and pump pressure, (*b*) governor pressure, and throttle pressure, (*c*) throttle pressure, and manifold pressure.
15. In reverse: (*a*) the front band and front clutch apply, (*b*) the rear band and rear clutch apply, (*c*) both the front and rear bands apply.

Purpose and Operation of Components In the following, you are asked to write the purpose and operation of various components of the Borg-Warner automatic transmission. If you have any trouble writing your explanations, turn back into the chapter and reread the pages that will give you the information you need. But don't copy. Try to tell it in your own words. This is a very good way to fix the explanations firmly in your mind.

1. Describe the power flow through the transmission in L, D, and R.
2. Explain the basic difference between the conditions in the transmission in L low and D low.
3. Explain the basic difference between the conditions in the transmission in D low and D second.
4. Explain the basic difference between the conditions in the transmission in D second and D, or direct drive.
5. Describe the operation of the governor.
6. Explain how the servo-orifice control valve times the release and application for the various operating conditions.
7. Explain how the primary regulator valve operates; how the secondary regulator valve operates.

8. Explain the purpose and operation of the throttle valve and vacuum diaphragm.
9. Explain the purpose and operation of the modulator plunger and valve.
10. Explain the purpose and operation of the downshift valve and kickdown solenoid.
11. Explain how the 1-2 shift valve and plunger work.
12. Explain how the 2-3 shift valve and plunger work.

SUGGESTIONS FOR FURTHER STUDY

If possible, examine the complete transmission, as well as its various components. Torque converters are welded units and cannot be opened for examination. However, your school may have a cutaway torque converter you can examine. Make a list of the clutches and brake bands in the transmission, and show which are applied or released during the various shift positions.

If you can get copies of the hydraulic circuit in black and white (similar to those in the plates but uncolored), use colored pencils to show the circuits under pressure. (Schools sometimes prepare these for student use.) Tracing out the circuits in this manner (following the plates) is a good way to see exactly what happens in the hydraulic system. Also, study any shop manuals you can find which cover the operation and service procedures for the Borg-Warner transmission. Be sure to write in your notebook any important facts you learn.

chapter 12

BORG-WARNER THREE-SPEED AUTOMATIC-TRANSMISSION SERVICE

This chapter covers the trouble-diagnosis and service procedures for the Borg-Warner three-speed automatic transmission with a compound planetary gearset. The transmission was described in Chap. 11. This chapter is divided into three parts: "Normal Maintenance and Adjustments," "Trouble Diagnosis," and "Transmission Overhaul."

Normal Maintenance and Adjustments

⊘ 12-1 Checking and Adding Transmission Fluid Use only Dexron A.T.F. in this transmission! If you use another fluid, it may cause transmission failure. Fluid level should be checked every 12,000 miles [19,312 km] and be changed every 24,000 miles [38,624 km]. If the car is used in heavy city traffic, with long idling periods, the oil should be changed more often.

1. *CHECKING FLUID LEVEL* The fluid level can be checked with the transmission hot or cold. Remove the fluid-level gauge (Fig. 12-1) to check the fluid level. The car must be on a level surface, and the engine should be running. If the transmission is cold, as when just after starting, check the COLD side of the gauge. If the transmission is hot, check the HOT side. The fluid should be between the LOW and FULL marks. Add Dexron A.T.F. if necessary to bring the level up to FULL. *Do not overfill!* Too much fluid in the transmission can cause high fuel consumption and reduced top car speed. Low fluid can cause transmission overheating and clutch slippage.

2. *CHANGING FLUID* As noted above, only Dexron A.T.F. must be used in the transmission. The oil pan should be drained and the strainer (where present) and pan cleaned every 24,000 miles [38,624 km].

a. Raise the car on a hoist and remove the oil pan and gasket. Throw away the gasket. Drain the fluid from the pan. Clean the pan with solvent and dry it with clean compressed air.
b. Remove and clean the strainer (where present).
c. Install a new gasket on the strainer (where used) and install it.
d. Install a new gasket on the oil pan and install the pan. Tighten the attaching bolts to the correct torque.
e. Lower the car and add the specified amount of fluid to bring the level up to FULL.
f. Start the engine. With the brake applied, move the selector lever through all positions. Put the selector lever in P and recheck the fluid level. Add more fluid if necessary. *Do not overfill!* The car must be on a level surface when the fluid is checked.

NOTE: Bands must be adjusted every time the transmission fluid is changed.

⊘ 12-2 Towing Instructions The car can be towed short distances at low speed on its rear wheels with the selector lever in N *provided that the transmission is not damaged and is operating properly.* If the car must be towed any distance, and if there is any doubt about the condition of the transmission, either disconnect the drive shaft or tow the car on

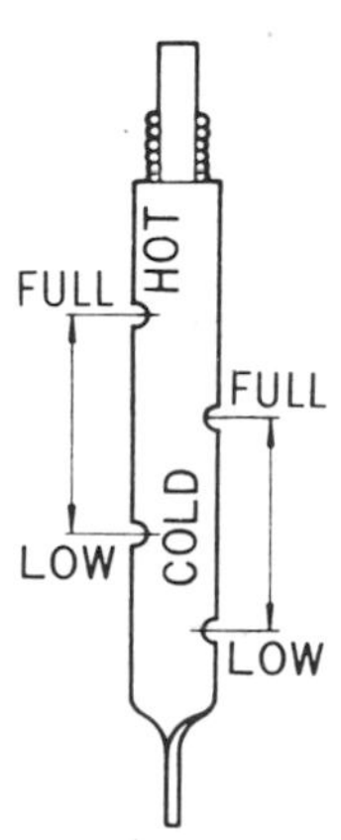

Fig. 12-1. Transmission oil-level gauge. (*Chrysler Corporation*)

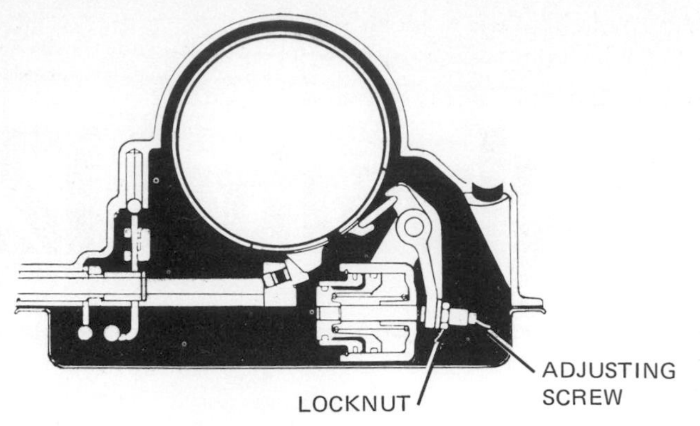

Fig. 12-2. Adjustment of front brake band. (*Chrysler Corporation*)

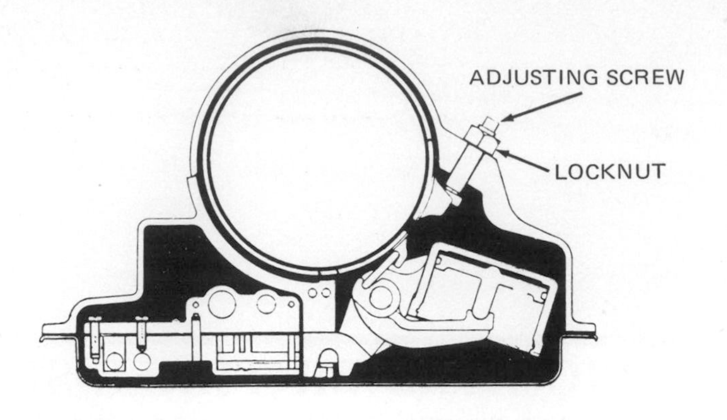

Fig. 12-3. Adjustment of rear brake band. (*Chrysler Corporation*)

its front wheels. Otherwise, the transmission could be seriously damaged.

⊘ 12-3 Band Adjustments The two brake bands must be adjusted as follows every time the fluid is changed, and also if transmission checks indicate the bands are not being applied correctly.

1. *FRONT-BAND ADJUSTMENT* The oil pan must be removed for front-band adjustment. Loosen the locknut (Fig. 12-2). Pivot the servo lever away from the servo and insert the special 0.250-inch [6.35 mm] gauge between the adjusting screw and the servo piston pin. Then, using the special adapter, spin the torque screwdriver to tighten the adjusting screw to the correct tension (10 pound-inches [11.5 kg-cm] torque for one model). Tighten the lock nut to the specified tension.

2. *REAR-BAND ADJUSTMENT* Rear-band adjustment (Fig. 12-3) is made by turning the adjusting screw located on the right-hand side of the outside wall of the transmission case. Loosen the locknut. Then, using the special socket tool, tighten the adjusting screw to the proper torque (10 pound-feet [1.38 kg-m]). Then back off the adjusting screw $\frac{3}{4}$ turn. Tighten the locknut to the specified tension.

Trouble Diagnosis

The chart and the list that follow (⊘ 12-4 and 12-5) will enable you to relate Borg-Warner-transmission troubles with possible causes. The chart in ⊘ 12-4 lists transmission troubles followed by numbers. The numbers refer to the list of possible trouble causes given in ⊘ 12-5.

⊘ 12-4 Transmission Troubles After each trouble given in the following chart you will find a series of numbers. The numbers refer to the list of possible trouble causes in ⊘ 12-5. For example, "Rough engagement in R, D, or L" could be caused by conditions 1, 2, 3, 4, 5, 6, or 7 listed in ⊘ 12-5.

BORG-WARNER TROUBLE-DIAGNOSIS CHART

TROUBLE	POSSIBLE-CAUSE NUMBER
ENGAGEMENT PROBLEMS	
1. Rough engagement in R, D, or L	1, 2, 3, 4, 5, 6, 7
2. Delayed engagement in D and L	1, 3, 5, 6, 8, 9, 10, 11, 12, 13, 14
3. Delayed engagement in R	1, 3, 5, 7, 8, 9, 10, 11, 12, 13, 14, 15
4. No engagement	3, 5, 8, 9, 10, 11, 13, 16, 17, 18
5. No drive in forward direction	6, 8, 9, 11, 13, 18, 19
6. Dragging D and L engagement	6, 8, 11
7. No drive in reverse	7, 8, 9, 11, 15
8. Transmission does not go into neutral	6
9. Car not held in P	20, 21
UPSHIFT PROBLEMS	
10. No 1-2 upshift	5, 9, 10, 11, 22, 23, 24, 25, 26
11. No 2-3 upshift	7, 9, 10, 11, 25, 26, 27, 28, 29, 30
12. Too high speed required to upshift	3, 5, 11, 22, 24, 25, 27, 28, 31
13. Upshift occurs at too low speed	2, 4, 5, 11, 22, 28

BORG-WARNER TROUBLE-DIAGNOSIS CHART (*Continued*)

TROUBLE	POSSIBLE-CAUSE NUMBER
ABNORMAL UPSHIFTS	
14. Slipping in 1-2 upshift	2, 3, 4, 5, 8, 9, 10, 11, 22
15. Slipping in 2-3 upshift	2, 3, 4, 5, 7, 8, 9, 10, 11, 22
16. Shock in 1-2 upshift	2, 3, 4, 5, 6, 19, 22, 26, 32
17. Shock in 2-3 upshift	2, 3, 4, 5, 7, 22
18. Tied up in 1-2 upshift	5, 7, 11, 15, 19, 32
19. Tied up in 2-3 upshift	5, 10, 11, 22
DOWNSHIFT PROBLEMS	
20. No 2-1 downshift	2, 15, 22, 24
21. No 3-2 downshift	2, 7, 22, 23, 24
22. Unexpected 3-2 downshift	7, 8, 10
23. Downshift at too high speed	2, 5, 11, 23, 25, 28
24. Downshift at too low speed	2, 5, 11, 24, 25, 27, 28
ABNORMAL DOWNSHIFTS	
25. Slipping in 2-1 downshifting	32
26. Slipping in 3-2 downshifting	3, 4, 5, 10, 11, 22, 28
27. Shock in 2-1 downshift	6, 11, 19
28. Shock in 3-2 downshift	3, 4, 5, 7, 22
KICKDOWN PROBLEMS	
29. No 3-2 kickdown	24, 27, 29, 30, 31
30. No 2-1 kickdown	24, 27, 29, 30
REVERSE PROBLEMS	
31. Slipping or dragging	6, 7, 11, 15
32. Tied up in R	6, 11
LINE-PRESSURE PROBLEMS	
33. Too low at idle	1, 2, 3, 5, 8, 9, 10, 11, 12, 13, 18
34. Too high at idle	1, 2, 3, 4, 33
35. Too low at stall	2, 3, 4, 5, 6, 8, 10, 11, 26
36. Too high at stall	3, 4, 5, 26
ENGINE SPEED AT STALL	
37. Too low	18
38. Too high in D position	3, 5, 6, 8, 9, 10, 11, 15, 16, 18, 19
39. Too high in R position	7, 8, 9, 15, 16, 18
NOISES	
40. In neutral	6, 7, 13, 18
41. In parking position	6, 18
42. In all speed positions	13, 18, 34
43. In 1 and 2 only	6, 34
OVERHEATING	
44. Overheating	8, 15, 18, 22
SHIFTING PROBLEMS	
45. Selector lever operates roughly or hard	35, 36
46. Starter will not operate in N or P	37, 38, 39

⊘ 12-5 Possible Trouble Causes The list that follows is related by numbers to the troubles listed in ⊘ 12-4. In every case, the cure of the trouble is obvious. You either make an adjustment, replace a part or an assembly, or tighten a screw or other part—whatever is required.

1. Engine speed wrong
2. Vacuum diaphragm maladjusted
3. Primary regulator inoperative
4. Throttle valve inoperative
5. Valve-body screw loose or missing
6. Front clutch malfunctioning—seized, slipping, etc.
7. Rear clutch malfunctioning—seized, slipping, etc.
8. Oil level low
9. Manual linkage improperly installed or out of adjustment
10. Oil tube not installed or installed incorrectly
11. Sealing ring not installed or damaged
12. Pump-check valve not installed or inoperative
13. Pump worn or damaged
14. Converter check valve not installed or inoperative
15. Rear band slipping, damaged, or worn, or servo defective
16. Input shaft defective
17. Converter-pump driving hub damaged
18. Converter defective
19. One-way clutch slipping or improperly installed
20. Parking linkage defective
21. Output shaft defective
22. Front band slipping, damaged, or worn, or servo defective
23. Governor valve inoperative
24. 1-2 shift valve inoperative
25. Throttle valve inoperative
26. Modulator valve inoperative
27. 2-3 shift valve inoperative
28. 2-3 shift-valve plunger inoperative
29. Kickdown switch faulty or connection defective
30. Kickdown solenoid, switch, or wiring faulty
31. Vacuum diaphragm maladjusted or faulty
32. One-way clutch seized
33. Secondary regulator valve inoperative
34. Planetary-gear assembly defective
35. Linkage maladjusted or worn
36. Detent worn
37. Neutral safety switch defective or maladjusted
38. Circuit defective
39. Interlock system inoperative

⊘ 12-6 Line-Pressure Check To check the line pressure, make sure the engine is in good operating condition and at operating temperature. Attach an oil-pressure gauge to the transmission case after removing the pipe plug (Fig. 12-4). Install a vacuum gauge in the line to the vacuum diaphragm.

Set the selector lever at R. Apply the brakes and place blocks under the wheels so that the car does not move. The oil-gauge readings should be as noted in the chart (Fig. 12-5). If the line pressure is low, remove the vacuum-diaphragm hose and turn the adjusting screw clockwise (Fig. 12-4). If the line pressure is high, turn the adjusting screw counterclockwise. One turn of the screw changes the adjustment about 10 psi [0.703 kg/cm^2]. If you cannot get the correct line pressures, check the vacuum diaphragm (⊘ 12-7).

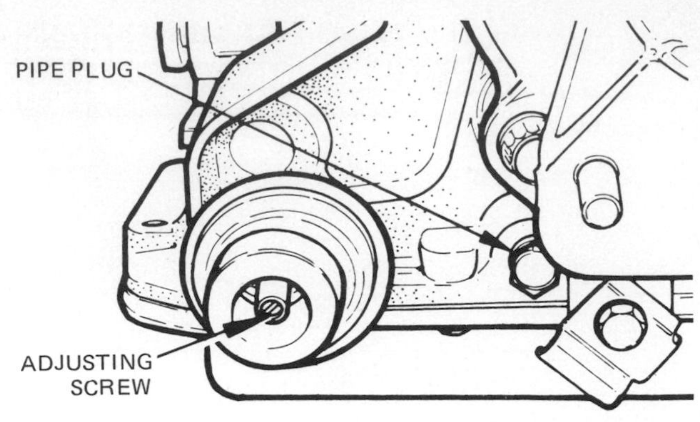

Fig. 12-4. Adjustment of vacuum diaphragm. (*Chrysler Corporation*)

If the oil pressure is normal in R, check the pressure in D and L with the throttle open just enough to give 12 inches [304.8 mm] of vacuum. If the oil pressure is low in D and L but normal in R, check for sticking or other troubles in the governor valve.

If the line pressure is normal in all ranges but no upshift takes place in D, the governor valve is probably sticking closed. Make a road test (or use a chassis dynamometer) to check the shifts listed in Fig. 11-5.

⊘ 12-7 Vacuum-Diaphragm Check The vacuum diaphragm and throttle valve are shown in Fig. 11-14. If the pressure cannot be corrected by adjusting the vacuum diaphragm as explained in ⊘ 12-6, check the diaphragm further. Look for external damage. Remove the vacuum hose and check the hose and tube for oil or gasoline. If there is visible damage or if oil or gas is found, replace the vacuum diaphragm.

If the vacuum diaphragm looks okay, remove it and test it with a source of variable vacuum. Put

CONDITION	MANIFOLD VACUUM, in	LINE PRESSURE, psi
1. Throttle partly open	12	92–98
2. Idling	17.7	55–68
3. Stall	2	160–195

Note: In test 1, open throttle enough to get 12 inches vacuum. Stall test should last not more than 10 seconds. Then return selector lever to N to cool transmission.

Fig. 12-5. Chart of manifold vacuums and line pressures with different operating conditions. (*Chrysler Corporation*)

the rod in place. Increase the vacuum slowly and note the movement of the rod. It must retract completely with a vacuum of 18 inches [457.2 mm]. Next, clamp the vacuum hose to prevent escape of the vacuum in the diaphragm. The rod should remain stationary. Failure of the vacuum diaphragm in either of these tests indicates a leak, which requires replacement of the unit.

⊘ 12-8 Kickdown-Switch and Solenoid Checks To check the kickdown switch, disconnect the lead wire from the solenoid terminal and connect a voltmeter from the lead clip to ground. With the ignition switch at ON and the engine not running, push the accelerator pedal to the floor. If the meter shows no voltage, then the switch requires adjusting or replacing.

To check the kickdown solenoid (Fig. 11-21), use the following procedures:

1. *NO 2-3 UPSHIFT* Disconnect the lead to the solenoid and test the car either on the road or on a chassis dynamometer. If the 2-3 upshift now occurs, the kickdown switch is faulty. If no 2-3 upshift takes place, oil is probably leaking into the solenoid. Replace it.
2. *NO FORCED DOWNSHIFT* Disconnect the wire from the solenoid. Connect a jumper lead from the battery insulated terminal to the solenoid terminal. If the solenoid does not click, it is defective. If the solenoid does click, the trouble is in the switch or circuit to the solenoid.

⊘ 12-9 Torque-Converter Check If the converter seems to be defective (⊘ 12-5, item 18), check it in the following manner.

1. With the selector lever in L or D and blocks under the rear wheels of the car, apply the brakes and open the throttle fully. Check the stall speed of the converter with a tachometer.

CAUTION: Limit the stall check to not more than 10 seconds. Then return the selector lever to N to cool the transmission.

If the stall speed is 300 rpm below the specified value, there is trouble in the torque converter.

2. If the car has poor acceleration or fails to start up a steep slope, either the stator or the stator one-way clutch may be defective. The stator is probably turning backward, and so no torque multiplication results.

Check the stall speed and if it is about 600 rpm below specifications, replace the torque converter.

3. If the car has poor acceleration over 30 mph [48.28 km/h] and low top speed, chances are the stator one-way clutch is locked up, preventing the stator from freewheeling. Replace the torque converter.
4. If the stall speed is too high, the converter fluid may be low or the clutches in the transmission may be slipping.

Transmission Overhaul

⊘ 12-10 Transmission Removal The following procedure applies to transmissions that have the selector lever in the console on the floor of the car (Fig. 11-4). If the selector lever is on the steering column, the procedure is somewhat different. That is, the linkages are different, and a different way of disconnecting them is required. See the applicable manufacturers' shop manuals for details.

1. *OPERATIONS INSIDE THE ENGINE COMPARTMENT* First disconnect the ground strap from the battery. Next disconnect the wire from the kickdown switch. Disconnect the vacuum hose from the vacuum diaphragm. Disconnect and remove the starting motor. Unclip the starting-motor cable from the transmission.

Remove the two upper bolts attaching the transmission to the engine. Remove the clips. Remove oil-filler-tube bracket and tube.

2. *OPERATIONS OUTSIDE THE ENGINE COMPARTMENT* Raise the car on a hoist and drain the transmission fluid. Then disconnect the speedometer cable. Disconnect the drive line from the differential yoke. Pull the shaft out of the transmission. Disconnect the exhaust pipe from the pipe bracket.

Remove the front and rear transmission oil pipes. Disconnect the control-rod assembly. Remove the bell housing cover and the four bolts while rotating the converter.

Place a transmission jack with a wide pad (to prevent damage to the oil pan) under the transmission. Remove two through bolts, ground cable, spacer, and insulator. Remove the support bracket by pulling it off sideways. Then remove the converter-housing-to-engine bolts and lower the transmission from the car.

⊘ 12-11 Transmission-Service Precautions Cleanliness is of the utmost importance in transmission work. The tiniest, almost invisible, piece of lint from a cleaning rag or a small piece of dirt can cause a valve to hang up and prevent normal transmission action. Under some conditions, this could cause fatal damage to the transmission.

1. Never mix parts of different transmissions. Keep all parts belonging to a transmission in one place so that they will not become mixed with other transmission parts.
2. Clean the outside of the transmission thoroughly before starting to disassemble it. Plug all openings and use steam if it is available. The workbench, tools, your hands, and all parts must be kept clean at all times. Do not allow dust to blow in from the outside or from other areas of the shop and settle on the transmission parts. Dirt can mean big trouble.
3. Before installing screws into aluminum parts, such as the case, dip the screws into transmission fluid to lubricate them. Lubrication will prevent the screws from galling the aluminum threads and seizing.

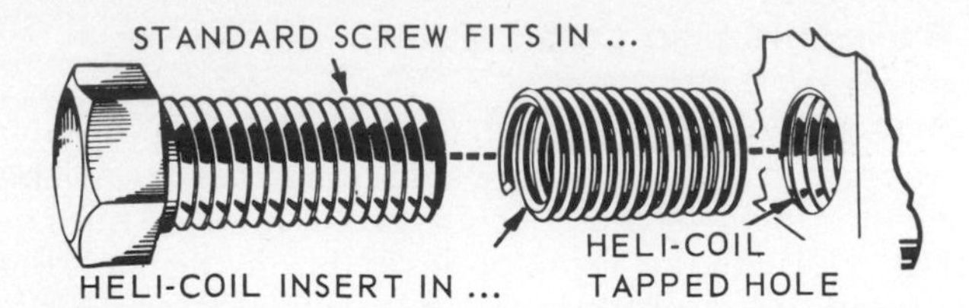

Fig. 12-6. Heli-Coil installation. (*Chrysler Corporation*)

4. If the threads in an aluminum part are stripped, repair them with a Heli-Coil, as shown in Fig. 12-6. To repair a defective thread, drill a hole and then tap it with a special Heli-Coil tap. Finally, install a Heli-Coil to bring the hole back to its original thread size.
5. Special care and special tools to protect the seals must be used during the assembly procedure to prevent damage to the seals. The slightest flaw in a seal or sealing surface can result in an oil leak.
6. The aluminum castings are relatively soft and can be easily nicked, scratched, or burred. Use great care in handling them.
7. Discard all O-ring seals, gaskets, and oil seals that are removed, and use new ones on reassembly.
8. During reassembly, lubricate all internal parts with transmission fluid.
9. During disassembly, clean and inspect all parts, as explained in ⊘ 12-12.

⊘ 12-12 Transmission Disassembly In the following order, remove:

1. Vacuum tube and hose.
2. Torque converter. Be careful not to damage the oil pump driving finger, gear slot, and oil seals.
3. Converter housing.
4. Speedometer driven-gear assembly and gasket.
5. The 15 oil-pan attaching bolts, loosening them in the numbered order shown in Fig. 12-7. Then remove the oil pan.

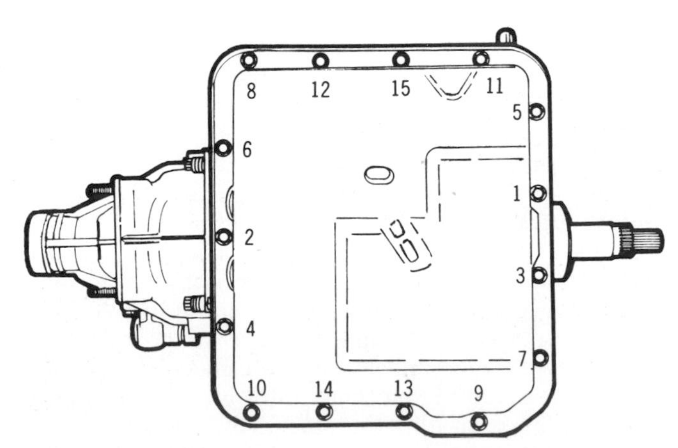

Fig. 12-7. Order in which oil-pan attaching bolts should be loosened when the oil pan is removed. (*Chrysler Corporation*)

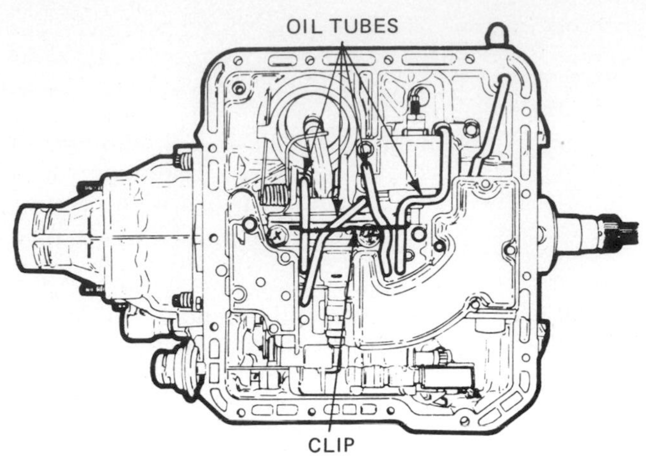

Fig. 12-8. Locations of oil tubes and clip. (*Chrysler Corporation*)

6. Oil-tube clip and four oil tubes (Fig. 12-8). If the tubes stick, pry them loose, being careful not to damage them or the transmission parts.
7. Magnet, vacuum diaphragm, and rod.
8. Kickdown wiring.
9. The three-valve-body bolts and valve body. Lift the valve body off carefully so that you do not damage the oil-feed tube.
10. Extension housing.
11. Oil pump. First measure the end play of the input shaft with a dial gauge or special tool and feeler gauge (Fig. 12-9). If the end play is not correct (0.01 to 0.03 inch [0.25 to 0.76 mm]), replace the old input-shaft thrust washer with a new one that will give the proper end play. Thrust washers come in various thicknesses. See A in Fig. 12-10.

Next, remove the oil tube from the pump adapter. Do not drop the O ring. Then remove the six oil-pump attaching bolts. Press the input shaft in and pull off the pump assembly. Remove the gasket.
12. Front clutch. Pull it out together with the input-

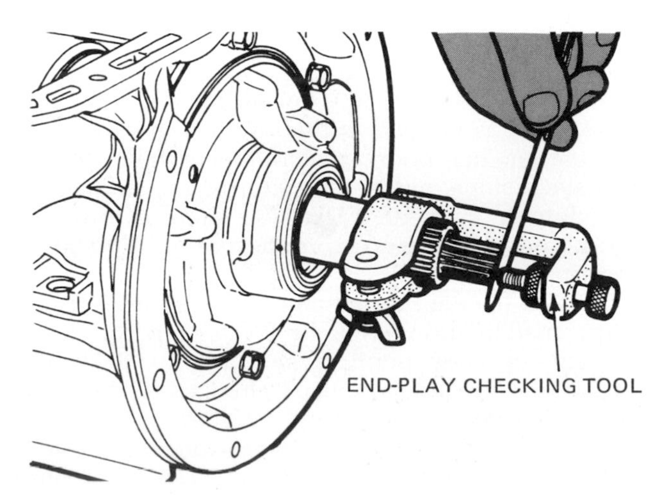

Fig. 12-9. Measuring input-shaft end play. (*Chrysler Corporation*)

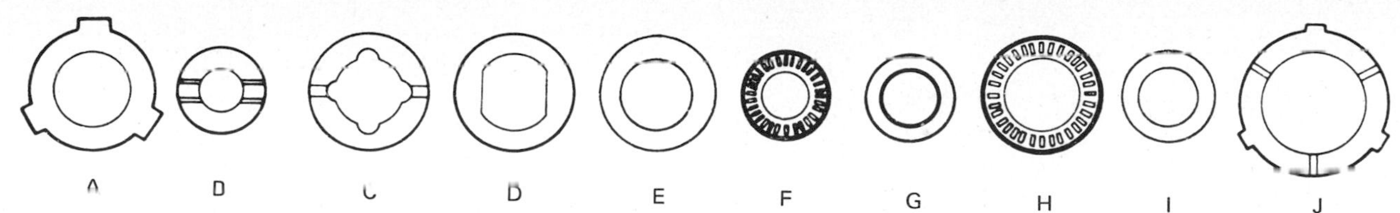

A. Input-shaft thrust washer (selective adjustment)—between pump adapter and converter support and input shaft
B. Front-clutch-hub thrust washer—between input shaft and front-clutch hub
C. Front-clutch-cylinder thrust washer (bronze)—between front-clutch cylinder and rear-clutch spring seat
D. Front-clutch-cylinder thrust washer (steel)—between front-clutch cylinder and rear-clutch spring seat
E. Thrust washer—between reverse sun gear and forward sun gear
F. Needle thrust bearing—between forward sun gear and planet cover
G. Needle-thrust-bearing plate—between forward sun gear and planet cover
H. Needle thrust bearing—between planet cover and output shaft
I. Needle-thrust-bearing plate—between planet cover and output shaft
J. Output-shaft thrust washer—between output shaft and transmission case

Fig. 12-10. Types of thrust washers, thrust bearings, and plates. (*Chrysler Corporation*)

shaft thrust washer. Take out the two front-clutch thrust washers (Fig. 12-11).

13. Rear clutch. Pull off the rear clutch with the sun gear (Fig. 12-12). Then remove the two sealing rings and separate the forward sun gear from the clutch (Fig. 12-13).

14. Front servo. Remove the two bolts (Fig. 12-14). They are of different lengths. The long bolt (to the rear) is a dowel bolt. Remove the front brake band by squeezing the ends together.

15. Rear servo. Remove the two bolts, rear servo, and strut (Fig. 12-15).

16. Planetary gear and center support. Remove two screws from the two sides of the transmission case. Pull off the center support, planetary gearset, needle bearing, and washers (Fig. 12-16). Remove the center support and clutch from the planetary gearset. Then remove the snap ring and one-way clutch outer race from the gearset.

17. Rear brake band. Squeeze the band together and take it out from the front of the case.

18. Speedometer drive gear and ball by removing the snap ring.

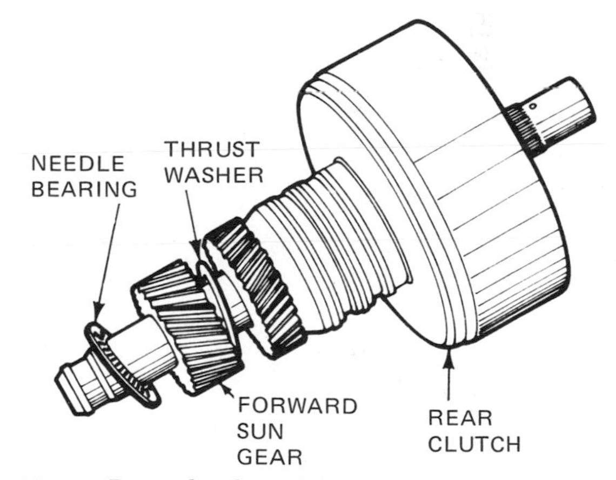

Fig. 12-12. Rear clutch and forward sun gear. (*Chrysler Corporation*)

19. Governor and ball by removing the snap ring.

20. Air-breather baffle plate and adapter.

21. Output shaft and ring gear with thrust washer (J in Fig. 12-10) from front of case.

CAUTION: Do not damage the white metal bushing at the rear of the case.

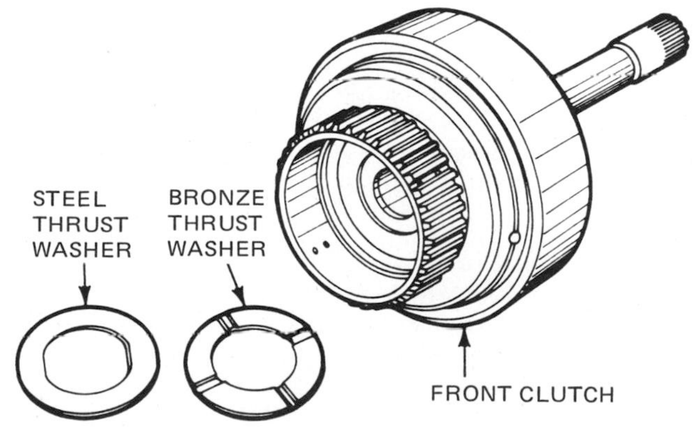

Fig. 12-11. Front clutch with thrust washers. (*Chrysler Corporation*)

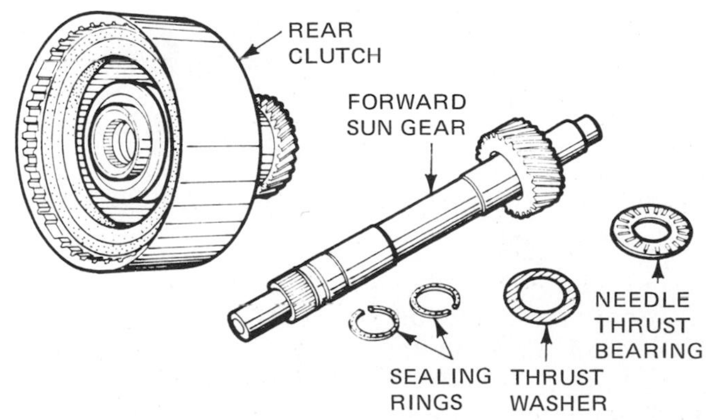

Fig. 12-13. Disassembled rear clutch and forward sun gear. (*Chrysler Corporation*)

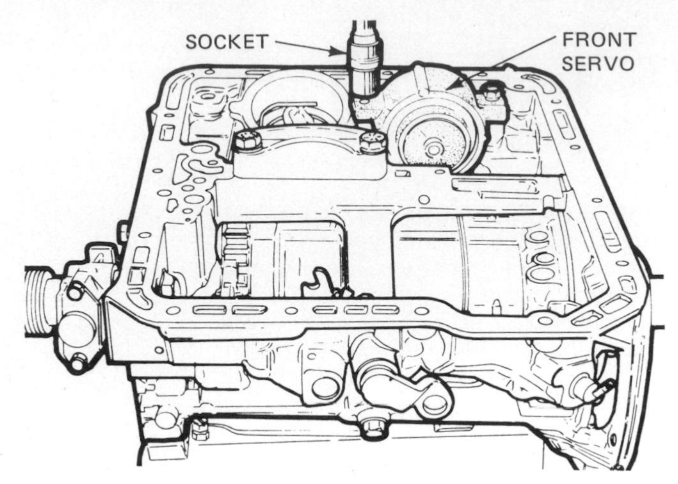

Fig. 12-14. Removing the front servo. (*Chrysler Corporation*)

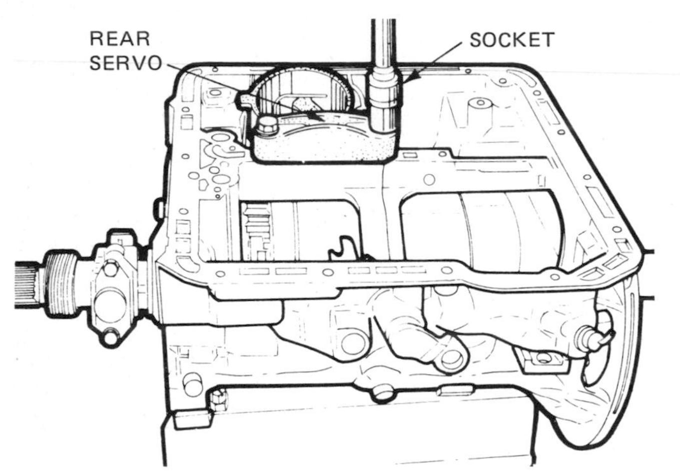

Fig. 12-15. Removing the rear servo. (*Chrysler Corporation*)

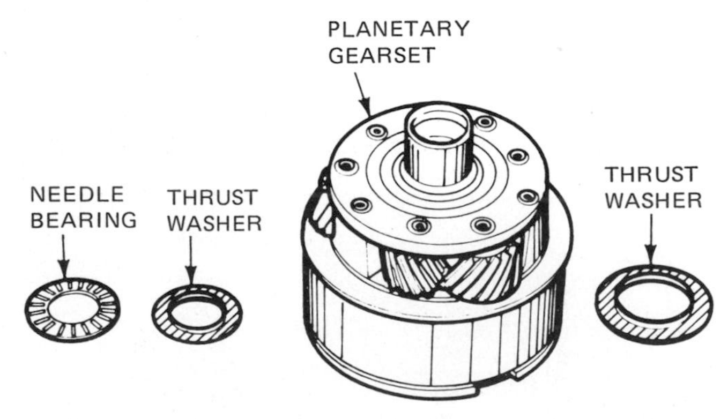

Fig. 12-16. Planetary gearset. (*Chrysler Corporation*)

22. Shaft lever and parking brake. See Figs. 12-17 and 12-18. Pull out the roll pin, anchor pin, and toggle pin. To remove the anchor pin, use a magnet or shake it out. Then pull the toggle pin to the rear.

Loosen the nut and remove the inside manual lever with the manual control rod. Pull out the taper

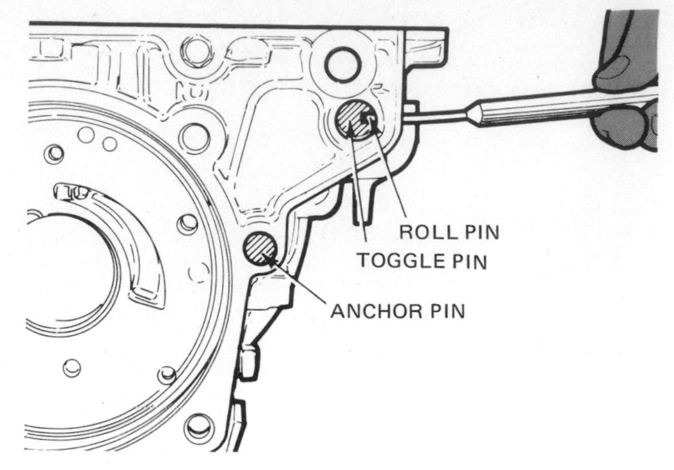

Fig. 12-17. Removing the shaft lever and parking brake, step 1. (*Chrysler Corporation*)

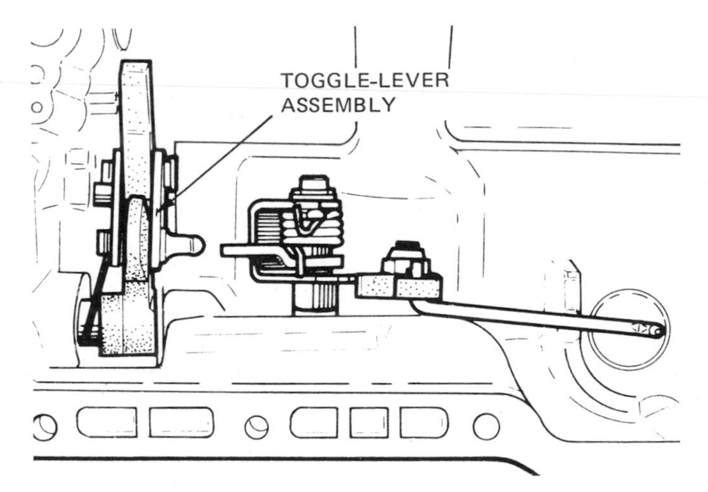

Fig. 12-18. Removing the shaft lever and parking brake, step 2. (*Chrysler Corporation*)

pin and draw the manual shaft out. Remove the retainer spring and washer from the torsion-lever pin and then remove the torsion lever, toggle lever, and toggle-lever assembly with spring. Then pull out the taper pin and pull the torsion-lever pin out from the case.

23. The two oil-cooler connecters, rear-band adjusting screw, and pipe plugs.

NOTE: The preceding list is the complete disassembly procedure. As a rule, the complete procedure is not necessary. Only the major parts need be removed for cleaning, inspection, and replacement.

⊘ 12-13 Front-Clutch Service Figure 12-19 shows the disassembled front clutch. To disassemble the clutch, remove the two snap rings (1 and 6). Check the plates and thrust washer for wear and damage. Also, check the sealing rings, O rings, and piston check valve for defects. Replace defective parts.

To reassemble the front clutch, install the sealing ring in the piston groove. Insert the O ring in

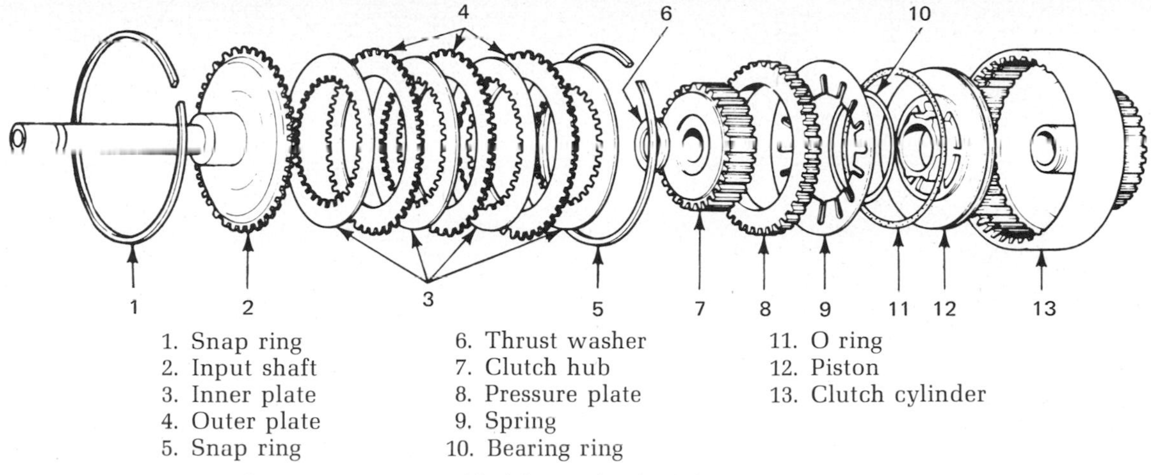

1. Snap ring
2. Input shaft
3. Inner plate
4. Outer plate
5. Snap ring
6. Thrust washer
7. Clutch hub
8. Pressure plate
9. Spring
10. Bearing ring
11. O ring
12. Piston
13. Clutch cylinder

Fig. 12-19. Disassembled front clutch. (*Chrysler Corporation*)

the cylinder-ring groove. Use the special piston tool to install the piston (Fig. 12-20). Then install the other parts in the order shown in Fig. 12-19.

⊘ **12-14 Rear-Clutch Service** Figure 12-21 shows the disassembled rear clutch. Note the key (11) that locks the gear to the clutch cylinder (10). To disassemble the clutch, remove the snap ring, spacer, and clutch plates. Note that the inner plates are identical to the inner plates of the front clutch. However, the outer plates are slightly conical and are different from the front-clutch outer plates. To remove the spring, use the spring compressor, as shown in Fig. 12-22, so that the snap ring can be removed. Then remove the other parts, as shown.

Inspect all parts for wear or damage. Replace defective parts.

To reassemble the rear clutch, use the spring compressor, as shown in Fig. 12-23, and then assemble all parts in the order shown in Fig. 12-21. When installing the slightly conical outer plates, be sure they all face the same direction.

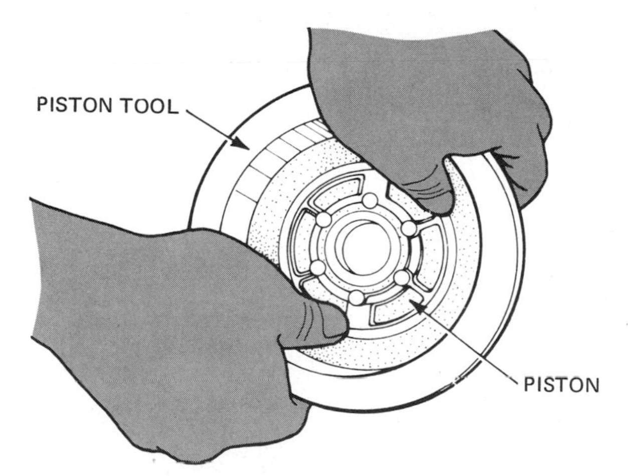

Fig. 12-20. Using a piston tool to install the piston in the front clutch. (*Chrysler Corporation*)

⊘ **12-15 Front Servo** Figure 12-24 shows the disassembled front servo. To disassemble it, remove the snap ring. Reassemble in the order illustrated in Fig. 12-24.

⊘ **12-16 Rear Servo** Figure 12-25 illustrates the disassembled rear servo. It is disassembled by re-

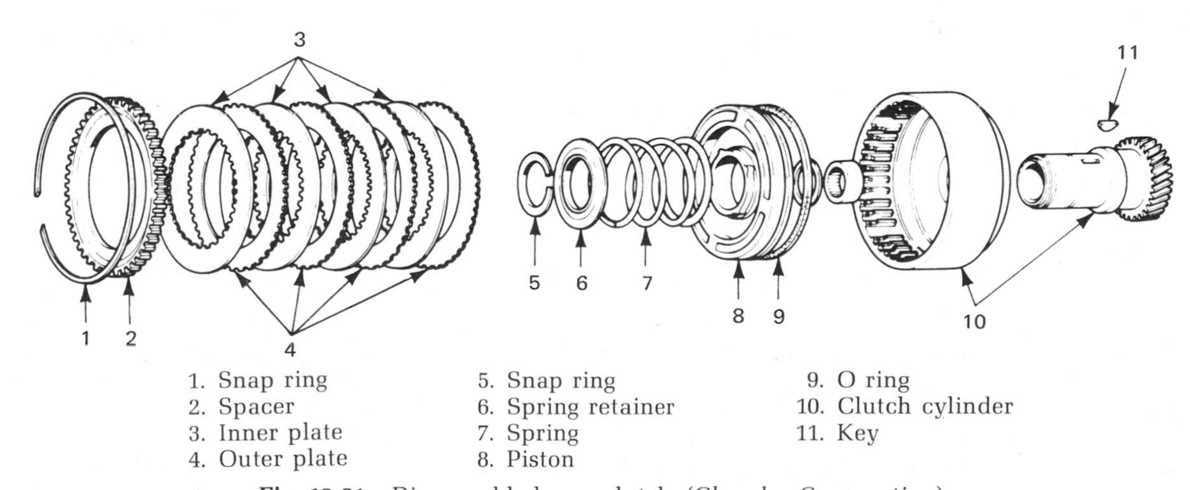

1. Snap ring
2. Spacer
3. Inner plate
4. Outer plate
5. Snap ring
6. Spring retainer
7. Spring
8. Piston
9. O ring
10. Clutch cylinder
11. Key

Fig. 12-21. Disassembled rear clutch. (*Chrysler Corporation*)

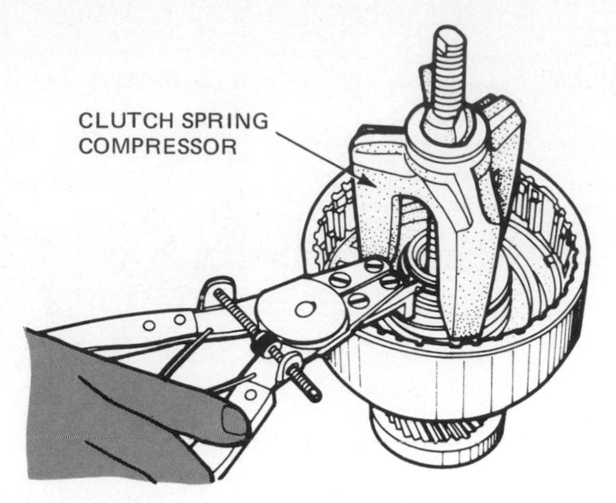

Fig. 12-22. Using a clutch-spring compressor to disassemble the rear clutch. (*Chrysler Corporation*)

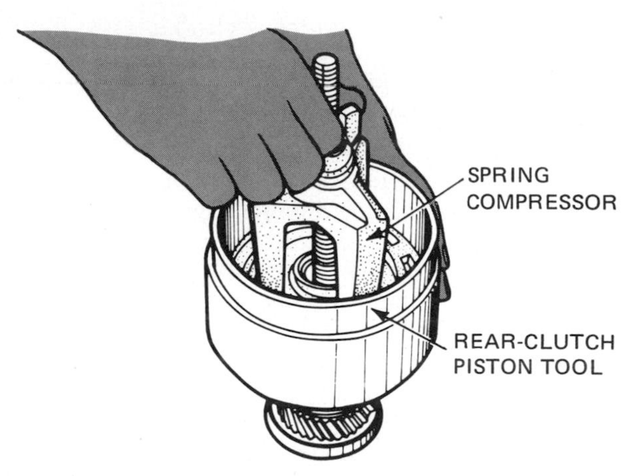

Fig. 12-23. Using a clutch-spring compressor to assemble the rear clutch. (*Chrysler Corporation*)

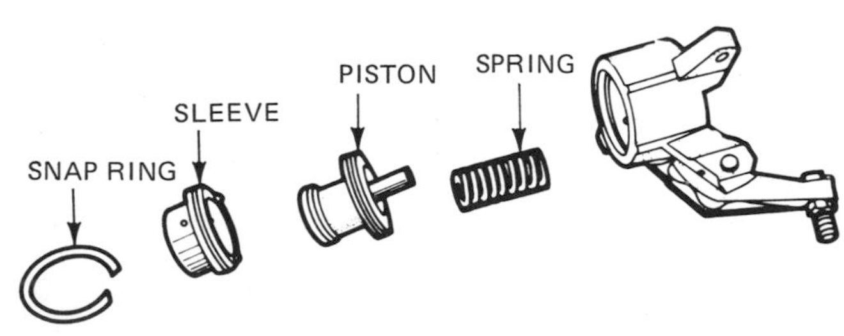

Fig. 12-24. Dissassembled front servo. (*Chrysler Corporation*)

moving the spring and pulling out the pivot pin. Reassemble as shown in Fig. 12-25.

⊘ **12-17 Governor** The disassembled governor is shown in Fig. 12-26. Removing the screws and retainer permits complete disassembly of the unit. Inspect the valve for free movement in the body. To reassemble, follow the order shown in Fig. 12-26.

⊘ **12-18 Oil Pump** Remove five bolts and one screw to separate the oil pump from the support (Fig. 12-27). Scribe a mark across the driven and drive gears before removing them. The mark will permit realignment of the parts in the original order on reassembly. Check the oil seal and O ring, and check the pump housing and gears for wear or damage.

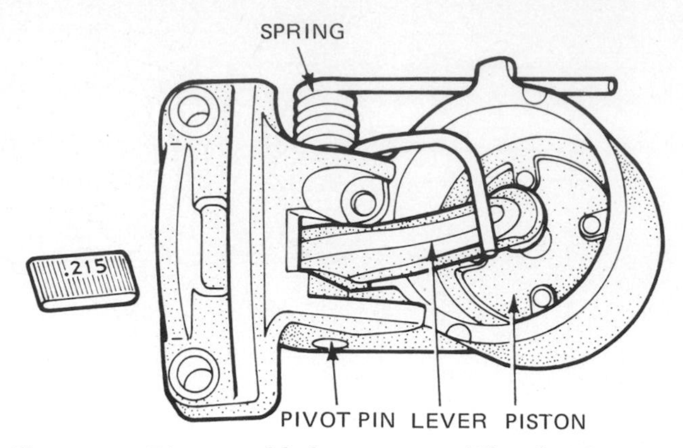

Fig. 12-25. Dissassembled rear servo. (*Chrysler Corporation*)

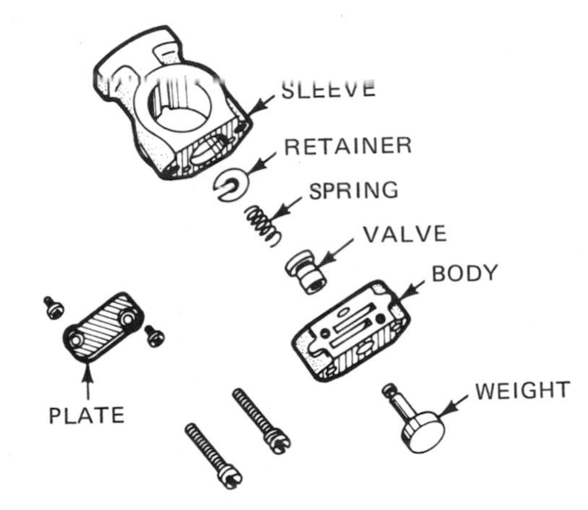

Fig. 12-26. Dissassembled governor. (*Chrysler Corporation*)

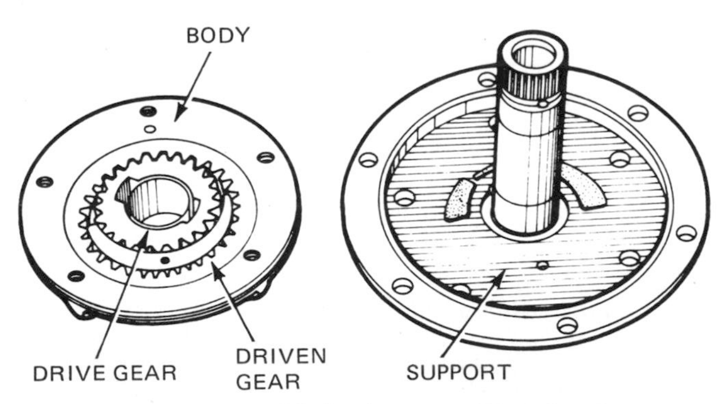

Fig. 12-27. Dissassembled oil pump. (*Chrysler Corporation*)

⊘ **12-19 Planetary Gearset** Figure 12-28 shows the planet-pinion carrier with the center support and one-way clutch. Note that the planet-pinion carrier, with the planet pinions, is serviced as a complete unit and should not be disassembled.

Inspect the bearings, gears, and one-way clutch for damage or wear. Replace defective parts.

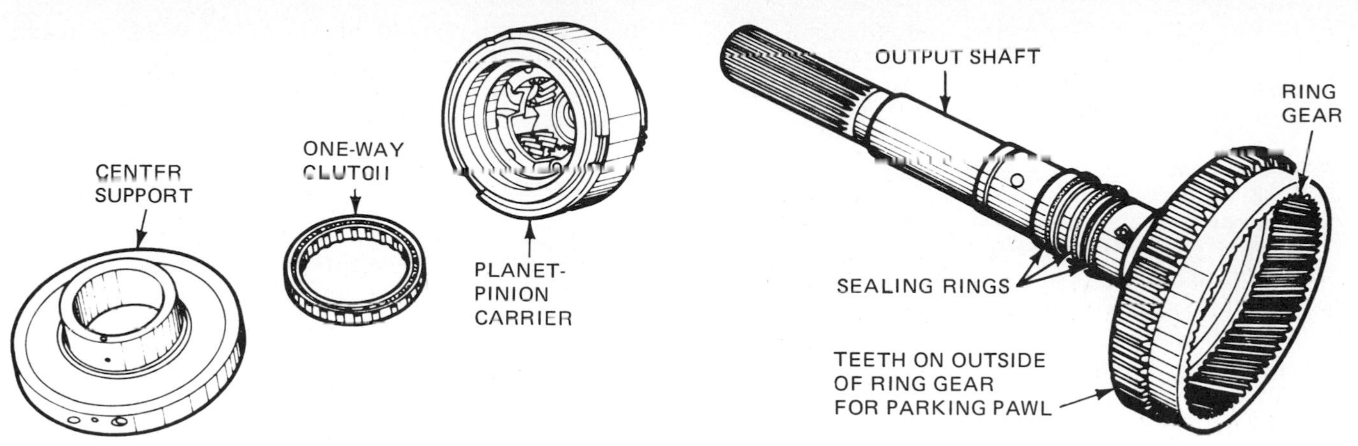

Fig. 12-28. Dissassembled planetary-gear carrier. (*Chrysler Corporation*)

Fig. 12-29. Output shaft and ring gear. (*Chrysler Corporation*)

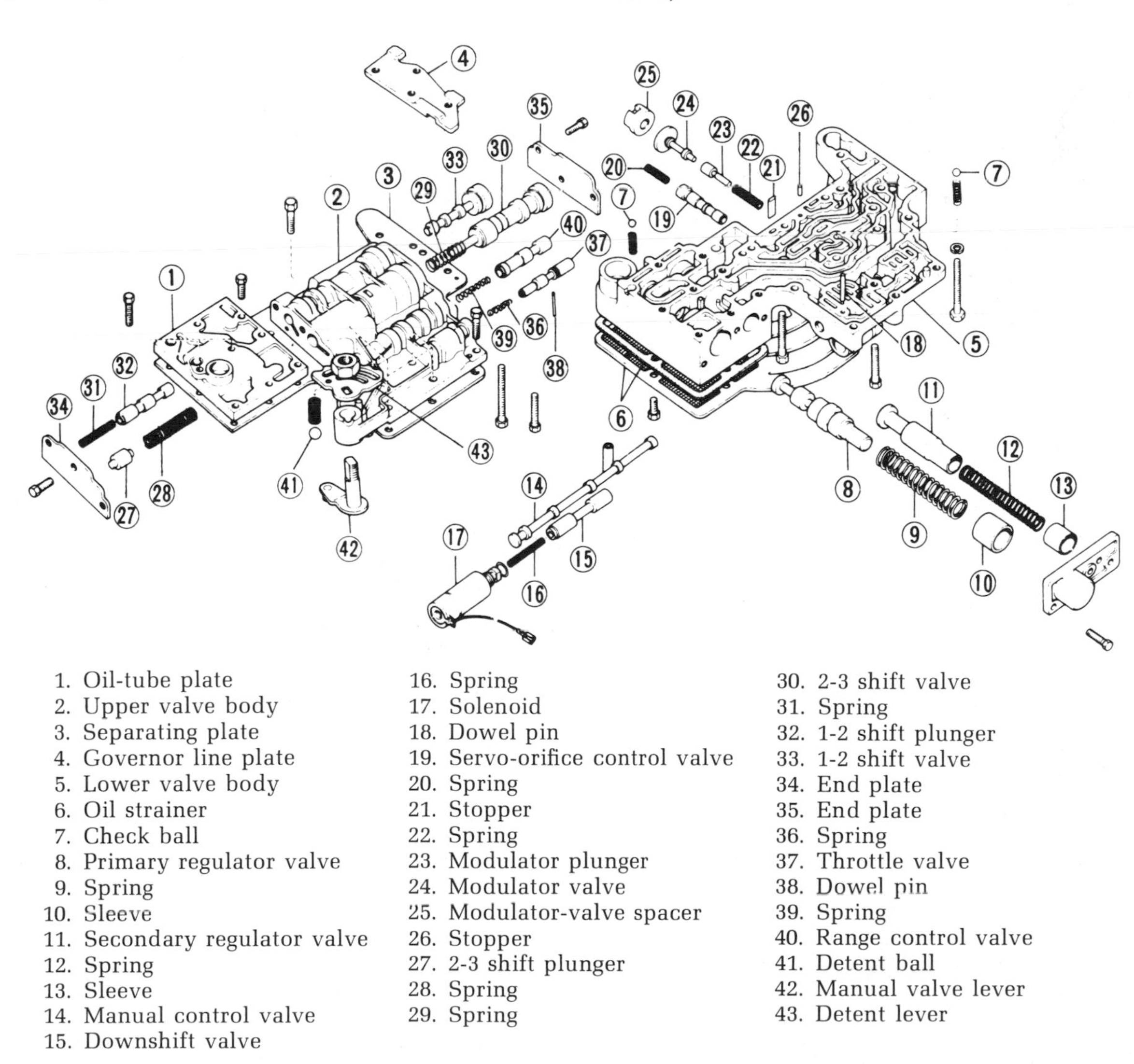

1. Oil-tube plate
2. Upper valve body
3. Separating plate
4. Governor line plate
5. Lower valve body
6. Oil strainer
7. Check ball
8. Primary regulator valve
9. Spring
10. Sleeve
11. Secondary regulator valve
12. Spring
13. Sleeve
14. Manual control valve
15. Downshift valve
16. Spring
17. Solenoid
18. Dowel pin
19. Servo-orifice control valve
20. Spring
21. Stopper
22. Spring
23. Modulator plunger
24. Modulator valve
25. Modulator-valve spacer
26. Stopper
27. 2-3 shift plunger
28. Spring
29. Spring
30. 2-3 shift valve
31. Spring
32. 1-2 shift plunger
33. 1-2 shift valve
34. End plate
35. End plate
36. Spring
37. Throttle valve
38. Dowel pin
39. Spring
40. Range control valve
41. Detent ball
42. Manual valve lever
43. Detent lever

Fig. 12-30. Dissassembled valve body. (*Chrysler Corporation*)

⊘ 12-20 Output Shaft and Ring Gear Figure 12-29 shows the output shaft and ring gear. Check the oil-seal ring and ring gear for damage. Replace defective parts.

⊘ 12-21 Control-Valve Body Assembly Figure 12-30 shows the completely disassembled valve body. If it is necessary to disassemble the body, proceed carefully so that you do not mix the parts. Handle parts with care because any nick or scratch on the valves or plungers can cause serious trouble in the transmission. Make sure the valves move freely in the valve body. With valves and valve body dry, put valves in place and turn the valve body vertically. Valves should fall out of their own weight. Check oil passages in the valve body to make sure they are clean. On reassembly, apply transmission fluid to all parts.

⊘ 12-22 Transmission Reassembly Note the transmission-service precautions listed in ⊘ 12-11. Lubricate parts with transmission fluid on reassembly. To hold thrust washers and other parts in place, coat them lightly with petroleum jelly—*not grease!* Tighten all screws, bolts, and nuts to the proper specifications. Handle parts with care because some are made of relatively soft material and can be nicked or scratched easily. Proceed as follows:

1. Install the parking pawl and lever (Fig. 12-31). Note the relationship between the inside manual lever and the toggle-lift lever. Secure the parts with pins.
2. Install the solenoid terminal if it was removed, being sure to use the O ring.
3. Install the oil-cooler connections, rear-band adjusting screw, and pipe plug if they were removed.
4. Install the output shaft (Fig. 12-29). Hold the thrust washer in place with petroleum jelly. A thrust washer of the right thickness (previously selected) must be used to ensure correct end play (Fig. 12-9).
5. Install the adapter (Fig. 12-32). Tighten five bolts to the correct tension.
6. Install the air-breather baffle plate.
7. Install the governor. Put a $\frac{1}{4}$-inch [6.35 mm] ball on the output shaft and slide the governor into place. The cover plate should face the rear. Secure with a snap ring.
8. Install the speedometer drive gear using a $\frac{3}{16}$-inch [4.76 mm] ball. Secure with a snap ring.
9. Install the rear brake band (Fig. 12-33). Make sure the band seat is lined up with the adjusting screw. Note that the rear brake band is narrower than the front brake band.
10. Install the planet-pinion carrier. Stick the needle thrust bearing on the planet-pinion carrier with petroleum jelly. Make sure the one-way-clutch flange faces the front when installing it. Then install the carrier (Fig. 12-34).
11. Install the center support. Make sure the one-way-clutch lubricating hole and rear-clutch feed hole are visible from underneath. Tighten the two bolts to the specified tension.

NOTE: The center-support-screw lock washers are also oil retainers. Make sure the screws are tightened enough so that the flat rims contact the case.

12. Install the rear servo (Fig. 12-35). Hold the strut in place with petroleum jelly while installing the servo. Install the magnet on the bolt head.
13. Install the front band, making sure the seat is in line with the case anchor pin.
14. Install the clutches. With thrust washers held in place by petroleum jelly and the sealing rings installed, bring the two clutches together (Fig. 12-36). Then install the assembly in the transmission case.
15. Install the front servo. Attach the strut to the servo lever with petroleum jelly. Then install the servo. Make sure the strut fits properly into the brake-band slot.
16. Install the oil pump. Use a new gasket between

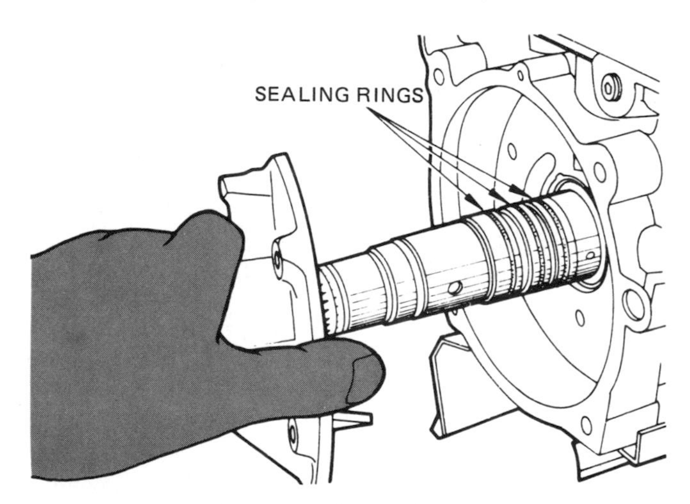

Fig. 12-32. Installing the adapter. (*Chrysler Corporation*)

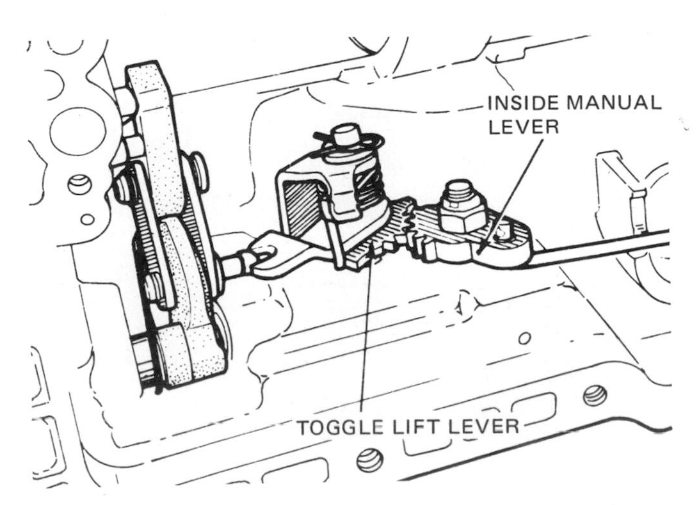

Fig. 12-31. Installing the parking pawl and lever. (*Chrysler Corporation*)

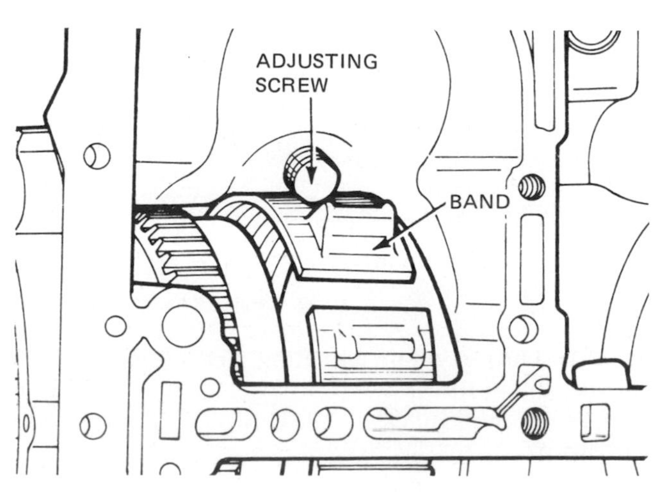

Fig. 12-33. Installing the rear brake band. (*Chrysler Corporation*)

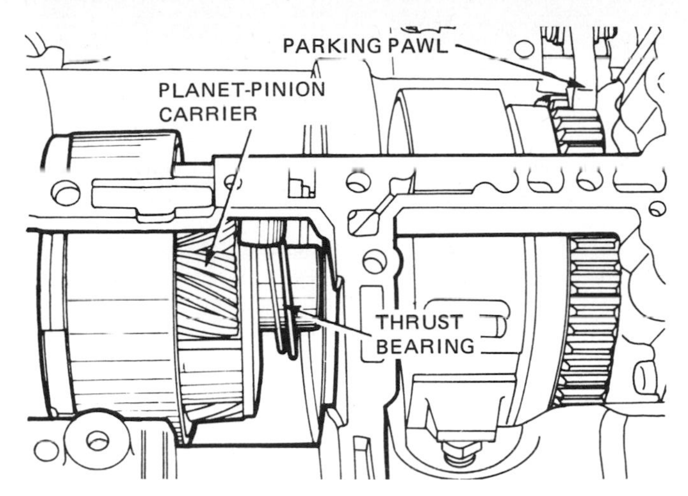

Fig. 12-34. Installing the planetary gearset. (*Chrysler Corporation*)

the oil pump and the transmission case. Stick the input-shaft thrust washer to the oil pump with petroleum jelly. Install the pump and tighten the six bolts to specifications. Recheck the input-shaft end play (Fig. 12-9). If end play is not correct, replace the thrust washer with a thrust washer of the correct thickness.
17. Install the extension housing. Use a new gasket.
18. Install the drive-gear assembly with a new gasket.
19. Install the oil tubes (Fig. 12-37).
20. Install the valve body. Make sure the manual control rod is fitted properly in the detent lever. Connect the solenoid wire to the terminal on the case. Tighten three bolts to specifications. Do not tighten too much!
21. Install the four oil tubes shown in Fig. 12-8. Install the clip.
22. Install the oil pan, tightening the screws in the sequence shown in Fig. 12-38.
23. Install the converter housing with six bolts.
24. Install the torque converter, making sure to properly align the converter-hub finger with the slot in the oil-pump drive gear.

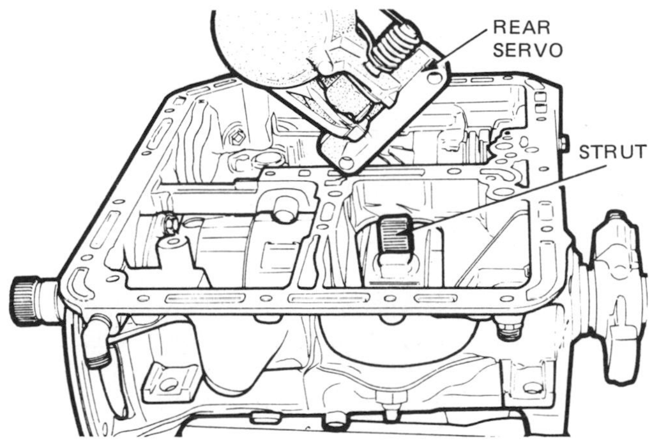

Fig. 12-35. Installing the rear servo. (*Chrysler Corporation*)

⊘ 12-23 Transmission Installation When installing the transmission, observe the following procedure:

1. Use a transmission jack to raise the transmission into place.
2. Tighten the insulator bolts and nuts to the specified torque.
3. With the selector lever in N or P, turn the ignition switch on and make sure the starting motor operates.
4. See ⊘ 12-24 for installation of the transmission control.

⊘ 12-24 Transmission Control Figure 12-39 shows the console-mounted selector lever and associated parts. Refer to this figure if disassembly is necessary. Figure 12-40 shows the selector-lever positions. Figure 12-41 shows how to adjust the rod-adjusting nut. Figure 12-42 shows how the neutral safety switch should be adjusted when it is installed. With the selector lever in N, there should be 0.059-inch [1.50 mm] clearance between the selector lever and the switch, as shown. The control rod should be adjusted as shown in Fig. 12-43.

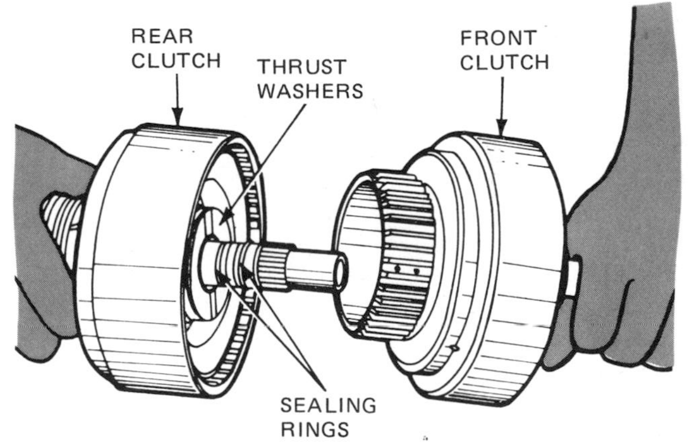

Fig. 12-36. Installing the front and rear clutches. (*Chrysler Corporation*)

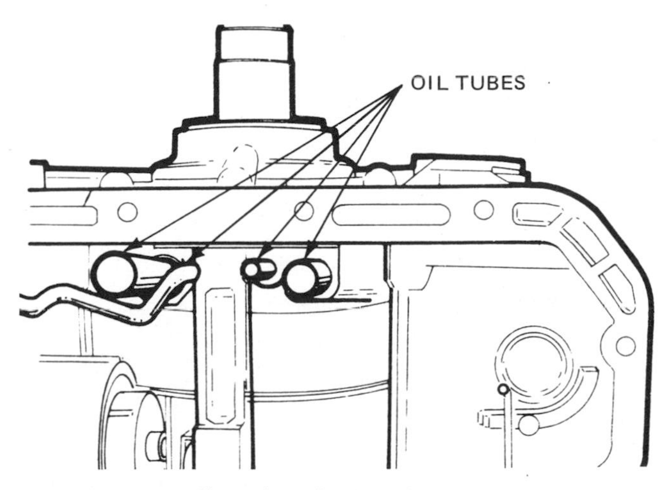

Fig. 12-37. Installing the oil tubes. (*Chrysler Corporation*)

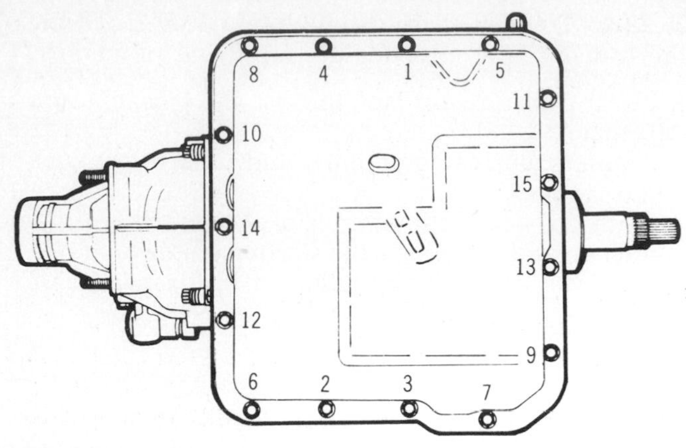

Fig. 12-38. Order in which the oil-pan attaching bolts should be tightened when the oil pan is installed. (*Chrysler Corporation*)

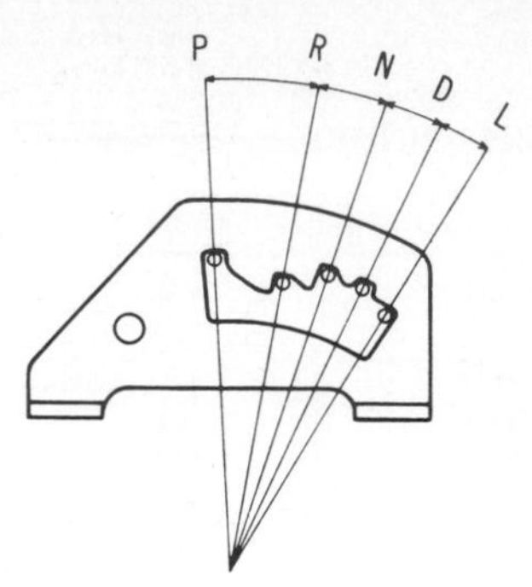

Fig. 12-40. Selector-lever positions. (*Chrysler Corporation*)

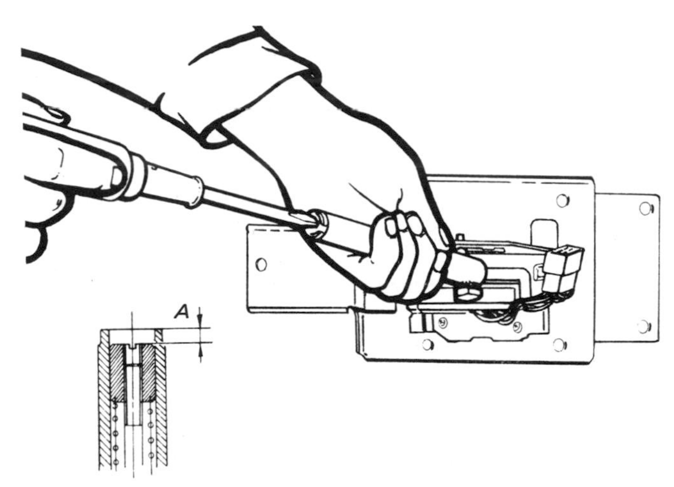

Fig. 12-41. Adjusting the rod adjusting nut. (*Chrysler Corporation*)

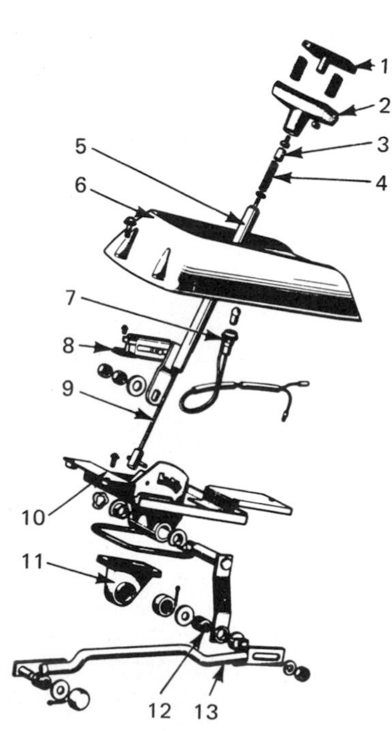

1. Push button
2. Shift handle
3. Rod adjusting nut
4. Rod-return spring
5. Selector-lever assembly
6. Position-indicator assembly
7. Indicator-light-socket assembly
8. Inhibitor switch
9. Shift-lever rod
10. Shift-lever-bracket assembly
11. Lever bracket cover
12. Transmission control arm
13. Transmission control rod

Fig. 12-39. Disassembled view of a console-type transmission selector lever and linkages. (*Chrysler Corporation*)

CHAPTER 12 CHECKUP

NOTE: Since the following is a chapter review test, you should review the chapter before taking it.

Chapter 11 described the operation and construction of the Borg-Warner automatic transmission which uses a compound planetary gearset to get three forward speeds and reverse. This chapter has explained how the transmission is maintained and serviced. To check up on yourself and see how well you remember the service information presented in the chapter, take the test that follows.

Completing the Sentences The sentences that follow are incomplete. After each sentence there are several words, or phrases, only one of which will correctly complete the sentence. Write each sentence in your notebook, selecting the proper word or phrase to complete it correctly.

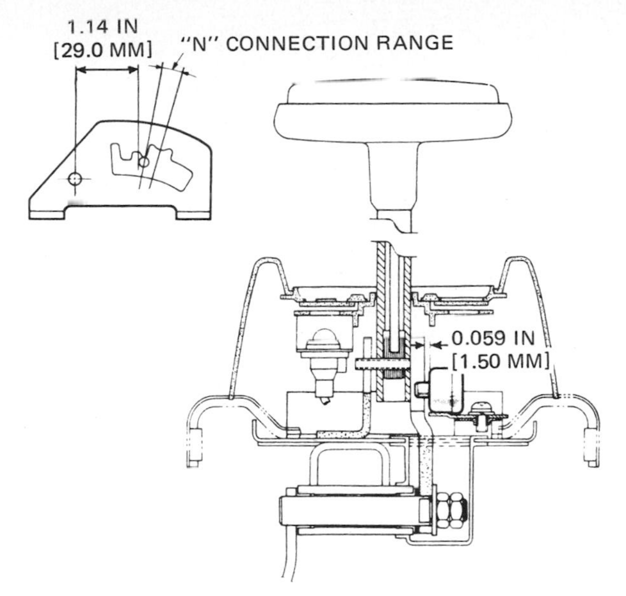

Fig. 12-42. Installation of the neutral safety switch. (*Chrysler Corporation*)

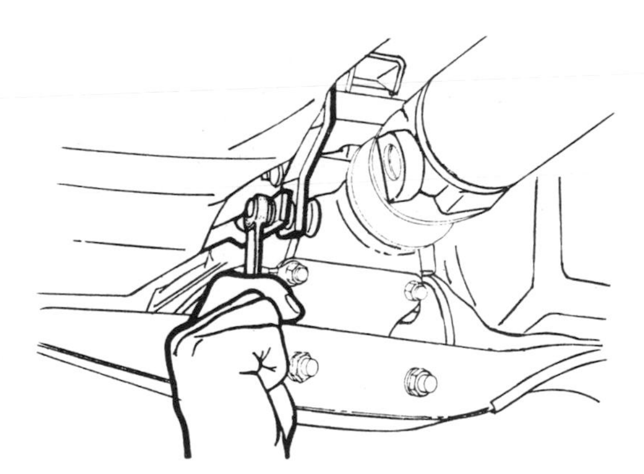

Fig. 12-43. Adjusting the control rod. (*Chrysler Corporation*)

1. The proper fluid to use when adding transmission fluid is: (*a*) W10-10 HD, (*b*) Dexron A.T.F., (*c*) 3-in-1 oil.
2. The fluid level can be checked with the transmission: (*a*) hot only, (*b*) cold only, (*c*) either hot or cold.
3. To adjust the front band, the: (*a*) transmission must be removed, (*b*) oil pan must be removed, (*c*) servo must be removed.
4. To adjust the rear band, the: (*a*) transmission must be removed, (*b*) oil pan must be removed, (*c*) adjusting screw on the outside wall of the case should be turned.
5. As a first step in checking line pressure, check the pressure in: (*a*) R, (*b*) D, (*c*) L.
6. To adjust line pressure, turn the adjusting screw in the: (*a*) primary regulator, (*b*) secondary regulator, (*c*) vacuum diaphragm.
7. The kickdown switch is checked with the ignition switch at ON, the accelerator pedal floored, and the: (*a*) engine not running, (*b*) engine running, (*c*) car on the road or dynamometer.
8. Poor acceleration and failure to start up a steep slope could be caused by: (*a*) a locked-up stator one-way clutch, (*b*) the stator spinning backward due to a defective one-way clutch, (*c*) excessive converter fluid.

Troubles and Service Procedures In the following, write the required trouble causes and service procedures in your notebook. Do not copy from the book. Try to write the trouble causes and procedures in your own words. Writing them in your own words will help you remember them.

1. Describe the procedure of adding oil; of changing oil.
2. Explain how to adjust the front band; the rear band.
3. List the causes of no 1-2 upshift; of no 2-3 upshift.
4. List the causes of slipping in the 1-2 upshift; in the 2-3 upshift.
5. List the causes of failure to downshift from 2 to 1; from 3 to 2.
6. List the causes of failure to kickdown.
7. List the causes of slipping or dragging in R.
8. List the causes of excessively low pressure at idle.
9. List the causes of excessively high pressure at idle.
10. Explain how to check line pressure.
11. Explain how to check and adjust the vacuum diaphragm.
12. Explain how to check the kickdown switch; the kickdown solenoid.
13. Explain how to check the converter and what the various results, such as poor acceleration over 30 mph [48.28 km/h], could mean as related to the converter.
14. Describe the procedure of removing the transmission from the car.
15. Describe the procedure of disassembling the transmission into its major components.
16. Explain how to check the end play of the input shaft.
17. Describe servicing of the front clutch.
18. Describe servicing of the rear clutch.
19. What are the main precautions to observe in overhauling the transmission?
20. Describe the transmission-reassembly procedure.

SUGGESTIONS FOR FURTHER STUDY

When you are in the automatic-transmission shop, pay special attention to the required service procedures. Notice the special tools and how they are used. Review the service procedures outlined in the appropriate manufacturers' service manuals. Write in your notebook any important facts you learn for reference later.

chapter 13

GENERAL MOTORS TYPE 350 TURBO HYDRA-MATIC TRANSMISSION

This chapter describes the construction and operation of the Type 350 Turbo Hydra-Matic transmission used on many General Motors cars. This transmission provides three forward speeds and reverse, and includes a torque converter. Chapter 14 discusses the servicing of this transmission, including trouble diagnosis, minor repairs, and complete disassembly and reassembly of the transmission.

⊘ 13-1 Type 350 Turbo Hydra-Matic Automatic Transmission This transmission is shown in cutaway view in Fig. 13-1. It includes a torque converter, two planetary gearsets, four multiple-disk clutches, two roller clutches, and a brake band. Also, there is a hydraulic system with several valves and servos to provide the pressures required to operate the clutches and brake band. The system provides three forward speeds and reverse. A selector lever and quadrant on the steering-wheel mast (Fig. 13-2) permit the driver to select the desired transmission operating conditions. For example, for normal town and country driving, a driver would select the D, or drive, position. In this position the transmission will start out in low gear and upshift through intermediate or second to high gear as car speed and accelerator position require. The transmission will also downshift for passing when the accelerator is pushed down to full open or when car speed drops off sufficiently.

In L1, or low, the transmission will stay in low gear. This position is used for pulling a heavy load or for going down a steep grade. The low-gear position permits the engine to help brake the car and thus saves the brakes from long periods of use when going down a steep hill.

There is another forward selector-lever position, variously called L2 or S (super range). In L2, the transmission will upshift from low to second when first starting out, after car speed has increased sufficiently. Also, when the selector is moved from D to L2, the transmission will downshift from high to second at speeds below about 70 mph [112.65 km/h]. This permits the engine to help brake the car.

The transmission described in this chapter is used on many different cars with many different types of engine. The transmission is basically the same, regardless of what engine it is used with. However, there are different models of the Type 350 Turbo Hydra-Matic transmission, and they vary in some details. For instance, some models may have more plates in the multiple-plate clutches than others. Also, valves and valve springs can be altered to produce different shifting patterns. For example, when the transmission is used with one engine, the upshift points might be designed to be on the high side. But with another engine having a different torque output curve, best car operation might require that the transmission upshift at somewhat lower speeds. However, the general service procedures for the Type 350 transmission covered in Chap. 14 will be found to be the same for all models.

⊘ 13-2 Oil Cooler The hydraulic oil used in the transmission becomes hot. Part of the heat comes from the engine because the transmission is so close to the engine. But much of the heat comes from the workings of the transmission itself. As the oil is tossed from torque-converter pump to turbine to pump again, it gets hot. Rapid stirring of any liquid will heat it up because of the frictional effects of the layers of liquid passing over each other. In addition to the heating effect in the torque converter, the oil is also heated as it serves as lubricant between the gears and between rotating shafts and bearings.

To prevent overheating of the oil and the transmission itself, many transmissions have an oil-cooling system such as the one shown in Fig. 13-3. The oil circulates through a small radiator that is part of the engine cooling system. In Figs. 13-3 and 13-4, for example, the transmission oil-cooler radiator is located inside the right-hand engine-cooling-system radiator tank. It is immersed in the coolant of the

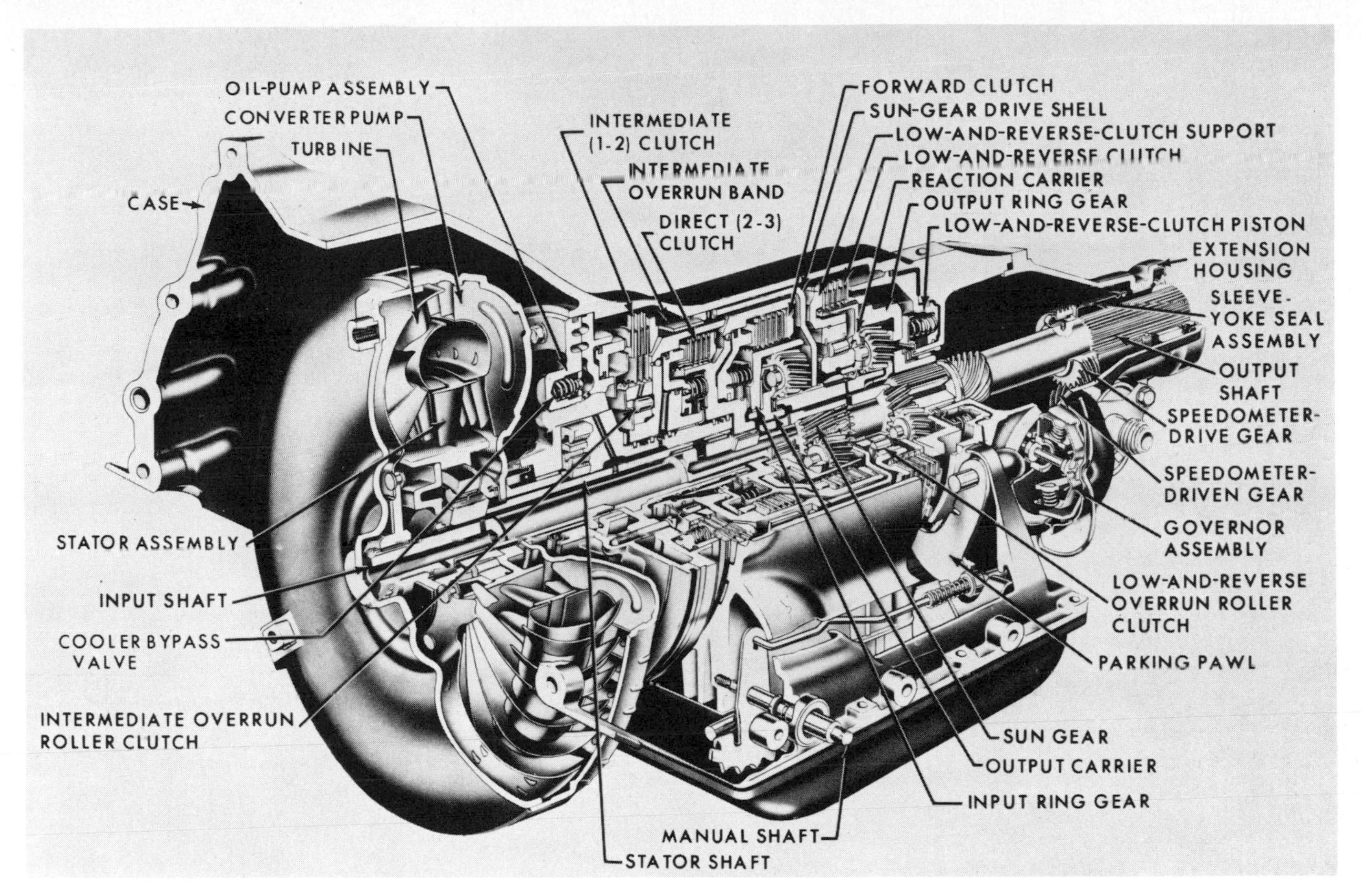

Fig. 13-1. Cutaway view of the Type 350 Turbo Hydra-Matic automatic transmission. (*Oldsmobile Division of General Motors Corporation*)

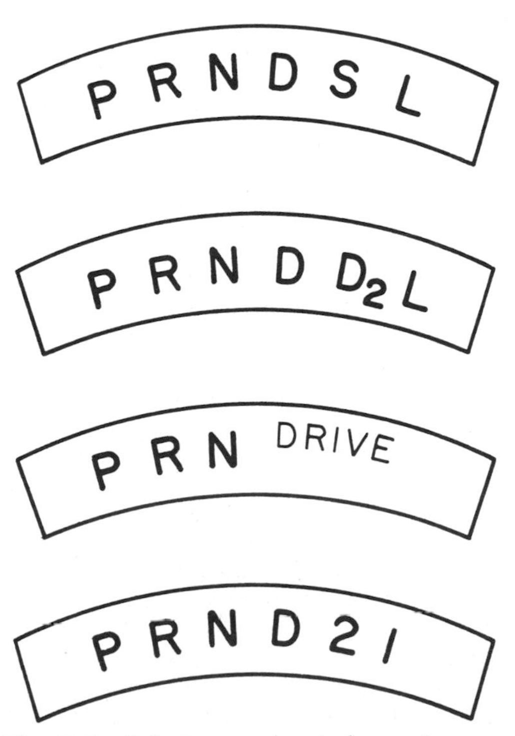

Fig. 13-2. Selector quadrants for various cars.

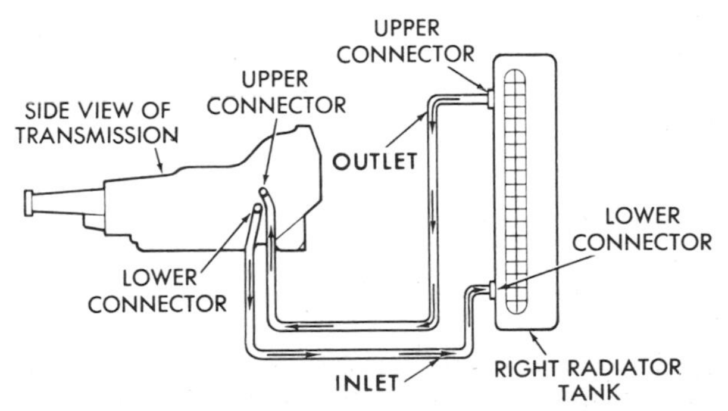

Fig. 13-3. Transmission oil-cooling system. (*Cadillac Motor Car Division of General Motors Corporation*)

engine cooling system. This arrangement permits the engine cooling system to keep the transmission oil and thus the transmission at a sufficiently cool operating temperature.

⊘ 13-3 Clutches, Brake Band, and Planetary Gearsets We shall now begin our study of the Type 350 Turbo Hydra-Matic automatic transmission by looking at the planetary gearsets and their controls.

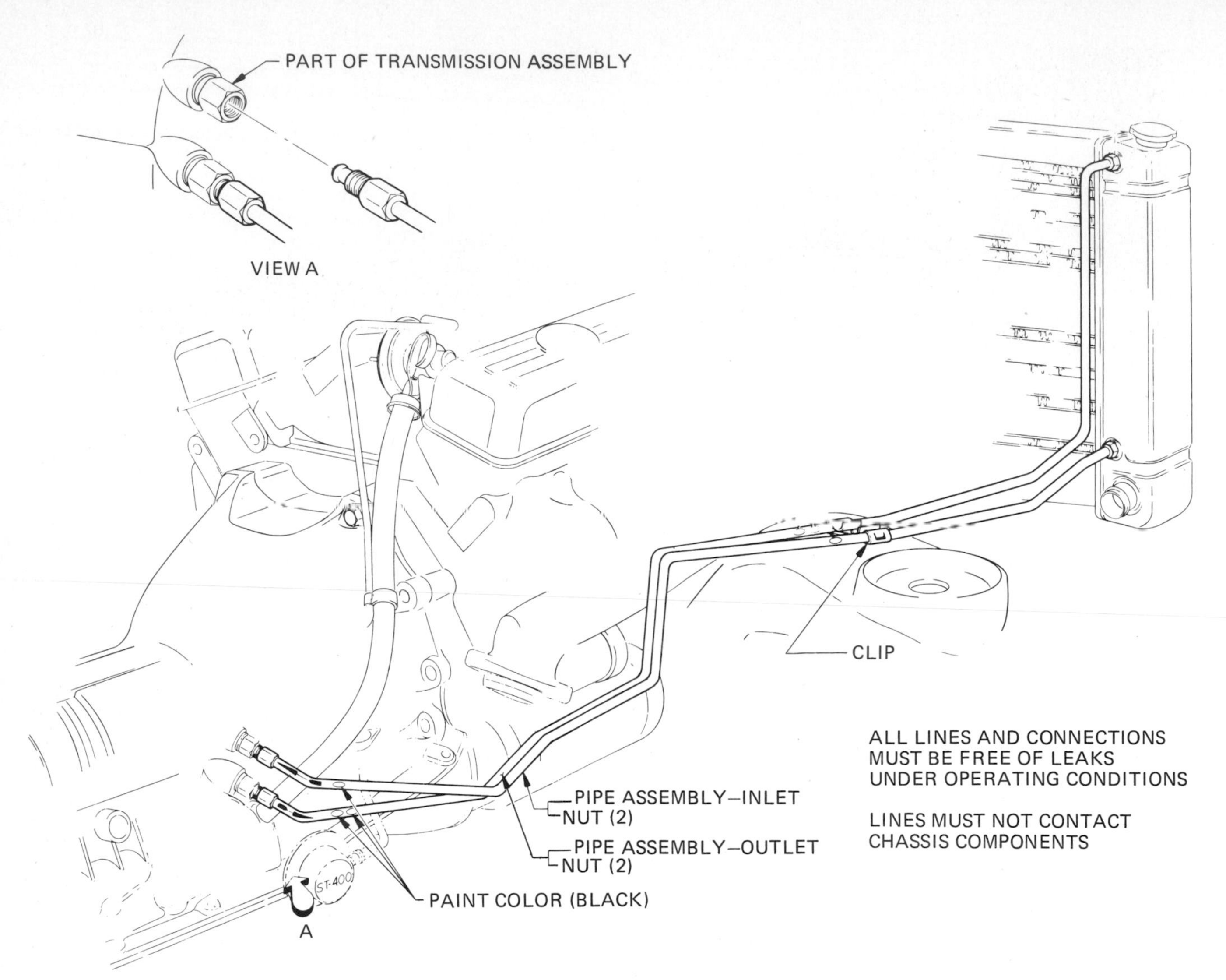

Fig. 13-4. Transmission oil-cooler lines between the transmission and the radiator. (*Buick Motor Division of General Motors Corporation*)

These controls include the clutches and brake band. The clutches and brake band are, in turn, controlled by the hydraulic control system. We shall discuss this system later.

Figure 13-1 is a cutaway view of the complete transmission showing the locations of the various components. Figure 13-5 will give you a better idea of their locations. Let us name the components, from left to right in Fig. 13-5. First is the intermediate overrun roller clutch. This clutch is an overrunning clutch that is closely associated with the intermediate clutch, as we shall explain soon. The intermediate clutch, next in line, is a multiple-disk clutch. We explained (Chap. 7) how overrunning and multiple-disk clutches work. The overrunning clutch permits rotation in one direction but not in the other. Also, it permits one member to run faster than, or overrun, another member. The multiple-disk clutch can lock up two members if the disks are pressed together by hydraulic pressure.

Next in line is the intermediate overrun band. It is positioned around the extension from the front-gearset sun gear. When the brake band is applied, it holds the sun gear stationary. We shall explain the effect of this action shortly.

Next is the direct clutch. This is a multiple-disk clutch. Then there are the front planetary gearset, forward clutch, the low-and-reverse multiple-disk clutch, the low-and-reverse roller clutch, and the rear planetary gearset.

We shall move very carefully and slowly in our explanations of how all these parts work together to achieve the various gear ratios and reverse. We want to be sure that you understand everything that is discussed because once you understand one automatic transmission, you will have little trouble understanding all the others.

⊘ 13-4 Power Flow in D, Low Gear When the driver moves the selector lever to D (drive), linkage to the valve body in the transmission moves a man-

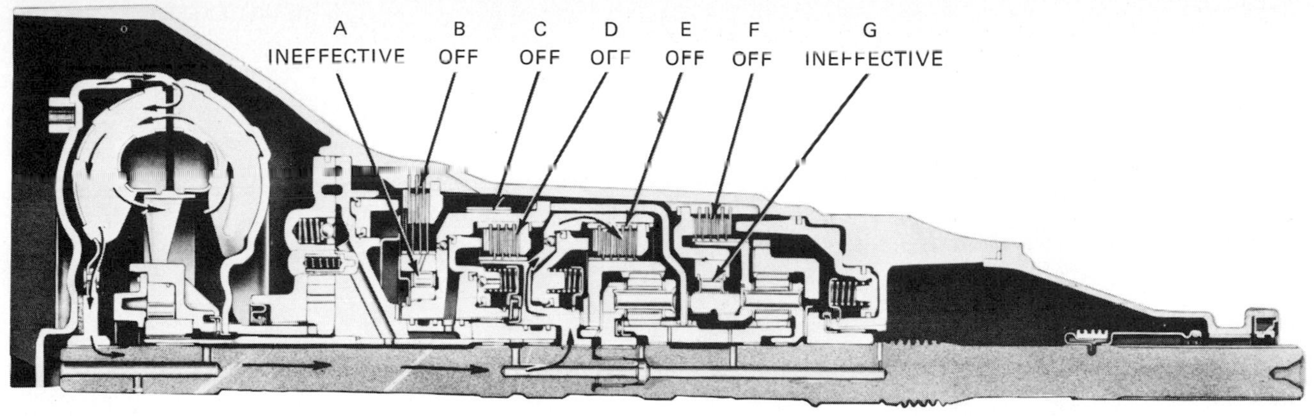

Fig. 13-5. Sectional view of the upper half of the Type 350 Turbo Hydra-Matic automatic transmission. Transmission is in neutral with engine running. (*Oldsmobile Division of General Motors Corporation*)

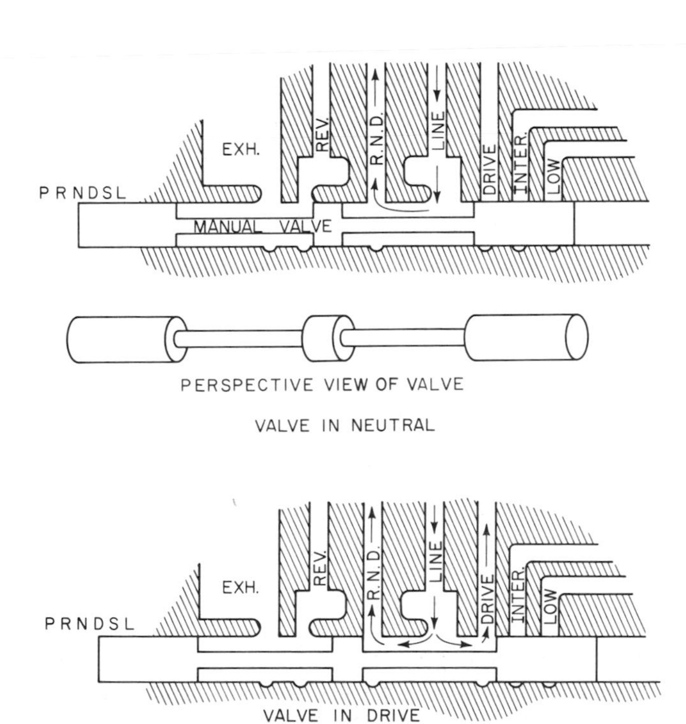

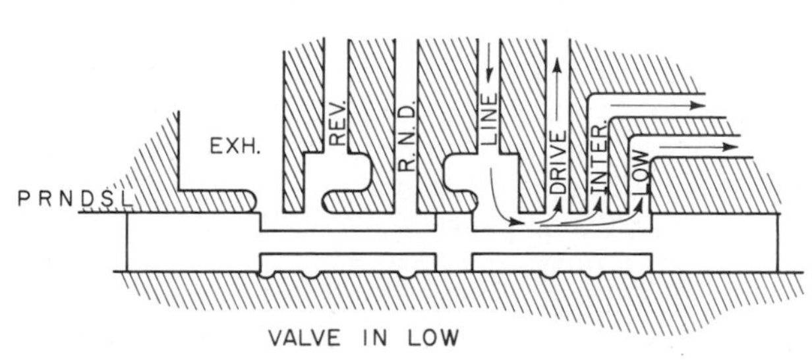

Fig. 13-6. Details of a manual valve, showing the valve in neutral (*top*) and in D (drive) and L (low) positions.

ual valve. Figure 13-6 shows the valve in the valve body. When the valve is pushed in from P to D, hydraulic pressure from the pump can flow into the D circuit of the valve body, as shown. This pressure can then flow from the valve body to other valves in the hydraulic system and from there to the disk clutches and brake-band servo. Now let us see what happens when the selector lever is moved to D and the car pulls away from the curb.

When first starting out in D, that is, in the drive range, the transmission is in low gear. Oil pressure is applied to the forward clutch, as shown in Fig. 13-7. The other disk clutches and brake band are off. With this condition, the power flow is as shown in Fig. 13-7. Power input is at the torque-converter pump (1). (Refer to Fig. 13-7 for locations of numbers and parts.) Power flows to the turbine (2) and to input shaft (3). The turbine hub is splined to the input shaft. The forward-clutch drum assembly (5) is splined to the input shaft at 4. Thus, this drum assembly must rotate with the input shaft. The forward clutch is on, which causes the ring gear of the front gearset to rotate as shown to the upper left in Fig. 13-7. The ring gear drives the planet gears clockwise, as shown, which causes the sun gear to turn counterclockwise. The sun gear of the front gearset (7) and the sun gear of the rear gearset (8) make up a single piece. That is, the two sun gears are at the two ends of a hollow shaft. This hollow shaft rides on the output shaft.

When the front sun gear turns, the rear sun gear must turn also because they are a single piece, as we already mentioned. This action causes the planet gears of the rear set to rotate in a clockwise direction, as shown. The planet-gear cage is held and prevented from rotating by the low-and-reverse

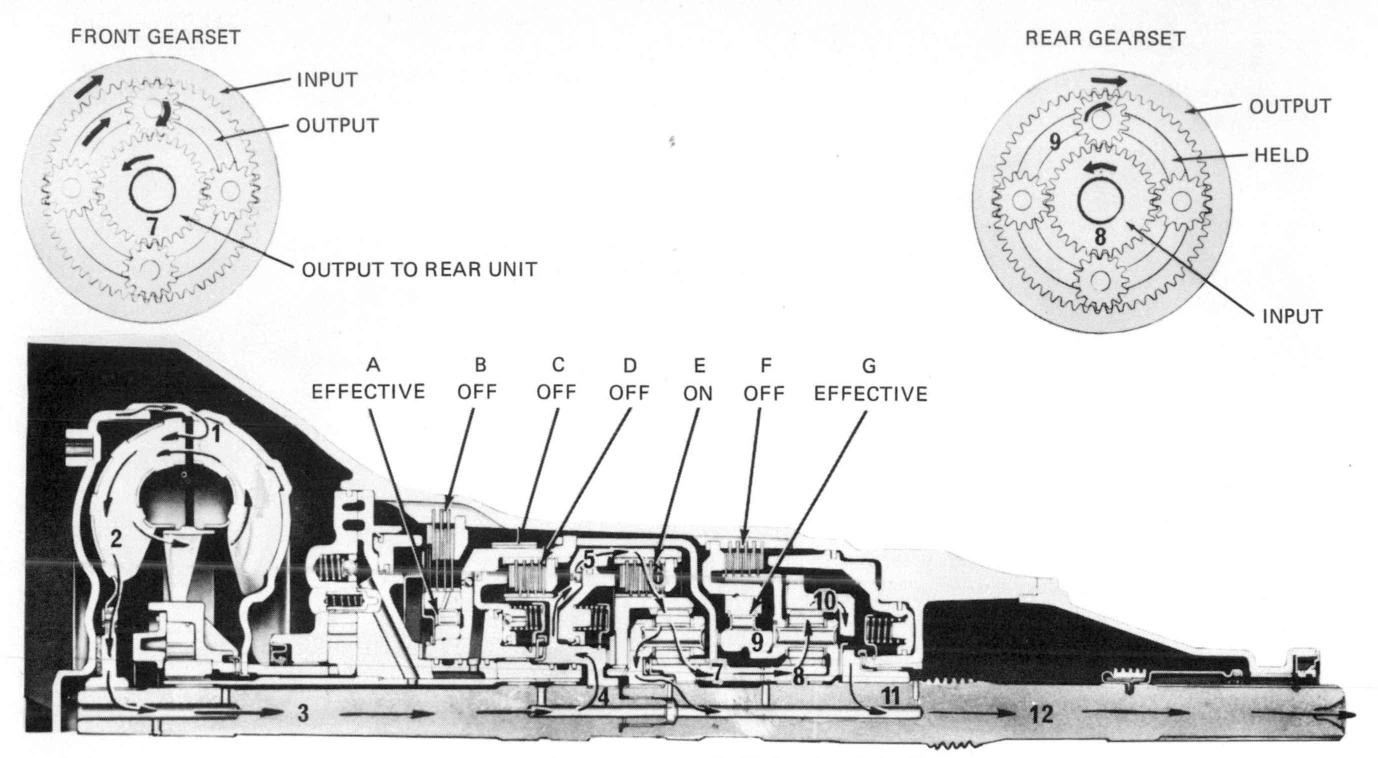

A. INTERMEDIATE-OVERRUN ROLLER CLUTCH—EFFECTIVE
B. INTERMEDIATE CLUTCH—OFF
C. INTERMEDIATE-OVERRUN BAND—OFF
D. DIRECT CLUTCH—OFF
E. FORWARD CLUTCH—ON
F. LOW-AND-REVERSE CLUTCH—OFF
G. LOW-AND-REVERSE ROLLER CLUTCH—EFFECTIVE

Fig. 13-7. Power flow, shown by arrows, through the transmission in drive range and low gear. (*Oldsmobile Division of General Motors Corporation*)

roller clutch (9). Note that the outer race of the roller clutch is attached to the transmission case so that it cannot turn. The inner race is part of the planet-pinion-cage assembly. Thus, when the sun gear rotates counterclockwise, as shown to the upper right, it attempts to cause the planet-pinion cage to turn counterclockwise also. However, the roller clutch (9) locks to prevent this counterclockwise rotation. As a result, the planet pinions act as idlers and drive the ring gear (10) of the rear gearset in a clockwise direction.

The ring gear (10) of the rear gearset is splined to the output shaft (12) at 11. Thus the output shaft turns in a clockwise direction to give the car forward motion. Note that there is gear reduction in both the front and rear gearsets. Their combined gear reduction is 2.52:1. That is, the input shaft must turn 2.52 times to turn the output shaft once. Remember, also, that there can be a "gear" reduction in the torque converter. In other words, the pump can turn several times faster than the turbine, thus producing a speed reduction. This also produces a torque increase, as noted previously.

Note that the intermediate overrun roller clutch is effective, or locked in D, low gear. This means that the sun gear, sun-gear drive shell, direct-clutch housing, intermediate roller clutch, and intermediate-clutch plates are all turning in a counterclockwise direction.

⊘ 13-5 Power Flow in D, Second Gear As engine and car speed increase, an upshift from low gear to second gear occurs. The manner in which the upshift occurs was explained in ⊘ 8.4. We shall examine further the part that the hydraulic system plays in upshifting in ⊘ 13-10.

When conditions are right for the upshift, the hydraulic system applies pressure to the intermediate clutch (4 in Fig. 13-8). This action changes the path of power flow through the transmission, as follows: The power enters through the torque converter and input shaft as before (1, 2, and 3 in Fig. 13-8). Application of the intermediate clutch locks the outer race of the intermediate overrun roller clutch (5) to the transmission case—through the clutch disks (4). The inner race of the intermediate overrun roller clutch (6) is assembled to the sun-gear shell (7) and sun gear (8). When the outer race is locked stationary and the roller clutch is effective, that is, also locked, then the sun gear is locked and cannot turn. It tries to turn in a counterclockwise direction just as it did in first gear. But the intermediate overrun roller clutch locks up to prevent this

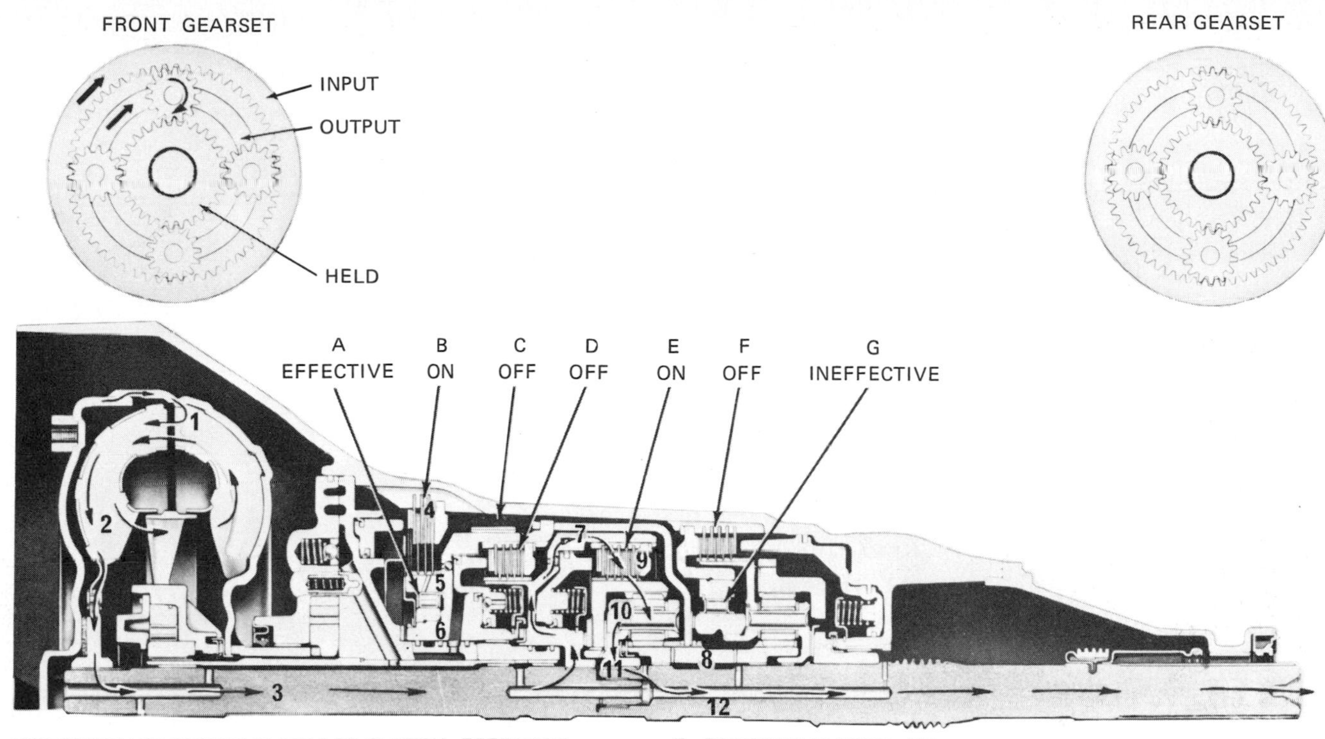

A. INTERMEDIATE-OVERRUN ROLLER CLUTCH—EFFECTIVE
B. INTERMEDIATE CLUTCH—ON
C. INTERMEDIATE-OVERRUN BAND—OFF
D. DIRECT CLUTCH—OFF
E. FORWARD CLUTCH—ON
F. LOW-AND-REVERSE CLUTCH—OFF
G. LOW-AND-REVERSE ROLLER CLUTCH—INEFFECTIVE

Fig. 13-8. Power flow, shown by arrows, through the transmission in drive range and second gear. (*Oldsmobile Division of General Motors Corporation*)

movement, and so the sun gear is held stationary.

The ring gear of the front gearset (9) is locked to the input shaft (3) through the forward clutch (9), which is still engaged. With the hydraulic system signaling for the upshift from low to second gear, both the intermediate clutch (4) and the forward clutch (9) are effective. That is, they are engaged, or locked up. When this happens, the ring gear of the front gearset must turn with the input shaft. We noted that the sun gear (8) is held as the intermediate clutch (4) engages. This means that the clockwise rotation of the ring gear rotates the planet pinions against the stationary sun gear, forcing the planet pinions to walk around the sun gear. As a result, the planet-pinion cage (10) is carried around in a clockwise direction. The planet-pinion cage is splined to the output shaft (12) at 11. Thus, when the planet-pinion cage rotates, it causes the output shaft to rotate.

Note that the rear gearset is ineffective. There is gear reduction through the transmission of 1.52:1. That is, the input shaft must turn 1.52 times to cause the output shaft to turn once. Additional speed reduction and torque increase are available through the torque converter, of course.

The reason that the rear gearset is ineffective is that the output shaft turns the ring gear of the rear gearset at the same speed as the output shaft is turning. The two are splined together, you will recall. The sun gear is held stationary. This combination causes the planet gears to walk around the stationary sun gear, carrying the planet-pinion cage with them. This effect is the same as in the front gearset. The planet-pinion cage of the rear gearset is assembled to the inner race of the low-and-reverse roller clutch. This clutch now permits the inner race to freewheel so the planet-pinion cage can revolve without restraint

⊘ 13-6 Power Flow in D, Direct Drive In D (direct drive, or third gear) we have the conditions shown in Fig. 13-9. The intermediate clutch (4) and forward clutch (5) remain engaged, but now the direct clutch (6) becomes engaged. The direct clutch thus locks the ring gear (7) of the front gearset to the sun gear. Note that the sun gear (8) is locked to the input shaft (3) through the direct clutch (6) and the sun-gear shell (9). At the same time, the ring gear (7) is locked to the input shaft (3) through the forward clutch (5). With both the sun gear and the ring gear locked to the input shaft, both gears must turn with the input shaft. This means that the planet gearset is locked up. The planet-pinion cage turns at the same speed as the input shaft. Since the cage is splined to the output shaft, the output shaft also turns at the same

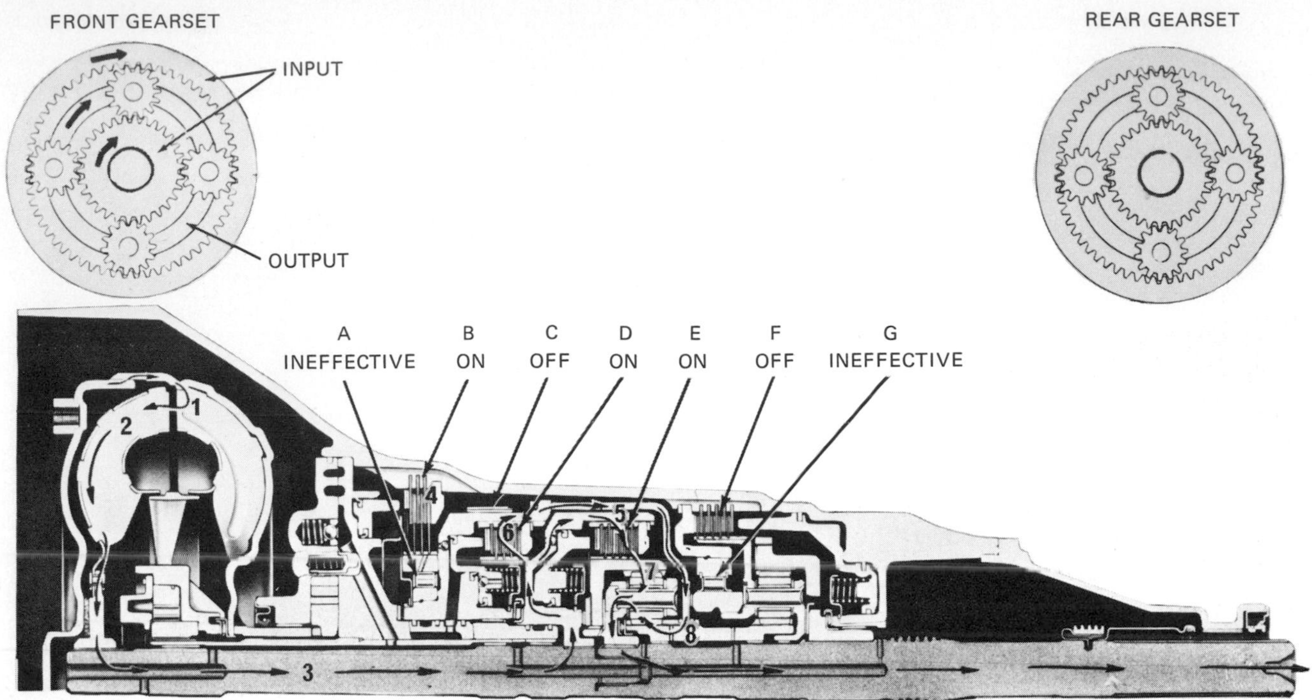

A. INTERMEDIATE-OVERRUN ROLLER CLUTCH–INEFFECTIVE
B. INTERMEDIATE CLUTCH–ON
C. INTERMEDIATE-OVERRUN BAND–OFF
D. DIRECT CLUTCH–ON
E. FORWARD CLUTCH–ON
F. LOW-AND-REVERSE-CLUTCH–OFF
G. LOW-AND-REVERSE ROLLER CLUTCH–INEFFECTIVE

Fig. 13-9. Power flow, shown by arrows, through the transmission in drive range and direct, or third, gear. (*Oldsmobile Division of General Motors Corporation*)

speed as the input shaft. The transmission is now in direct drive, with a 1:1 gear ratio.

⊘ 13-7 Power Flow in L, First Gear If the driver moves the selector lever to L—the low range—the transmission will shift to low gear and remain in that gear. The internal conditions are shown in Fig. 13-10. The manual valve is positioned by the shift to L so as to produce the following: The intermediate clutch is off; the intermediate overrun band is off; the direct clutch is off; the forward clutch is on; and the low-and-reverse clutch is on. Now, the power flow is as shown in Fig. 13-10: from the input shaft through the forward clutch to the front-gearset ring gear. The ring gear is turned in a clockwise direction. This causes the planet pinions of the front gearset to turn in a clockwise direction, thus driving the sun gear in a counterclockwise direction. As the sun gear turns counterclockwise, it causes the planet pinions of the rear gearset to turn clockwise, thus driving the ring gear of the rear gearset clockwise. This ring gear is splined to the output shaft, and so the output shaft turns. With gear reduction in both the front and rear gearsets, the total gear reduction through the transmission is 2.52:1.

⊘ 13-8 Power Flow in D2 or S, Second Gear If the driver selects D2 or S, the hydraulic system will shift the transmission into second gear and hold it there, except at very low car speed, which will cause the transmission to downshift into low. The conditions will be as shown in Fig. 13-11. The intermediate clutch is applied to allow the intermediate overrun roller clutch to hold the shell and sun gear stationary (against counterclockwise rotation). Power flows through the applied forward clutch to the ring gear of the front gearset. The sun gear is stationary, and so the planet gears of the front gearset walk around the sun gear. This movement turns the planet-pinion cage of the front gearset and output shaft in a clockwise direction. The reaction of the planet pinions against the sun gear is taken either by the intermediate overrun roller clutch or by the intermediate overrun band. The intermediate overrun band provides overrun braking as it holds the sun gear stationary. The gear ratio is 1.52:1.

⊘ 13-9 Power Flow in Reverse When the driver moves the selector lever to R (reverse), the hydraulic system will shift the transmission into reverse. The conditions will be as shown in Fig. 13-12. Only the direct clutch and the low-and-reverse clutch will be engaged. Power flows from the input shaft through the direct clutch to the sun-gear shell and sun gear. The sun gear rotates clockwise. This causes the planet pinions of the rear gearset to rotate counter-

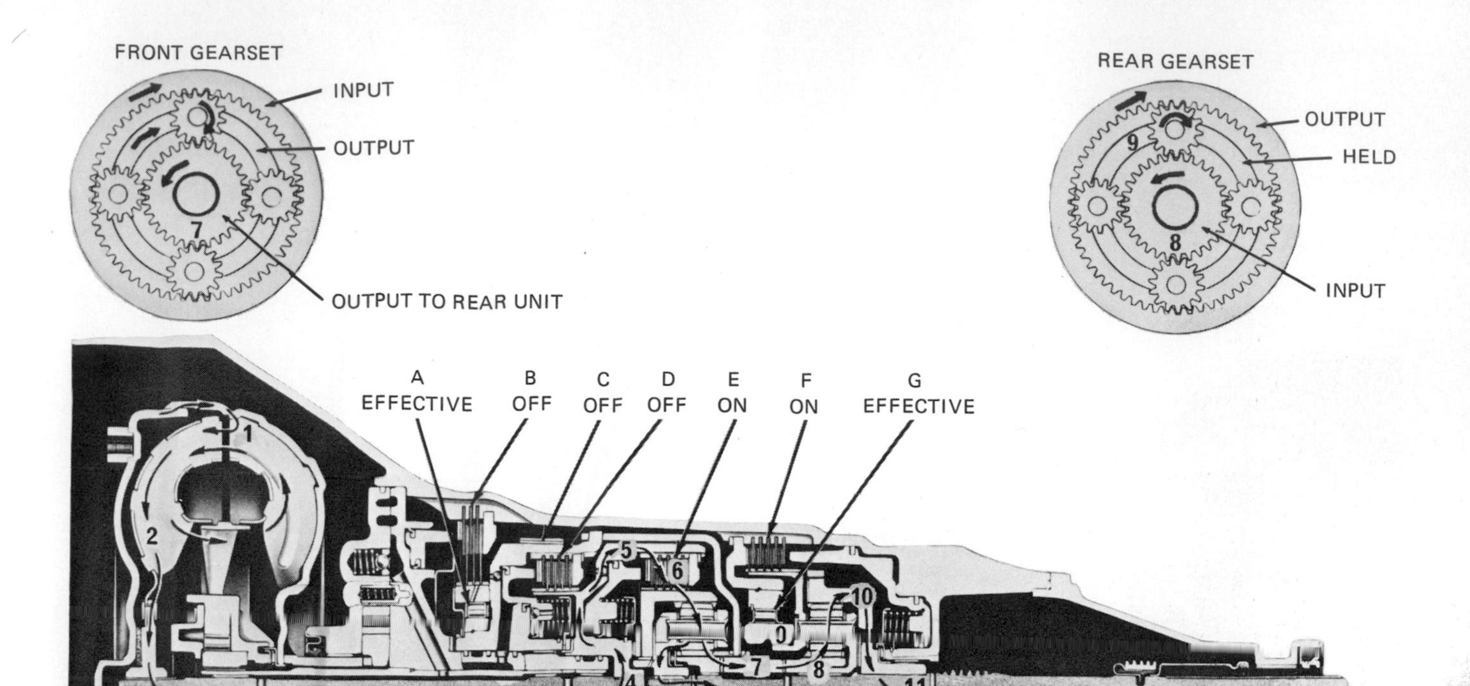

A. INTERMEDIATE-OVERRUN ROLLER CLUTCH—EFFECTIVE
B. INTERMEDIATE CLUTCH—OFF
C. INTERMEDIATE-OVERRUN BAND—OFF
D. DIRECT CLUTCH—OFF
E. FORWARD CLUTCH—ON
F. LOW-AND-REVERSE CLUTCH—ON
G. LOW-AND-REVERSE ROLLER CLUTCH—EFFECTIVE

Fig. 13-10. Power flow, shown by arrows, through the transmission in low range and low, or first, gear. (*Oldsmobile Division of General Motors Corporation*)

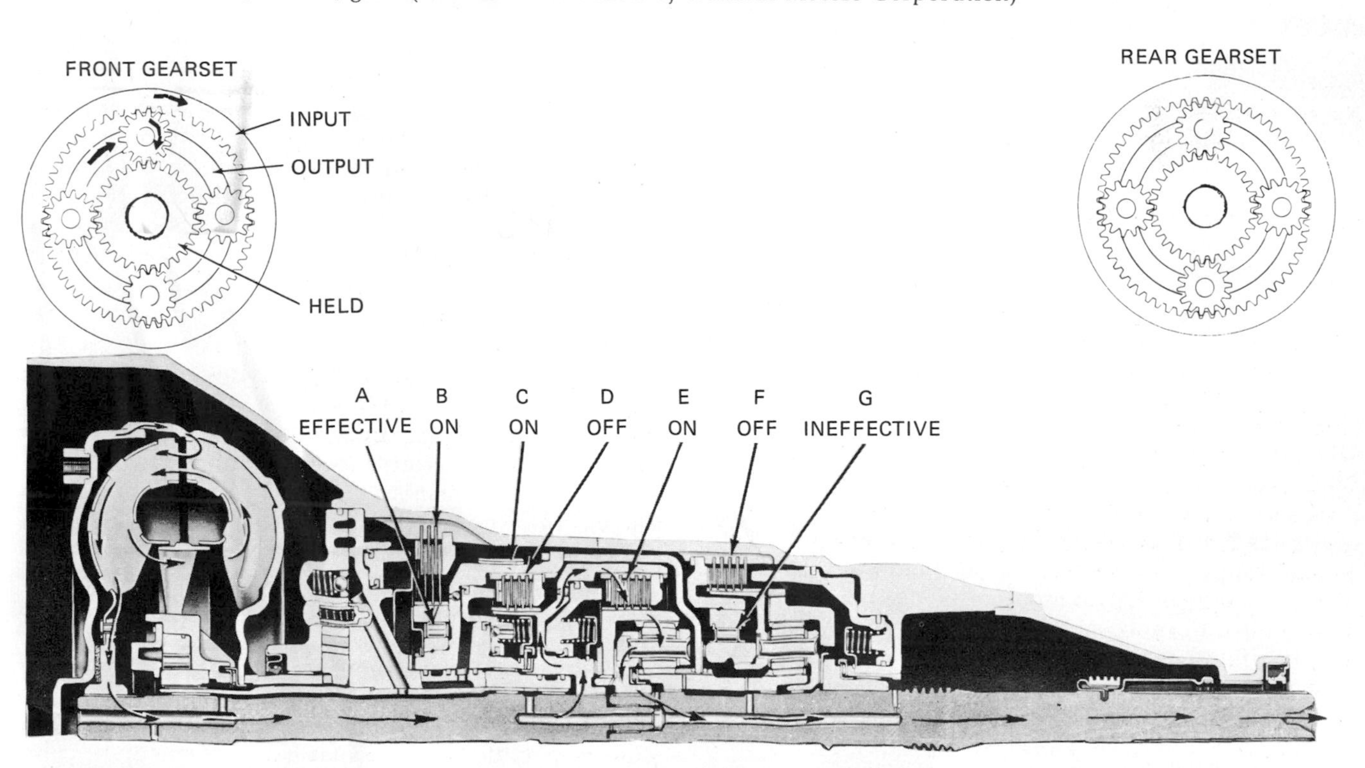

A. INTERMEDIATE-OVERRUN ROLLER CLUTCH—EFFECTIVE
B. INTERMEDIATE CLUTCH—ON
C. INTERMEDIATE-OVERRUN BAND—ON
D. DIRECT CLUTCH—OFF
E. FORWARD CLUTCH—ON
F. LOW-AND-REVERSE CLUTCH—OFF
G. LOW-AND-REVERSE ROLLER CLUTCH—INEFFECTIVE

Fig. 13-11. Power flow, shown by arrows, through the transmission in super range (D2) and second gear. (*Oldsmobile Division of General Motors Corporation*)

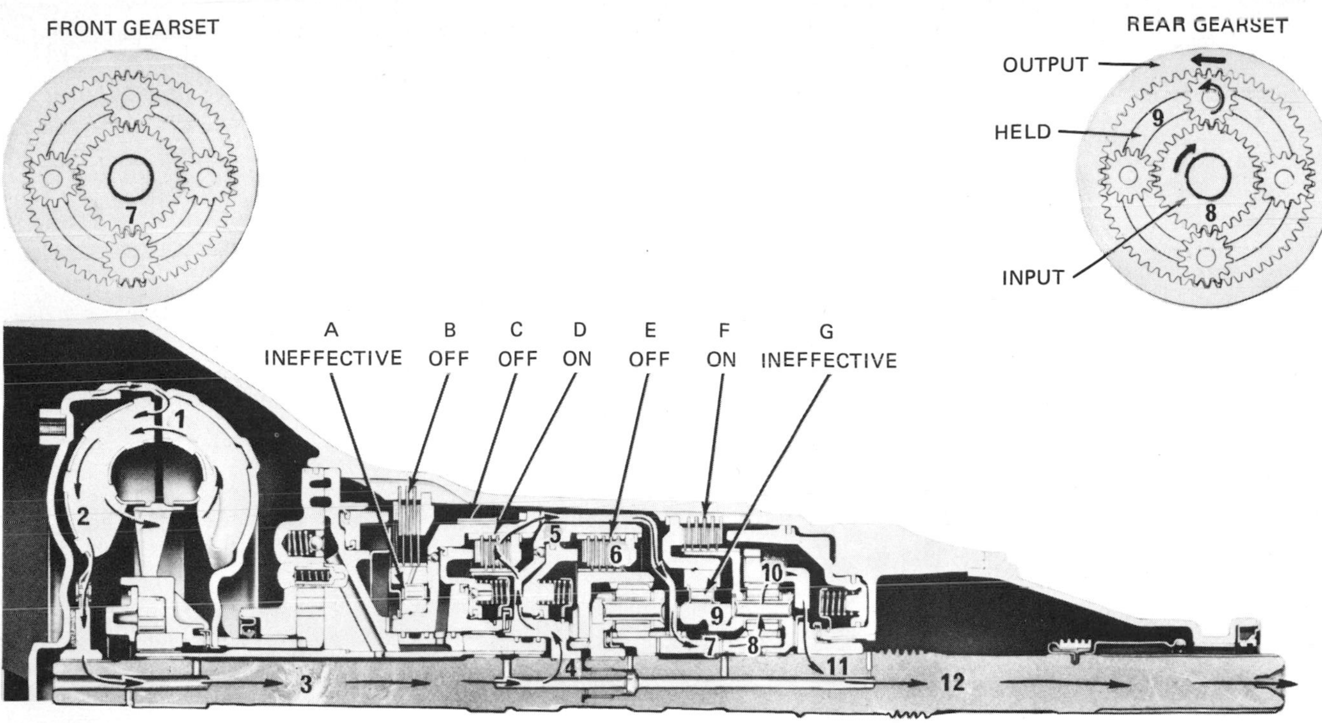

Fig. 13-12. Power flow, shown by arrows, through the transmission in reverse. (*Oldsmobile Division of General Motors Corporation*)

clockwise. The planet-pinion cage of the rear gearset is held by the low-and-reverse clutch. Thus, the planet pinions act as idlers and cause the ring gear of the rear gearset to rotate in a counterclockwise direction. Since this ring gear is splined to the output shaft, the output shaft will also turn counterclockwise. This causes the car to back up. The gear ratio through the transmission is 1.93:1.

Check Your Progress

Progress Quiz 13-1 The Type 350 Turbo Hydra-Matic transmission is one of the more complicated transmissions. If you understand it, you will be able to understand every other type of automatic transmission in automobiles. The quiz that follows covers the mechanical aspects of the transmission. Make sure you understand the gearing and how it works by taking the quiz.

Completing the Sentences The sentences that follow are incomplete. After each sentence there are several words or phrases, only one of which will correctly complete the sentence. Write each sentence in your notebook, selecting the proper word or phrase to complete it correctly.

1. The 350 transmission includes a torque converter, two planetary gearsets, two roller clutches, a brake band, and: (*a*) two multiple-disk clutches, (*b*) three multiple-disk clutches, (*c*) four multiple-disk clutches.
2. The 350 transmission provides reverse and: (*a*) two forward speeds, (*b*) three forward speeds, (*c*) four forward speeds.
3. When the transmission is in low gear in drive range, both roller clutches are effective and the following is on: (*a*) forward clutch, (*b*) low-and-reverse clutch, (*c*) intermediate clutch.
4. When the transmission is in second gear in drive range, the intermediate overrun roller clutch is effective and the following are on: (*a*) low-and-reverse clutch and forward clutch, (*b*) intermediate clutch and intermediate band, (*c*) intermediate clutch and forward clutch.
5. With the transmission in direct drive, neither roller clutch is effective and the following are on: (*a*) intermediate band and direct clutch; (*b*) forward and low clutches; (*c*) forward, intermediate, and direct clutches.
6. With the transmission in low gear in L range, both the roller clutches are effective and the following are on: (*a*) forward clutch and low-and-reverse clutch; (*b*) forward and direct clutches; (*c*) low, forward, and direct clutches.

7. With the transmission in super range (D2) and second gear, the intermediate overrun roller clutch is effective and the following are on: (*a*) intermediate and forward clutches and intermediate band; (*b*) intermediate and low clutches; (*c*) direct, forward, and intermediate clutches.
8. With the transmission in reverse, neither roller clutch is effective and the following are on: (*a*) low-and-reverse clutch and forward clutch, (*b*) low-and-reverse clutch and direct clutch, (*c*) low-and-reverse clutch and brake band.
9. When the intermediate overrun band is applied, it locks the: (*a*) ring gear, (*b*) sun gear, (*c*) planet-pinion carrier.
10. When the forward clutch is on, the ring gear of the front gearset is locked to the: (*a*) sun gear, (*b*) output shaft, (*c*) input shaft.

⊘ 13-10 Hydraulic System The hydraulic system includes a pump, driven by the engine, which produces hydraulic pressure. Also, there are a number of valves that control the pressure and direct it to the proper clutch or brake-band servo to produce the proper gearshift. In addition, there are devices for smoothing shifts and modifying them to suit engine and highway operating conditions. We shall now study the hydraulic system in its entirety, including the various valves.

Almost all the valves in the transmission are located in a valve body. The valve body is a die casting, a little less than 1 foot [0.305 m] square, which is located at the bottom of the transmission in the oil pan. The valve body contains most of the valves, as shown in Fig. 13-13. We shall describe each of these in following sections. After we explain the function of each valve, we shall find out how the complete system works. Plates 9 to 16 show the complete hydraulics system during different operating conditions.

⊘ 13-11 Oil Pump The oil pump is located at the front of the transmission just behind the torque converter, as shown in Fig. 13-1. The oil pump is driven by the engine and is the gear type, as shown in Fig. 13-14. As the gears rotate, the spaces between the teeth first become larger so that oil is "sucked"

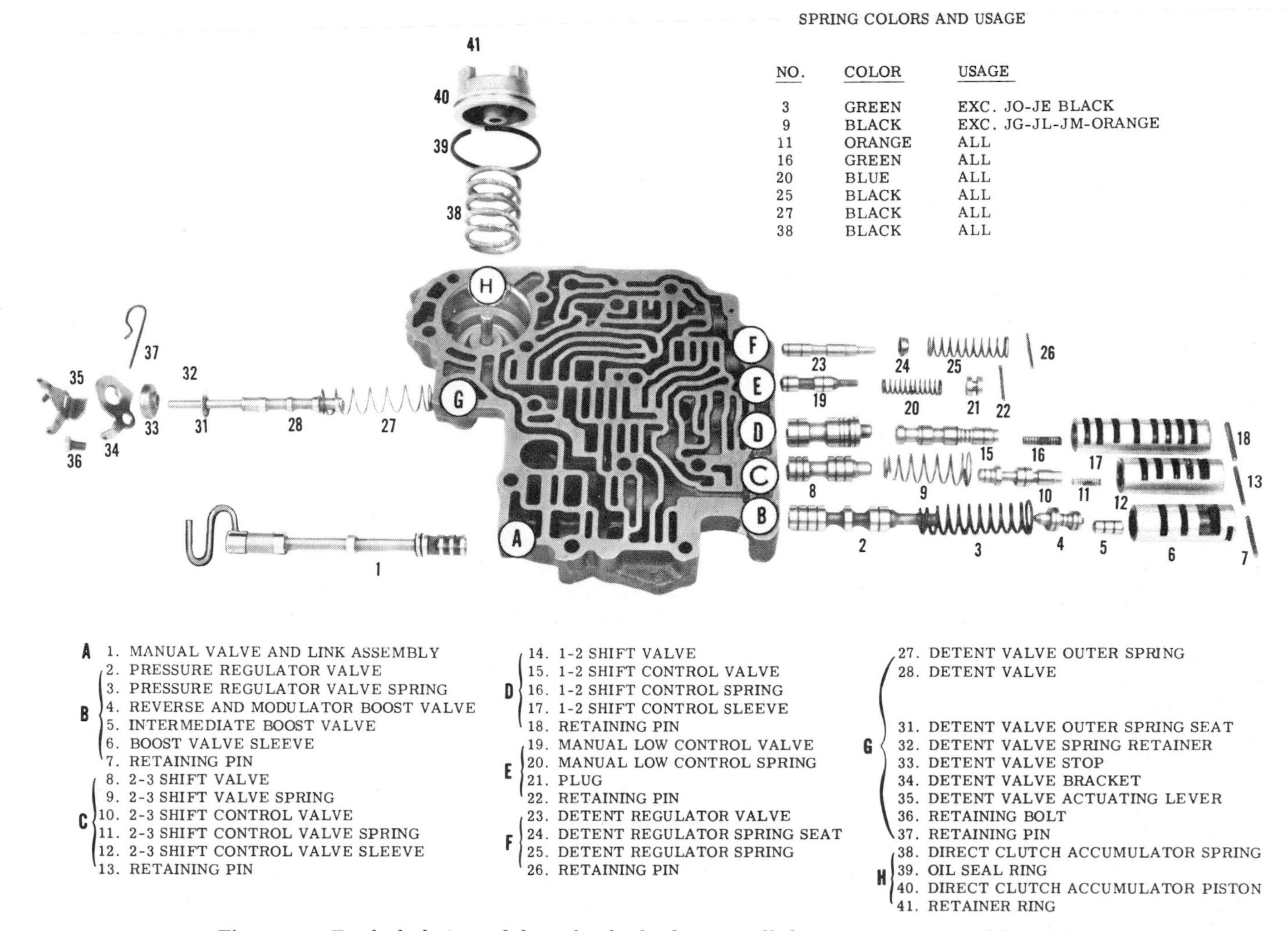

SPRING COLORS AND USAGE

NO.	COLOR	USAGE
3	GREEN	EXC. JO-JE BLACK
9	BLACK	EXC. JG-JL-JM-ORANGE
11	ORANGE	ALL
16	GREEN	ALL
20	BLUE	ALL
25	BLACK	ALL
27	BLACK	ALL
38	BLACK	ALL

Fig. 13-13. Exploded view of the valve body showing all the interior parts. (*Oldsmobile Division of General Motors Corporation*)

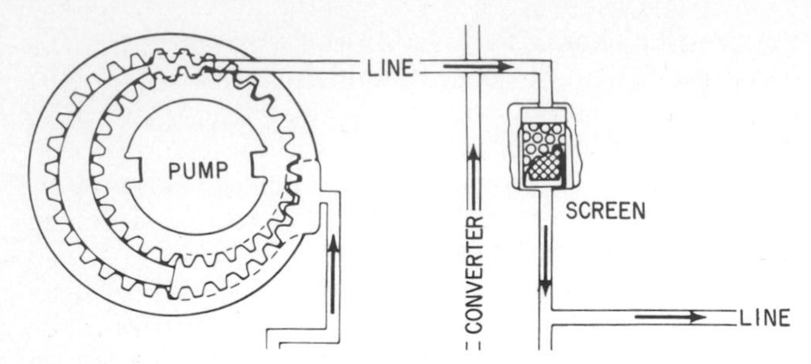

Fig. 13-14. Schematic view of the oil pump. (*Oldsmobile Division of General Motors Corporation*)

into the spaces. Actually, atmospheric pressure pushes the oil into the increasing spaces between the teeth. Then, when the teeth approach each other and mesh, the oil between them is forced out into the line. Now this pump pressure varies greatly according to engine speed. At high engine speed, the pump pressure would be very high if it were not for the pressure-regulating valve train. This valve reduces the pressure as necessary to meet the operating conditions. We shall discuss this valve in more detail in ⊘ 13-14.

⊘ 13-12 The Vacuum-Modulator Assembly This assembly, shown in sectional view in Fig. 13-15, contains a metal bellows, a diaphragm, and two springs. Its purpose is to modulate, or modify, the line pressure with changing engine output and also with changes in altitude. It does this by changing the valve position in its cylinder. This cylinder has several openings, or ports, to admit oil and allow it to flow out. You will recall from our previous discussions of spool valves (in Chap. 7) that as the valve moves it can partly shut off or increase the valve-port openings. If the port that admits pump pressure is partly shut off, the effect is to lower the pressure that exits from the valve assembly. If you study the valve in Fig. 13-15, you will note that there are four ports: line, modulator, detent, and governor. We are interested at the moment in the first two, line and modulator. We discuss the other two ports later.

The metal bellows is evacuated. It contains no air and thus has a vacuum inside it. Atmospheric pressure acting on the bellows tends to collapse it, that is, shorten its length. Opposing this pressure is the heavy spring inside the bellows. However, changes in atmospheric pressure will change the length of the bellows. High altitudes, where atmospheric pressure is lower, will allow the bellows to lengthen. This effect counterbalances the loss of engine power due to the lower atmospheric pressure, as we shall explain soon.

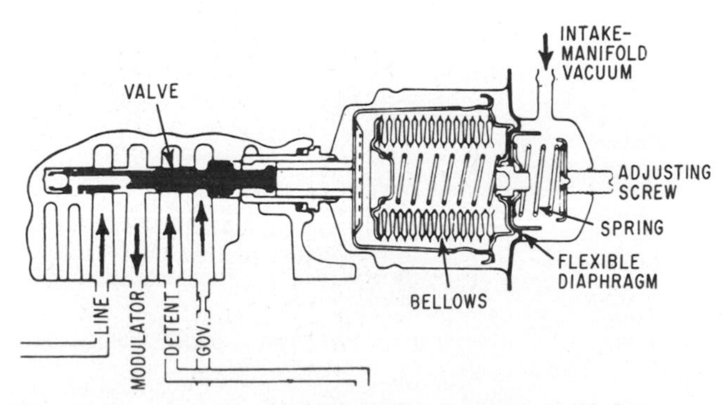

Fig. 13-15. Sectional view of the vacuum-modulator assembly. (*Oldsmobile Division of General Motors Corporation*)

The flexible diaphragm is open to atmospheric pressure on one side and to intake-manifold vacuum on the other side. Thus, as this vacuum changes, the diaphragm moves, thereby moving the modulator valve.

Now, let us add up these two actions and see what we have: The bellows and the external spring (to the right in Fig. 13-15) act to move the modulator valve so as to increase modulator pressure. The flexible diaphragm acts on the modulator valve so as to reduce modulator pressure as engine intake-manifold pressure increases. For example, suppose the driver is cruising along at part throttle but gaining speed so that the hydraulic system signals for an upshift. The intake-manifold vacuum is high so that the flexible diaphragm acts to reduce modulator pressure. With lower modulator pressure, the shift takes place more slowly and is thus less noticeable. On the other hand, if the driver opens the throttle to demand full engine power, intake-manifold vacuum drops and the flexible diaphragm has less effect. This means that the modulator pressure increases and the shift takes place more quickly. With an open throttle and high engine speed, the shift must take place fast to prevent slippage and destructive heat buildup in the clutch. But a quick shift under the circumstances is least noticeable and most desirable.

The combination has another effect. Engine power is lost at high altitudes due to the lower atmospheric pressure. With reduced engine power, fast shifts would be harsh and noticeable. So shifts should slow down. This slowing-down action is produced by the combined effect of the bellows and the diaphragm. The diaphragm, being larger in effective area than the bellows, acts to reduce modulator pressure. Remember that the bellows expands at higher altitudes, and that this tends to increase modulator pressure. But the lowered atmospheric pressure, acting on the flexible diaphragm, tends to reduce modulator pressure. Because the diaphragm is larger, it overcomes the bellows action and does cause a decrease in modulator pressure.

NOTE: We have talked about "fast" and "slow" shifts. Actually, the critical part of the shift is over in a second or so. This critical time includes only the moment when the brake band or clutch is actually taking hold. This might be compared to your ramming one fist into the palm of your other hand, as opposed to moving the fist to the palm more gently. The event is over very quickly either way, but the fist-ramming is faster.

⊘ 13-13 Governor

⊘ 13-13 Governor Governor pressure is critical to the operation of the transmission. The governor is driven by the output shaft of the transmission and is thus directly related to car speed (not engine speed). The governor contains two pairs of weights, as shown in Fig. 13-16. These weights are held in place by springs. When car speed, and governor speed, increases, centrifugal force acts on the weights, forcing them to move outward against the spring pressure. This action pushes the governor valve upward in the assembly, tending to close off the exhaust port and partly open the drive port to line pressure. The effect is that as car speed increases, governor pressure increases. Thus, governor pressure is a function of car speed alone. As governor pressure increases, it acts on other valves, getting them ready to produce shifts when the proper engine and car speed and throttle opening are reached. We explain how this happens later.

Governor pressure also has an effect on the vacuum-modulator valve. Look at Fig. 13-15. Note that governor pressure is introduced into a port that is located at an undercut in the modulator valve. The land on one side of the valve undercut is larger than the land on the other side. That means that the governor pressure tends to move the valve in the direction of the larger land (or to the right in Fig. 13-15). The result is that as governor pressure increases, signaling higher car speed, the modulator valve tends to move so as to decrease modulator pressure. The effect is to slightly modify the shifting speed at high car speeds, thus preventing harsh shifts.

Note that the governor has two sets of weights: The primary weights are effective at low speeds and allow the governor pressure to build up rapidly until about 20 mph [32.19 km/h]. At this point, the primary weights have moved out as far as possible. Now, the secondary weights take over as car speed further increases. They produce a slower increase in governor pressure. Thus, we have a rapid increase in governor pressure at low car speeds and a slower increase at higher speeds. This two-step curve of pressure increase allows pressure to build up rapidly at first but prevents it from going too high in the high-car-speed ranges.

⊘ 13-14 Pressure-Regulator Valve The pressure-regulator valve, shown in Fig. 13-17 regulates the pressure from the pump so as to avoid excessive pressures in the lines going to the other valves and to the clutches. The valve is shown in Fig. 13-13 and is numbered 2. The pump pressure is applied to one

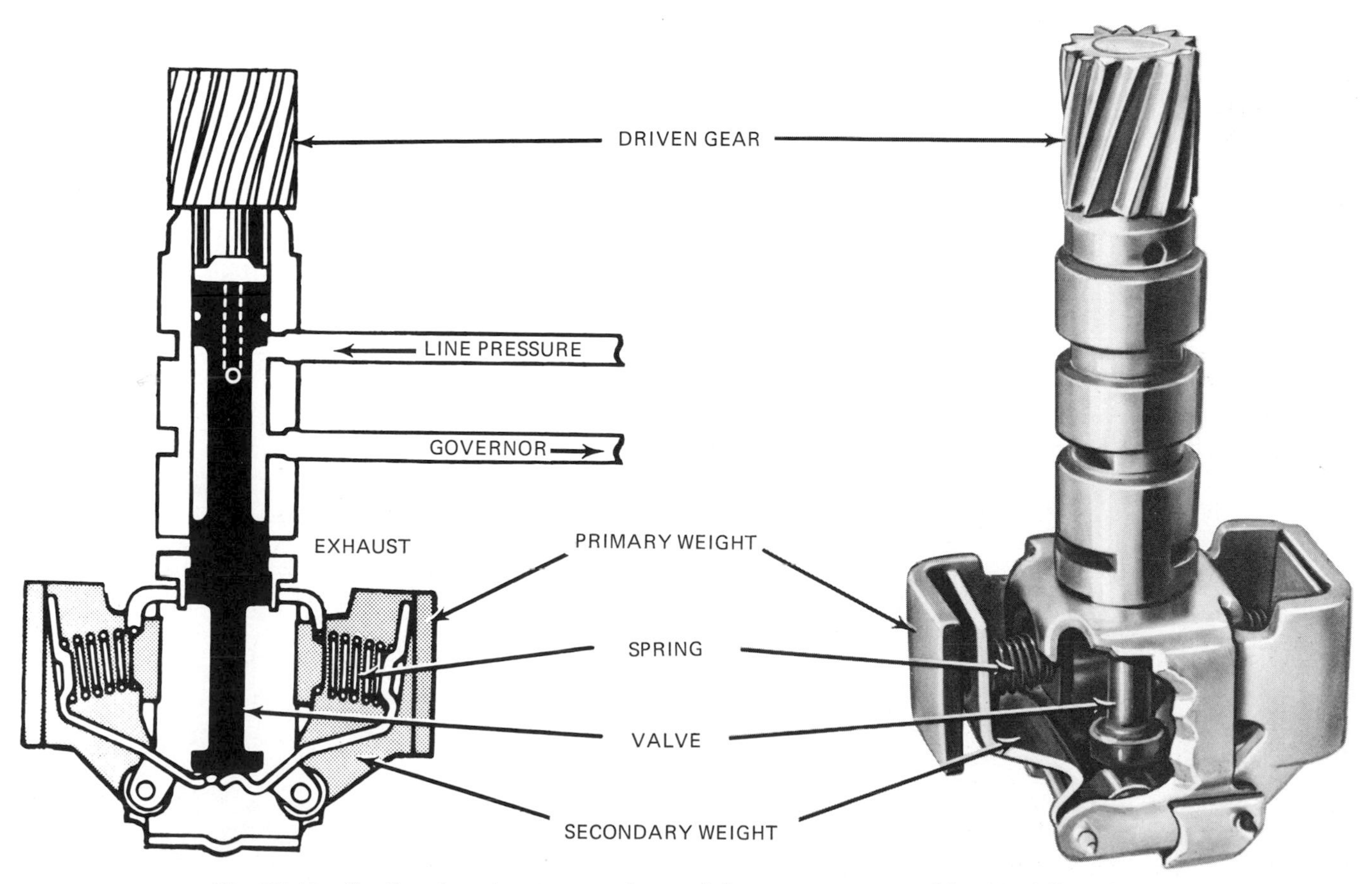

Fig. 13-16. Sectional and cutaway views of the governor assembly. (*Cadillac Motor Car Division of General Motors Corporation*)

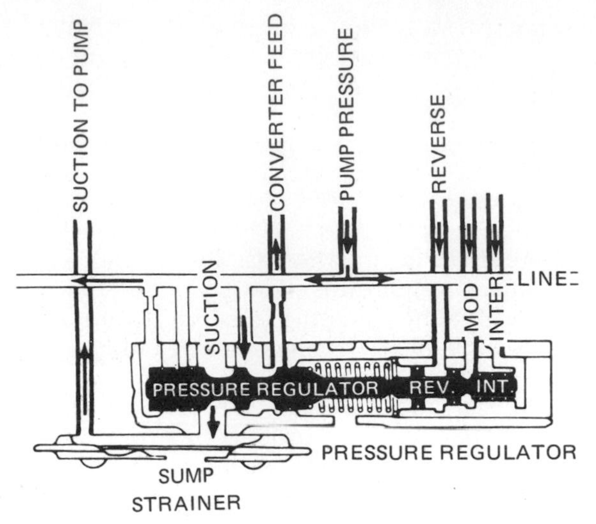

Fig. 13-17. Sectional view of the pressure-regulator valve train. There are three parts to the train, aside from the spring: the pressure-regulator valve, the reverse-and-modulator boost valve (labeled REV), and the intermediate boost valve (labeled INT). (*Oldsmobile Division of General Motors Corporation*)

end (left in Fig. 13-17) of the valve. Opposing this is the spring force at the other end of the valve. The valve therefore positions itself so as to give the pressure designated by the spring force. Note that the pump pressure is reduced as necessary by the valve position, which allows some of the oil to flow to the oil sump, or oil pan. Dumping oil in this manner reduces the pressure.

The pressure is also affected by the modulator pressure during intermediate-, or second-, gear operation. The modulator pressure tends to raise the line pressure so as to provide a more rapid shift from second to high. This prevents excessive slippage of the clutch during this shift. Also, pressure is affected when the transmission is shifted into reverse. During reverse operation, very firm application of the low-and-reverse clutch is necessary. This is produced by raising the line pressure. When the shift is made, oil pressure from the line is introduced into the reverse section of the pressure-regulator-valve train. This pressure acts on the larger land of the reverse boost valve (labeled REV in Fig. 13-17 and numbered 4 in Fig. 13-13), adding to the spring force to obtain an increase in line pressure.

⊘ 13-15 Manual Valve We have already introduced the manual valve (⊘ 8-4) and learned that this valve directs oil pressure to different parts of the hydraulic circuit, according to where it has been set by the driver. The valve is shown in Fig. 13-18 in sectional view, and the actual valve and link can be seen in Fig. 13-13, where it is numbered 1. Line pressure enters the valve body and flows to drive, intermediate, low, or reverse, according to the position of the valve. Figure 13-6 shows the valve in neutral, drive, and low.

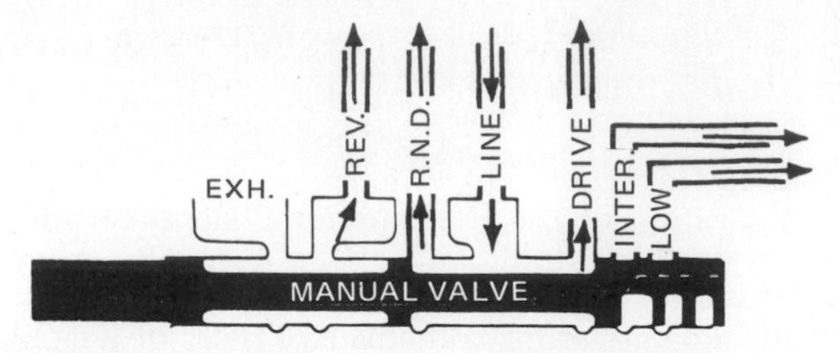

Fig. 13-18. Sectional view of the manual valve. (*Oldsmobile Division of General Motors Corporation*)

⊘ 13-16 Intermediate-Clutch Accumulator We've already described and illustrated (see Fig. 7-15) one version of an accumulator. Its purpose is to cushion the application of a clutch or brake band. The accumulator shown in Fig. 13-19 cushions the application of the intermediate clutch during shifts from first to second. It is therefore known as the *1-2 accumulator*.

When the hydraulic system calls for a shift from first to second, oil pressure is applied to the intermediate clutch. This pressure enters the accumulator through the line marked "1-2 CL" in Fig. 13-19. This stands for intermediate, or 1-2, clutch. This pressure is applied to one side of the piston. Line pressure is applied to the other side of the piston. Before the shift is called for, and without 1-2 clutch pressure, line pressure pushes the piston far to the right in the accumulator body. This compresses the spring. But when 1-2 clutch pressure comes on, it combines with the spring pressure to push the piston back against line pressure. This action gives the 1-2 clutch oil a place to go besides the intermediate clutch. The result is to allow the clutch to apply smoothly. The accumulator provides a cushion, in effect, so that all the 1-2 clutch oil does not suddenly ram into the clutch and cause it to apply suddenly. Instead, the oil pressure builds up slowly so that the clutch application is smooth.

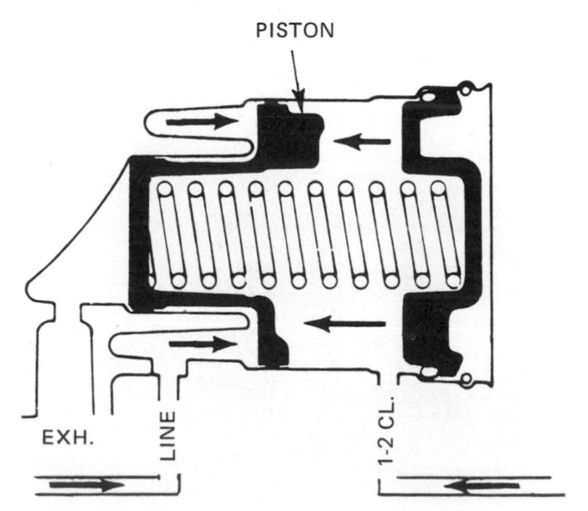

Fig. 13-19. Sectional view of the 1-2 accumulator. (*Oldsmobile Division of General Motors Corporation*)

⊘ 13-17 Direct-Clutch Accumulator The direct-clutch, or 2-3, accumulator does the same job for the direct clutch that the intermediate-clutch accumulator does for the intermediate clutch. It cushions the application of the direct clutch during the shift from 1-2 to 2-3. Figure 13-20 is a sectional view of the direct-clutch accumulator. Number 40 in Fig. 13-13 shows the piston.

The accumulator works this way: Before the shift to direct is called for, there is no 2-3 clutch pressure. The RND (reverse, neutral, drive) pressure pushes down on the piston. Also, 1-2 clutch pressure acts between the two parts of the piston. These pressures push the piston down and compress the lower spring. At the same time, 1-2 pressure tends to separate the two parts of the piston, acting almost like an auxiliary spring. Then, when the hydraulic system signals for a shift from second to high, or direct, the 2-3 clutch pressure is applied to the lower part of the piston. This pressure forces the piston up against the RND pressure. At the same time, 1-2 clutch pressure is released. Thus, there is a combined upward pressure that allows the piston to move up. This action produces a relatively gradual buildup of 2-3 clutch pressure so that the direct clutch applies smoothly.

Note that the 2-3 accumulator also includes the servo which applies the intermediate overrun band under certain conditions, as we shall explain later.

⊘ 13-18 1-2 Shift Valve The 1-2 shift valve, shown in Fig. 13-21, directs oil pressure to the transmission to cause it to shift from first to second or from second to first. It is numbered 14 in Fig. 13-13. The operation of this valve is controlled by governor pressure, detent pressure, modulator pressure, and spring force. Governor pressure, we know, is directly related to car speed. It acts on the left end

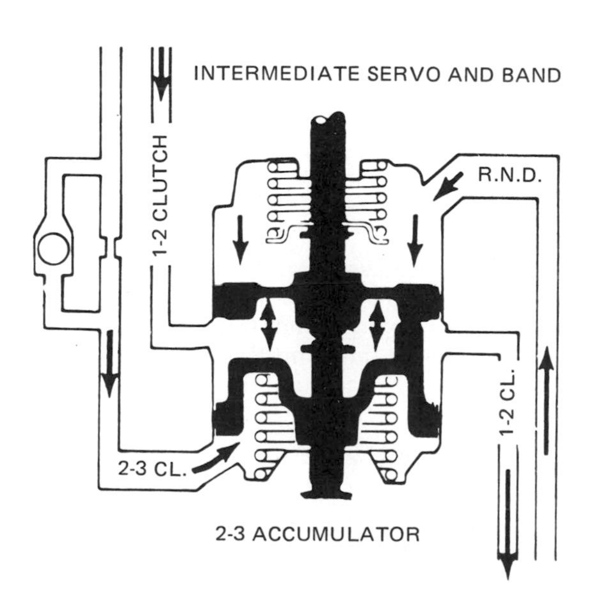

Fig. 13-20. Sectional view of the 2-3 accumulator. (*Oldsmobile Division of General Motors Corporation*)

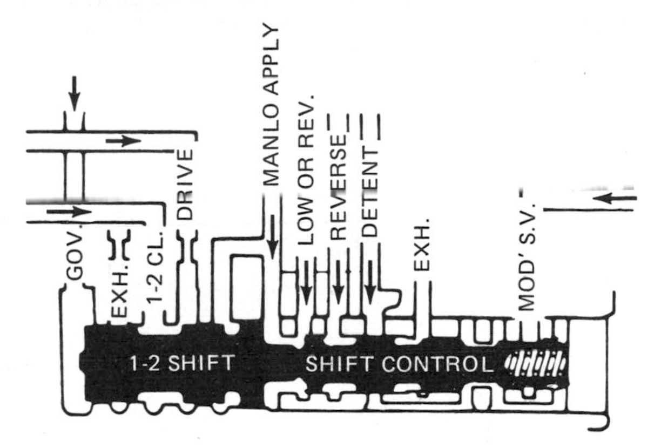

Fig. 13-21. Sectional view of the 1-2 shift valve train. There are two parts to the train, aside from the spring: the shift valve and the shift control valve. (*Oldsmobile Division of General Motors Corporation*)

(Fig. 13-21) of the 1-2 shift valve and increases with car speed. Opposing this pressure is a spring, the modulator pressure, and the detent pressure. We shall see how the opposing forces operate to allow upshifting—or downshifting under certain conditions—when we study the complete hydraulic system.

⊘ 13-19 2-3 Shift Valve The 2-3 shift valve, shown in Fig. 13-22, is numbered 8 in Fig. 13-13. It routes oil pressure in the transmission to cause a shift from second to direct or from direct to second. Its operation is controlled by several pressures—governor, modulator, and detent—as well as by spring force. We shall explain how these pressures can produce the upshift or downshift when we examine the complete hydraulic system.

⊘ 13-20 Detent Valve The detent valve, shown in sectional view in Fig. 13-23 and numbered 28 in Fig. 13-13, is actuated by the downshift cable which is

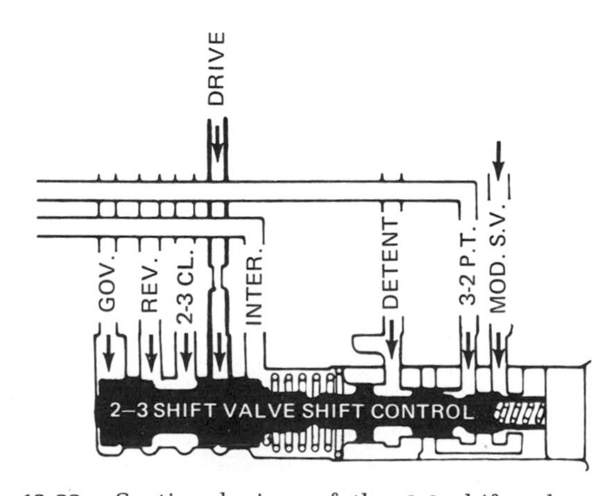

Fig. 13-22. Sectional view of the 2-3 shift-valve train. There are two parts to the train, aside from the spring: the shift valve and the shift control valve. (*Oldsmobile Division of General Motors Corporation*)

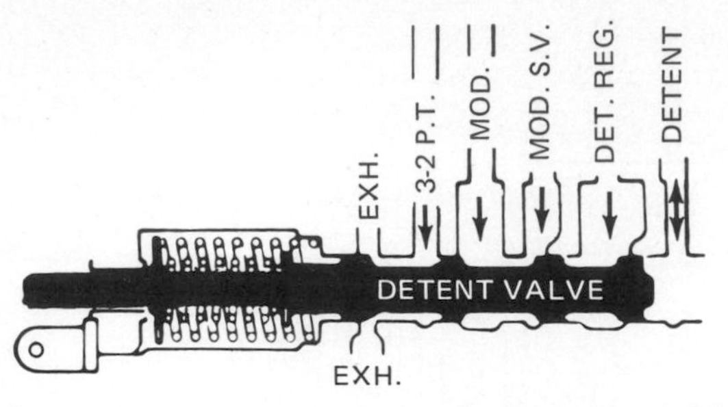

Fig. 13-23. Sectional view of the detent valve. (*Oldsmobile Division of General Motors Corporation*)

linked to the accelerator. When the accelerator pedal is pushed all the way down, the linkage moves the detent valve so that it changes the routing of the oil through the transmission. If the transmission is in direct drive and the car is traveling at a high speed, the new oil routes cause the transmission to downshift from high to second. If the transmission is in either second or direct and the car is traveling at a low speed, the new oil routes cause the transmission to shift to first. That is, the shift is either 2-1 or 3-1. More on this when we examine the complete hydraulic circuits.

⊘ 13-21 Detent Regulator Valve The detent regulator valve and spring are shown in sectional view in Fig. 13-24. They are numbered 23 and 25 in Fig. 13-13. This valve changes line pressure into detent regulator oil pressure, which is then used to control the car speed at which 1-2 and 2-3 upshifts will occur with full throttle.

⊘ 13-22 Oil-Cooler Bypass Valve The oil-cooler bypass valve is a spring-loaded valve which permits oil to be fed from the torque converter directly into the hydraulic system when the oil is cold or if there is a restriction in the cooler or cooler lines. When the oil is cold, it is thicker and flows more slowly. Higher pressures result, which unseat the valve ball so that the oil can bypass the cooler. The same actions occur if the cooler circuit is restricted. This ensures adequate oil circulation to transmission parts.

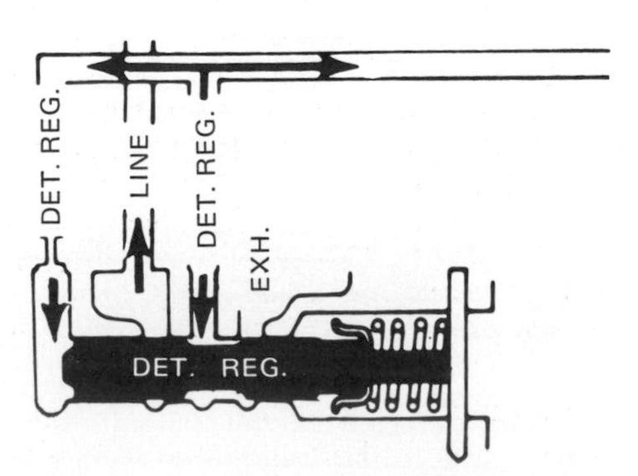

Fig. 13-24. Sectional view of the detent regulator valve. (*Oldsmobile Division of General Motors Corporation*)

⊘ 13-23 Manual-Low-Control Valve The manual-low-control valve is numbered 19 in Fig. 13-13. When the manual valve is positioned by the driver in low and the car is traveling below 45 mph [72.42 km/h], oil is directed to the 1-2 shift valve so that it moves to the downshifted position. At the same time, it moves the 1-2 shift control valve to the upshifted position, which sends low apply oil to the low-and-reverse clutch to cause this clutch to engage. Now, the transmission is shifted to low and will remain there. It cannot upshift regardless of engine or car speed.

At speeds above 45 mph [72.42 km/h], shifting to low is prevented by the manual-low-control valve because the high governor pressure prevents the valve from moving.

⊘ 13-24 Hydraulic System We have looked at the transmission gearing with its related clutches and brake band and the various valves in the hydraulic system. Now we shall put everything together and find out just what takes place during the different phases of transmission operation.

We previously discussed (⊘ 8-4) a simplified version of the shifting action and learned that shifts are controlled by two factors: car speed, which provides a signal through governor pressure, and engine speed, which provides a signal through intake-manifold vacuum. In the following sections, we shall examine the eight conditions in the transmission:

1. Neutral, engine running
2. Drive range, first gear
3. Drive range, second gear
4. Drive range; third, or direct, gear
5. Low range, first gear
6. Super range, second gear
7. Drive-range detent downshift
8. Reverse

As we have mentioned, the terms *neutral, drive range, low range, super range* (also called *low two* or *drive two*), and *reverse* refer to the manual-valve positions. These are indicated on the selector quadrant on the steering column as N, D (or Dr), L, S (or L2 or D2), and R.

⊘ 13-25 Neutral, Engine Running The conditions in the transmission are as shown in Plate 9 when the manual valve is in neutral and the engine is running. These are the conditions shown in Fig. 13-5. The conditions of the clutches and band are as follows:

Intermediate clutch—off
Direct clutch—off
Forward clutch—off
Low-and-reverse clutch—off
Intermediate overrun roller clutch—freewheeling
Low-and-reverse roller clutch—freewheeling
Intermediate overrun band—off

When the engine starts, with the manual valve in neutral, oil flows from the pump to the pressure-regulator valve, where it is controlled at line pressure and is called simply LINE in Plate 9. Oil flows from the pressure-regulator valve to the converter to fill it. Oil begins to circulate through the oil cooler and flows from the oil cooler to the transmission lubricating lines so that all bearing surfaces receive an immediate supply of oil. Line oil also flows from the pressure-regulator valve to the manual valve, which directs the oil, now called *reverse-neutral-drive* (RND) *oil,* to the servo part of the 2-3 accumulator.

At the same time, *line oil* flows to the 1-2 accumulator and strokes it in preparation for the 1-2 shift. By "stroke" we mean the piston is moved by the oil pressure. *Line oil* also flows to the vacuum-modulator valve. It emerges from this valve as *modulator oil,* and this modulator oil flows to the pressure-regulator-valve train and to the detent valve. It emerges from this valve as *modulator shift-valve oil* and flows to the 1-2 and 2-3 shift valves.

None of the valves reacts at this time. However, the valves are in readiness to react just as soon as the manual valve is shifted from neutral.

⊘ 13-26 Drive Range, First Gear The conditions in the transmission are as shown in Plate 10 with the manual valve in drive range and the transmission in first gear. These conditions are shown in Fig. 13-7. The *line oil* entering the manual valve emerges as *drive oil.* You will recall that the manual valve has already passed oil, called *reverse-neutral-drive oil,* to the 2-3 accumulator, stroking it in readiness for an upshift. The drive oil emerging from the manual valve flows to the 1-2 and 2-3 shift valves, where it has no immediate effect. Drive oil also flows to the governor and forward clutch, causing the forward clutch to apply. The conditions of the clutches and band are then as follows:

Intermediate clutch—off
Direct clutch—off
Forward clutch—on
Low-and-reverse clutch—off
Intermediate overrun roller clutch—locked
Low-and-reverse roller clutch—locked
Intermediate overrun band—off

Drive oil enters the governor and emerges as a variable-pressure oil, called *governor oil.* This variable pressure is applied to the ends of the 1-2 and 2-3 shift valves as well as to the modulator valve.

In summary, line oil is fed to the modulator valve, 1-2 accumulator, detent pressure-regulator valve, and manual valve. Line oil emerges from the manual valve as reverse-neutral-drive oil and as drive oil. This drive oil flows to the forward clutch, engaging it, so that the transmission is in first. The drive oil also flows to the governor and the 1-2 and 2-3 shift valves. RND oil emerges from the manual valve and flows to the servo part of the 2-3 accumulator. This oil strokes the 2-3 accumulator in preparation for it to cushion the 2-3 clutch (the direct clutch) when the shift is made from second to high. The 1-2 accumulator has already been stroked by line pressure to prepare it to cushion the 1-2 clutch (the intermediate clutch) when the shift is made from first to second.

⊘ 13-27 Drive Range, Second Gear The conditions in the transmission are as shown in Plate 11 with the manual valve in drive range and the transmission in second gear. These are the conditions shown in Fig. 13-8. As vehicle speed and governor pressure increase, the force of the governor pressure acting on the end of the 1-2 shift valve overcomes the valve-spring force. This moves the 1-2 shift valve so that drive oil is admitted to the 1-2, or intermediate, clutch, causing it to engage. As a result, the transmission upshifts to second gear. The conditions of the clutches and band are then as follows:

Intermediate clutch—on
Direct clutch—off
Forward clutch—on
Low-and-reverse clutch—off
Intermediate overrun roller clutch—locked
Low-and-reverse clutch—freewheeling
Intermediate overrun band—off

Note that modulator shift-valve pressure also helps the 1-2 shift-valve spring so that, in actuality, governor pressure must overcome both the spring and modulator shift-valve pressures before it can move the shift valve. Modulator shift-valve pressure starts from the modulator valve as simply modulator pressure. Modulator pressure, as we have already explained (⊘ 13-12), is related to engine intake-manifold vacuum and atmospheric pressure. The major influence is from intake-manifold vacuum. With part-throttle, high-vacuum conditions, modulator pressure is low. This low pressure passes through the detent valve and emerges as modulator shift-valve pressure, which in this case is the same pressure. The low modulator (or modulator shift valve) pressure adds little force to the 1-2 shift-valve spring. Therefore, a relatively low car speed, meaning a relatively low governor pressure, can produce the upshift from first to second.

On the other hand, if the throttle is wide open so that the intake-manifold vacuum is low, then the modulator pressure will be high. This high pressure, added to the 1-2 shift-valve spring, makes it necessary for the car to reach a higher speed before the shift can take place. This higher car speed is required so that the higher governor pressure needed to make the shift is reached. The governor pressure must go higher in order to overcome both the spring and higher modulator pressure.

Notice that the 1-2, or intermediate, clutch oil also flows to the 2-3 accumulator and from there to the 1-2 accumulator. This oil cushions the intermediate-clutch application. A typical application has these shift points: When in drive range with a full throttle, that is, with the throttle wide open, the upshift will occur at about 50 mph [80.47 km/h]. However, with a minimum throttle, the upshift will occur at about 12 mph [19.31 km/h]. So you can see that modulator and governor pressures play a very important role in controlling the shift point.

⊘ 13-28 Drive Range; Third, or Direct, Gear

The conditions in the transmission are as shown in Plate 12 with the manual valve in drive range and the transmission in third, or direct, gear. These conditions are shown in Fig. 13-9. As car speed and governor pressure increase, the increasing pressure, acting on the end of the 2-3 valve, overcomes the force of the 2-3 shift-valve spring and modulator pressure to move the shift valve. This movement allows drive oil to emerge from the 2-3 shift valve—as 2-3 clutch oil. This action causes the 2-3, or direct, clutch to engage so that the transmission goes into direct drive. This produces the following conditions of the clutches and band:

Intermediate clutch—on
Direct clutch—on
Forward clutch—on
Low-and-reverse clutch—off
Intermediate overrun roller clutch—freewheeling
Low-and-reverse roller clutch—freewheeling
Intermediate overrun band—off

Note that modulator pressure (modulator shift-valve pressure) is also helping the 2-3 shift-valve spring so that, in actuality, governor pressure must overcome both the spring force and modulator pressure before it can move the 2-3 shift valve. With part-throttle, high-intake-manifold-vacuum conditions, the modulator pressure is low. With a wide-open throttle and low intake-manifold vacuum, the modulator pressure is high. This variation in pressure makes a great difference in the shift point in shifting from second to direct. For example, on one application, the upshift to direct will occur at about 85 mph [136.79 km/h] with a wide-open throttle. It will occur at about 20 mph [32.19 km/h] with a minimum throttle opening.

You will note that it is the increasing governor pressure that produces the upshift from first to second and from second to third. There must be a considerable difference in the two shifting pressures. Pressure must be relatively low to produce the 1-2 shift and relatively high to produce the 2-3 shift. In a typical application, a governor pressure of 46 psi [3.234 kg/cm^2] is required to produce the 1-2 shift with wide-open throttle. A governor pressure of 83 psi [5.835 kg/cm^2] is required to produce the 2-3 shift with a wide-open throttle.

Now look at what happens at the 2-3 accumulator. When the 2-3 clutch oil flows to the direct clutch, it also flows to the lower part of the 2-3 clutch, forcing the piston upward against the RND pressure. This movement cushions the engagement of the direct clutch.

⊘ 13-29 Low Range, First Gear

When the driver moves the selector lever to L, or low, the conditions in the transmission are as shown in Plate 13. These are the conditions shown in Fig. 13-10. The transmission is in low, or first, gear, with the following conditions of the clutches and band:

Intermediate clutch—off
Direct clutch—off
Forward clutch—on
Low-and-reverse clutch—on
Intermediate overrun roller clutch—locked
Low-and-reverse roller clutch—locked
Intermediate overrun band—off

Now let us see what happens in the hydraulic circuit when the driver shifts to L, or low. Line oil entering the manual valve emerges as low oil, which passes through the manual-low-control valve. The oil then passes through the 1-2 shift valve, which directs it, as low or reverse oil, to the low-and-reverse clutch so that it engages. Note that the manual valve also directs line oil, in this case called *intermediate oil,* to the 2-3 shift valve. In both the 1-2 and 2-3 shift valves, the valves are moved against governor pressure to shut off the oil pressure to the intermediate and direct clutches so that they can release.

Governor pressure and throttle opening also play a part in the actual timing of the downshift. For instance, if the car speed is above about 45 mph [72.42 km/h] or if engine speed is above about 3,600 rpm, then the governor or modulator pressure will not permit the downshift. In other words, governor pressure acts against the ends of the 1-2 and 2-3 shift valves. Modulator pressure acts against the other ends, that is, the ends of the shift-control valves. Both of these pressures must be overcome by the low, or the manual-low-apply, pressure before downshift will occur. This condition protects the engine. If it were not for this protection, and if a downshift did occur at high car speed, then the transmission would spin the engine so fast that it would be severely damaged.

⊘ 13-30 Super Range, Second Gear

Super range, second gear, is also known as low two (L2) or drive two (D2). When the driver moves the selector lever to this position, the conditions in the transmission are as shown in Plate 14. These are the conditions shown in Fig. 13-11. The transmission is in second gear, with the following clutch and band conditions:

Intermediate clutch—on
Direct clutch—off
Forward clutch—on
Low-and-reverse clutch—off
Intermediate overrun roller clutch—locked
Low-and-reverse roller clutch—freewheeling
Intermediate overrun band—on

Now let us see what happens in the hydraulic system when the driver moves the selector lever to this position. Line oil entering the manual valve now exists as intermediate oil, which is then directed to two places: to the 2-3 shift valve and to the end of the intermediate boost valve. This valve is marked INT in Plate 14 and is to the right of the pressure-regulator valve in the illustration. It is numbered 5 in Fig. 13-13. This action increases minimum line pressure to 95 psi [6.679 kg/cm^2]. The intermediate oil moves the 2-3 shift valve to the downshifted position, and this releases the direct valve. Note that when the manual valve is moved to the intermediate position, the RND oil is exhausted from the top of the servo, which is the upper part of the 2-3 accumulator. Then, 1-2 oil, acting through the 1-2 accumulator, flows to the lower part of the servo and forces the servo piston up. This movement applies the intermediate overrun band so that the shell and sun gear of the front gearset are locked (through the intermediate overrun roller clutch). Now, as shown in Fig. 13-11, power flows through the forward clutch to the ring gear of the front gearset. The planet gears walk around the sun gear to produce the gear reduction.

Note that the situation here is different from the conditions when the manual valve is in D, or drive, and the transmission has upshifted from 1 to 2. Here, the sun gear is being held by the intermediate overrun band, and it cannot rotate. When the manual valve is in D with the transmission in intermediate or second gear, an upshift will occur, as previously explained, when the 2-3, or direct, clutch applies. But when the manual valve is in intermediate, an upshift cannot occur because the 2-3 oil is exhausted through the 2-3 shift valve. The 2-3 oil drains back through the reverse line.

The transmission will remain in second as long as the manual valve remains in intermediate unless the car slows down to about 9 mph [14.48 km/h]. If this should happen, then the governor pressure will drop so low that the 1-2 shift valve will move. This movement allows the 1-2 clutch oil to exhaust so that the clutch disengages and the transmission drops back into first.

⊘ 13-31 Drive-Range Detent Downshift When operating below 75 mph [120.70 km/h], a forced downshift can be made from direct to second if the accelerator is fully depressed. When this happens, the detent valve is moved by the cable that connects it to the throttle linkage. The detent valve is moved to its extreme inner position (to the right in Plate 15). The hydraulic system is thus set up in the situation shown in Plate 15. Modulator oil now goes to the 3-2 part-throttle line, and detent regulator oil is routed to the modulator shift valve and detent passages, as shown. Detent regulator oil therefore acts on both the 1-2 and 2-3 shift control valves. Also, modulator pressure acts on the 2-3 shift control valve through the 3-2 part-throttle passage. Detent regulator oil is also routed to the modulator valve through the detent passage.

Modulator oil, detent regulator oil, and the force of the 2-3-shift-control-valve spring move the 2-3 shift valve to the downshifted position below approximately 75 mph [120.70 km/h] so that the transmission downshifts into second gear.

The transmission can also be downshifted from second to first (2-1) or from third to first (3-1) below about 35 mph [56.33 km/h]. This is because detent regulator oil is directed to the 1-2 shift control valve. This action allows detent regulator oil plus the 1-2 shift control spring to move the 1-2 shift valve to the downshifted position to place the transmission into first gear.

In the detent downshifted position, these are the conditions of the clutches and band:

Intermediate clutch—on
Direct clutch—off
Forward clutch—on
Low-and-reverse clutch—off
Intermediate overrun roller clutch—locked
Low-and-reverse roller clutch—freewheeling
Intermediate overrun band—off

⊘ 13-32 Reverse When the driver moves the selector lever to R, or reverse, the conditions in the transmission are as pictured in Plate 16. These are the conditions shown in Fig. 13-12. The conditions of the clutches and band are as follows:

Intermediate clutch—off
Direct clutch—on
Forward clutch—off
Low-and-reverse clutch—on
Intermediate overrun roller clutch—freewheeling
Low-and-reverse roller clutch—freewheeling
Intermediate overrun band—off

With the manual valve in the reverse position, line pressure enters the reverse circuit. Reverse oil then flows to the following:

Direct clutch
Low-and-reverse clutch
1-2 shift valve
2-3 shift valve
Reverse boost valve

The reverse boost valve, numbered 4 in Fig. 13-13, acts to increase the line pressure to about 250 psi [17.58 kg/cm^2]. This higher pressure is needed to hold the clutches during reverse operation. Strong holding power is required at this time.

CHAPTER 13 CHECKUP

NOTE: Since the following is a chapter review test, you should review the chapter before taking it.

With a good understanding of the Type 350 Turbo Hydra-Matic transmission, you will be equipped to understand any automatic transmission. Furthermore, you will be able to move ahead to service these transmissions with confidence because you will understand how the transmission is constructed and how it operates. Now, check your memory on how well you absorbed everything you read in the chapter by taking the test that follows.

Completing the Sentences The sentences that follow are incomplete. After each sentence there are several words or phrases, only one of which will correctly complete the sentence. Write each sentence in your notebook, selecting the proper word or phrase to complete it correctly.

1. Counting all the clutches in the 350 transmission, including the roller clutches and the multiple-disk clutches, there are a total of: (*a*) four, (*b*) five, (*c*) six.
2. In the Powerglide, the sun gears are two separate parts. In the Type 350 Turbo Hydra-Matic: (*a*) they are also two separate parts, (*b*) they are a single piece, (*c*) there are three sun gears.
3. A major purpose of the vacuum-modulator assembly is to produce a modulated oil pressure correlated with: (*a*) engine speed, (*b*) car speed, (*c*) intake-manifold vacuum.
4. The purpose of the governor is to produce an oil pressure correlated with: (*a*) engine speed, (*b*) car speed, (*c*) intake-manifold vacuum.
5. The purpose of the pressure-regulator valve is to: (*a*) reduce pump pressure to line pressure, (*b*) maintain pump pressure, (*c*) vary pressure with car speed.
6. The purpose of the reverse boost valve is to: (*a*) aid shifting into reverse, (*b*) increase line pressure in reverse, (*c*) boost the low-reverse valve.
7. The purpose of the accumulator is to cushion the application of the: (*a*) selector lever, (*b*) manual valve, (*c*) clutch.
8. The purpose of the detent valve is to: (*a*) reroute oil pressure to produce a downshift, (*b*) reroute oil pressure to prevent a downshift, (*c*) hold the manual valve in detent.
9. The purpose of the 1-2 shift valve is to route oil pressure so as to: (*a*) hold gears in 1-2 position, (*b*) produce an upshift, (*c*) engage the direct and intermediate clutches.
10. The purpose of the 2-3 shift valve is to route oil pressures so as to: (*a*) engage the roller clutches, (*b*) hold gears in 2-3 position, (*c*) produce an upshift.

Purpose and Operation of Components In the following, you are asked to write the purpose and operation of the various components used in the Type 350 Turbo Hydra-Matic transmission. If you have any difficulty with any of the questions, reread the pages in the chapter that will give you the information you need. Then write your explanation. Don't copy; try to tell it in your own words. This is a good way to fix the information in your mind. Write in your notebook.

1. Describe the relationship of the ring gears, planet pinions, and sun gears to each other.
2. Describe the operation of the overrun roller clutch.
3. Describe the power flow in D, low gear.
4. Describe the power flow in D, second gear.
5. Describe the power flow in D, direct drive.
6. Describe the power flow in L, first gear.
7. Describe the power flow in D2, second gear.
8. Describe the power flow in reverse.
9. Explain how the oil pump works.
10. Explain the purpose and operation of the vacuum-modulator assembly.
11. Explain the purpose and operation of the governor.
12. Explain how the pressure-regulator valve works.
13. Explain the purpose and operation of the accumulator.
14. Explain how the 1-2 shift valve works.
15. Explain how the detent valve works.
16. Make a drawing of the part of the hydraulic system that is active when the transmission is in first gear, drive range.
17. Make a drawing of the part of the hydraulic system that is active when the transmission is in second gear, drive range.
18. Make a drawing of the part of the hydraulic system that is active when the transmission is in direct drive.
19. Make a drawing of the part of the hydraulic system that is active when the transmission is in reverse.
20. Explain how a drive-range detent downshift takes place.

SUGGESTIONS FOR FURTHER STUDY

Whenever you have a chance, examine the 350 transmission and its component parts. These are available in school shops and in automotive shops that handle automatic-transmission work. Note especially the construction and relationship of the gearing and the clutches and bands that control the gears. Study the car shop manuals covering the transmission. Be sure to write everything of importance you learn in your notebook.

chapter 14

TYPE 350 TURBO HYDRA-MATIC TRANSMISSION SERVICE

This chapter discusses the trouble-diagnosis and service procedures for the Type 350 Turbo Hydra-Matic automatic transmission. The chapter is divided into four parts: "Normal Maintenance and Adjustments," "Trouble Diagnosis," "On-the-Car Repairs," and "Transmission Overhaul."

⊘ **14-1 Type 350 Transmission Variations** The Type 350 transmission is used on many cars with a wide variety of engines, both sixes and eights. Some minor modifications of the transmission are made so that it will cooperate properly with the engine with which it is used. These modifications have no effect on the service procedures, but they do give the characteristics the transmission needs for engine torque and output curves. For example, the transmission used on Buick six-cylinder engines has fewer plates in the clutches than the transmission used on Buick eight-cylinder engines. Fewer plates are required to handle the lower torque of the smaller engine. Also, springs and valves may be changed so as to provide different shift points. But, as we said, the service procedures are essentially the same for all models of the Type 350 Turbo Hydra-Matic transmission.

Normal Maintenance and Adjustments

⊘ **14-2 Checking and Adding Transmission Fluid** The recommendation is to use only Dexron transmission fluid (also called *transmission oil*). Using another fluid could ruin the transmission. Fluid level should be checked at every engine-oil change. If oil is needed, it should be added as explained below. Every 24,000 miles [38,624 km] the oil pan should be drained and the strainer cleaned. Then fresh oil should be added, as we shall explain. If the car is used in heavy city traffic, where the engine is regularly idled for long periods, the oil pan should be drained and the strainer cleaned every 12,000 miles [19,312 km].

Fluid level can be checked with the engine cold or hot. But, it is better to check it with the engine and transmission hot (180 degrees Fahrenheit) [82.2°C]. Both methods are described below.

1. *CHECKING FLUID LEVEL AND ADDING FLUID WITH TRANSMISSION HOT* Proceed as follows:

a. Block the car wheels and apply the parking brake. Start the engine. *Do not race the engine!* Move the selector lever through each range. If the parking brake releases, hold the car with the foot brake. Reapply the parking brake when you return the selector lever to park.

b. Immediately check the fluid level with the selector lever in park. The engine should be running at slow idle and the car should be on a level surface. The fluid level should be at the FULL mark on the dipstick.

c. If the fluid level is low, add only enough to bring it up to FULL. *Do not overfill!* Overfilling will cause foaming and loss of fluid through the vent pipe. If any amount of fluid is lost in this manner, complete loss of drive through the transmission could occur.

2. *CHECKING FLUID LEVEL AND ADDING FLUID WITH TRANSMISSION COLD*[1] Proceed as you did for checking with the transmission hot with this exception: With the transmission cold, the fluid level should be 1/4 inch [6.35 mm] below the ADD mark on the dipstick. If it is lower, add fluid to bring the fluid level up to 1/4 inch [6.35 mm] below the ADD mark, no more. When the transmission heats up, the fluid will expand and rise to the FULL mark. *Do not overfill!*

NOTE: Buick, Chevrolet, and Pontiac recommend that the fluid level be 1/4 inch [6.35 mm] below the ADD mark. Oldsmobile, however, recommends that the fluid level be brought up to 3/4 inch [19.05 mm] below the FULL mark with the transmission cold.

[1] 80 degrees Fahrenheit [26.7°C].

3. *DRAINING THE OIL PAN AND CLEANING STRAINER* The oil pan should be drained and the strainer cleaned every 24,000 miles [38,624 km] (12,000 miles [19,312 km] in severe city service).

a. Raise the car on a hoist and remove the oil pan and gasket. Throw the gasket away. Drain oil from the pan. Clean the pan with solvent and dry it thoroughly with clean compressed air.

b. Remove the strainer assembly and clean it. Discard the oil-strainer-to-valve-body gasket.

c. Install a new gasket on the strainer and reinstall the strainer.

d. Install a new gasket on the oil pan and install the oil pan. Tighten attaching bolts to 13 pound-feet [1.80 kg-m].

e. Lower the car and add $1\frac{1}{2}$ quarts [1.42 l] of transmission fluid.

f Block the car wheels, set the parking brake, and start the engine. *Do not race the engine.* Move the selector lever through each range.

g. Immediately check the fluid level with the selector lever in P (park) and with engine idling. The car must be on a level surface.

h. Add fluid to bring the fluid level to $\frac{1}{4}$ inch [6.35 mm] below the ADD mark. *Do not overfill!*

⊘ 14-3 Towing Instructions Usually, the car may be safely towed on its rear wheels with the selector lever in neutral for short distances at speeds no greater than 35 mph [56.33 km/h]. But there are exceptions: If the transmission is not operating properly, if higher speeds are necessary, or if the car must be towed more than 50 miles [80.47 km], the *drive shaft must be disconnected.*

As an alternative, the car can be towed on its front wheels. In this case, the steering wheel must be secured to keep the front wheels in the straight-ahead position.

⊘ 14-4 Adjustments Only two adjustments are required with the transmission in the car: linkage to the selector lever and linkage to the accelerator (for the detent downshift). The adjustments are different for different cars, and so you should always refer to the appropriate shop manual for the car you are working on. There, you will find the instructions that apply. Figures 14-1 and 14-2 show typical adjustment procedures for the shifter linkage to the selector lever and for the detent cable.

Trouble Diagnosis

Although the transmission is fairly complex and packs a considerable number of moving parts in a small space, causes of troubles are usually easy to spot. A specific trouble can have only certain specific causes.

All the various car manufacturers who use the Type 350 automatic transmission recommend that oil-pressure checks be made with the transmission in operation to help locate causes of trouble. The procedures recommended, as well as the pressures specified under different operating conditions, vary somewhat from manufacturer to manufacturer. Thus, always refer to the shop manual that specifically covers the car and transmission you are checking. A typical diagnosis procedure follows.

⊘ 14-5 Trouble Diagnosis The chart shown in Fig. 14-3 will help you to locate causes of trouble in the Type 350 transmission. Before we look at the chart, however, here is a typical diagnosis procedure as outlined by one of the car manufacturers using the transmission:

1. Check and correct the oil level (see ⊘ 14-2).
2. Check and correct the detent-cable adjustment.
3. Check the vacuum line and fittings for leakage.
4. Check and adjust the manual linkage as necessary.
5. Shop-test and road-test the car, as follows. Install the oil-pressure gauge by connecting it to the transmission line-pressure tap located to the left of the 1-2 accumulator cover (shown being removed in Fig. 14-4). Then test the car in the shop and on the road, using all selector-lever positions and noting transmission operation and oil pressures under different conditions.

NOTE: If the engine is not performing satisfactorily, it should be tuned up and given whatever service is necessary to bring it up to normal operating conditions. Poor engine performance can result in faulty

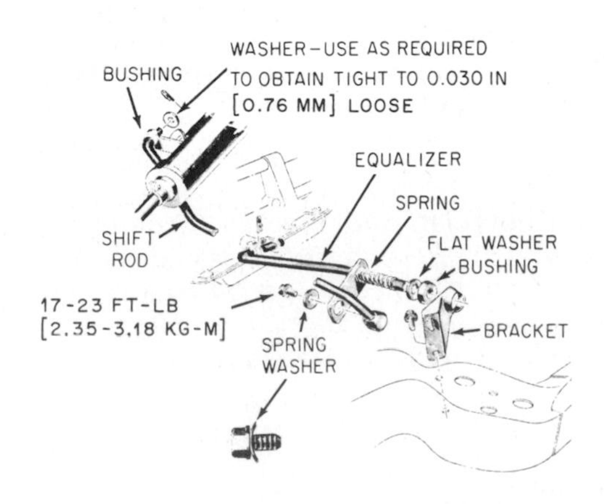

SHIFT-ROD ADJUSTMENT

1. SET TRANSMISSION OUTER LEVER IN DRIVE POSITION
2. HOLD UPPER SHIFT LEVER AGAINST DRIVE-POSITION STOP IN UPPER STEERING COLUMN (DO NOT RAISE LEVER)
3. TIGHTEN SCREW IN CLAMP ON LOWER END OF SHIFT ROD TO SPECIFIED TORQUE
4. CHECK OPERATION:
   a. WITH KEY IN RUN POSITION AND TRANSMISSION IN REVERSE, BE SURE THAT KEY CANNOT BE REMOVED AND THAT STEERING WHEEL IS NOT LOCKED
   b. WITH KEY IN LOCK POSITION AND SHIFT LEVER IN PARK, BE SURE THAT KEY CAN BE REMOVED, THAT STEERING WHEEL IS LOCKED, AND THAT THE TRANSMISSION REMAINS IN PARK WHEN THE STEERING COLUMN IS LOCKED

Fig. 14-1. Shift-rod adjustment. (*Oldsmobile Division of General Motors Corporation*)

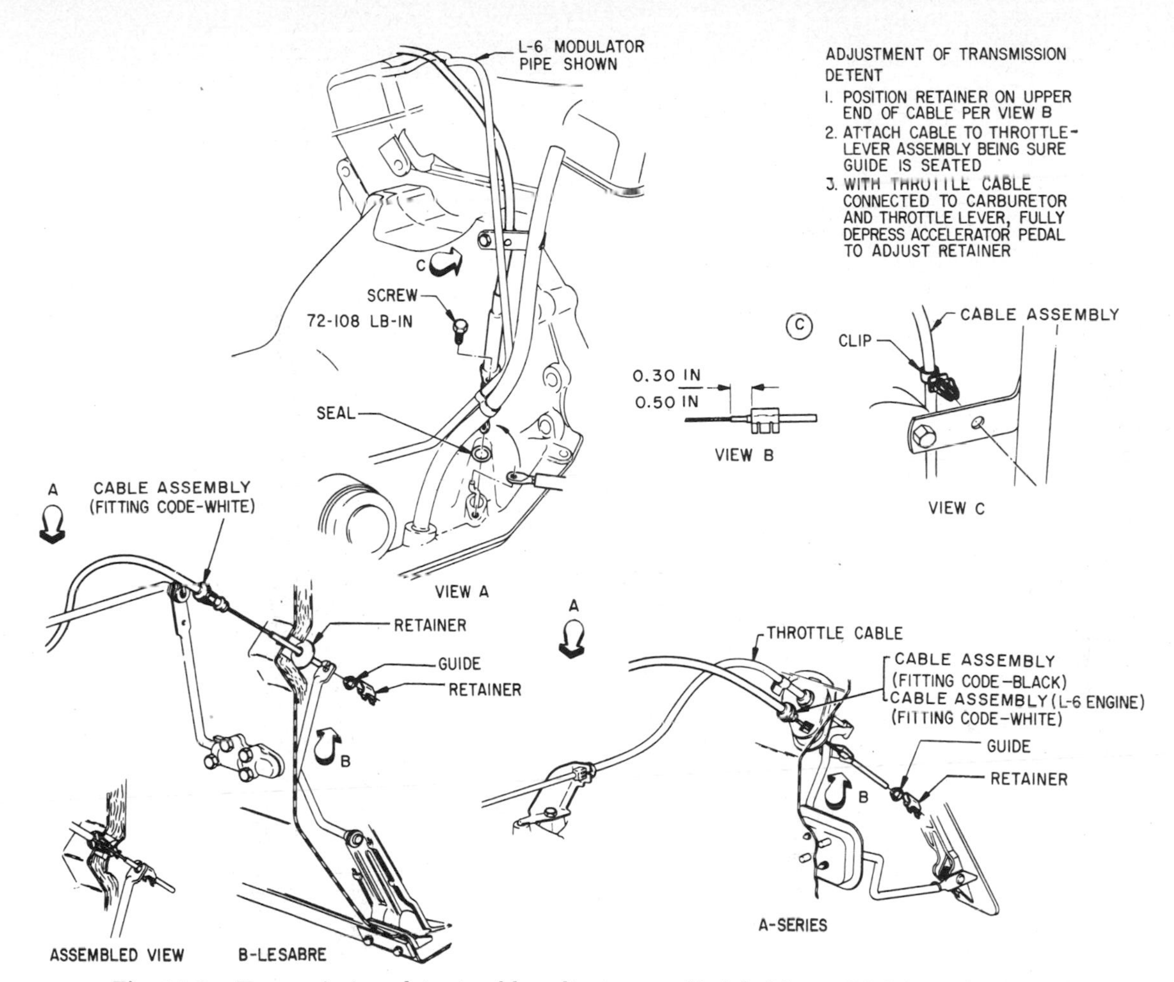

Fig. 14-2. Transmission detent-cable adjustment. (*Buick Motor Division of General Motors Corporation*)

transmission shifts and can make the transmission look bad even though it is in normal condition.

CAUTION: When road-testing a car, obey all traffic laws. Drive safely. Use a chassis dynamometer (⊘ 1-9) if it is available so that you can make all the tests in the shop and will not have to take the car out on the road.

6. The tabulation that follows is the recommendation of one car manufacturer for the test procedure and pressure readings, in pounds per square inch, that should be obtained.

CAR STATIONARY

Parking brake applied—wheels blocked

RANGE	MIN. psi AT 1,000 rpm WITH VACUUM LINE CONNECTED	MAX. psi AT 1,200 rpm WITH VACUUM LINE DISCONNECTED
Park and neutral	60	156
Reverse	85	240
Drive	60	156
Super	85	156
Low	85	156

ROAD OR DYNAMOMETER TEST

Vacuum line and pressure gauge installed

THROTTLE DRIVE RANGE	IDLE	WIDE OPEN
First	60	153
Second	60	153
Third	60	
Reverse	85	239
Low (coast at 3 mph)	85	
Super (coast at 30 mph)	85	

7. Refer to the trouble-diagnosis chart (Fig. 14-3) to determine causes of possible troubles that show up during the tests. In most cases, the chart will help you locate the causes so that you can make the necessary repairs. Sometimes, it will be necessary to remove the transmission from the car for disassembly so that defective parts can be replaced. Other troubles, however, may be solved by replacement of parts without removal of the transmission.

On-the-Car Repairs

⊘ 14-6 Repairs That Do Not Require Transmission Removal Some parts can be removed and replaced

PROBLEM / CAR ROAD TEST LEGEND X–PROBLEM AREA VS. CAUSE * –@ "O" VACUUM ONLY O–BALLS/#2/3/4 ONLY L –LOCKED S–STUCK POSSIBLE CAUSE	ALL RANGES–SLIPS	DRIVE–SLIPS–NO 1ST GEAR	LINE PRESSURE–ALL LOW	LINE PRESSURE–ALL HIGH	1-2 INTERM. CL. PRES. HIGH	1-2 INTERM. CL. PRES. LOW	2-3 DIRECT CL. PRES. LOW	2-3 DIRECT CL. PRES. HIGH	NO 1-2 UPSHIFT	1-2 U.S. EARLY/LATE	1-2 UPSHIFT @ W.O.T. ONLY	SLIPS–1-2 UPSHIFT	ROUGH–1-2 UPSHIFT	NO 2-3 UPSHIFT	2-3 U.S.–EARLY/LATE	SLIPS–2-3 UPSHIFT	ROUGH–2-3 UPSHIFT	NO WOT. 1-2 UPSHIFT	NO–PART TH. DOWN SHIFT	NO–FULL TH. DOWN SHIFT	2-3 UPSHIFT–W.O.T. ONLY	HARSH–DOWN SHIFT	L1 RANGE–NO ENG. BRAKING	L2 RANGE–NO ENG. BRAKING	NEUTRAL–DRIVES IN NEUTRAL	REVERSE–NO REVERSE	SLIPS IN REVERSE	PARK–NO PARK–RATCHETS	NOISY–ALL RANGES	1-2 2-3 SHIFT NOISY	REV. & D. L1 & L2 NOISY	LOW COOLER–FLOW	SPEWS OIL OUT BREATHER	HUNTS BETWEEN 2 & 3 AND 3 & 2
LOW OIL LEVEL/WATER IN OIL	X	X	X			X		X				X				X											X		X	X		X	X	
VACUUM LEAK				X	X		X		X	X	X		X		X		X				X													
MODULATOR &/OR VALVE	X	X	*	X	X							X	X		X	X	X										X							
STRAINER &/OR GASKET	X	X	X			X		X																			X		X			X	X	
GOVERNOR–VALVE/SCREEN			X	X					X	X					X																			
VALVE BODY–GASKET/PLATE	X	X	X			X		X	X	X		X		X	X	X											X		X					
PRES. REG. &/OR BOOST VALVE	X	X	X	X	X	X	X	X	X			X	X			X	X						X	X		X	X							
BALL (#1) SHY	X	X	*			X				X		X			X	X						O				X	X							
1-2 SHIFT VALVE						X			X	X	X	X											X				X							
2-3 SHIFT VALVE								X						X	X	X	X		X															
MANUAL LOW [illegible] VALVE																							X											
DETENT VALVE & LINKAGE				X							[illegible]				[illegible]				X	X	X													S
DETENT REG. VALVE																		X																
2-3 ACCUMULATOR						X						X					X																	
MANUAL VALVE/LINKAGE	X	X	X																				X	X		X	X							
POROSITY/CROSS LEAK	X	X	X	X	X	X	X	X	X	X	X	X	X	X	X	X	X	X					X	X		X	X							
PUMP–GEARS	X	X	X			X		X				X				X													X					
PRIMING VALVE SHY	X	X	X																														S	
COOLER VALVE LEAK																																X	X	
CLUTCH SEAL RINGS	X	X	X			X		X	X			X		X		X							X	X		X	X							
POROUS/CROSS LEAK	X	X	X			X		X	X			X	X	X		X		X								X	X					X	X	
GASKET SCREEN–PRESSURE	X	X	X																							X	X		X			X		
BAND–INTERM. O.R.																								X										
CASE–POROUS/X LEAK	X	X	X	X	X	X	X	X	X	X	X	X	X	X	X	X	X						X			X	X					X		
1-2 ACCUMULATOR	X	X	X			X			X			X	X																					
INTERMED. SERVO		X	X			X						X												X										
FORWARD CLUTCH ASS'Y		X																					X	X	X	L								
DIRECT CLUTCH ASS'Y								X						X		X										X	X			X				
INTERMED. CL. ASS'Y						X			X			X																		X				
L & REV. CL. ASS'Y																							X			X	X							
INT. ROLLER CL. ASS'Y									X			X												X										
L. & R. ROLLER CL. ASS'Y		X																																
PARK PAWL/LINKAGE																												X						
CONVERTER ASS'Y																													X		X			
GEAR SET & BEARINGS																													X	X				

Fig. 14-3. Type 350 Turbo Hydra-Matic transmission trouble-diagnosis chart. (*Chevrolet Motor Division of General Motors Corporation*)

Fig. 14-4. Using a special tool to remove the intermediate-clutch-accumulator retaining ring.

without removing the transmission from the car. These include:

Oil pan and gasket
Oil screen
Valve-body assembly
Direct-clutch accumulator and servo (after valve body is removed)
Manual control linkage and parking pawl
Extension housing and seal
Vacuum modulator
Speedometer drive gear
Transmission-control-spark switch
Governor assembly
Intermediate-clutch accumulator

These parts, with the exception of the last two, require the same procedure used during complete disassembly and reassembly of the transmission. Thus, these removal-and-replacement procedures are described on later pages where complete transmission disassembly and reassembly are described. Governor assembly and intermediate-clutch-accumulator removal and replacement require different procedures if the transmission is not removed from the car, and these procedures are described next.

⊘ 14-7 Governor-Assembly Removal and Replacement The governor assembly can be removed and replaced with the transmission on the car, as follows:

1. Put the car on a hoist and disconnect the speedometer cable at the transmission.
2. Remove the governor cover retainer and cover. Do not damage the O-ring seal.
3. Remove the governor. Check weights and valve to make sure they move freely. If they do not, repair or replace the parts as necessary.
4. To replace the governor, slip it into place and install the cover, using a brass drift around the outside flange of the cover. Do not distort the cover, and be sure the O-ring seal is not cut or damaged.
5. Replace the retainer and connect the speedometer cable. Lower the car from the hoist and check the fluid level.

⊘ 14-8 Intermediate-Clutch-Accumulator Removal and Replacement The intermediate-clutch accumulator can be removed and replaced with the transmission on the car, as follows:

1. Remove the two transmission oil-pan bolts below the intermediate-clutch cover. Install a special tool, as shown in Fig. 14-4, in place of the bolts removed.
2. Press in on the tool handle and use an awl to remove the retaining ring, as shown in Fig. 14-4.
3. Remove the cover O-ring seal, spring, and intermediate-clutch accumulator.
4. To replace put the intermediate-clutch accumulator into place. Lubricate the rings. Rotating the piston slightly will help start the rings into the piston bore.
5. Put the spring, O-ring seal, and cover into place. Press in on the cover and replace the retaining ring.
6. Remove the tool and replace the oil-pan bolts.

Transmission Overhaul

⊘ 14-9 Transmission Removal and Replacement Individual variations in engine-compartment arrangements require different removal and replacement procedures. Always check the car shop manual for the car you are working on for the specific instructions that apply. The following is a typical removal-and-replacement procedure, which applies to late-model Chevrolet vehicles.

Before raising the car, disconnect the negative battery cable and release the parking brake. Then proceed as follows:

1. Raise the car on a hoist and remove the drive shaft. Scribe marks on the shaft and flange so that the shaft can be reinstalled in its original position.
2. Disconnect the speedometer cable, detent cable, modulator vacuum line, and oil cooler pipes.
3. Disconnect the shift-control linkage.
4. Support the transmission with a special transmission jack.
5. Disconnect the rear mount from the frame cross member.
6. Remove two bolts at each end of the frame cross member and remove the cross member.
7. Remove the oil-cooler lines, vacuum-modulator line, speedometer cable, and detent cable at the transmission.
8. Remove the pan from under the converter. Mark the converter and flywheel so that they can be reattached in their original position. Then remove the converter-to-flywheel bolts.
9. Loosen the exhaust-pipe-to-manifold bolts about $\frac{1}{4}$ inch [6.35 mm] (not required on Nova). Lower the transmission until the jack is barely supporting it.

CAUTION: On V-8 engines, have an assistant on the top side to observe the clearance of all engine components while the transmission rear end is being lowered. There is a chance that the distributor may be forced against the dash and damaged.

10. Remove the transmission-to-engine mounting bolts and remove the oil-filler tube from the transmission.
11. Raise the transmission to its normal position, support the engine with a jack, and slide the transmission rearward from the engine and lower it away from the vehicle.

CAUTION: Use a converter-holding tool, or keep the rear of the transmission lower than the front. Otherwise, the converter is likely to slide off and drop to the floor. This would ruin it or hurt anyone standing underneath.

12. Replacement is essentially the reverse of removal except for the following added steps: Before replacing the flex-plate-to-converter bolts, make sure that the attaching lugs on the converter are flush with the flex plate and that the converter rotates freely by hand in this position. Then, hand-start all three bolts and tighten them finger tight before torquing to specifications. This ensures proper alignment of the converter. After replacement is complete, remove the car from the hoist and check the linkage for proper adjustment. Check the fluid level, as explained in ⊘ 14-2, and add fluid as necessary.

⊘ 14-10 Transmission-Service Precautions Cleanliness is of the utmost importance in transmission work. The tiniest, almost invisible piece of lint from a cleaning rag or a small piece of dirt can cause a valve to hang up and prevent normal transmission action. Under some conditions, this could cause fatal damage to the transmission.

1. Never mix parts between transmissions. Keep all parts belonging to a transmission in one place so that they will not get mixed with parts of other transmissions.
2. Clean the outside of the transmission thoroughly before starting to disassemble it. Plug all openings and use steam if it is available. The workbench, tools, your hands, and all parts must be kept clean at all times. Do not allow dust to blow in from outside or from other areas of the shop and settle on the transmission parts. This could mean big trouble.
3. Before installing screws into aluminum parts such as the case, dip the screws into transmission fluid to lubricate them. Lubrication will prevent their galling the aluminum threads and will prevent the screws from seizing.
4. If the threads in an aluminum part are stripped, repair can be made with a Heli-Coil, as shown in Fig. 14-5. To repair a defective thread, drill a hole and then tap it with a special Heli-Coil tap. Finally, install a Heli-Coil to bring the hole back to its original thread size.
5. Special care and special tools to protect the seals must be used during assembly to prevent damage to the seals. The slightest flaw in a seal or sealing surface can result in an oil leak.
6. The aluminum castings are relatively soft and can be easily nicked, scratched, or burred. Use great care in handling them.
7. Discard all O-ring seals, gaskets, and oil seals that are removed, and use new ones on reassembly.
8. During reassembly, lubricate all internal parts with transmission fluid.
9. During disassembly, clean and inspect all parts, as explained in ⊘ 14-11.

⊘ 14-11 Cleaning and Inspecting Transmission Parts After disassembly, clean and inspect all parts, as follows: Wash all metal parts in solvent and blow them dry with compressed air. Do not use cleaning solvents that could damage rubber seals or clutch facings. Make sure all oil passages are clean. Check small passages with small wire such as tag wire. Inspect parts as follows:

1. Check linkage and pivot points for excessive wear.
2. Check bearing and thrust faces for wear and scoring.
3. Check mating surfaces of castings and end plates for burrs that could cause poor seating and oil leaks. To remove burrs or irregularities, lay a piece of crocus cloth on a very flat surface such as a piece of plate glass. Then lay the part on the crocus cloth and move it back and forth. This process is called *lapping* and will bring the metal surface to a smooth finish.
4. Check for damaged grooves or lands where O rings seat. Irregularities here can cause serious oil leaks.
5. Check castings for cracks and sand holes and for damaged threads. Thread repairs can often be made with Heli-Coils, as already noted (see Fig. 14-5).
6. Check gear teeth for chipping, scoring, or wear.
7. Make sure that valves are free of burrs and that the shoulders are square and sharp. Burrs can be removed from bores by honing. Valves must slide freely in the bores.
8. Inspect the composition clutch plates for damaged surfaces and loose facings. If you can remove flakes of facing with your thumbnail, replace the plates. Discoloration, however, is normal and is not a sign of failure.
9. Inspect steel clutch plates for scored surfaces.
10. Inspect springs for distortion or collapsed coils.

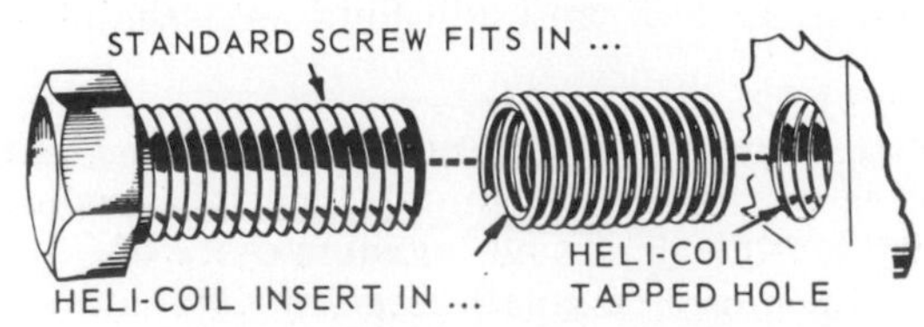

Fig. 14-5. Heli-Coil installation. (*Chrylser Corporation*)

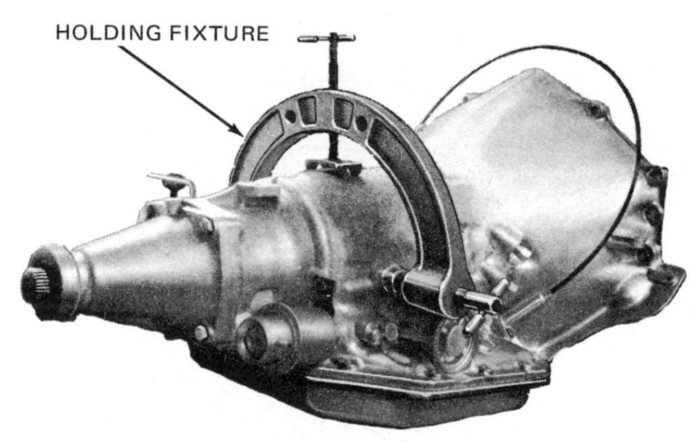

Fig. 14-6. Transmission in a holding fixture. (*Chevrolet Motor Division of General Motors Corporation*)

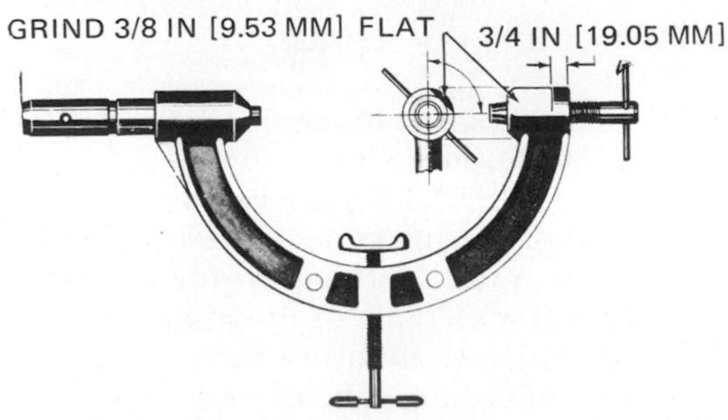

Fig. 14-7. Modifying the holding fixture to accept the transmission. (*Oldsmobile Division of General Motors Corporation*)

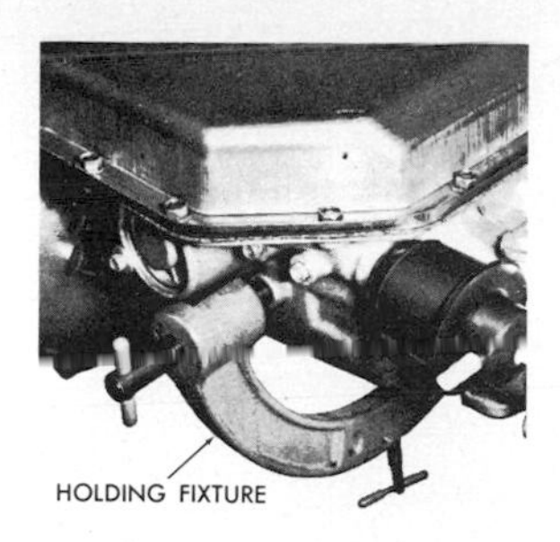

Fig. 14-8. Transmission in a holding fixture. (*Buick Motor Division of General Motors Corporation*)

Fig. 14-9. Removing the torque converter from the transmission. (*Oldsmobile Division of General Motors Corporation*)

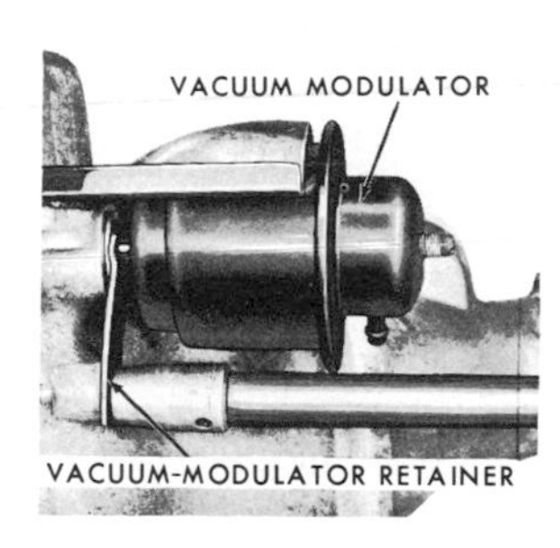

Fig. 14-10. Vacuum-modulator assembly and retainer. (*Buick Motor Division of General Motors Corporation*)

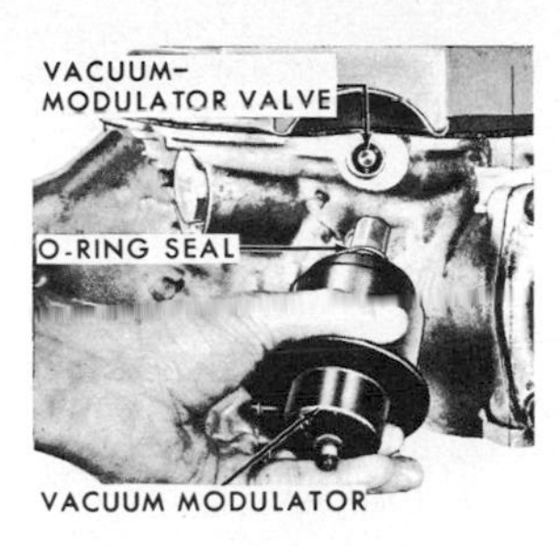

Fig. 14-11. Removing the vacuum-modulator assembly. (*Buick Motor Division of General Motors Corporation*)

Fig. 14-12. Using a wrench to remove the bolts that attach the extension housing to the transmission case. (*Buick Motor Division of General Motors Corporation*)

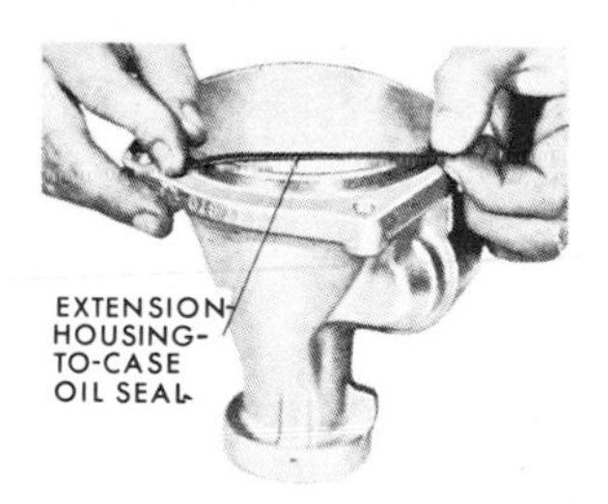

Fig. 14-13. Removing the square-cut O-ring seal from the extension housing. (*Buick Motor Division of General Motors Corporation*)

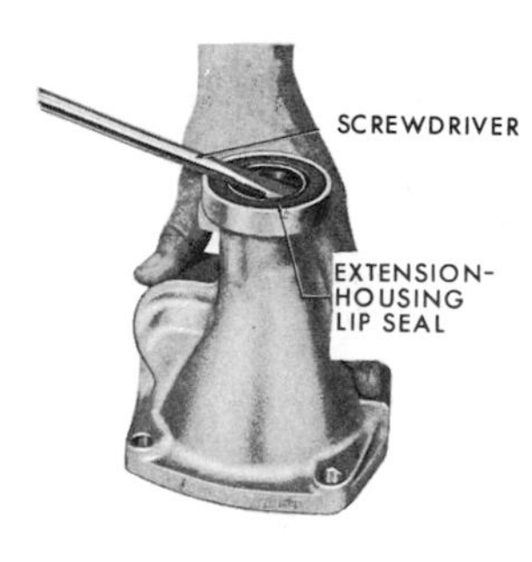

Fig. 14-14. Removing the extension-housing lip seal with a screwdriver. (*Buick Motor Division of General Motors Corporation*)

11. Check the bushings and test for wear by inserting the mating part into the bushing and noting the amount of looseness. According to one shop manual, wear is excessive if the clearance is more than 0.008 inch [0.20 mm]—checked with a wire feeler gauge.
12. If the transmission has internal damage to clutches or other parts, or if it shows evidence that foreign material has been circulating in the oil, install a new oil filter when reassembling the transmission. Also, flush out the oil cooler and lines and the torque converter.
13. Be sure the lubrication holes in the converter turbine shaft are open.

⊘ 14-12 Transmission Disassembly None of the moving parts requires forcing when disassembling or reassembling the transmission. Bushing removal and replacement do require driving or pressing, of course. The cases might fit tightly, and you can loosen them with a rawhide or plastic mallet. Never use a hard hammer.

As a first step, the transmission must be placed in a holding fixture, as shown in Fig. 14-6. If the fixture has not been modified to accept the transmission, grind a flat on it, as shown in Fig. 14-7. This flat will allow the fixture to clear the 1-2 accumulator, as shown in Fig. 14-8. Proceed as follows:

1. Turn the transmission so that the oil pan is up. If you have used a torque-converter holding tool, remove it. Then slip off the torque converter, as shown in Fig. 14-9.
2. Remove the vacuum-modulator-assembly attaching bolt and retainer, as shown in Fig. 14-10. Now, remove the vacuum-modulator assembly, O-ring seal, and modulator valve, shown in Fig. 14-11. The vacuum modulator is checked as explained in ⊘ 14-21.
3. Remove the extension housing by taking out the four attaching bolts, using a wrench, as shown in Fig. 14-12. Remove the square-cut O-ring seal, as shown in Fig. 14-13. Use a screwdriver to remove the extension-housing lip seal from the housing, as shown in Fig. 14-14. If it is necessary to replace the

housing bushing, use a screwdriver to collapse it, as shown in Fig. 14-15. Clean the housing carefully and inspect it for any damage. Then install a new housing bushing with special tools, as shown in Fig. 14-16. A new housing lip seal can then be installed with the special tool, as shown in Fig. 14-17. The extension housing is now ready for reinstallation on the transmission case and can be set aside until the rest of the transmission has been disassembled and reassembled.

4. To remove the yoke seal from the end of the output shaft, install the special tool, as shown in Fig. 14-18, to keep the seal from cocking, and then tap the seal off with a screwdriver, as shown. Next, take off the speedometer-drive gear by depressing the retaining clip, as shown in Fig. 14-19. The gear can be slid off the shaft with the clip depressed. Remove the clip.

5. Next, remove the governor by first prying off the governor retaining wire with a screwdriver, as shown in Fig. 14-20. Remove the governor cover and O-ring seal from the case (Fig. 14-21). Use a screwdriver to pry between the cover flange and the case. But be extremely careful to avoid denting or bending the cover or gouging the transmission case. The dimple in the end of the cover provides the proper end play for the governor. If the cover is bent or damaged in any way, it must be discarded and a new cover installed. Remove the O-ring seal from the cover. Now, the governor assembly can be withdrawn from the case, as shown in Fig. 14-22.

6. Now, remove the oil pan, strainer, and valve body. First, take out 13 attaching-screw-and-washer assemblies and lift off the oil pan and gasket, as shown in Fig. 14-23. Then remove the two screws attaching the oil-pump strainer to the valve body, as shown

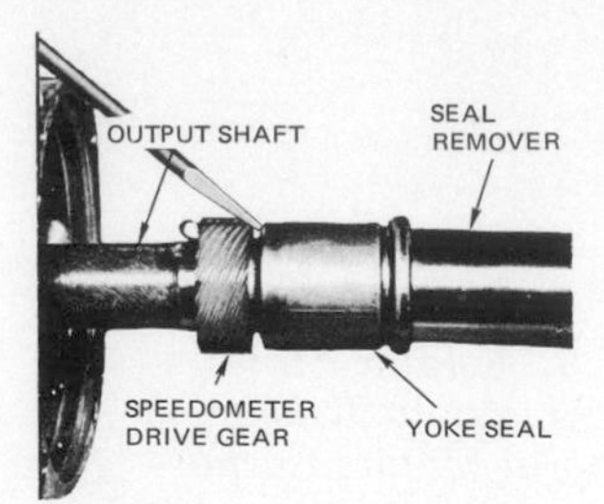

Fig. 14-18. Removing the yoke seal from the end of the transmission output shaft. (*Oldsmobile Division of General Motors Corporation*)

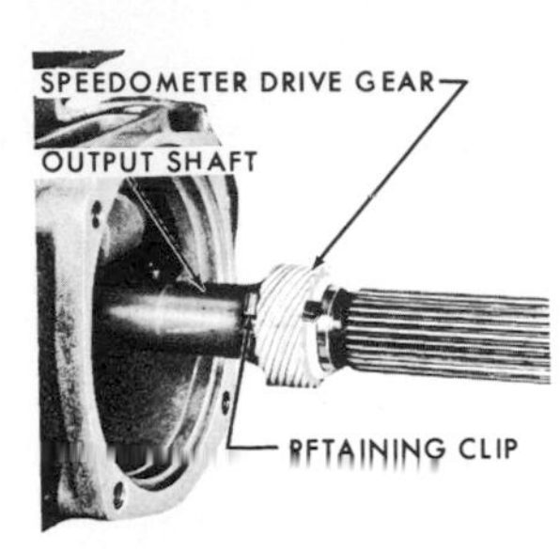

Fig. 14-19. Speedometer drive gear and retaining clip. (*Buick Motor Division of General Motors Corporation*)

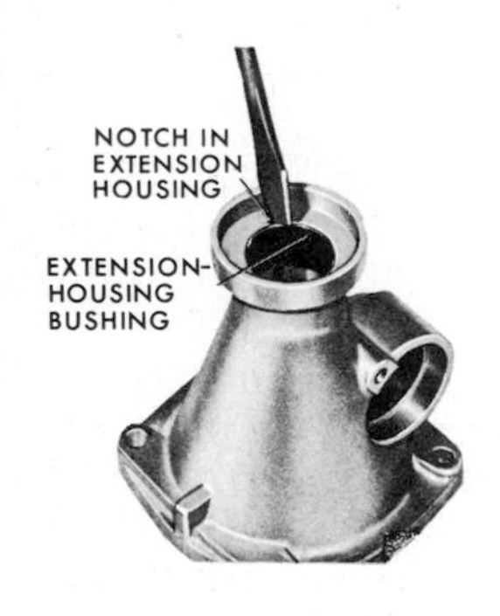

Fig. 14-15. Removing the extension-housing bushing with a screwdriver. (*Buick Motor Division of General Motors Corporation*)

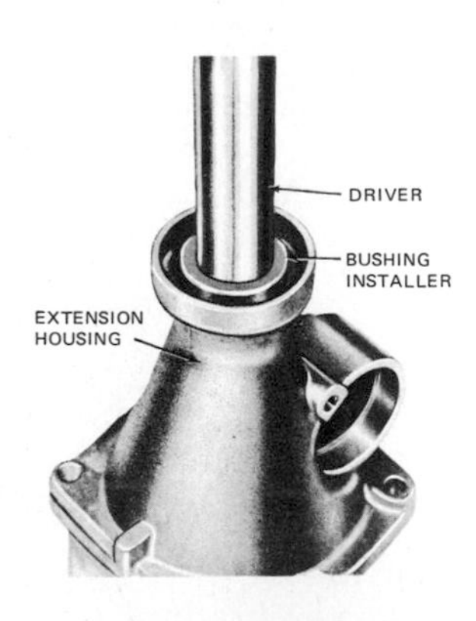

Fig. 14-16. Installing a new extension-housing bushing. (*Buick Motor Division of General Motors Corporation*)

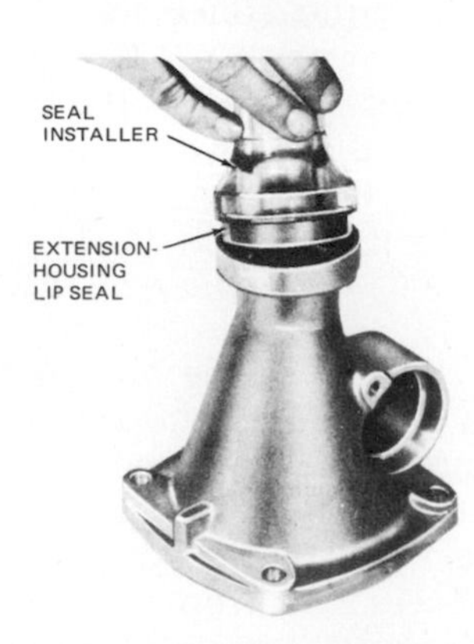

Fig. 14-17. Installing a new extension-housing lip seal. (*Buick Motor Division of General Motors Corporation*)

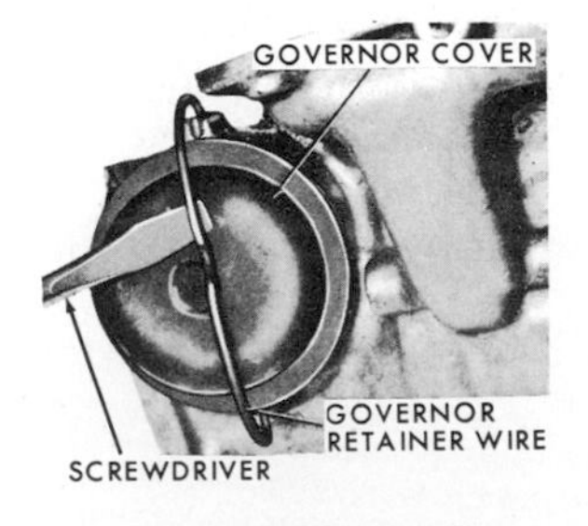

Fig. 14-20. Prying off the governor-cover retainer wire with a screwdriver. (*Buick Motor Division of General Motors Corporation*)

Fig. 14-21. Removing the governor cover. (*Buick Motor Division of General Motors Corporation*)

Fig. 14-22. Removing the governor assembly from the case. (*Buick Motor Division of General Motors Corporation*)

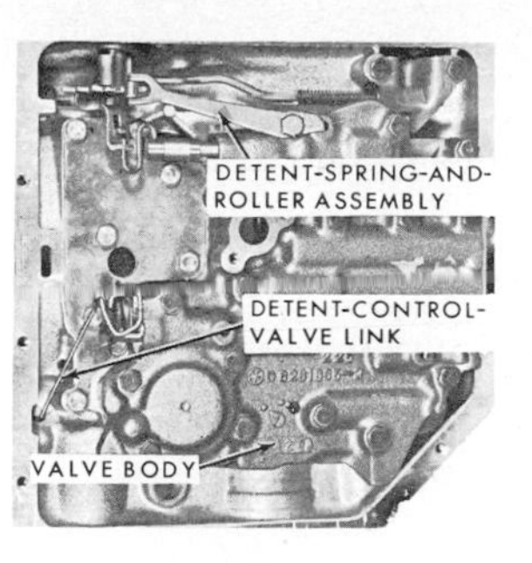

Fig. 14-25. Locations of the detent-spring-and-roller assembly and the detent-control-valve link on the valve body. (*Buick Motor Division of General Motors Corporation*)

Fig. 14-23. Removing the oil pan and gasket. (*Buick Motor Division of General Motors Corporation*)

Fig. 14-26. Removing the valve-body-to-spacer plate gastet. (*Buick Motor Division of General Motors Corporation*)

in Fig. 14-24. Take the oil-pump-strainer gasket off the valve body.

Next, remove the detent-spring-and-roller assembly from the valve body. See Fig. 14-25. Remove the bolts attaching the valve body to the case. Now, lift the valve body from the case, carefully guiding the manual-valve link from the range-selector inner lever. Remove the detent-control-valve link from the detent actuating lever. Lay the valve body aside on a clean surface in preparation for further disassembly.

Remove, from the case, the following: valve-body-to-spacer-plate gasket, as shown in Fig. 14-26, spacer-support-plate bolts, and the spacer support plate, as shown in Fig. 14-27. Then remove the valve-body spacer plate and gasket (Fig. 14-28).

7. From the case, remove the oil-pump pressure screen from the pressure hole, as shown in Fig. 14-29, and the governor screens, as shown in Fig. 14-30. Clean the screens and lay them aside in a clean, safe place. Remove the four check balls from the case face. Their locations are shown in Fig. 14-31.

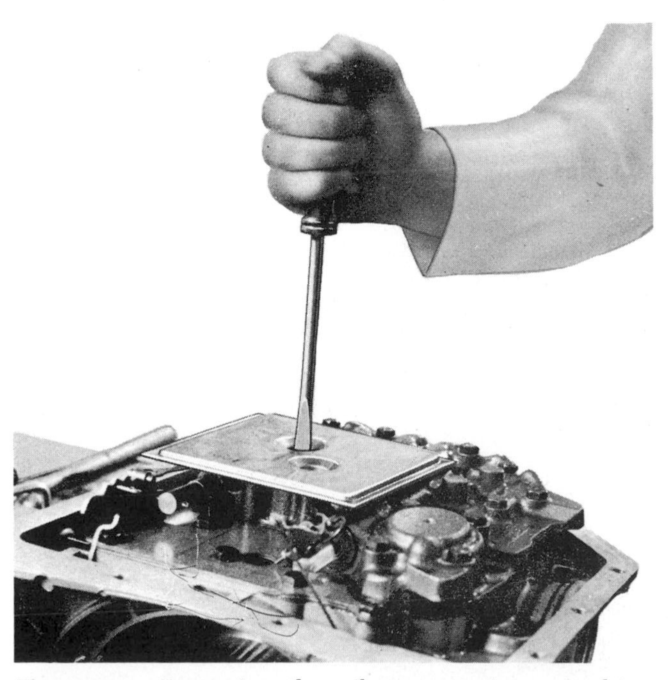

Fig. 14-24. Removing the oil-pump-strainer attaching screws. (*Chevrolet Motor Division of General Motors Corporation*)

Fig. 14-27. Removing the spacer support plate. (*Buick Motor Division of General Motors Corporation*)

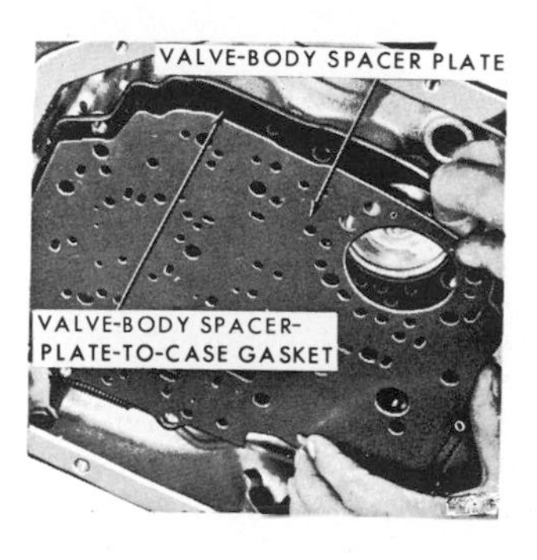

Fig. 14-28. Removing the valve-body spacer plate and gasket. (*Buick Motor Division of General Motors Corporation*)

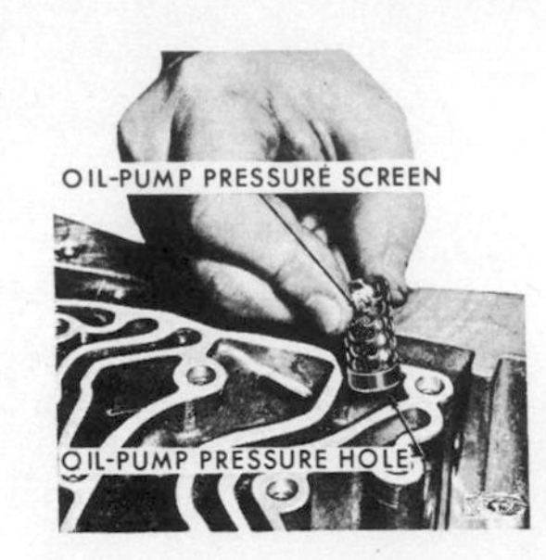

Fig. 14-29. Removing the oil-pump pressure screen. (*Buick Motor Division of General Motors Corporation*)

Fig. 14-30. Removing the governor screens. (*Buick Motor Division of General Motors Corporation*)

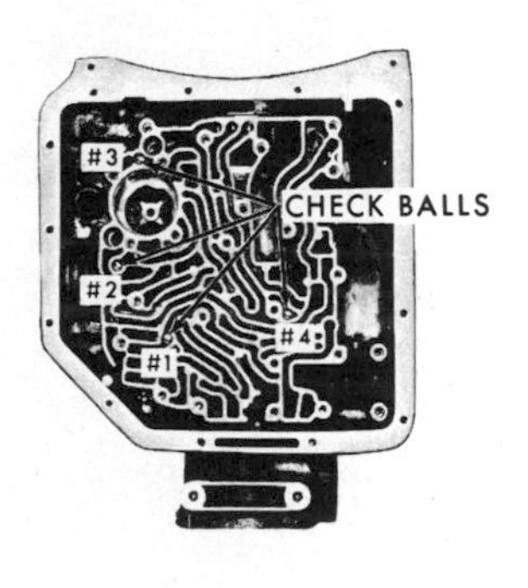

Fig. 14-31. Locations of the four check balls in case. (*Buick Motor Division of General Motors Corporation*)

8. Next, remove the manual shaft, inner lever, and parking pawl. First, take off the manual-control-valve-link retainer from the range-selector inner lever. Then use a screwdriver, as shown in Fig. 14-32, to remove the manual-shaft-to-case retainer. Next, use a wrench, as shown in Fig. 14-33, to remove the jam nut holding the range-selector inner lever to the manual shaft. Now, remove the manual shaft from the case. Remove the range-selector inner lever and parking-pawl actuating rod, as shown in Fig. 14-34. If the manual-shaft-to-case lip seal is damaged, remove it with a screwdriver, as shown in Fig. 14-35.

If necessary to remove the parking pawl or shaft, proceed as follows: Remove the parking-lock bracket, shown in Fig. 14-36. Then remove the retaining plug, parking-pawl shaft, parking pawl, and disengaging spring. These parts are shown in Fig. 14-37.

9. Remove the intermediate-servo piston and metal oil-seal ring, as shown in Fig. 14-38. Then remove the washer, spring seat, and apply pin. Now, check to see if a new long or short apply pin is required on reassembly. Do this by using a special apply-pin gauge and straightedge, as shown in Fig. 14-39. Press down on the gauge and note whether the upper end of the gauge is above or below the straightedge. There are two replacement apply pins: long and short. If the gauge is below the straightedge, the long

Fig. 14-32. Removing the manual-shaft-to-case retainer. (*Buick Motor Division of General Motors Corporation*)

Fig. 14-33. Removing the jam nut from the manual shaft. (*Buick Motor Division of General Motors Corporation*)

Fig. 14-34. Removing the range-selector inner lever and parking-pawl actuating rod. (*Buick Motor Division of General Motors Corporation*)

Fig. 14-35. Removing the manual-shaft lip seal with a screwdriver. (*Buick Motor Division of General Motors Corporation*)

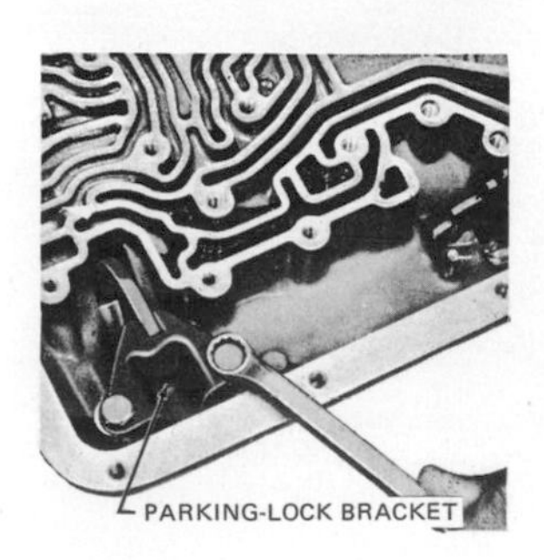

Fig. 14-36. Removing the parking-lock bracket. (*Buick Motor Division of General Motors Corporation*)

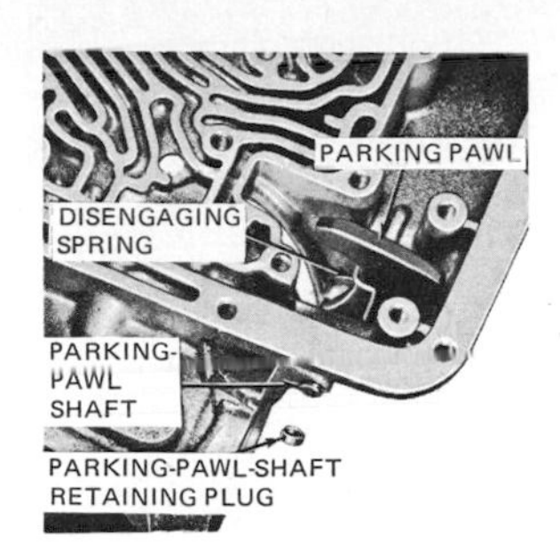

Fig. 14-37. Locations of the parking pawl, parking-pawl shaft, parking-pawl-shaft retaining plug, and disengaging spring. (*Buick Motor Division of General Motors Corporation*)

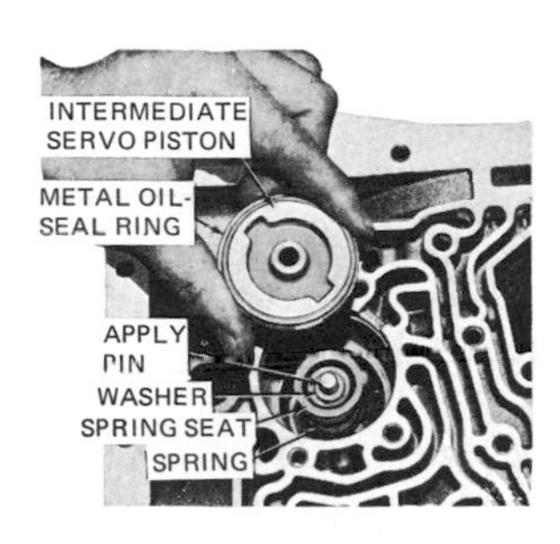

Fig. 14-38. Removing the intermediate servo piston and metal oil-seal ring. (*Buick Motor Division of General Motors Corporation*)

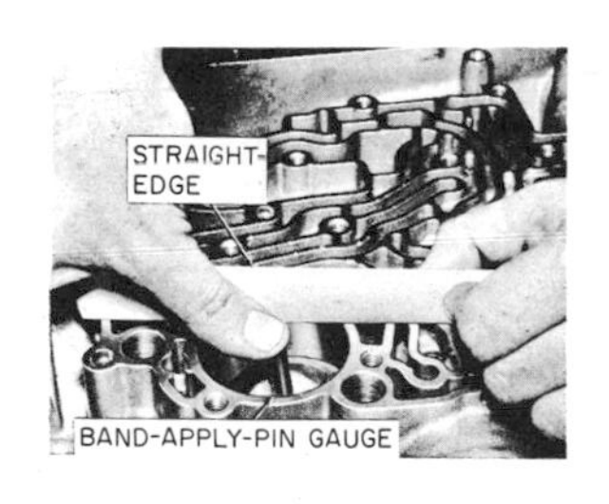

Fig. 14-39. Using a special gauge and straightedge to determine what apply pin to use on reassembly. (*Buick Motor Division of General Motors Corporation*)

pin should be used; if above, the short pin should be used. Selecting the proper length of pin is equivalent to adjusting the brake band. Make a note of which pin is required so that it can be installed on reassembly.

10. Remove the pump as follows: With the transmission turned, pump end up, remove the eight pump attaching bolts and washer-type seals, as shown in Fig. 14-40. Discard the seals. Install two threaded slide hammers into the threaded holes in the pump body and tighten the jam nuts, as shown in Fig. 14-41. Carefully use the slide hammers to draw the pump up and out of the case. Remove and discard the pump-assembly-to-case gasket, shown in Fig. 14-42. Lay the pump aside on a clean surface for later disassembly.

11. Remove the intermediate-clutch plates and band as follows: First take out the intermediate-clutch cushion spring, as shown in Fig. 14-43, and then remove the intermediate-clutch plates, as shown in Fig. 14-44. Inspect the plates as already explained (in ⊘ 14-11). Next, remove the intermediate-clutch pressure plate, as shown in Fig. 14-45, and the overrun brake band, shown in Fig. 14-46.

12. Next, remove the forward- and direct-clutch assemblies from the case, as shown in Fig. 14-47. Lay

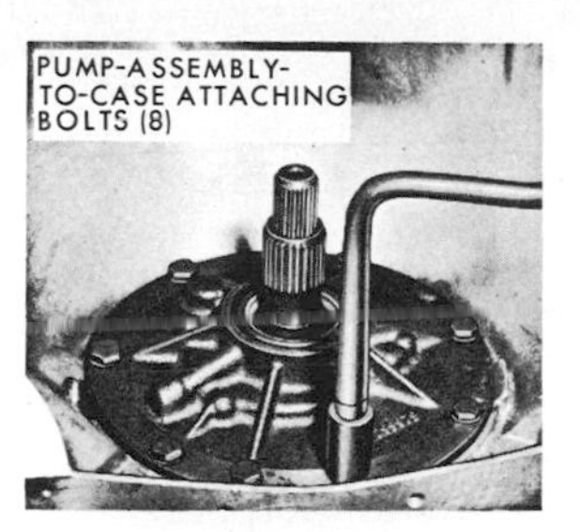

Fig. 14-40. Removing the pump-assembly attaching bolts. (*Buick Motor Division of General Motors Corporation*)

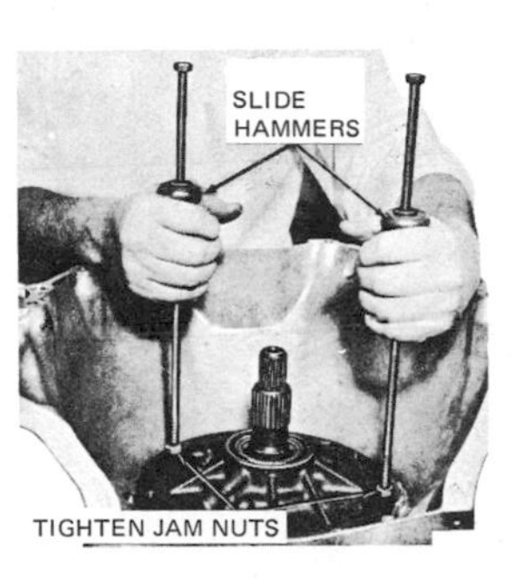

Fig. 14-41. Removing the pump assembly with slide hammers. (*Buick Motor Division of General Motors Corporation*)

Fig. 14-42. Location of the pump-assembly-to-case gasket. (*Buick Motor Divison of General Motors Corporation*)

Fig. 14-43. Removing the intermediate-clutch cushion spring. (*Buick Motor Division of General Motors Corporation*)

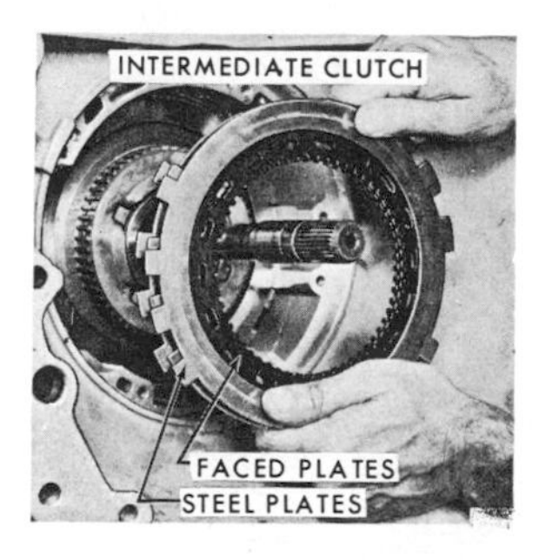

Fig. 14-44. Removing the intermediate-clutch plates. The faced plates are also known as *composition plates.* (*Buick Motor Division of General Motors Corporation*)

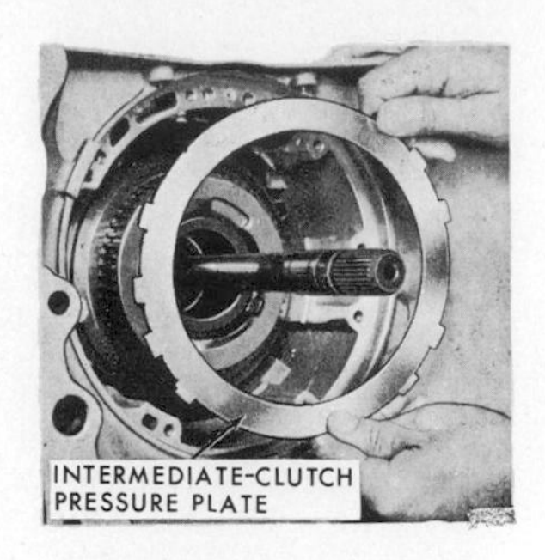

Fig. 14-45. Removing the intermediate-clutch pressure plate. (*Buick Motor Division of General Motors Corporation*)

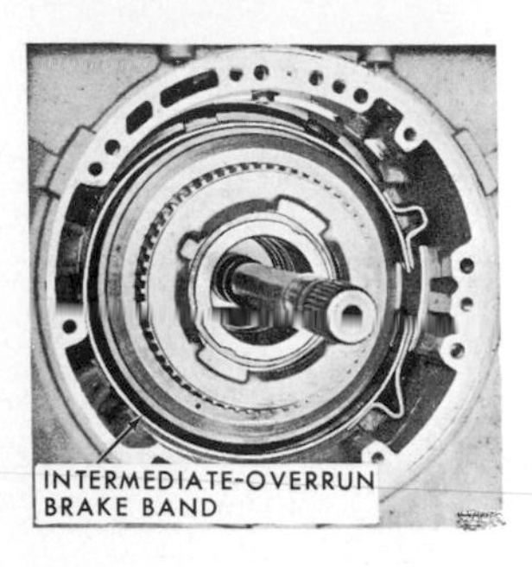

Fig. 14-46. Location of the intermediate-overrun brake band. (*Buick Motor Division of General Motors Corporation*)

Fig. 14-47. Removing the forward- and direct-clutch assemblies. (*Buick Motor Division of General Motors Corporation*)

Fig. 14-49. Removing the input ring gear. (*Buick Motor Division of General Motors Corporation*)

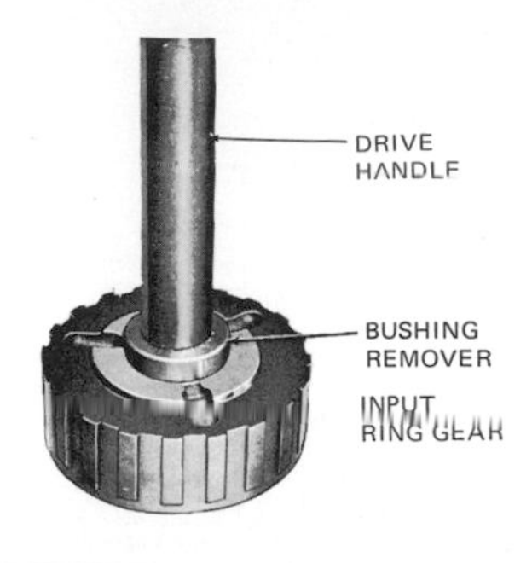

Fig. 14-50. Removing the bushing from the input ring gear with special tools. (*Buick Motor Division of General Motors Corporation*)

Fig. 14-51. Removing the output-carrier thrust washer. (*Buick Motor Division of General Motors Corporation*)

them aside in a clean place to await further disassembly.

13. Remove the front input-ring-gear thrust washer, as shown in Fig. 14-48, and then remove the input ring gear, as shown in Fig. 14-49. Check the bushing in the ring gear. If it is worn or galled, use a special tool to remove and replace it. Thread the tool onto the drive handle and use the tool, as shown in Fig. 14-50, to remove the bushing. The same tool is used to install the new bushing.

14. To remove the output-carrier assembly, first remove the input-ring-gear-to-output-carrier thrust washer, as shown in Fig. 14-51. Then remove and discard the output-carrier-to-output-shaft snap ring, as shown in Fig. 14-52. Now the output-carrier assembly will slide out, as shown in Fig. 14-53. Lay it aside on a clean surface to await further disassembly.

15. Next, remove the sun-gear-and-drive-shell assembly by pulling it out, as shown in Fig. 14-54.

16. Remove the low-and-reverse-clutch-support assembly by first prying out the retaining ring, as shown in Fig. 14-55. Now, grasp the output shaft, as shown in Fig. 14-56, and pull up until the low-and-reverse-roller-clutch-support assembly clears the low-and-reverse-clutch-support retainer spring. The support assembly will then slide out. Now, use pliers (Fig. 14-57) to remove the retainer spring.

Fig. 14-48. Removing the front input-ring-gear thrust washer. (*Buick Motor Division of General Motors Corporation*)

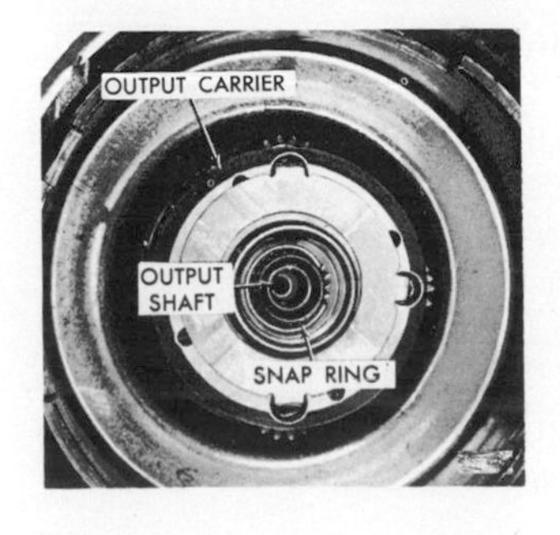

Fig. 14-52. Location of the output-carrier-to-output-shaft snap ring. (*Buick Motor Division of General Motors Corporation*)

Fig. 14-53. Removing the output-carrier assembly. (*Buick Motor Division of General Motors Corporation*)

Fig. 14-54. Removing the sun-gear-drive-shell assembly. (*Buick Motor Division of General Motors Corporation*)

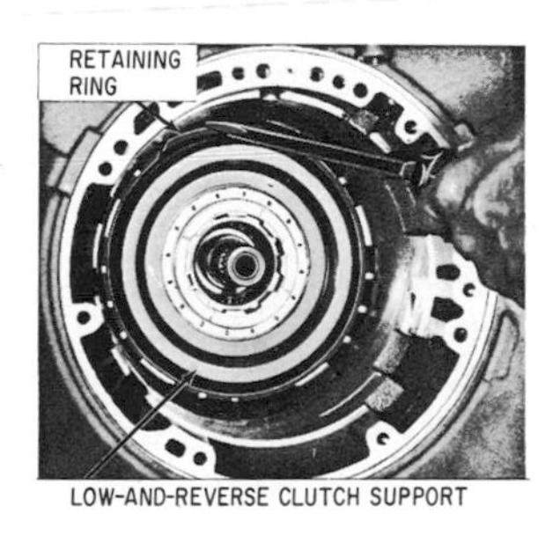

Fig. 14-55. Removing the low-and-reverse-clutch-support retaining ring. (*Buick Motor Division of General Motors Corporation*)

Fig. 14-56. Removing the low-and-reverse-clutch support. (*Buick Motor Division of General Motors Corporation*)

Fig. 14-57. Removing the low-and-reverse-clutch-support retainer spring. (*Buick Motor Division of General Motors Corporation*)

17. Now reach into the case and remove the low-and-reverse-clutch plates, as shown in Fig. 14-58, followed by the reaction carrier, as shown in Fig. 14-59. Inspect the clutch plates, as explained previously (in ⊘ 14-11). If the reaction-carrier bushing is worn or galled, replace it with the special tools, as shown in Fig. 14-60.

18. Remove the output-ring-gear-and-shaft assembly from the case, as shown in Fig. 14-61. Remove the output-ring-gear-to-case needle bearing from either the case or the output shaft. Then take the tanged thrust washer from the assembly, as shown in Fig.

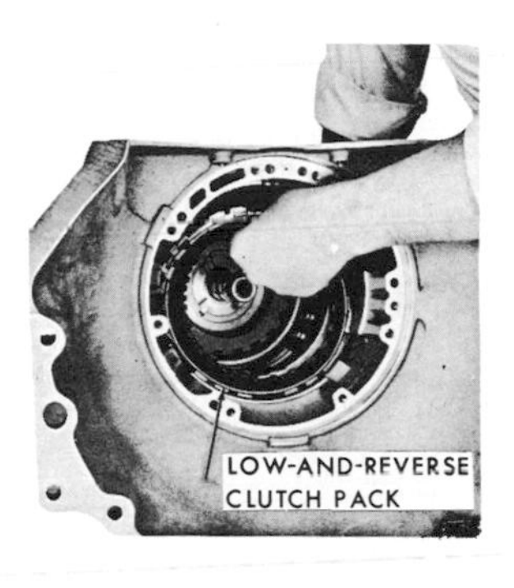

Fig. 14-58. Removing the low-and-reverse-clutch plates. The plates together are called the *clutch pack*. (*Buick Motor Division of General Motors Corporation*)

Fig. 14-59. Removing the reaction carrier. (*Buick Motor Division of General Motors Corporation*)

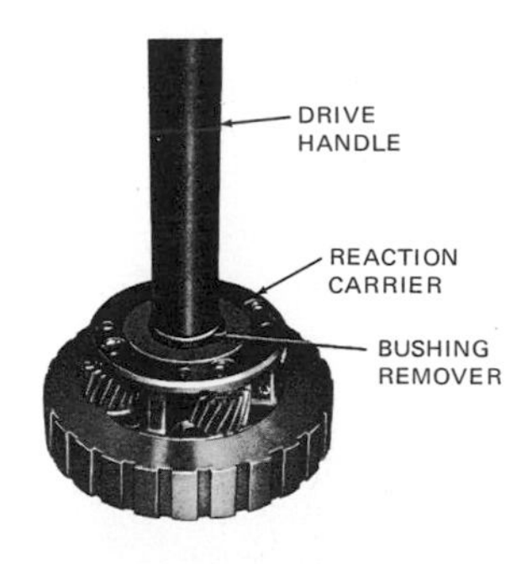

Fig. 14-60. Removing the bushing from the reaction carrier with special tools. (*Buick Motor Division of General Motors Corporation*)

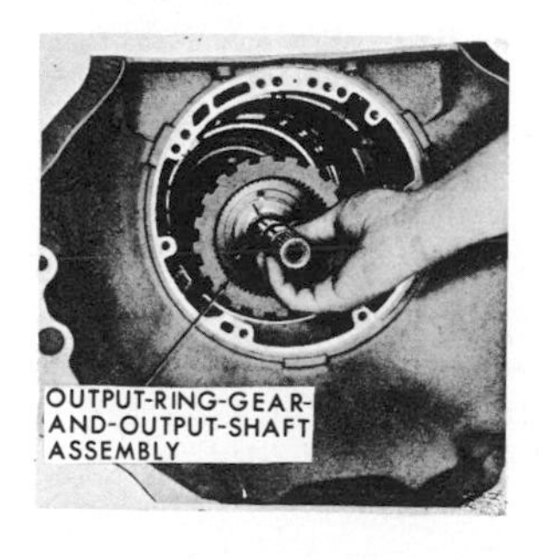

Fig. 14-61. Removing the output-ring-gear-and-output-shaft assembly. (*Buick Motor Division of General Motors Corporation*)

14-62. Next, take off the output-ring-gear-to-output-shaft snap ring and discard it. The snap-ring location is shown in Fig. 14-63. Slip the output shaft from the output ring gear.

Remove the output-ring-gear-to-case needle-bearing assembly, as shown in Fig. 14-64.

Check the output-shaft bushing. If it is worn or galled, replace it with special tools, as shown in Figs. 14-65 and 14-66. The procedure is to assemble tool J-9534-01 into adapter J-2619-4 and then assemble them to the slide hammer, as shown. Thread the assembly into the bushing and clamp the slide hammer in a vise. Then grasp the output shaft to remove the bushing. Then use tool J-23062-7, assembled into drive handle J-8092, as shown in Fig. 14-66, and press the new bushing into place 0.140 inch [3.56 mm] below the end surface of the output shaft.

19. To remove the low-and-reverse-clutch piston, the springs behind the piston must be compressed by installing a special compressor tool, as shown in Fig. 14-67. Then the piston retaining ring and spring retainer can be removed. Figure 14-68 shows in sectional view how the tool is installed to compress the springs. Remove the 17 piston return springs, shown in Fig. 14-69. Now, the low-and-reverse-clutch-piston assembly can be removed. Use compressed air to aid in the removal of the piston assembly. Apply compressed air in the passage shown pointed out by a pencil in Fig. 14-70. The low-and-reverse-clutch-piston assembly has three seals: the outer seal, center seal, and inner seal. All three seals should be removed and discarded.

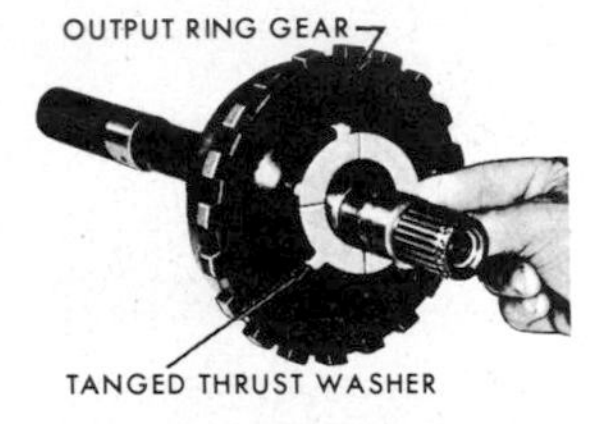

Fig. 14-62. Removing the tanged thrust washer from the output ring gear. (*Buick Motor Division of General Motors Corporation*)

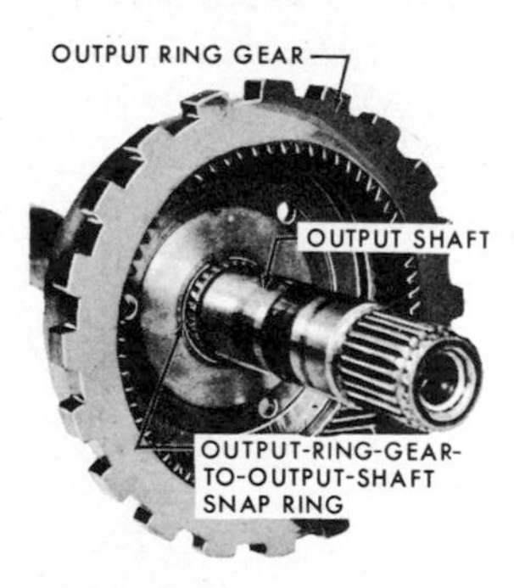

Fig. 14-63. Location of the output-ring-gear-to-output-shaft snap ring. (*Buick Motor Division of General Motors Corporation*)

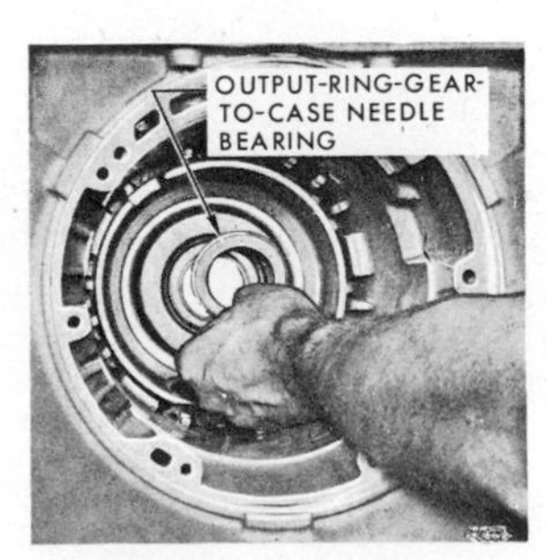

Fig. 14-64. Removing the output-ring-gear-to-case needle bearing. (*Buick Motor Division of General Motors Corporation*)

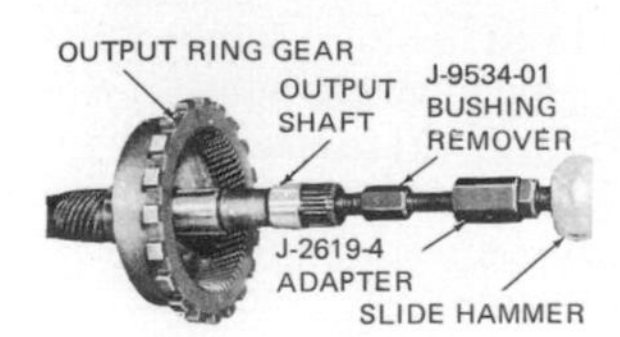

Fig. 14-65. Removing the output-shaft bushing. (*Buick Motor Division of General Motors Corporation*)

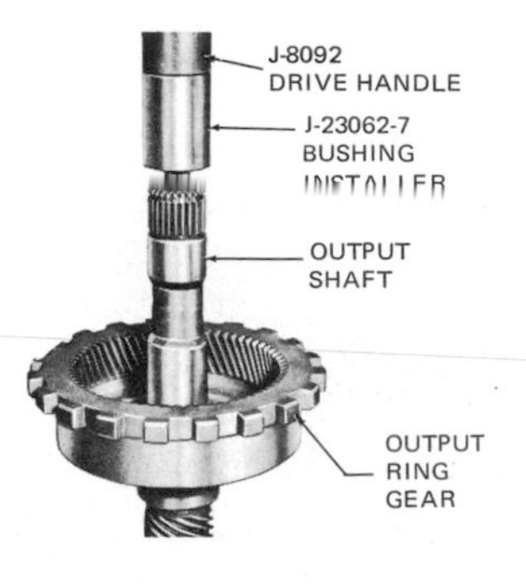

Fig. 14-66. Installing the output-shaft bushing. (*Buick Motor Division of General Motors Corporation*)

Fig. 14-67. Using a special tool to compress the low-and-reverse-clutch-piston springs. (*Chevrolet Motor Division of General Motors Corporation*)

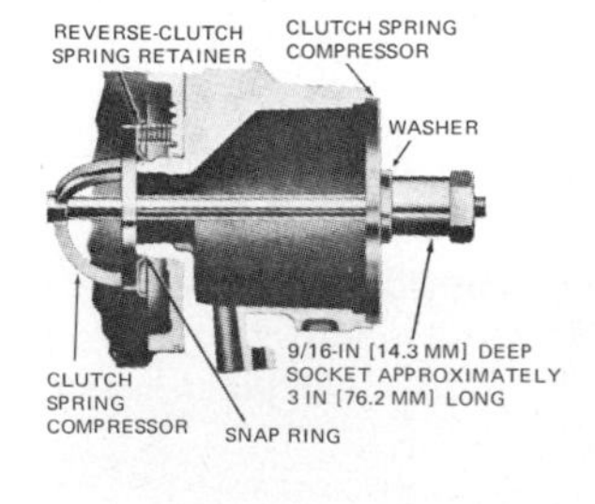

Fig. 14-68. Sectional view of the special tool compressing the springs. (*Oldsmobile Division of General Motors Corporation*)

Fig. 14-69. Location of the low-and-reverse-clutch piston and piston-return springs. (*Buick Motor Division of General Motors Corporation*)

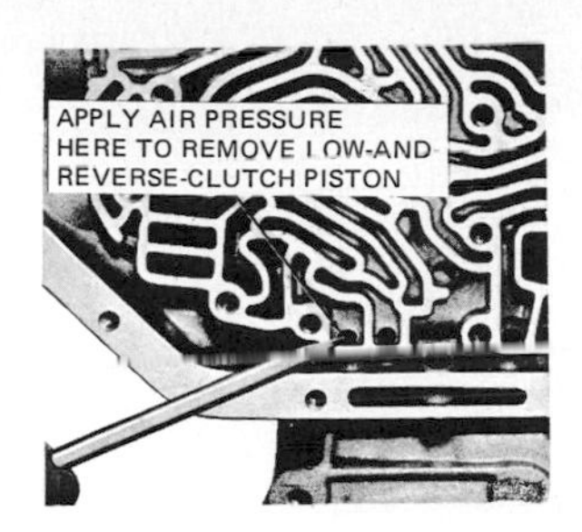

Fig. 14-70. Pencil points to the passage at which compressed air is applied to aid in removal of the low-and-reverse-clutch piston. (*Buick Motor Division of General Motors Corporation*)

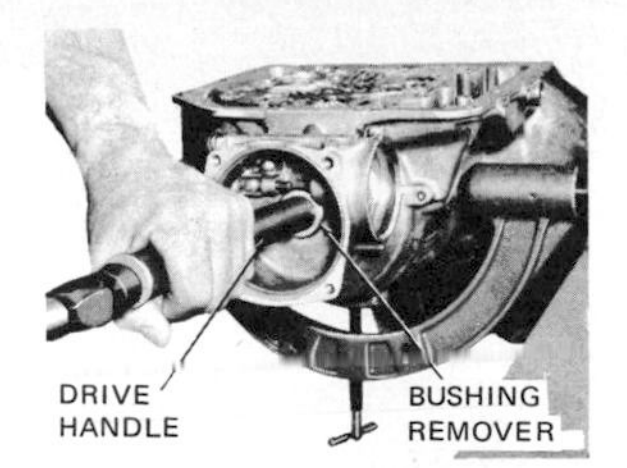

Fig. 14-71. Removing the case bushing. (*Buick Motor Division of General Motors Corporation*)

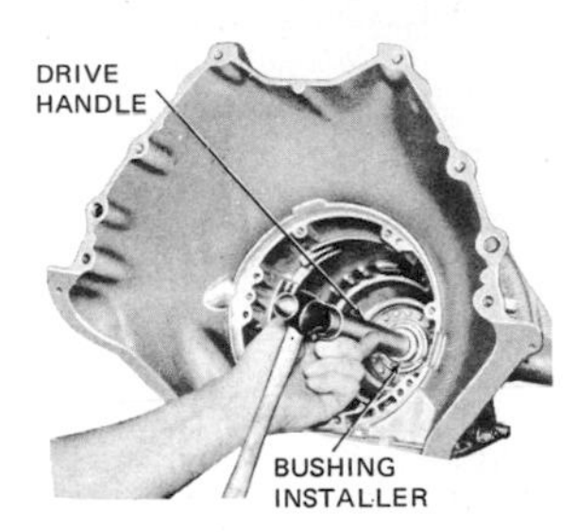

Fig. 14-72. Installing a new case bushing. (*Buick Motor Division of General Motors Corporation*)

20. If the case bushing is worn or otherwise damaged, remove it by using the special removing tool installed on the drive handle, as shown in Fig. 14-71. Note that the bushing is driven inward. The same tools are then used to drive the new bushing into place, as shown in Fig. 14-72. The bushing should be pressed in to 0.195 inch [4.95 mm] below the chamfered edge of the case. Make sure that the split on the bushing is opposite the notch on the case.

21. Removal and replacement of the intermediate-clutch accumulator have already been discussed (⊘ 14-8). The tool used is shown in Fig. 14-4. A disassembled view of the intermediate-clutch accumulator with retaining ring, cover and seal, spring, and piston is shown in Fig. 14-73.

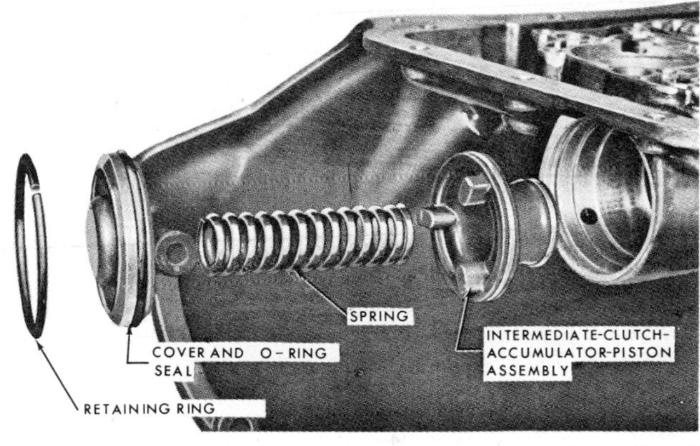

Fig. 14-73. Disassembled view of the intermediate clutch (*Buick Motor Division of General Motors Corporation*)

⊘ 14-13 Oil-Pump-Assembly Service The oil-pump assembly is shown in exploded view in Fig. 14-74. Note that the assembly includes the intermediate-clutch-piston assembly and related parts.

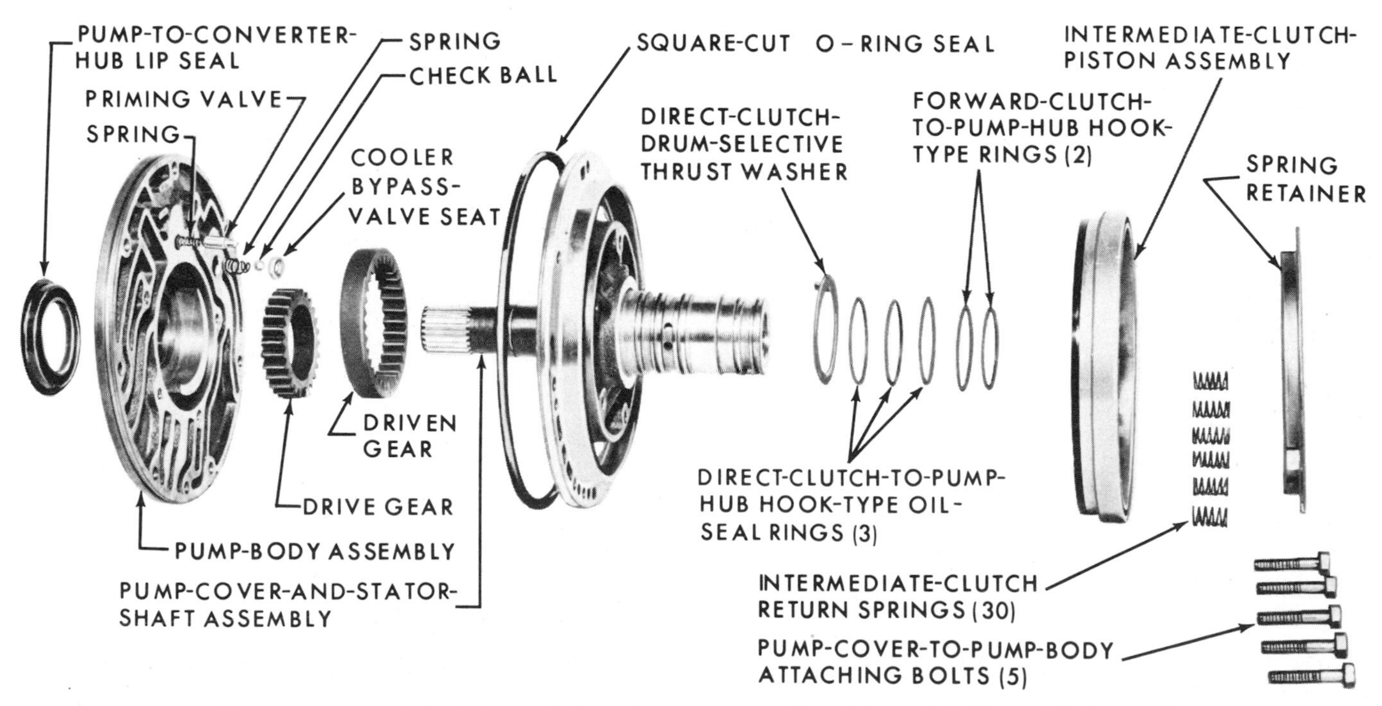

Fig. 14-74. Exploded view of the oil-pump assembly. (*Chevrolet Motor Division of General Motors Corporation*)

Disassembly

To disassemble the oil-pump assembly, put it on the bench with the shaft through a hole in the bench.

1. Remove the five pump-cover-to-pump-body bolts, as shown in Fig. 14-75. This permits removal of the spring retainer, 30 springs, and intermediate-clutch-piston assembly, all shown in Fig. 14-74. The intermediate-clutch piston has two seals: an inner seal and an outer seal. Both seals should be removed and discarded so that new seals can be installed.
2. There are five hook-type oil-seal rings on the pump hub, as shown in Fig. 14-74. These are removed by unhooking the ring ends and lifting them up and off the hub. The removal of one ring is shown in Fig. 14-76. The selective thrust washer, shown in Fig. 14-74, can now be removed. Note that this thrust washer comes in various thicknesses, and the correct thickness must be selected to produce the proper end play in the transmission.
3. Lift the pump-cover-and-stator-shaft assembly from the pump body, as shown in Fig. 14-77. Remove the pump drive and driven gears from the pump body. These are shown in Fig. 14-74. Remove and discard the large square-cut O-ring seal.
4. Turn the pump body over on two wood blocks to prevent damage to the surface, and remove the pump-to-converter-hub lip seal, as shown in Fig. 14-78. Turn the pump body inner face up, and remove the priming valve and spring. These are shown in Fig. 14-74.
5. Then remove the cooler bypass-valve seat, check ball, and spring. Two methods for removal are given. In one method, a bolt extractor is used, as shown in Fig. 14-79. In the other, the bypass passage is filled with grease and a special tool is driven into the passage to force out the valve seat, check ball, and spring, as shown in Fig. 14-80.

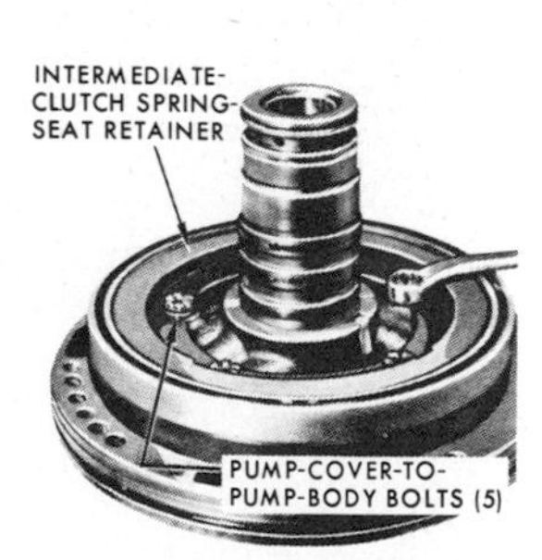

Fig. 14-75. Removing pump-cover-to-pump-body bolts. (*Chevrolet Motor Division of General Motors Corporation*)

Fig. 14-76. Removing the top hook-type oil-seal ring from the pump hub. (*Chevrolet Motor Division of General Motors Corporation*)

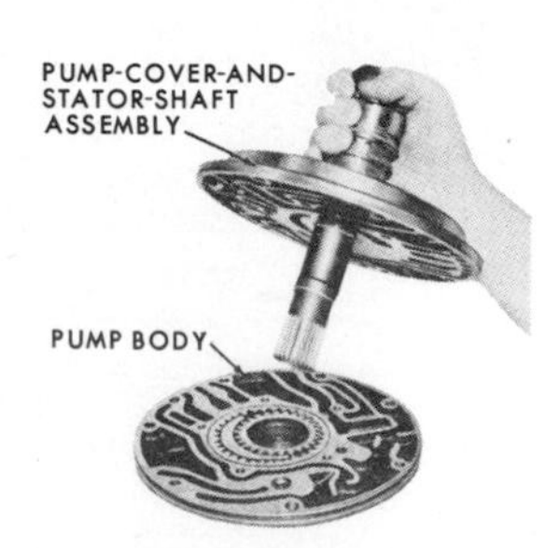

Fig. 14-77. Lifting the pump-cover-and-stator-shaft assembly from the pump body. (*Chevrolet Motor Division of General motors Corporation*)

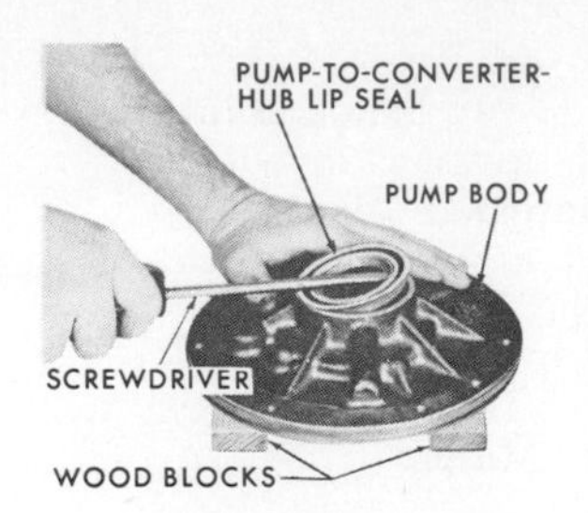

Fig. 14-78. Removing the pump-to-converter-hub lip seal. (*Chevrolet Motor Division of General Motors Corporation*)

Inspection of Parts

Wash all parts in cleaning solvent and blow out all oil passages. Do not use rags to dry parts. Traces of lint can cause valves to hang up and ruin transmission performance. Check all parts for nicks, scoring, or other damage. If the pump-body bushing is worn or galled, replace the bushing or pump body. Chevrolet recommends replacing the body. Buick recommends replacing the bushing in the body.

With parts clean and dry, install pump gears and check the clearance between the pump-body face and the gear faces with a straightedge and feeler gauges, as shown in Fig. 14-81. The clearance should be between 0.0005 and 0.0015 inch [0.013 and 0.043 mm]. If the clearance is excessive, the gears and body should be replaced.

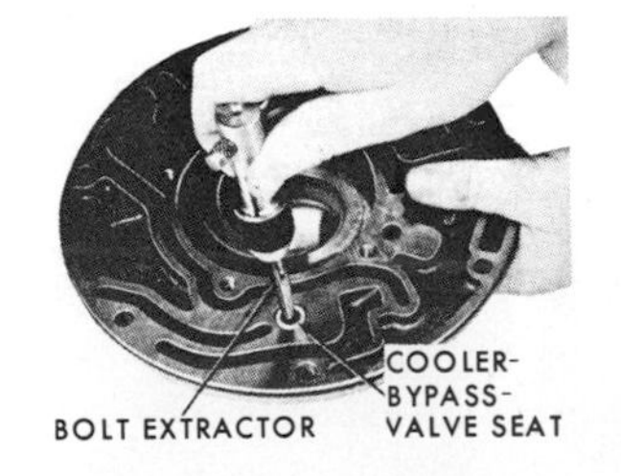

Fig. 14-79. Removing the cooler bypass-valve seat by means of a bolt extractor. (*Chevrolet Motor Division of General Motors Corporation*)

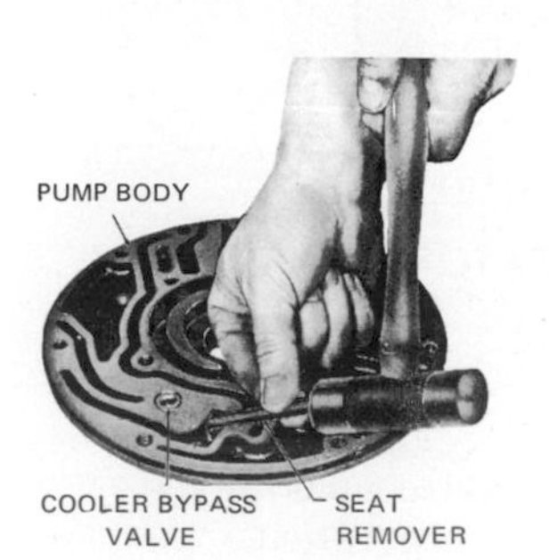

Fig. 14-80. Removing the cooler bypass-valve seat with grease and an extractor. (*Buick Motor Division of General Motors Corporation*)

Fig. 14-81. Checking gear clearance with a straightedge and feeler gauge. (*Chevrolet Motor Division of General Motors Corporation*)

Reassembly

1. If the pump-to-converter-hub lip seal has been removed, as shown in Fig. 14-78, replace it by using a special driver tool, as shown in Fig. 14-82.

2. Turn the gear body over and install the gears, making sure that the driving tang on the driving gear is up and that the marks on the gears, where present, align, as shown in Fig. 14-83.

3. Install the priming valve and spring, cooler bypass-valve seat, check ball, and spring. These are all shown in Fig. 14-74. When installing the valve seat, tap it down with a soft hammer or brass drift until it is flush to 0.010 inch [0.25 mm] below the surface.

4. To the hub on the pump cover, install five hook-type oil-seal rings. These are shown in Fig. 14-74 and are shown installed in Fig. 14-76.

5. Install the inner seal and outer seal on the intermediate-clutch piston. Now carefully install the intermediate-clutch-piston assembly in the pump cover, being careful not to damage the seals. Install 30 clutch return springs. Put the intermediate-clutch-spring seat retainer into place and install five bolts.

6. Bring the pump-cover-and-stator-shaft assembly down onto the pump body, as shown in Fig. 14-77. Be sure to align them properly. Figures 14-84 and 14-85 show the various passages in the faces of the pump body and cover. One convenient alignment

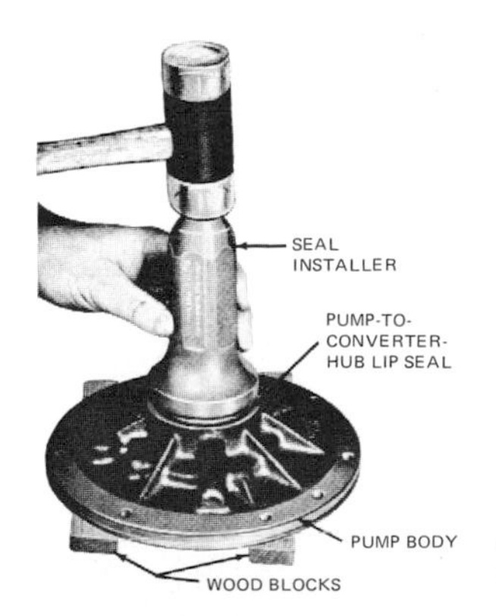

Fig. 14-82. Installing the pump-to-converter-hub lip seal. (*Chevrolet Motor Division of General Motors Corporation*)

Fig. 14-83. Screwdriver points to aligning marks on gears. (*Chevrolet Motor Division of General Motors Corporation*)

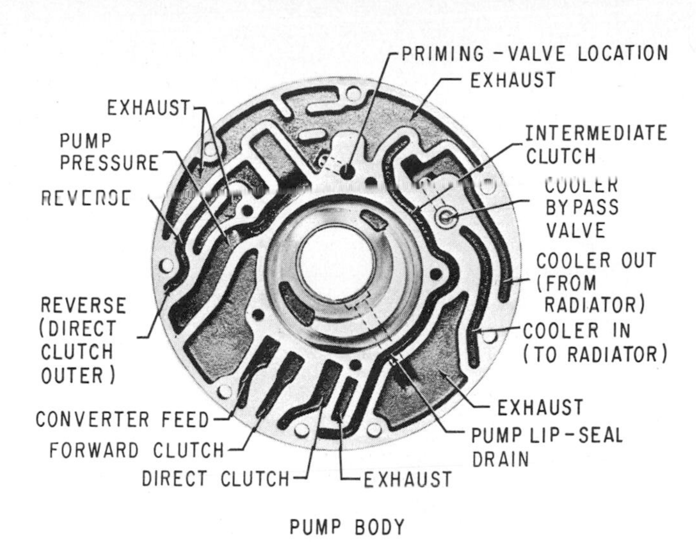

Fig. 14-84. Pump-body oil passages. (*Chevrolet Motor Division of General Motors Corporation*)

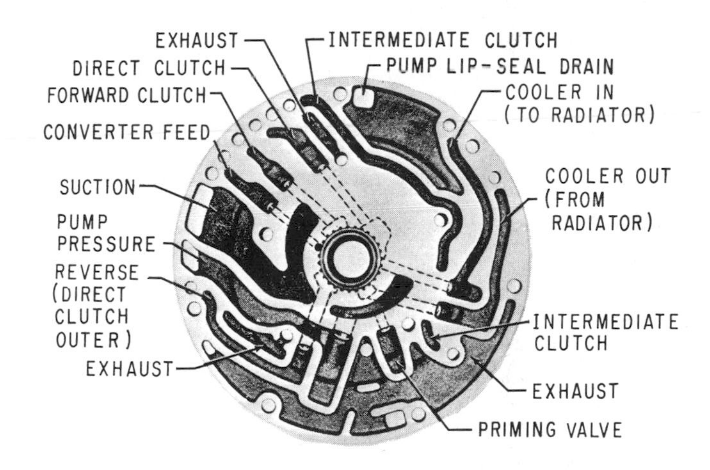

Fig. 14-85. Pump-cover oil passages. (*Chevrolet Motor Division of General Motors Corporation*)

procedure is to align the priming valve in the pump body with the priming-valve cavity in the pump cover. See Figs. 14-84 and 14-85. Install the square-cut O-ring seal. Tighten the five attaching bolts to 18 pound-feet [2.487 kg-m] torque.

⊘ 14-14 Direct-Clutch Service The direct-clutch assembly, with overrun-clutch parts, is shown in Fig. 14-86. To remove the overrun parts, pry out the retaining ring with a screwdriver, as shown in Fig. 14-87. The retainer, outer race, and roller-clutch assembly can then be lifted off. Next, the direct-clutch assembly itself can be disassembled. It is shown in exploded view in Fig. 14-88.

Disassembly

1. Start by removing the clutch-drum-to-forward-clutch-housing special thrust washer.

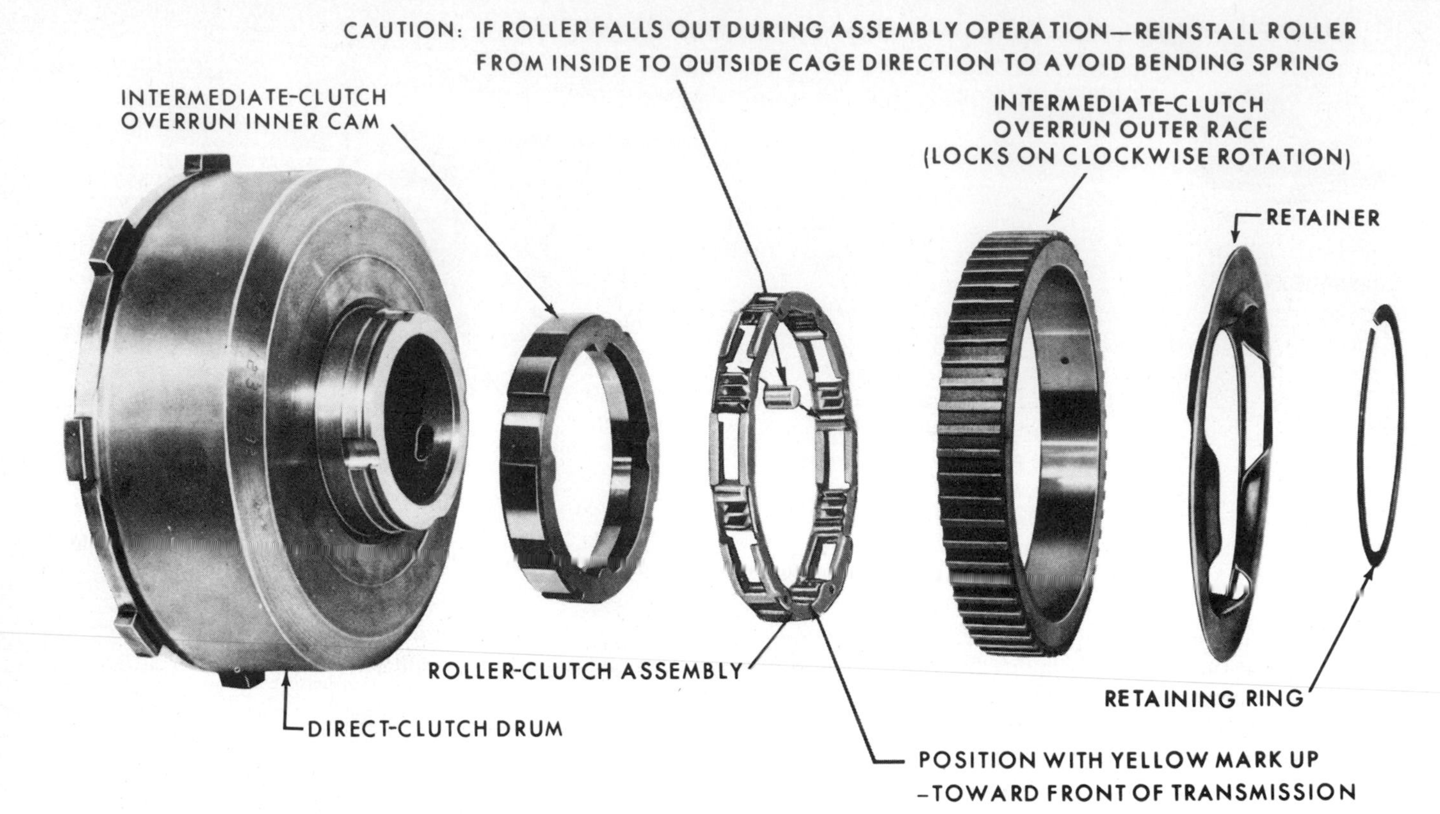

Fig. 14-86. Direct-clutch assembly with intermediate-overrun-clutch parts. (*Chevrolet Motor Division of General Motors Corporation*)

2. Use a screwdriver, as shown in Fig. 14-89, to remove the retaining ring, pressure plate, and clutch plates.
3. Use the special tool, as shown in Fig. 14-90, to compress the springs. Then remove the retaining ring, spring seat, 17 clutch return springs, and direct-clutch-piston assembly.

Inspection
Check clutch plates for signs of burning, wear, or scoring. Check springs for collapsed coils or distortion. Check the piston and clutch housing for wear and scores. Make sure all oil passages are open and that the ball check works freely. Make sure roller-clutch inner and outer races are free of scratches, indentations, or other signs of wear. The roller cage should be free of excessive wear, and roller springs should be in good condition.

RETAINING RING
OVERRUN CLUTCH
DIRECT-CLUTCH ASSEMBLY

Fig. 14-87. Removing the overrun-clutch retaining ring. (*Chevrolet Motor Division of General Motors Corporation*)

Reassembly
1. Install new inner and outer seals on the direct-clutch piston.
2. Install the direct-clutch-piston center seal on the drum with the lip facing upward, as shown in Fig. 14-91.
3. Install the direct-clutch piston in the drum using a piece of 0.020 inch [0.51 mm] piano wire crimped into copper tubing, as shown in Fig. 14-92, to help get the seal into place.
4. Install 17 clutch return springs and put the spring retainer in position. Compress springs with a special tool, as shown in Fig. 14-90, and install the retaining ring.
5. Lubricate the clutch plates with transmission fluid and install them, alternating faced and steel plates.
6. Install the pressure plate and retaining ring.
7. Install the roller-clutch assembly, outer race, retainer, and retainer ring (all shown in Fig. 14-86). Make sure that the outer race can freewheel in a counterclockwise direction only.

⊘ **14-15 Forward-Clutch Service** Figure 14-93 shows the forward-clutch assembly in exploded view.

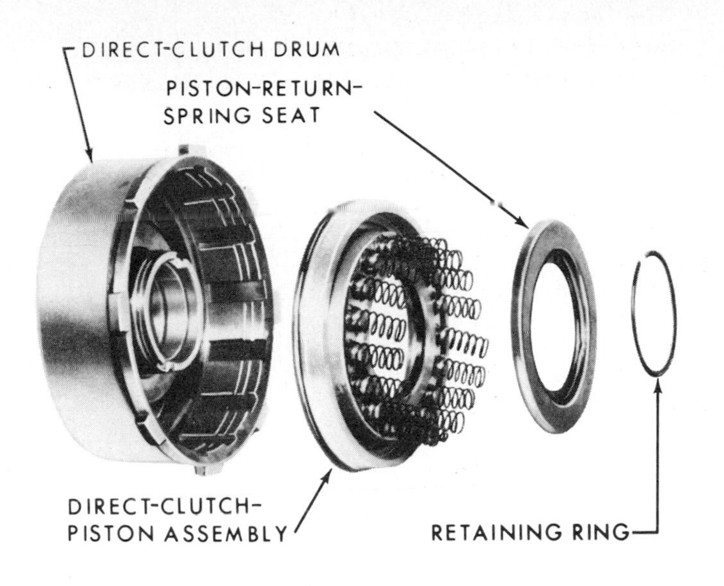

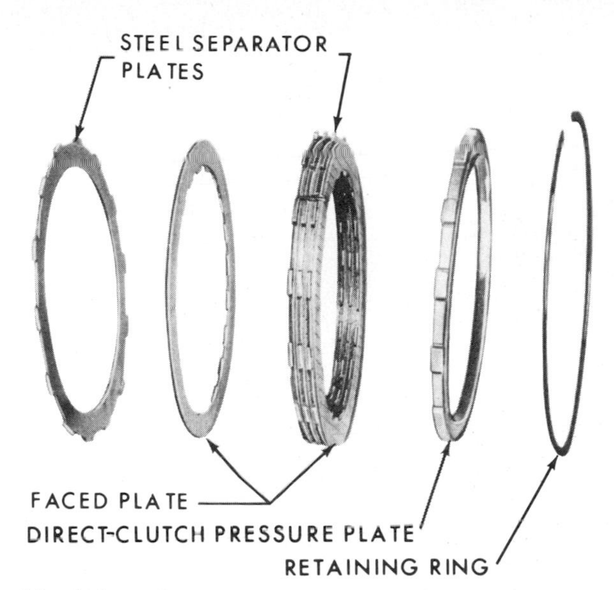

Fig. 14-88. Exploded view of a direct-clutch assembly. (*Chevrolet Motor Division of General Motors Corporation*)

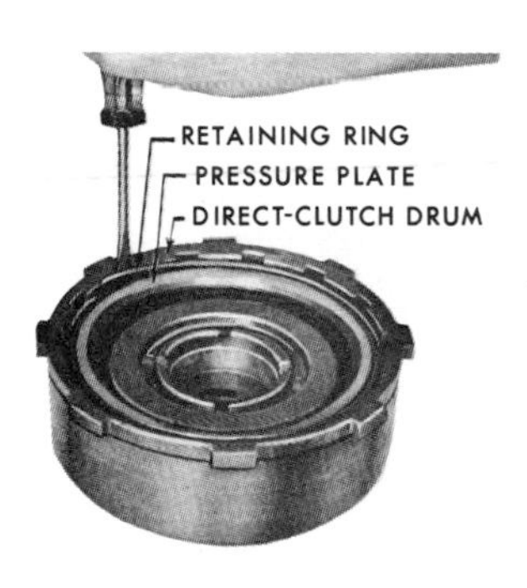

Fig. 14-89. Removing the pressure-plate retaining ring. (*Chevrolet Motor Division of General Motors Corporation*)

Fig. 14-90. Using a special tool to compress the springs so that the retaining ring can be removed. (*Chevrolet Motor Division of General Motors Corporation*)

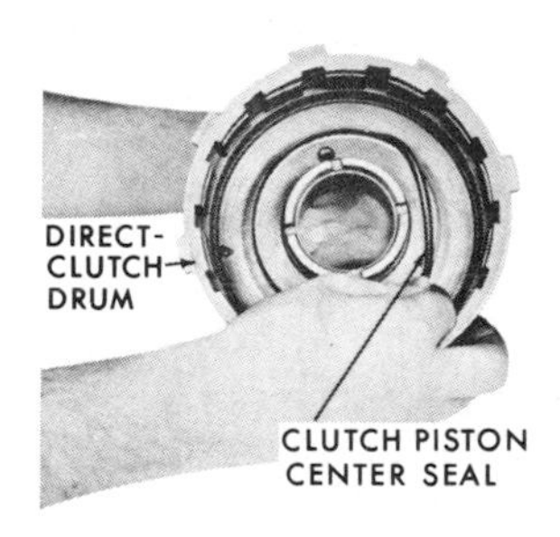

Fig. 14-91. Installing the direct-clutch-piston center seal on the drum. (*Chevrolet Motor Division of General motors Corporation*)

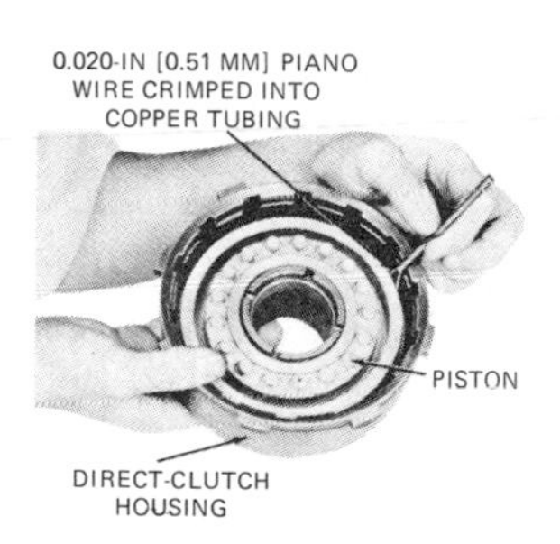

Fig. 14-92. Using a piece of 0.020-inch [0.51 mm] piano wire to install the direct-clutch piston in the drum. (*Chevrolet Motor Division of General Motors Corporation*)

Disassembly

1. Remove the pressure-plate-to-drum retaining ring, as shown in Fig. 14-94.

2. Remove the pressure plate, clutch plates, and cushion spring. These are all shown in Fig. 14-93.

3. Use the special compressing tool, as shown in Fig. 14-95, to compress springs so that the retaining ring can be removed. Now, the piston return seat, 21 springs, and forward-clutch-piston assembly can be removed.

4. Remove the piston inner and outer seals.

Inspection

Check clutch plates for signs of wear, burning, or scoring. Check springs for collapsed coils or signs of distortion. Check the piston and clutch drum for signs of wear or other damage. Make sure oil passages are open. Inspect the input shaft for damaged splines, worn bushing journals, cracks, or other damage, and make sure the oil passages are open. Check the ball-check exhaust, as shown in Fig. 14-96, to make sure it is free.

Reassembly

1. Install the inner and outer seals on the piston. Then install the forward-clutch-piston assembly in the clutch drum, using a piece of 0.02-inch [0.51 mm]

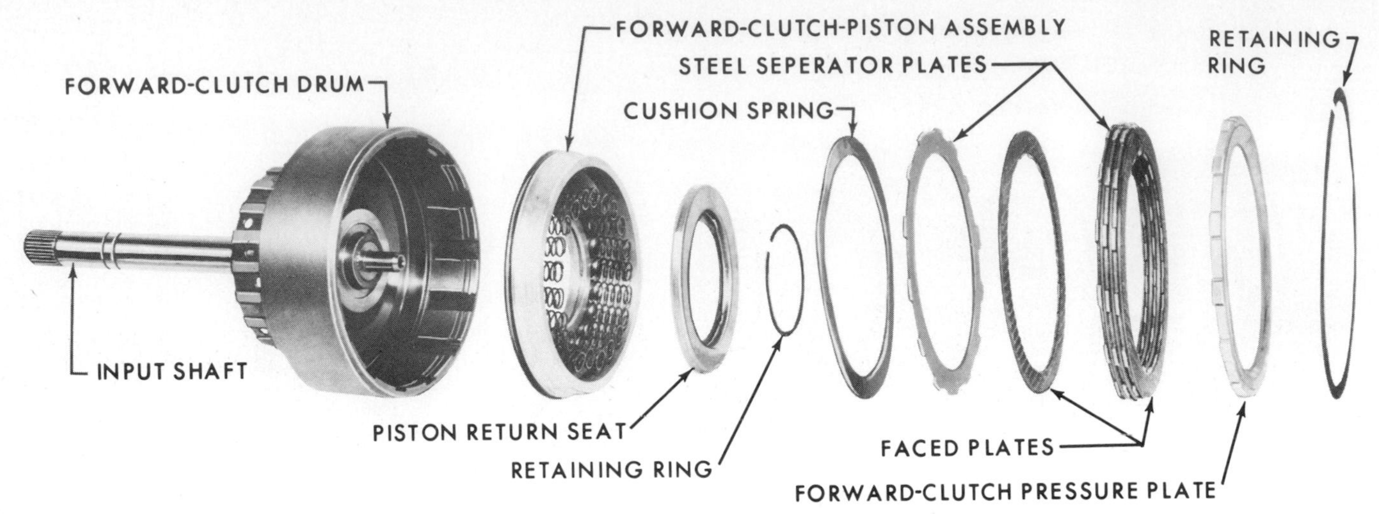

Fig. 14-93. Exploded view of the forward-clutch assembly. (*Chevrolet Motor Division of General Motors Corporation*)

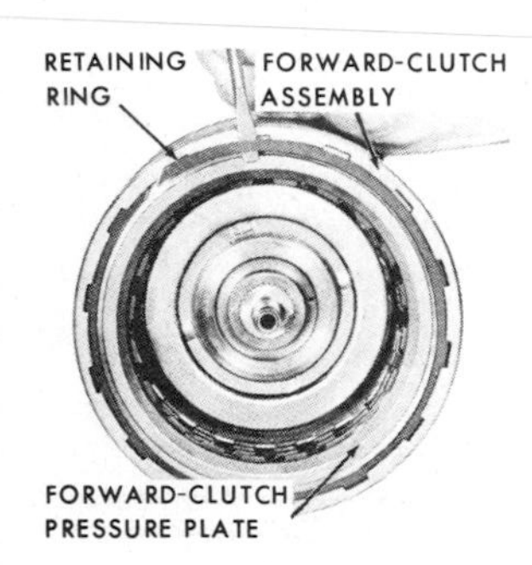

Fig. 14-94. Removing the pressure-plate-to-drum retaining ring. (*Chevrolet Motor Division of General Motors Corporation*)

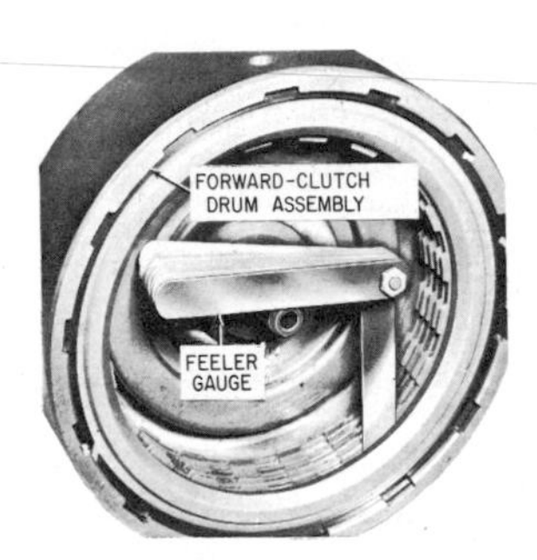

Fig. 14-97. Measuring clearance between the pressure plates and the nearest clutch plate. (*Buick Motor Division of General Motors Corporation*)

Fig. 14-95. Compressing the piston return spring so that the retaining ring can be removed. (*Chevrolet Motor Division of General Motors Corporation*)

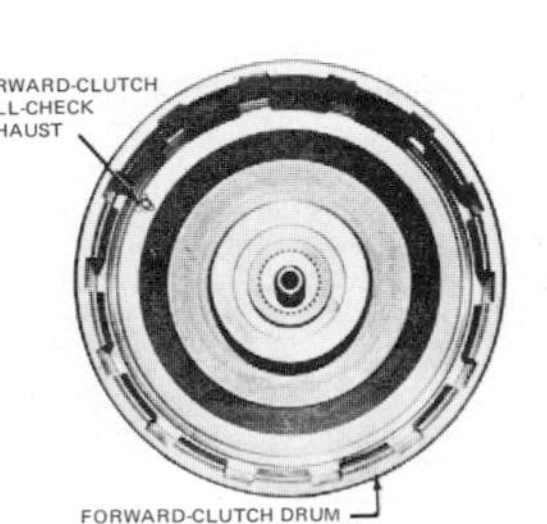

Fig. 14-96. Location of the ball-check exhaust in the forward-clutch drum (*Chevrolet Motor Division of General Motors Corporation*)

piano wire crimped into copper tubing. This procedure is similar to the procedure for installing the direct-clutch piston, shown in Fig. 14-92.

2. Install the 21 springs and the spring retainer. Compress the springs with a special tool, as shown in Fig. 14-95, and install the retaining ring.
3. Lubricate the clutch plates with transmission oil and install the cushion spring, clutch plates, forward-clutch pressure plate, and retaining ring.
4. Use a feeler gauge to check the clearance between the pressure plate and the clutch plate, as shown in Fig. 14-97. If the clearance is less than 0.0105 inch [0.263 mm], a thinner pressure plate should be used. If the clearance is greater than 0.082 inch [2.08 mm], a thicker pressure plate should be used. The correct clearance is between 0.0105 and 0.082 inch [0.263 and 2.08 mm]. Pressure plates of three thicknesses are available.

⊘ 14-16 Sun-Gear and Drive-Shell Service The sun gear and drive shell, along with the other components of the planetary-gear train, are shown in exploded view in Fig. 14-98.

1. Remove the sun-gear-to-drive-shell rear retaining ring, as shown in Fig. 14-99. Then lift off the rear flat-steel thrust washer. Then remove the front retaining ring, as shown in Fig. 14-100.
2. Check the gear and shell for wear.
3. On reassembly, use new retaining rings.

NOTE: If sun-gear bushings require replacement, they may be removed with a cape chisel, as shown in Fig. 14-101, and the new bushings pressed in with the special tool required.

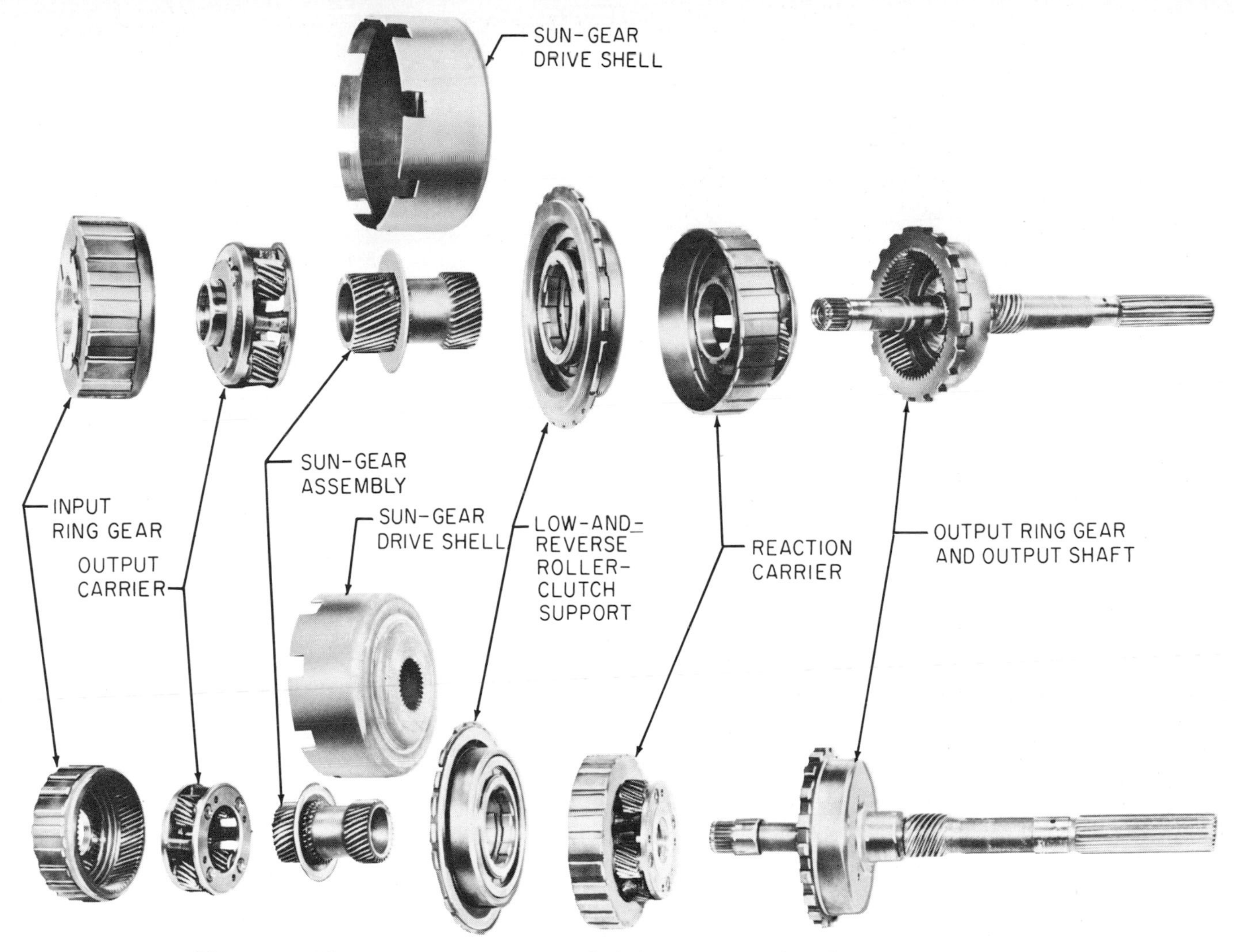

Fig. 14-98. Planetary-gear train in exploded view. Note that there are two views of each part. (*Chevrolet Motor Division of General Motors Corporation*)

Fig. 14-99. Removing the rear retaining ring from the sun gear. (*Buick Motor Division of General Motors Corporation*)

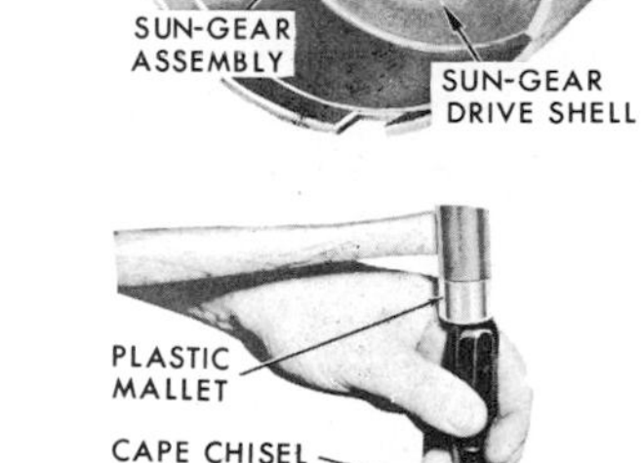

Fig. 14-100. Location of the front retaining ring in the sun gear. (*Buick Motor Division of General Motors Corporation*)

⊘ 14-17 Low-and-Reverse-Roller-Clutch Service This assembly is shown in exploded view in Fig. 14-102. The parts are separated by removing the thrust washer and retaining ring. Inspect the roller races and rollers for wear, scratches, or other damage. Check the springs for distortion.

When reassembling, be sure that the roller-clutch assembly is installed in the support with the

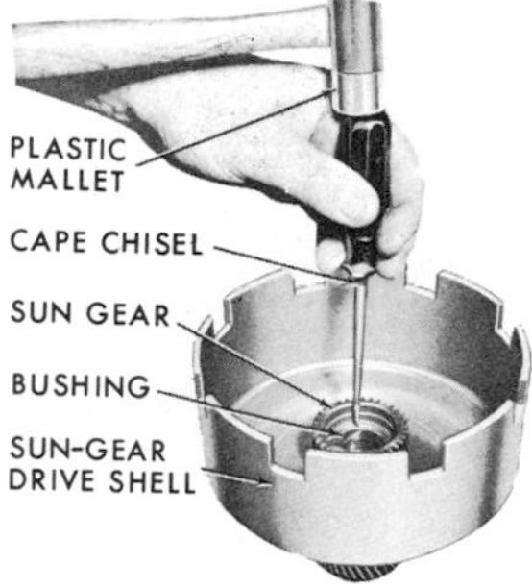

Fig. 14-101. Removing sun-gear bushings with a cape chisel and plastic hammer. (*Buick Motor Division of General Motors Corporation*)

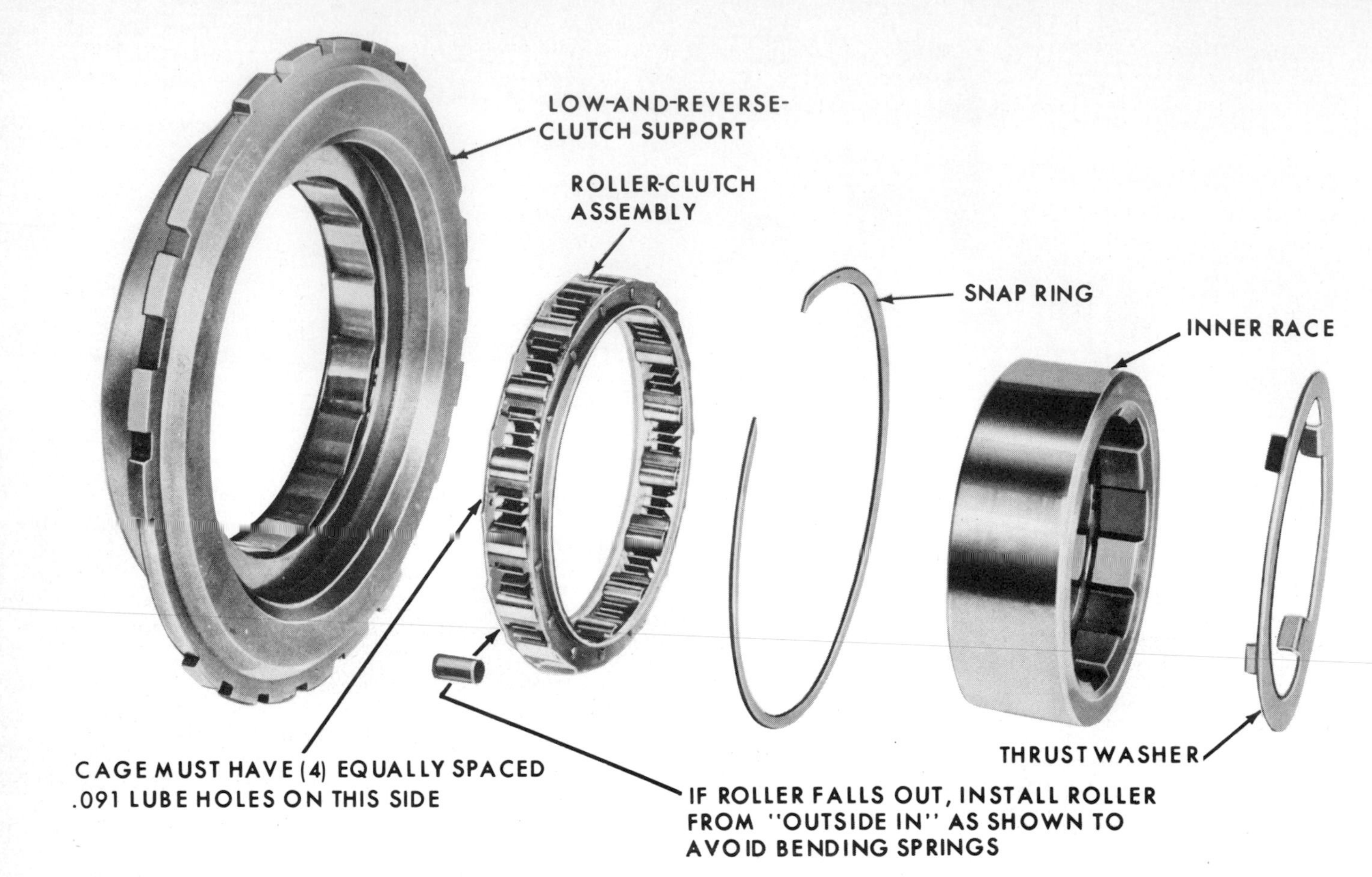

Fig. 14-102. Exploded view of the low-and-reverse-clutch assembly. (*Chevrolet Motor Division of General Motors Corporation*)

oilholes to the rear of the transmission. Make sure that the inner race freewheels in a clockwise direction only.

⊘ 14-18 Governor Service The governor is serviced by replacing it completely, with the exception of the driven gear, which can be replaced. The reason for this is that the parts are selectively fitted and each assembly is carefully calibrated to give the required performance. If driven-gear replacement is required, or if foreign matter has gotten into the governor so that it does not perform properly, the governor can be disassembled, as follows.

Refer to Fig. 14-103. Cut off one end of each governor weight pin and remove the pins, thrust cap, weights, and springs. Remove the valve from the governor sleeve.

Wash all parts in cleaning solvent and air-dry. Blow out all passages. Inspect the sleeve and valve for nicks, burrs, or other damage. Check the governor sleeve for free operation in the bore of the transmission case. The governor valve should slide freely in the bore of the sleeve. Springs and weights should be in good condition. If the driven gear is damaged, it may be replaced as follows.

A special service package is available, containing a new driven gear, two weight retaining pins,

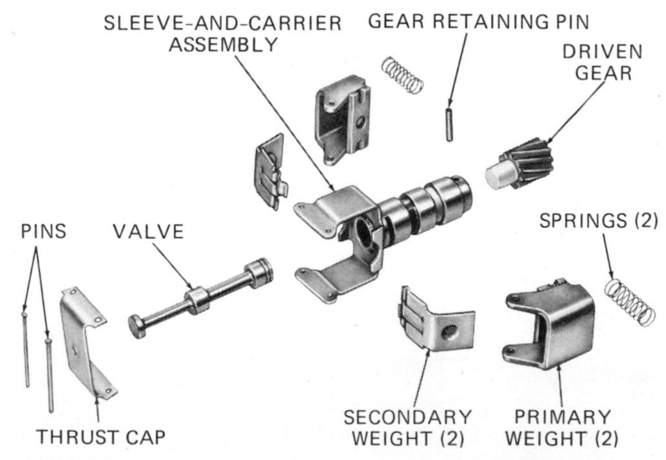

Fig. 14-103. Disassembled view of the governor. (*Chevrolet Motor Division of General Motors Corporation*)

and a gear-retainer split pin. To replace the driven gear, drive out the old split pin with a small punch. Support the sleeve on $\frac{3}{16}$-inch [4.762 mm] plates installed in the exhaust slots of the sleeve, and press the gear out of the sleeve in an arbor press. Clean the governor sleeve. Press the new driven gear into the sleeve until it is nearly seated. Remove any chips that may have been shaved off the gear, and press the gear on in until it bottoms on the shoulder. Drill a new pinhole 90 degrees from the old one. Install the split retaining pin. Wash the sleeve off to remove any chips or dirt.

Reassemble the governor by installing the valve in the sleeve, large land end first. Install the weights, springs, and thrust cap and secure with new pins. Crimp both ends of the pins to keep them from falling out. Check for free operation of weights and valve in the sleeve.

⊘ 14-19 Valve-Body Service The valve body is shown in disassembled view in Fig. 13-13. There is nothing especially difficult about disassembling the body, although extreme care must be used to avoid dirt and damage to finished surfaces. The various valves and springs are held in place by retaining pins. The direct-clutch accumulator piston is held in place by a retaining ring. Extreme cleanliness is essential in working on the valve body and valves. The slightest piece of lint or dirt can cause a valve to hold up, and this could cause the transmission to malfunction and even be severely damaged. Be sure to identify all springs as they are removed so that they can be replaced in the proper positions.

Handle the valves with great care in order to avoid damaging the operating surfaces. Here, again, any dent or burr can cause the valve to hang up with possible serious consequences.

Clean all parts in cleaning solvent and air-dry them. Make sure all oil passages in the valve body are clear. Follow Fig. 13-13 carefully when replacing the valves.

⊘ 14-20 Transmission Reassembly Again, we stress the extreme importance of cleanliness when handling any transmission parts. Hands, tools, and working area must be clean. If work is stopped before reassembly is complete, cover all openings with clean cloths.

During reassembly, lubricate all bushings with transmission oil. Coat thrust washers, both sides, with petroleum jelly.

Use all new seals and gaskets. Don't take chances on old ones.

Tighten all parts evenly and in the proper sequence when replacing screws or bolts. Use new retaining rings, as called for, and be careful not to overstress retaining rings because this could cause them to loosen in service with possible disastrous results. The reassembly procedure follows.

1. Install the low-and-reverse-clutch piston, with new seals in place, with the notch in the piston adjacent to the parking pawl.
2. Install 17 piston return springs and the spring retainer. Use the compressor tool shown in Figs. 14-67 and 14-68 to compress the springs so that the retaining ring can be installed.
3. Install the output ring gear on the output shaft. Install the retaining ring. See Fig. 14-63. Install the reaction-carrier-to-output-ring-gear thrust washer into the output-ring-gear support. This thrust washer has three tangs. See Fig. 14-62.
4. Put the output-ring-gear-to-case needle bearing into position and install the output-shaft assembly in the case.
5. Install the reaction-carrier assembly, as shown in Fig. 14-59.
6. Oil and install the low-and-reverse-clutch plates, as shown in Fig. 14-58. Start with a steel plate and alternate with faced plates. Secure with a retainer spring, as shown in Fig. 14-57.
7. Install the low-and-reverse-clutch-support assembly, as shown in Fig. 14-56. The notch of the retainer spring should be located as shown in Fig. 14-104.

NOTE: The splines in the inner race of the roller clutch must align with the splines on the reaction carrier.

8. Install the low-and-reverse-roller-clutch-inner-race-to-sun-gear-shell thrust washer. Install the retaining ring, as shown in Fig. 14-55.
9. Install the sun-gear-and-drive-shell assembly. See Fig. 14-54.
10. Install the output-carrier assembly, Fig. 14-53. Secure with a retainer ring, as shown in Fig. 14-52, and install the thrust washer, Fig. 14-51.
11. Install the input ring gear, Fig. 14-49, and the thrust washer, Fig. 14-48.
12. Install the direct-clutch assembly and special thrust washer to the forward-clutch assembly.

There are two designs here. One design uses a thrust washer between the forward and direct clutches. The second design uses a Torrington needle bearing. When replacing the first design, use the new design parts. That is, use a new direct clutch, forward-clutch housing, and needle bearing.
13. Install the clutch assemblies in the case, as shown in Fig. 14-47. Make sure that the forward-clutch face plates are positioned over the input ring gear and that the tangs on the direct-clutch housing are installed into the slots on the sun-gear drive shell.
14. Install the intermediate-overrun brake band, as shown in Fig. 14-46.
15. Install the intermediate-clutch pressure plate, as shown in Fig. 14-45.
16. Oil and install the intermediate-clutch plates, starting with a face plate and alternating steel and

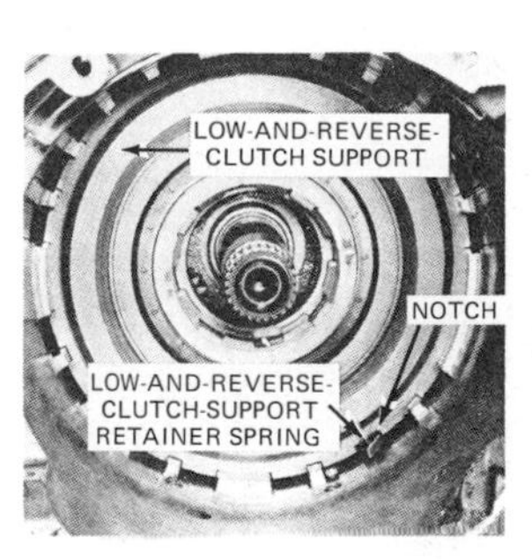

Fig. 14-104. Location of notch in the retainer spring. (*Chevrolet Motor Division of General Motors Corporation*)

face plates. See Fig. 14-44. The notch in the steel reaction plates is installed toward the selector-lever inner bracket.

17. Install the intermediate-clutch cushion spring, as shown in Fig. 14-43.

18. The pump is installed next, but during this procedure, the end play of the input shaft must be checked. If the end play is incorrect, the pump must be removed so that a selective thrust washer of the proper thickness can be installed. The procedure is as follows:

Install a selective-fit thrust washer, oil-pump gasket, and oil pump. Use two guide pins in two opposing holes in the case to align the oil pump as it is brought into position. Install and tighten the pump-to-case bolts. Mount a dial indicator, as shown in Fig. 14-105. Pull out on the input shaft and set the dial indicator to zero. Push in on the input shaft to read the end play. The end play should be between 0.032 and 0.064 inch [0.81 and 1.63 mm]. If it is not, then a different selective-fit thrust washer should be used. This means removal and reinstallation of the pump. Selective-fit washers are available in three thicknesses: 0.066, 0.083, and 0.100 inch [1.68, 2.11, and 2.54 mm].

19. After the proper selective-fit thrust washer is determined, the complete pump-installation procedure is as follows:

a. Install a new pump-to-case gasket, as shown in Fig. 14-42.

b. Install the correct thrust washer, first covering both sides with petroleum jelly.

c. Install a new square-cut oil-seal ring.

d. Install the pump, using two guide pins, as in item *18a,* to align it. Attach with bolts, using new washer-type seals.

CAUTION: Check the rotation of the input shaft as the pump is pulled down into place. If the input shaft cannot be rotated freely, the direct- and forward-clutch housings have not been properly installed. That is, the clutch plates are not indexing. Remove the necessary parts to secure proper indexing. Otherwise, parts will break as the pump is pulled down into place.

20. Put the drive-gear retainer clip into the hole in the output shaft, align the slot in the speedometer-drive gear with a retainer clip, and install the gear on the shaft. See Fig. 14-19.

21. Install the extension-housing-to-case square-cut ring seal, and attach the extension housing to the case with attaching bolts. Torque to 25 pound-feet [3.46 kg-m].

22. Install the parking pawl, with the tooth toward the inside of the case. Then install the disengaging spring on the pawl and slide the shaft into place. See Fig. 14-37. Drive the retainer plug flush to 0.010 inch [0.25 mm] below the case, using a $\frac{3}{8}$-inch [9.53 mm] rod. Stake the plug in three places.

23. Install the park-lock bracket, as shown in Fig. 14-36, and torque bolts to 29 pound-feet [4.01 kg-m].

24. With the actuating rod attached to the range-selector inner lever, put the assembly into place, as shown in Fig. 14-34.

25. Install the manual shaft through the case and the range-selector inner lever. Install the retaining nut on the manual shaft and torque to 25 pound-feet [3.46 kg-m]. Install the manual-shaft-to-case spacer clip.

26. Next, install the intermediate-servo piston, washer, spring seat, and apply pin. See Fig. 14-38. We have already explained how to check the band apply pin for correct length. See Fig. 14-39. When installing the piston, use a new metal oil-seal ring.

27. Install the four check balls in the transmission-case pockets, as shown in Fig. 14-31.

CAUTION: If check balls are missing, complete transmission failure may occur.

28. Install the oil-pump pressure screen and governor screens in the case, as shown in Figs. 14-29 and 14-30.

29. Install the valve-body spacer-plate-to-case gasket and spacer plate, as shown in Fig. 14-28.

30. Install the valve-body-to-spacer-plate gasket and spacer support plate. See Figs. 14-27 and 14-28. Torque the support-plate bolts to 13 pound-feet [1.80 kg-m].

31. Connect the detent-control-valve link to the actuating lever. See Fig. 14-25.

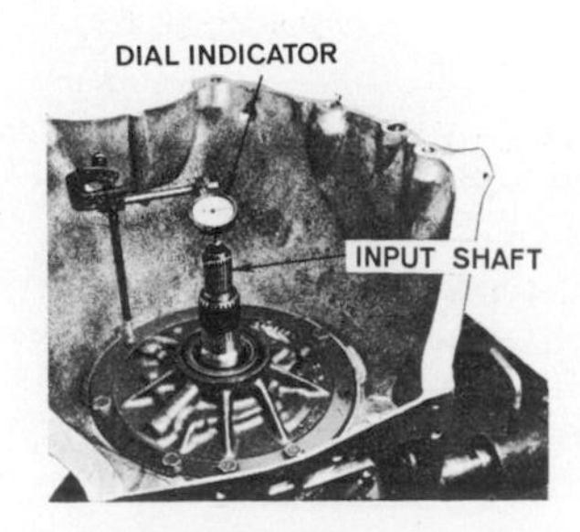

Fig. 14-105. Dial indicator in place to check input-shaft end play. (*Oldsmobile Division of General Motors Corporation*)

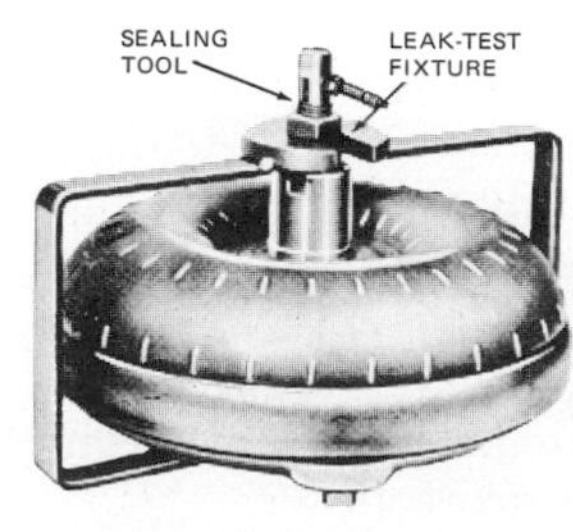

Fig. 14-106. Setup to check the converter for leaks. (*Oldsmobile Division of General Motors Corporation*)

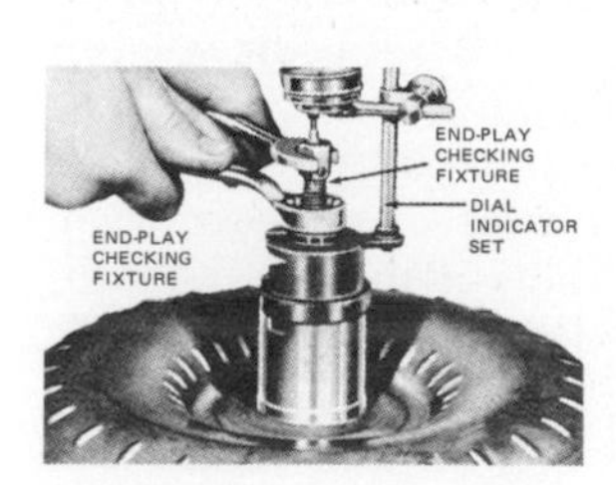

Fig. 14-107. Test setup to check converter end play. (*Oldsmobile Division of General Motors Corporation*)

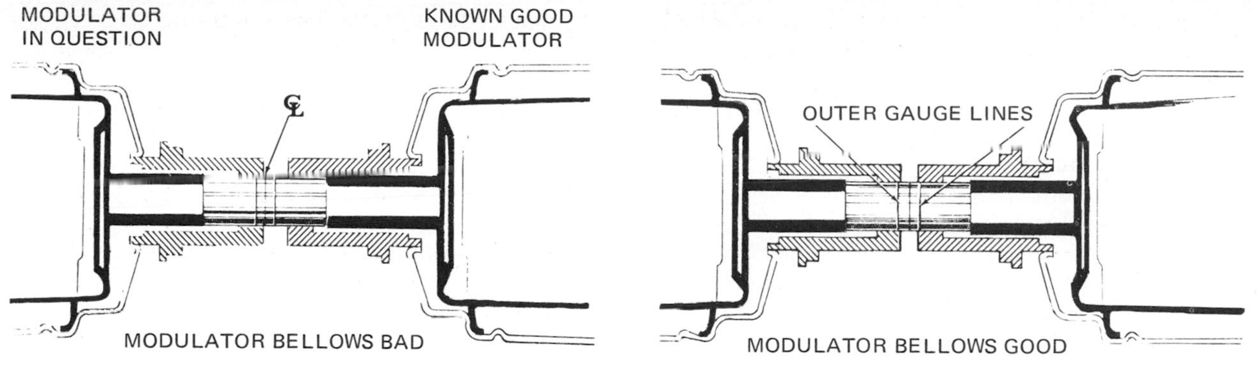

Fig. 14-108. Checking a possible defective modulator bellows against a modulator known to be good. (*Chevrolet Motor Division of General Motors Corporation*)

32. Install the valve body. Connect the manual-control-valve link to the range-selector inner lever. Torque bolts in random sequence to 13 pound-feet [1.80 kg-m]. Leave a bolt loose for the detent-spring-and-roller assembly. See Fig. 14-25.

CAUTION: When handling the valve body, be careful that the retainer pins do not fall out.

33. Install the manual-shaft-to-case retainer ring and detent-spring-and-roller assembly, as shown in Fig. 14-25.
34. Install the strainer-assembly gasket and strainer. See Fig. 14-24.
35. Install the new gasket and oil pan. See Fig. 14-23.

NOTE: The intermediate-clutch-accumulator assembly must be installed, as explained in ⊘ 14-8, before the oil pan is installed.

36. Install the governor assembly, cover, seal, and retainer wire. See Figs. 14-20 to 14-22.
37. Install the vacuum-modulator valve and retainer clip. See Figs. 14-10 and 14-11. Lubricate the O-ring seal to prevent damaging it. Torque the clip bolt to 12 pound-feet [1.66 kg-m].
38. Before installing the converter, check for leaks by installing the sealing tool, as shown in Fig. 14-106. Fill the converter with air to 80 psi [5.625 kg/cm^2], and submerge it in water to check for leaks.
39. Check converter end play with special tools and dial indicator, as shown in Fig. 14-107. Tighten the hex nut, and zero the dial indicator. Loosen the hex nut and note indicator reading. This is the end clearance. It should not exceed 0.050 inch [1.27 mm].
40. Converter leakage or excessive end play requires replacement of the assembly.
41. Refer to ⊘ 14-9 for the transmission-installation procedure.

⊘ 14-21 Vacuum-Modulator Check Turn the vacuum modulator so that the vacuum side points down. If oil comes out, the diaphragm is defective and the modulator must be replaced. Make a bellows-comparison check with a vacuum modulator known to be good, as shown in Fig. 14-108. Using the gauge as shown, install the two modulators on either end of the gauge. Hold the modulators horizontal, and push them toward each other until either modulator sleeve end just touches the line in the center of the gauge. The gap between the other modulator sleeve end and the line should be $\frac{1}{16}$ inch [1.59 mm] or less. If it is more, the modulator in question should be discarded.

CHAPTER 14 CHECKUP

NOTE: Since the following is a chapter review test, you should review the chapter before taking it.

The service procedures for the Type 350 Turbo Hydra-Matic transmission, as detailed in the chapter, are rather lengthy and not the sort of procedures you memorize. They are included to give you a good idea of how to go about servicing the transmission. Furthermore, the illustrations will help you visualize the sizes and shapes of the various component parts so that you can readily recognize them when you see them in the shop. When you overhaul a transmission, you will have the appropriate shop manual before you, and you will follow that manual. Of course, by the time you have serviced many transmissions, you will know the step-by-step procedures by heart and will need to refer to the shop manual only to check on special points or look up specifications. The checkup below focuses on the general instructions, with some questions directed specifically to the Type 350 transmission.

Completing the Sentences The sentences that follow are incomplete. After each sentence there are several words or phrases, only one of which will correctly complete the sentence. Write each sentence in your notebook, selecting the proper word or phrase to complete it correctly.

1. In transmissions used on the big engines, the clutches have: (*a*) fewer plates, (*b*) same number of plates, (*c*) more plates.

2. The Type 350 transmission should have the oil changed every: (*a*) year, (*b*) 12,000 miles [19,312 km], (*c*) 24,000 miles [38,624 km].
3. Only two in-car adjustments are required on the Type 350 transmission: (*a*) selector-lever linkage and throttle linkage, (*b*) linkage and brake band, (*c*) brake band and throttle-valve linkage.
4. In making the oil-pressure checks on the Type 350 transmission, what you are actually checking is: (*a*) pump pressure, (*b*) TV (throttle-valve) pressure, (*c*) line pressure.
5. Several parts can be removed from the transmission while it is still in the car, including the: (*a*) vacuum modulator and direct clutch, (*b*) valve-body assembly and extension housing, (*c*) governor assembly and front brake band.
6. Low line pressures could be due to low oil level, governor or valve-body trouble, or to: (*a*) pump trouble, (*b*) clutch slippage, (*c*) band slippage.
7. High line pressure in all ranges could be due to vacuum-modulator or governor trouble, stuck pressure regulator or boost valve, or to a: (*a*) stuck shift valve, (*b*) stuck detent valve, (*c*) faulty accumulator.
8. Slipping in reverse could be caused by low oil pressure, a stuck pressure regulator or boost valve, or by a: (*a*) defective governor, (*b*) slipping direct or low-and-reverse clutch, (*c*) detent valve sticking.
9. Slipping in all ranges could be caused by low oil pressure, trouble in the valve body or 1-2 accumulator, or by: (*a*) incorrect manual-valve linkage adjustment, (*b*) defective governor action, (*c*) shift valves stuck.
10. A rough 2-3 upshift could be caused by trouble in the valve body, a stuck 2-3 shift valve, faulty 2-3 accumulator action, or by: (*a*) a defective governor, (*b*) faulty vacuum-modulator action, (*c*) band not releasing smoothly.

Troubles and Service Procedures In the following, you should write in your notebook the required trouble causes and service procedures. Do not copy from the book; try to write the trouble causes and service procedures in your own words. Writing them in your own words will be of great help to you because it will enable you to remember them better; you will thereby be greatly benefited when you go into the automotive shop.

1. Describe the procedure of adding oil; of changing oil.
2. Explain how to make the two required adjustments on the Type 350 transmission in a car.
3. Explain how to make a pressure check of the transmission.
4. List possible causes if the transmission slips in all ranges.
5. List possible causes if line pressure is low.
6. List possible causes if line pressure is high.
7. List possible causes if there is no 1-2 upshift.
8. List possible causes if the 1-2 upshift is rough.
9. List possible causes if the 2-3 upshift is rough.
10. List possible causes if the transmission slips in reverse.
11. List possible causes if the transmission has harsh downshifts.
12. List possible causes if the transmission drives in neutral.
13. List the service precautions to take when working on transmissions.
14. List the procedures required to clean and inspect transmission parts.
15. List the transmission parts in the order in which they are removed from the transmission.
16. Refer to a shop manual of a car using the 350 transmission, and prepare a transmission removal story outlining, step by step, how to take the transmission out of the car.

SUGGESTIONS FOR FURTHER STUDY

If you can be in the transmission shop when a 350 transmission is removed from a car and overhauled, you will learn a great deal about the procedures. Reading the procedures in the book is one thing, but watching them being done is something else again. You really need both to get the whole story—read how it is done and watch it being done—and then, of course, do it yourself. Naturally, you must have a good deal of basic experience on the job before you can tackle the overhaul of a transmission all by yourself. Meantime, study the shop manuals, watch the procedures, and sooner or later you will be doing the job yourself. Write everything of importance you learn in your notebook so that you will have it for ready reference.

chapter 15

GENERAL MOTORS TYPE 400 TURBO HYDRA-MATIC TRANSMISSION

The General Motors Type 400 Turbo Hydra-Matic transmission provides three forward speeds and reverse. It includes a torque converter and has three multiple-plate clutches, a sprag overrun clutch, a roller overrun clutch, and two brake bands. These components are actuated by hydraulic pressure to produce automatic upshifts or downshifts in the drive range as required by engine output and road speed. This transmission is used on Buick, Cadillac, Chevrolet, Oldsmobile, and Pontiac cars. A highly modified version of the transmission is used on the front-drive Cadillac Eldorado and the Oldsmobile Toronado. This chapter describes the construction and operation of the transmission. Chapter 16 gives details of the service procedures.

⊘ 15-1 Type 400 Turbo Hydra-Matic Transmission Figure 15-1 is a cutaway view of the Type 400 Turbo Hydra-Matic transmission. The transmission functions in a manner very similar to that of the Type 350 transmission, described in Chap. 13. Even though the internal construction is different, with somewhat different gearing arrangements and clutch and brake-band setups, the two transmissions are alike in operation. There is a selector lever and quadrant on the steering-wheel mast, as shown in Fig. 15-2. The selector lever permits the driver to select the desired transmission operating conditions.

For normal town and country driving, the driver would select the D, or drive, position. This causes the transmission to start out in low gear and upshift through intermediate to high gear as car speed and acceleration position dictate. The transmission will also downshift for passing when the accelerator is pushed down to full open or when car speed drops off sufficiently.

When the selector lever is placed in L1, the transmission will stay in low gear. This position is for pulling a heavy load or for going down a steep grade. The low-gear position permits the engine to help brake the car and thus saves the brakes from long periods of use when going down a steep hill.

When the selector lever is placed in L2, also known as the super range in some cars, the transmission will upshift to second but not to high or third gear. The transmission can be downshifted by opening the throttle wide open, as for passing another car. It will also downshift of its own accord if car speed drops low enough.

There is also reverse, as in all transmissions, for backing the car.

In this transmission, just as in all other automatic transmissions, the shifting is produced by hydraulic pressures which actuate various valves, servos, and multiple-plate clutches. The servos apply or release brake bands; the clutches release or lock up planetary-gear components. The result is the shifts that take place in the transmission.

The transmission also includes an oil cooler, as shown in Figs. 13-3 and 13-4. The oil circulates through a small radiator that is part of the engine cooling system.

We shall now study the conditions in the transmission when it is in the different gear positions. A chart listing these conditions is given in Fig. 15-2.

⊘ 15-2 Low Gear, Drive Range In drive range, when first starting out, the transmission is in low gear. This condition is shown in Plate 17. Take a moment to study this illustration. Locate these components: front clutch, direct clutch, front band, intermediate clutch, intermediate sprag clutch, roller clutch, and rear band. Now notice the oil flow in the hydraulic circuits. Movement of the manual valve to D allows line pressure from the oil pump to enter the following: forward clutch, 1-2 shift valve, governor assembly, 1-2 accumulator valve, and detent regulator valve.

In Chaps. 9, 13, and 14, where we studied the Type 300 and Type 350 automatic transmissions, we described in detail the construction and operation

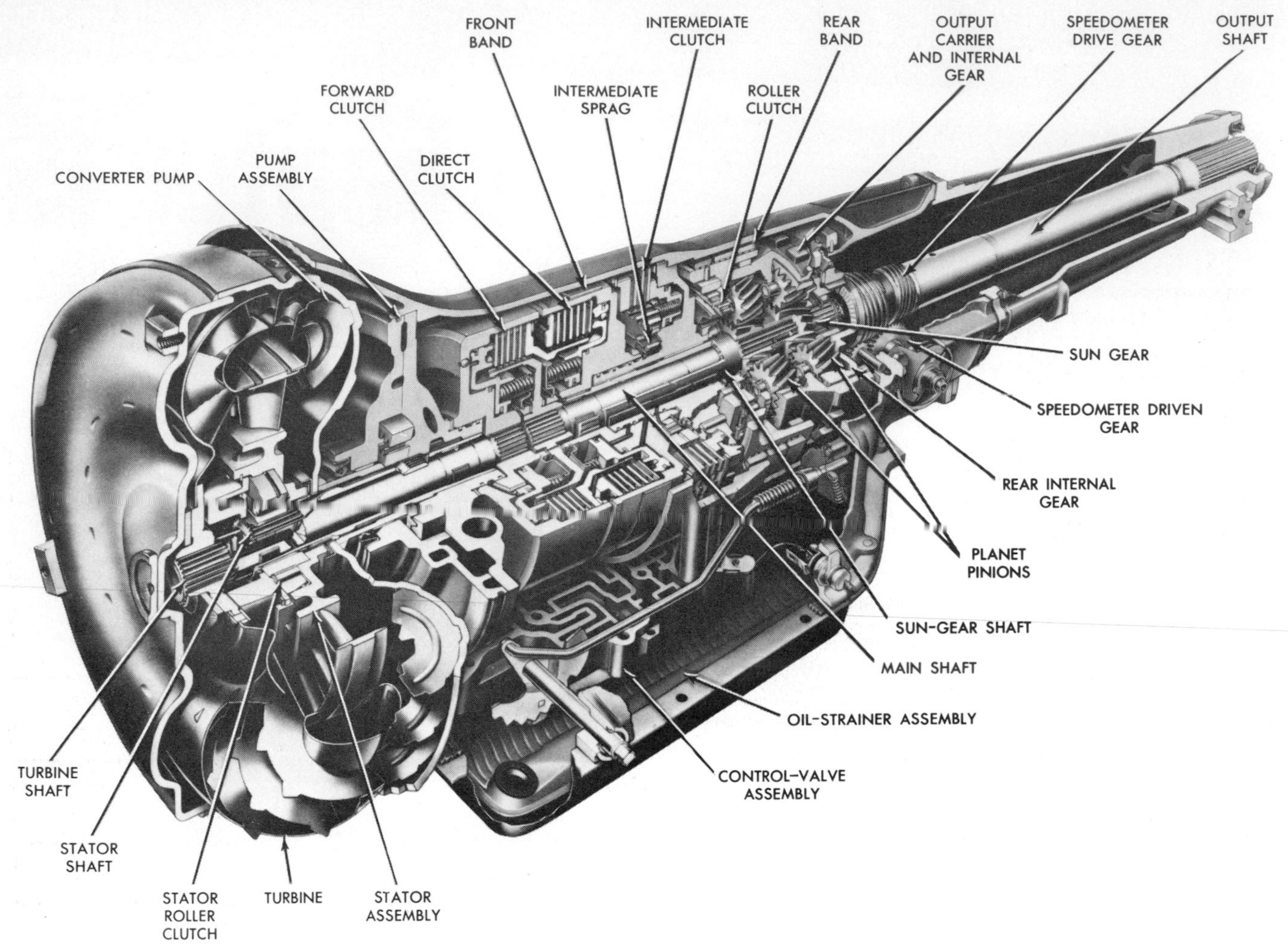

Fig. 15-1. Cutaway view of the Type 400 Turbo Hydra-Matic automatic transmission. (*Cadillac Motor Car Division of General Motors Corporation*)

of the following hydraulic-circuit components: manual valve, pressure-regulator valve, boost valve, governor, servos, accumulators, vacuum-modulator assembly, 1-2 shift valve, 2-3 shift valve, and detent valve. If you are unsure of the operation of any of these components, refer to these chapters, where they are discussed in detail.

There is one difference in the detent valve that should be noted. In the Type 350 automatic transmission, the detent valve is operated by a mechanical linkage from the throttle. When the throttle is opened wide, it causes the detent valve to move so that a downshift occurs. In the Type 400 automatic transmission, the action is caused by a detent solenoid. When the throttle is opened wide, a switch in the throttle linkage is operated to connect the detent solenoid to the battery. This action causes the solenoid to dump detent oil, thus causing the detent valve to produce the downshift. We shall discuss this action later.

Now, back to the low gear in drive range. Drive oil is directed to the forward clutch, as already noted. This causes the forward clutch to apply. The power flow, shown by the arrows in Plate 17, is through the forward clutch to the main shaft and from there to the rear planetary internal gear. The internal gear drives the rear planetary planet pinions, which then drive the sun gear. The front end of the sun gear, meshed with the front planetary-planet pinions, drives these front pinions, which, in turn, drive the front internal gear. The front internal gear drives the output shaft. There is gear reduction in both the front and rear planetary gearsets, giving a combined gear reduction of 2.5:1. Additional "gear" reduction is available in the torque converter, giving a total gear reduction of 5:1 through the transmission.

The reaction of the front pinions against the front internal gear is taken by the front pinion carrier and the roller-clutch assembly. The roller clutch holds the pinion carrier stationary so that the pinions can act as idler gears and cause the front internal gear to be driven.

Drive oil entering the governor exits at a variable pressure, depending on car speed. This pressure, called *governor pressure,* acts against the ends of the

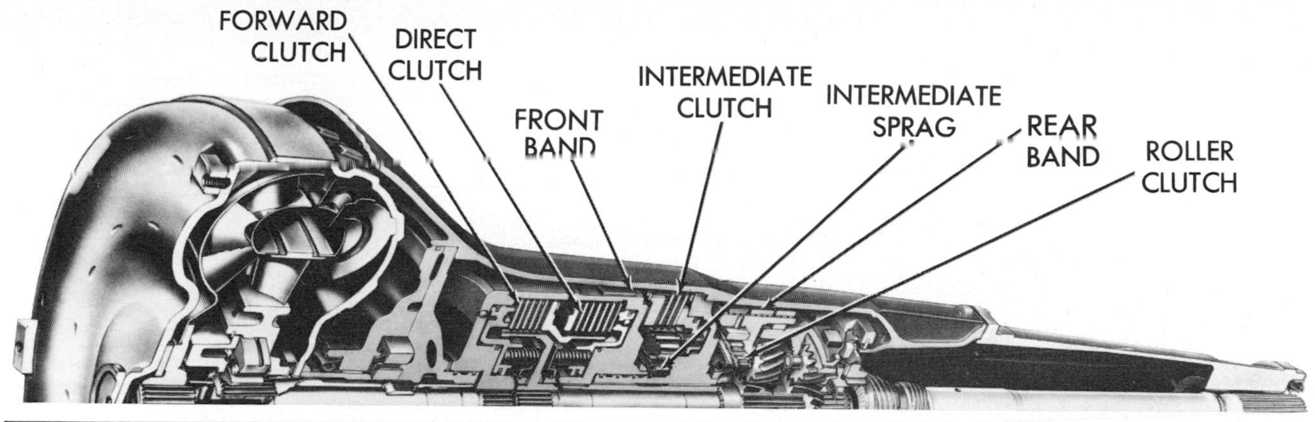

SELECTOR POSITION	PUMP PRESSURE	FORWARD CLUTCH	DIRECT CLUTCH	FRONT BAND	INT. CLUTCH	INT. SPRAG	ROLLER CLUTCH	REAR BAND
PARK–NEUT.	60-150	OFF	OFF	OFF	OFF	OFF	OFF	OFF
DRIVE 1 LEFT 2 3	60-150 60-150 60-150	ON ON ON	OFF OFF ON	OFF OFF OFF	OFF ON ON	OFF ON OFF	ON OFF OFF	OFF OFF OFF
DRIVE 1 RIGHT 2	150 150	ON ON	OFF OFF	OFF ON	OFF ON	OFF ON	ON OFF	OFF OFF
LO 1 2	150 150	ON ON	OFF OFF	OFF ON	OFF ON	OFF ON	ON OFF	ON OFF
REV.	95-240	OFF	ON	OFF	OFF	OFF	OFF	ON

Fig. 15-2. Cutaway view of the Type 400 Turbo Hydra-Matic automatic transmission with clutch-, band-, and sprag-application chart. (*Cadillac Motor Car Division of General Motors Corporation*)

1-2 and 2-3 shift valves. It also acts on an area of the modulator valve. All this activity prepares the transmission for an upshift to second gear.

⊘ 15-3 Second Gear, Drive Range As both car speed and governor pressure increase, the pressure of the governor oil acting on the 1-2 shift valve overcomes the force of the 1-2 valve spring and the modulator oil pressure. The modulator pressure varies according to throttle-valve opening, as explained in ⊘ 13-12. In effect, car speed, or governor pressure, overcomes engine speed, or throttle-valve opening, so that the 1-2 shift valve is forced to move. When it does, line pressure is admitted to the intermediate clutch. The intermediate clutch is applied, and the conditions shown in Plate 18 result.

The arrows in Plate 18 show the power flow through the transmission. Power flows through the forward clutch to the main shaft and from there to the rear planetary internal gear. The internal gear drives the rear planetary planet pinions. This is the same thing that happens when the transmission is in low gear in the drive range. From here on, however, the situation is different.

For now, the intermediate clutch is applied to hold the sun gear stationary. The stationary disks of the intermediate clutch are splined to the transmission case. The other disks are splined to the outer race of the sprag clutch. Thus, when the intermediate clutch is applied, the outer race is held stationary. Now, the inner race of the sprag clutch, which is attached to the sun gear, is locked up. This means the sun gear is locked and cannot turn.

With the sun gear locked, the planet pinions can still rotate, but they must walk around the sun gear; and, in doing so, they carry the planet-pinion carrier around. The carrier, attached to the output shaft, thus causes the output shaft to turn. The gear reduction is through only the rear planetary gearset and is approximately 1.5:1.

Intermediate-clutch pressure from the 1-2 shift valve not only flows to the intermediate clutch. It also flows to the rear servo, front servo and accumulator pistons, and 2-3 shift valve. The intermediate-clutch oil moves the rear-servo accumulator piston against the 1-2 accumulator oil and accumulator spring. This action allows a relatively soft application of the intermediate clutch. We have already explained how the accumulator works in ⊘ 7-9 and 9-16. It cushions clutch application.

Intermediate-clutch oil also flows to the front servo, preparing it for an upshift into third, or direct, gear. At the same time, intermediate-clutch oil flows to a land on the 2-3 shift valve, where it is stopped.

⊘ 15-4 Third Gear, Drive Range As both car speed and governor pressure increase further, the pressure of the governor oil acting on the 2-3 shift valve overcomes the force of the 2-3 shift-valve spring and modulator oil pressure. Modulator pressure varies according to throttle-valve opening, as explained in ⊘ 13-12. In effect, car speed, or governor pressure, overcomes engine speed, or throttle-valve opening. The 2-3 shift valve is forced to move. Now, line pressure is directed to the direct clutch. This pressure is called *direct-clutch oil pressure.* The direct clutch applies, and the situation shown in Plate 19 results. At the same time, direct-clutch oil flows to the front servo. This servo also serves as the accumulator for the upshift from second to third. The direct-clutch oil causes the accumulator piston to move upward against the spring and intermediate-clutch pressure. This cushions the application of the direct clutch.

Now let us see what we have: Power flows through the forward clutch, which is still applied, to the main shaft. At the same time, the direct clutch, which is also applied, directs power to the sun-gear shaft and sun gear. In other words, the rear internal gear, which is driven by the main shaft, and the sun gear, driven through the direct clutch and sun-gear shaft, are turning at the same speed. In effect, the planetary gears are locked up, all turning as a unit. Thus, the gear ratio through the transmission is 1:1.

A part-throttle downshift can be produced below about 33 mph [53.11 km/h] by depressing the accelerator pedal sufficiently. This action reduces intake-manifold vacuum, which, in turn, causes the vacuum-modulator valve to move. When this happens, modulator-valve pressure causes the 3-2 valve to move. (This is the 3-2 valve, not the 2-3 valve.) The 3-2 valve moves against direct-clutch oil. Now, the 3-2 valve passes modulator oil to act on the 2-3 modulator valve, which is to the right of the 2-3 valve in Plate 19. The 2-3 modulator valve then moves the 2-3 valve, cutting off direct-clutch oil. The direct clutch releases, and the transmission drops back into second gear.

⊘ 15-5 Detent Downshift While the car is operating below about 70 mph [112.6 km/h], a forced downshift into second gear, called a *detent downshift,* can be made by depressing the accelerator pedal all the way. This action operates a detent switch, which then connects the detent solenoid in the transmission to the battery. The detent solenoid then opens an orifice, which causes the line oil acting on the detent valve to exhaust, as shown in Plate 20. When this happens, the detent valve is moved to the left (Plate 20). Now drive oil entering the detent valve exits at a regulated pressure of approximately 70 psi [4.921 kg/cm^2]. This oil is called *detent oil.* Detent oil acts on the following:

1. 1-2 regulator valve
2. 2-3 modulator valve
3. 3-2 valve
4. 1-2 primary accumulator valve
5. Vacuum-modulator valve

The detent oil, acting on the 2-3 modulator valve, causes the 2-3 modulator valve to move the 2-3 valve to cut off the direct-clutch oil. This action causes the direct clutch to disengage so that the transmission downshifts to second. If the car is moving below about 20 mph [32.2 km/h], the downshift will be to first. This downshift occurs because detent oil is also directed to the 1-2 regulator valve. This causes the detent oil to shift the 1-2 valve to cut off intermediate-clutch oil so that the downshift is made to first.

Detent oil also acts on the modulator valve to prevent modulator pressure from regulating below 70 psi [4.92 kg/cm^2] at high speeds or altitudes.

⊘ 15-6 Super Range Super range is also known as L2 and is indicated on the selector quadrant as either S or L2. In this range, the transmission is in second gear and cannot upshift to high, or third, gear. Hydraulic-circuit conditions are as shown in Plate 21. The manual valve has been moved to allow drive oil to flow in the S, or super-range, line. Now, the oil in the S line, called *super oil,* acts at the 2-3 valve and at the pressure-boost valve, which is part of the pressure-regulator-valve train. The super oil, acting on the boost valve, causes the pressure regulator to raise the line pressure to 150 psi [10.546 kg/cm^2]. This pressure is the line pressure, the drive pressure, and also the super pressure. All are connected at the manual valve. The super pressure, acting on the 2-3 valve, forces the 2-3 valve to close, shutting off direct-clutch oil. If the transmission is in high, this action releases the direct clutch so that the transmission drops down into second gear.

If the transmission is in low, it will upshift into second. The intermediate-clutch oil acts on the front servo to apply the front brake band. The brake band must be applied to hold the sun gear stationary. If the band were not applied, the sun gear would overrun the intermediate sprag clutch when the car goes down a hill and engine braking is required. The clutch and band conditions are as follows:

Forward clutch applied
Direct clutch released
Intermediate clutch applied
Roller clutch ineffective
Intermediate sprag clutch effective
Front band applied
Rear band released

⊘ 15-7 Low Range If the selector lever and manual valve are moved to L (low) when first starting out, the forward clutch is applied but all other clutches are off. This condition is the same as when first starting out in the drive range. However, there are differences, as you will see. The conditions are as shown in Plate 22. The low selector-lever position is used mainly to get maximum downhill engine braking at speeds below about 40 mph [64.37 km/h]. The low oil from the manual valve is directed to the rear servo, 1-2 accumulator valve, detent regulator valve, and 1-2 shift valve.

If the car speed is above 40 mph [64.37 km/h], the governor pressure, acting on the 1-2 valve, will prevent the 1-2 valve from moving to cause oil cutoff from the intermediate clutch. Thus, downshift cannot be made. However, if the speed drops to less than 40 mph [64.37 km/h], then modulator and detent oil at line pressure, acting on the 1-2 regulator valve and the 1-2 detent valve, overcome governor pressure. They combine to cause the 1-2 shift valve to move so as to cut off intermediate-clutch oil. The intermediate clutch releases, leaving only the forward clutch engaged.

The power flow is as shown in Plate 22. This situation is the same, insofar as the power flow is concerned, as the one shown in Plate 17, which shows power flow in low gear in drive range. There is a difference, however: Low oil from the manual valve flows to the rear servo to apply the rear band. This band must apply to prevent the front planet-pinion carrier from rotating during engine braking. If the front planet-pinion carrier is not held, it will overrun the roller clutch when the car is going down a hill in low; no engine braking will result. The clutch and band conditions are as follows:

Forward clutch applied
Intermediate clutch released
Direct clutch released
Roller clutch effective
Sprag clutch ineffective
Front band released
Rear band applied

⊘ 15-8 Reverse When the selector lever is moved to reverse, the manual lever moves to the reverse position and the conditions shown in Plate 23 result. The manual valve directs line pressure into the reverse circuit so that it acts on the following:

1. Direct clutch
2. 2-3 shift valve
3. Rear-servo piston
4. Pressure-boost valve

The direct clutch applies, but the other clutches are disengaged. There is no governor pressure, nor is there any pressure on the 1-2 or 2-3 valves. Reverse oil enters the rear servo and causes it to apply the rear brake band. Reverse oil also acts on the pressure boost valve, causing it to increase the line and thus reverse pressure. With these conditions—the direct clutch applied and the rear band applied—the transmission goes into reverse.

The power flow is to the sun-gear shaft and sun gear. This action causes the front pinions and front internal gear to turn counterclockwise in reverse rotation. As a result, the front internal gear, being directly attached to the output shaft, causes the output shaft to turn in a reverse direction so that the car is backed. The clutch and band conditions are as follows:

Forward clutch released
Intermediate clutch released
Direct clutch applied
Roller clutch ineffective
Sprag clutch ineffective
Front band released
Rear band applied

CHAPTER 15 CHECKUP

NOTE: Since the following is a chapter review test, you should review the chapter before taking it.

Previous chapters described in detail the operation of all the valves, clutches, servos, brake bands, and planetary-gear systems found in automatic transmissions. This chapter has not repeated these explanations. Instead, it has concentrated on the special arrangements of these elements in the Type 400 Turbo Hydra-Matic transmission and explained how they interact to produce the three forward speeds—L2, L1, and detent downshift—and reverse. Now, test your knowledge of the material covered in this chapter by taking the quiz that follows.

Completing the Sentences The sentences that follow are incomplete. After each sentence there are several words or phrases, only one of which will correctly complete the sentence. Write each sentence in your notebook, selecting the proper word or phrase to complete it correctly.

1. In first speed in drive range, the overrun roller clutch is effective and the following is on: (*a*) low clutch, (*b*) forward clutch, (*c*) intermediate clutch.
2. In second speed in drive range, the intermediate sprag overrun clutch is effective and the following are on: (*a*) low and intermediate clutches, (*b*) forward and direct clutches, (*c*) intermediate and forward clutches.
3. In third speed, or direct drive, neither overrun clutch is effective and the following are on: (*a*) forward, direct, and intermediate clutches; (*b*) front band and direct clutch; (*c*) both bands and direct clutch.
4. In a detent downshift, the intermediate sprag overrun clutch is effective and the following are on: (*a*) forward and direct clutches, (*b*) forward and intermediate clutches, (*c*) front band and forward clutch.

5. In super range, or L2, the intermediate sprag overrun clutch is effective and the following are on: (*a*) front and rear bands and forward clutch, (*b*) front band and direct clutch, (*c*) front band and forward and intermediate clutches.
6. In low range, first speed, the roller overrun clutch is effective and the following are on: (*a*) rear band and forward clutch, (*b*) forward band and rear clutch, (*c*) forward clutch and front band.
7. In reverse, neither overrun clutch is effective and the following are on: (*a*) direct clutch and front band, (*b*) direct clutch and rear band, (*c*) rear clutch and rear band.
8. In the 400 transmission, there are two overrun clutches, two brake bands, and: (*a*) one clutch, (*b*) two clutches, (*c*) three clutches.

Purpose and Operation of Components In the test below, you are asked to write the purpose and operation of the various components used in the transmission described in this chapter. If you have any difficulty with any of the questions, reread the pages in the chapter that will give you the information you need. Then write your explanation. Don't copy; try to tell it in your own words. This is a good way to fix the information in your mind. Write in your notebook.

1. Describe the power flow in the transmission in first speed in drive range.
2. Describe the power flow in second speed in drive range.
3. Describe the power flow in direct drive.
4. Describe the power flow in a detent downshift.
5. Describe the power flow in super range, or L2.
6. Describe the power flow in low range, first speed.
7. Describe the power flow in reverse.
8. What is the major difference in the method of operating the detent valve in the Type 350 and Type 400 transmissions?
9. Make a drawing of the part of the hydraulic system that is active with the transmission in first speed in drive range.
10. Make a drawing of the part of the hydraulic system that is active with the transmission in second speed in drive range.
11. Make a drawing of the part of the hydraulic system that is active with the transmission in direct drive.
12. Make a drawing of the part of the hydraulic system that is active with the transmission in a detent downshift.
13. Make a drawing of the part of the hydraulic system that is active with the transmission in super range, or L2.
14. Make a drawing of the part of the hydraulic system that is active with the transmission in low range, first speed.
15. Make a drawing of the part of the hydraulic system that is active with the transmission in reverse.

SUGGESTIONS FOR FURTHER STUDY

As we have suggested for the transmissions described in previous chapters, examine transmission parts to help you understand how they are made, how they go together, and how they work. Also, study the shop manuals applying to cars using the transmission. Be sure to write any interesting facts you learn in your notebook.

chapter 16

TYPE 400 TURBO HYDRA-MATIC TRANSMISSION SERVICE

This chapter discusses the trouble-diagnosis and service procedures for the Type 400 Turbo Hydra-Matic automatic transmission. The chapter is divided into four parts: "Normal Maintenance and Adjustments," "Trouble Diagnosis," "On-the-Car Repairs," and "Transmission Overhaul."

⊘ 16-1 Type 400 Transmission Variations The Type 400 transmission is used on many cars and with a variety of engine sizes and performance characteristics. To cooperate properly with the engine it is used with, some minor modifications of the transmission are made. These do not affect the servicing procedures because the basic design and construction are the same for all Type 400 transmissions.

Normal Maintenance and Adjustments

⊘ 16-2 Checking and Adding Transmission Fluid All car manufacturers specify that only Dexron transmission fluid, also called *transmission oil,* be used in the Type 400 transmission. Using any other fluid, with other specifications, could ruin the transmission. Fluid level should be checked at every engine oil change. If oil is needed, it should be added, as explained in Chap. 14, ⊘ 14-2, which covers the addition of transmission fluid to the Type 350 transmission. The procedures are the same for both transmissions.

Every 24,000 miles [38,624 km], the oil pan should be drained and the filter replaced. Then fresh transmission fluid should be added, as explained in ⊘ 16-3. If the car is used in heavy city traffic, where the engine is regularly idled for long periods, the oil pan should be drained and the filter replaced every 12,000 miles [19,312 km].

⊘ 16-3 Draining Oil Pan and Replacing Filter The oil pan should be drained and the filter replaced every 24,000 miles [38,624 km] (every 12,000 miles [19,312 km] in severe city service). The procedure is as follows:

1. Raise the car on a hoist or place on jack stands and provide a container to catch draining fluid.
2. Remove the oil pan and gasket. Throw the gasket away.
3. Drain fluid from the oil pan. Clean the pan with solvent and dry with compressed air.
4. Remove the oil-filter retainer bolt, oil filter, intake-pipe assembly, and intake-pipe-to-case O-ring seal (see Fig. 16-9). Discard the old filter and O-ring seal.
5. Install a new O-ring seal on the intake pipe and put the pipe assembly into the grommet on the new filter assembly. Install the pipe-and-filter assembly in the case, attaching the filter to the valve body with the retainer bolt torqued to 10 pound-feet [1.38 kg-m].
6. Install the new gasket on the oil pan and install the oil pan, tightening the attaching bolts to 12 pound-feet [1.66 kg-m].
7. Lower the car and add specified quantity of transmission fluid through the filler tube.

NOTE: Recommendations vary with different car manufacturers. Refer to the shop manual for the car being serviced.

8. With the manual lever in park, start the engine. *Do not race the engine!* Move the manual lever through each range.
9. Check fluid level with the selector lever in park and the vehicle on a level surface. Add more fluid to bring the fluid up to the specified level.

CAUTION: Do not overfill!

⊘ 16-4 Towing Instructions The car may be safely towed on its rear wheels, with the selector lever in neutral, for short distances at speeds no greater than 35 mph [56.33 km/h] under most conditions. However, if the transmission is not operating properly, if higher speeds are necessary, or if the car must be towed more than 50 miles [80.47 km], *the drive shaft must be disconnected!* Otherwise, the transmission could be completely ruined.

As an alternative, the car can be towed on its front wheels. In this case, the steering wheel must be secured to keep the front wheels in the straight-ahead position.

⊘ **16-5 Adjustments** There are two linkage adjustments, one to the detent switch, the second between the selector lever and the manual valve in the transmission. A considerable variety of linkages and detent switches are found on cars using the Type 400 transmission. Therefore, always refer to the shop manual covering the car you are servicing when you check and adjust linkages.

Typical linkage adjustments are illustrated in Figs. 16-1 to 16-3. Figure 16-1 shows one detent-switch adjustment. Figure 16-2 shows a typical shift-linkage adjustment on a car with the selector lever on the steering column. Figure 16-3 shows the shift-cable adjustment on a console-equipped car. On these, the shift lever is on the floor-mounted console to the right of the driver.

Trouble Diagnosis

Although the transmission is fairly complex and packs a great number of parts in a small space, it is usually easy to spot causes of trouble. A specific trouble can have only certain specific causes. All the various car manufactures who install the Type 400 automatic transmission on their cars recommend that an oil-pressure check be made with the transmission in operation to help locate causes of trouble. The procedures recommended, as well as the pressures specified under different operating conditions, vary somewhat from manufacturer to manufacturer. Therefore, always check the shop manual that specifically covers the car and transmission you are working on before proceeding to trouble-diagnose a transmission. A typical procedure follows.

⊘ **16-6 Trouble Diagnosis** The procedure that follows is for any Type 400 automatic transmission.

1. Check and correct the oil level, as explained in ⊘ 16-2.
2. Check the detent switch and linkage.
3. Check the manual-valve linkage.
4. Check the vacuum lines and fittings for leaks.
5. Install an oil-pressure gauge and test the car on the road and in the shop. The oil-pressure gauge is connected to the transmission line-pressure tap. When testing the car, use all selector-lever positions and note transmission operation and oil pressures under different conditions.

CAUTION: When road-testing the car, obey all traffic laws. Drive safely. If a chassis dynamometer

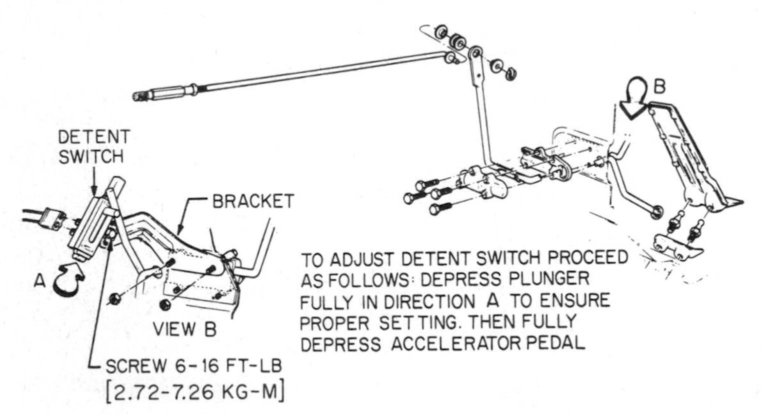

Fig. 16-1. Detent-switch adjustment. (*Buick Motor Division of General Motors Corporation*)

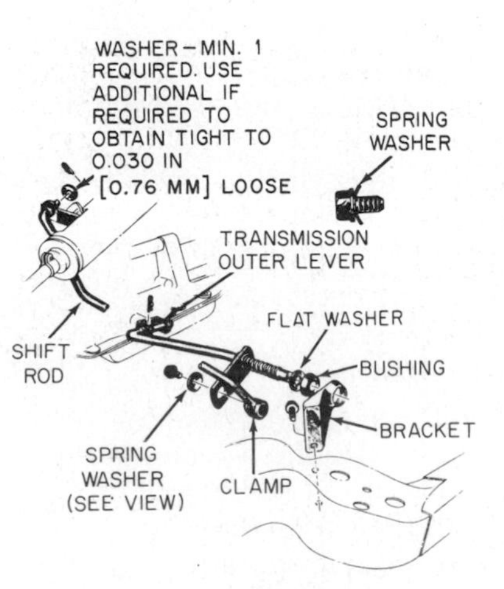

SHIFT-ROD ADJUSTMENT

1. SET TRANSMISSION OUTER LEVER IN DRIVE POSITION
2. HOLD UPPER SHIFT LEVER AGAINST DRIVE POSITION STOP IN UPPER STEERING COLUMN (DO NOT RAISE LEVER)
3. TIGHTEN SCREW IN CLAMP ON LOWER END OF SHIFT ROD TO SPECIFIED TORQUE
4. CHECK OPERATION:
   a. WITH KEY IN "RUN" POSITION AND TRANSMISSION IN REVERSE, BE SURE THAT KEY CANNOT BE REMOVED AND THAT STEERING WHEEL IS NOT LOCKED
   b. WITH KEY IN "LOCK" POSITION AND SHIFT LEVER IN PARK, BE SURE THAT KEY CAN BE REMOVED, THAT STEERING WHEEL IS LOCKED, AND THAT THE TRANSMISSION REMAINS IN PARK WHEN THE STEERING COLUMN IS LOCKED

Fig. 16-2. Shift-linkage adjustment. (*Oldsmobile Division of General Motors Corporation*)

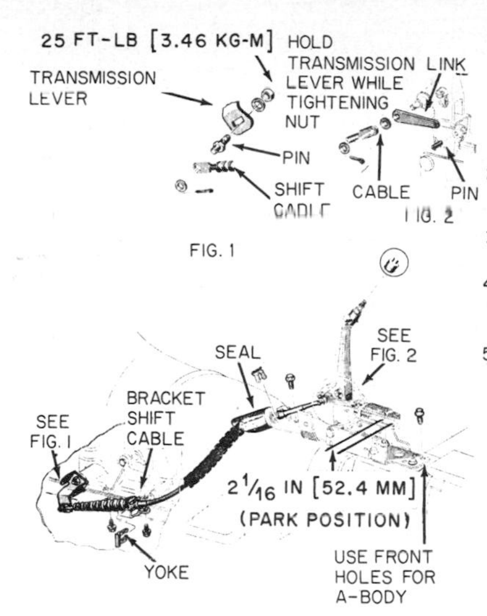

SHIFT-CABLE ADJUSTMENT

1. LOOSEN SHIFT-ROD CLAMP SCREW, LOOSEN PIN IN TRANSMISSION MANUAL LEVER
2. PLACE SHIFT LEVER IN P POSITION, PLACE TRANSMISSION MANUAL LEVER IN P POSITION AND IGNITION KEY IN LOCK POSITION
3. PULL SHIFT ROD LIGHTLY AGAINST LOCK STOP AND TIGHTEN CLAMP SCREW
4. MOVE PIN IN MANUAL TRANSMISSION LEVER TO GIVE "FREE PIN" FIT AND TIGHTEN ATTACHING NUT
5. CHECK OPERATION:
   a. MOVE SHIFT HANDLE INTO EACH GEAR POSITION AND SEE THAT TRANSMISSION MANUAL LEVER IS ALSO IN DETENT POSITION
   b. WITH KEY IN "RUN" POSITION AND TRANSMISSION IN REVERSE, BE SURE THAT KEY CANNOT BE REMOVED AND THAT STEERING WHEEL IS NOT LOCKED
   c. WITH KEY IN "LOCK" POSITION AND TRANSMISSION IN PARK, BE SURE THAT KEY CAN BE REMOVED AND THAT STEERING WHEEL IS LOCKED

Fig. 16-3. Shift-cable adjustment on a console-equipped car. (*Oldsmobile Division of General Motors Corporation*)

(⊘ 1-9) is available, use it to perform all the tests so that you don't have to put the car on the road.

NOTE: An engine that is not performing satisfactorily may affect transmission operation and produce such signs as faulty shifts, making the transmission look bad. The engine should be tuned up and given whatever other service is necessary to bring it up to normal operating condition before the transmission tests are performed.

Following are the specifications supplied by one car manufacturer for the oil pressures under different operating conditions:

a. In super range, second gear, at a steady road speed of 25 mph [40.23 km/h], the oil pressure should be between 145 and 155 psi [10.195 and 10.898 kg/cm^2].
b. In drive range, zero to full throttle, the oil pressure should vary between 60 and 150 psi [4.219 and 10.546 kg/cm^2].
c. In drive range, third gear, zero throttle at 30 mph [48.28 km/h], the oil pressure should be 60 psi [4.219 kg/cm^2].
d. In reverse, zero to full throttle, the oil pressure should be between 95 and 260 psi [6.679 and 18.280 kg/cm^2].

With the vacuum tube disconnected from the vacuum-modulator valve, the car stationary (parking brake on and wheels blocked), the engine running at 1,200 rpm, and at various altitudes, the oil pressures should be as follows.

FEET ABOVE SEA LEVEL	D, N, P	S OR L	R
0	150	150	244
2,000	150	150	233
4,000	145	150	222
6,000	138	150	212
8,000	132	150	203

CAUTION: Total running time for these tests must be not more than 2 minutes!

6. Refer to the trouble-diagnosis chart that follows to determine possible causes of troubles that show up during the tests. In most cases, the chart will help you track down the trouble cause quickly so that you can make necessary repairs. Some troubles may require removal of the transmission for complete overhaul. Other troubles, however, may require only replacement of some parts without the necessity of a complete teardown.

⊘ 16-7 Type 400 Transmission Trouble-Diagnosis Chart The chart that follows tabulates transmission complaints, possible causes, and checks or corrections to make.

TYPE 400 TRANSMISSION TROUBLE-DIAGNOSIS CHART

COMPLAINT	POSSIBLE CAUSE	CHECK OR CORRECTION
1. No drive in drive range	a. Low oil level	Leaks: porous casting, damaged seals, defective diaphragm in vacuum modulator
	b. Manual linkage maladjusted	Readjust
	c. Low oil pressure	Check filter O ring, filter for blockage, pump, pressure regulator, case porosity
	d. Forward clutch not applying	Piston cracked, seals damaged, plates burned, leak in feed circuit, clutch ball check stuck
	e. Roller clutch damaged	Check for broken spring or damaged cage

TYPE 400 TRANSMISSION TROUBLE-DIAGNOSIS CHART (*Continued*)

COMPLAINT	POSSIBLE CAUSE	CHECK OR CORRECTION
2. 1-2 shift at full throttle only	a. Detent switch sticking or defective	Replace
	b. Detent solenoid loose or sticking	Tighten, check gasket; replace if defective
	c. Valve body	Check for loose mount, defective gasket, stuck detent valve train, 3-2 valve sticking
	d. Case porosity	Repair, replace
3. First speed only; no upshift	a. Governor assembly	Governor valve sticking, drive gear loose
	b. Valve body	1-2 valve stuck, feed channel blocked, valve body leaking due to loose bolts, gasket defect
	c. Case	Intermediate-clutch plug leaking, porosity, governor feed channel blocked or damaged to allow cross feeding
	d. Intermediate clutch	Piston seals, oil rings, orifice plug damaged or missing
4. No 2-3 upshift in drive range	a. Detent solenoid stuck open	Replace
	b. Detent switch	Adjust or replace as required
	c. Valve body	2-3 shift train stuck, body gaskets leaking
	d. Direct clutch	Piston seals, oil rings damaged or missing, ball check stuck or missing
5. Drive in neutral	a. Manual linkage maladjusted	Readjust
	b. Forward clutch does not release	This also causes no reverse; piston stuck or feed line clogged
6. No drive in reverse or slips in reverse	a. Low oil level	Refill, check for leakage, damaged seals, etc.
	b. Manual linkage maladjusted	Readjust
	c. Low oil pressure	Vacuum modulator defective or valve sticking, pump assembly defective, pressure or boost valve sticking
	d. Valve body	Leakage due to bad or missing gaskets, low-and-reverse ball check missing, 2-3 valve stuck open, reverse feed passage clogged
	e. Rear servo	Servo-piston seal ring damaged or missing; short band apply pin
	f. Rear band	Burned or loose lining, apply pin or anchor pin not engaged, band broken
	g. Direct clutch	Outer seal damaged or missing, plates burned (can be caused by stuck ball check in piston)
	h. Forward clutch does not release	Will also cause drive in neutral; piston stuck or line clogged
7. Slips in all ranges and on start	a. Low oil level	Refill, check for leaks, damaged seals, etc.
	b. Low oil pressure	Vacuum modulator defective or valve stuck, pump assembly defective or gasket damaged, pressure regulator or boost valve stuck
	c. Clutches slipping	Plates burned due to low oil pressure, worn plates, damaged piston seals
8. Slips on 1-2 shift	a. Low oil level	Add oil, check for leaks, damaged seals, etc.
	b. Low oil pressure	Vacuum modulator defective or valve stuck, pump assembly defective or gasket damaged, pressure regulator or boost valve stuck
	c. Valve body	1-2 accumulator valve train sticking, porosity causing leakage, body loose or gasket missing
	d. Front accumulator	Oil ring damaged or missing
	e. Rear accumulator	Oil ring damaged or missing, case bore damaged
	f. Case	Intermediate-clutch plug leaking, leakage between circuits
	g. Intermediate clutch	Piston seals missing or damaged, plates burned; center-support oil rings leaking
9. Rough 1-2 shift	a. Oil pressure incorrect	Vacuum modulator defective or valve stuck, pump or gasket defective, regulator or boost valve stuck

TYPE 400 TRANSMISSION TROUBLE-DIAGNOSIS CHART *(Continued)*

COMPLAINT	POSSIBLE CAUSE	CHECK OR CORRECTION
9. Rough 1-2 shift (Cont.)	b. Valve body	1-2 accumulator valve train stuck, valve body loose or gaskets defective
	c. Case	Intermediate-clutch ball missing or not sealing, leakage between circuits
	d. Rear-servo accumulator	Oil rings damaged, piston stuck, broken spring, bore damaged
10. Slips on 2-3 shift	a. Oil level low	Check as for 8a above
	b. Oil pressure low	Check as for 8b above
	c. Valve body	Accumulator piston pin leaking at swedge end
	d. Case porosity	Repair or replace
	e. Direct clutch	Piston seals leaking, ball check leaking, center-support oil-seal rings damaged
11. Rough 2-3 shift	a. High oil pressure	Modulator assembly defective or valve sticking, pressure regulator or boost valve inoperative
	b. Front accumulator	Spring missing or piston stuck
12. No engine braking L2, second gear	a. Front servo and accumulator	Spring broken or piston stuck
	b. Front band	Broken, burned, not engaged on anchor pin or servo pin
13. No engine braking L1	a. Valve body	Low-and-reverse check ball missing
	b. Rear servo	Oil seal, bore, or piston damaged; apply pin short
	c. Rear band	Broken, burned, not engaged on pins
14. No part-throttle downshift	a. Oil pressure	Check as for item 8b above
	b. Valve body	3-2 valve stuck, spring broken
15. No detent downshifts	a. Valve body	3-2 valve stuck, spring broken, detent valve train stuck
	b. Detent switch	Out of adjustment, circuit open
	c. Detent solenoid	Inoperative, circuit open
16. Low or high shift points	a. Oil pressure incorrect	Engine trouble—check intake-manifold vacuum at transmission; vacuum modulator defective or valve stuck, pressure-regulator valve train
	b. Governor defective	Valve sticking, feed lines leaking or plugged
	c. Detent solenoid	Stuck open, loose, etc.; this can cause late shifts
	d. Valve body	Detent valve train, 3-2 valve, 1-2 shift valve train (if 1-2 regulator valve stuck, it would cause constant 1-2 shift point regardless of throttle opening) Leakage between channels or from valve body
	e. Porous case	Repair or replace; also, check for missing or leaking intermediate plug
17. Won't hold in park	a. Manual linkage maladjusted	Readjust
	b. Internal linkage	Parking-brake lever and actuator defective, parking pawl broken
18. Transmission noisy*	a. Pump noise	Oil high or low, plugged filter, O ring damaged, air or water in oil, pump gears damaged, crescent interference with gears
	b. Gear noises	Will occur in first or second; probable cause, damaged gears in planetary systems
	c. Clutch noise	Will occur during shifts; check which shift to pinpoint clutch; cause can be burned clutch plates
	d. Converter noise	Converter damage, bearings worn, loose converter attaching bolts

*Make sure the noise is not coming from water pump, alternator, air conditioner, power steering, etc., before looking into transmission. These components can be halted by removing the drive belt and running the engine. *Do not run engine more than 2 minutes without engine fanbelt to avoid engine overheating and damage!*

Check Your Progress

Progress Quiz 16-1 The ability to spot the causes of troubles is a mark of the automatic-transmission expert. If you know how the transmission works, and know what could cause the different troubles, you can easily solve problems that could keep the nonexpert guessing for hours. Find out how well you remember the Type 400 transmission troubles and causes by taking the following quiz.

Correcting Troubles Lists The purpose of this exercise is to help you spot related and unrelated troubles in a list. For example, in the list "No drive in drive range: low oil level, manual linkage maladjusted, forward clutch not applying, governor valve stuck," you can see that "governor valve stuck" does not apply because it is the only condition listed that could not cause the trouble.

In each of the following lists, you will find one item that does not belong. Write each list in your notebook, *but do not write* the item that does not belong.

1. 1-2 shift at full throttle only: detent switch sticking, detent solenoid not functioning, valve-body troubles, porous case, forward clutch not releasing.
2. First speed only, no upshift: intermediate clutch not applying, case porosity, forward clutch not applying, intermediate clutch not applying, 1-2 valve stuck, governor inoperative.
3. No 2-3 upshift in drive range: detent solenoid stuck open, detent switch malfunctioning, 2-3 shift train stuck, intermediate clutch not applying, direct clutch not applying.
4. Drive in neutral: manual linkage maladjusted, forward clutch not releasing, governor valve stuck.
5. No drive in reverse or slips in reverse: low oil pressure, manual linkage maladjusted, rear band not applying, front band not applying, direct clutch not applying, forward clutch not releasing.
6. Slips in all ranges and on start: low oil level, low oil pressure, clutches slipping, governor valve stuck.
7. Slips on 1-2 shift: low oil level, 1-2 accumulator valve train sticking, accumulators malfunctioning, direct clutch not releasing, intermediate clutch slipping.
8. Low or high shift points: oil pressure incorrect, governor defective, detent solenoid stuck or loose, trouble in the valve body, bands not applying.
9. No engine braking in L1: low-and-reverse check ball missing, rear servo malfunctioning, front band not applying, rear band not applying.
10. Rough 1-2 shift: oil pressure incorrect, 1-2 accumulator valve train stuck, intermediate-clutch ball missing, rear-servo accumulator malfunctioning, direct clutch not applying.

On-the-Car Repairs

⊘ 16-8 Repairs That Do Not Require Transmission Removal Some parts can be removed from the transmission and replaced without removing the transmission from the car. These parts are listed below. (Removal-and-replacement procedures are detailed in ⊘ 16-9 to 16-23, which cover the complete transmission overhaul.)

1. Parts that can be removed from the transmission on the car include:
   Output-shaft O ring
   Oil pan and gasket
   Extension-housing gasket and seal
   Governor
   Vacuum modulator and valve
   Rear seal
   Oil-cooler lines
   Speedometer drive and driven gears
2. Parts that can be removed after the oil pan is removed include:
   Filter assembly and O ring
   Manual detent lever, shaft, and seal
   Parking-lock actuator
   Pressure-regulator valve and boost valve
   Parking pawl and shaft and cup plug
   Electric connector
   Rear servo
   Detent solenoid
   Valve body and spacer plate
   Pressure-switch assembly
   Six check balls from valve body (see Fig. 16-4)
   Front servo

Transmission Overhaul

⊘ 16-9 Transmission Removal Individual variations in engine-compartment arrangements require somewhat different removal-and-replacement procedures on different cars. Always check the car shop manual for the car you are working on for the specific instructions that apply. The following is a typical removal-and-replacement procedure applying to late-model Chevrolets.

Fig. 16-4. Location of six check balls in the valve body. *(Chevrolet Motor Division of General Motors Corporation)*

Before raising the car, disconnect the negative battery cable and release the parking brake. Then, proceed as follows:

1. Place the car on a hoist and remove the drive shaft. Scribe marks on the shaft and flange so that the shaft can be reinstalled in its original position.
2. Disconnect the speedometer cable, electric leads, vacuum line, oil-cooler pipes, and shift-control linkage.
3. Support the transmission with a suitable transmission jack.
4. Remove the frame cross member as follows. Disconnect the rear transmission mount from the cross member. Then remove two bolts from each end of the cross member. On some models, such as Corvette, a through bolt at the inside of the frame and the parking-brake pulley must also be removed.
5. On some cars, such as late-model Oldsmobiles, an engine-support bar must be installed, as shown in Fig. 16-5.
6. Remove the converter underpan. Mark the flywheel and converter so that they can be reattached in their original position. Remove the attaching bolts.
7. Loosen the exhaust-pipe-to-manifold bolts about $\frac{1}{4}$ inch [6.35 mm].
8. Lower the jack and transmission until the jack is barely supporting the transmission. Then remove the transmission-to-engine mounting bolts and slide the transmission away from the engine.

NOTE: The oil-filler tube may require removal from the transmission if it interferes with transmission removal. Also, the catalytic converter and converter support bracket may have to be disconnected to provide clearance for transmission removal.

CAUTION: When lowering the jack, be careful not to allow the distributor or distributor cap to hit the firewall.

Fig. 16-5. Engine-support bar installed in preparation for removal of transmission. (*Oldsmobile Division of General Motors Corporation*)

9. Install a converter-holding tool so that the converter does not slip off.
10. With the transmission clear of the engine, lower the transmission from the car.
11. Replacement is essentially the reverse of removal except for the following additional step: Before installing the flex-plate-to-converter bolts, make sure the converter is properly aligned in its original position. Make sure the weld nuts on the converter are flush with the flex plate and that the converter rotates freely by hand in this position. Hand-start all bolts and tighten finger tight. Then torque to the specified tightness.
12. Add oil as necessary.
13. Adjust linkages as necessary.

⊘ 16-10 Transmission-Service Precautions Remember that cleanliness is of the utmost importance in transmission work. The tiniest bit of lint from a cleaning rag or a particle of dirt can cause a valve to hang up and prevent normal transmission action. Under some conditions, this could ruin the transmission. The outside of the transmission must be cleaned before it is opened. The workbench, the tools, your hands, and all parts must be kept clean at all times. Cover the transmission with a clean cloth if you have to leave it in a disassembled condition for a while.

For other cautions regarding transmission work, reread ⊘ 14-10.

⊘ 16-11 Cleaning and Inspecting Transmission Parts During and after the disassembly, clean and inspect all parts as covered in ⊘ 14-11. If the transmission shows signs of internal damage and circulation of foreign material in the oil, the oil filter should be discarded and a new one installed on reassembly. Furthermore, the oil-cooler lines and oil cooler should be flushed out to make sure all dirty oil is removed. During disassembly, all oil will, of course, be removed from the transmission. After reassembly and reinstallation on the car, the transmission should be filled with the correct type of transmission oil.

⊘ 16-12 Transmission Disassembly None of the parts requires forcing when disassembling or reassembling the transmission, except for the bushings. These are all press fits and must be pressed out if replacement is required. As a first step in disassembling the transmission, put it in a holding fixture, such as the one shown in Fig. 14-7. Then proceed as follows:

1. With the transmission turned so that the oil pan is up, remove the converter-holding tool and converter.
2. Remove the attaching bolt, retainer, and modulator assembly, as shown in Fig. 16-6. Remove the modulator valve from the case.
3. Remove the governor, as shown in Fig. 16-7, after

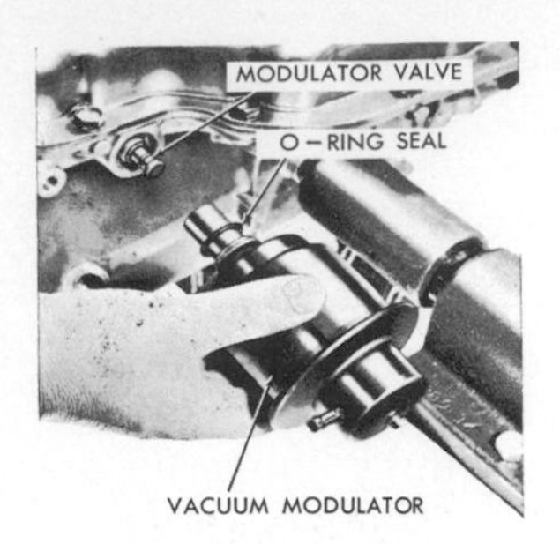

Fig. 16-6. Removing the vacuum-modulator assembly. (*Buick Motor Division of General Motors Corporation*)

Fig. 16-8. Removing the speedometer-driven-gear assembly from the case. (*Buick Motor Division of General Motors Corporation*)

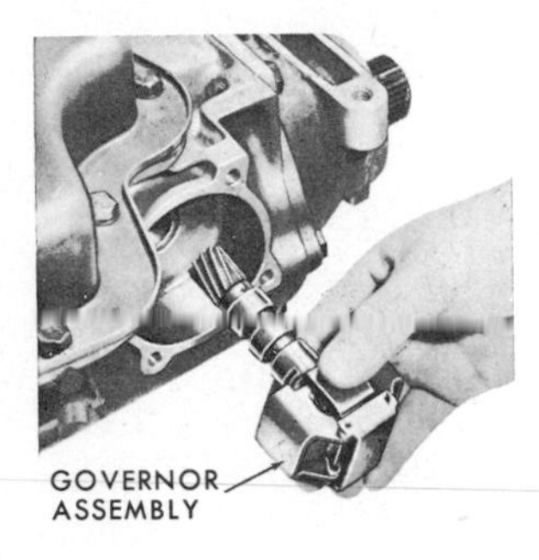

Fig. 16-7. Removing the governor assembly. (*Buick Motor Division of General Motors Corporation*)

removing attaching bolts, cover, and gasket. Next, remove the speedometer-driven-gear assembly from the case, as shown in Fig. 16-8, after the attaching bolt and retainer have been removed.

4. Remove the oil-pan attaching bolts, oil pan, and gasket. Discard the gasket.

5. Remove the filter-assembly retainer bolt and lift out the filter assembly with the intake pipe, as shown in Fig. 16-9.

6. Remove the valve-body attaching bolts and the detent-roller-and-spring assembly. Do not remove the solenoid attaching bolts at this time. Lift off the valve-body assembly with the governor pipes, as shown in Fig. 16-10. Detach the pipes from the valve body.

CAUTION: Do not let the manual valve fall out of the valve body.

7. Disconnect the detent-solenoid wire from the connector in the case.

8. Remove the valve-body-to-spacer gasket and discard it.

9. Remove the rear-servo-cover attaching bolts, servo cover, and gasket. Discard the gasket. Lift out the rear-servo assembly from the case, as shown in Fig. 16-11. Then remove the accumulator spring and check the band apply pin, as explained in step 10.

10. Attach the band-apply-pin selection gauge, as shown in Fig. 16-12, to the transmission case with attaching bolts. Torque to 6 to 10 pound-feet [0.829 to 1.382 kg-m]. Apply 25 pound-feet [3.455 kg-m] torque to the nut on the gauge and note the position of the steps on the gauge in relation to the gauge surface. If both steps are above the gauge surface, a short pin must be used on reassembly. If the gauge surface is between the steps, use the medium pin. If both steps are below the surface, use a long pin. The three lengths are identified by rings on them, as in the table in Fig. 16-12. Selecting the proper pin is equivalent to adjusting the band. If a new pin is required, make note of the pin size to be used on

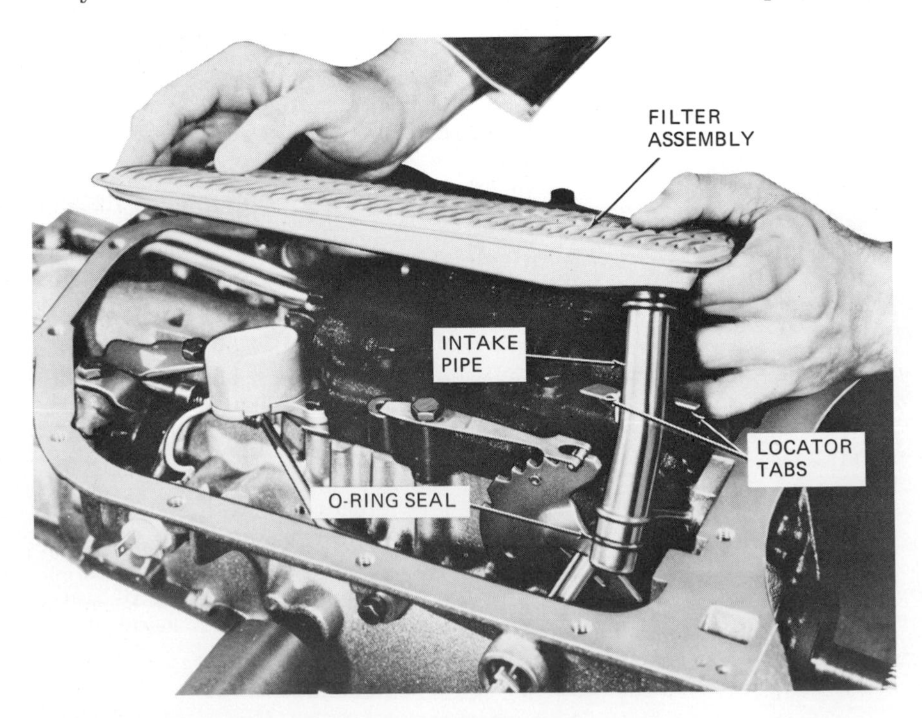

Fig. 16-9. Removing the filter assembly. (*Oldsmobile Division of General Motors Corporation*)

Fig. 16-10. Lifting off the valve-body assembly. This is also called the *control-valve assembly.* (*Cadillac Motor Car Division of General Motors Corporation*)

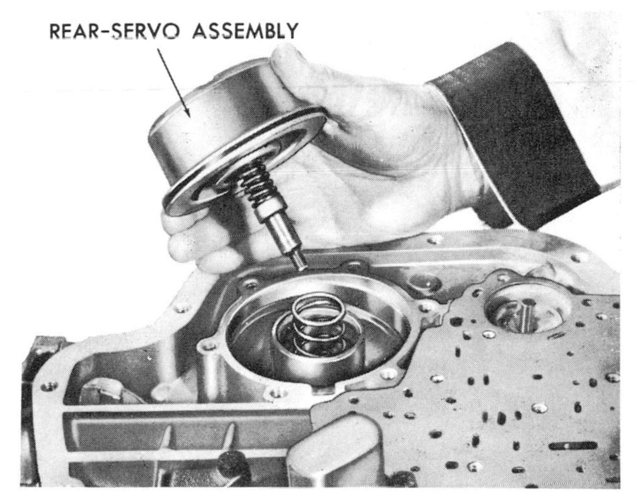

Fig. 16-11. Removing the rear servo from the case. (*Cadillac Motor Car Division of General Motors Corporation*)

IDENTIFICATION	PIN LENGTH
Three rings	Long
Two rings	Medium
One ring	Short

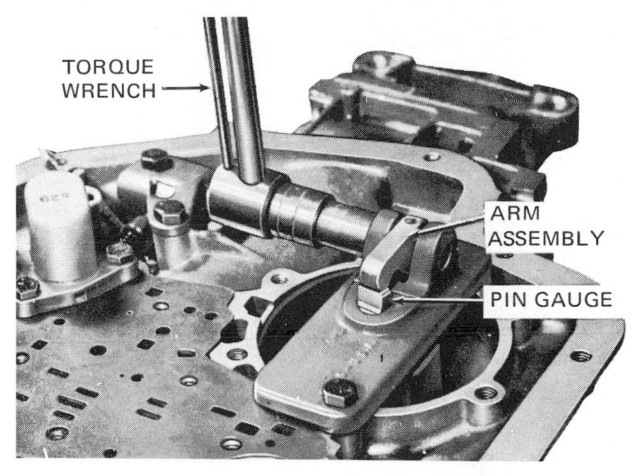

Fig. 16-12. *Top,* pin-length identification table; *bottom,* using special gauges and a torque wrench to determine which band apply pin to use on reassembly. (*Cadillac Motor Car Division of General Motors Corporation*)

reassembly and remove the gauge from the transmission case.

11. Remove the electric-connector sleeve from the case by compressing the fingers on the connector so that it can be pulled out of the case.

12. Remove the detent-solenoid attaching bolts, solenoid, and gasket.

13. Remove the valve-body-assembly spacer and gasket. Remove the six check balls from the cored passages in the transmission case. See Fig. 16-4.

14. Remove the front-servo piston, washer, pin, retainer, and spring from the transmission case, as shown in Fig. 16-13.

15. If necessary to remove the internal manual linkage, refer to Fig. 16-14. First, unthread the jam nut holding the detent lever to the manual shaft. Remove the manual-shaft retaining pin from the case. Remove the detent lever from the manual shaft. Then remove the manual shaft and jam nut from the case.

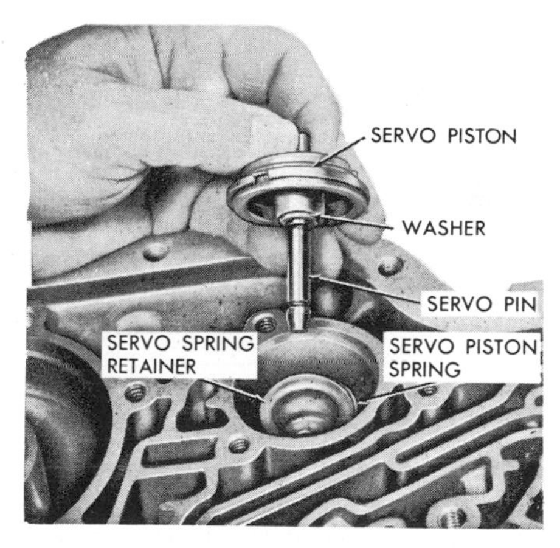

Fig. 16-13. Removing the front-servo piston and related parts from the transmission case. (*Buick Motor Division of General Motors Corporation*)

If the manual-shaft lip seal is damaged, remove it from the case. Next, remove the parking-pawl-actuator-rod-and-detent-lever assembly. Remove the attaching bolts and the parking-lock bracket. Then remove the parking-pawl return spring, shaft retainer, and cup plug. The plug can be pried out with a screwdriver inserted between the parking-pawl shaft and the rib in the case. Now, remove the parking-pawl shaft and parking pawl.

16. Examine the extension-housing rear oil seal, and if it needs replacement, pry it out. Then remove the extension housing and gasket by removing attaching bolts, as shown in Fig. 16-15.

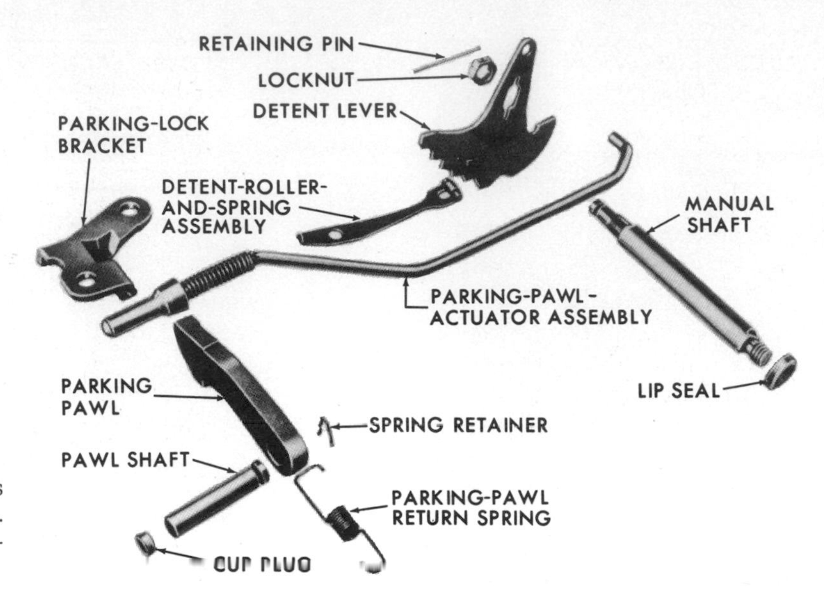

Fig. 16-14. Internal linkages found in the transmission. (*Buick Motor Division of General Motors Corporation*)

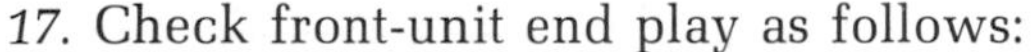

Fig. 16-15. Removing the extension-housing attaching bolts. (*Buick Motor Division of General Motors Corporation*)

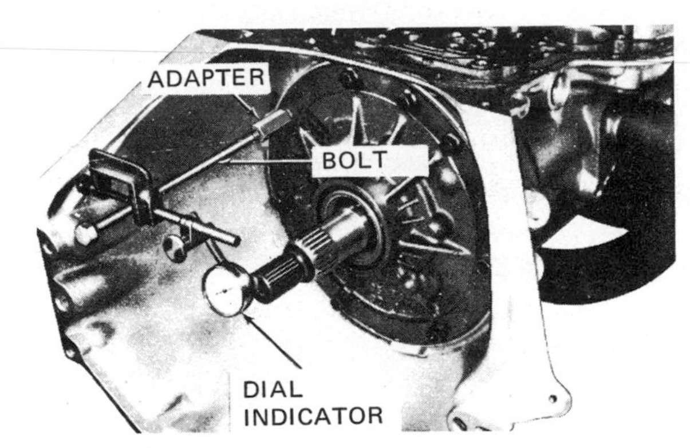

Fig. 16-16. Dial indicator mounted in readiness to check front unit end play. (*Cadillac Motor Car Division of General Motors Corporation*)

17. Check front-unit end play as follows:

a. Remove one front-pump attaching bolt and sealing washer. Install a slide-hammer bolt into the bolt hole and mount a dial indicator on the bolt, as shown in Fig. 16-16. Pull the output shaft forward while pushing the turbine shaft in as far as it will go.

b. Set the dial indicator to zero and pull the turbine shaft forward as far as it will go. Note the indicator reading, which should be 0.003 to 0.024 inch [0.076 to 0.610 mm].

c. This end play is controlled by a selective washer located between the pump cover and the forward-clutch housing. If the end play is not correct, a new selective washer will be required on reassembly. The various thicknesses of washers available and their identifying colors are listed in Fig. 16-17.

d. Note the amount the end play is off, and when you remove the old selective washer from the transmission, identify it by color or mike it to determine its thickness. You will then know what thickness the new washer must have to correct the end play. Suppose, for example, that the end play was 0.032 inch [0.813 mm], which is 0.008 inch [0.203 mm] too much. Suppose, then, that when you measure the old selective washer you find it is 0.071 inch [1.803 mm]. You would use the next larger size, which measures 0.082 to 0.086 inch [2.083 to 2.184 mm] in thickness. This would bring the end play down to between 0.017 and 0.021 inch [0.432 and 0.533 mm]. You see, the new

THICKNESS, in	COLOR
0.060–0.064	Yellow
0.071–0.075	Blue
0.082–0.086	Red
0.093–0.097	Brown
0.104–0.108	Green
0.115–0.119	Black
0.126–0.130	Purple

Fig. 16-17. Color identifications of selective washers of various thicknesses.

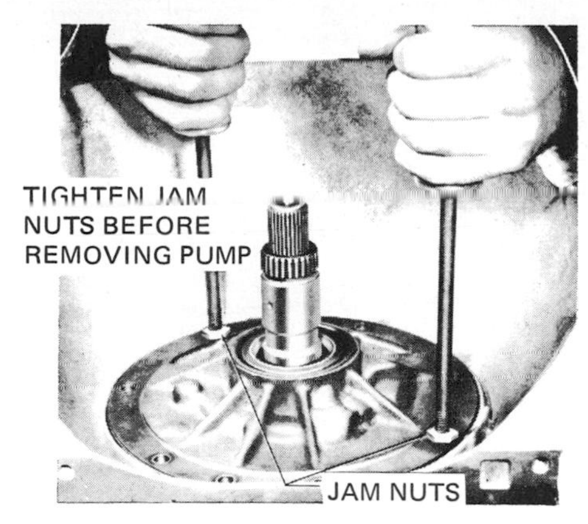

Fig. 16-18. Removing the pump with slide hammers. (*Buick Motor Division of General Motors Corporation*)

Fig. 16-19. Removing the forward-clutch-and-turbine-shaft assembly. (*Buick Motor Division of General Motors Corporation*)

washer is 0.011 to 0.015 inch [0.279 to 0.381 mm] thicker than the old one. These dimensions subtract from the old end play to give the new end play, as mentioned above.

18. If the pump-to-shaft seal requires replacement, pry the seal from the pump.

19. Next, remove the oil pump, using a pair of slide hammers, as shown in Fig. 16-18. All attaching bolts must be removed, and the two slide hammers should be installed in the bolt holes opposite each other.

20. Remove and discard the pump-to-case seal ring and gasket.

21. Remove the forward-clutch-and-turbine-shaft assembly, as shown in Fig. 16-19. If the thrust washer did not come out with the assembly, remove it from the case.

22. Remove the direct-clutch assembly, as shown in Fig. 16-20, followed by the front band assembly and the sun-gear shaft.

23. Now, with the rear end of the transmission up, mount the dial indicator on the slide-hammer bolt, as shown in Fig. 16-21. Move the output shaft up and down to determine the end play. It should be between 0.007 and 0.019 inch [0.177 and 0.483 mm]. The selective washer controlling this end play is the steel

Fig. 16-20. Removing the direct-clutch assembly. (*Buick Motor Division of General Motors Corporation*)

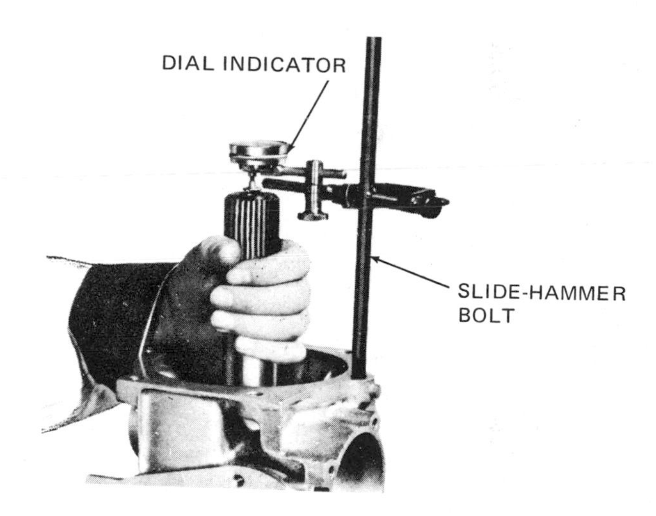

Fig. 16-21. Checking rear end play with a dial indicator. (*Chevrolet Motor Division of General Motors Corporation*)

THICKNESS, in	IDENTIFICATION NOTCH AND NUMERAL	
0.074–0.078	None	1
0.082–0.086	On side of tab 1	2
0.090–0.094	On side of tab 2	3
0.098–0.102	On end of tab 1	4
0.106–0.110	On end of tab 2	5
0.114–0.118	On end of tab 3	6

Fig. 16-22. Identification of selective washers to adjust rear end play.

washer having three lugs that is located between the thrust washer and the rear face of the transmission case. If a different washer thickness is required to bring the end play within specifications, select it from the table in Fig. 16-22.

24. Remove the center-support-to-case bolt with a $\frac{3}{8}$-inch [9.525 mm] 12-point thin-wall deep socket, as shown in Fig. 16-23.

Fig. 16-23. Removing the center-support bolt. (*Chevrolet Motor Division of General Motors Corporation*)

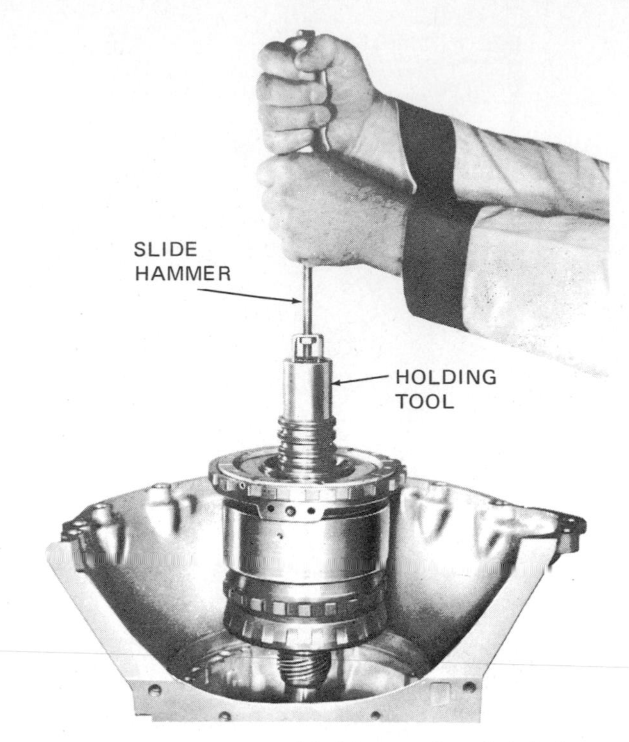

Fig. 16-25. Using a special holding tool and slide hammer to remove the center-support-and-gear assembly. (*Chevrolet Motor Division of General Motors Corporation*)

25. Remove the snap ring that retains the intermediate-clutch backing plate and clutch plates, as shown in Fig. 16-24. Remove the clutch backing plate and clutch plates. Examine the plates for defects, as noted in ⊘ 14-11.
26. Remove the snap ring that retains the center support in the case so that the complete gear assembly can be removed from the case. Use the special holding tool J-21795 and a slide hammer, as shown in Fig. 16-25. Remove the thrust washer from the rear of the output shaft or from the transmission case.
27. Put the gear assembly on the workbench, output shaft facing down through a hole in the workbench. Or mount it in a special holding fixture, output shaft down.
28. Now remove from the transmission case the three-tanged selective washer, support-to-case spacer, and rear-band assembly. Figure 16-26 shows the proper position of the support-to-case spacer.

⊘ 16-13 Gear-Assembly Service We shall now describe the service procedures for the subassemblies removed from the transmission case, starting with the gear assembly.

1. Remove the center-support assembly by lifting straight up. Then remove the center-support-to-reaction-carrier thrust washer, as shown in Fig. 16-27.
2. Remove the center-support-to-sun-gear races and thrust bearing, as shown in Fig. 16-28.

Fig. 16-24. Removing the intermediate-clutch-plate snap ring. (*Buick Motor Division of General Motors Corporation*)

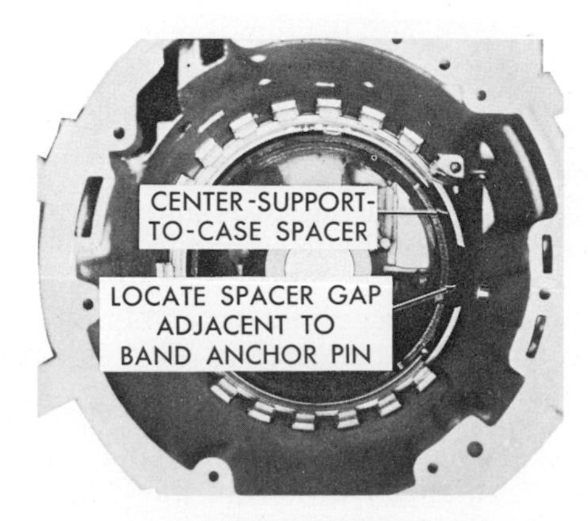

Fig. 16-26. Proper position of the center-support-to-case spacer. (*Oldsmobile Division of General Motors Corporation*)

Fig. 16-27. Removing the center-support-to-reaction-carrier thrust washer. (*Chevrolet Motor Division of General Motors Corporation*)

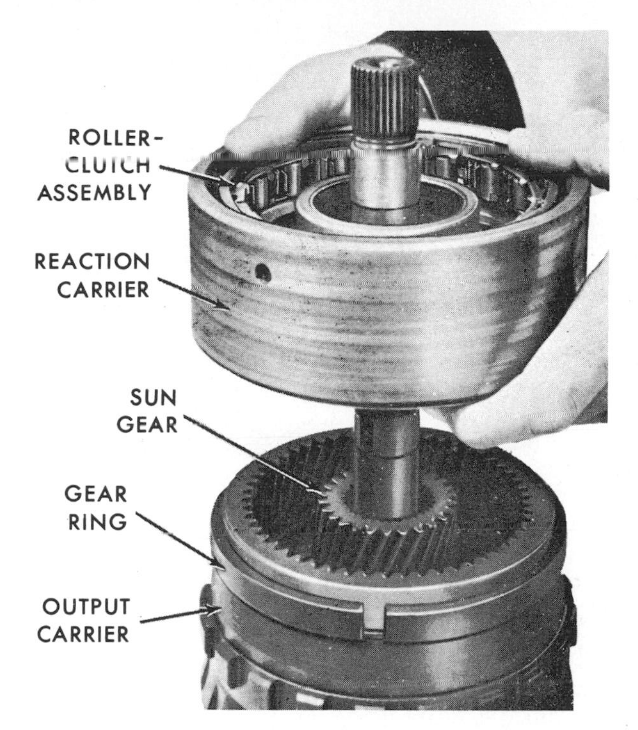

Fig. 16-29. Lifting off the roller-clutch assembly and reaction carrier. (*Cadillac Motor Car Division of General Motors Corporation*)

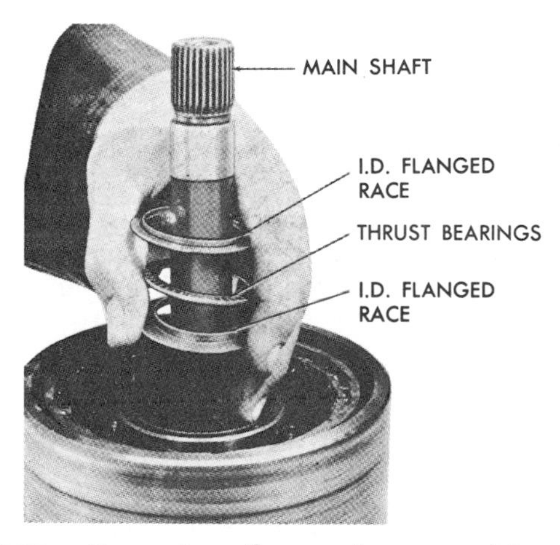

Fig. 16-28. Removing the center-support-to-sun-gear thrust bearing and races. (*Oldsmobile Division of General Motors Corporation*)

Fig. 16-30. Using a gear-puller to remove a speedometer steel drive gear. (*Chevrolet Motor Division of General Motors Corporation*)

3. Lift off the roller-clutch assembly and reaction carrier, as shown in Fig. 16-29. Remove the front-internal-gear ring, shown in Fig. 16-29, from the output-carrier assembly. Lift out the sun gear and the tanged thrust washer.
4. Turn the assembly over so that the output shaft is up. Remove the output shaft by first taking out the snap ring and lifting up on the output shaft.
5. If the speedometer drive gear needs replacement, note whether it is nylon or steel. If it is nylon, remove it by depressing the retaining clip so that the gear can be slid off the shaft. If it is steel, it must be pulled with a special puller, as shown in Fig. 16-30. Then press the new steel drive gear down into position, as shown in Fig. 16-31. The distance to press the gear down on the shaft varies with different cars. Refer to the car shop manual that applies to the car you are working on.
6. Remove the thrust bearing and two races from the top of the rear internal gear. This thrust bearing is

Fig. 16-31. Installing a new speedometer steel drive gear. (*Chevrolet Motor Division of General Motors Corporation*)

located between the rear internal gear and the output shaft in the transmission assembly.

7. You can now remove the rear internal gear and main shaft, as shown in Fig. 16-32.

8. Remove the thrust bearing and two races from inside the internal gear. To separate the internal gear from the main shaft, remove the snap ring.

9. If planet pinions or bearings require replacement, refer to Fig. 16-33. The removal must be done carefully in order not to damage or weaken the carrier. First, remove the stake marks with a ½-inch [12.7 mm] drill. Then, press out the pinion pin with a punch. On reassembly, hold the needle bearings and washers in place with petroleum jelly. Stake the new pin in three places.

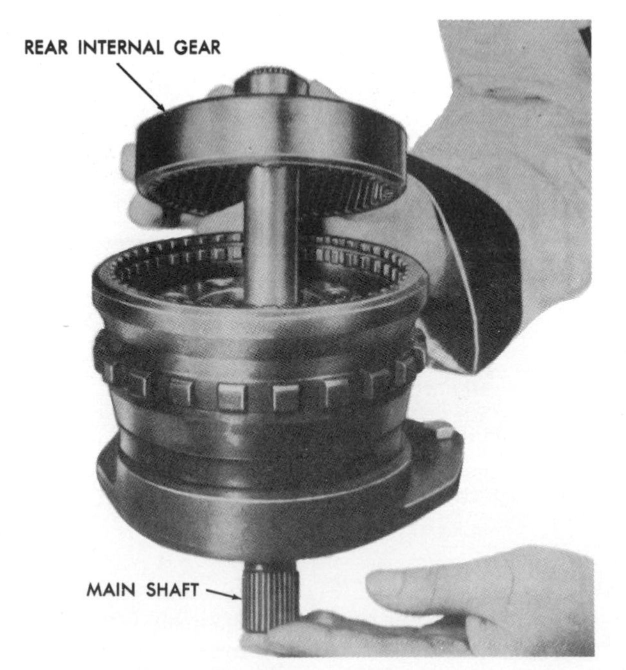

Fig. 16-32. Removing the rear internal gear and main shaft. (*Chevrolet Motor Division of General Motors Corporation*)

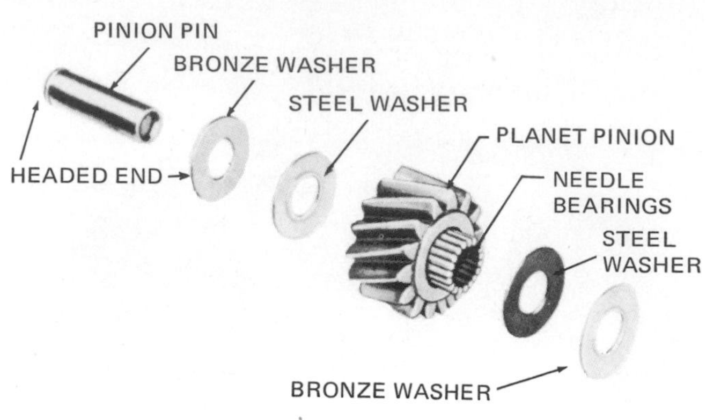

Fig. 16-33. Planet pinion and related parts. (*Buick Motor Division of General Motors Corporation*)

10. If the bushing in the output shaft is worn or galled, use special tools and a slide hammer to remove it and a special installing tool to press the new bushing into place.

11. To start the reassembly, install the rear internal gear on the end of the main shaft having the snap-ring groove. Then install the internal-gear thrust races and thrust bearing against the inner face of the rear internal gear, retaining them in place with petroleum jelly, as follows:

a. Put the large race against the internal gear with the flange facing forward or up.

b. Put the thrust bearing against the race.

c. Put the small race against the bearing with the inner flange facing into the gearing or down.

12. Install the output carrier over the main shaft so that the pinions mesh with the rear internal gear.

13. Now put the partial assembly through a hole in the bench so that the main shaft hangs down.

14. Next install the rear-internal-gear-to-output-shaft thrust races and thrust bearing, retaining them with petroleum jelly. Put the small race on first, center flange facing up, followed by the thrust bearing and larger race with outer flange cupped over the bearing.

15. Install the output shaft into the output-carrier assembly and secure it with the output-carrier snap ring.

16. Lubricate the output-shaft-to-case thrust washer with petroleum jelly and then install it.

17. Lubricate the reaction-carrier-to-output-carrier thrust washer with petroleum jelly, and install it with tabs facing down into pockets.

18. Install the sun gear, inner chamfer down. Install the ring gear over the output carrier. Then install the sun-gear shaft.

19. Install the reaction carrier, as shown in Fig. 16-29. Lubricate the center-support-to-sun-gear thrust races and thrust bearing with petroleum jelly, and install them as shown in Fig. 16-28. The large race, center flange up, goes over the sun-gear shaft first, followed by the thrust bearing and the second race, center flange up.

20. Lubricate the center-support-to-reaction-carrier thrust washer with petroleum jelly, and install it into the recess in the center support, as shown in Fig. 16-27.

21. If the rollers have been removed from the cage, reinstall them by compressing the energizing spring with a forefinger and inserting the roller from the outside. Do not distort the springs, and make sure the curved ends of leaves of springs are positioned against the rollers.

22. Install the roller-clutch assembly into the reaction carrier in the position shown in Fig. 16-29. Install the center support into the roller clutch in the reaction carrier.

NOTE: Check the roller clutch by holding the reaction carrier and turning the center support. It should turn counterclockwise only.

⊘ 16-14 Governor Service The governor used in the Type 400 Turbo Hydra-Matic transmission is similar to the one used in the Type 350 Turbo Hydra-Matic transmission. Therefore, the service procedure for the Type 350 governor described in ⊘ 14-18 also applies to the Type 400 governor.

⊘ 16-15 Valve-Body Service The valve body is shown in disassembled view in Fig. 16-34. There is nothing especially difficult about disassembling the body. However, extreme care must be used to keep everything absolutely clean and to avoid damaging any operating surfaces. The slightest piece of lint or dirt can cause a valve to hold up, which could cause the transmission to malfunction and possibly be severely damaged. The same thing could happen if a valve were dropped or dented so that it tended to hang up in the bore in the valve body.

When disassembling the valve body, make sure that all the springs are properly identified so that they can be put back into the bores from which they were removed. You will need a special tool to compress the accumulator piston, as shown in Fig. 16-35, so that the E-ring retainer can be removed to permit removal of the piston and its components, as shown in Fig. 16-36.

Clean all parts in cleaning solvent and air-dry them. Make sure all oil passages in the valve body are clear. Follow Fig. 16-34 carefully when reinstalling valves and springs.

⊘ 16-16 Rear-Servo and Accumulator Service Figure 16-36 shows the disassembled rear servo and accumulator. Disassembly begins by removing the retaining E ring. If the check made in ⊘ 16-12, item *10,* indicates that a pin of a different length is required, substitute the new pin when reassembling the servo. Make sure oil-seal rings are in good condition. If they are not, install new ones.

⊘ 16-17 Pump Service The pump cover is shown in disassembled view in Fig. 16-37. The boost valve and pressure-regulator valve are removed by remov-

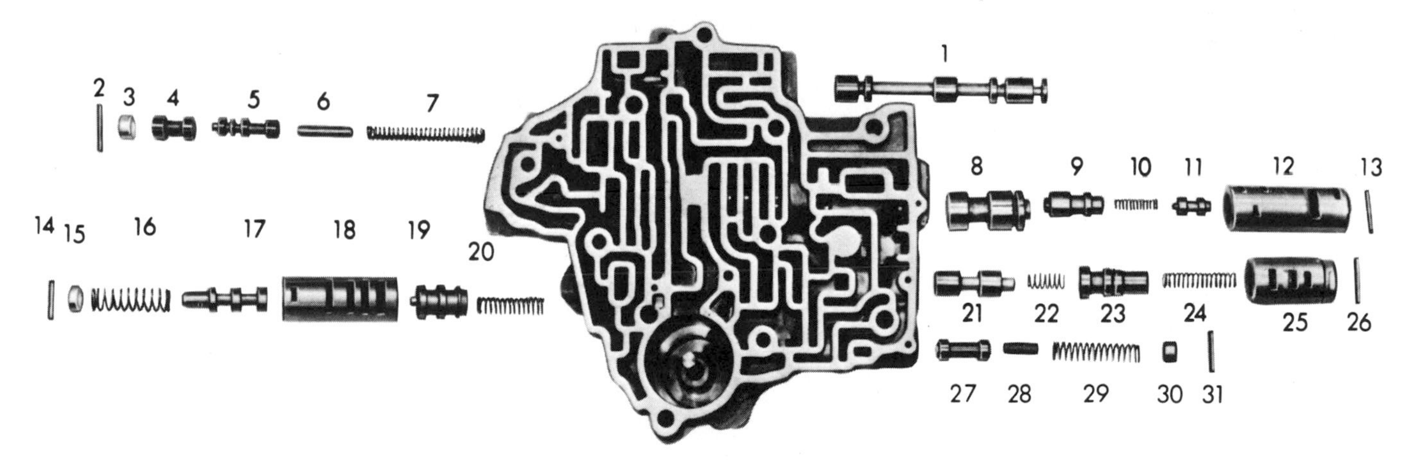

1. Manual Valve
2. Retaining pin
3. Bore plug
4. Detent valve
5. Detent regulator
6. Spacer
7. Detent spring
8. 1-2 valve
9. 1-2 detent valve
10. 1-2 regulator spring
11. 1-2 regulator valve
12. 1-2 bushing
13. Retaining pin
14. Retaining pin
15. Bore Plug
16. 1-2 accumulator secondary spring
17. 1-2 accumulator valve
18. 1-2 accumulator bushing
19. 1-2 accumulator primary valve
20. 1-2 accumulator primary spring
21. 2-3 valve
22. 3-2 intermediate spring
23. 2-3 modulator valve
24. 2-3 valve spring
25. 2-3 bushing
26. Retaining pin
27. 3-2 valve
28. Spacer
29. 3-2 spring
30. Bore plug
31. Retaining pin

Fig. 16-34. Disassembled view of a valve body. (*Cadillac Motor Car Division of General Motors*)

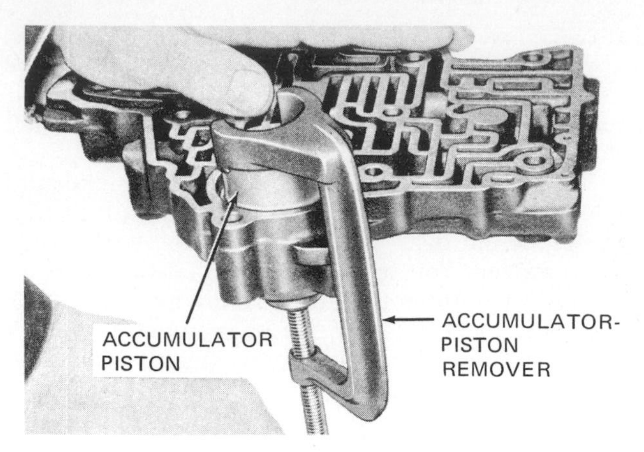

Fig. 16-35. Special accumulator piston remover in place to compress the piston so that the E ring can be removed. (*Chevrolet Motor Division of General Motors Corporation*)

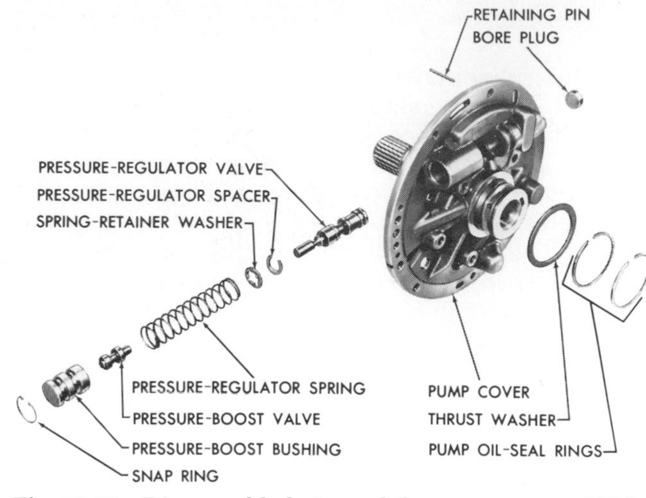

Fig. 16-37. Disassembled view of the pump cover. (*Oldsmobile Division of General Motors Corporation*)

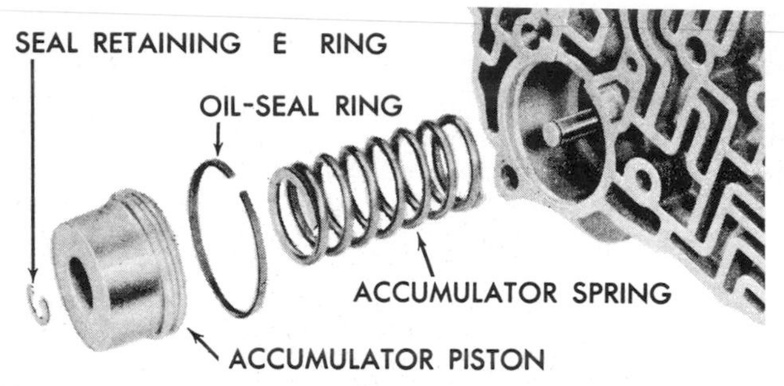

Fig. 16-36. Front-accumulator piston and related parts. (*Chevrolet Motor Division of General Motors Corporation*)

Fig. 16-38. Checking gear clearance with a feeler gauge and straightedge. (*Oldsmobile Division of General Motors Corporation*)

ing the snap ring with special snap-ring pliers. Then the pump cover is removed from the pump body by taking out the attaching bolts. Remove the hook-type oil rings from the pump cover.

Next, mark the drive and driven gears so that they can be replaced in the same relative positions. Remove them for inspection. Check the gear pocket and crescent for scoring, galling, or other damage that would require replacement of the pump body. Install the gears and check end clearance, as shown in Fig. 16-38. The clearance should be between 0.0008 and 0.0035 inch [0.0203 and 0.0889 mm].

On reassembly, be sure to install the drive gear with the tangs up, as shown in Fig. 16-38. Follow Fig. 16-37 when installing the valves and valve parts. To install the snap ring, use a small screwdriver between a spring coil and the bore of the boost valve to keep the spring compressed.

⊘ 16-18 Forward-Clutch Service To disassemble the forward clutch, refer to Fig. 16-39, which shows partly disassembled forward and direct clutches. Place the forward clutch, with the turbine shaft in a hole, on the work bench. Remove the large direct-clutch snap ring so that the direct-clutch hub can be lifted off the assembly, as shown in Fig. 16-40.

Next, take out the forward-clutch hub, thrust washers, clutch plates, and waved steel plate, all shown in Fig. 16-39.

Now, the forward clutch can be disassembled. If the turbine shaft must be removed from the forward clutch, press it out in an arbor press, as shown in Fig. 16-41.

Use a spring compressor, as shown in Fig. 16-42, to compress the clutch springs. This permits removal of the snap ring, as shown. Now, remove the tool, spring retainer, and 16 clutch release springs. Remove the clutch piston and take off the piston inner and outer seals. Remove the center piston seal from the forward-clutch housing. Use new seals on reassembly.

CAUTION: The forward and direct pistons have identical inside and outside diameters. It is possible to reverse the two during reassembly; so study Fig. 16-43 to make sure you identify them correctly. The piston with the check ball is the direct-clutch piston.

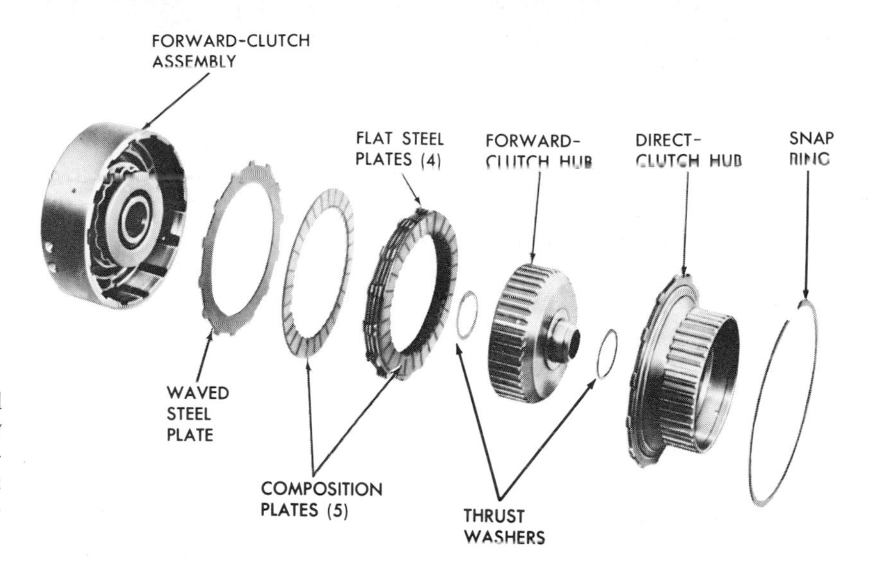

Fig. 16-39. Forward and direct clutches in partly disassembled view. (*Cadillac Motor Car Division of General Motors Corporation*)

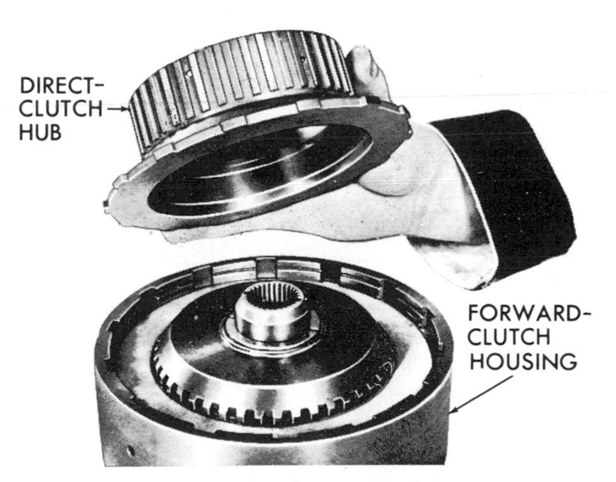

Fig. 16-40. Removing the direct-clutch hub. (*Oldsmobile Division of General Motors Corporation*)

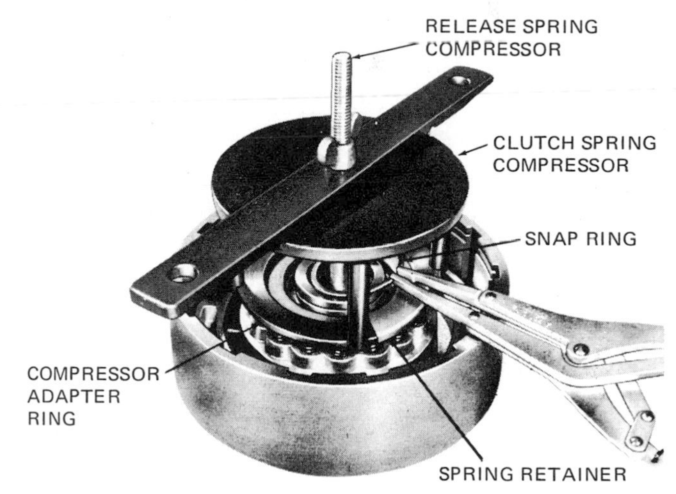

Fig. 16-42. Using a spring compressor so that the clutch snap ring can be removed. (*Pontiac Motor Division of General Motors Corporation*)

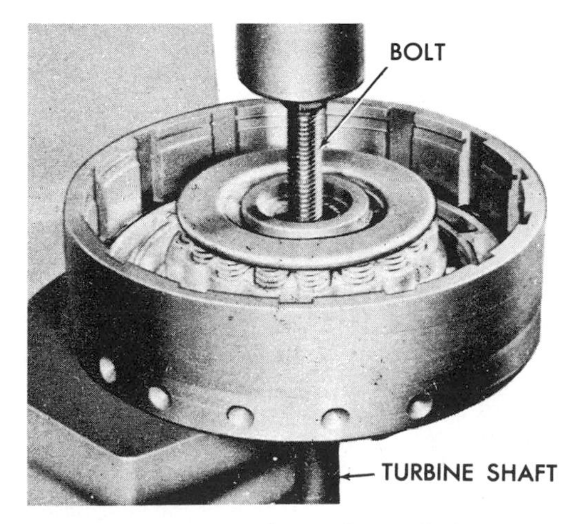

Fig. 16-41. Pressing out the turbine shaft, using a bolt to press on the shaft. (*Chevrolet Motor Division of General Motors Corporation*)

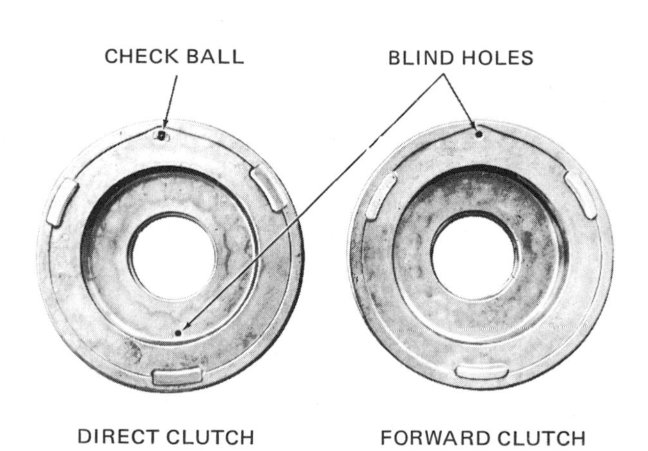

Fig. 16-43. Identification of clutch pistons. (*Pontiac Motor Division of General Motors Corporation*)

To reassemble the forward clutch, install a new center seal in the housing, lip face up. Then put seal protectors over the clutch hub and around the inner face of the clutch housing, as shown in Fig. 16-44, and install the piston.

Put 16 release springs into the pockets in the piston. Put the spring retainer and snap ring on springs and use the compressor tool, as shown in Fig. 16-42, to compress springs so that the snap ring can be installed.

If the turbine shaft was removed, press it into place in the forward-clutch drum. Put the thrust washers on the forward-clutch hub, retaining them in place with petroleum jelly. Then put the hub into the clutch housing. Oil and install the waved steel plate and clutch plates. See Fig. 16-39 for the lineup. The composition and flat steel plates alternate. Install the direct-clutch hub and retaining snap ring. Put the forward-clutch housing on the pump delivery sleeve and air-check the clutch operation, as shown in Fig. 16-45.

⊘ 16-19 Direct-Clutch and Intermediate-Sprag-Clutch Service Figure 16-46 is an exploded view of the direct-clutch-and-intermediate-sprag-clutch assembly. Remove the sprag assembly by removing the snap ring, retainer, and outer race along with the sprag bushings. Then turn the direct clutch over and remove the large snap ring so that the backing plate and clutch plates can be taken out, as shown in Fig. 16-47.

NOTE: Different models of the Type 400 transmission use different numbers of clutch plates.

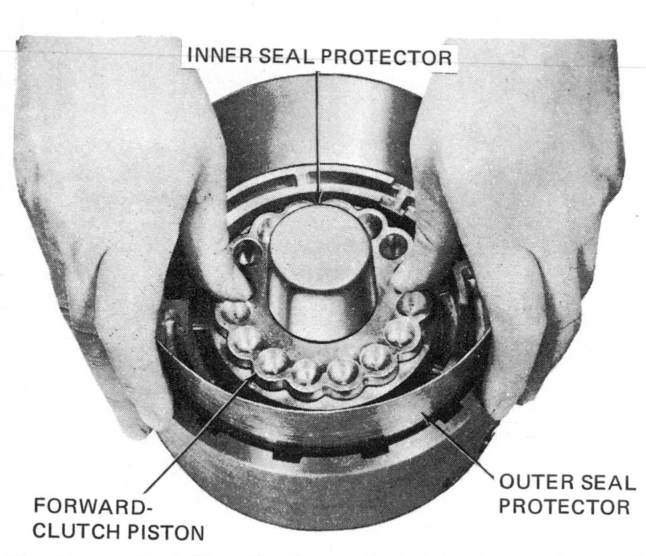

Fig. 16-44. Installing the forward-clutch piston after seal protectors have been temporarily put in place. (*Pontiac Motor Division of General Motors Corporation*)

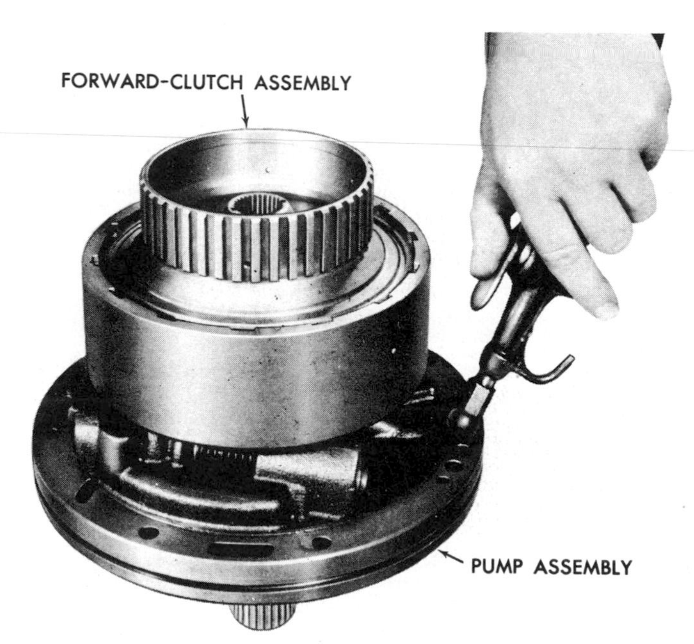

Fig. 16-45. Air-checking the forward clutch. (*Chevrolet Motor Division of General Motors Corporation*)

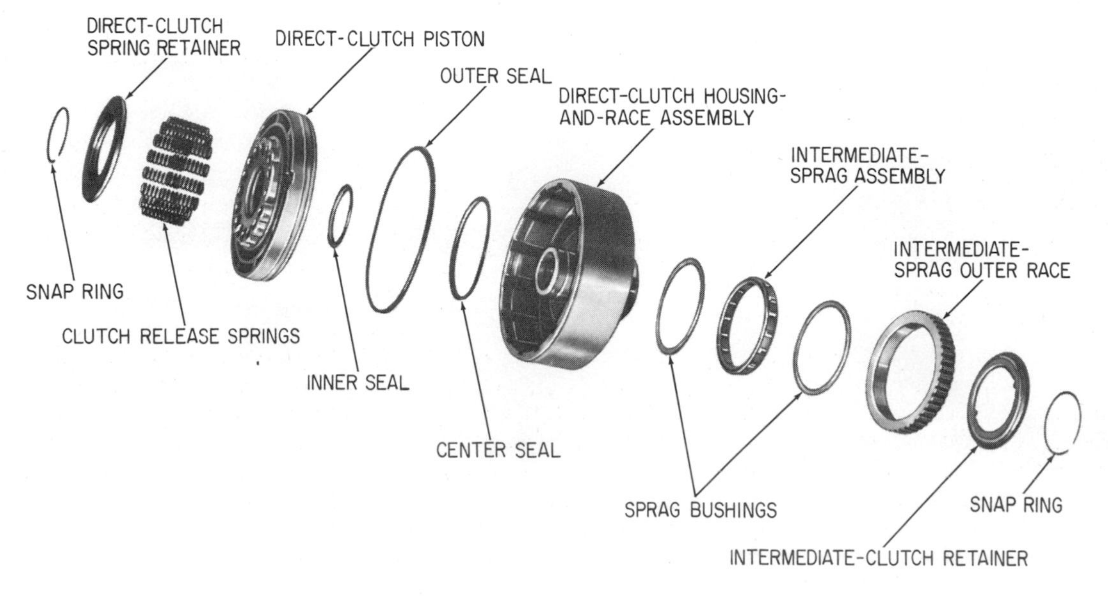

Fig. 16-46. Exploded view of the direct-and-intermediate-sprag-clutch assembly. (*Chevrolet Motor Division of General Motors Corporation*)

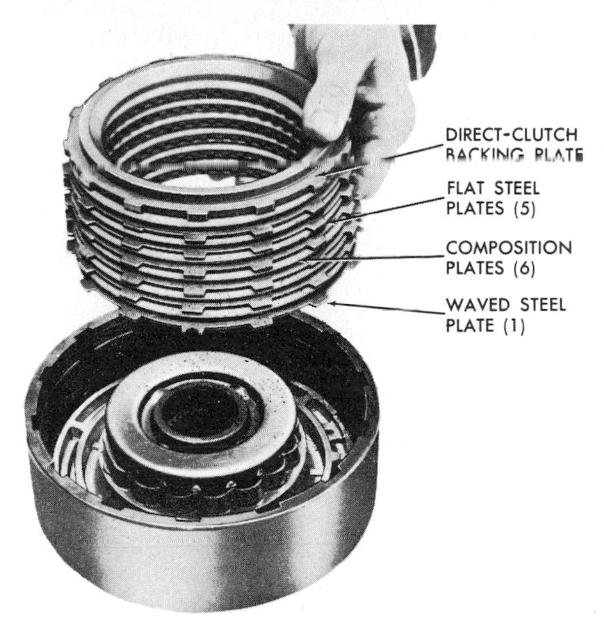

Fig. 16-47. Removing the backing plate and clutch plates from the direct clutch. (*Cadillac Motor Car Division of General Motors Corporation*)

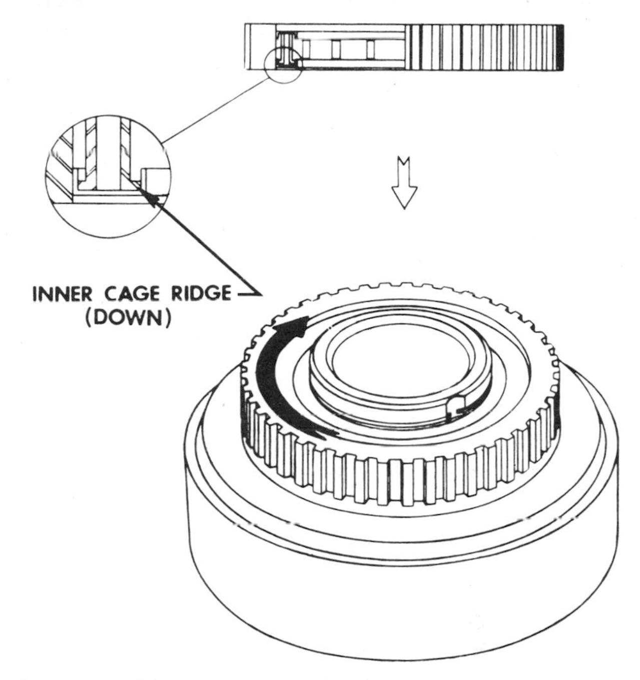

Fig. 16-48. The sprag assembly must be installed so that it rotates in a clockwise direction. (*Cadillac Motor Car Division of General Motors Corporation*)

Next, use a spring compressor similar to the one used on the forward clutch in Fig. 16-42, to compress the clutch springs so that the snap ring, 16 springs, and piston can be removed.

CAUTION: Do not mix the forward-clutch and direct-clutch springs!

Remove the three piston seals; two are the inner and outer seals on the piston itself, and the third is the center seal in the direct-clutch housing. Discard the seals.

On reassembly, make sure the correct piston is used. See Fig. 16-43 for identification. Install the three new seals, two on the piston and one in the direct-clutch housing. Use inner and outer seal protectors similar to those used for the forward-clutch piston in Fig. 16-44. Then install the direct-clutch piston. Install the 16 springs, spring retainer, and snap ring. Use a spring compressor to compress the springs so that the snap ring can be installed. Lubricate the clutch plates and install them, alternating steel and composition plates, followed by the direct-clutch backing plate. Then install the backing-plate snap ring.

Turn the unit over and install one sprag bushing, cup side up, over the inner race. Put the sprag assembly into the outer race. The sprag assembly must be installed one way and one way only so that it will rotate clockwise, as shown in Fig. 16-48. The ridge on the inner cage must face down when bringing the sprag assembly, with the outer race, down over the inner race. Put the other sprag bushing over the sprag, cup side down. Secure with the sprag retainer, cup side down, followed by the snap ring.

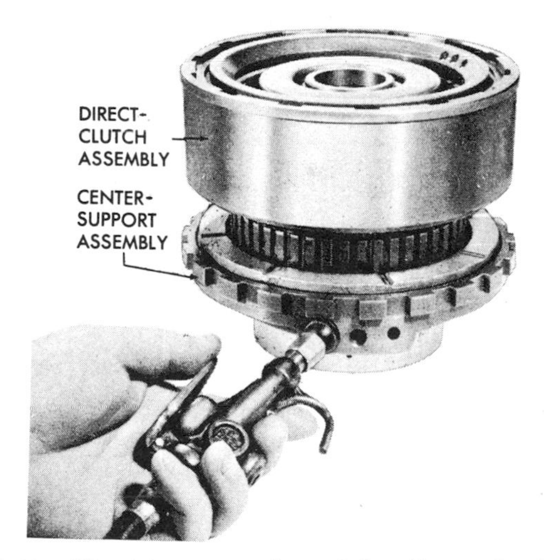

Fig. 16-49. Checking operation of the direct clutch with air. (*Chevrolet Motor Division of General Motors Corporation*)

Put the direct-clutch assembly over the center support, as shown in Fig. 16-49, and check the clutch operation with compressed air, as shown.

⊘ 16-20 Center-Support and Intermediate-Clutch Service Figure 16-50 shows the center support and intermediate clutch in disassembled view. To disassemble, remove the four hook-type oil-seal rings from the center-support assembly and then use the intermediate-clutch compressor to compress the springs so that the snap ring can be removed. Now, the spring retainer, three springs, and clutch piston

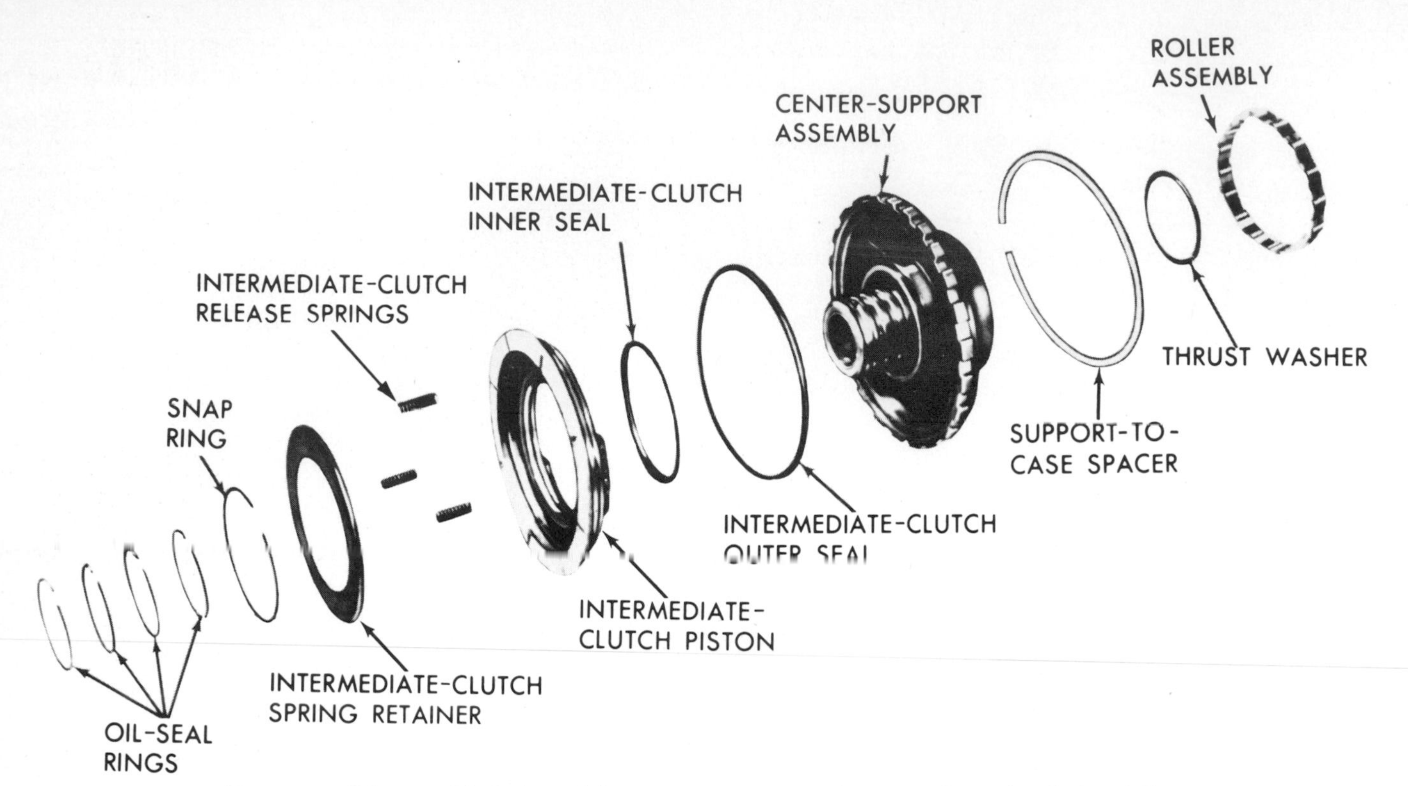

Fig. 16-50. Disassembled view of the center support and intermediate clutch. (*Cadillac Motor Car Division of General Motors Corporation*)

can be removed. Remove the inner and outer piston seals. Do not remove the three bolts attaching the roller-clutch inner race to the center support.

NOTE: On some models, there are 12 intermediate-clutch springs.

If the bushing in the center support is worn or galled, use special tools to remove the old bushing and install a new one. Make sure the elongated hole in the bushing aligns with the drilled hole in the oil-delivery sleeve closest to the piston. Drive the bushing to flush to 0.010 inch [0.254 mm] below the top of the oil delivery sleeve.

To reassemble, install new seals on the piston and then use an inner seal protector to install the piston on the center-support hub. Put the release springs in place on the piston, followed by the spring retainer and snap ring. Then use the spring compressor to compress the springs so that the snap ring can be installed to hold the retainer in place.

Install the four hook-type oil rings. Air-check the operation of the intermediate clutch by applying air to the center hole in the center support. Then install the center support into the roller clutch in the reaction carrier and check the roller-clutch action, as explained at the end of ⊘ 16-13.

⊘ 16-21 Torque-Converter Inspection Two checks of the torque converter should be made for leakage and for end play. The leakage test is made by installing the special tool, as shown in Fig. 14-106, and filling the converter with air at 80 psi [5.624 kg/cm^2]. Then, submerge the converter in water and check for air leaks. The end clearance is checked by mounting a dial indicator, as shown in Fig. 14-107, and tightening the hex nut. Then, the dial indicator should be set to zero and the nut loosened. This will allow the shaft to drop so that the end clearance will register on the dial indicator. If the end clearance is over 0.050 inch [1.27 mm], the converter is defective and must be replaced.

⊘ 16-22 Vacuum-Modulator Check The vacuum modulator is checked as explained in ⊘ 14-21. The modulators for the Type 350 and the Type 400 transmissions are similar.

⊘ 16-23 Transmission Reassembly If the parking pawl, shaft, and retainer have been removed, replace them. See Fig. 16-14. The cup should be driven in with a $\frac{3}{8}$-inch [9.525 mm] rod until the shaft bottoms on the case rib. Then install the parking-lock bracket, guides, and two attaching bolts torqued to 15 to 20 pound-feet [2.07 to 2.76 kg-m].

1. Install the center-support-to-case spacer, as pictured in Fig. 16-26. Do not confuse this spacer, which is 0.040 inch [1.016 mm] thick and flat on both sides, with the center-support-to-case snap ring, which is beveled on one side, or the backing-plate-to-case snap ring, which is flat on both sides and 0.093 inch [23.622 mm] thick.
2. Install the rear-band assembly so that the two lugs

index with the two anchor pins. Make sure the ends of the bands are seated on the lugs.
3. Lubricate and install the selective washer in the slots provided inside the rear of the transmission case.
4. Install the complete gear assembly, as shown in Fig. 16-25. Make certain the center-support bolt aligns with the hole in the case. Make sure the tangs on the output-shaft-to-case thrust washer are positioned in the pockets.
5. Install the center-support-to-case retaining snap ring with bevel side up. Locate the gap in the ring adjacent to the band anchor pin. Make sure the ring is seated in the case.
6. Install the case-to-center-support bolt by first using the locating tool, inserted into the direct-clutch passage, as shown in Fig. 16-51. Apply pressure downward to force the center support counterclockwise viewed from the front of the case. While holding the center support firmly against the case splines, torque the bolt to 20 to 25 pound-feet [2.764 to 3.455 kg-m] with a ⅜-inch [9.525 mm] 12-point thin-wall deep socket.

CAUTION: Do not raise burrs on the valve-body mounting face of the case.

7. Lubricate with transmission oil and install the intermediate-clutch plates, alternating the steel and composition plates, starting with the waved steel and composition plate. Then install the intermediate-clutch backing plate, ridge up. Secure with a snap ring. The gap in the snap ring should be opposite the band anchor pin.
8. Recheck the rear end play as described in ⊘ 16-12, item 23.
9. Install the front band with the band anchor hole over the band anchor pin and apply lug, facing the servo hole, as shown in Fig. 16-52.
10. Install the direct-clutch-and-sprag assembly, as shown in Fig. 16-20. It will be necessary to twist the housing so that the sprag outer race indexes with the composition clutch plates.
11. Install the forward-clutch-hub-to-direct-clutch-

Fig. 16-51. Using a special tool to properly locate the center support. (*Buick Motor Division of General Motors Corporation*)

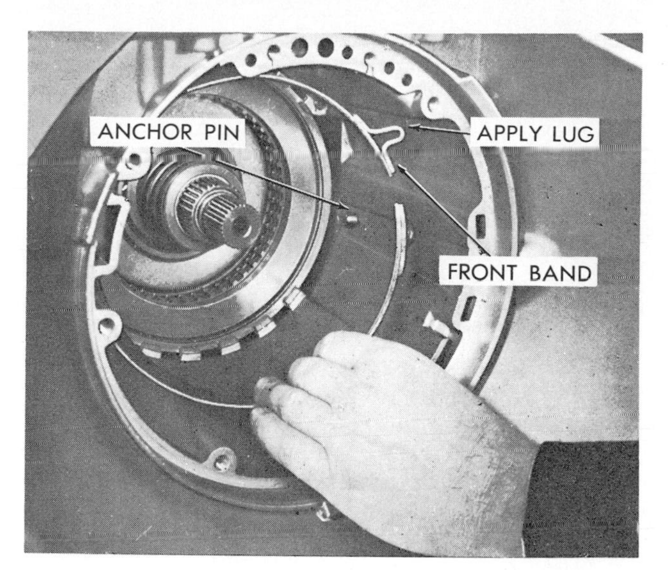

Fig. 16-52. Proper position for installing the front band. (*Cadillac Motor Car Division of General Motors Corporation*)

housing thrust washer, retaining it on the clutch hub with petroleum jelly.
12. Install the forward-clutch-and-turbine-shaft assembly, as shown in Fig. 16-19. Index the direct-clutch hub so that the end of the main shaft will bottom on the end of the forward-clutch hub. When seated, the forward clutch will be approximately 1¼ inch [31.75 mm] from the pump face in the case.
13. Install the guide pins in the transmission case and slide the pump gasket and pump down into place, installing *all but one* attaching bolt and seal. Remove the guide pins during bolt installation. Torque to 15 to 20 pound-feet [2.073 to 2.764 kg-m].

CAUTION: Always use new seals on the bolts.

NOTE: If the turbine shaft cannot be rotated as the pump is being pulled down into place, the forward- or direct-clutch housing has not been properly installed to index with all the clutch plates. This condition must be corrected before the pump is pulled down into place or something will be broken.

14. Install a new front seal if necessary, using a special tool to drive it into place.
15. Recheck front-unit end play, as explained in ⊘ 16-12, item 17.
16. Remove the dial indicator and install the last pump-attaching bolt.
17. Install the rear-extension housing, using a new gasket and torquing the bolts to 20 to 25 pound-feet [2.764 to 3.455 kg-m]. Install a new seal, if necessary, using special tools.
18. Install the remainder of the manual linkage, referring to Fig. 16-14 for the proper relationship.
19. Install the front servo and parts, as shown in Fig. 16-13. The pin and washer should be installed so that the tapered end is contacting the band. Install a new

seal on the piston, if necessary, and put the piston on the pin with identification numbers facing the bottom pan.
20. Check the freeness of the piston by moving it back and forth in the bore.
21. Install six check balls in the transmission-case pockets, as shown in Fig. 16-4.
22. Install the valve-body-spacer-plate-to-case gasket. This gasket has an extension for the solenoid.
23. Install the spacer plate, solenoid gasket, and solenoid, leaving solenoid attaching bolts loose.
24. Install the O-ring seal on the solenoid-case connector. Install the connector with lock tabs facing into the case, positioning the locator tab in the notch on the side of the case. Connect the detent wire and lead wire to the case connector. Check by pulling on the solenoid connector wire. If it pulls out of the connector, turn the wire clip over and reinsert it into the connector.
25. Install the rear-servo assembly, as shown in Fig. 16-11. Put the gasket and servo cover in place and secure with attaching bolts tightened to 15 to 20 pound-feet [2.073 to 2.764 kg-m].
26. Put the valve-body-to-spacer-plate in place and then install two guide pins, as shown in Fig. 16-53. Install the gasket and put the valve body into position. Governor pipes should be installed on the valve body before the valve body is put into position. Make sure the pipes fit properly into the transmission case and that the manual valve is properly indexed with the pin on the detent lever.
27. Install attaching bolts and the detent-roller-and-spring assembly. Make sure the lead-wire clip is installed. Center the roller on the detent lever. Remove the two guide pins and install the attaching bolts. Tighten the valve body and the solenoid attaching bolts to 6 to 10 pound-feet [0.829 to 1.382 kg-m]. Connect the lead wire to the pressure switch.
28. Install a new oil filter and O ring on the intake pipe and install the assembly, as shown in Fig. 16-9. Use a new gasket and install the oil pan, torquing

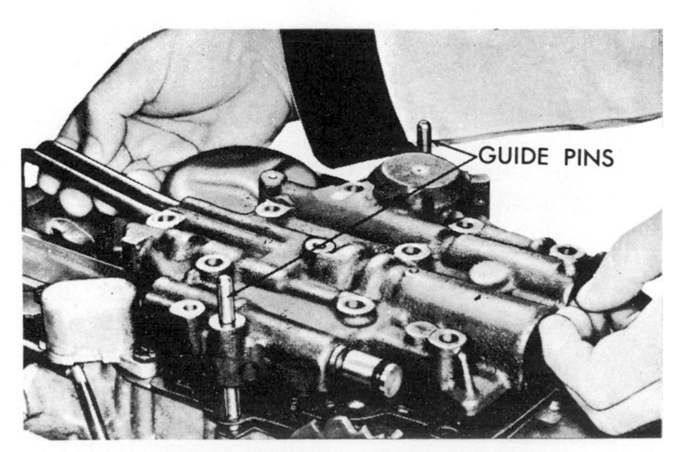

Fig. 16-53. Guide pins in place to guide the valve body to the proper position for installation. *(Oldsmobile Division of General Motors Corporation)*

the bolts to 10 to 13 pound-feet [1.382 to 1.796 kg-m].
29. Install the modulator valve in the case, stem end out. With a new O ring on the vacuum modulator, install the assembly into the case, as shown in Fig. 16-6. Secure with retainer and attaching bolt. Torque to 15 to 20 pound-feet [2.073 to 2.764 kg-m].
30. Install the governor, as shown in Fig. 16-7, and attach the cover with a new gasket and four attaching bolts. Torque to 15 to 20 pound-feet [2.07 to 2.76 kg-m].
31. Install the speedometer-driven-gear assembly, as shown in Fig. 16-8, and secure with retainer and attaching bolt.
32. Install the torque-converter assembly on the oil pump, making sure you do not damage the seal. Make certain the converter-hub drive tangs are fully engaged with the pump drive-gear tang. Install the converter-holding tool to keep the converter from falling off when it is installed on the car.
33. Install the transmission and test it, as explained in ⊘ 14-9 and 14-21.

CHAPTER 16 CHECKUP

NOTE: Since the following is a chapter review test, you should review the chapter before taking it.

By now, having studied several automatic transmissions, you should begin to see that there are many similarities insofar as their operation, component parts, and service procedures are concerned. There are still individual variations that could fool you if you didn't check the shop manual that applies. Yet you do know what troubles low oil level or pressure or slipping clutches can cause. The checkup that follows concentrates on the Type 400 transmission.

Completing the Sentences The sentences that follow are incomplete. After each sentence there are several words or phrases, only one of which will correctly complete the sentence. Write each sentence in your notebook, selecting the proper word or phrase to complete it correctly.

1. The 400 transmission should have the oil changed every: (*a*) 12,000 miles [19,312 km], (*b*) 24,000 miles [38,624 km], (*c*) year.
2. Only two in-car adjustments are required on the 400 transmission: (*a*) detent switch and selector-lever linkage, (*b*) brake band and selector-lever linkage, (*c*) selector-lever and throttle-valve linkage.
3. When making the oil-pressure checks on the 400 transmission, you actually check: (*a*) pump pressure, (*b*) TV (throttle-valve) pressure, (*c*) line pressure.
4. The rear band can be adjusted by: (*a*) turning a screw, (*b*) changing the band apply pin, (*c*) changing a spring.
5. Comparing the direct-clutch and the forward-clutch piston, you will find that the direct-clutch

piston has: (*a*) an extra seal ring, (*b*) a check ball, (*c*) a larger outside diameter.

Troubles and Service Procedures In the following, you should write in your notebook the required trouble causes and service procedures. Do not copy from the book; try to write them in your own words. Writing the trouble causes and procedures in your own words will help you to remember them; you will thereby be greatly benefited when you go into the automotive shop.

1. Describe the procedure of adding oil; of changing oil.
2. Explain how to make the two adjustments called for on the 400 transmission in a car.
3. Explain how to make the pressure check of the 400 transmission.
4. List possible causes if there is no drive in drive range.
5. List possible causes if the 1-2 shift will occur only at full throttle.
6. List possible causes if there is no upshift from first.
7. List possible causes if there is no 2-3 upshift in drive range.
8. List possible causes if there is drive in neutral.
9. List possible causes if there is no drive in reverse or if transmission slips in reverse.
10. List possible causes if the transmission slips in all ranges.
11. List possible causes if the transmission slips on the 1-2 shift.
12. List possible causes if the 1-2 shift is rough.
13. List possible causes if the transmission slips on the 2-3 shift.
14. List possible causes of a rough 2-3 shift.
15. List possible causes of no engine braking in L1; in L2.
16. List possible causes if there is no part-throttle downshift.
17. List possible causes if there is no detent downshift.
18. List possible causes of low or high shift points.
19. List possible causes of a noisy transmission.
20. List the service precautions to take when working on transmissions.
21. Make a list of the procedures required to clean and inspect transmission parts.
22. Make a list of the transmission parts in the order in which they are removed from the Type 400 transmission.
23. Refer to a shop manual of a car using the Type 400 transmission and prepare a transmission-removal story outlining the step-by-step procedure of taking the transmission out of the car.

SUGGESTIONS FOR FURTHER STUDY

If you can be in a shop while a 400 transmission is being removed from a car and overhauled, you will find it a fascinating experience. You will learn a great deal. It is always a pleasure to watch an expert doing a job, and someday you will give other newcomers pleasure as they watch you work.

To become an expert in the field of automatic-transmission work, you need to study and learn how transmissions are constructed, how they work, and how to trouble-diagnose and overhaul them. You are well on your way to becoming an expert. Study shop manuals. Watch the service procedures. Sooner or later you will be doing the job yourself. Write everything of importance you learn in your notebook so that you will be able to refer to it quickly.

chapter 17

CHRYSLER TORQUEFLITE AUTOMATIC TRANSMISSION

In past chapters, we have looked at a variety of automatic transmissions, including the two-speed Powerglide and the three-speed Types 350 and 400 Turbo Hydra-Matic transmissions. Now, we shall look at another three-speed automatic transmission—the TorqueFlite—which has only five active control members in the gearing system: the front clutch, the rear clutch, two brake bands, and an overrunning clutch. The TorqueFlite transmission, with minor variations, has been supplied for the full line of Chrysler Corporation cars, including Chrysler, Imperial, Dodge, and Plymouth.

⊘ 17-1 TorqueFlite Gearsets Figure 17-1 is a sectional view of a late-model TorqueFlite automatic transmission. It combines a torque converter and a fully automatic three-speed transmission. The transmission contains two multiple-disk clutches, an overrunning clutch, two brake bands operated by servos, and a compound planetary gearset. In addition, it has a hydraulic system containing several valves which provide control of the multiple-disk clutches and brake-band servos.

Figure 17-2 shows the essential operating parts of the planetary gearsets, clutches, bands, and overrunning clutch. This planetary gearset contains one long sun gear, two sets of planetary pinions in their carriers, and two internal, or ring, gears. The long sun gear meshes with both sets of planetary pinions. The sun gear is connected to the front clutch by a driving shell which is splined to the sun gear. This driving shell extends around and in front of the front clutch. Study Figs. 17-1 and 17-2 to establish this relationship in your mind.

⊘ 17-2 TorqueFlite Gearshifts Figure 17-3 shows, in chart form, the various patterns of clutch engagement and band application to obtain the various gear ratios through the gearsets. Figures 17-2 to 17-8 show these arrangements. Figure 17-2, for example, shows the power flow through the transmission with the selector lever in D (drive) when first breaking away from a standing start. The power flows from the input shaft through splines to the hub on which the inner drum of the front clutch and the outer drum of the rear clutch are mounted. These parts all turn as a unit at all times. When the front clutch is disengaged and the rear clutch is engaged, as shown in Fig. 17-2, then the ring gear of the front gearset must turn also. This movement drives the planet pinions of the front gearset, and they, in turn, drive the sun gear. Note that there is gear reduction through the front gearset.

As the sun gear turns, it drives the planet pinions of the rear gearset, and they, in turn, drive the ring gear. The planet-pinion carrier is mounted on an overrunning clutch, and this clutch holds the carrier stationary. The arrangement produces an additional gear reduction so that the total gear reduction through the two gearsets is 2.45:1. In other words, the input shaft must turn 2.45 times to cause the output shaft to turn once.

Now note the situation when the transmission upshifts to second, as shown in Fig. 17-4. The front, or kickdown, band and the rear clutch are applied. Applying the kickdown band locks the sun gear in a stationary position. Now, the power flow is through the rear clutch, planet pinions, and carrier to the output shaft. Gear reduction is achieved in the front gearset only. The total gear reduction is 1.45:1.

In direct drive, shown in Fig. 17-5, both front and rear clutches are engaged, locking the front gearset together (sun gear to ring gear). Thus, the assembly turns as a unit to give direct drive.

Figure 17-6 shows the situation with the selector lever in second. The transmission is held in the downshifted position. Figure 17-7 shows the conditions when the selector lever is in L (low). They are similar to the conditions shown in Fig. 17-2, which is low with the selector lever in D when first starting out. However, with the selector lever in L, the low-

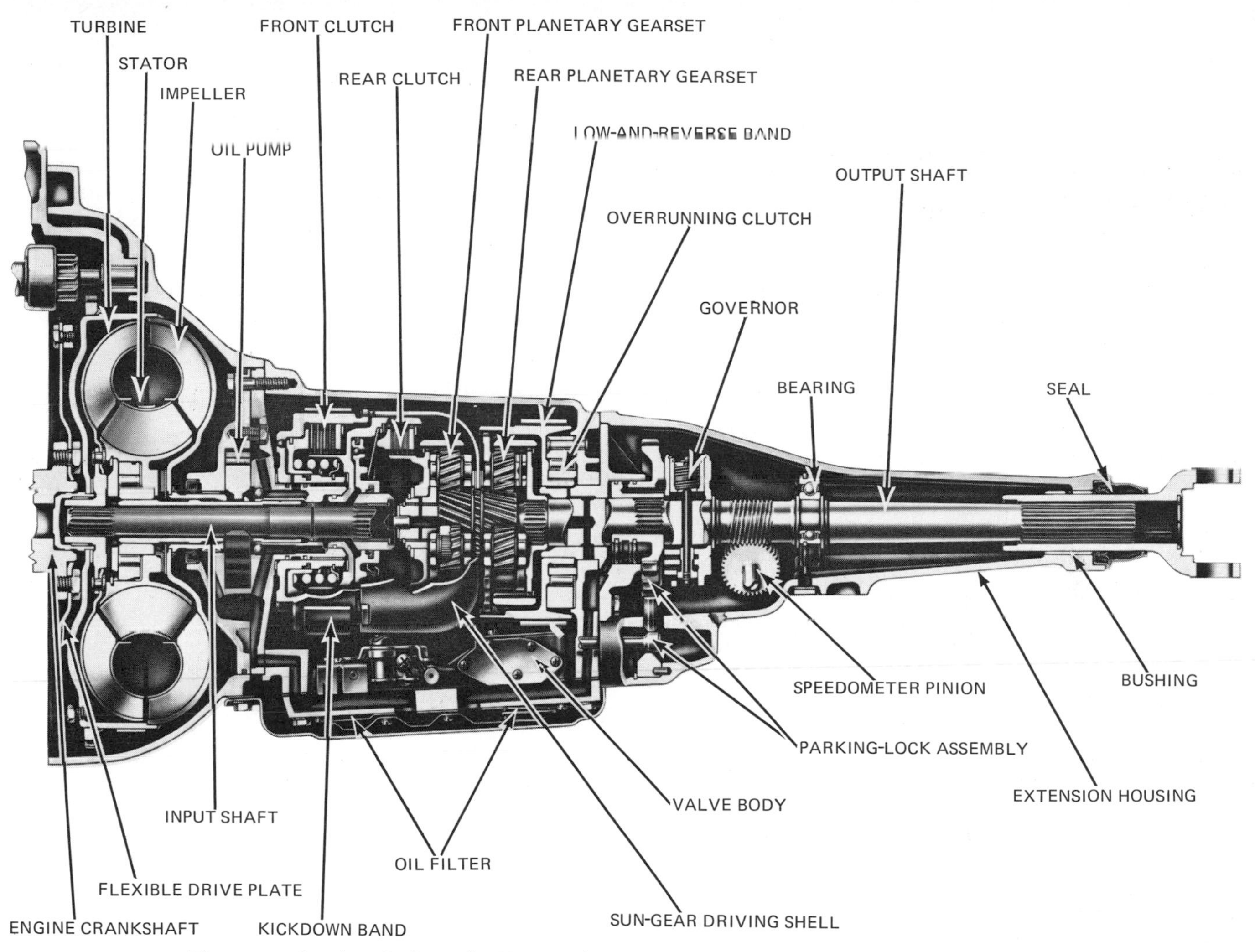

Fig. 17-1. Sectional view of a TorqueFlite automatic transmission. (*Chrysler Corporation*)

and-reverse band is applied to lock the rear planet carrier (instead of the overrunning clutch holding it). The reason for this change is that the low-and-reverse band will remain applied as long as the selector lever is in L. The 2.45:1 gear reduction will therefore remain in effect. But if the selector lever were in D, then the overrunning clutch would hold the rear planet carrier only until upshifting occurred. The overrunning clutch would then permit the carrier to overrun, as pictured in Figs. 17-4 and 17-5.

In reverse, shown in Fig. 17-8, the front clutch is engaged and the low-and-reverse band applied. Now the power flow is through the front clutch, driving shell, sun gear, and rear planetary gearset to the output shaft. Note that the low-and-reverse band, when applied, holds the rear-planetary-gearset carrier stationary. The planetary gears therefore act as idlers and cause the output shaft to turn in the reverse direction so that the car is backed. The gear ratio is 2.20:1.

⊘ 17-3 TorqueFlite Hydraulic Circuits Let us now study the hydraulic circuits, with the various valves, that control the shifting of the automatic transmission (Plates 24 to 30).

Plate 24 shows the hydraulic circuit and the four active gearset controls, which are the two clutches and the two servos. First, note that Chrysler has somewhat different names for some of the transmission components. The KD servo means *kickdown servo;* it applies the front band (also called the *KD band*) under certain conditions. The shuttle valve is a valve we have not run into before; it cushions certain shifts, as we shall explain later. The TC (torque-converter) control valve (lower left) prevents excessive pressures in the torque converter. The kickdown valve is also different. All valves are described in following sections.

⊘ 17-4 Drive Range, Low Gear, or Breakway Chrysler calls low gear in drive range the "break-

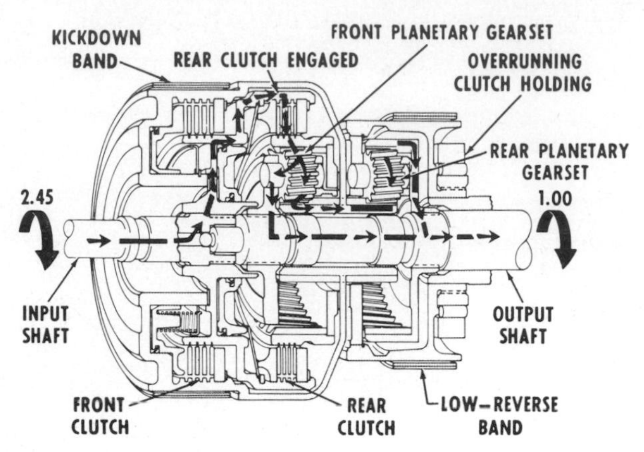

Fig. 17-2. Power flow in D (drive range) when first breaking away from a standing start with gears in first, or low. (*Chrysler Corporation*)

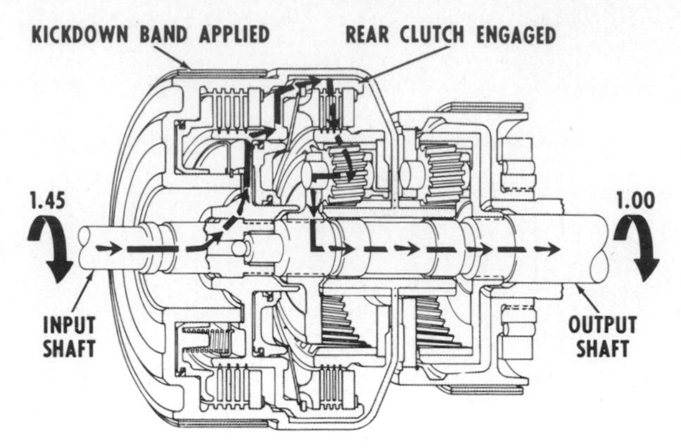

Fig. 17-4. Power flow in D after an upshift to second gear or after a kickdown from direct. (*Chrysler Corporation*)

away" gear. Transmission conditions are shown in Fig. 17-2 and Plate 24. The regulator valve, at the lower left in Plate 24, controls line pressure at a value dependent on throttle opening. It does this by the interaction between the kickdown valve (to middle right) and the throttle valve. The kickdown valve is mechanically linked to the carburetor throttle linkage.

As the driver pushes down on the accelerator pedal, the throttle opens and the kickdown valve is moved inward (to left in Plate 24). This movement increases the spring pressure on the throttle valve, forcing it to move. The throttle valve then passes more line pressure so that the pressure exiting from the throttle valve, called *throttle-valve pressure,* increases. This increasing pressure acts on the plug to the right of the regulator valve. The plug, in turn, reacts on the regulator valve so that it tends to cut off the dump, or "suction," line to the pump. The regulated pressure therefore increases. This provides greater holding power for the clutches and brake bands at higher speeds and with higher torques passing through the transmission.

As you can see, in breakaway—low gear in drive range—the manual valve has been moved so that

CLUTCH-ENGAGEMENT AND BAND-APPLICATION CHART

LEVER POSITION Drive Ratio	FRONT CLUTCH	REAR CLUTCH	FRONT (KICKDOWN) BAND	REAR (LOW-REVERSE) BAND	OVERRUNNING CLUTCH
N-NEUTRAL	Disengaged	Disengaged	Released	Released	No movement
D-DRIVE					
(Breakaway) 2.45:1	Disengaged	Engaged	Released	Released	Holds
(Second) 1.45:1	Disengaged	Engaged	Applied	Released	Overruns
(Direct) 1.00:1	Engaged	Engaged	Released	Released	Overruns
KICKDOWN					
(To Second) 1.45:1	Disengaged	Engaged	Applied	Released	Overruns
(To Low) 2.45:1	Disengaged	Engaged	Released	Released	Holds
2-SECOND 1.45:1	Disengaged	Engaged	Applied	Released	Overruns
1-LOW 2.45:1	Disengaged	Engaged	Released	Applied	Partial hold
R-REVERSE 2.20:1	Engaged	Disengaged	Released	Applied	No movement

Fig. 17-3. Chart of clutch engagements and band applications for all transmission operating conditions. (*Chrysler Corporation*)

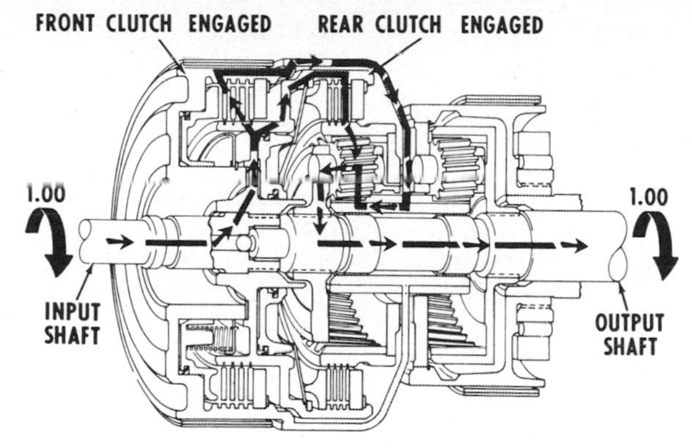

Fig. 17-5. Power flow in D after an upshift to direct drive. (*Chrysler Corporation*)

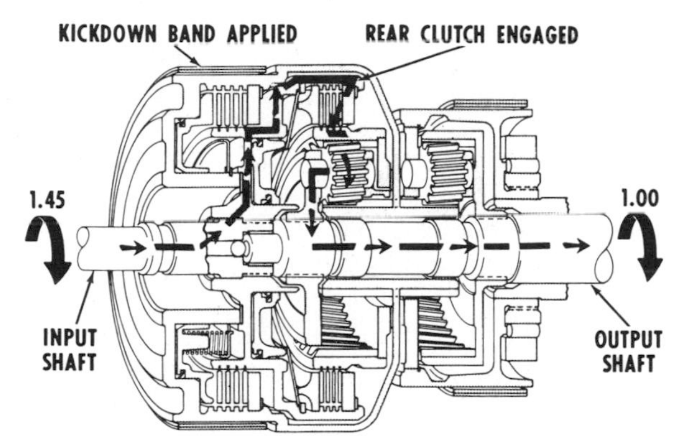

Fig. 17-6. Power flow in D2, or with the selector lever moved to 2. (*Chrysler Corporation*)

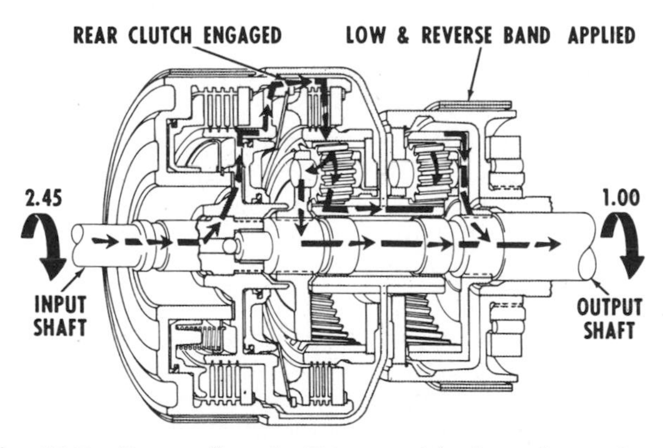

Fig. 17-7. Power flow in D1, or with the selector lever moved to 1. (*Chrysler Corporation*)

line pressure is directed to the rear clutch, engaging it. The power flow is shown in Fig. 17-2. Note that governor pressure is applied to the governor plugs behind the 1-2 and 2-3 shift valves. Opposing this governor pressure, at the opposite ends of the 1-2 and 2-3 shift valves, is spring pressure plus throttle-valve pressure. You will recall that throttle-valve pressure increases with increased throttle opening. This increase in pressure is due to the movement of the kickdown valve, which is linked to the throttle linkage.

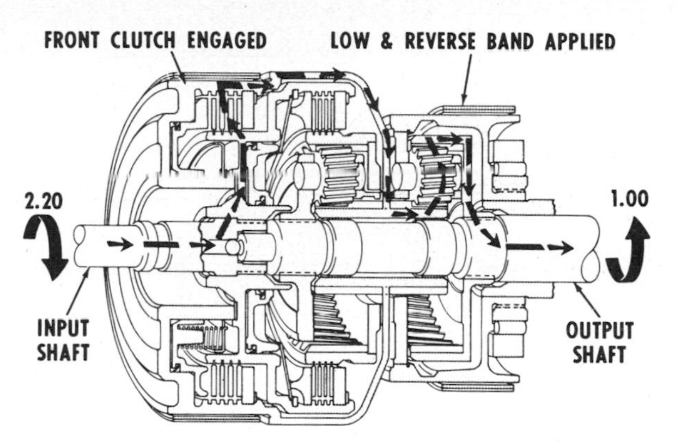

Fig. 17-8. Power flow in R, or reverse. (*Chrysler Corporation*)

As car speed and governor pressure increase, the increasing governor pressure will overcome the 1-2 shift-valve spring and throttle-valve pressure. Thus, the transmission will upshift to second.

⊘ 17-5 Drive Range, Second Gear When the upshift occurs, the governor plug, with governor pressure behind it, has pushed the 1-2 shift valve to the right, as shown in Plate 25. Note that, for simplicity, we have shown only the upper part of the hydraulic circuit in Plate 25. The lower part remains essentially unchanged. As the 1-2 shift valve moves, it allows line pressure to flow to the KD servo so that the front, or kickdown, band is applied, as shown in Fig. 17-4. Now, the power flow is as can be seen in Fig. 17-4.

Note that the 1-2 shift-valve movement also allows line pressure to flow to the accumulator. The accumulator cushions the application of the kickdown band. We have explained how accumulators work in ⊘ 7-10, 13-16, and 13-17.

The point at which the upshift occurs depends on two factors: car speed and throttle opening. With a small throttle opening, there will be a relatively small throttle-valve pressure; thus a relatively low car speed and governor pressure can produce the upshift. The actual speed at which upshifts occur with closed and wide-open throttle varies greatly with different cars and engines. The Chrysler engineers get this variation in different transmission models by using different valve springs, clutch plates, and so on. A typical pattern would be for a closed-throttle 1-2 upshift to take place at about 12 mph [19.31 km/h]. With wide-open throttle, the 1-2 upshift would then take place at about 40 mph [64.37 km/h]. For this particular application, the upshift would vary between those two extremes, depending on the amount of throttle opening.

⊘ 17-6 Drive Range; Direct, or Third, Gear With further increase in car speed and thus in governor pressure, the governor pressure reaches a value sufficient to overcome the spring and throttle-valve pressures acting on the 2-3 shift valve, and it is forced to move to the right, as shown in Plate 26. Now, line pressure is directed through the 2-3 shift valve to the front clutch, causing it to engage. At the same time, line pressure is admitted to the upper part of the kickdown servo, forcing the servo piston to move down and release the kickdown brake band. The situation is then as shown in Fig. 17-5.

The shuttle valve, to the upper right in Plate 26, acts to smooth out the application of the front clutch and the release of the kickdown brake band. It does this by modulating the line pressure going to the release side of the kickdown piston and to the front clutch. The modulation is related to throttle-valve pressure, which acts on the throttle plug to the left of the shuttle valve. This action positions the shuttle valve according to throttle-valve pressure, and the shuttle valve then modulates the pressure going to the release side of the kickdown piston and to the front clutch.

For example, consider a "lift-foot" upshift, which is made by accelerating the car in second and then lifting the foot to move the throttle toward a closed position. When this happens, throttle-valve pressure on the 2-3 shift valve is reduced so that governor pressure can move the 2-3 valve to produce the upshift. The reduced throttle-valve pressure on the throttle plug eases the pressure on the shuttle valve, allowing it to reduce line pressure to the kickdown piston and front clutch. As a result, the band releases more slowly and the front clutch engages more softly. Thus, the upshift is mild and without any noticeable jerk. On the other hand, if the throttle opening is maintained, the kickdown band must release rapidly and the front clutch must engage quickly to produce a smooth 2-3 upshift.

The point at which a 2-3 upshift takes place depends upon car speed and throttle opening. The greater the throttle opening, the higher the car speed at which the upshift will take place. For a typical example, with a closed throttle, the upshift could take place at 18 mph [28.968 km/h]. But on the same application, with a wide-open throttle, the upshift would not take place until 70 mph [112.65 km/h].

⊘ 17-7 Kickdown in Drive Range If quick acceleration is required to pass another car, the driver pushes the accelerator down all the way to wide-open throttle. The linkage from the throttle to the kickdown valve, shown to the lower right in Plate 22, pushes the kickdown valve to the left. This movement allows throttle-valve pressure to pass through the kickdown valve, and this pressure, now termed *kickdown pressure,* passes to the 2-3 and 1-2 shift valves. If the kickdown pressure is great enough to overcome governor pressure, a downshift will occur. To say it another way, throttle-valve pressure must overcome governor pressure, or car speed. When the downshift occurs, the oil is dumped from the front clutch and the upper, or release, side of the kickdown-servo piston. The front clutch releases, and the kickdown servo applies the kickdown band; thus, the situation is as shown in Fig. 17-6. The transmission has dropped down to second gear.

The shuttle valve comes into play here to smooth the downshift. It permits the kickdown to take place rapidly at low car speed but slows the kickdown at higher car speed. The reason is this. At low car speed, very little time is required to complete the shift because there is a comparatively small change in engine speed between direct drive and kickdown, or second gear. However, at high car speed, the engine speed must change considerably to pick up the difference between direct gear and second. Therefore, the shuttle valve slows the downshift at higher car speeds. If it were not for this, there would be a very noticeable jerk as the downshift occurred.

If the car is in second gear and moving slowly, the kickdown will shift the gears from second to first. The throttle-valve pressure will be sufficient to overcome governor pressure at the 1-2 shift valve, causing it to move and produce the downshift to first. As the 1-2 shift valve moves, it cuts off the pressure to the kickdown servo so that the band releases, thus producing the situation illustrated in Fig. 17-2.

⊘ 17-8 Kickdown in Drive Range, Eight-Cylinder Engines The transmission used with eight-cylinder engines has extra valves to control the kickdown band on 1-2 upshifts and 3-2 downshifts. See Fig. 17-9. The added valves include the limit valve (to upper right), the 1-2 shift control valve (just above the throttle valve), and the throttle plug in the 2-3 shift-valve assembly.

In Fig. 17-9, the conditions shown are for a full-throttle kickdown in D (drive). With this condition, the kickdown valve releases pressure to the 2-3 shift valve, forcing it to the left, as shown. This is the same action as in the transmission for six-cylinder engines (⊘ 17-7). However, at the same time, governor pressure lifts the limit valve (to upper right) and cuts off the flow of pressure from B to C. This, in turn, cuts off pressure from the throttle plug in the 2-3 shift assembly. This action controls the 3-2 downshift according to car speed (governor pressure). That is, it prevents downshifts if the car speed is too high.

⊘ 17-9 Reverse When the driver shifts the selector lever to R (reverse), the hydraulic circuit is as shown in Plate 28 and the power flow is as shown in Fig. 17-8. Note that everything in the hydraulic circuit becomes ineffective except the pump, regulator valve, torque-converter control valve, and manual valve. The manual valve feeds line pressure to the pressure-regulator valve so that the pressure regulator regulates to a higher pressure—as much as 230 to 260 psi [16.171 to 18.280 kg/cm^2]. This high pressure ensures good firm application of the low-and-reverse band and the front clutch.

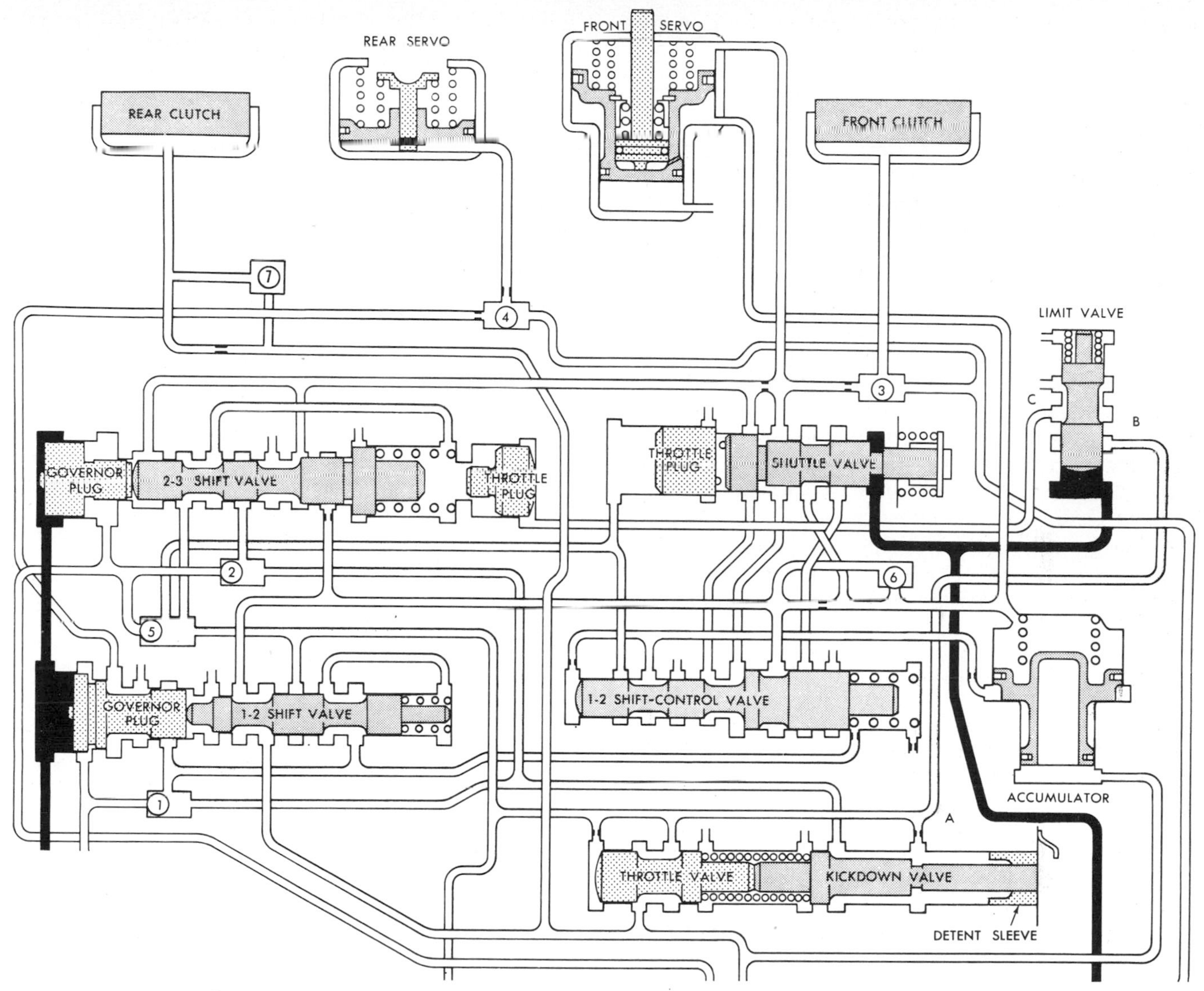

Fig. 17-9. Hydraulic circuit (upper part) showing the added valves for the transmission used with eight-cylinder engines. (Compare with Plates 24 and 27.)

⊘ 17-10 Drive 2 Range This is the condition that results if the driver places the selector lever in drive 2 position. The manual valve is moved so that line pressure is applied to the rear clutch and the apply side of the kickdown-servo piston. (See Plate 29.) With the rear clutch and kickdown band applied, the situation shown in Fig. 17-6 results.

At high car speed, with the transmission in direct, the downshift will not occur until governor pressure (car speed) drops enough to permit line pressure to overcome governor pressure and move the governor plug (to the left in Plate 29). This action allows the spring to move the 2-3 shift valve so that the line pressure is cut off from the front clutch and the release side of the kickdown-servo piston.

At low car speed, the transmission can downshift from second to first. It can then upshift to second with increasing car speed. But it will not upshift to third as long as the selector lever and the manual valve are in the drive 2 range.

⊘ 17-11 Drive 1 Range This is the condition that results if the driver places the selector lever in drive 1 position. The transmission stays in low gear. The hydraulic circuit is as shown in Plate 30, and the power flow is as shown in Fig. 17-7. The manual valve moves so that line pressure is admitted back of the 1-2 governor plug and the 2-3 governor plug. This action prevents any upshift.

If car speed is above a certain minimum and the transmission is in second or direct gear, the downshift to first will not take place until the car speed, and governor pressure, drops to the required minimum. When this happens, governor pressure, working against the governor plugs, is overcome by line pressure directed to the shift valves. The downshift takes place.

Drive 1 and 2 ranges are for pulling heavy loads up long hills where the accelerator is kept more or less wide open for $\frac{1}{2}$ mile [0.804 km] or more. These lower gears prevent transmission overheating on the hard pull. They also permit engine braking of the car when going down long hills.

CHAPTER 17 CHECKUP

NOTE: Since the following is a chapter review test, you should review the chapter before taking it.

You are making great progress in your studies of automatic transmissions and have covered all the essentials leading to an expert understanding of how they are constructed and how they operate. Studies of the TorqueFlite plus the Ford transmissions, which we have yet to cover, add to your knowledge and prepare you to understand any transmission that may be brought out in the future. Now check your memory of the details of the TorqueFlite transmission by taking the test that follows.

Completing the Sentences The sentences that follow are incomplete. After each sentence there are several words or phrases, only one of which will correctly complete the sentence. Write each sentence in your notebook, selecting the proper word or phrase to complete it correctly.

1. Counting all the active control members in the TorqueFlite transmission, including the clutches and bands, there are: (*a*) four, (*b*) five, (*c*) six.
2. The compound planetary-gear system has two sets of planetary pinions, two internal or ring gears, and: (*a*) three sun gears, (*b*) two sun gears, (*c*) one sun gear.
3. The front clutch is connected by a driving shell to the: (*a*) output shaft, (*b*) rear ring gear, (*c*) sun gear.
4. In D range on breakaway in first gear, the overrun clutch holds and the: (*a*) front clutch is engaged, (*b*) rear clutch is engaged, (*c*) kickdown band is applied.
5. In D range in second gear, the overrun clutch overruns, the rear clutch is engaged, and the: (*a*) front clutch is engaged, (*b*) rear band is applied, (*c*) front band is applied.
6. In direct drive, the overrun clutch overruns, both clutches are engaged, and: (*a*) the front band is applied, (*b*) the rear band is applied, (*c*) both bands are released.
7. In second range, second gear, the overrun clutch overruns, the kickdown or front band is applied, and the: (*a*) rear clutch is engaged, (*b*) rear band is applied, (*c*) front clutch is engaged.
8. In low range, low gear, the rear band is applied and the: (*a*) front clutch is engaged, (*b*) front band is applied, (*c*) rear clutch is engaged.
9. In reverse, the rear band is applied and the: (*a*) rear clutch is engaged, (*b*) front clutch is engaged, (*c*) front band is applied.
10. The sun gear is locked in a stationary position by: (*a*) front-clutch engagement, (*b*) front-band application, (*c*) rear-band application.

Purpose and Operation of Components In the following, you are asked to write the purpose and operation of the various components in the transmission. If you have any difficulty with any of the questions, reread the pages in the chapter that will give you the information you need. Then write your explanation or make your drawing. Don't copy; try to tell it in your own words. This is a good way to remember the information. Write in your notebook.

1. Make a chart showing the band application and clutch engagements for the different operating conditions similar to the one in the text.
2. Describe the power flow in D, breakaway, or low gear.
3. Describe the power flow in D, second gear.
4. Describe the power flow in direct drive.
5. Describe the power flow in second gear with the selector lever in D2.
6. Describe the power flow in first gear with the selector lever in L.
7. Describe the power flow in reverse.
8. Explain how the shuttle valve works.
9. Draw the part of the hydraulic circuit that is active in D in low gear.
10. Draw the part of the hydraulic circuit that is active in D, second gear.
11. Draw the part of the hydraulic circuit that is active in direct drive.
12. Draw the part of the hydraulic circuit that is active in second gear, D2.
13. Draw the part of the hydraulic circuit that is active in first gear, L.
14. Draw the part of the hydraulic circuit that is active in reverse.

SUGGESTIONS FOR FURTHER STUDY

As we have suggested in past chapters, your best bet to learn more about transmissions is to go to a shop that does automatic-transmission work and watch the experts on the job. Note how they remove, overhaul, and install transmissions. Study the tools, shop manuals, and working procedures. Write all important facts you learn in your notebook.

chapter 18

TORQUEFLITE TRANSMISSION SERVICE

This chapter covers the trouble-diagnosis and service procedures for the Chrysler TorqueFlite transmission used on Chrysler, Imperial, Dodge, and Plymouth automobiles. Chapter 17 covers the construction and operation of this transmission. There are two basic models of the TorqueFlite, the A-727 and the A-904, as explained below.

⊘ 18-1 Basic Models of the TorqueFlite There are two basic models of the TorqueFlite transmission: the A-727 and the A-904. They are used with the engines shown in the chart in Fig. 18-1. Note that the A-904 is used with small-output engines while the A-727 is used with high-output engines. There are a number of internal differences that we shall illustrate and describe when we discuss the transmissions. For example, the A-904 has one large spring in its front clutch while the A-727 has a number of small springs. There are also differences in the rear clutches, planetary-gear trains, and overrunning clutches. We discuss these differences later.

⊘ 18-2 Operating Cautions For mountain driving, either with heavy loads or when pulling a trailer, use the 2 (second) or 1 (low) selector-lever position when on upgrades requiring a heavy throttle for ½ mile [0.804 km] or more. Using these positions reduces the possibility that the transmission and convertor will overheat.

The TorqueFlite transmission will not permit starting the engine by pushing or towing.

When towing the vehicle, use a rear-end pickup or remove the drive shaft. The car can be towed with the rear wheels on the ground and the drive shaft not removed for short distances at low speeds of 30 mph [48.28 km] or less. However, remember that the transmission receives lubrication only when the engine is running. Therefore, it is always best to use the rear-end pickup or remove the drive shaft.

TORQUEFLITE TRANSMISSION APPLICATION AND STALL SPEED

TRANSMISSION ASSEMBLY NO.*	ENGINE in^3	TRANSMISSION TYPE	CONVERTER DIAMETER, in	STALL rpm	APPLICATION
3681841	198	A-904	10¾	1,625–1,925	Any models
3681841	225	A-904	10¾	1,800–2,100	equipped with
3681843	318	A-904-LA	10¾	2,125–2,425	these engines
3681861	225	A-727	11¾	1,400–1,700	Police & taxi
3681862	318	A-727	11¾	1,725–2,025	Police & taxi
3681863	360–4 hp	A-727	10¾	2,200–2,500	(VLBJRW)
3681862	360–2 360–4	A-727	10¾	2,300–2,600	(PDC)
3681864	400–2 400–4	A-727	11¾	1,875–2,175	(BJRWPDC)
3681865	400–4 hp	A-727	10¾	2,400–2,700	Hi. Perf. (BJRWPDC)
3681866	440–4	A-727	11¾	1,975–2,275	(PDCY)
3681867	440–4	A-727	10¾	2,600–2,900	Hi. Perf. (RW)
3681867	440–4	A-727	11¾	2,100–2,400	Hi. Perf. (PDC)

*Part numbers subject to change during model year. Number is found on left side of transmission oil-pan flange.

Fig. 18-1. Table of TorqueFlite transmission applications and stall speeds. (*Chrysler Corporation*)

Trouble Diagnosis

⊘ 18-3 Transmission Trouble Diagnosis Figures 18-2 to 18-5 are trouble-diagnosis charts for the TorqueFlite transmission. Figure 18-2 is a general chart, listing possible trouble conditions in the left-hand column. Possible causes are listed horizontally at the top of the chart. To find the causes of a trouble, locate the trouble in the column to the left. Then run across the chart horizontally, noting the Xs on the horizontal line. Each X indicates a possible cause.

Figure 18-3 is a diagnosis guide to follow if the car will not move in any selector-lever position. Figure 18-4 is a diagnosis guide to follow if there is abnormal noise that seems to come from the transmission. Figure 18-5 is a diagnosis guide to follow if the transmission leaks fluid.

The chart in Fig. 18-6 (p. 258) lists the shift patterns of the TorqueFlite transmission as used on different engines. Note that the figures vary somewhat, but this is not important as long as the shifts are smooth and responsive and there is no noticeable engine runaway.

Three tests are made on the TorqueFlite transmission for diagnostic purposes: the *stall test, hydraulic-control pressure tests,* and *air-pressure tests.* These are covered in following sections.

⊘ 18-4 Stall Test The stall test consists in determining engine speed obtained at full throttle in the D (drive) range. This test checks the torque-converter stator clutch and the holding ability of the transmission clutches. Check the transmission oil level and bring the engine up to operating temperature before beginning the stall test.

CAUTION: Both the parking and service brakes must be fully applied and both front wheels blocked while making this test.

SPECIAL CAUTION: Do not allow anyone to stand in front of the car during the test!

Do not hold the throttle wide open for more than 5 seconds at a time! If you have to make more than one stall check, run the engine at 1,000 rpm in neutral for 20 seconds between runs to cool the transmission. If engine speed exceeds the maximum limits shown in the chart in Fig. 18-1, release the throttle at once since this condition indicates clutch slippage.

To make the stall test, connect a tachometer that can be read from the driver's seat, block the front wheels, apply both brakes, and open the throttle wide. Read the top engine speed and compare it with the reading in the chart (Fig. 18-1) for the model being tested.

CAUTION: Make the test in 5 seconds or less!

1. *STALL SPEED ABOVE SPECIFICATIONS* If engine speed increases by more than 200 rpm above the specification in the chart, clutch slippage is indicated. Release the throttle at once to avoid damage to the transmission. Clutch slippage requires further checking, and the transmission hydraulic-control- and air-pressure checks should be made, as explained in ⊘ 18-5 to 18-7.
2. *STALL SPEED BELOW SPECIFICATIONS* Low stall speed with a *properly tuned engine* indicates torque-converter stator-clutch problems, and a road test is necessary to determine what is wrong. If stall speeds are 250 to 350 rpm below specifications and the vehicle operates properly at highway speeds but has poor through-gear acceleration, the stator clutch is slipping. If the stall speed and acceleration are normal, but abnormally high throttle opening is required to maintain highway speed, the stator clutch has seized. With either of these stator defects, the torque converter must be replaced.

⊘ 18-5 Hydraulic-Control Pressure Tests These tests check the pressures in the hydraulic system during different operating conditions and can reveal certain defects, as we shall now explain.

1. *LINE PRESSURE AND KICKDOWN-SERVO RELEASE PRESSURE* This check is made in D with the rear wheels free to turn. Transmission fluid must be at operating temperature (150 to 200 degrees Fahrenheit [65.6 to 93.8°C]). Install an engine tachometer, raise the car on a hoist that leaves the rear wheels free, and position the tachometer so it can be read from under the car.

a. Connect two 0- to 100-psi [0 to 7.031 kg/cm^2] pressure gauges to the pressure-takeoff plugs to check line pressure and kickdown-servo release pressure. See Fig. 18-7 (p. 258) for connection points.

b. With the control in D, speed up the engine slightly so that the transmission shifts into direct gear. Reduce the engine slowly to 1,000 rpm. Now, line pressure must be 54 to 60 psi [3.797 to 4.219 kg/cm^2], and kickdown-servo release pressure must not be more than 3 psi [0.21 kg/cm^2] below line pressure.

c. Disconnect the throttle linkage from the transmission throttle lever and move the transmission throttle lever gradually to the full-throttle position. Line pressure must rise to a maximum of 90 to 96 psi [6.328 to 6.750 kg/cm^2] just before or at kickdown into low gear. Kickdown-servo release pressure must follow line pressure up and should be not more than 3 psi [0.021 kg/cm^2] below it.

If the line pressure is not 54 to 60 psi [3.797 to 4.219 kg/cm^2] at 1,000 rpm, make adjustments, as explained in ⊘ 18-6. This requires removal of the valve body. If the kickdown-servo release pressures are below specifications, with line pressures okay, there is excessive leakage in the front clutch or in the kickdown-servo circuit.

CONDITION	1 Engine idle speed too high	2 Hydraulic pressures too low	3 Low-reverse band out of adjustment	4 Valve-body malfunction or leakage	5 Low-reverse servo, band or linkage malfunction	6 Low fluid level	7 Incorrect gearshift-control-linkage adjustment	8 Oil filter clogged	9 Faulty oil pump	10 Worn or broken input-shaft seal rings	11 Aerated fluid	12 Engine idle speed too low	13 Incorrect throttle-linkage adjustment	14 Kickdown band out of adjustment	15 Overrunning clutch not holding	16 Output-shaft bearing and/or bushing damaged	17 Governor-support seal rings broken or worn	18 Worn or broken reaction-shaft-support seal rings	19 Governor malfunction	20 Kickdown servo band or linkage malfunction	21 Worn or faulty front clutch	22 High fluid level	23 Breather clogged	24 Hydraulic pressure too high	25 Kickdown-band adjustment too tight	26 Faulty cooling system	27 Insufficient clutch-plate clearance	28 Worn or faulty rear clutch	29 Rear clutch dragging	30 Planetary gear sets broken or seized	31 Overrunning clutch worn, broken or seized	32 Overrunning clutch inner race damaged
HARSH ENGAGEMENT FROM NEUTRAL TO D OR R	X			X																				X				X				
DELAYED ENGAGEMENT FROM NEUTRAL TO D OR R		X		X	X	X	X	X	X	X	X	X						X			X							X				
RUNAWAY UPSHIFT		X		X		X		X			X		X					X		X	X											
NO UPSHIFT		X		X		X	X						X				X	X	X	X	X											
3-2 KICKDOWN RUNAWAY		X		X		X					X		X	X				X		X	X											
NO KICKDOWN OR NORMAL DOWNSHIFT				X									X						X	X												
SHIFTS ERRATIC		X		X		X	X	X	X		X		X				X	X	X	X	X											
SLIPS IN FORWARD DRIVE POSITIONS		X		X		X	X	X	X	X	X		X		X													X			X	
SLIPS IN REVERSE ONLY		X	X	X	X	X	X		X		X							X			X											
SLIPS IN ALL POSITIONS		X		X		X		X	X	X	X																					
NO DRIVE IN ANY POSITION		X		X		X		X	X																					X		
NO DRIVE IN FORWARD DRIVE POSITIONS		X		X		X				X					X													X		X	X	
NO DRIVE IN REVERSE		X	X	X	X		X											X			X							X		X		
DRIVES IN NEUTRAL				X			X																				X	X	X			
DRAGS OR LOCKS			X																						X					X	X	
GRATING, SCRAPING, GROWLING NOISE			X											X		X														X	X	
BUZZING NOISE				X		X					X																					X
HARD TO FILL, OIL BLOWS OUT FILLER TUBE								X			X											X	X									
TRANSMISSION OVERHEATS	X	X				X	X		X																X	X	X					
HARSH UPSHIFT		X											X	X										X								
DELAYED UPSHIFT													X	X			X	X	X	X	X											

Fig. 18-2. Diagnosis chart for the TorqueFlite transmission. (*Chrysler Corporation*)

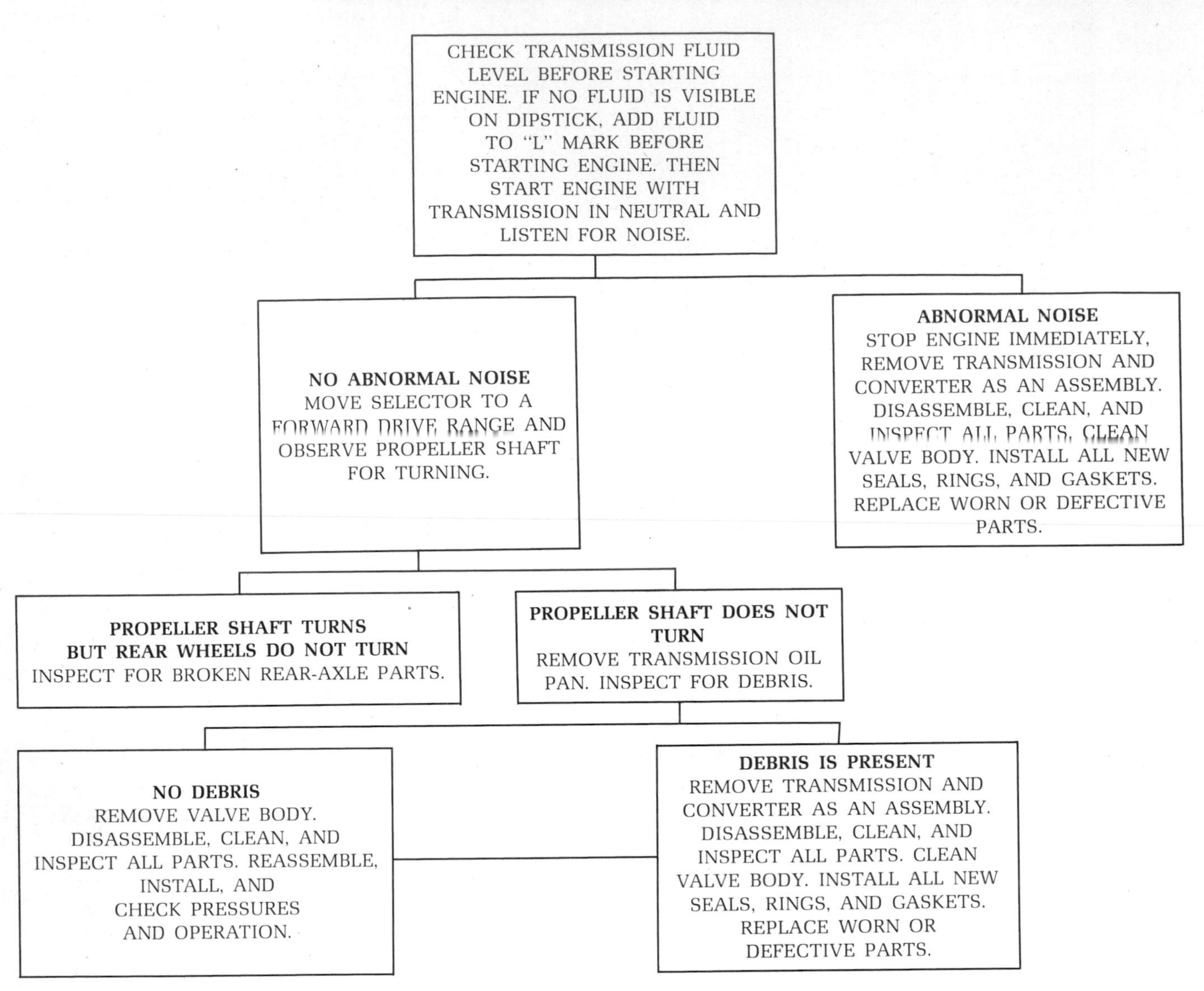

Fig. 18-3. Diagnosis guide to find trouble if the vehicle will not move. (*Chrysler Corporation*)

NOTE: Always check the external transmission throttle lever for looseness on the valve-body shaft when making the pressure check.

2. *LUBRICATION PRESSURE* This check is made by installing a "tee" fitting at the cooler return line, shown in Fig. 18-8. Connect a 0- to 100-psi [0 to 7.031 kg/cm^2] pressure gauge to the "tee" fitting. At 1,000 engine rpm, with the throttle closed and the transmission in D, lubrication pressure should be 5 to 15 psi [0.352 to 1.055 kg/cm^2]. It will just about double as the throttle is opened to maximum line pressure.

3. *REAR-SERVO APPLY PRESSURE* Connect a 0- to 300-psi [0 to 21.09 kg/cm^2] gauge to the rear-servo apply-pressure takeoff, shown in Fig. 18-8. With the transmission in R and the engine running at 1,600 rpm, the pressure should be 230 to 300 psi [16.171 to 21.090 kg/cm^2].

4. *GOVERNOR PRESSURE* Connect a 0- to 100 psi [0 to 7.031 kg/cm^2] gauge to the governor-pressure takeoff point, shown in Fig. 18-8. Governor pressure should fall within the limits shown in the chart in Fig. 18-6. If it does not, or if pressure does not fall to 0 to 1.5 psi [0 to 0.106 kg/cm^2] when the car stops, the governor valve or weights are sticking and the valve assembly must be removed for service.

⊘ 18-6 Hydraulic-Control Pressure Adjustments

There are two pressure adjustments that can be made: *line pressure* and *throttle pressure*. We explained how to check line pressure in ⊘ 18-5. If it is not correct, then the throttle pressure must be adjusted first, followed by an adjustment of line pressure. These adjustments require removal of the valve body, as we shall explain. The valve body should be mounted in a valve-body repair stand,

COLOR PLATES OF AUTOMATIC-TRANSMISSION HYDRAULIC CIRCUITS

William H. Crouse
Donald L. Anglin

The color plates included here show the hydraulic circuits of the most widely used automatic transmissions under varying operating conditions. The circuits themselves are largely self-explanatory. However, the student should also refer to the text pages covering the transmission in order to correlate the actions of the various valves, clutches, and brake bands with the hydraulic circuits.

McGraw-Hill

CONTENTS

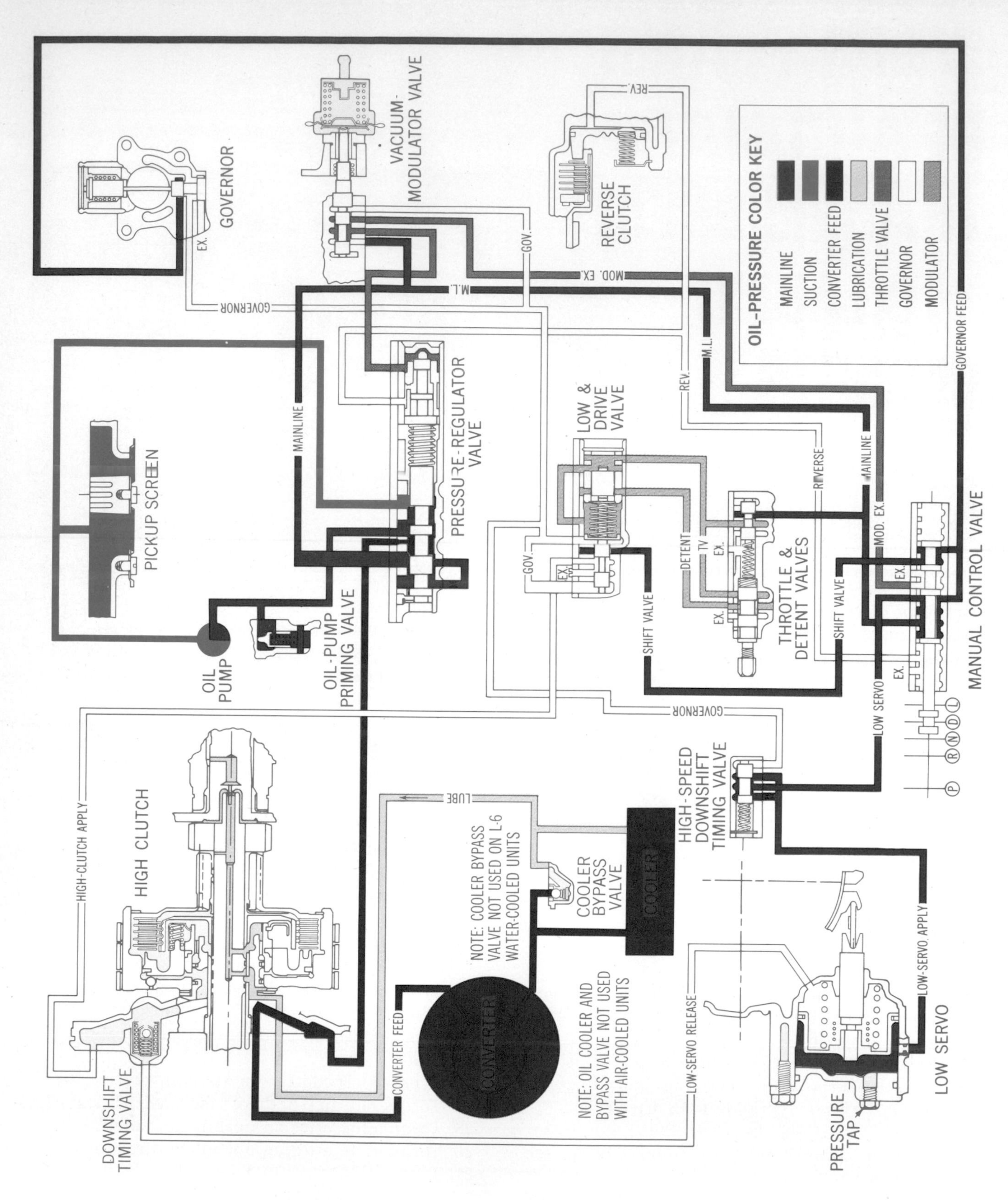

Plate 1. Complete hydraulic circuit of the Powerglide with the transmission in low gear, drive range, before an upshift. The clutches are released and the low band is applied. (*Chevrolet Motor Division of General Motors Corporation*)

3668630

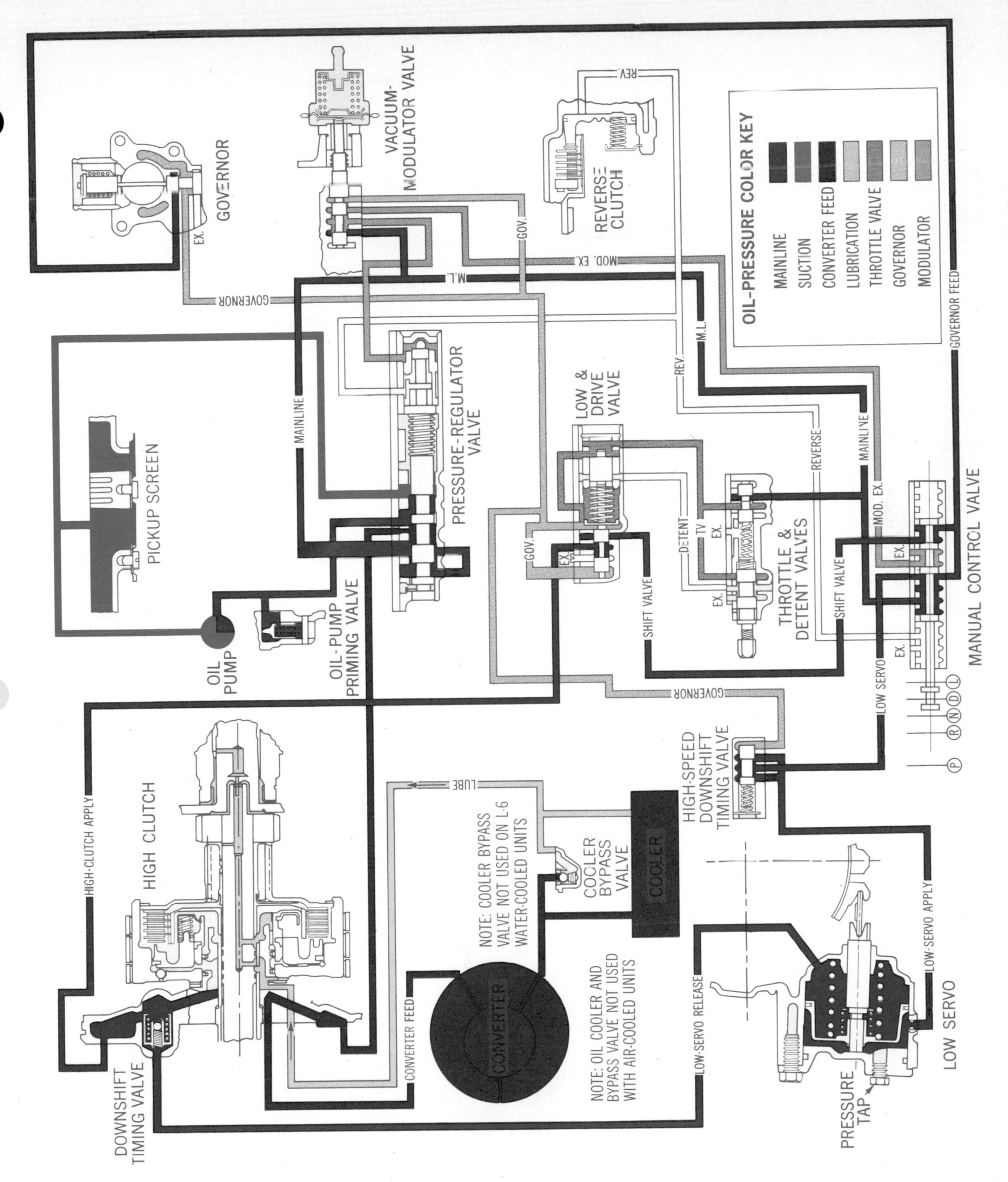

Plate 2. Complete hydraulic circuit of the Powerglide after the transmission has upshifted into direct drive. The low band is released and the high clutch is applied. (*Chevrolet Motor Division of General Motors Corporation*)

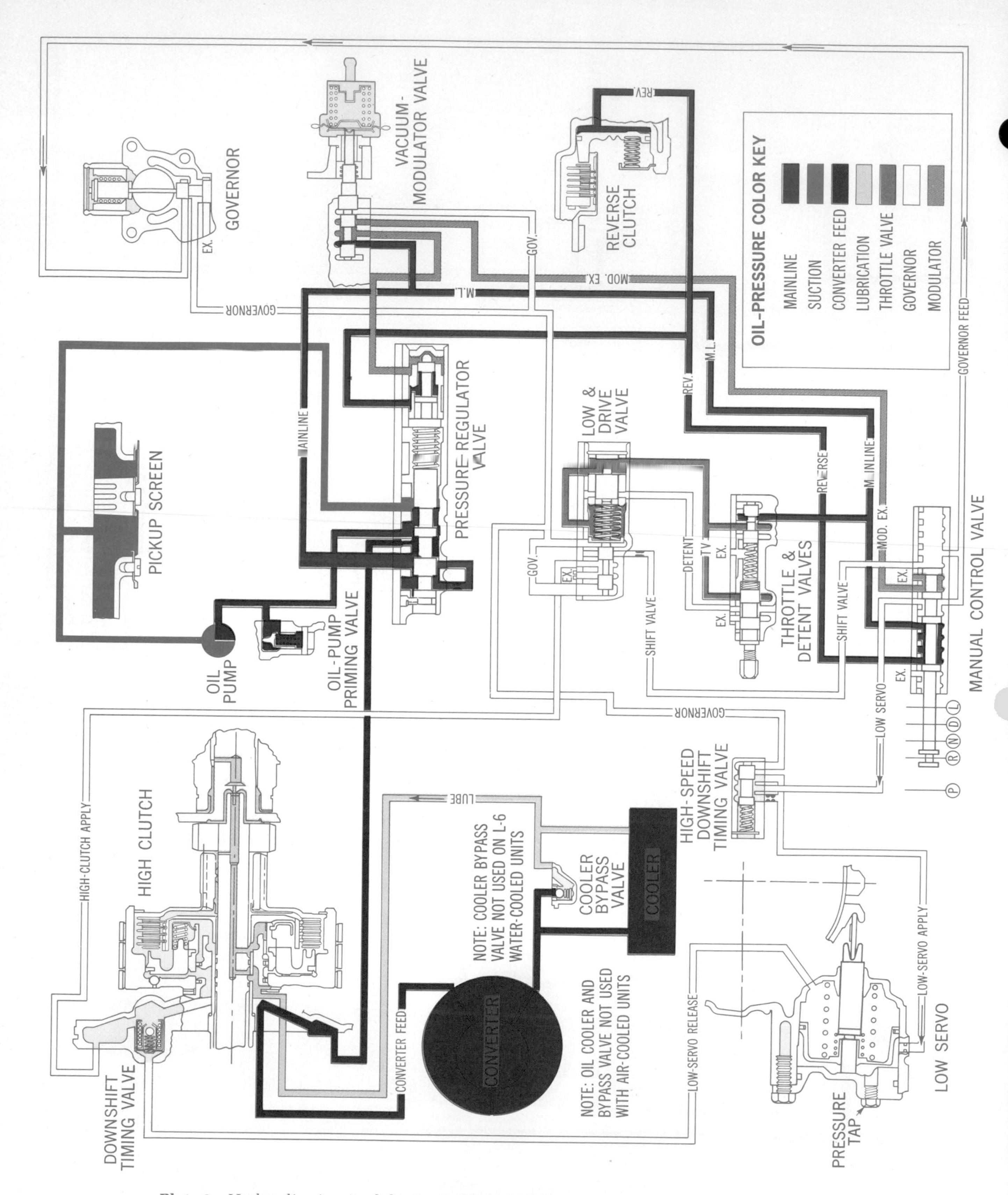

Plate 3. Hydraulic circuit of the Powerglide with the transmission in reverse. The high clutch and low band are released, and the reverse clutch is applied. (*Chevrolet Motor Division of General Motors Corporation*)

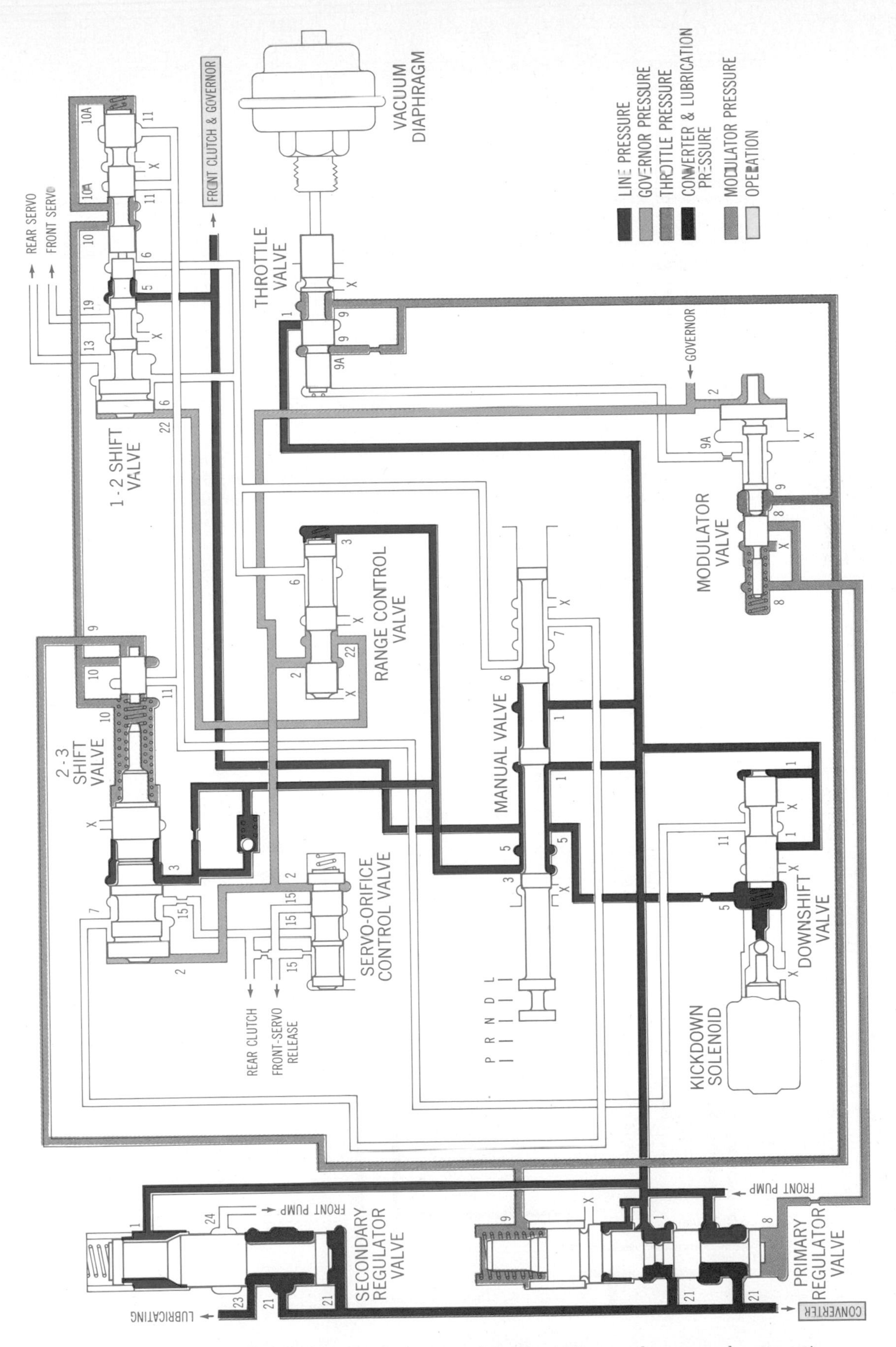

Plate 4. Conditions in the hydraulic circuit of the Borg-Warner three-speed automatic transmission with the compound planetary gearset with the selector lever in D and the transmission in low. Line pressure is applying the front clutch. The planet-pinion carrier is prevented from rotating backward by the one-way clutch. (*Chrysler Corporation*)

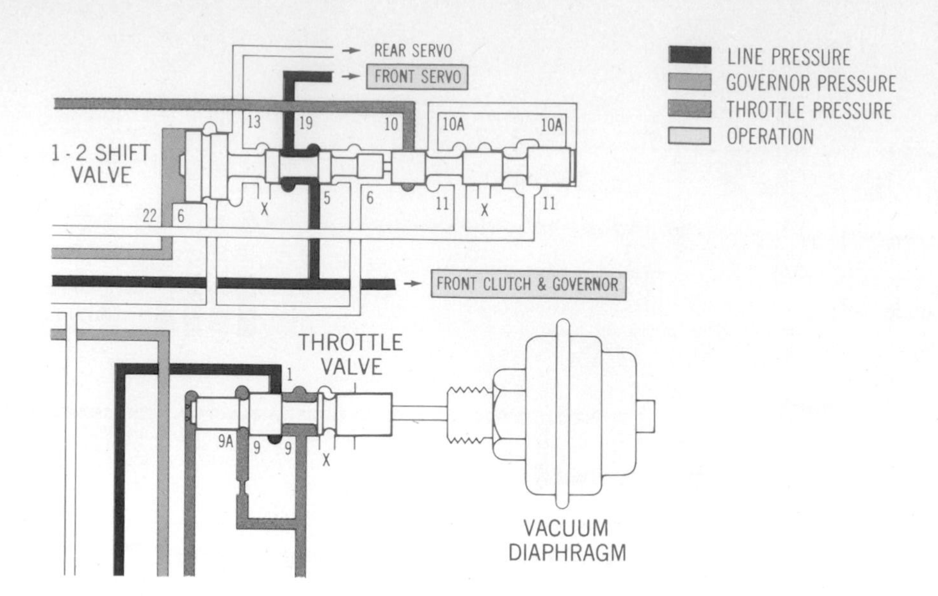

Plate 5. Conditions at the 1-2 shift valve in the hydraulic circuit of the Borg-Warner three-speed automatic transmission. Here, the selector lever is in D and the tranmission has upshifted to second. Line pressure is applying the front clutch and the front brake band. (*Chrysler Corporation*) Refer next to Plate 7.

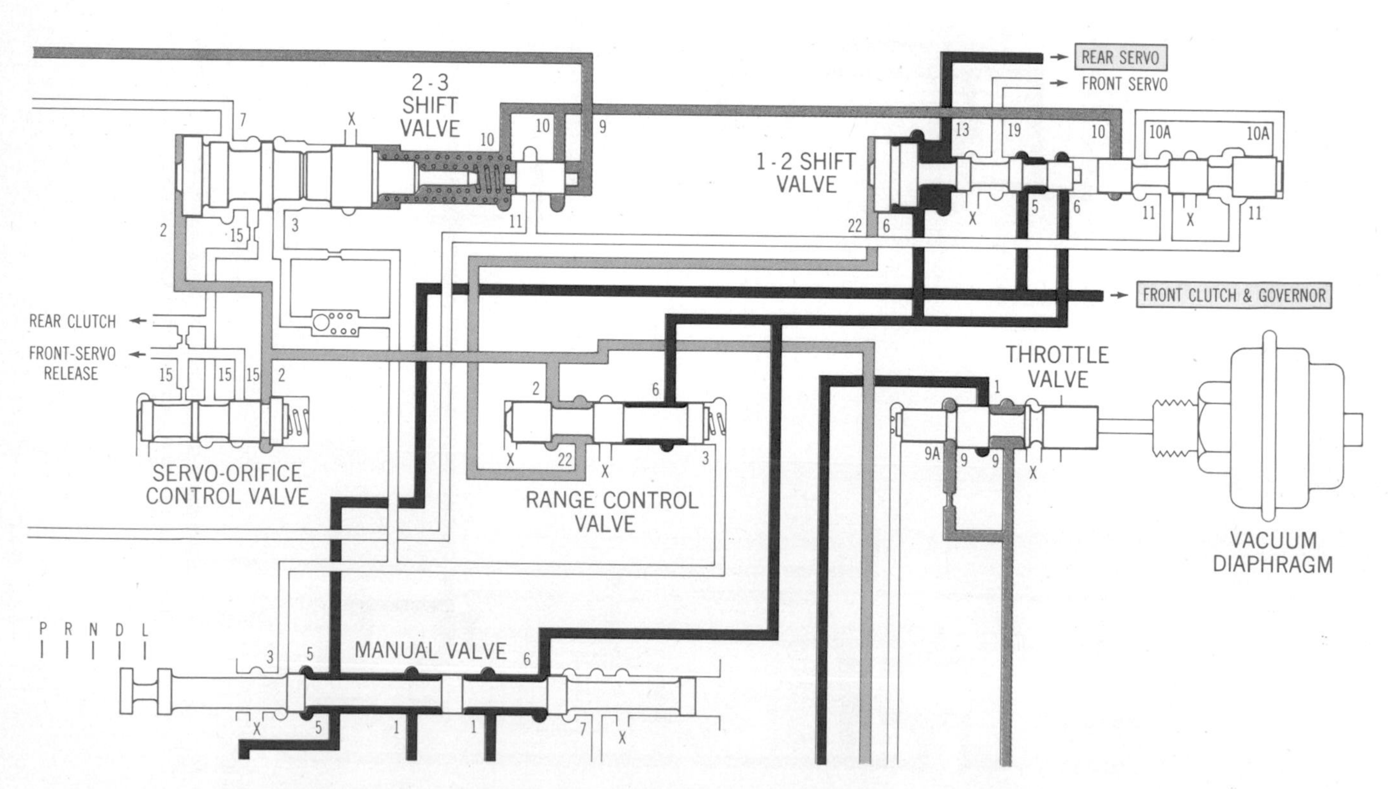

Plate 6. Conditions in the hydraulic circuit of the Borg-Warner three-speed automatic transmission with the compound planetary gearset with the transmission in low, L. Line pressure is cut off from the 2-3 shift valve. The 1-2 shift valve is feeding line pressure to the rear servo so the rear brake band is on as well as the front clutch. (*Chrysler Corporation*) Refer next to Plate 8.

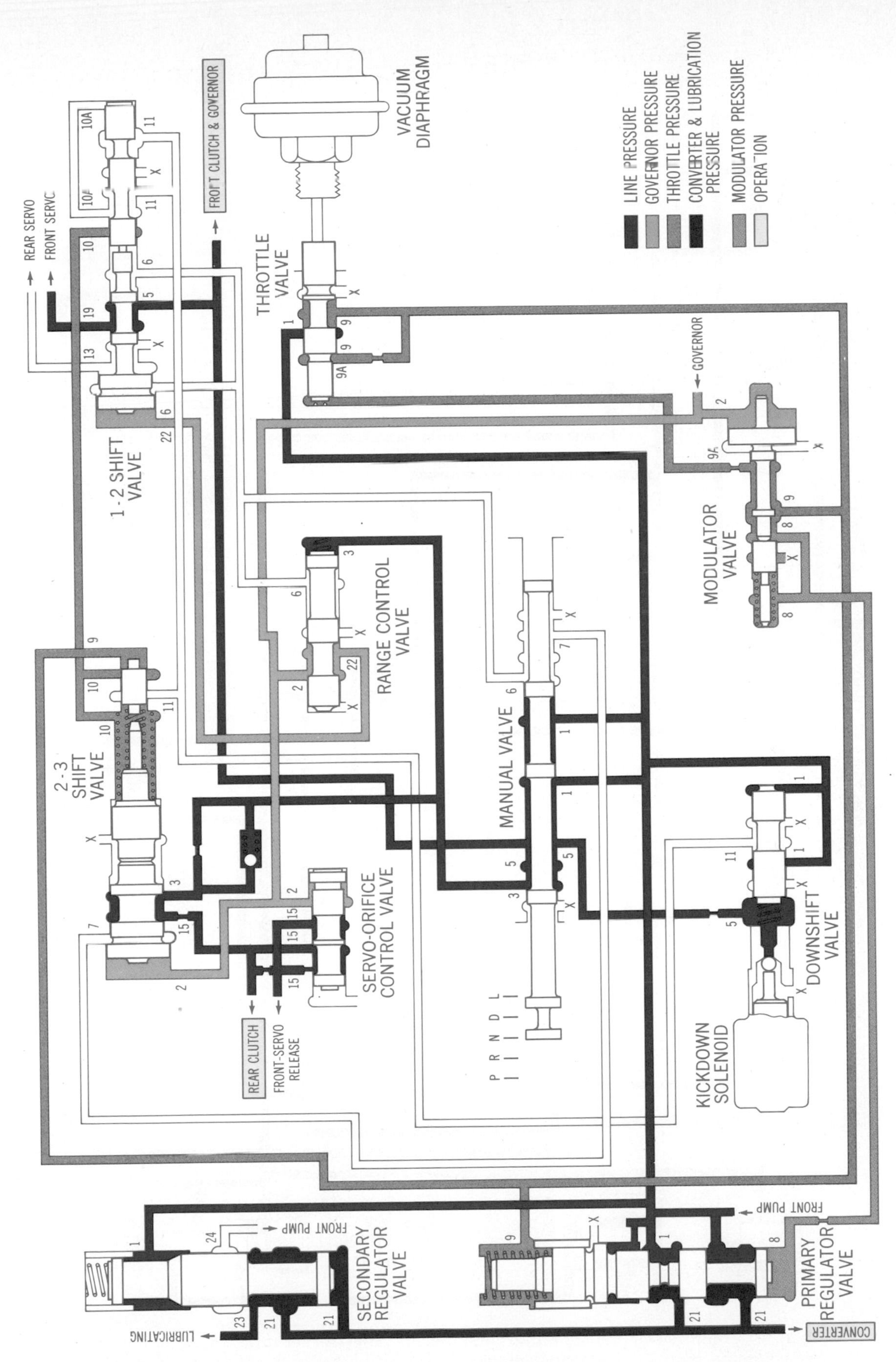

Plate 7. Conditions in the hydraulic circuit of the Borg-Warner three-speed automatic transmission with the compound planetary gearset with the transmission in direct drive. (*Chrysler Corporation*) Refer next to Plate 6.

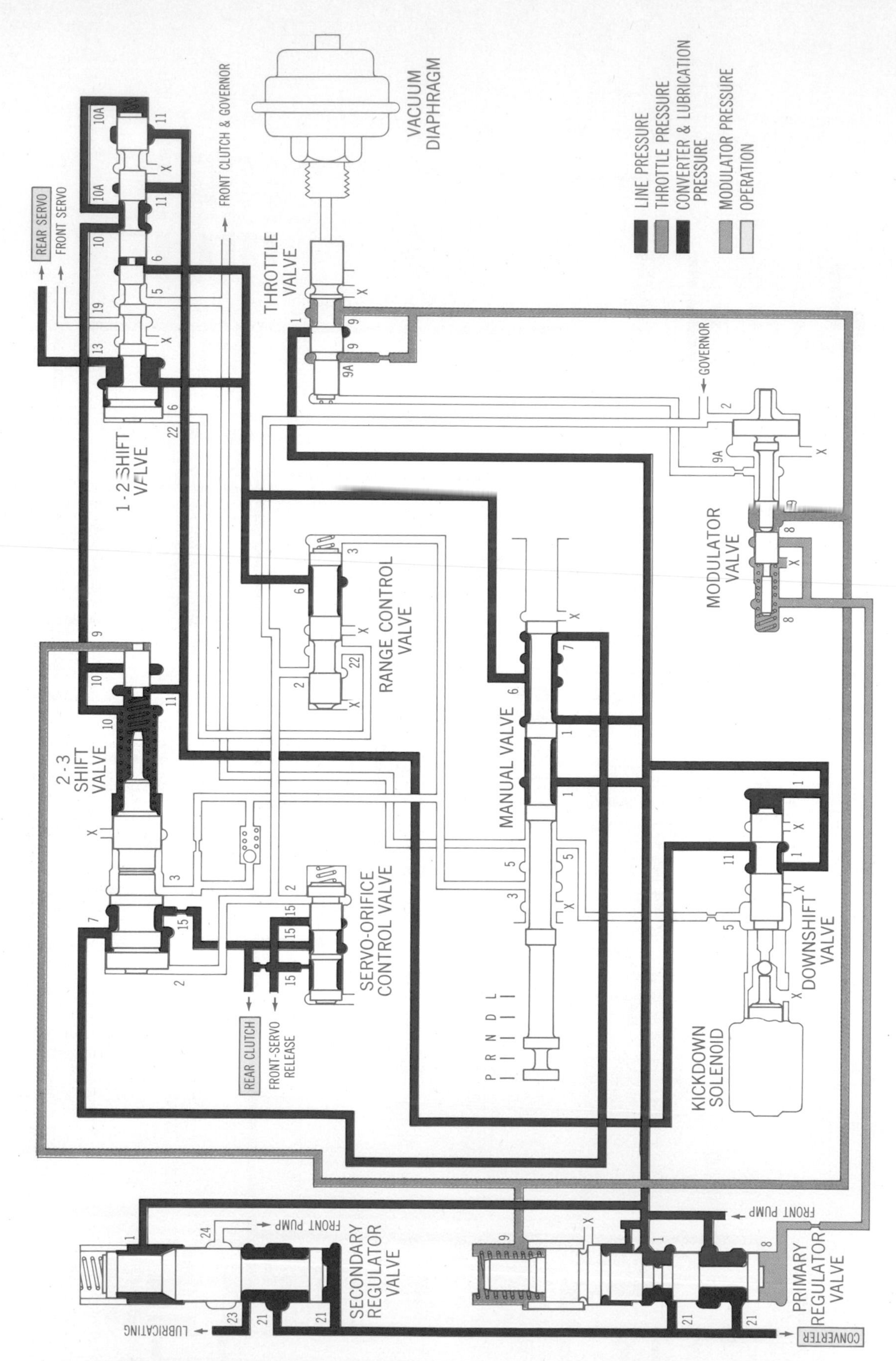

Plate 8. Conditions in the hydraulic circuit of the Borg-Warner three-speed automatic transmission with the compound planetary gearset with the transmission in R. The front brake band and front clutch are released. The rear clutch and rear brake band are applied. (*Chrysler Corporation*)

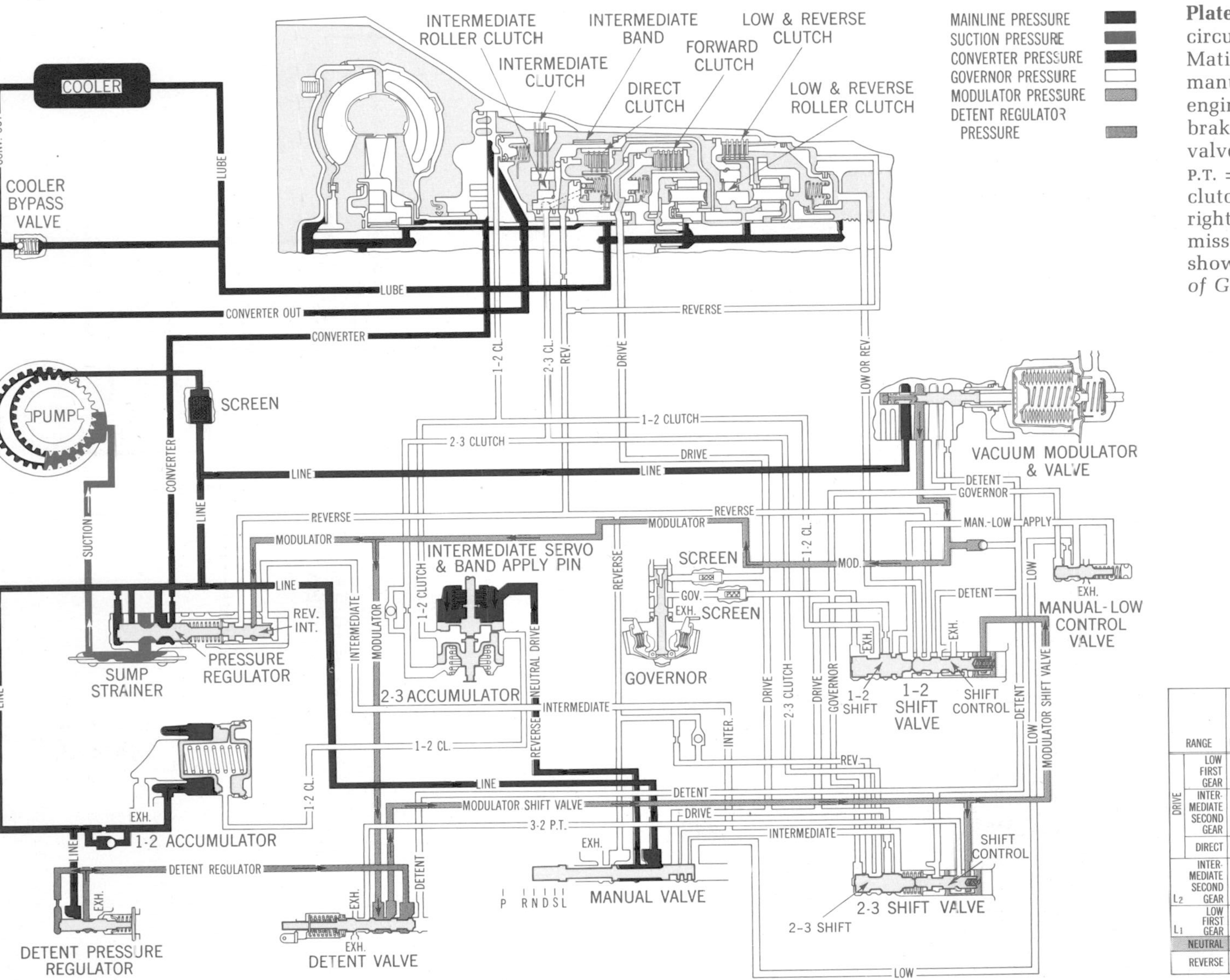

Plate 9. Conditions in the hydraulic circuit of the Type 350 Turbo Hydra-Matic automatic transmission with the manual valve in neutral and with the engine running. All clutches and the brake band are off. REV. = reverse boost valve, INT. = intermediate boost valve, P.T. = part throttle, and 1-2 CL = 1-2 clutch. The red band in the table (lower right) indicates the action of each transmission element for the range and gear shown in the plate. (*Oldsmobile Division of General Motors Corporation*)

	RANGE	RATIO	INTER-MEDIATE CLUTCH	DIRECT CLUTCH	FORWARD CLUTCH	LOW-AND-REVERSE CLUTCH	INTER-MEDIATE OVERRUN BAND	LOW-AND-REVERSE ROLLER CLUTCH	INTER-MEDIATE OVERRUN ROLLER CLUTCH
DRIVE	LOW FIRST GEAR	2.52			ON			LOCKED	LOCKED
	INTER-MEDIATE SECOND GEAR	1.52	ON		ON			FREE-WHEEL	LOCKED
	DIRECT	1.00	ON	ON	ON			FREE-WHEEL	FREE-WHEEL
L_2	INTER-MEDIATE SECOND GEAR	1.52	ON		ON			FREE-WHEEL	LOCKED
L_1	LOW FIRST GEAR	2.52			ON	ON		LOCKED	LOCKED
	NEUTRAL		OFF	OFF	OFF	OFF	OFF		
	REVERSE	1.93		ON		ON		FREE-WHEEL	FREE-WHEEL

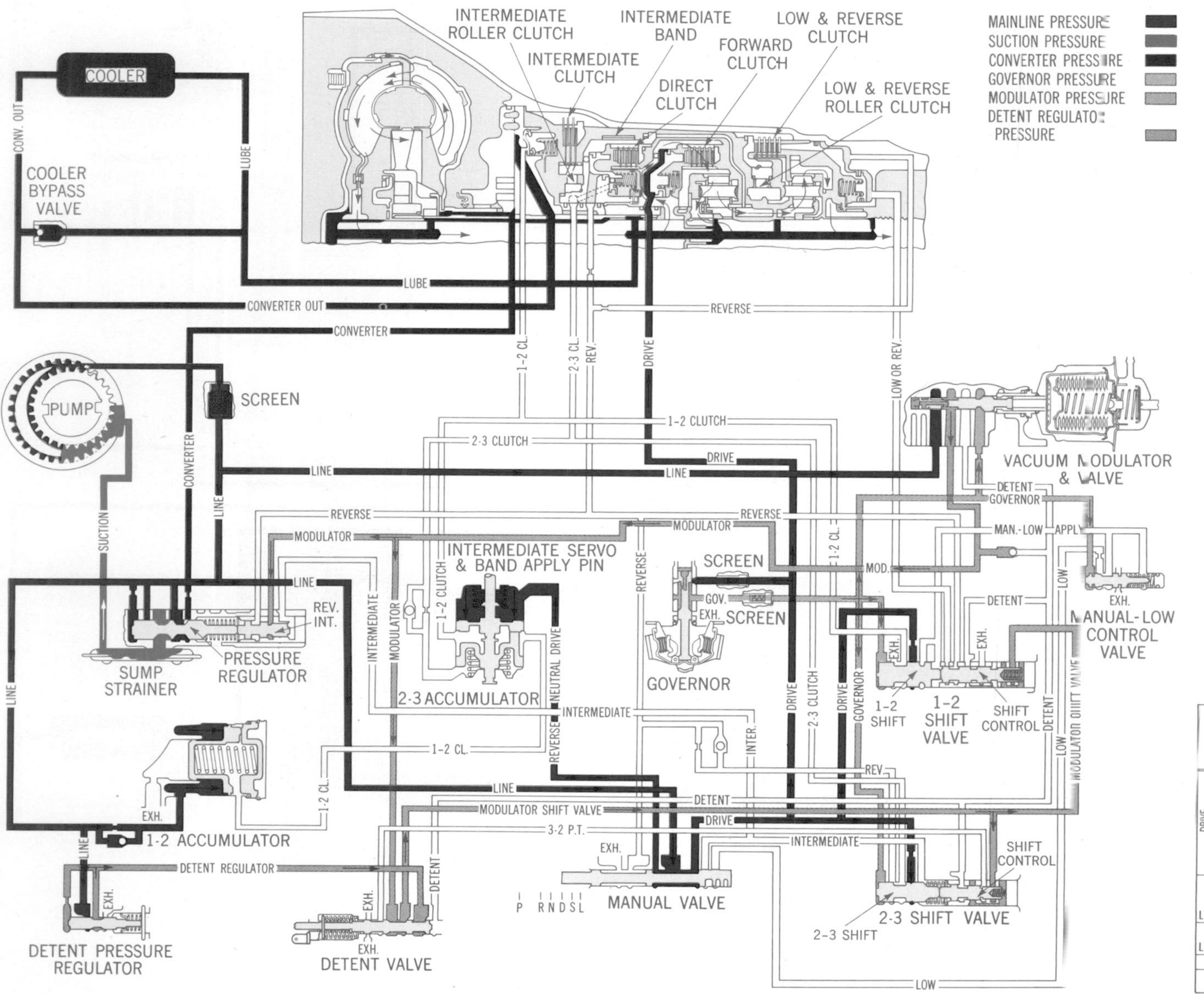

Plate 10. Conditions in the hydraulic circuit of the Type 350 Turbo Hydra-Matic automatic transmission with the manual valve in the drive range and with the transmission in first gear. Both roller clutches are effective, that is, locked, and the forward clutch is applied. Arrows in the transmission show power flow. The red band in the table (lower right) indicates the action of each transmission element for the range and gear shown in the plate. (*Buick Motor Division of General Motors Corporation*)

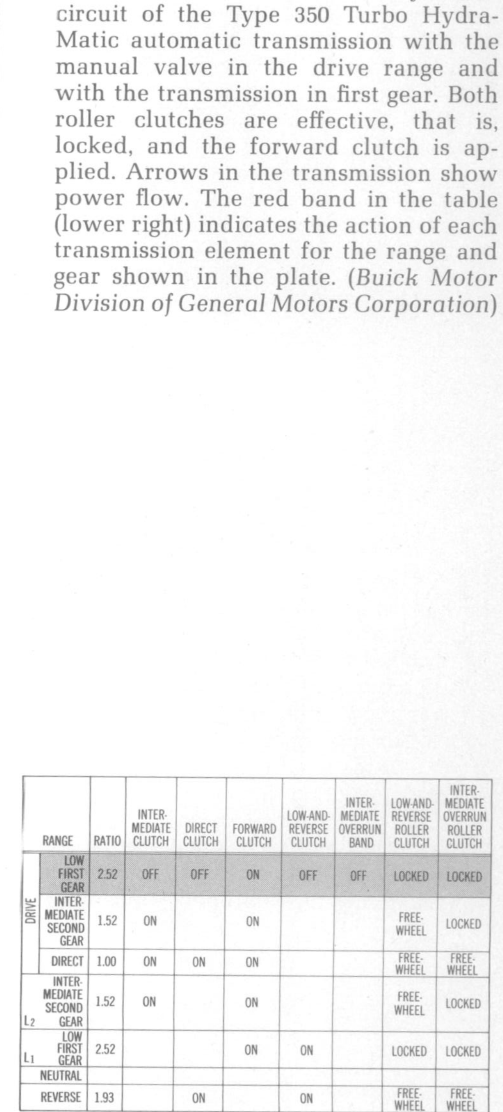

RANGE		RATIO	INTERMEDIATE CLUTCH	DIRECT CLUTCH	FORWARD CLUTCH	LOW-AND-REVERSE CLUTCH	INTERMEDIATE OVERRUN BAND	LOW-AND-REVERSE ROLLER CLUTCH	INTERMEDIATE OVERRUN ROLLER CLUTCH
DRIVE	LOW FIRST GEAR	2.52	OFF	OFF	ON	OFF	OFF	LOCKED	LOCKED
	INTERMEDIATE SECOND GEAR	1.52	ON		ON			FREE-WHEEL	LOCKED
	DIRECT	1.00	ON	ON	ON			FREE-WHEEL	FREE-WHEEL
L2	INTERMEDIATE SECOND GEAR	1.52	ON		ON			FREE-WHEEL	LOCKED
L1	LOW FIRST GEAR	2.52			ON	ON		LOCKED	LOCKED
NEUTRAL									
REVERSE		1.93		ON		ON		FREE-WHEEL	FREE-WHEEL

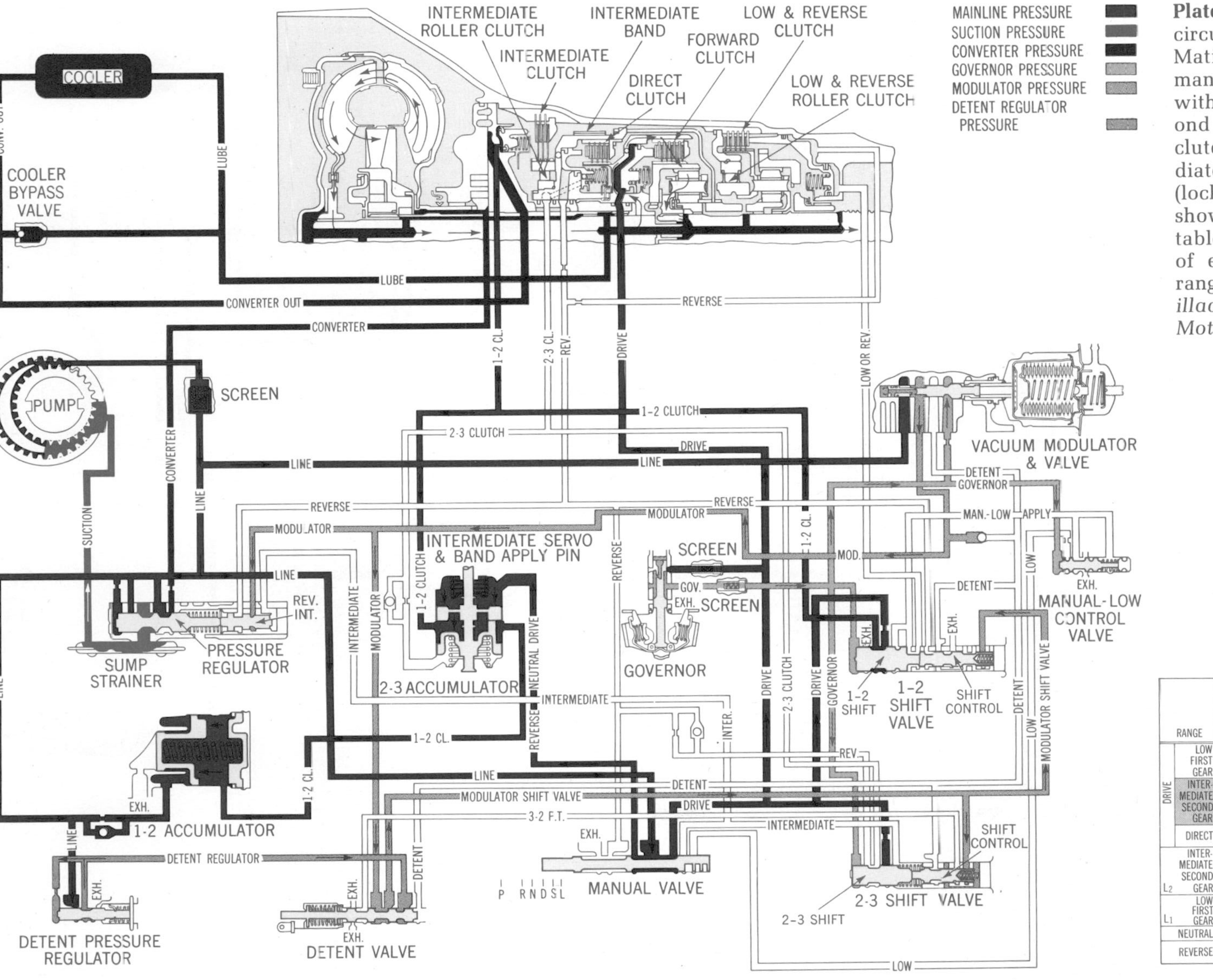

RANGE		RATIO	INTER-MEDIATE CLUTCH	DIRECT CLUTCH	FORWARD CLUTCH	LOW-AND-REVERSE CLUTCH	INTER-MEDIATE OVERRUN BAND	LOW-AND-REVERSE ROLLER CLUTCH	INTER-MEDIATE OVERRUN ROLLER CLUTCH
DRIVE	LOW FIRST GEAR	2.52			ON			LOCKED	LOCKED
DRIVE	INTER-MEDIATE SECOND GEAR	1.52	ON	OFF	ON	OFF	OFF	FREE-WHEEL	LOCKED
DRIVE	DIRECT	1.00	ON	ON	ON			FREE-WHEEL	FREE-WHEEL
L2	INTER-MEDIATE SECOND GEAR	1.52	ON		ON			FREE-WHEEL	LOCKED
L1	LOW FIRST GEAR	2.52			ON	ON		LOCKED	LOCKED
NEUTRAL									
REVERSE		1.93		ON		ON		FREE-WHEEL	FREE-WHEEL

Plate 11. Conditions in the hydraulic circuit of the Type 350 Turbo Hydra-Matic automatic transmission with the manual valve in the drive range and with the transmission upshifted to second gear. The intermediate and forward clutches are applied, and the intermediate overrun roller clutch is effective (locked). Arrows in the transmission show power flow. The red band in the table (lower right) indicates the action of each transmission element for the range and gear shown in the plate.(*Cadillac Motor Car Division of General Motors Corporation*)

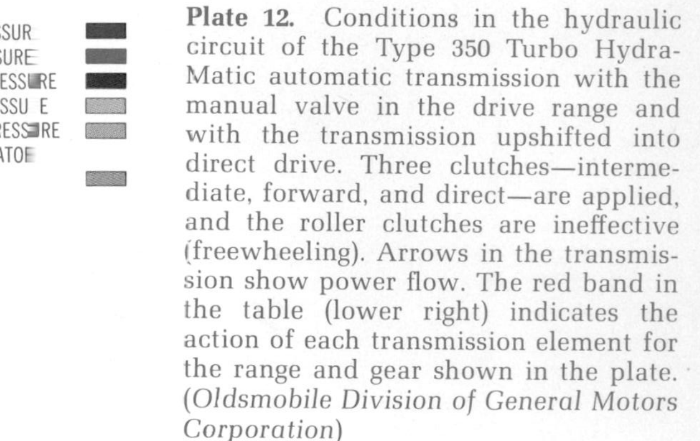

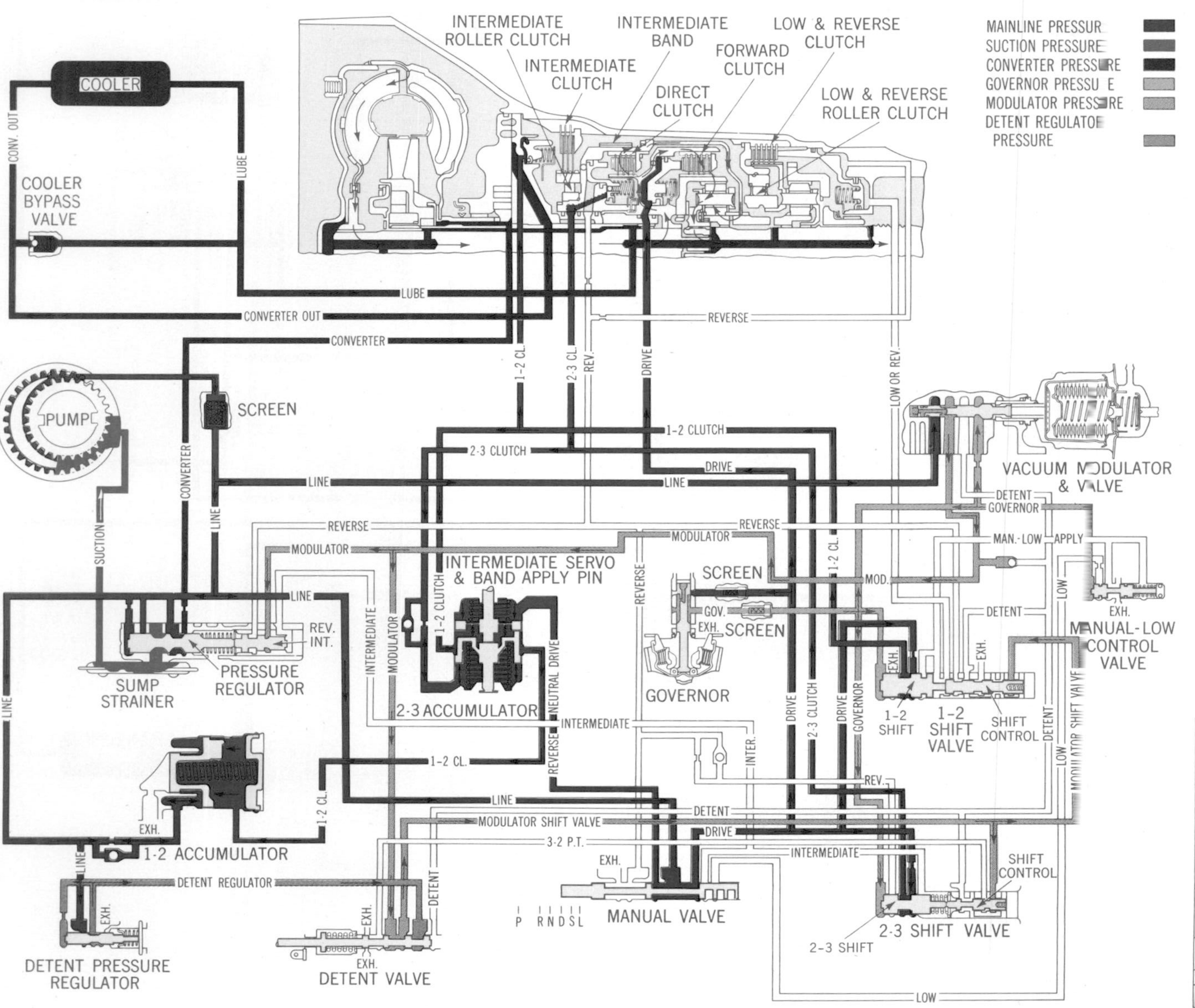

Plate 12. Conditions in the hydraulic circuit of the Type 350 Turbo Hydra-Matic automatic transmission with the manual valve in the drive range and with the transmission upshifted into direct drive. Three clutches—intermediate, forward, and direct—are applied, and the roller clutches are ineffective (freewheeling). Arrows in the transmission show power flow. The red band in the table (lower right) indicates the action of each transmission element for the range and gear shown in the plate. (*Oldsmobile Division of General Motors Corporation*)

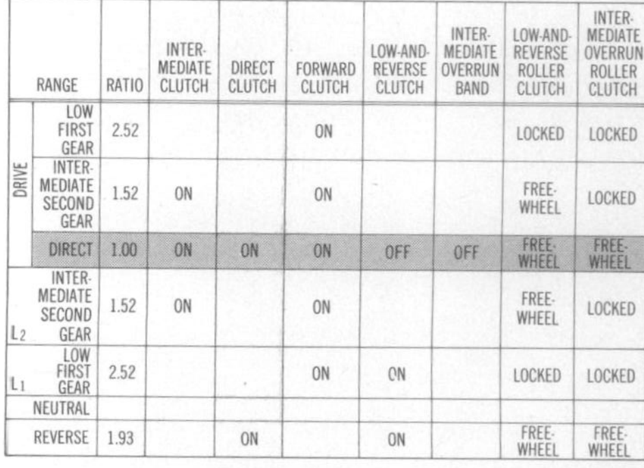

	RANGE	RATIO	INTERMEDIATE CLUTCH	DIRECT CLUTCH	FORWARD CLUTCH	LOW-AND-REVERSE CLUTCH	INTERMEDIATE OVERRUN BAND	LOW-AND-REVERSE ROLLER CLUTCH	INTERMEDIATE OVERRUN ROLLER CLUTCH
DRIVE	LOW FIRST GEAR	2.52			ON			LOCKED	LOCKED
	INTERMEDIATE SECOND GEAR	1.52	ON		ON			FREE-WHEEL	LOCKED
	DIRECT	1.00	ON	ON	ON	OFF	OFF	FREE-WHEEL	FREE-WHEEL
L2	INTERMEDIATE SECOND GEAR	1.52	ON		ON			FREE-WHEEL	LOCKED
L1	LOW FIRST GEAR	2.52			ON	ON		LOCKED	LOCKED
	NEUTRAL								
	REVERSE	1.93		ON		ON		FREE-WHEEL	FREE-WHEEL

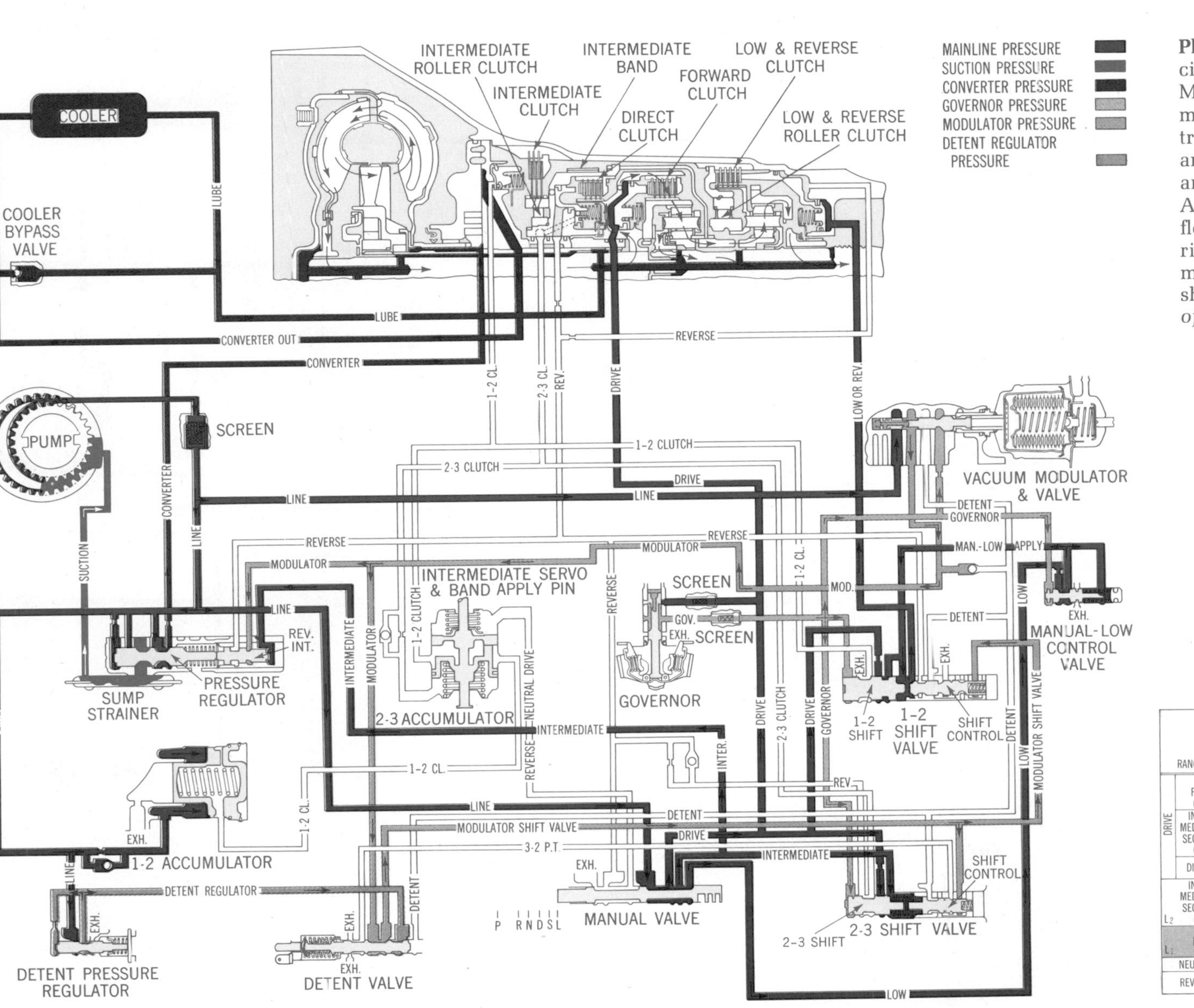

Plate 13. Conditions in the hydraulic circuit of the Type 350 Turbo Hydra-Matic automatic transmission with the manual valve in the low range and the transmission in first gear. The forward and low-and-reverse clutches are on, and both roller clutches are effective. Arrows in the transmission show power flow. The red band in the table (lower right) indicates the action of each transmission element for the range and gear shown in the plate. (*Oldsmobile Division of General Motors Corporation*)

	RANGE	RATIO	INTER-MEDIATE CLUTCH	DIRECT CLUTCH	FORWARD CLUTCH	LOW-AND-REVERSE CLUTCH	INTER-MEDIATE OVERRUN BAND	LOW-AND-REVERSE ROLLER CLUTCH	INTER-MEDIATE OVERRUN ROLLER CLUTCH
DRIVE	LOW FIRST GEAR	2.52			ON			LOCKED	LOCKED
DRIVE	INTER-MEDIATE SECOND GEAR	1.52	ON		ON			FREE-WHEEL	LOCKED
DRIVE	DIRECT	1.00	ON	ON	ON			FREE-WHEEL	FREE-WHEEL
L2	INTER-MEDIATE SECOND GEAR	1.52	ON		ON			FREE-WHEEL	LOCKED
L1	LOW FIRST GEAR	2.52	OFF	OFF	ON	ON	OFF	LOCKED	LOCKED
	NEUTRAL								
	REVERSE	1.93		ON		ON		FREE-WHEEL	FREE-WHEEL

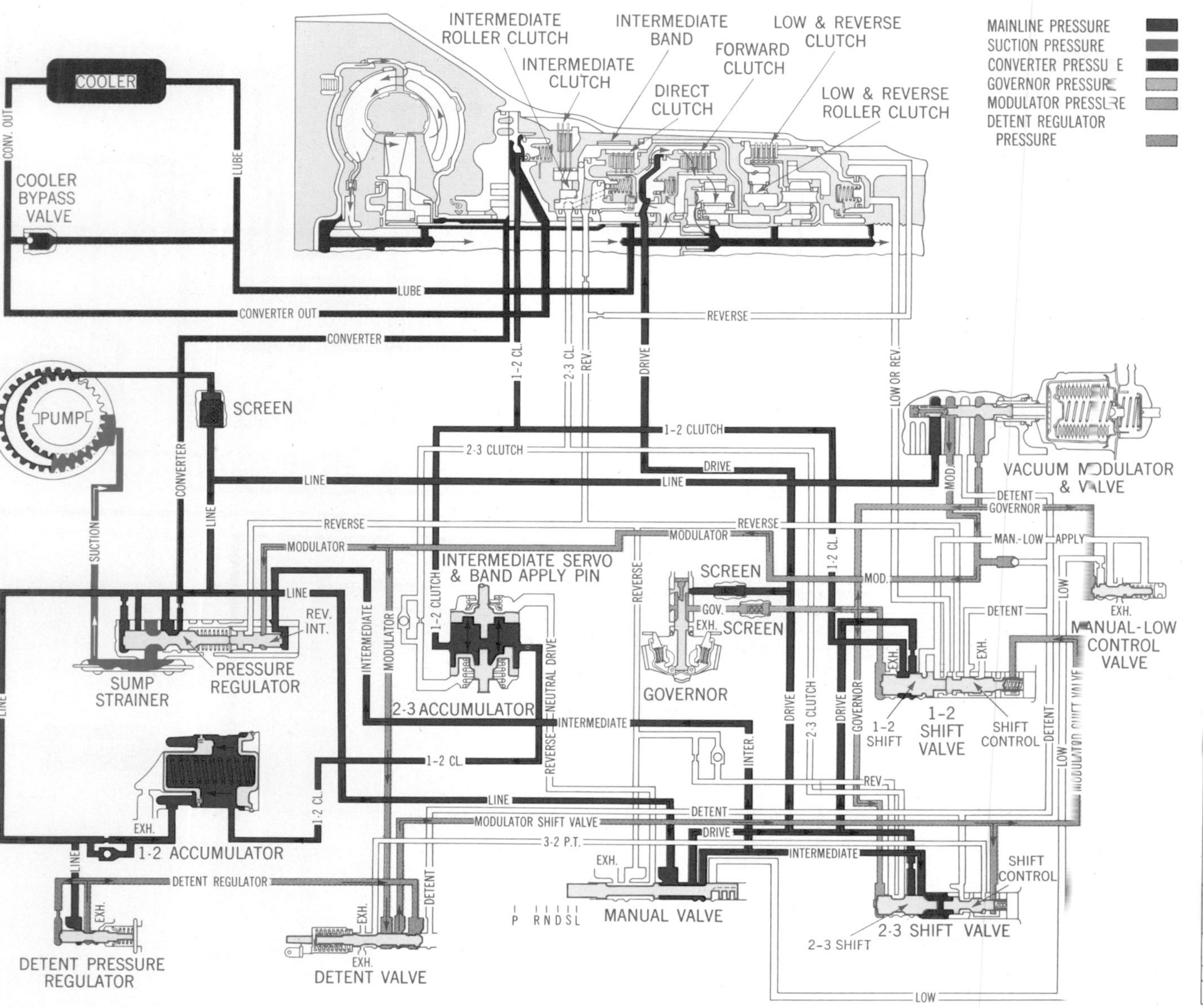

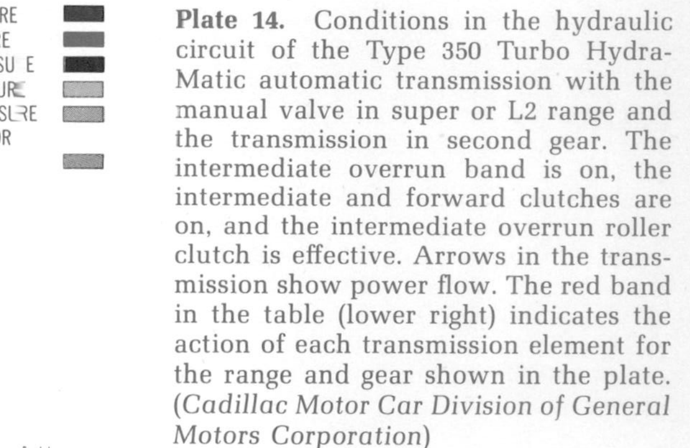

Plate 14. Conditions in the hydraulic circuit of the Type 350 Turbo Hydra-Matic automatic transmission with the manual valve in super or L2 range and the transmission in second gear. The intermediate overrun band is on, the intermediate and forward clutches are on, and the intermediate overrun roller clutch is effective. Arrows in the transmission show power flow. The red band in the table (lower right) indicates the action of each transmission element for the range and gear shown in the plate. (*Cadillac Motor Car Division of General Motors Corporation*)

	RANGE	RATIO	INTER-MEDIATE CLUTCH	DIRECT CLUTCH	FORWARD CLUTCH	LOW-AND-REVERSE CLUTCH	INTER-MEDIATE OVERRUN BAND	LOW-AND-REVERSE ROLLER CLUTCH	INTER-MEDIATE OVERRUN ROLLER CLUTCH
DRIVE	LOW FIRST GEAR	2.52			ON			LOCKED	LOCKED
DRIVE	INTER-MEDIATE SECOND GEAR	1.52	ON		ON			FREE-WHEEL	LOCKED
DRIVE	DIRECT	1.00	ON	ON	ON			FREE-WHEEL	FREE-WHEEL
L2	INTER-MEDIATE SECOND GEAR	1.52	ON	OFF	ON	OFF	ON	FREE-WHEEL	LOCKED
L1	LOW FIRST GEAR	2.52			ON	ON		LOCKED	LOCKED
	NEUTRAL								
	REVERSE	1.93		ON		ON		FREE-WHEEL	FREE-WHEEL

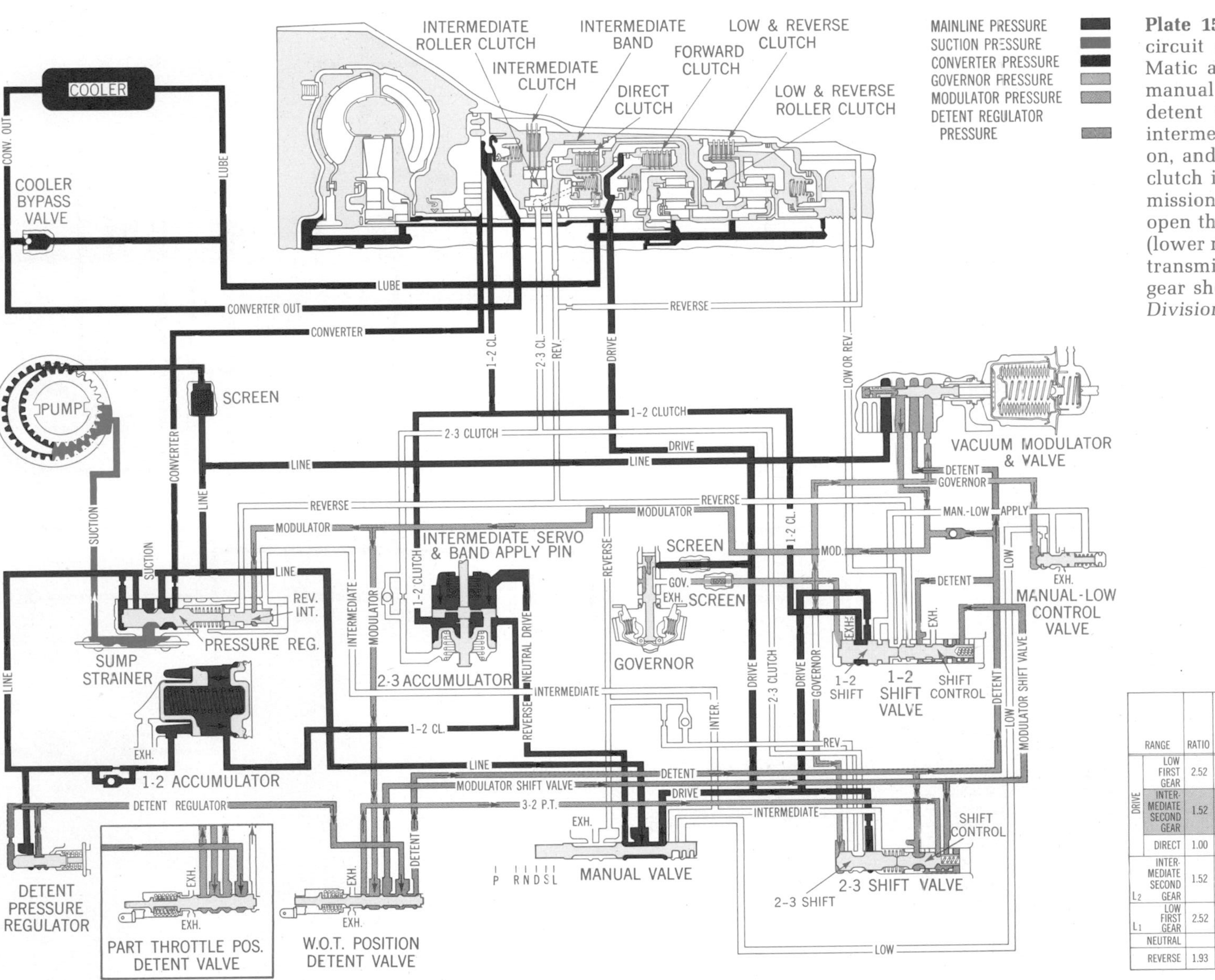

Plate 15. Conditions in the hydraulic circuit of the Type 350 Turbo Hydra-Matic automatic transmission with the manual valve in the drive range after a detent downshift to second gear. The intermediate and forward clutches are on, and the intermediate overrun roller clutch is effective. Arrows in the transmission show power flow. W.O.T. = wide-open throttle. The red band in the table (lower right) indicates the action of each transmission element for the range and gear shown in the plate. (*Pontiac Motor Division of General Motors Corporation*)

	Range	Ratio	Intermediate clutch	Direct clutch	Forward clutch	Low-and-reverse clutch	Intermediate overrun band	Low-and-reverse roller clutch	Intermediate overrun roller clutch
Drive	Low first gear	2.52			ON			LOCKED	LOCKED
Drive	Intermediate second gear	1.52	ON	OFF	ON	OFF	OFF	FREE-WHEEL	LOCKED
Drive	Direct	1.00	ON	ON	ON			FREE-WHEEL	FREE-WHEEL
L2	Intermediate second gear	1.52	ON		ON			FREE-WHEEL	LOCKED
L1	Low first gear	2.52			ON	ON		LOCKED	LOCKED
	Neutral								
	Reverse	1.93		ON		ON		FREE-WHEEL	FREE-WHEEL

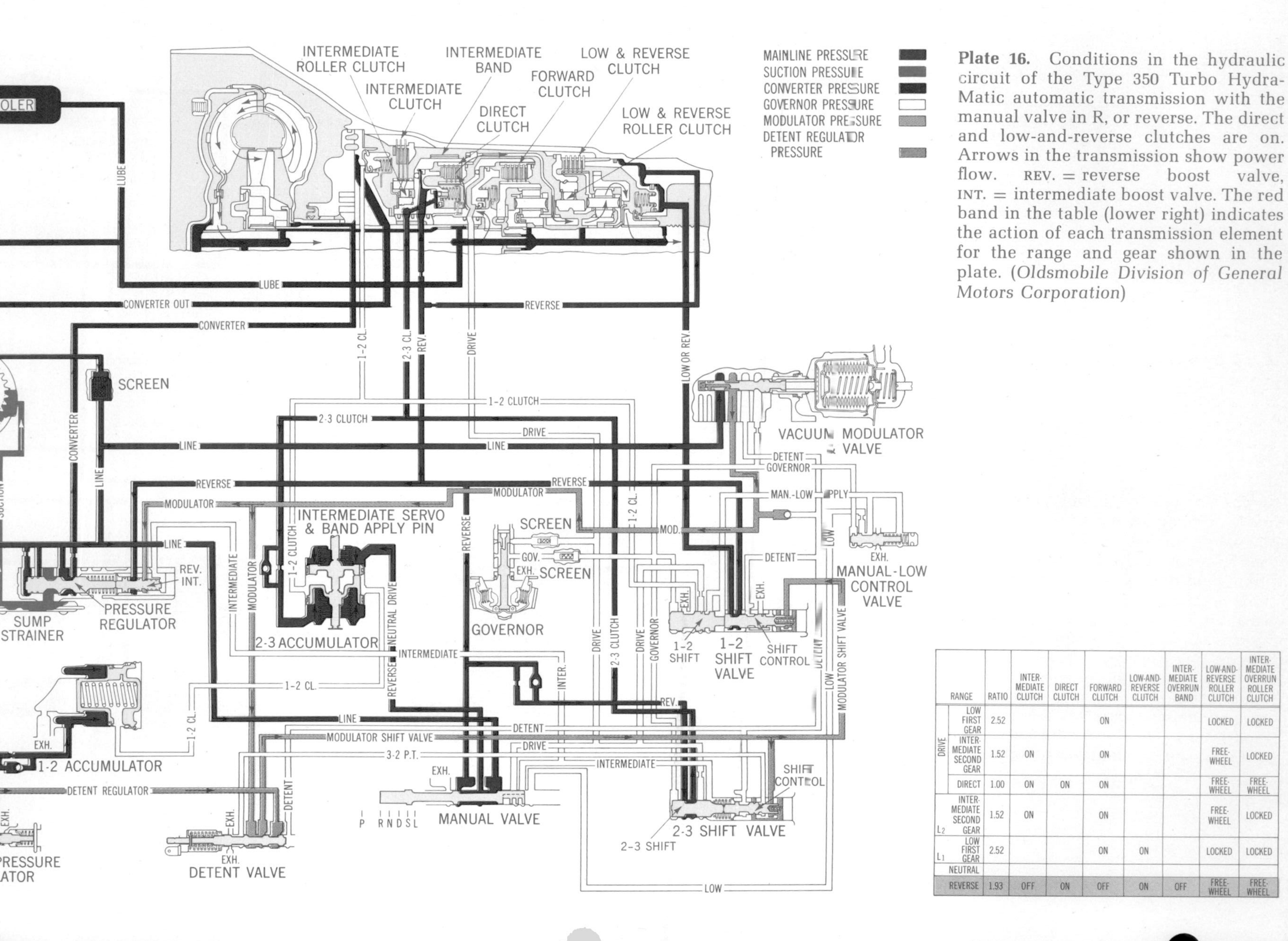

Plate 16. Conditions in the hydraulic circuit of the Type 350 Turbo Hydra-Matic automatic transmission with the manual valve in R, or reverse. The direct and low-and-reverse clutches are on. Arrows in the transmission show power flow. REV. = reverse boost valve, INT. = intermediate boost valve. The red band in the table (lower right) indicates the action of each transmission element for the range and gear shown in the plate. (*Oldsmobile Division of General Motors Corporation*)

	RANGE	RATIO	INTERMEDIATE CLUTCH	DIRECT CLUTCH	FORWARD CLUTCH	LOW-AND-REVERSE CLUTCH	INTERMEDIATE OVERRUN BAND	LOW-AND-REVERSE ROLLER CLUTCH	INTERMEDIATE OVERRUN ROLLER CLUTCH
DRIVE	LOW FIRST GEAR	2.52			ON			LOCKED	LOCKED
DRIVE	INTERMEDIATE SECOND GEAR	1.52	ON		ON			FREE-WHEEL	LOCKED
DRIVE	DIRECT	1.00	ON	ON	ON			FREE-WHEEL	FREE-WHEEL
L_2	INTERMEDIATE SECOND GEAR	1.52	ON		ON			FREE-WHEEL	LOCKED
L_1	LOW FIRST GEAR	2.52			ON	ON		LOCKED	LOCKED
	NEUTRAL								
	REVERSE	1.93	OFF	ON	OFF	ON	OFF	FREE-WHEEL	FREE-WHEEL

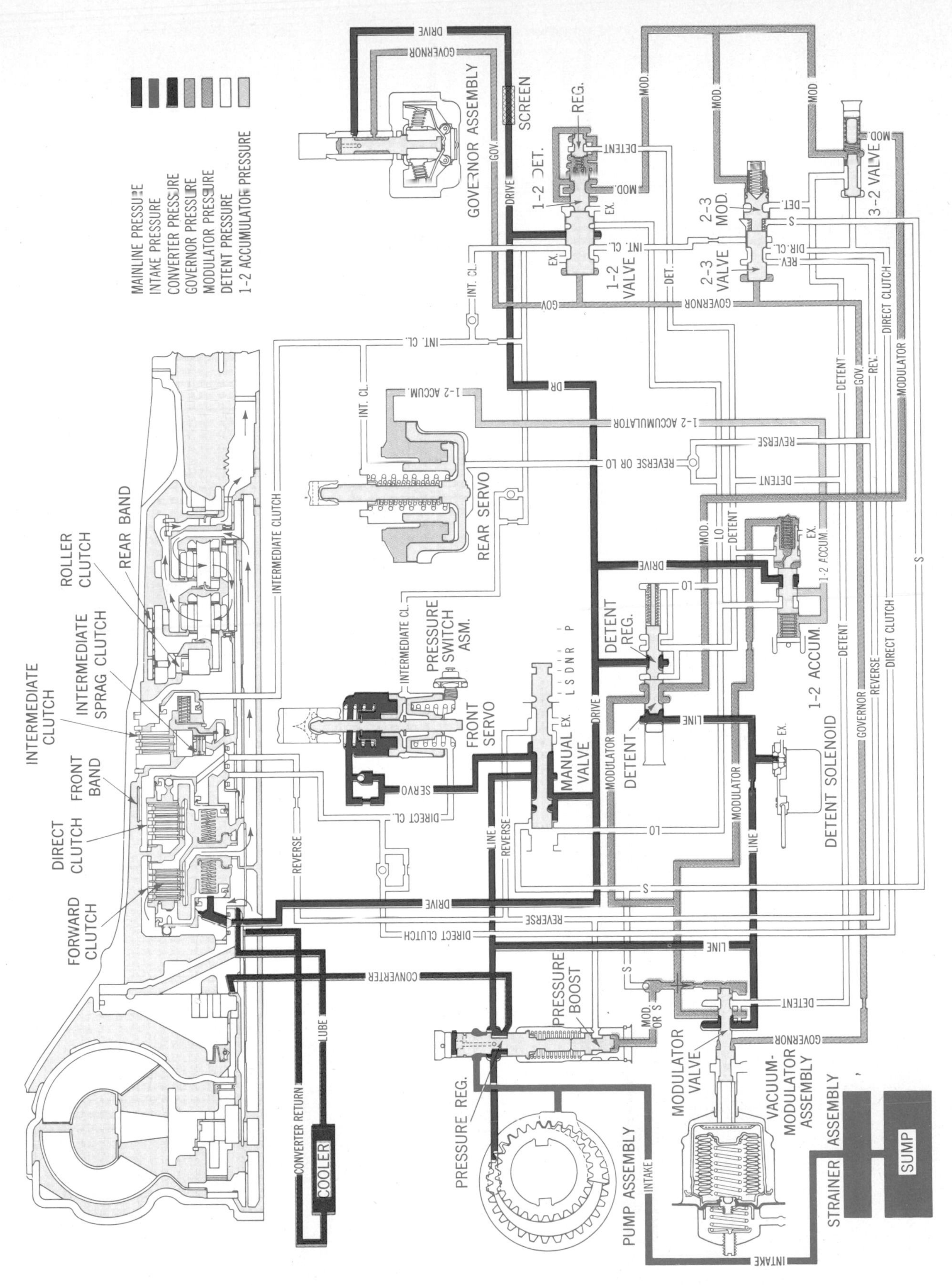

Plate 17. Conditions in the hydraulic circuit of the Type 400 Turbo Hydra-Matic automatic transmission with the transmission in drive range, first gear. The forward clutch (also called the front clutch) is applied, and the roller clutch is effective. Power flow through the transmission is shown by arrows. (*Buick Motor Division of General Motors Corporation*)

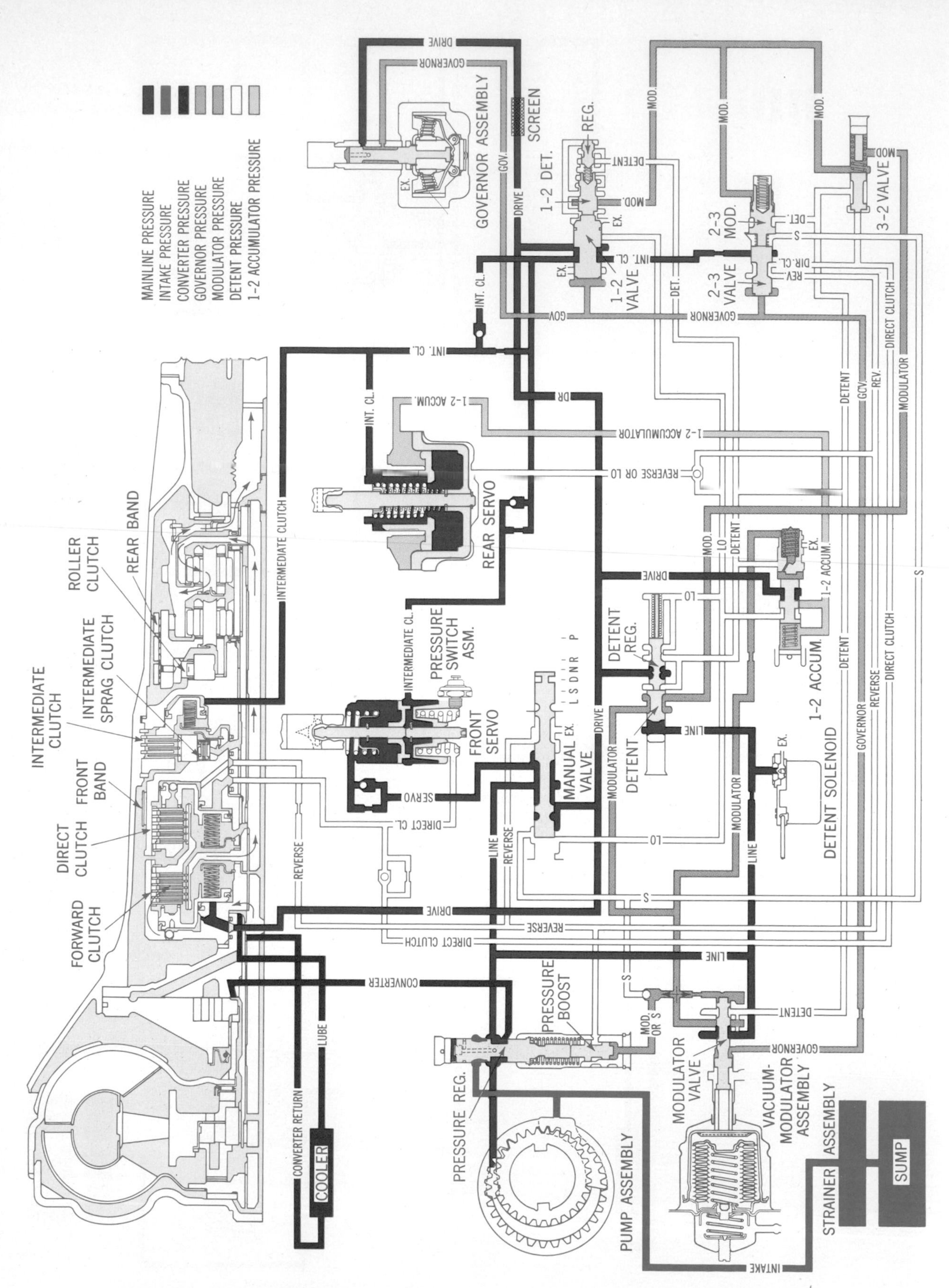

Plate 18. Conditions in the hydraulic circuit of the Type 400 Turbo Hydra-Matic automatic transmission with the transmission in drive range, second gear. The forward and intermediate clutches are applied, and the intermediate sprag clutch is effective. Power flow through the transmission is shown by arrows. (*Pontiac Motor Division of General Motors Corporation*)

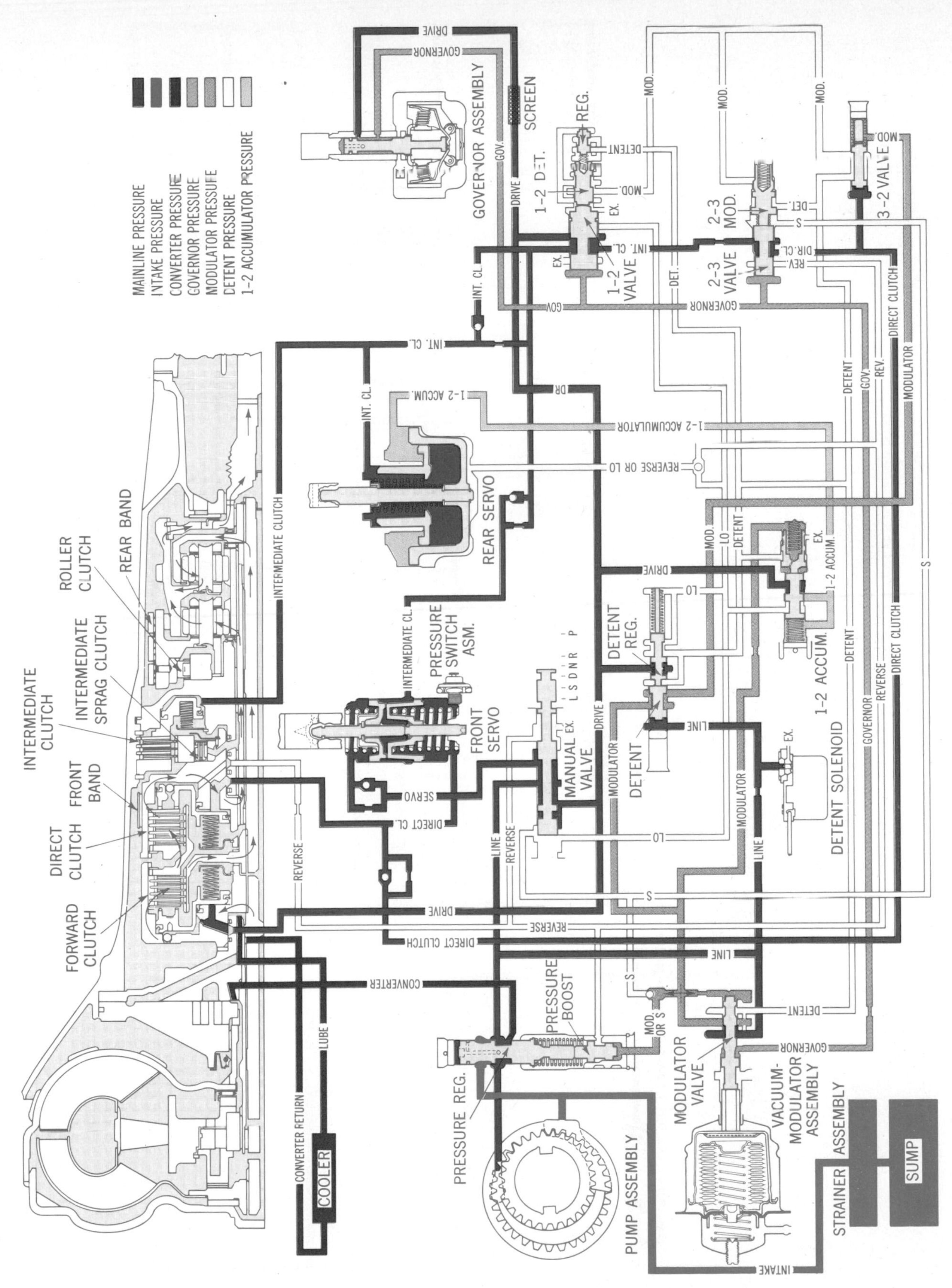

Plate 19. Conditions in the hydraulic circuit of the Type 400 Turbo Hydra-Matic automatic transmission with the transmission in drive range, third gear or direct drive. The forward, direct, and intermediate clutches are applied. Power flow through the transmission is shown by arrows. (*Pontiac Motor Division of General Motors Corporation*)

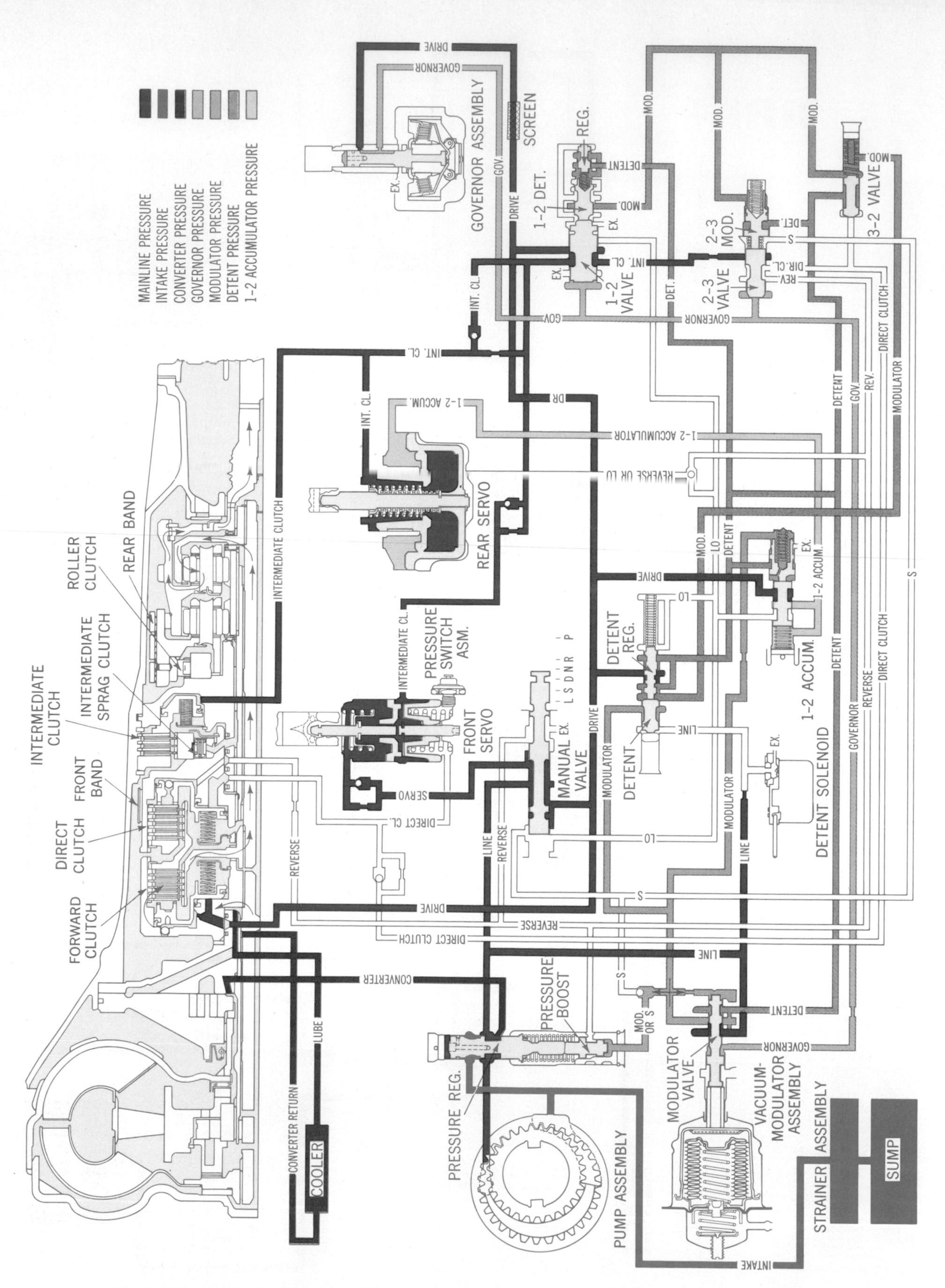

Plate 20. Conditions in the hydraulic circuit of the Type 400 Turbo Hydra-Matic automatic transmission after a detent downshift to second. The forward and intermediate clutches are applied, and the intermediate sprag clutch is effective. Power flow through the transmission is shown by arrows. (*Oldsmobile Division of General Motors Corporation*)

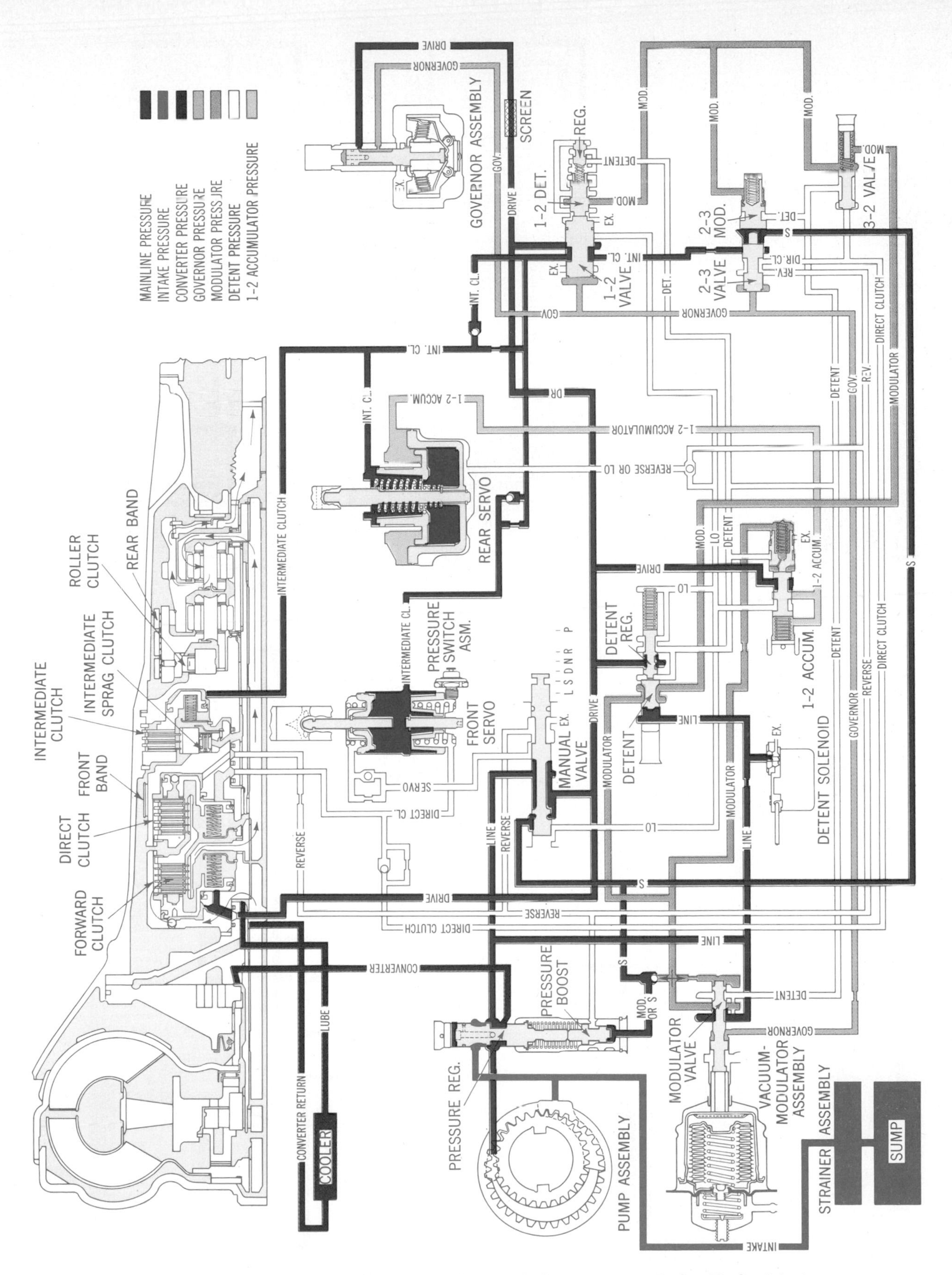

Plate 21. Conditions in the hydraulic circuit of the Type 400 Turbo Hydra-Matic automatic transmission with the transmission in L2 or super range, second gear. The forward and intermediate clutches are applied. The front band is applied to give engine braking going down hills. The intermediate sprag clutch is effective. Power flow through the transmission is shown by arrows. (*Buick Motor Division of General Motors Corporation*)

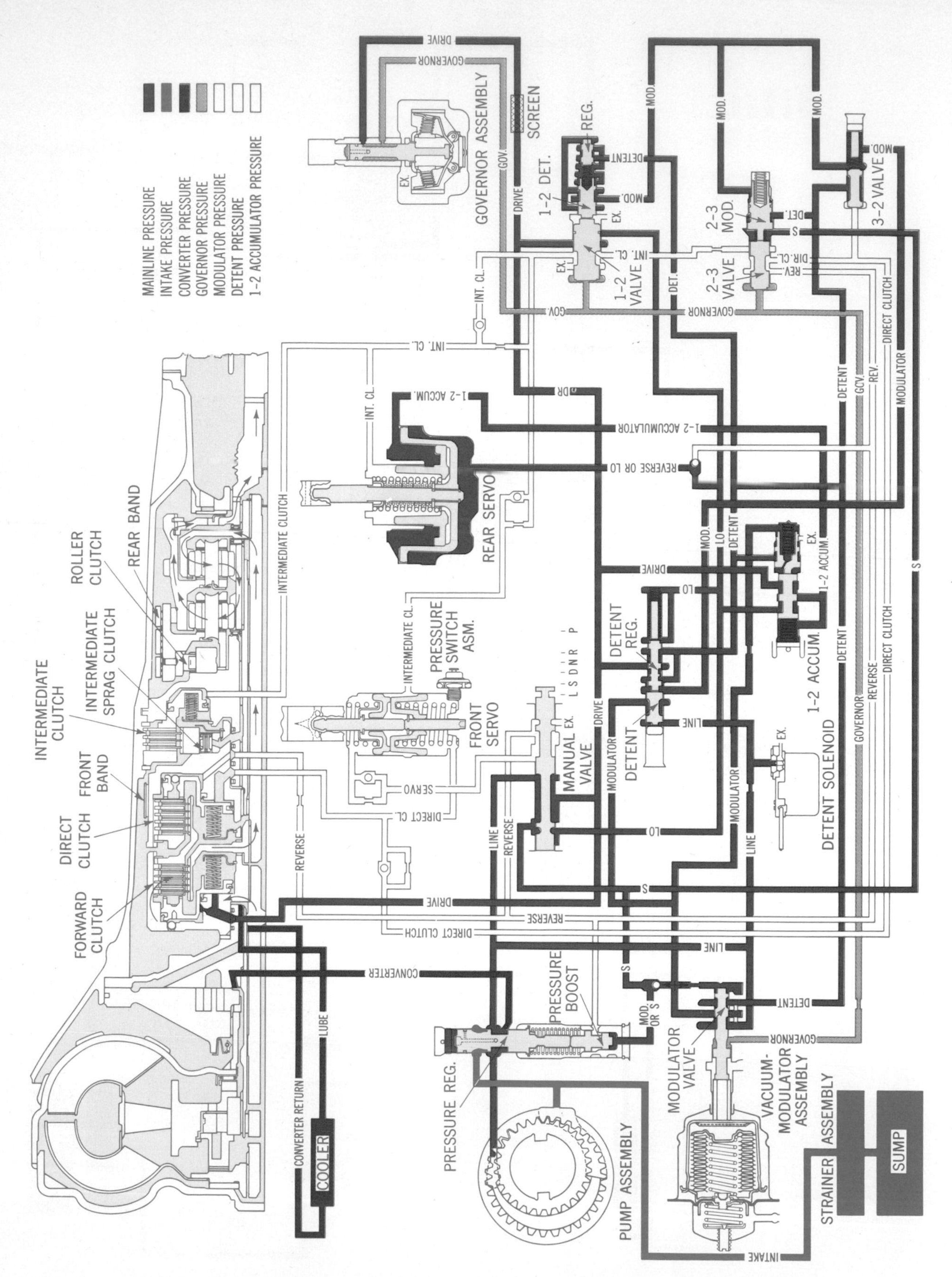

Plate 22. Conditions in the hydraulic circuit of the Type 400 Turbo Hydra-Matic automatic transmission with the transmission in low range, first gear. The forward clutch is applied and the roller clutch is effective. The rear band is applied to give engine braking going down hills. Power flow through the transmission is shown by arrows. (*Buick Motor Division of General Motors Corporation*)

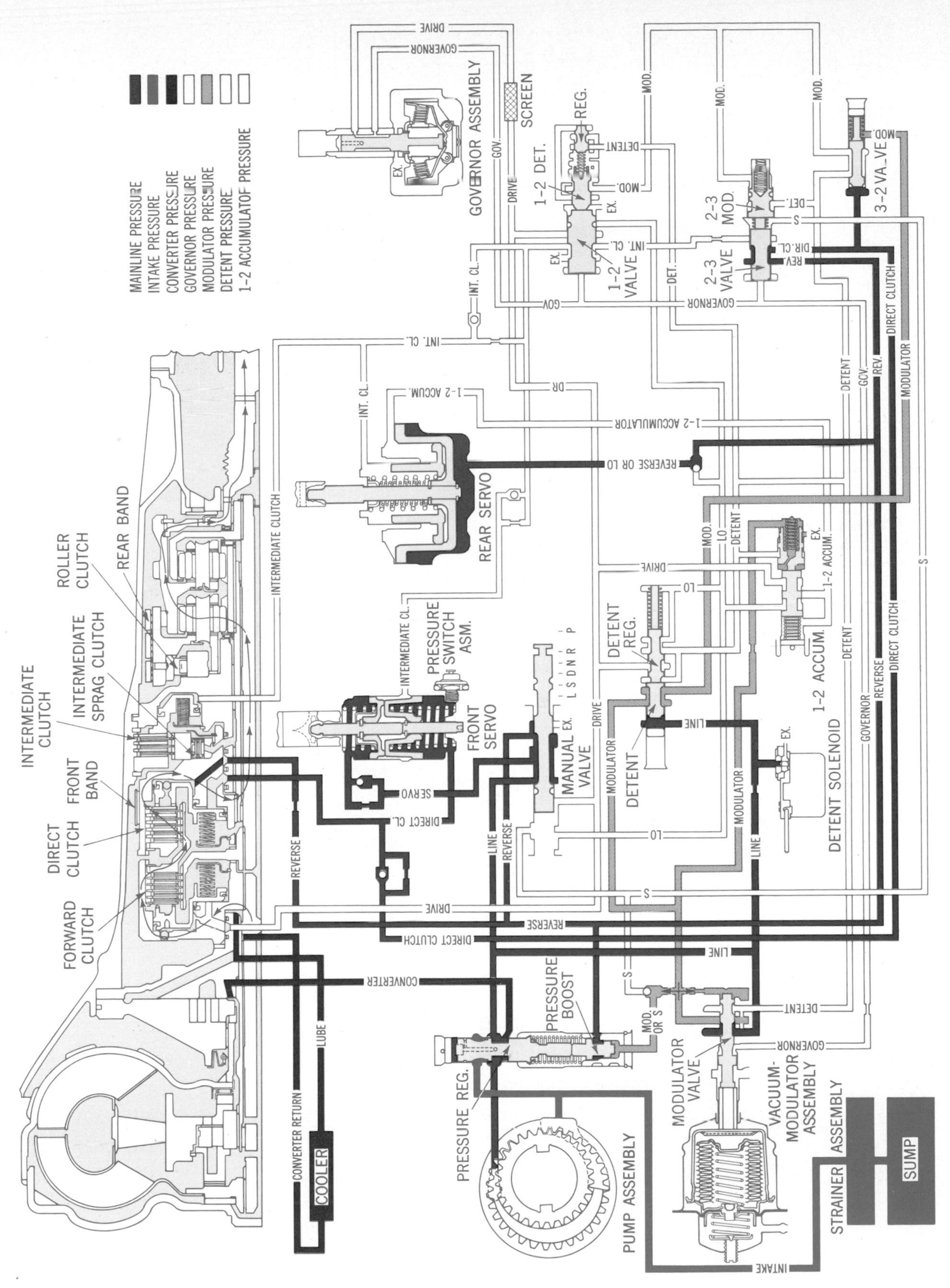

Plate 23. Conditions in the hydraulic circuit of the Type 400 Turbo Hydra-Matic automatic transmission with the transmission in reverse. The direct clutch is applied and the rear band is applied. Power flow through the transmission is shown by arrows. (*Cadillac Motor Car Division of General Motors Corporation*)

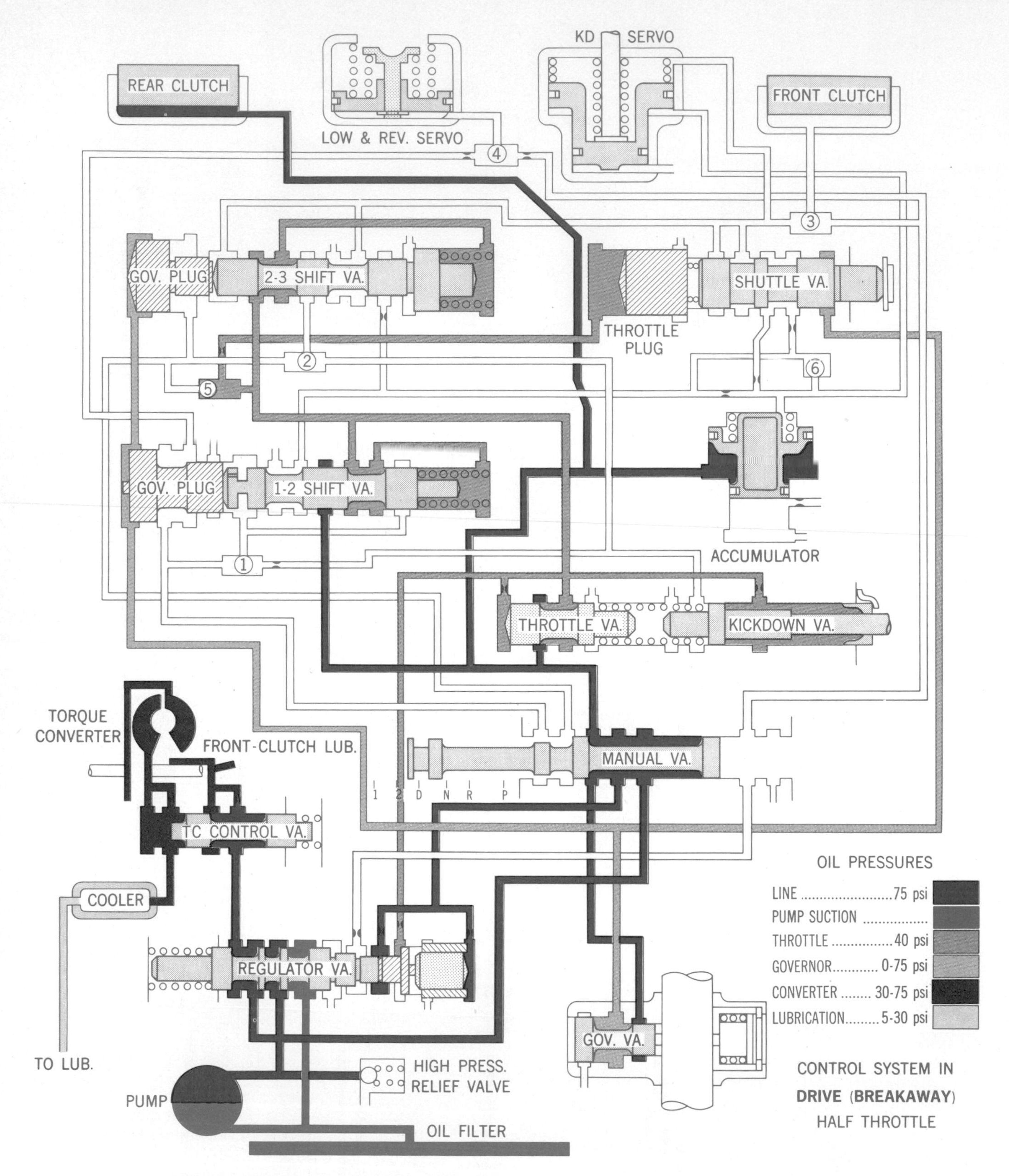

Plate 24. Hydraulic circuit of the TorqueFlite automatic transmission in D when first breaking away from a standing start, with the gears in first, or low. TC CONTROL VA. is the torque-converter control valve, which prevents high pressure in the torque converter. This circuit is for a transmission used with a six-cylinder engine. The transmission for the eight-cylinder engine includes a limit valve and a throttle plug to provide added control of the kickdown band. (*Chrysler Corporation*)

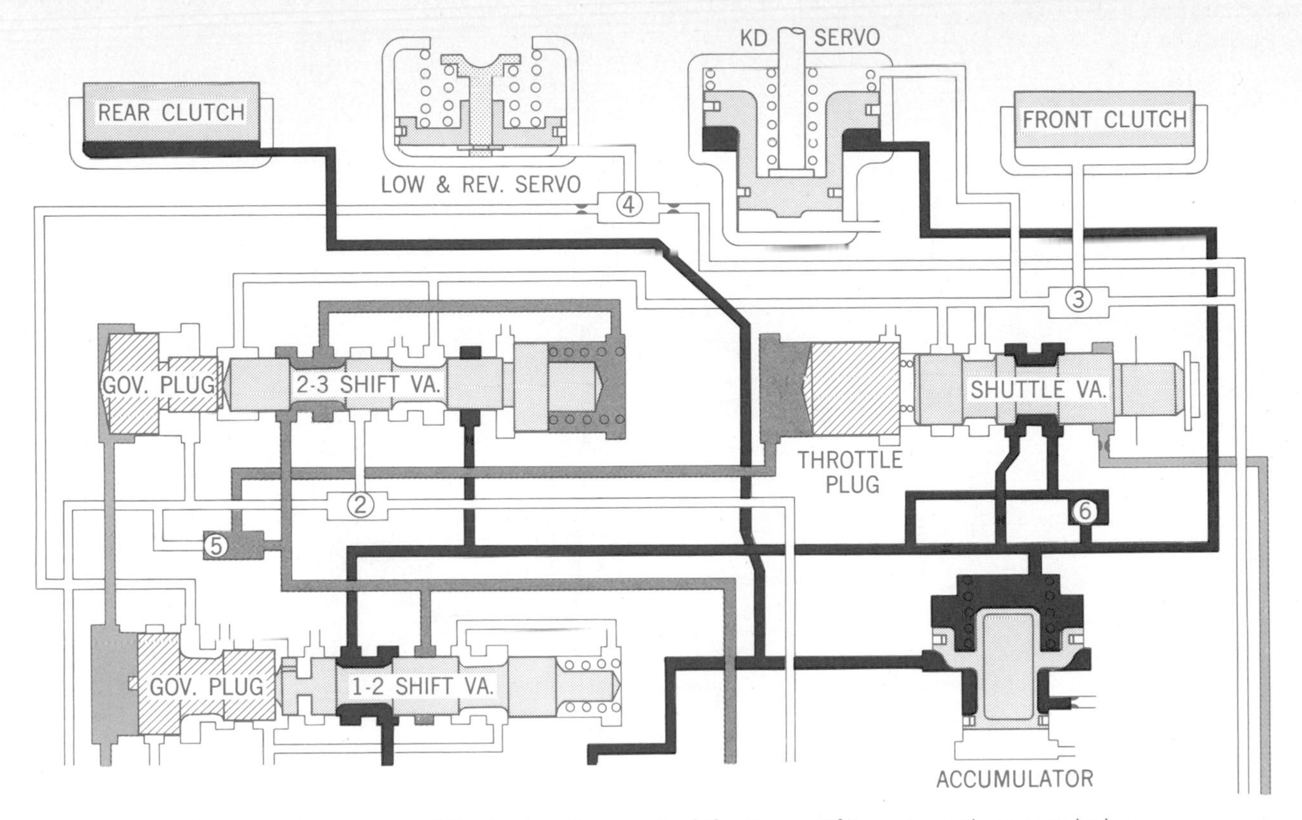

Plate 25. Upper part of the hydraulic circuit of the TorqueFlite automatic transmission in D after an upshift to second gear. Here, only the upper part of the circuit, where conditions have changed from Plate 24, is shown. (*Chrysler Corporation*)

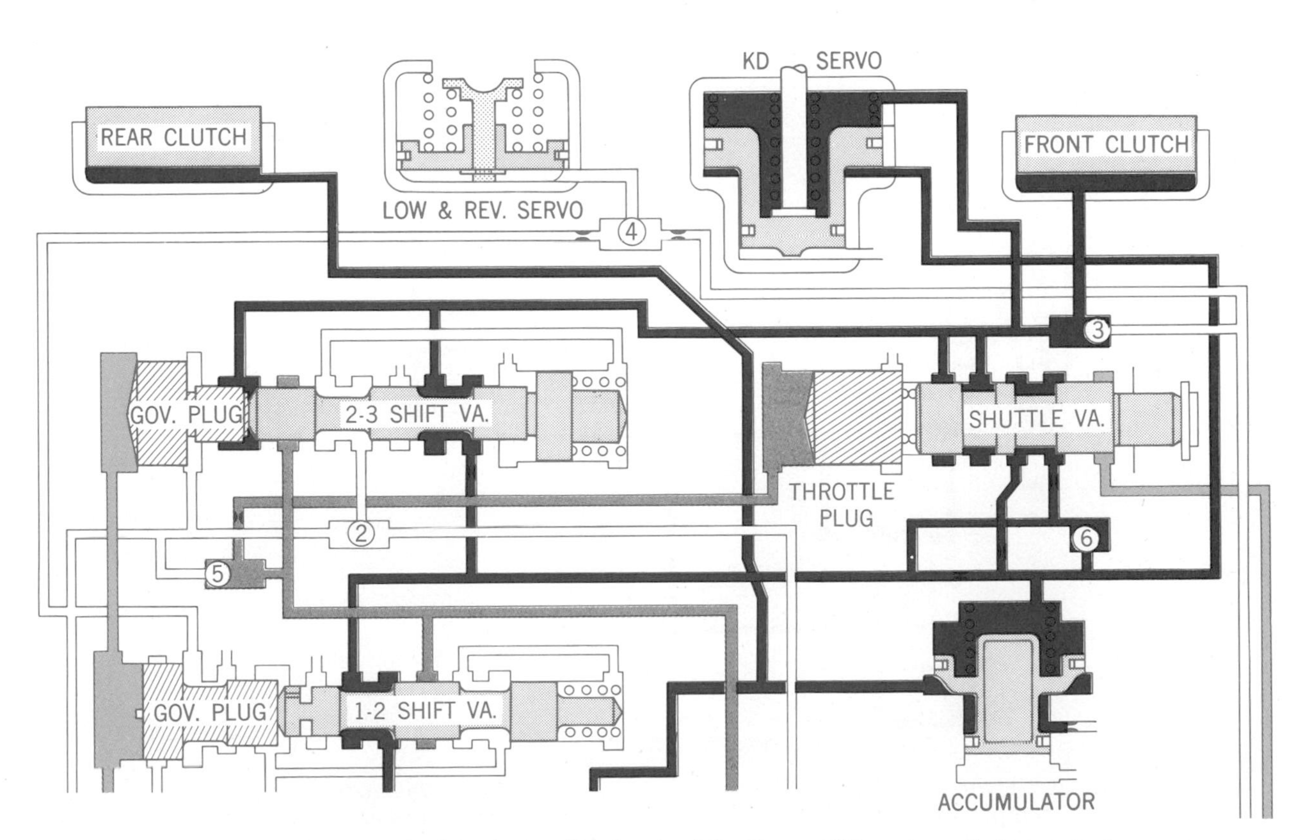

Plate 26. Upper part of the hydraulic circuit of the TorqueFlite automatic transmission in D after an upshift to direct drive. (*Chrysler Corporation*)

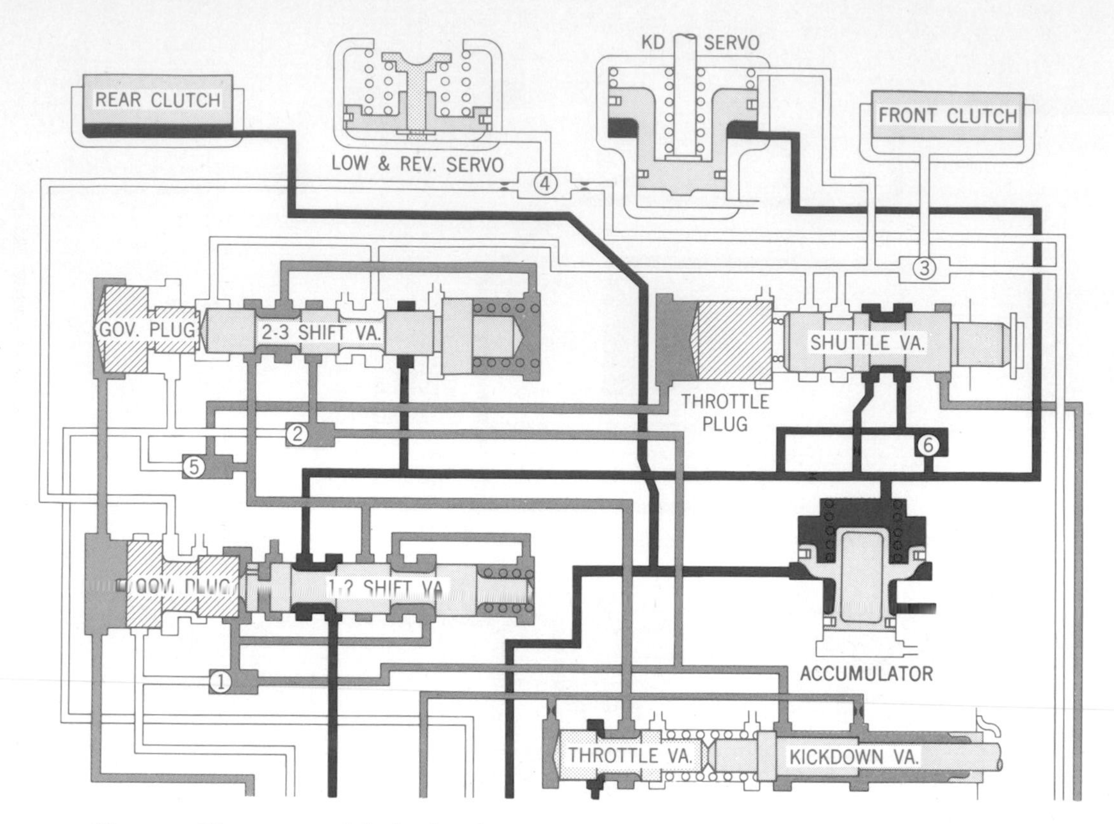

Plate 27. Upper part of the hydraulic circuit of the TorqueFlite automatic transmission after a kickdown downshift to second. (*Chrysler Corporation*)

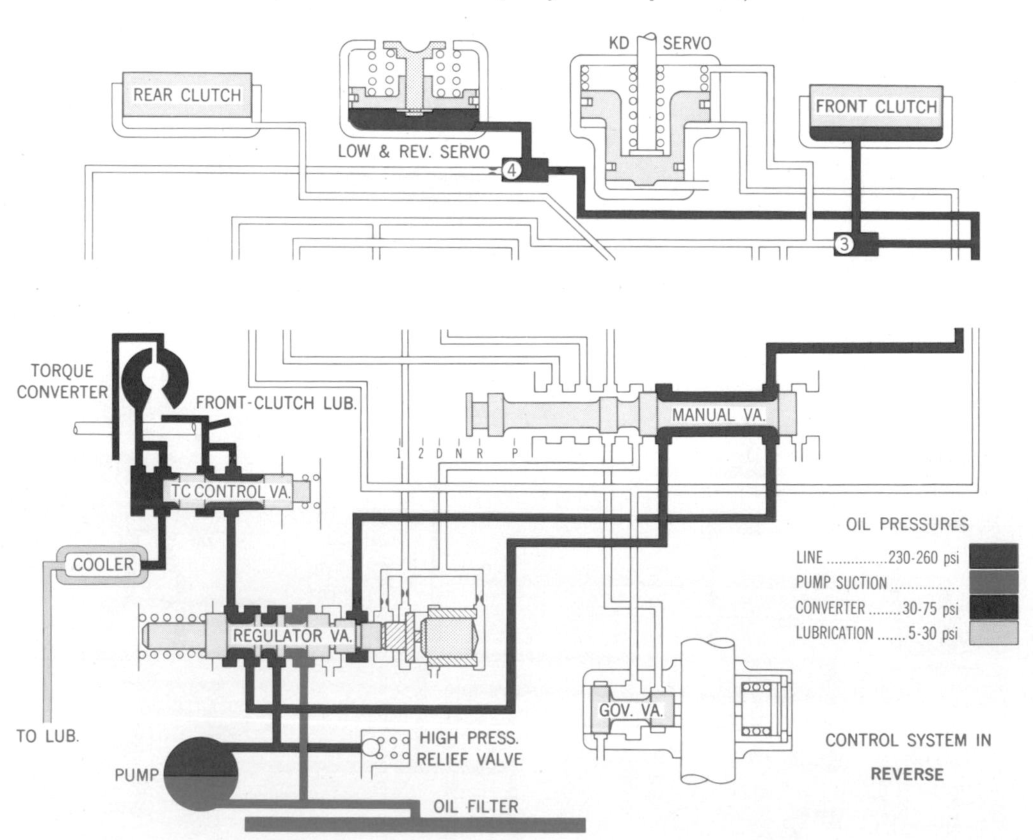

Plate 28. Hydraulic circuit with the TorqueFlite automatic transmission in reverse. Only the upper and lower parts of the circuit are shown because the central part, shown in an earlier plate, does not enter into the action in reverse. (*Chrysler Corporation*)

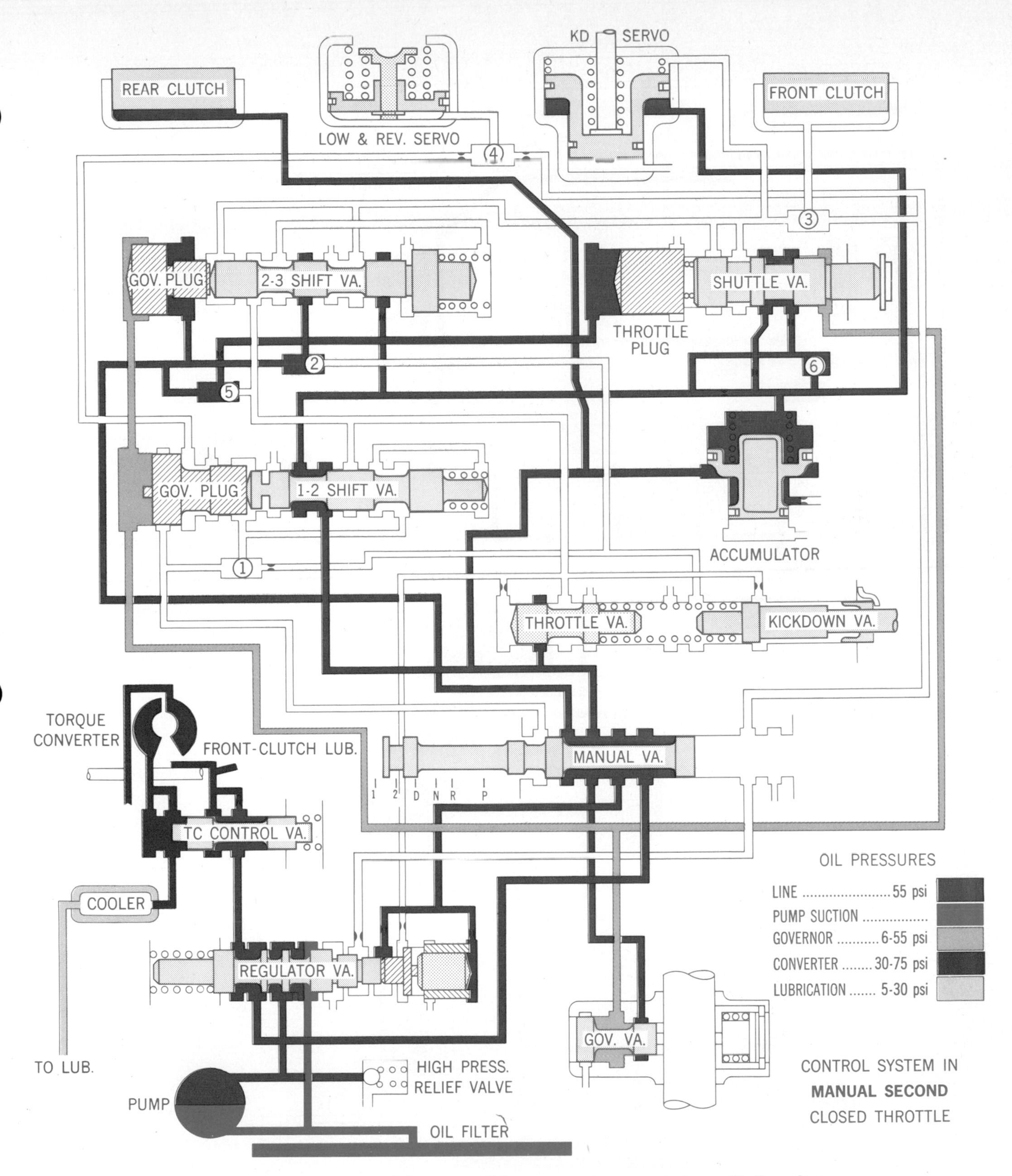

Plate 29. Hydraulic circuit of the TorqueFlite automatic transmission with the selector lever in 2 range and the transmission in second gear. (*Chrysler Corporation*)

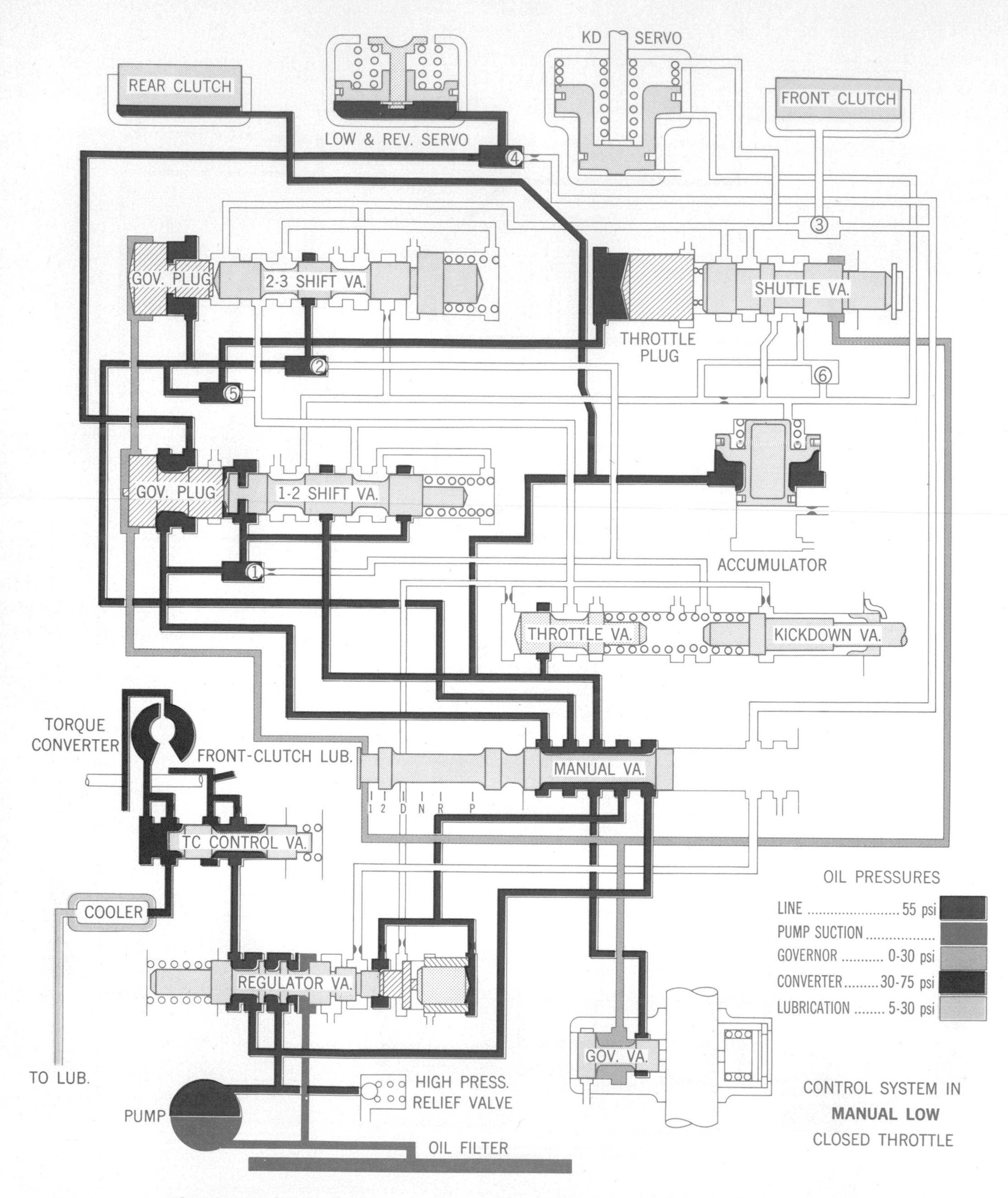

Plate 30. Hydraulic circuit of the TorqueFlite automatic transmission with the selector lever in 1 range and the transmission in first gear. (*Chrysler Corporation*)

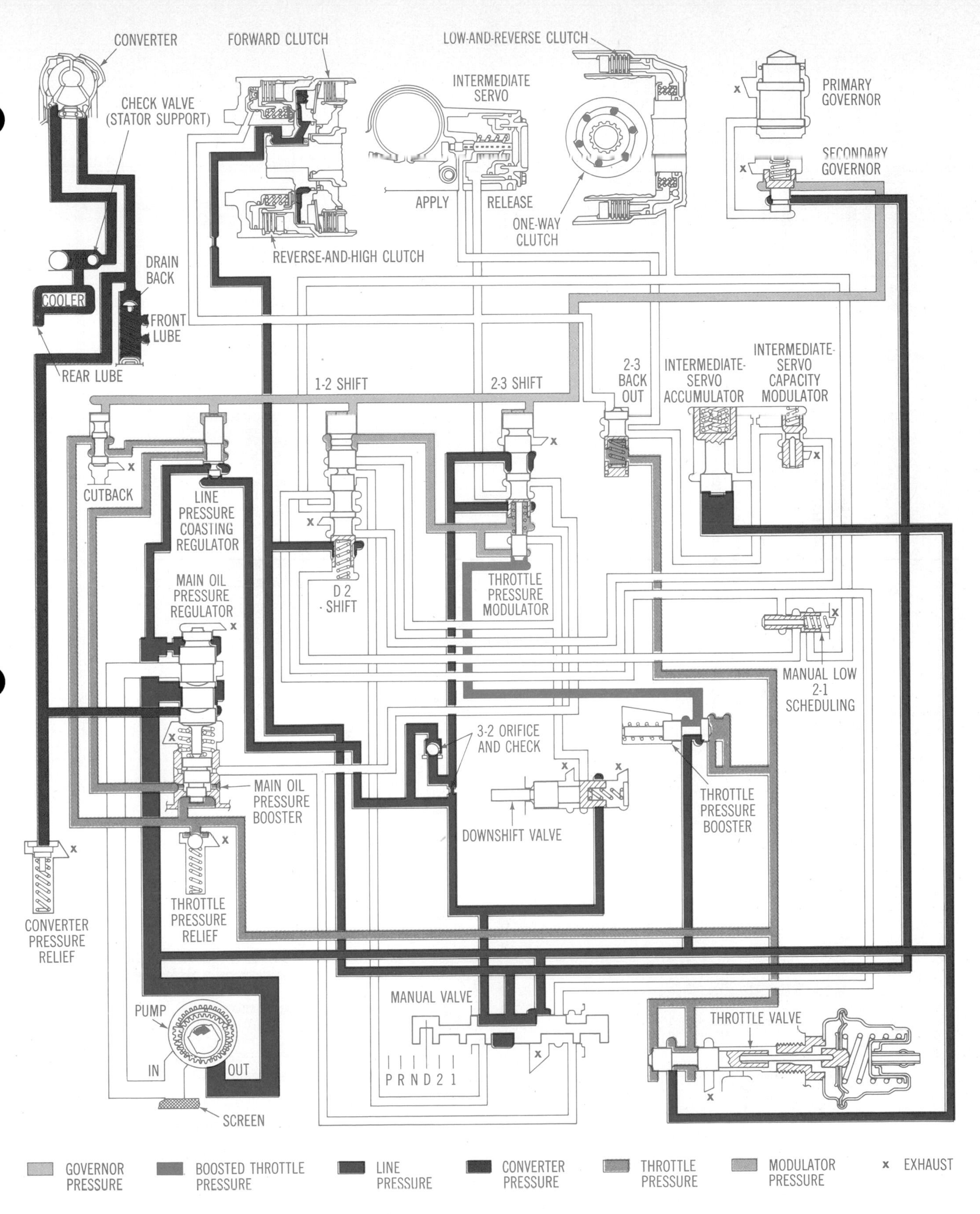

Plate 31. Conditions in the hydraulic circuit of the C6 automatic transmission with the transmission in first gear, drive range. (*Ford Motor Company*)

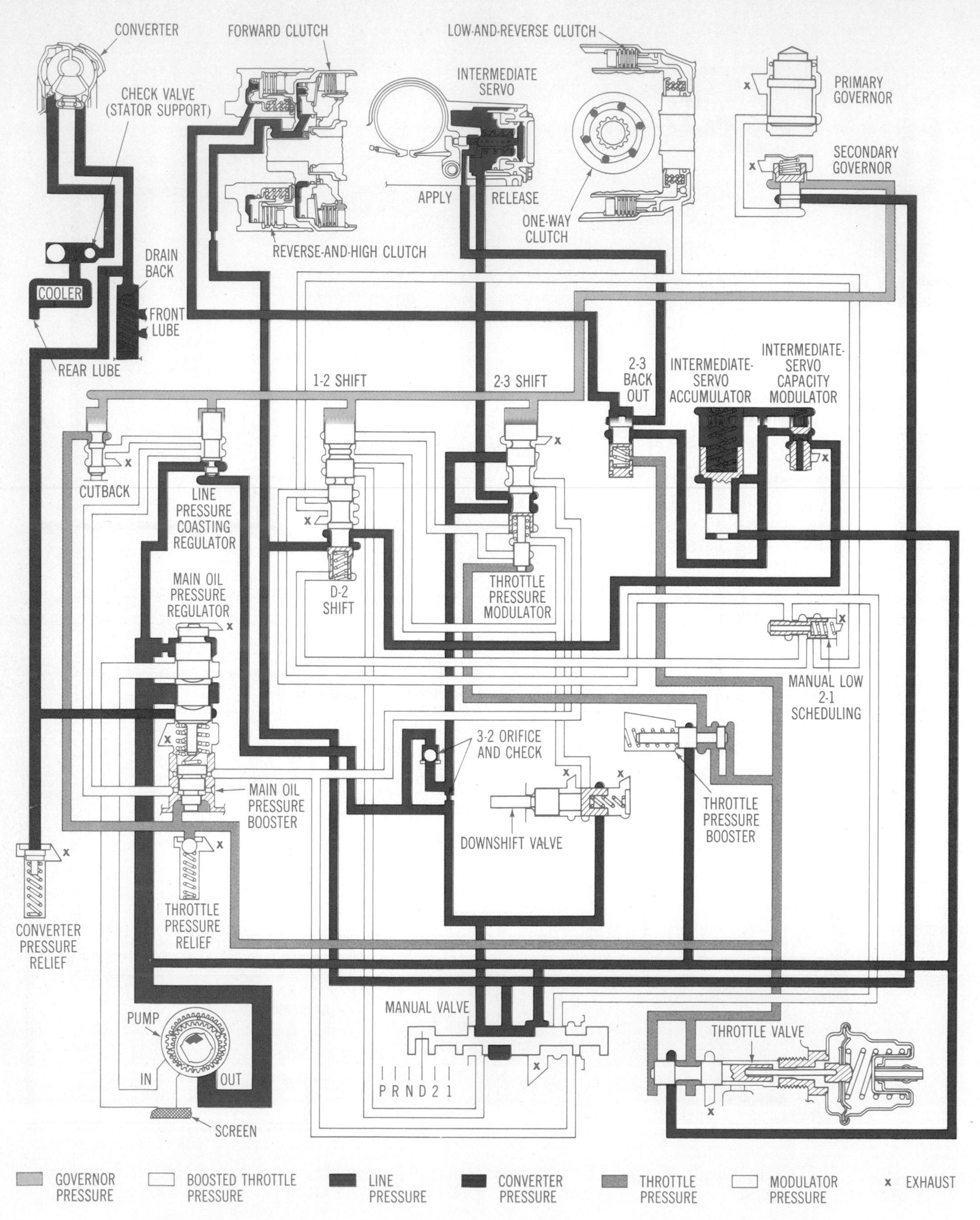

Plate 32. Conditions in the hydraulic circuit of the C6 automatic transmission with the transmission in direct drive. (*Ford Motor Company*)

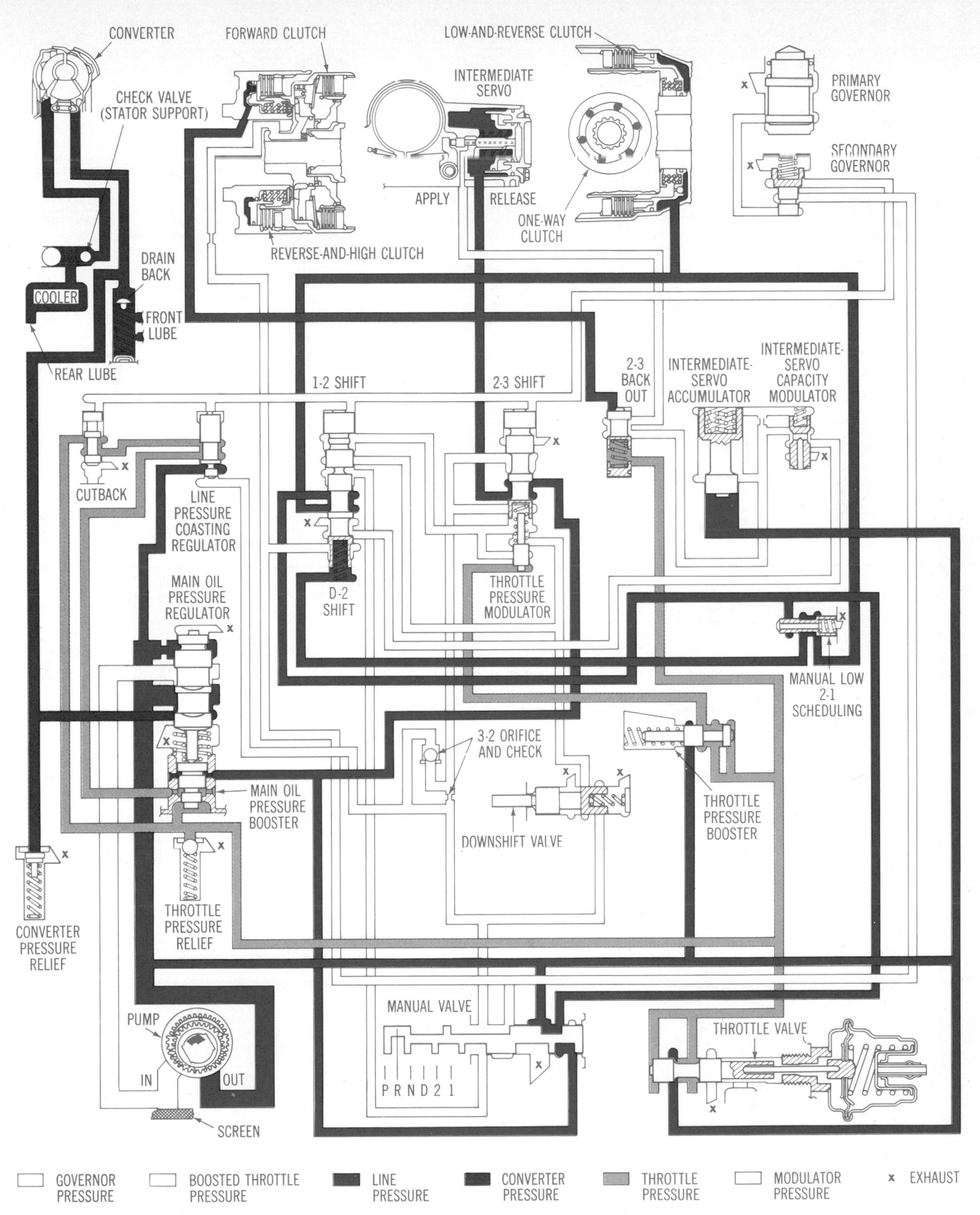

Plate 33. Conditions in the hydraulic circuit of the C6 automatic transmission with the transmission in reverse. (*Ford Motor Company*)

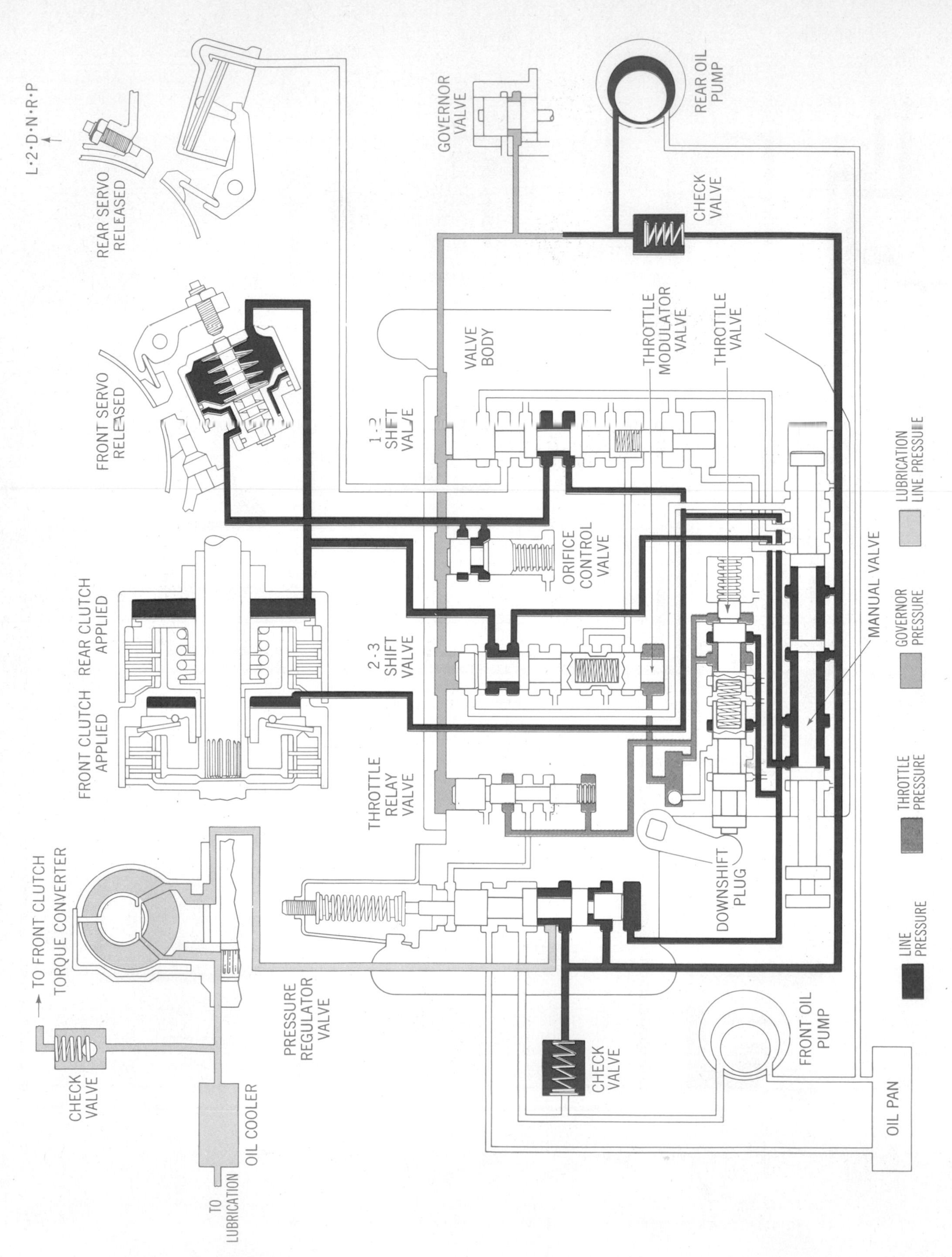

Plate 34. Hydraulic circuit in the Toyoglide automatic transmission with the transmission in drive range, third gear. (*Toyota Motor Sales Company, Ltd.*)

0-07-014637-3

DIAGNOSIS GUIDE—ABNORMAL NOISE

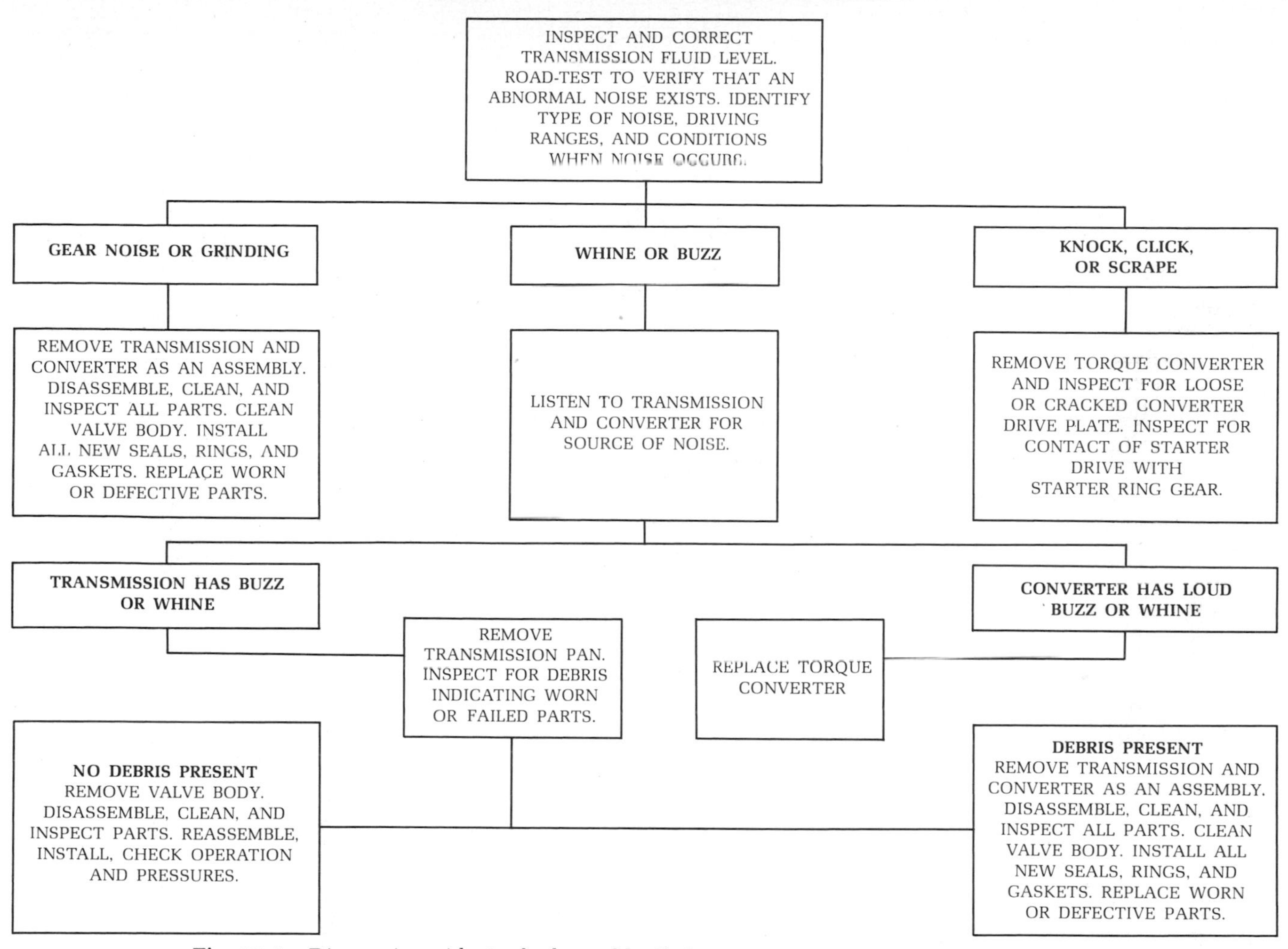

Fig. 18-4. Diagnosis guide to find trouble if the transmission has abnormal noises. (*Chrysler Corporation*)

DIAGNOSIS GUIDE—FLUID LEAKS

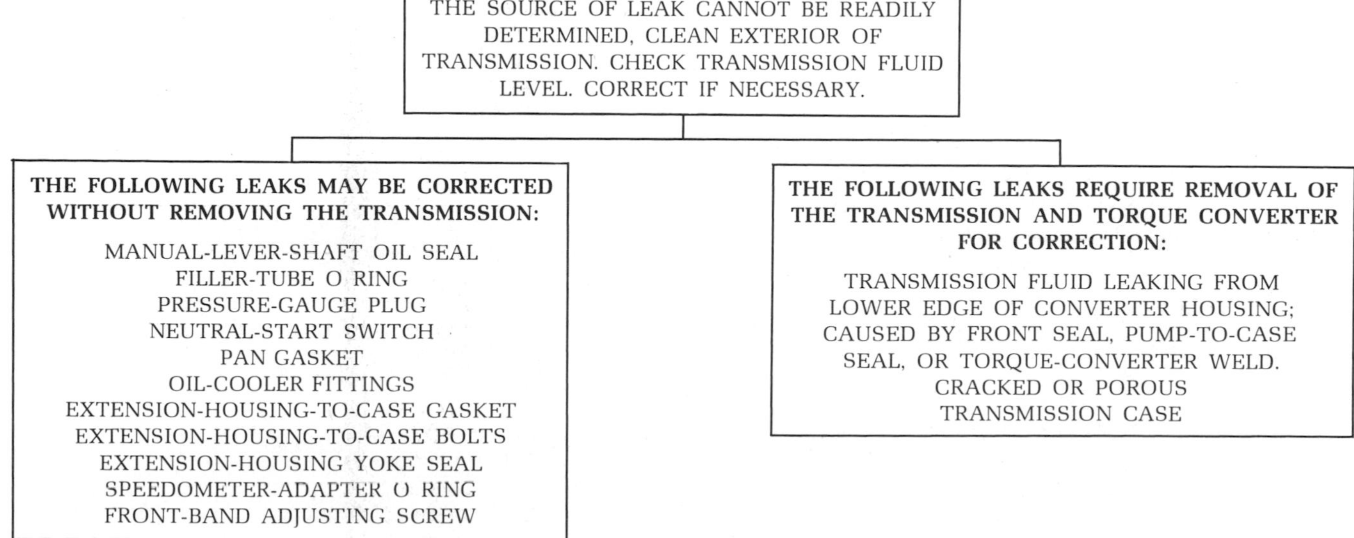

Fig. 18-5. Diagnosis chart to find trouble if the transmission leaks. (*Chrysler Corporation*)

AUTOMATIC-SHIFT SPEEDS AND GOVERNOR-PRESSURE CHART
(Approximte mph)

CARLINE	VL	RW	BJ	PDC	PDC	PDC	Y
Engine in³	198 225	318	360-4 Hi. Perf.	360-2 360-4	400-2 400-4	400-4 & 440 Hi. Perf.	440
Axle ratio	2.76	2.71	3.23	2.71	2.76	3.23	3.23
Tire size	6.95 X 14	E78 X 14	F70 X 14	F78 X 15	G78 X 15	G78 X 15	L84 X 15
Throttle minimum							
1-2 upshift	9–16	8–16	8–15	9–16	9–16	8–15	8–16
2-3 upshift	15–25	15–25	15–23	17–25	15–25	15–23	16–25
3-1 downshift	9–13	8–12	7-10	8–11	9–12	8–13	8–11
Throttle wide open							
1-2 upshift	31–43	35–48	33–44	37–51	37–51	36–47	33–46
2-3 upshift	65–75	72–89	62–71	77–89	76–88	62–77	68–80
Kickdown limit							
3-2 W.O.T. downshift	62–72	70–79	60–68	73–83	73–83	60–72	66–75
3-2 part-throttle downshift	47–57	26–52	26–46	[illegible]	[illegible]	27–50	25–45
3-1 W.O.T. downshift	28–33	27–37	28–35	28–38	28–38	28–38	25–34
Governor pressure*							
15 psi	20–22	20–22	17–18	20–22	21–22	16–19	18–20
40 psi	38–43	44–50	40–45	46–50	46–50	42–46	41–46
60 psi	57–62	66–71	56-61	68–73	68–73	59–63	60–65

*Governor pressure should be from 0 to 1.5 psi at standstill, or downshift may not occur.

Note: Figures given are typical for other models. Changes in tire size or axle ratio will cause shift points to occur at corresponding higher or lower vehicle speeds.

Fig. 18-6. Chart showing automatic-transmission shift speeds and governor pressures. Miles per hour are only approximate and will vary somewhat from one model to another. (*Chrysler Corporation*)

such as the one shown in Fig. 18-9. Also, note the cautions about cleanliness in the transmission-repair sections on later pages.

1. *THROTTLE PRESSURE* Loosen the locknut on the throttle-lever stop screw. Back the screw off 5 turns. Use the gauge pin, as shown in Fig. 18-10, between the throttle-lever cam and the kickdown valve. Push in on the tool to push the kickdown valve all the way in so that the throttle valve will bottom in the valve body. Now tighten the stop screw finger tight against the tang with the throttle-lever cam touching the gauge pin, as shown. Be sure the kickdown valve is pushed in as far as it will go. Remove the tool and tighten the locknut without turning the stop screw.

2. *LINE PRESSURE* First, check the distance be-

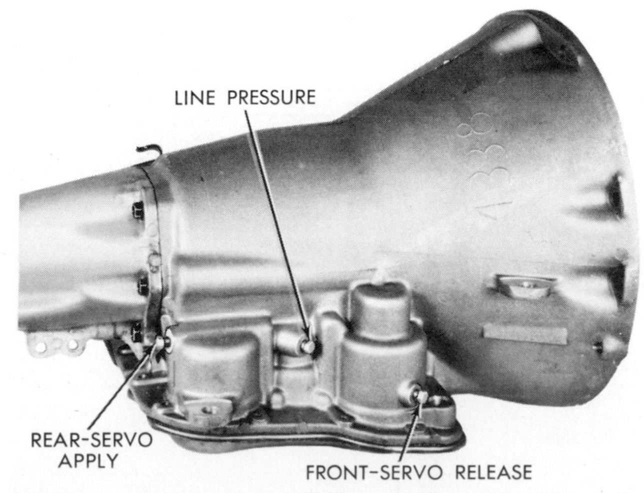

Fig. 18-7. Pressure-test locations on the right side of the transmission case. (*Chrysler Corporation*)

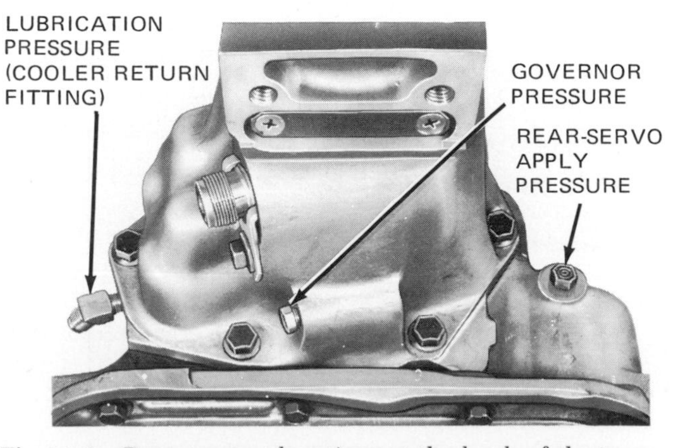

Fig. 18-8. Pressure-test locations at the back of the transmission case. (*Chrysler Corporation*)

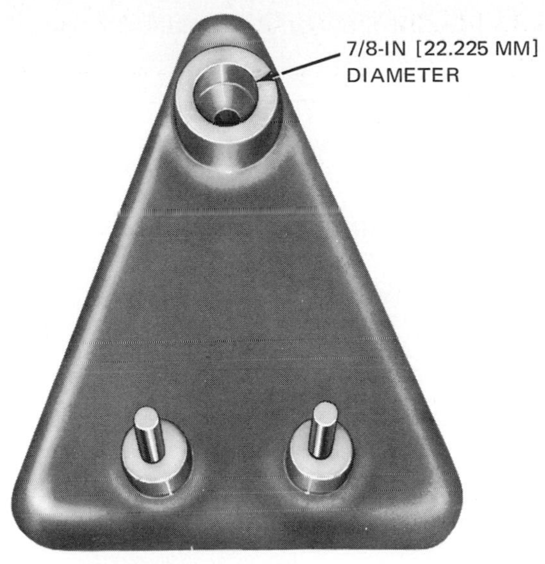

Fig. 18-9. Valve-body repair stand. (*Chrysler Corporation*)

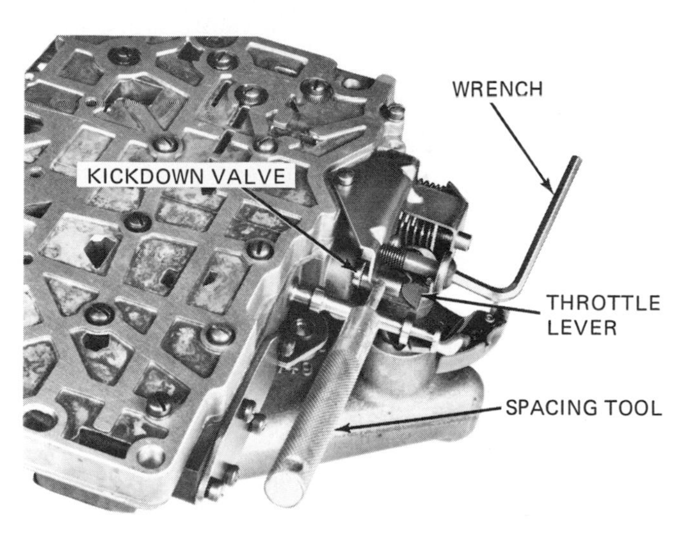

Fig. 18-10. Throttle-pressure adjustment. (*Chrysler Corporation*)

tween the manual valve and the line-pressure adjusting screw, as shown in Fig. 18-11. This distance should measure $1\frac{7}{8}$ inch [47.625 mm] with the manual valve in 1 (low). If the measurement is off, it means the spring retainer is not square with the valve body; the spring, hitting the pressure-regulator valve at an angle, may cause it to cock and hang up, thus preventing normal pressure regulation. You can correct the measurement by loosening the spring-retainer screws and moving the retainer as necessary. Tighten the screws and recheck the measurement to make sure it is correct.

Next, measure the distance between the valve body and the inner edge of the adjusting nut, as shown in Fig. 18-12. This distance should be approximately $1\frac{5}{16}$ inch [33.3 mm]. Adjustment can be made to obtain the proper line pressure by turning the adjusting screw with an Allen wrench. One

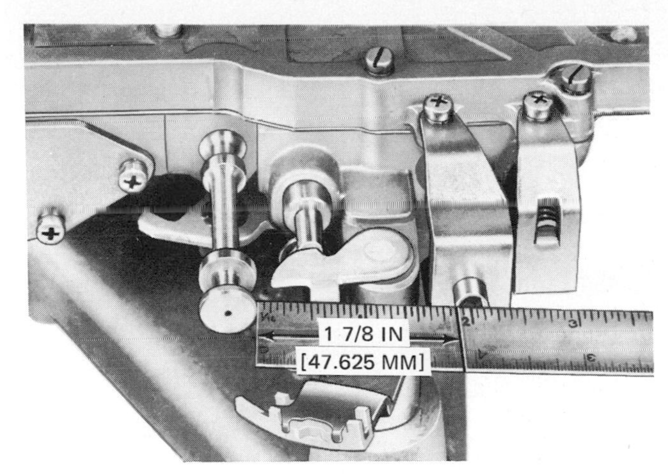

Fig. 18-11. Measuring spring-retainer location. (*Chrysler Corporation*)

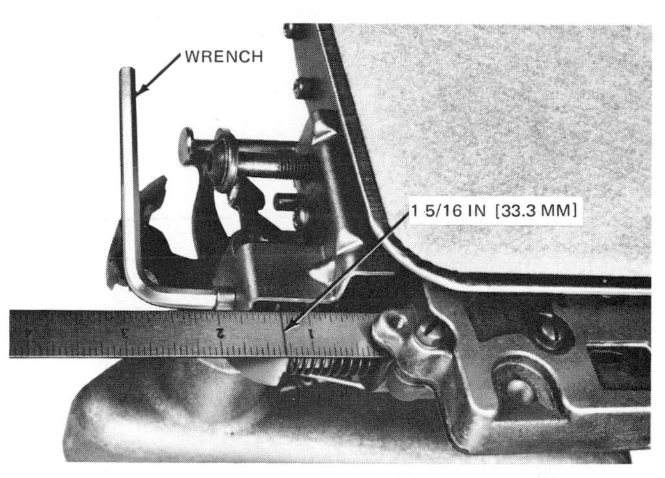

Fig. 18-12. Line-pressure adjustment. (*Chrysler Corporation*)

complete turn of the adjusting screw represents about $1\frac{2}{3}$ psi [0.117 kg/cm^2]. Turn the screw clockwise to decrease pressure, and counterclockwise to increase pressure.

⊘ 18-7 Air-Pressure Tests The clutches, bands, and servos can be checked with air pressure to determine whether they are working. First, remove the valve body, and then apply air pressure at 30 to 100 psi [2.109 to 7.031 kg/cm^2] to the proper oil passages in the transmission. These passages are shown in Fig. 18-13.

CAUTION: Compressed air must be clean and dry!

1. *FRONT CLUTCH* Apply air pressure to the front-clutch apply passage and listen for a dull thud, which indicates that the clutch has engaged. Hold pressure for a few seconds and check for oil leakage.
2. *REAR CLUTCH* Apply air pressure to the rear-clutch apply passage and listen for a dull thud, which indicates clutch engagement. Inspect for oil leaks.

Fig. 18-13. Air-pressure tests. (*Chrysler Corporation*)

NOTE: If you cannot hear the thud, touch the clutch housing with your fingertips as you apply air pressure. If the clutch engages, you can feel it.

3. *KICKDOWN SERVO AND LOW-AND-REVERSE SERVO* Apply air pressure to the kickdown-servo and low-and-reverse-servo passages, in turn, and note whether the related brake band tightens. It should tighten, and when air pressure is released, spring pressure should release the band.

NOTE: If clutches and servos operate properly, but shifting is incorrect, the trouble is probably in the valve body itself.

Transmission Adjustments

In addition to the hydraulic-control pressure adjustments described in ⊘ 18-6, the other adjustments to be made include *gearshift linkage, throttle-rod linkage,* and *brake band.*

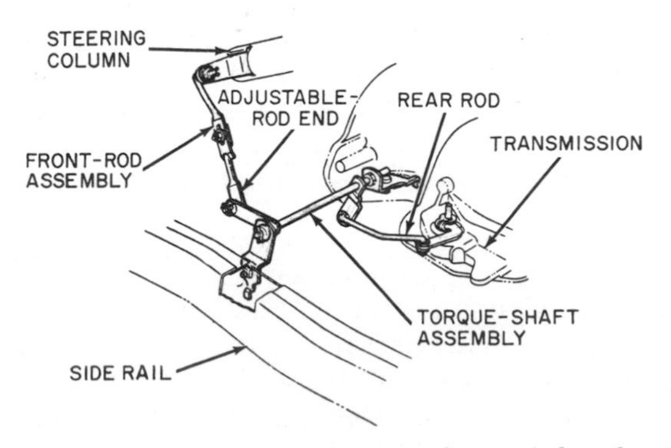

Fig. 18-14. Linkage to the steering column. (*Chrysler Corporation*)

⊘ 18-8 Gearshift-Linkage Adjustment Typical column and console gearshift linkages are shown in Figs. 18-14 and 18-15.

1. *COLUMN GEARSHIFT LINKAGE (FIG. 18-14)* With all parts assembled but with the adjustable rod end free, put the gearshift selector lever in park and lock the steering column with the ignition key. Then move the shift-control lever on the transmission all the way to the rear—to park. Set the adjustable rod to the proper length so that it installs freely.

Check to make sure the shift effort is free and the detents feel crisp. All stops must be positive. The detent position must be close enough to the gate stops in neutral and drive to ensure that the selector lever will not remain out of the detent position when placed against the gate and released. With the shift lever held down against the park gate, the engine must start when the ignition switch is turned to the START position.

2. *CONSOLE GEARSHIFT LINKAGE (FIG. 18-15)* Assemble all parts, but leave adjustable rod ends free. At the steering-column upper end, line up the locating slots in the bottom of the shift housing and bearing housing. Install a tool to hold this alignment, and lock the column with the ignition key. Put the console lever in park. Move the shift-control lever on the transmission all the way to the rear—to park. Set the adjustable rods to the proper length and attach, making sure that they install freely.

Check operation as noted above under column gearshift linkage.

NOTE: The console linkage, shown in Fig. 18-15, is for cars without the steering-column-mounted ignition switch. On these cars, the ignition switch not only turns the ignition on and off, but also locks the steering wheel when it is turned off. This is an added safeguard against theft of the car. The earlier system, shown in Fig. 18-15, has a linkage to the steering column which locks the steering shaft when the selector lever is moved to P (park). On the later models with the locking ignition switch, the linkage is simpler (Fig. 18-16).

⊘ 18-9 Throttle-Rod Adjustment Adjustment of the throttle rod varies widely according to the type of linkage used. The purpose of the adjustment, of course, is to make sure that the kickdown valve moves in proper synchronization with the throttle valve. A typical adjustment procedure follows. For other procedures, refer to the appropriate car shop manual. The model we shall discuss has a single-section throttle rod, as shown in Fig. 18-17.

1. Make sure the throttle-rod linkage is properly lubricated.
2. Disconnect the choke at the carburetor, or block the choke wide open. Open the throttle slightly to release the fast-idle cam; then release the throttle.
3. Loosen the transmission-throttle-rod-adjustment lock screw.

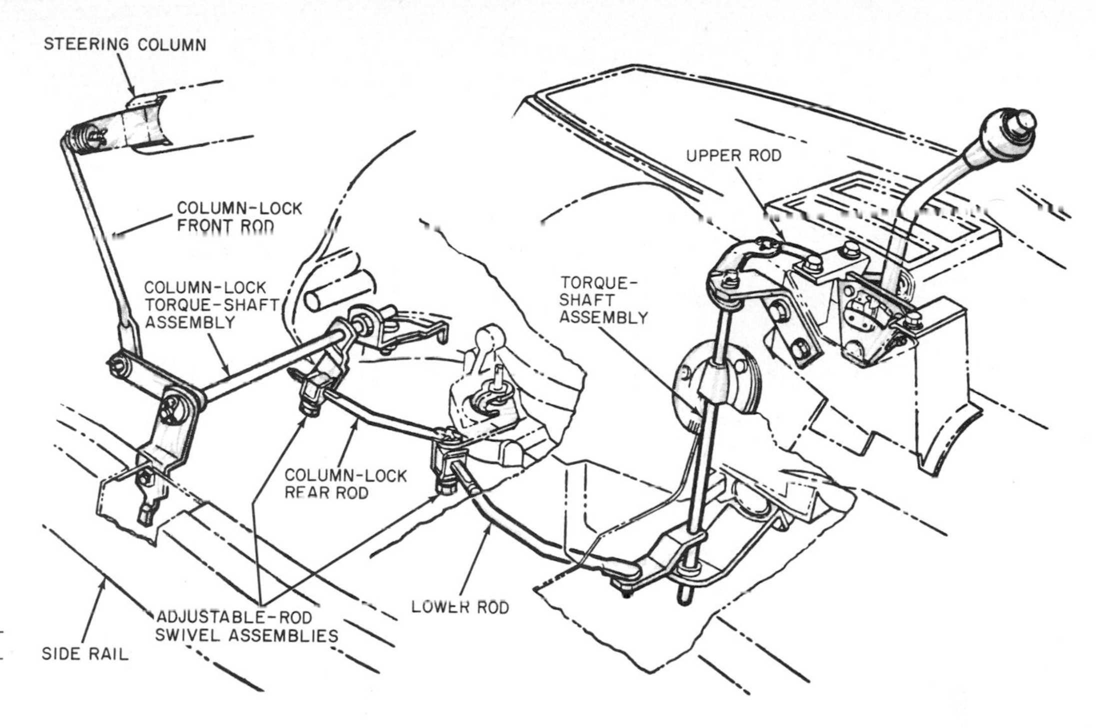

Fig. 18-15. Console linkage. (*Chrysler Corporation*)

4. Hold the transmission lever forward against its stop while adjusting the transmission linkage.
5. Adjust the transmission rod by lightly pulling forward on the slotted link so that the rear edge of the slot is against the carburetor lever pin. Tighten the transmission-rod-adjustment lock screw.

NOTE: Be sure to hold the slotted link and transmission lever in the forward position while tightening the screw.

6. Check transmission free movement by moving the slotted link to the full rear position, and then allow it to return slowly, making sure it comes to the full forward position.
7. Loosen the carburetor-cable-clamp nut. Adjust the position of the cable-housing ferrule in the clamp so that all slack is removed from the cable with the throttle at curb idle. Then back off the ferrule $\frac{1}{4}$ inch [6.35 mm] to provide free play. Tighten the cable clamp nut to 45 pound-inches [0.81 kg-mm]. Connect the choke rod or remove the block.

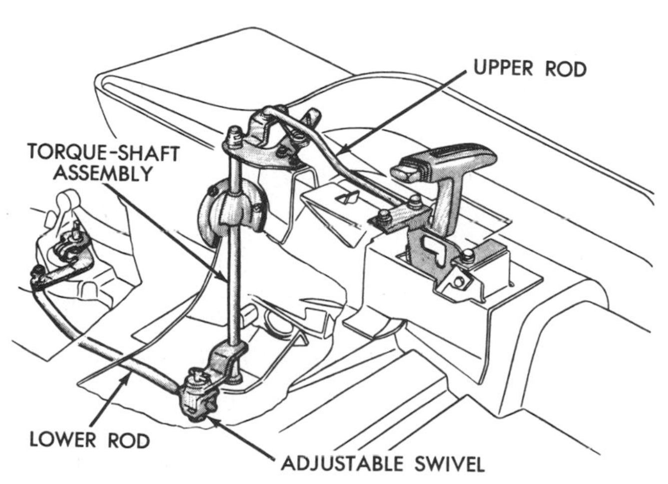

Fig. 18-16. Console linkage on a car with a steering-column ignition switch and steering lock. (*Chrysler Corporation*)

⊘ 18-10 Band Adjustments The two bands are adjusted as follows:

1. *KICKDOWN BAND* The kickdown-band adjusting screw is on the left side of the transmission case, as shown in Fig. 18-18. It is adjusted by loosening the locknut and backing off the screw about 5 turns, making sure it is backed off enough to be free of the band. Next, take a torque wrench and tighten the band adjusting screw to 72 pound-inches [2.29 kg-mm]. Then back off the adjusting screw 2 turns, hold the screw, and tighten the locknut to 35 pound-feet [4.837 kg-m]. (On the A-727 model used on the 426 engine, back off $1\frac{1}{2}$ turns only.)
2. *LOW-AND-REVERSE BAND* Raise the car, drain the transmission fluid, and remove the oil pan. Note the position of the adjusting screw in Fig. 18-19. Loosen the locknut and back off the screw about 5 turns. Make sure the screw is backed off enough to be free of the band. Tighten the adjusting screw to 72 pound-inches [2.29 kg-mm]. Then, on the A-904, back off the adjusting screw $3\frac{1}{4}$ turns, or 4 turns for the 318 cubic inch [5,211 cm^3 (cubic centimeter)] engine. On the A-727, back off 2 turns. Hold the adjusting screw and tighten the locknut to 30 pound-feet [4.146 kg-m].

Reinstall the oil pan, using a new gasket. Tighten the pan bolts to 150 pound-inches [2.70 kg-mm]. Fill the transmission with automatic-transmission fluid AQ-ATF Suffix A or Dexron, as explained in ⊘ 18-13.

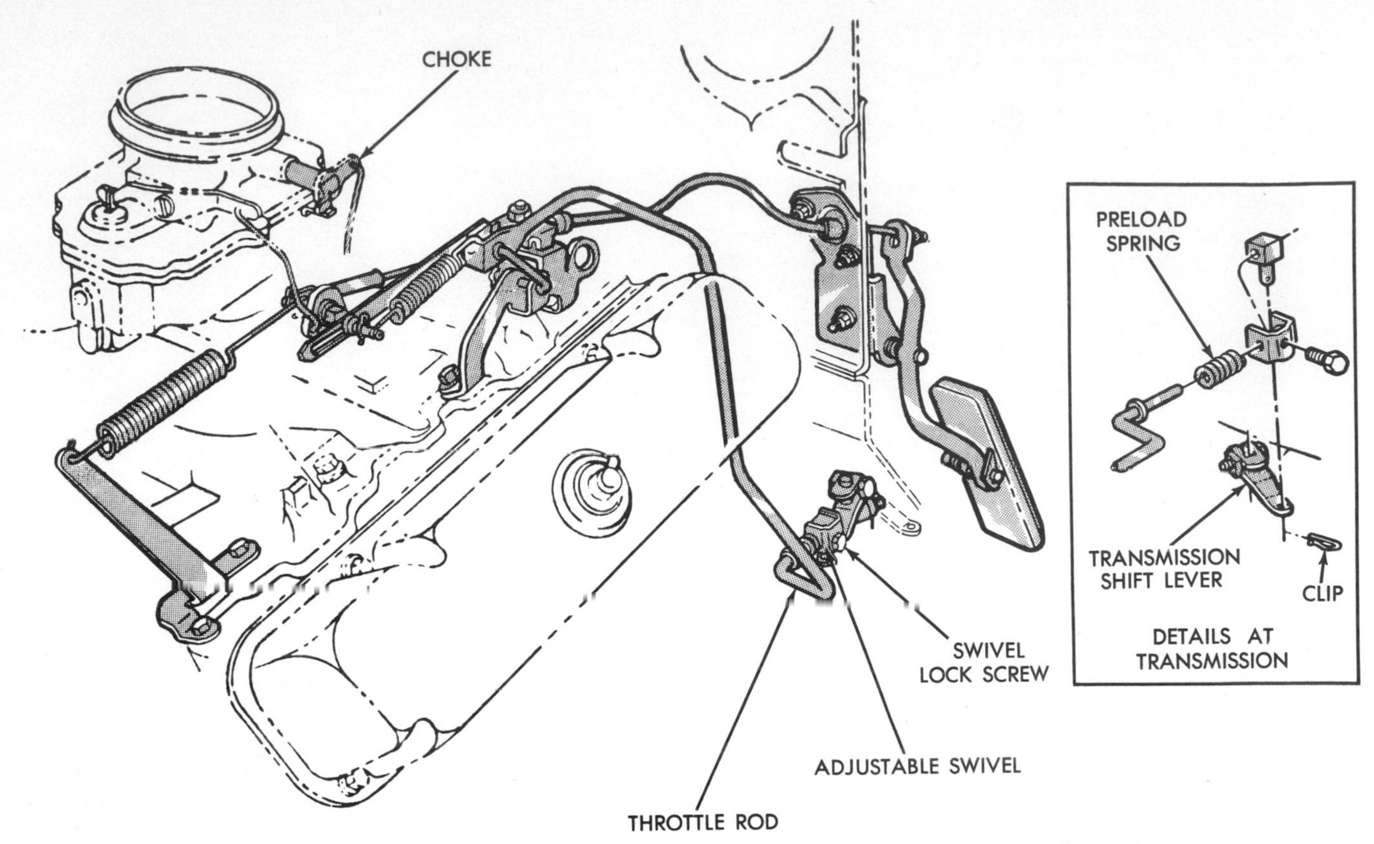

Fig. 18-17. Throttle-rod-adjustment on an eight-cylinder model with a single-section rod. (*Chrysler Corporation*)

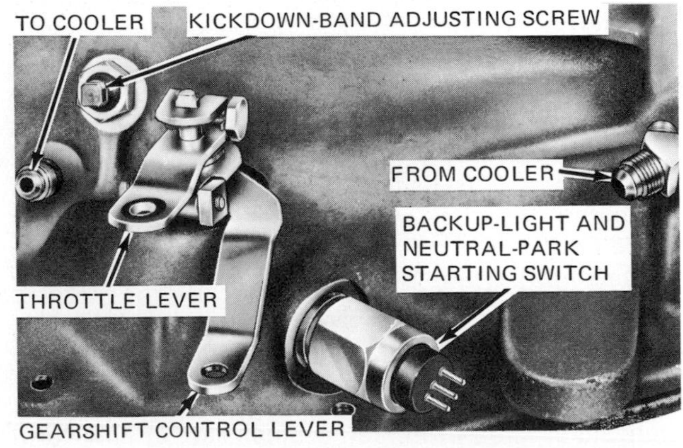

Fig. 18-18. External controls and adjustments. (*Chrysler Corporation*)

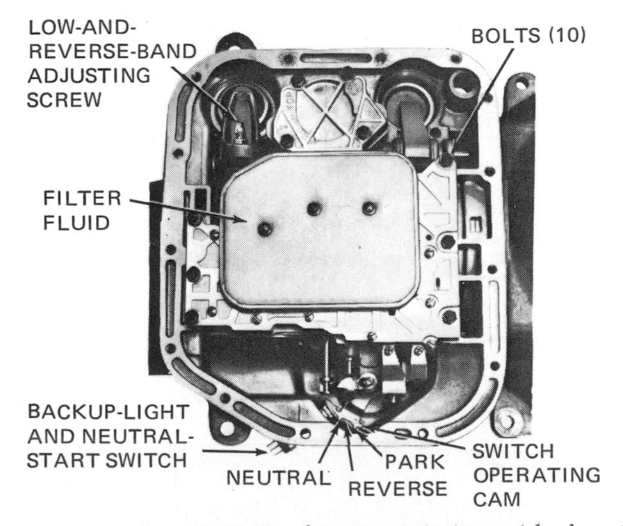

Fig. 18-19. Bottom view of a transmission with the oil pan removed. (*Chrysler Corporation*)

Check Your Progress

Progress Quiz 18-1 By now, you realize that similar troubles in different models of automatic transmissions have a family resemblance. For example, low oil pressure in any transmission will cause the same troubles: slipping, shift problems, burned clutch plates, and so on. Yet each model of transmission has its own adjustment procedures and problems, and the chapters in the book are designed to give you information on these. To find out how well you remember the TorqueFlite adjustments and checking procedures, take the following quiz.

Completing the Sentences The sentences that follow are incomplete. After each sentence there are several words or phrases, only one of which will correctly complete the sentence. Write each sentence in your notebook, selecting the proper word or phrase to complete it correctly.

1. Three diagnostic tests are to be made on the TorqueFlite: oil pressure, stall, and: (*a*) hydraulic control, (*b*) shift point, (*c*) air pressure.
2. To check the performance of the torque-convertcr stator clutch and transmission clutches, give the transmission the: (*a*) pressure test, (*b*) air-pressure test, (*c*) stall test.
3. Stall speed more than 200 rpm above specifications indicates: (*a*) poorly tuned engine, (*b*) clutch slippage, (*c*) band slippage.
4. Stall speed 250 to 300 rpm below specifications indicates: (*a*) torque-converter clutch slippage, (*b*) transmission clutch slippage, (*c*) brake-band slippage.
5. Rear-servo apply pressure, as compared with line pressure in D, should be: (*a*) higher, (*b*) about the same, (*c*) lower.
6. The hydraulic-control-pressure checks to be made include line pressure and kickdown-servo-release pressure, lubrication pressure, and: (*a*) clutch-release pressure, (*b*) rear-servo-apply pressure, (*c*) rear-servo-release pressure.
7. The two hydraulic-control-pressure adjustments that can be made are to the line pressure and the: (*a*) servo-apply pressure, (*b*) throttle pressure, (*c*) clutch-apply pressure.
8. In the air-pressure tests, you check the actions of the: (*a*) servos and clutches, (*b*) valves and pump, (*c*) servos and linkages.
9. In addition to the pressure adjustments, other adjustments to be made include throttle-rod linkage: (*a*) gearshift linkage and clutches, (*b*) throttle opening and gearshift linkage, (*c*) gearshift linkage and brake band.
10. If clutches and servos work properly but shifting is incorrect, the trouble is probably in the: (*a*) pump, (*b*) valve body, (*c*) linkages.

In-Car Service

A number of services can be performed in the car without removing the complete transmission. These include lubrication, replacing the backup and neutral starting switch, replacing the speedometer pinion, some seals, the governor, and so on. We shall discuss each of these services in following sections.

⊘ 18-11 Transmission-Service Precautions Remember: Cleanliness is of the utmost importance in transmission work. The smallest bit of lint from a cleaning rag or a particle of dirt can cause a valve to hang up and prevent normal transmission action. Under some conditions, this could ruin the transmission. The outside of the transmission must be cleaned before it is opened. The workbench, the tools, your hands, and all parts must be kept clean at all times. Cover the transmission with a clean cloth if you have to leave it for a while in a disassembled condition.

For other precautions, refer to ⊘ 14-10.

⊘ 18-12 Cleaning and Inspecting Transmission Parts During and after disassembly, clean and inspect all parts for wear or other damage. Refer to ⊘ 14-11, which covers the details. If the transmission shows signs of internal damage and it appears that foreign material has been circulating in the oil, all oil must be removed from the torque converter and transmission. Everything must be cleaned of the old oil, including the oil lines and oil cooler. A new oil filter should be installed on reassembly, and the transmission should be filled with fresh transmission oil, as explained in a following section.

For other details of cleaning and inspecting transmission parts, refer to ⊘ 14-11.

⊘ 18-13 Lubrication The Plymouth shop manual states that the transmission fluid and filter should provide satisfactory lubrication and protection to the transmission, and no change is recommended in normal use. If the vehicle is in severe service (towing a trailer, for example), then the recommendation is that the fluid be drained after 24,000 miles [38,624 km]. Fresh fluid should be put in after the old fluid is drained, along with a new filter. The bands should be adjusted after the fluid is changed.

EXCEPTION: The hemi-engine car should have the fluid and filter changed after the first 24,000 miles [38,624 km] or 2 years and every 12,000 miles [19,312 km] or 12 months after that.

The oil-and-filter-change procedure follows:

1. Raise the car on a hoist. Put a drain container with a large opening under the transmission oil pan.
2. Loosen pan bolts at one corner, tap the pan to break it loose, and allow oil to drain.
3. Remove the access plate and drain plug from the front of the converter, and allow converter fluid to drain.

NOTE: If the oil has been contaminated by a transmission failure, the torque converter must be flushed after draining. This procedure requires removal of the torque converter so that it can be laid on its side and 2 quarts [1.892 l] of solvent or kerosene can be poured into it. Then the converter must be shaken while the turbine and stator are rotated. Finally, the solvent or kerosene is drained out. The procedure should be repeated at least once. There are special tools to assist in this procedure.

4. Install a new filter and tighten the retaining screws to 35 pound-inches [0.63 kg-mm].
5. Clean the oil pan and reinstall with a new gasket; tighten bolts to 150 pound-inches [2.7 kg-mm].

6. Pour 6 quarts [5.676 l] of automatic-transmission fluid AQ-ATF Suffix A or Dexron through the filler tube.
7. Start and idle the engine for 2 minutes. Then, with the parking brake on or holding the foot brake down, move the selector lever through each position, returning it to the neutral position.
8. Add sufficient fluid to bring the fluid level to the ADD ONE PINT mark. After the transmission has reached operating temperature, recheck. The level should be between FULL and ADD ONE PINT.
9. Replace the dipstick and cap so that the cap is fully seated to keep dirt from entering the transmission.

⊘ 18-14 Backup and Neutral Starting Switch The location of the neutral starting switch is shown in Fig. 18-27. To check the switch, remove the wiring connector and check for continuity with a low-voltage test light between the center pin of the switch and the transmission case. The light should go on only when the transmission is in P or N. If the switch tests bad, check the gearshift-linkage adjustment before replacing the switch.

To replace the switch, unscrew it and allow fluid to drain into a container. Move the selector lever to P and then N and check to see that the switch-operating-lever fingers are centered in the switch opening in the case. Install a new switch, tightening it to 24 pound-feet [3.316 kg-m]. Retest the switch and add fluid to replace the fluid lost. See ⊘ 18-13.

The backup-light-switch circuit extends through the two outside terminals of the switch. To test the switch, remove the wiring connector from the switch and check for continuity between the two outside pins. Continuity should exist only with the transmission in R. No continuity should exist from either pin to the transmission case. If the switch is defective, replace it, as previously explained.

⊘ 18-15 Speedometer Pinion Speedometer-pinion parts are shown in Fig. 18-20. To replace the pinion, remove the bolt and retainer and carefully work the adapter and pinion out of the extension housing. If you find transmission fluid in the housing, replace the small oil seal in the adapter. The seal and retainer ring must be pushed in with a tool, as shown in Fig. 18-21 until the tool bottoms.

CAUTION: All parts must be clean!

Count the number of teeth on the pinion. Then rotate the adapter so that the number on the adapter which corresponds with the number of teeth will be at the bottom, or 6 o'clock. For example, if the number of teeth were 34, then you would locate the 32-38 at 6 o'clock, as shown in Fig. 18-22.

Install the retainer and bolt, with the retaining tangs in the adapter-positioning slots. Tap the adapter firmly into the extension housing, and

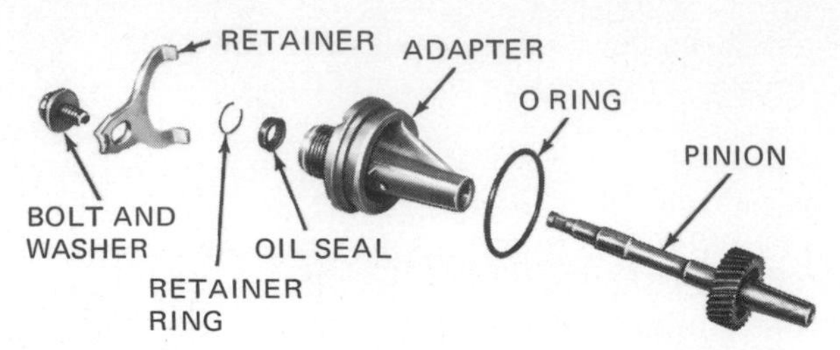

Fig. 18-20. Speedometer drive parts. (*Chrysler Corporation*)

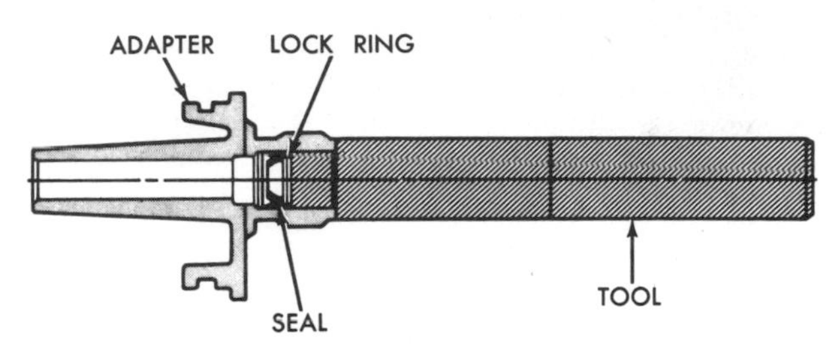

Fig. 18-21. Installing the speedometer-pinion seal with a special tool. (*Chrysler Corporation*)

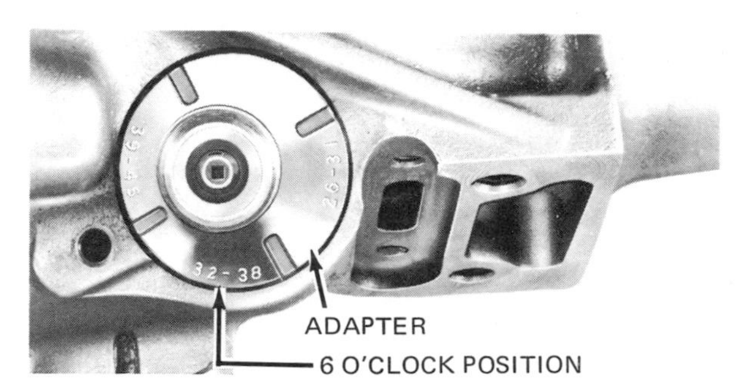

Fig. 18-22. Speedometer pinion and adapter installed but with retainer removed. (*Chrysler Corporation*)

tighten the retainer bolt to 100 pound-inches [1.8 kg-mm].

⊘ 18-16 Extension-Housing Yoke Seal To replace this seal, mark the parts for reassembly in the same relative positions and disconnect the drive shaft at the rear universal joint. Pull the shaft yoke out of the extension housing. Do not nick or scratch finished surfaces!

Then use special tools to remove the old seal and install a new one. Reattach the drive shaft after pushing the shaft yoke carefully into the extension housing and onto the main-shaft splines.

⊘ 18-17 Extension Housing and Output-Shaft Bearing To replace this bearing, the extension housing must be removed from the transmission, and this requires detachment of the drive shaft. Details of extension-housing removal are discussed in ⊘ 18-23.

With the extension housing off, use snap-ring pliers to remove the rear snap ring, shown in Fig. 18-23. Slide the old bearing off the output shaft and

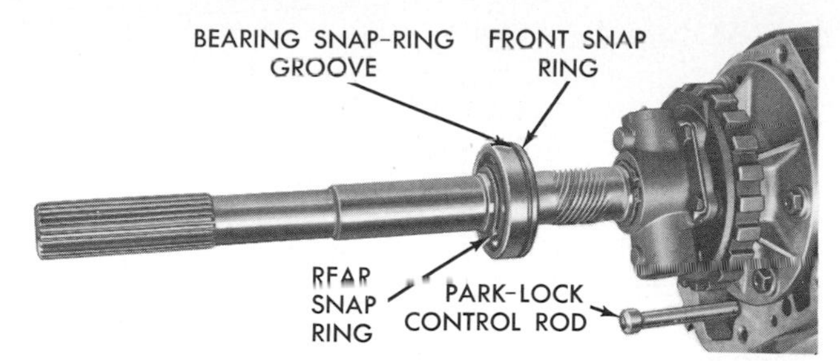

Fig. 18-23. Output-shaft bearing. (*Chrysler Corporation*)

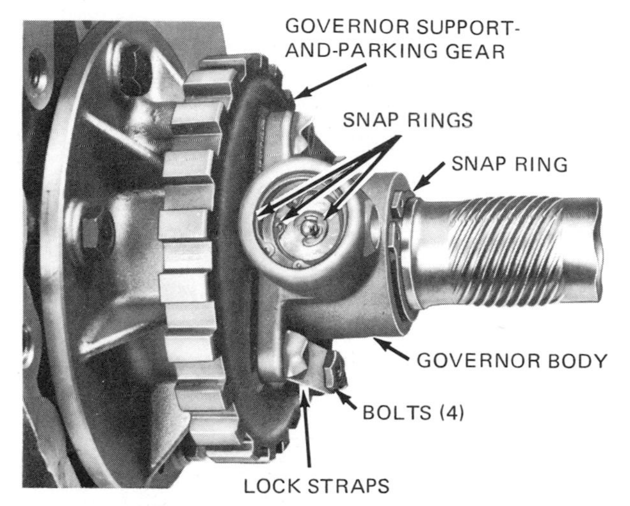

Fig. 18-24. Governor shaft-and-weight snap rings. (*Chrysler Corporation*)

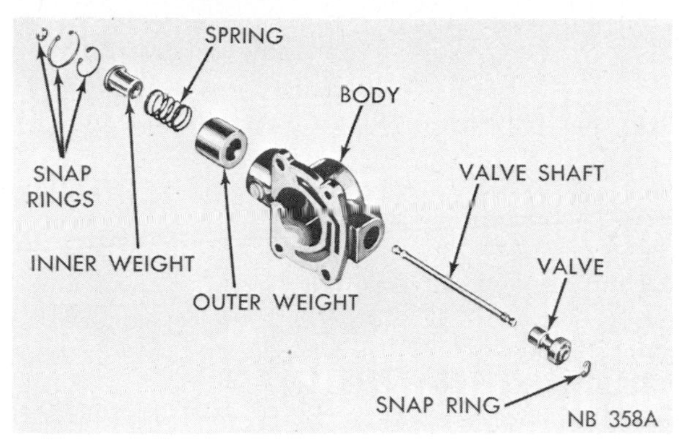

Fig. 18-25. Disassembled governor. (*Chrysler Corporation*)

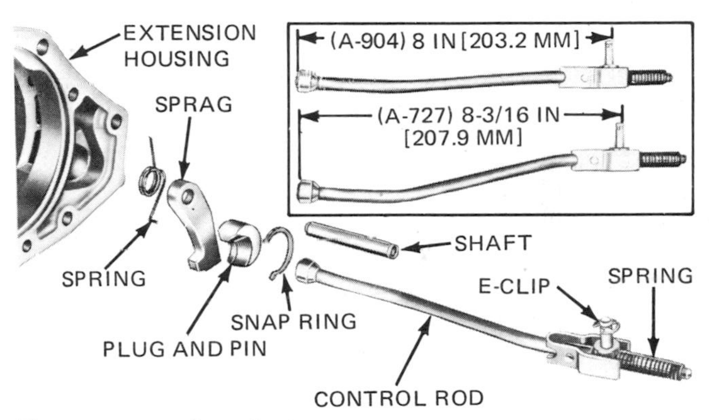

Fig. 18-26. Parking-lock components. (*Chrysler Corporation*)

install a new bearing. Secure it with the rear snap ring. Then replace the housing, as explained in ⊘ 18-23, and reconnect the drive shaft.

NOTE: With the transmission in the car, the transmission must be raised slightly with a jack and the center cross-member-and-support assembly must be removed before the extension housing can be removed.

⊘ 18-18 Governor Service To remove the governor, the extension-housing-and-output-shaft bearing must be removed, as already noted. Governor parts are held together by snap rings, as shown in Figs. 18-24 and 18-25. Remove the small snap rings to slide the valve and valve shaft from the governor body. Remove the other snap rings to complete disassembly. The governor body is held to the governor-support-and-parking gear by bolts. Both can be slid off the output shaft by removing the shaft snap ring.

Thoroughly clean and inspect all governor parts. The major trouble in governors is sticking valves or weights. You can remove rough surfaces with a crocus cloth. Clean parts before reassembly.

When reinstalling the governor body and support on the output shaft, make sure the valve-shaft hole in the output shaft aligns with the hole in the governor body. Assemble all parts in their original places, following Fig. 18-25, and secure with snap rings.

⊘ 18-19 Parking-Lock Components The parking-lock components are shown in Fig. 18-26. To replace them, the extension housing must be removed, as already explained. Follow Fig. 18-26 in removing and replacing parts.

⊘ 18-20 Valve-Body Assembly and Accumulator Piston To remove the valve body, raise the vehicle on a hoist and remove the transmission oil pan, as explained in ⊘ 18-13. Then proceed as follows:

1. Disconnect the throttle and gearshift linkage from the transmission levers. Loosen the clamp bolts and remove the levers. They are shown in Fig. 18-18.
2. Remove the E clip, shown in Fig. 18-27. Remove backup-light and neutral starting switch.
3. With the drain pan to catch oil, remove the 10 valve-body-to-transmission-case bolts, and lower the valve body from the case. Disconnect the parking-lock rod from the lever while lowering the valve body.
4. Withdraw the accumulator piston from the transmission case, and inspect the piston for scoring and

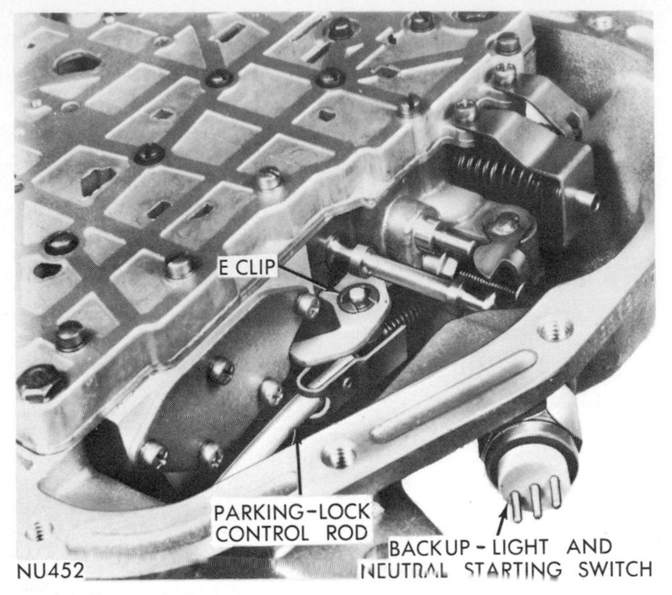

Fig. 18-27. Location of the parking-lock control-rod retaining E clip. (*Chrysler Corporation*)

the rings for wear or damage. Replace as necessary.
5. If the valve-body manual-lever-shaft seal requires replacement, drive it out of the case with a punch and use a special tool to install a new seal.

NOTE: Service the valve body and internal parts as explained in ⊘ 18-25.

To reinstall the valve body:

1. If the parking-lock rod was removed, insert it through the opening in the rear of the case with the knob positioned against the plug and sprag (see Fig. 18-26). Force the plug past the sprag, rotating the drive shaft if necessary.
2. Install the accumulator piston and position the accumulator spring in the valve body.
3. Put the manual lever in L and lift the valve body into approximate position. Connect the parking-lock rod to the manual lever and secure with an E clip. Then install the valve-body bolts finger tight.
4. Install the neutral starting switch. Put the manual lever in N. Shift the valve body as necessary to center the neutral finger over the neutral-switch plunger. Then snug bolts down evenly and tighten to 100 pound-inches [1.8 kg-mm].
5. Install the gearshift lever and tighten the clamp bolt. Make sure it is free to move in all lever positions. If binding exists, loosen the valve-body bolts and realign the body.
6. Make sure the throttle-shaft seal is in place. Install the flat washer and throttle lever, and tighten the clamp bolt. Reconnect the linkage and check linkage adjustments, as already explained.
7. Install the oil pan, using a new gasket. Add transmission fluid, as explained in ⊘ 18-13, to bring fluid up to the proper level.

Out-of-Car Service

The following sections describe transmission removal, disassembly, servicing of subassemblies, reassembly, and installation.

CAUTION: The transmission and torque converter must be removed and replaced as an assembly. Otherwise, the converter drive plate, pump bushing, and oil seal will be damaged. The drive plate cannot support a load. None of the transmission weight must be allowed to rest on the plate at any time.

⊘ 18-21 Transmission Removal First, connect a remote-control starter switch to the starting-motor solenoid so that the engine can be rotated from under the car. Disconnect the lead from the negative terminal of the ignition coil. This is the lead that goes to the distributor. Disconnecting this lead prevents the engine from starting and also protects the ignition coil from high voltage. Raise the car on a hoist. Proceed as follows:

1. Remove the cover plate from in front of the converter. Rotate the engine with the remote-control switch to bring the drain plug to the 6 o'clock position. With the container in place, remove the drain plug and drain converter and transmission.
2. Mark the converter and drive plate to aid in reassembly. One bolt hole is offset so that parts must go back in original position. This arrangement retains the original balance of the converter and engine.
3. Rotate the engine to locate two converter-to-drive-plate bolts at 5 and 7 o'clock. Remove the two bolts, rotate the engine again, and remove the other two bolts.

CAUTION: Do not rotate the converter or drive plate by prying with a screwdriver or other tool, as this could ruin the drive plate. Also, do not use the starting motor if the drive plate is not attached to the converter by at least one bolt. Do not use the starting motor if the transmission-case-to-engine bolts have been loosened. Either condition could cause transmission damage.

4. Disconnect the ground cable from the battery and remove the starting motor.
5. Disconnect the wire from the neutral starting switch.
6. Disconnect the gearshift rod from the transmission lever. Remove the gearshift-torque shaft from the transmission housing and left side rail.
7. Disconnect the throttle rod from the throttle lever on the transmission.
8. Disconnect oil-cooler lines from the transmission and remove the oil-filler tube. Disconnect the speedometer cable.
9. Mark parts for reassembly and disconnect the drive shaft at the rear universal joint. Pull the shaft assembly from the extension housing.

10. Remove rear-mount-to-extension-housing bolts.
11. Install the engine-support fixture and raise the engine slightly. See Figs. 18-28 and 18-29 for procedures on six- and eight-cylinder engines.

NOTE: Some models have exhaust systems that will interfere and will have to be partly removed for adequate clearance.

12. Remove the frame cross member.
13. Place the transmission-service jack under the transmission to support it.
14. Attach a small C clamp to the edge of the converter housing to hold the converter in place during transmission removal.
15. Remove converter-housing-to-engine retaining bolts. Carefully work the transmission rearward off the engine-block dowels and disengage the converter hub from the end of the crankshaft.

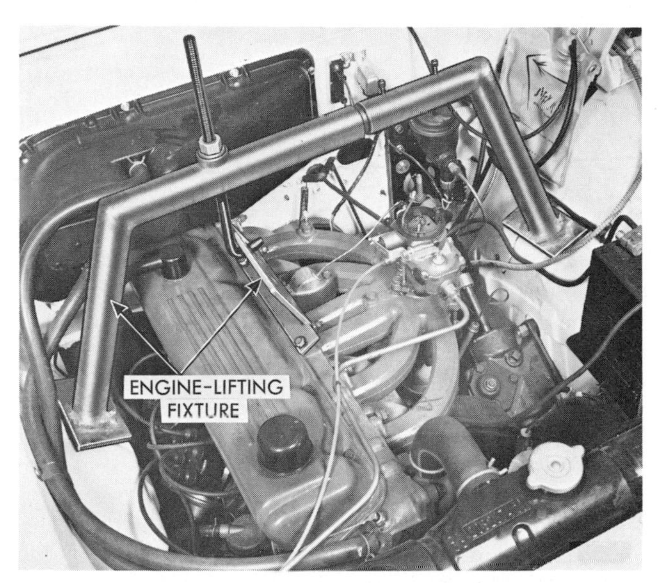

Fig. 18-28. Engine-lifting fixture for a six-cylinder engine. (*Chrysler Corporation*)

16. Lower the transmission jack and remove the converter assembly by taking off the C clamp. Then slide the converter off.
17. Mount the transmission in a repair stand for further service.

⊘ 18-22 Starting-Motor Ring-Gear The starting-motor ring gear is welded in four places to the outer diameter of the torque converter. If the ring gear must be replaced, these welds must be cut with a hacksaw or grinding wheel. Then the ring gear must be driven off with a brass drift and hammer.

CAUTION: Do not allow the converter to rest on its hub during this operation. This could ruin the converter.

The new ring gear must be heated so that it expands slightly, and then it can be driven onto the torque converter. Finally, the ring gear must be welded in place, with the new welds in the same places as the old welds.

⊘ 18-23 Transmission Disassembly Before disassembling the transmission, plug all openings and thoroughly clean the exterior, preferably by steam. We have already emphasized, in ⊘ 14-10 and 18-11 and elsewhere, the extreme importance of cleanliness when servicing automatic transmissions. Do not wipe parts with a shop towel. Instead, wash them in solvent and air-dry them. Do not scratch or nick finished surfaces; this could cause valve hangups or fluid leakage and faulty transmission operation. Proceed with disassembly as follows:

1. First, measure the drive-train end play by mounting a dial indicator, as shown in Fig. 18-30. Move the input shaft in and out to check end play. Specifications are 0.030 to 0.089 inch [0.762 to 2.260 mm] for the A-904 transmission and 0.037 to 0.084 inch [0.939 to 2.134 mm] for the A-727 transmission. Write

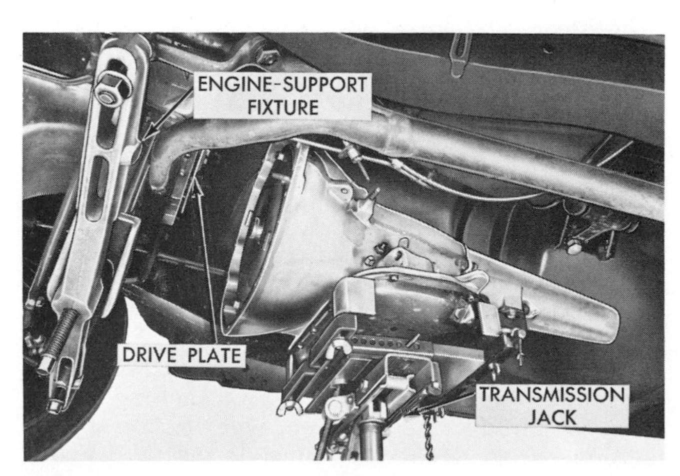

Fig. 18-29. Engine-support fixture for an eight-cylinder engine. (*Chrysler Corporation*)

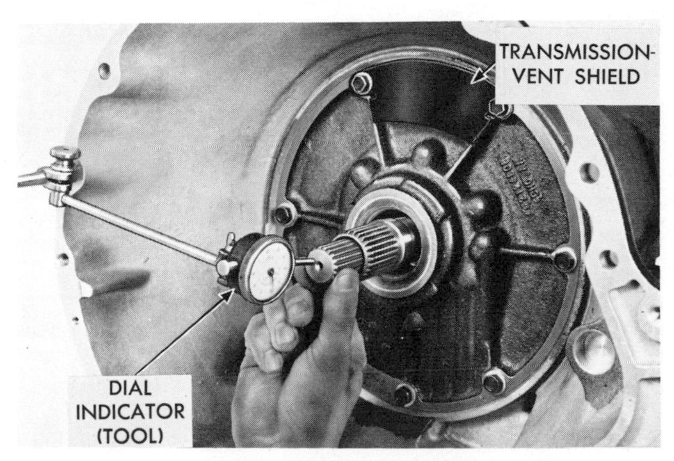

Fig. 18-30. Measuring drive-train end play with a dial indicator. (*Chrysler Corporation*)

down the indicator reading for reference when reassembling the transmission.
2. Remove the oil pan and gasket.
3. Remove the valve body and accumulator piston and spring, as explained in ⊘ 18-20.
4. To remove the extension housing, pull the parking-lock rod forward out of the case. Rotate the output shaft if necessary to align the parking gear and sprag to permit the knob on the end of the control rod to pass the sprag. Then proceed as follows:
a. Remove the speedometer and adapter assembly.
b. Remove extension-housing-to-transmission bolts.
c. Remove two screws, plate, and gasket from the bottom of the extension-housing mounting pad. Spread the large snap ring on the output-shaft bearing with a special tool, as shown in Fig. 18-31. With the snap ring spread as much as possible, carefully tap the extension housing off the output shaft and bearing.
5. We have explained how to remove the governor and support in ⊘ 18-18. This is the next assembly to come off the transmission.
6. The pump and reaction-shaft support come out next. First, tighten the front-band adjusting screw until the band is tight on the front-clutch retainer. This adjustment will hold the retainer when the pump is removed and thereby prevent damage to the clutch.

Next, remove the oil-pump-housing retaining bolts and use two slide hammers, as shown in Fig. 18-32, to remove the pump.
7. Loosen the front-band adjuster, remove the band strut, and slide the band from the case.
8. Slide the front-clutch assembly from the case.
9. Pull on the input shaft to slide the input-shaft-and-rear-clutch assembly out of the case.

CAUTION: Be careful not to lose the thrust washer located between the rear end of the input shaft and the forward end of the output shaft.

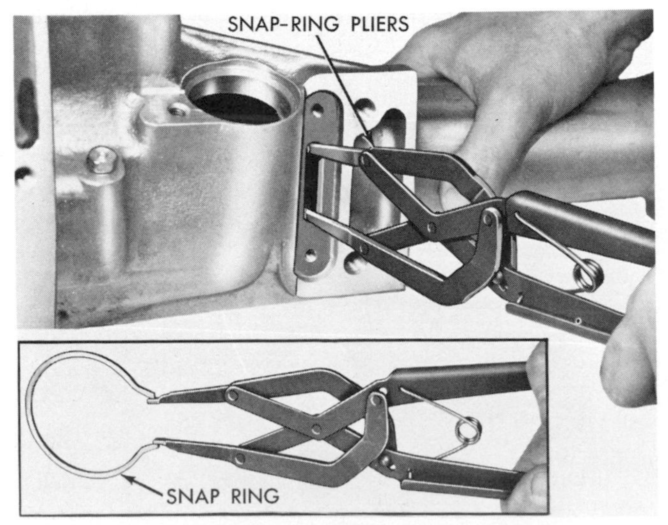

Fig. 18-31. Using snap-ring pliers to remove the snap ring from the output-shaft bearing. (*Chrysler Corporation*)

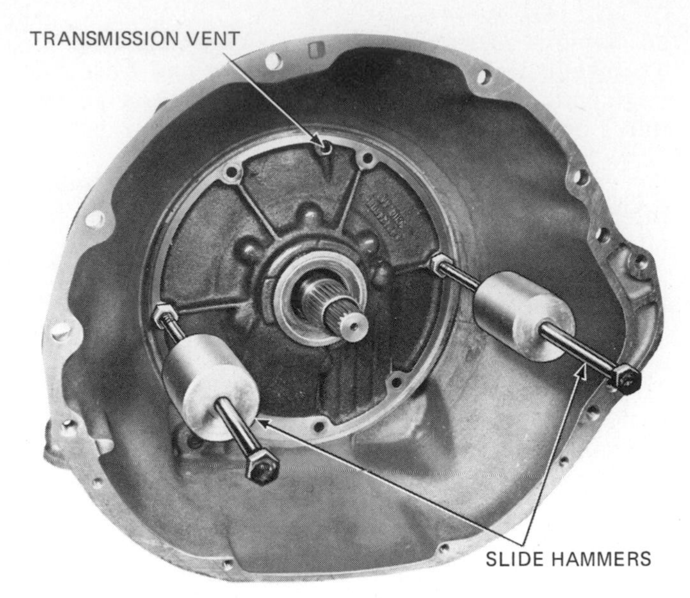

Fig. 18-32. Removing the pump and reaction-shaft support with slide hammers. (*Chrysler Corporation*)

10. Support the output shaft and driving shell with your hands, and slide the assembly forward and out through the case.

CAUTION: Be very careful to avoid damaging the ground surfaces on the output shaft.

11. Remove the low-and-reverse drum. Then loosen the rear-band adjuster, and remove the band strut and link and the band. On the double-wrap band (A-904LA), loosen the band adjusting screw to remove the band and then the low-and-reverse drum.
12. Note the position of the overrunning-clutch rollers and springs before disassembly so that you can reassemble everything in the original positions. Slide out the clutch hub and remove the rollers and springs. If the overrunning-clutch cam or spring retainer is damaged, refer to ⊘ 18-31 for the repair procedure.
13. Next, remove the kickdown servo. The servo spring must be compressed by an engine-valve-spring compressor, as shown in Fig. 18-33, so that the snap ring can be removed. This allows removal of the rod guide, springs, and piston rod. Do not damage these parts during removal. Now, the piston can be withdrawn from the transmission case.
14. Remove the low-and-reverse servo by compressing the piston spring with the engine-valve-spring compressor so that the snap ring can be removed. Then remove the spring retainer, spring, and servo piston and plug from the transmission case.

⊘ 18-24 Reconditioning Subassemblies We have already mentioned the necessity of keeping everything absolutely clean when working on transmissions. Also, great care must be used to avoid scratching or nicking finished surfaces. Crocus cloth can be used, within reason, to remove burrs and

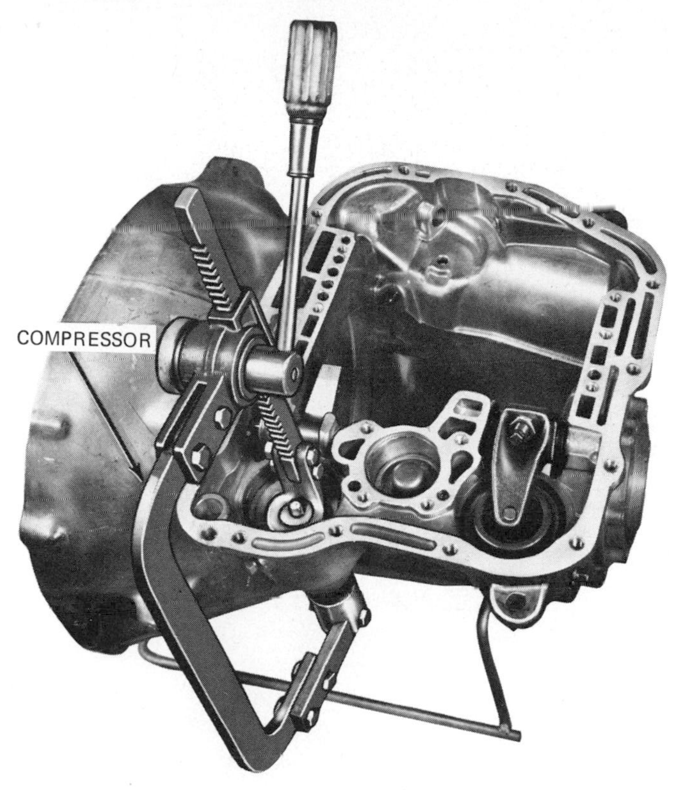

Fig. 18-33. Compressing the kickdown-servo spring with an engine-valve-spring compressor. (*Chrysler Corporation*)

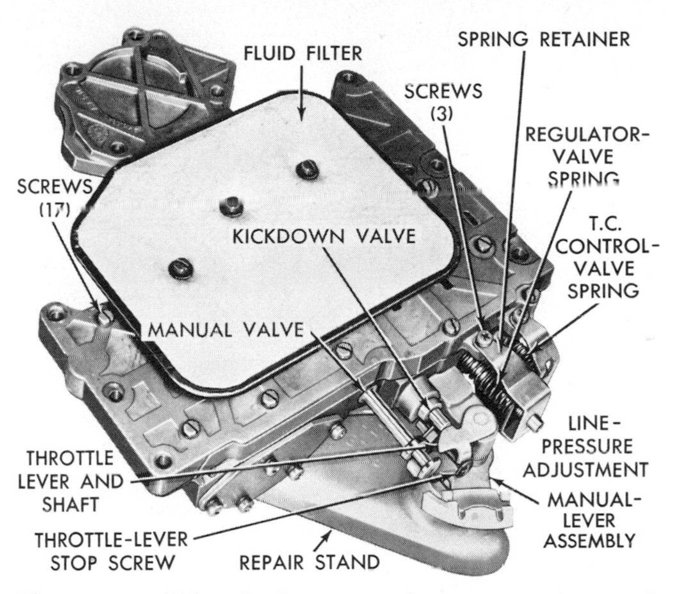

Fig. 18-34. Valve body mounted on a repair stand. (*Chrysler Corporation*)

rough spots. When using it on valves, take great care to avoid rounding off sharp edges. These sharp edges on valves are essential to proper valve operation. They help prevent dirt particles from getting between the valve and body and possibly causing the valve to stick.

Use new seal rings when reconditioning an automatic transmission. Also, check all bushings and replace those that are worn or galled. Use the special bushing removal-and-replacement tools supplied by Plymouth.

The following sections describe the service procedure for the various subassemblies removed from the TorqueFlite transmission.

⊘ **18-25 Valve Body** Use the special valve-body repair stand shown in Fig. 18-9 to work on the valve body. Never clamp any part of the body in a vise. Doing so can distort the body and cause leakage, valve hangups, and serious transmission trouble. Slide valves and plugs in and out of the valve body with great care to avoid damage. Proceed as follows:

1. With the valve body on the repair stand, as shown in Fig. 18-34, remove three screws and the filter.
2. Hold the spring retainer against the spring pressure and remove the three screws so that the spring retainer can be removed.
3. With the spring retainer off, remove the torque-converter valve spring and valve and the line-pressure adjusting screw, spring, and regulator valve. See Fig. 18-35.

CAUTION: Do not change the setting of the line-pressure adjusting screw.

4. Remove the transfer-plate attaching screws and lift the transfer plate and steel separator plate off the valve body. Invert the transfer plate and take off the small stiffener plate.
5. Remove and note locations of the seven steel balls and one spring from the valve body. (See Fig. 18-36.)

CAUTION: Do not mix the balls. The five smaller balls are the same size ($^1/_4$ inch [6.35 mm]), but there is one $^{11}/_{32}$-inch [8.73 mm] ball and one $^3/_8$-inch [9.53 mm] ball with a spring behind it.

6. Invert the valve body and lay it on a clean cloth. Remove the E clip and washer from the throttle-lever shaft. See Fig. 18-37. Hold the detent ball in its bore and slide the manual-lever assembly off the shaft. Remove the detent ball and spring.
7. Remove the manual valve by sliding it carefully out of the valve body with a rotating motion.
8. Remove the throttle lever and shaft.
9. Remove the shuttle-valve cover plate. See Fig. 18-37 for location. Then remove the E clip from the exposed end of the shuttle valve.
10. Remove the throttle-lever-stop-screw assembly, being careful not to disturb its setting. Then remove the kickdown detent, kickdown valve, throttle-valve spring, and throttle valve.
11. Remove the governor-plug end plate, and tilt the valve body to allow the throttle plug, spring, shuttle valve, and two governor plugs to slide out in your hand. Note that the 1-2 shift-valve governor plug has a longer stem.

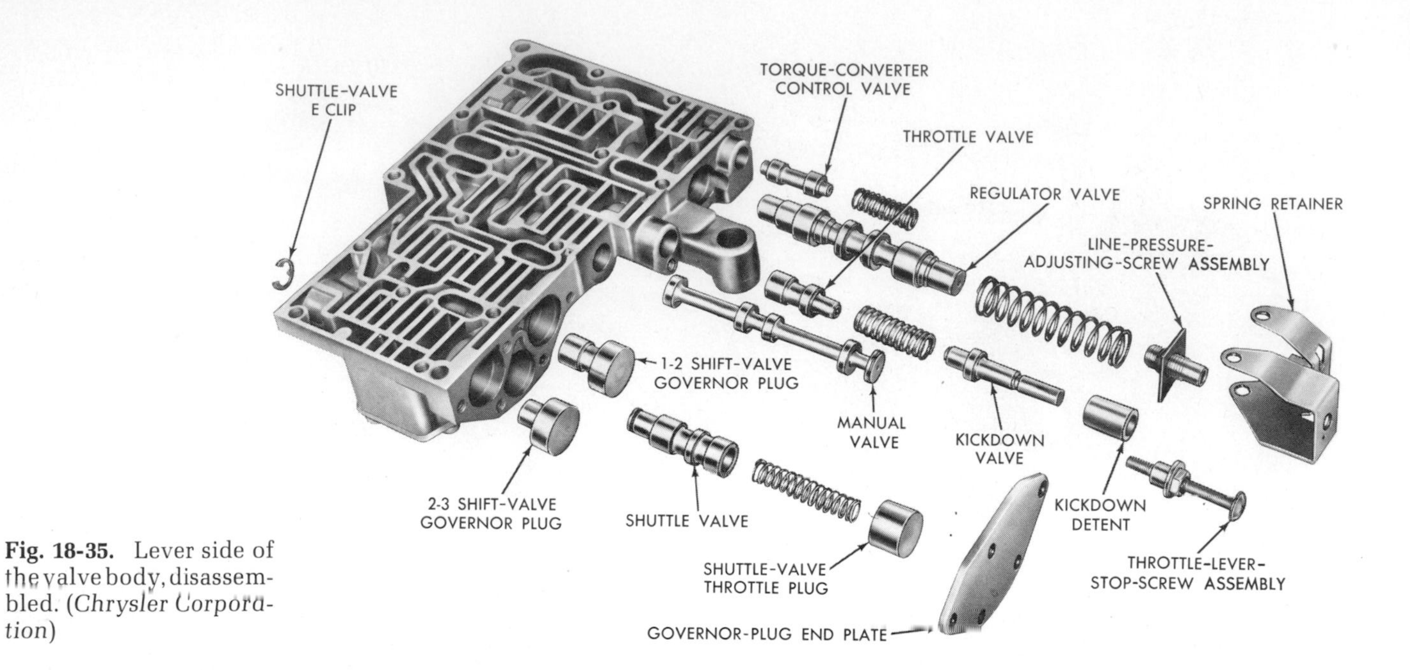

Fig. 18-35. Lever side of the valve body, disassembled. (*Chrysler Corporation*)

Fig. 18-36. Locations of steel balls in the valve body. (*Chrysler Corporation*)

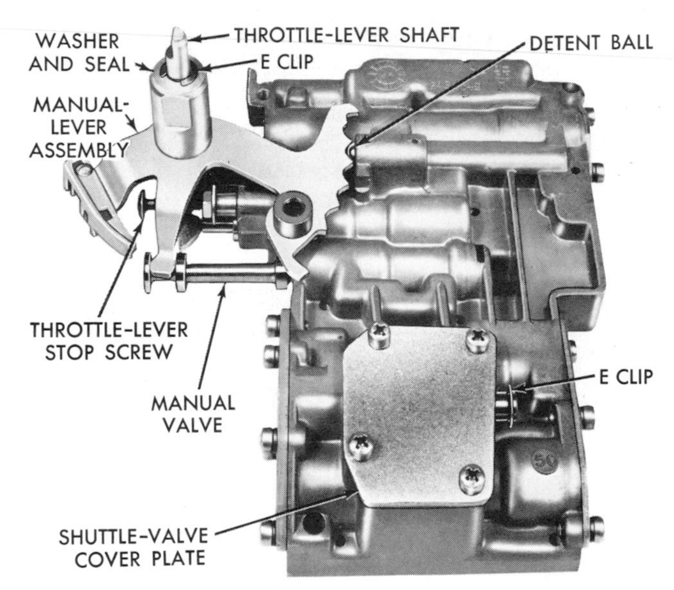

Fig. 18-37. Valve-body controls. (*Chrysler Corporation*)

12. Remove the shift-valve end plate or downshift-plug cover (on six-cylinder engines), and slide out the springs and valves (Fig. 18-38). Figure 18-39 shows a similar valve body with the same basic parts. It also shows parts that are included in the valve body for the A-727 and A-904-LA transmissions used with eight-cylinder engines. These parts are shown inside the dashed line. They include the limit valve and the throttle plug, which were discussed in ⊘ 17-8.

13. Clean and inspect parts. Soak parts in a suitable solvent for several minutes. Air-dry. Make sure all passages are clean. Check operating levers and shafts for wear, looseness, or bending. Check mating

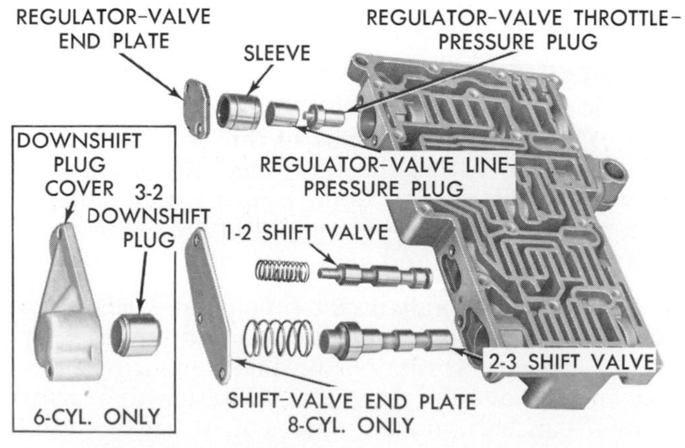

Fig. 18-38. Shift-valve side of the valve body, disassembled. (*Chrysler Corporation*)

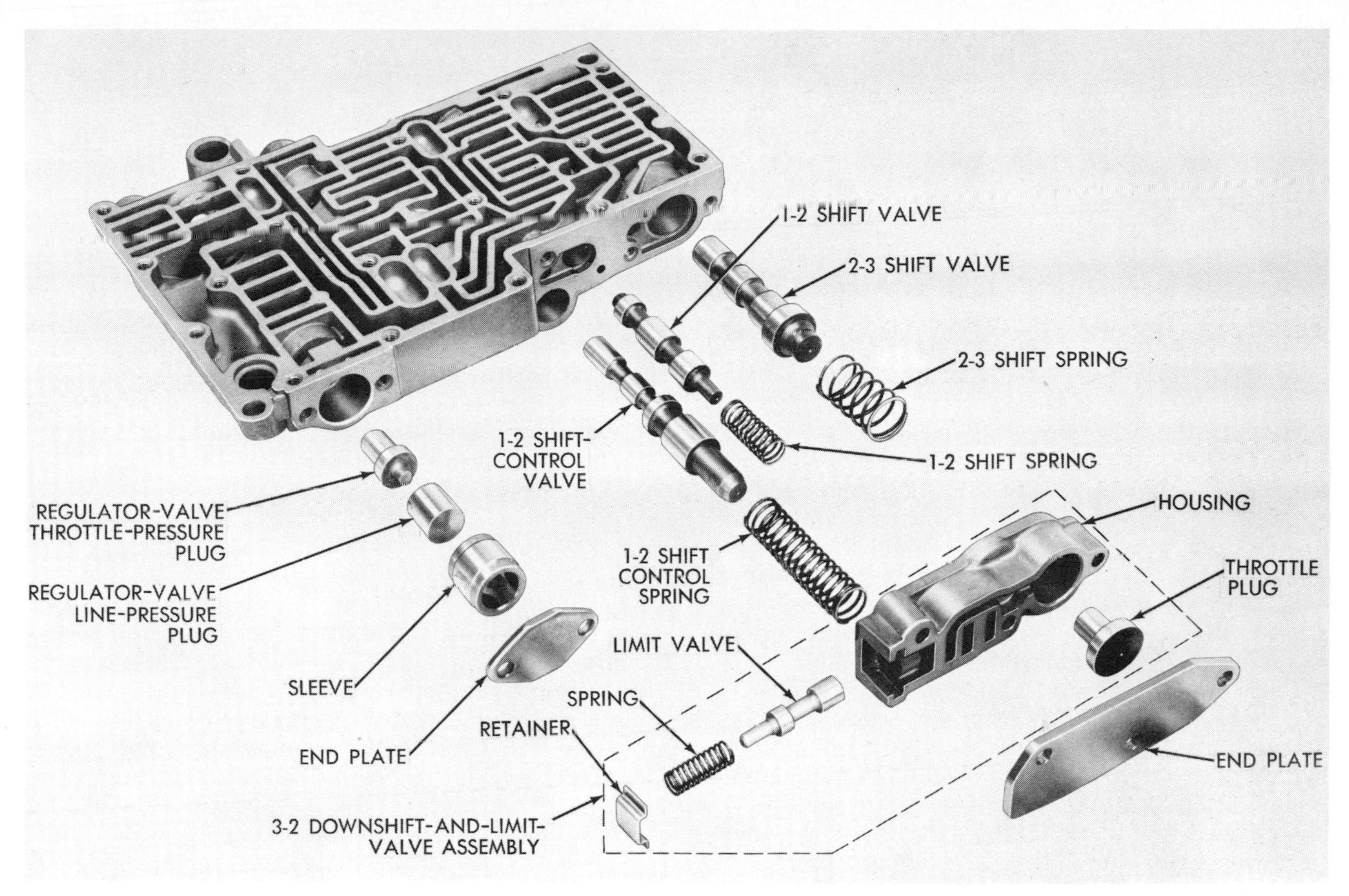

Fig. 18-39. Shift valves and pressure-regulator valve plugs for A-727 and A-904LA. (*Chrysler Corporation*)

surfaces for burrs, nicks, and scratches. Check valve springs for distortion and collapsed coils. Make sure all valves and plugs fit their bores freely. They should fall freely in the bores.

14. Reassembly is essentially the reverse of disassembly. Refer to Figs. 18-37 and 18-38. Stiffener-plate screws should be tightened to 28 pound-inches [0.504 kg-mm]. Also, the governor-plug-end-plate screws, shuttle-valve-cover-plate screws, shift-valve-end-plate (eight-cylinder) or downshift-plug-cover (six-cylinder) screws, and regulator-valve-end-plate screws should also be tightened to 28 pound-inches [0.504 kg-mm].

To install the spring retainer, tighten the three retaining screws to 28 pound-inches [0.504 kg-mm]. Measure and correct spring-retainer alignment, as shown in Fig. 18-11. Then adjust throttle and line pressures, as shown in Figs. 18-10 and 18-12.

⊘ 18-26 Other Subassemblies In previous sections, we have described the removal and replacement of the accumulator piston, extension-housing oil seal, parking-lock sprag, and governor-and-support assembly. Other subassemblies that we will look into in following sections are the oil pump, clutches, and planetary-gear train.

⊘ 18-27 Oil Pump and Reaction-shaft Support These are different for the A-904 and A-727 transmissions. See Figs. 18-40 and 18-41, which are disassembled views of the two oil pumps. Both are disassembled as follows:

1. Remove bolts from the rear side of the reaction-shaft support and lift the support off the pump.
2. Remove the rubber seal ring from the pump-body flange.
3. Drive out the oil seal with a blunt punch.
4. Inspect parts as follows:

a. Check the interlocking seal rings on the reaction-shaft support for wear or broken locks. Replace if necessary.

b. On the A-904, check the thickness of the thrust washer, which should be 0.043 to 0.045 inch [0.992 to 1.133 mm]. Replace if worn.

c. Check the machined surfaces for nicks and burrs, and check pump rotors for scores or pitting. Clean the rotors and pump body, and install the rotors to check clearances. Put a straightedge across the surface of the pump face and measure with a feeler gauge to the rotors. Clearance limits are 0.0015 to 0.003 inch [0.0381 to 0.0762 mm]. Rotor-tip clearance between rotor teeth should be 0.005 to 0.010 inch [0.127 to 0.254 mm]. Clearance be-

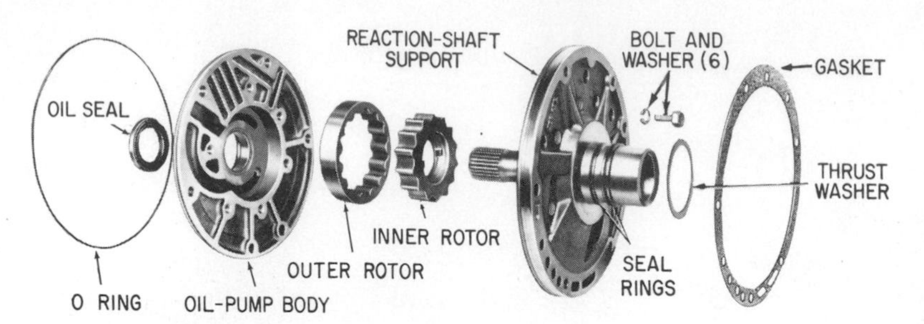

Fig. 18-40. Disassembled view of the oil pump and reaction-shaft support from the A-904. (*Chrysler Corporation*)

tween the outer rotor and its bore should be 0.004 to 0.008 inch [0.916 to 0.203 mm]. Replace the rotors and body if worn.

5. Replace the oil-pump-body bushing, if worn, as shown in Fig. 18-42 for the A-904 and as shown in Fig. 18-43 for the A-727. Be careful not to cock the removal or replacement tools because this could damage the bore in the oil-pump body. Stake the new bushing in place, as shown in Fig. 18-44. A gentle tap at each stake is all that is necessary. Use a narrow-bladed knife to remove high points around the stake that could interfere with the shaft.
6. Replace the reaction-shaft bushing if necessary. See Figs. 18-45 and 18-46.
7. Assemble the A-904 pump, as shown in Fig. 18-47. Note that two pilot studs are used to align the pump body and that the rotors are installed in the support and aligned with the aligning tool. Rotate the aligning tool as the pump is brought down into place. Tighten pump-body bolts to 160 pound-inches [2.88 kg-mm]. Install a new pump-body oil seal, lip facing in, using a special installing tool.
8. Assemble the A-727 pump by putting the rotors in the housing and installing the reaction-shaft support with the vent baffle over the vent opening. Tighten the retaining bolts to 160 pound-inches [2.88 kg-mm]. Install a new pump-body oil seal, lip facing in, with a special installing tool.

⊘ 18-28 Front Clutch The front clutches are different for the A-904 and A-727 transmissions. See Figs. 18-48 and 18-49. The disassembly procedures are similar, however, except for the piston-spring compressor that must be used. First, remove the large selective snap ring and remove the pressure plate and clutch plates. Next, use the spring compressor to remove the small snap ring, spring retainer, and spring or springs. Finally, bump the piston retainer on a wood block to remove the piston. Remove the seal rings from the piston.

Inspect the clutch plates and facings. Facings that are charred, pitted, glazed, or flaking require plate replacement. Steel plates that are burned, scored, or damaged should be replaced. Steel-plate-lug grooves in the clutch retainer must be smooth

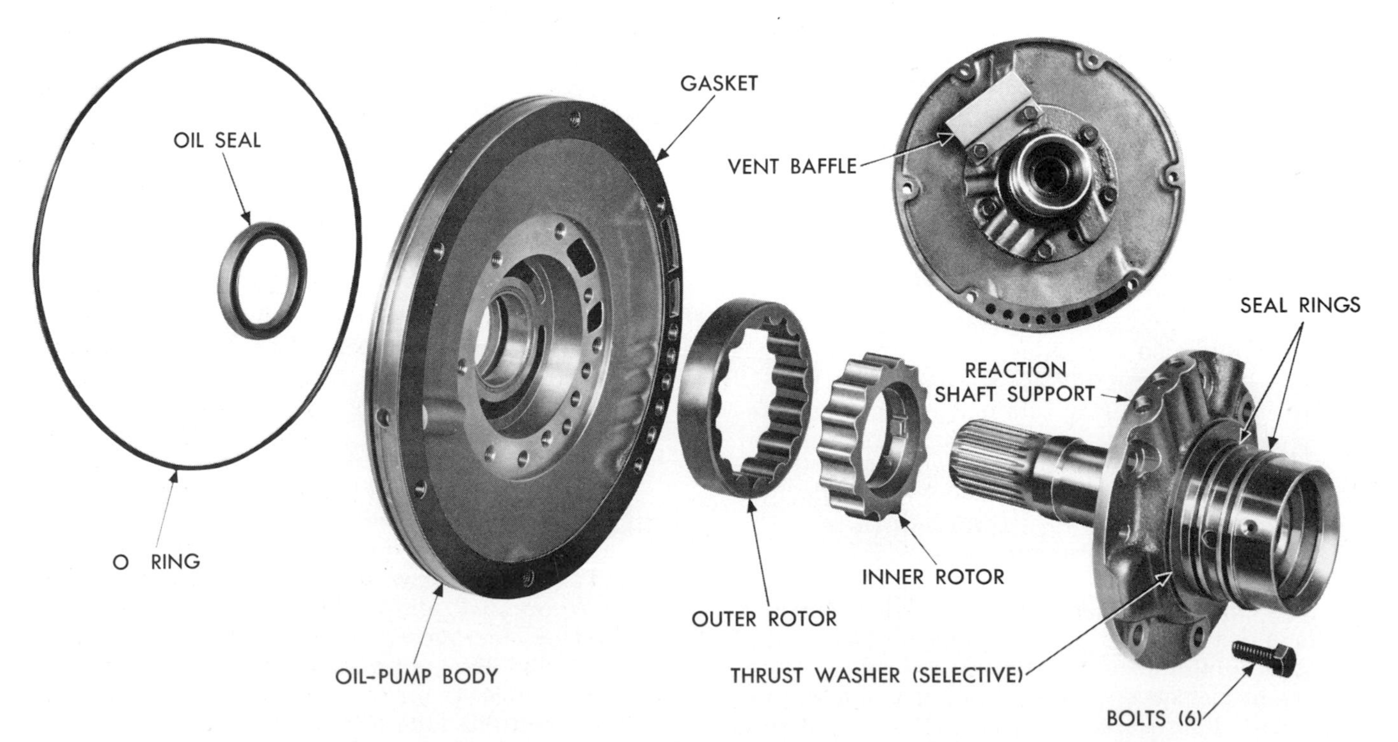

Fig. 18-41. Disassembled view of the oil pump and reaction-shaft support from the A-727. (*Chrysler Corporation*)

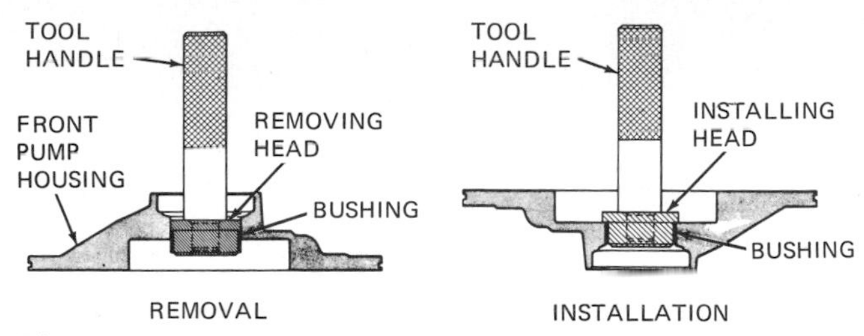

Fig. 18-42. Removal and installation of the pump-body bushing on the A-904. (*Chrylser Corporation*)

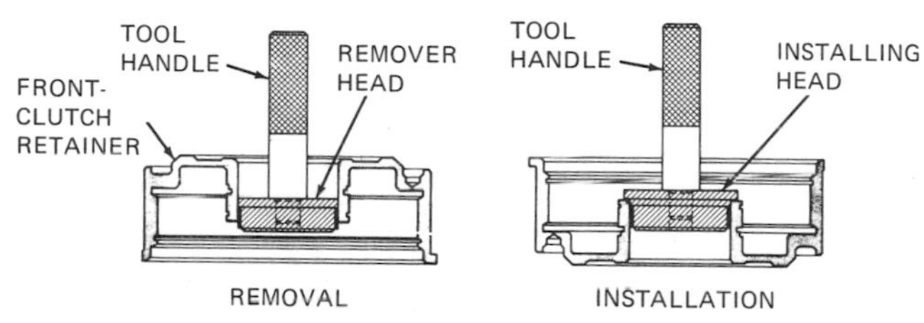

Fig. 18-43. Removal and installation of the pump-body bushing on the A-727. (*Chrysler Corporation*)

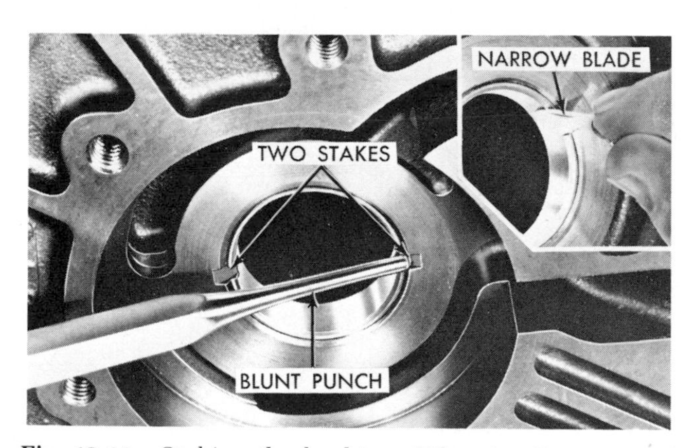

Fig. 18-44. Staking the bushing. (*Chrysler Corporation*)

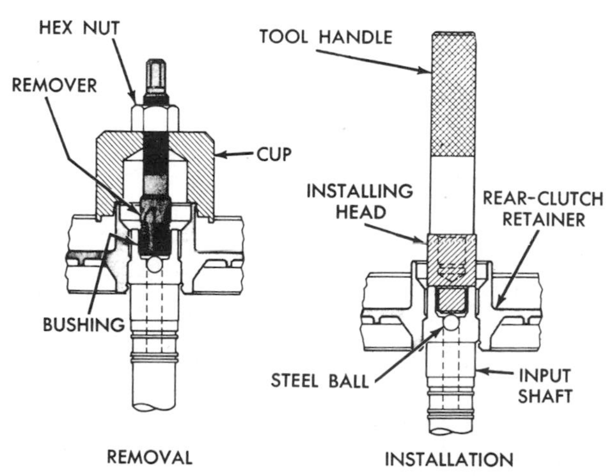

Fig. 18-45. Replacing the reaction-shaft bushing on the A-904. Note that the removal tool has threads which can be threaded into the bushing so that it can be pulled out. (*Chrysler Corporation*)

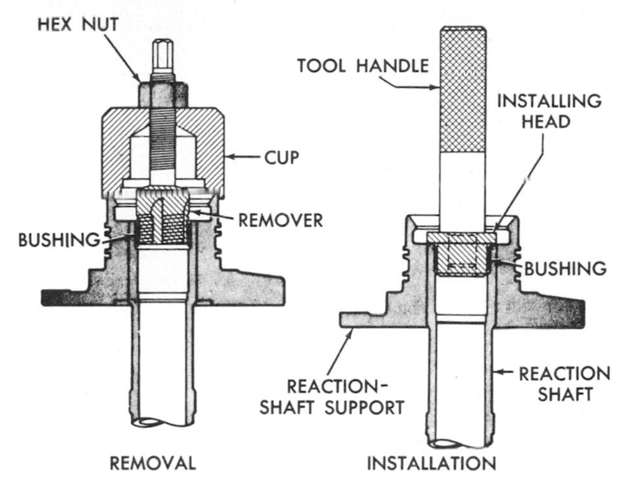

Fig. 18-46. Replacing the reaction-shaft bushing on the A-727. Note that the removal tool has threads which can be threaded into the bushing so that it can be pulled out. (*Chrysler Corporation*)

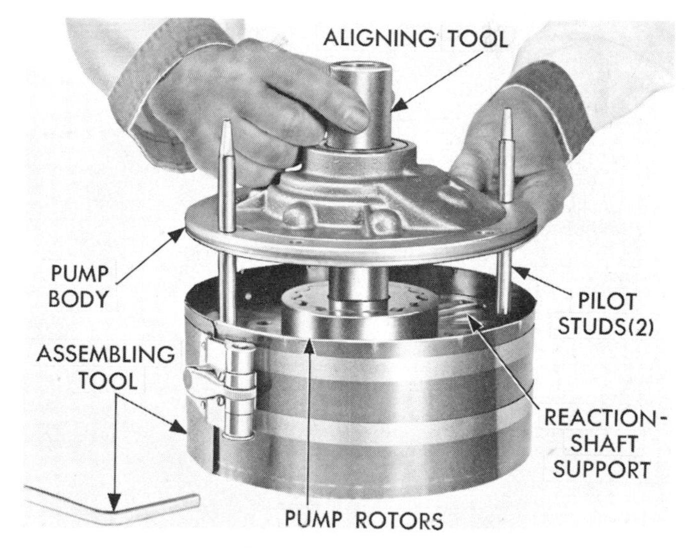

Fig. 18-47. Assembling the pump and reaction-shaft support for the A-904. (*Chrysler Corporation*)

so that plates can move freely. The brake-band surface on the outside of the piston retainer must be smooth, and the ball check must be free. Surfaces in the retainer on which the piston seals slide must be smooth. Use new seals on reassembly.

If the retainer bushing needs replacement, see Fig. 18-50. Special tools are needed to remove and replace the bushing. The procedure is similar for the A-904 and the A-727.

On reassembly, make sure the lips on the seal rings face down, or into the piston retainer, when installed on the piston. Lubricate all parts with transmission fluid for easier installation. Use the spring compressor to compress the spring or springs so that the snap ring can be installed. Make sure the spring or springs are in the same location as on the original assembly. Note that on the A-727, there may be from six to twelve springs.

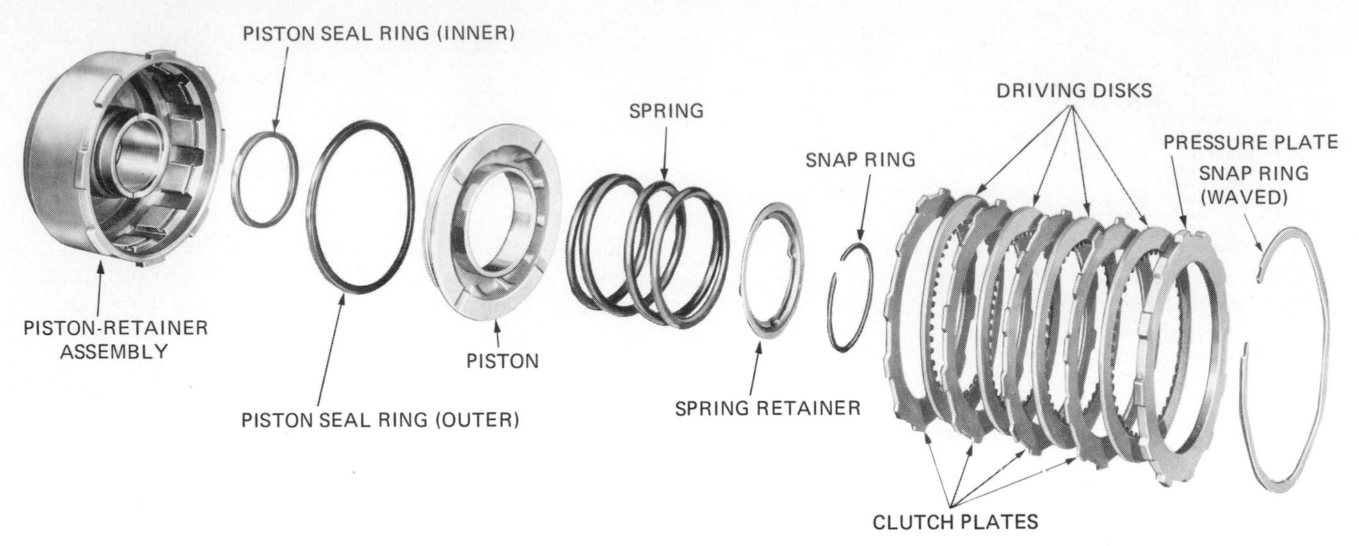

Fig. 18-48. Disassembled view of the A-904 front clutch. (*Chrysler Corporation*)

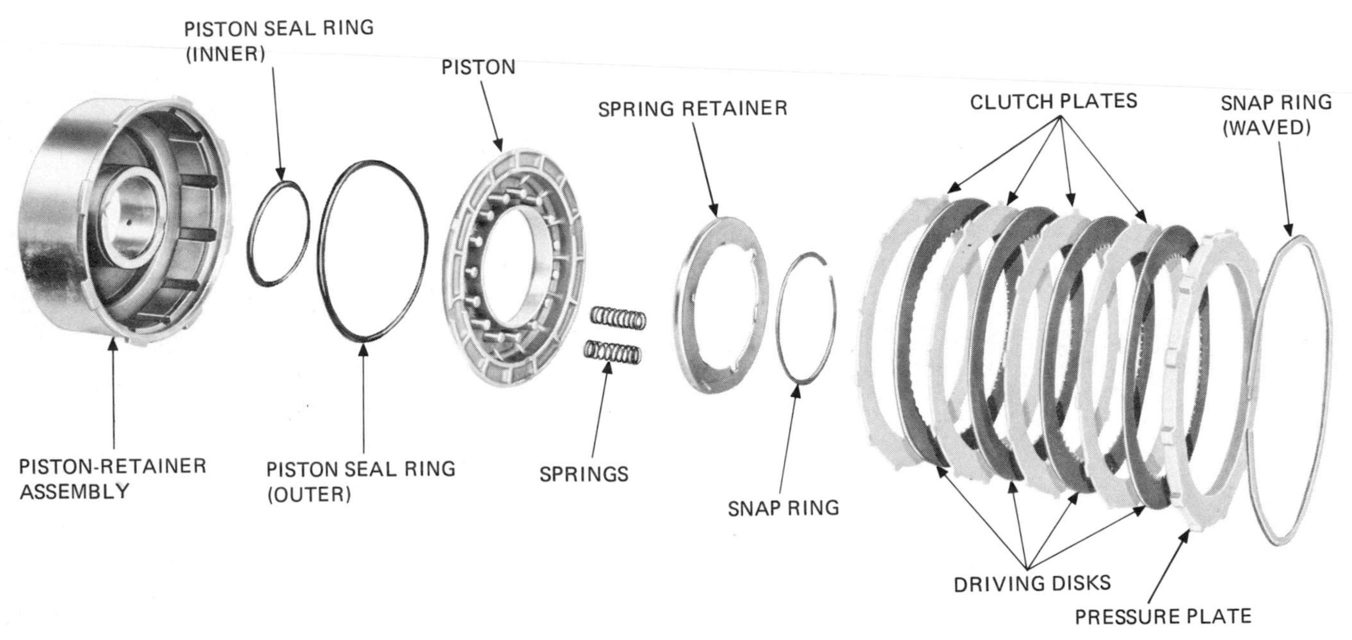

Fig. 18-49. Disassembled view of the A-727 front clutch. Only 2 of the 15 springs required are shown. (*Chrysler Corporation*)

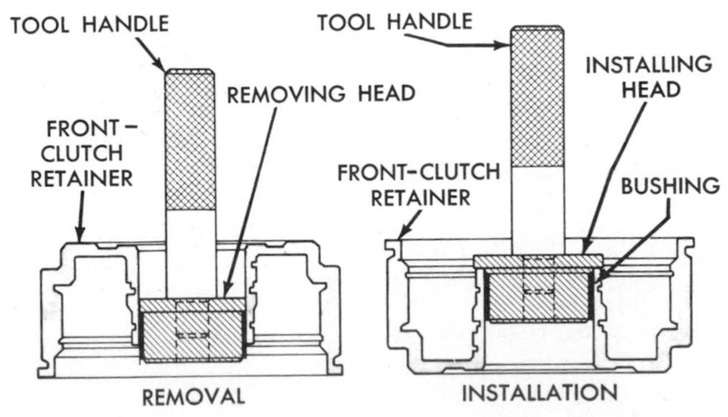

Fig. 18-50. Replacing the bushing in the A-904 front clutch piston retainer. (*Chrysler Corporation*)

After reassembly, check the clearance between the pressure plate and the selective snap ring, as shown in Fig. 18-51. The clearance varies with different engines. Check the shop manual for the proper specifications. There are snap rings of different thicknesses which can be installed to produce the correct clearance.

⊘ **18-29 Rear Clutch** Figures 18-52 and 18-53 show the rear clutches for the A-904 and A-727 transmissions in disassembled views. To disassemble the rear clutch, remove the large selective snap ring and lift the pressure plate, clutch plates, and inner pressure plate out of the retainer. Then carefully pry one

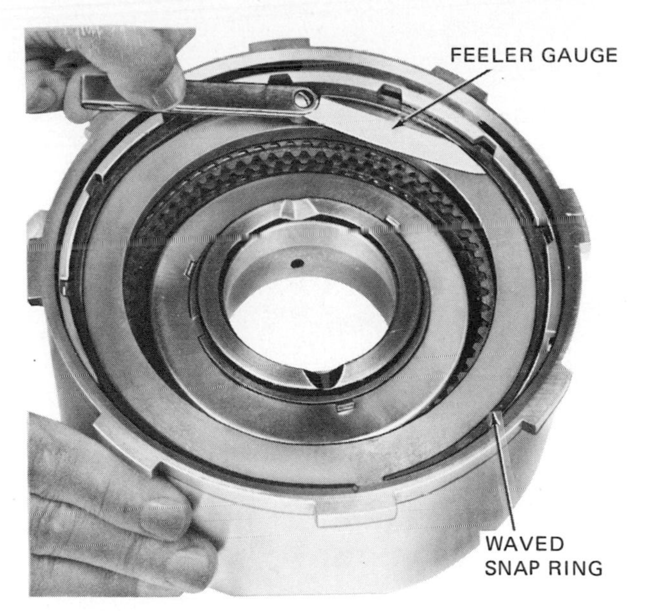

Fig. 18-51. Measuring front-clutch plate clearance. (*Chrysler Corporation*)

end of the wave spring out of its groove in the clutch retainer and remove the wave spring, spacer ring, and clutch-piston ring. Tap the retainer on a wood block to remove the piston. Take off the piston seals, noting the direction of the seal lips.

Inspect parts as for the front clutch, as explained in ⊘ 18-28.

On reassembly, lubricate parts with transmission fluid. To install the wave spring, start one end in the groove and then work the ring into the groove from that end to the other end. Check the clearance on the clutch by having an assistant press down firmly on the outer pressure plate and then measure with a feeler gauge between the selective snap ring and the pressure plate. If the clearance is not correct, install a snap ring of the correct thickness that will give the proper clearance. See the appropriate shop manual for specifications.

⊘ 18-30 Planetary-Gear Trains Figures 18-54 and 18-55 are disassembled views of the planetary-gear trains used in the A-904 and A-727 transmissions. With both, disassembly is essentially a process of removing the snap rings and sliding parts off the shaft. Before disassembly, however, measure the clearance between the shoulder on the output shaft and the rear-ring-gear-support hub, as shown in Fig. 18-56. This clearance is the gearset end play. If it is excessive, another selective snap ring of the proper thickness must be used to correct the end play.

See Figs. 18-54 and 18-55 for proper relationship of parts during reassembly. Make sure all parts are in good condition and not burred, nicked, or worn.

⊘ 18-31 Overrunning-Clutch Cam If the cam is worn, a condition which is rare, it can be replaced. On the A-904, it is held in place with rivets, which must be driven out. The new cam is then installed with retaining bolts. On the A-727, there is a setscrew that must be removed, followed by four bolts that hold the output-shaft support in place. The cam is then driven out of the transmission case by a punch inserted through the bolt holes.

⊘ 18-32 Servos Figures 18-57 and 18-58 show the kickdown and low-and-reverse servos in disassembled views. Disassembly is accomplished by removing the snap rings.

Transmission Reassembly

Keep everything clean. Use transmission fluid to lubricate parts during reassembly of the transmission. Do not use force to install mating parts. Everything should slide into place easily. If force is

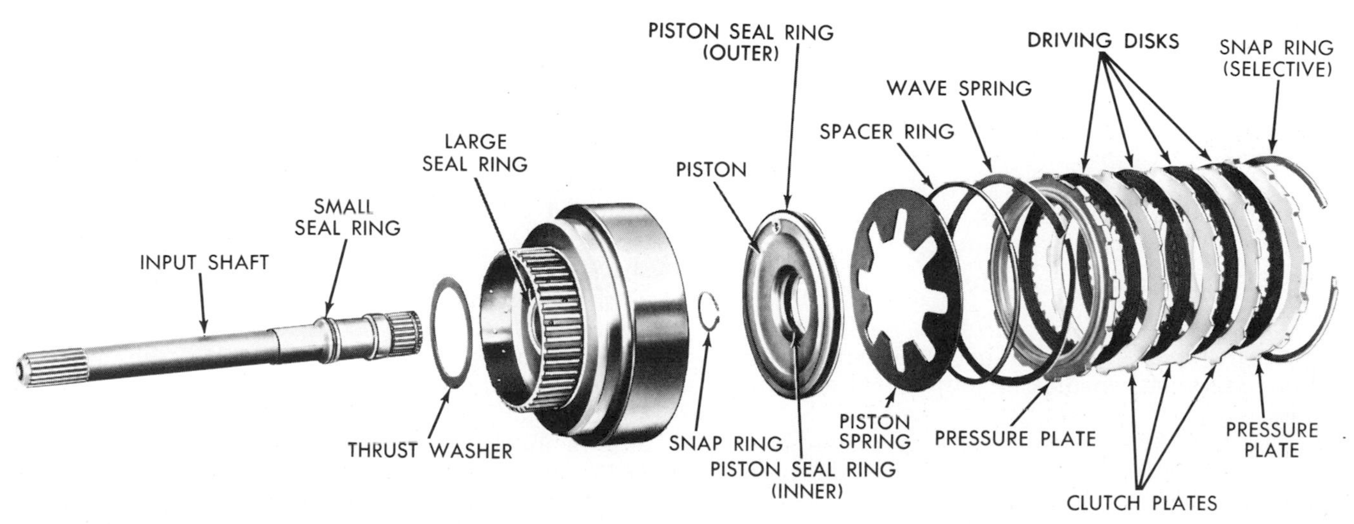

Fig. 18-52. Disassembled A-904 rear clutch. (*Chrysler Corporation*)

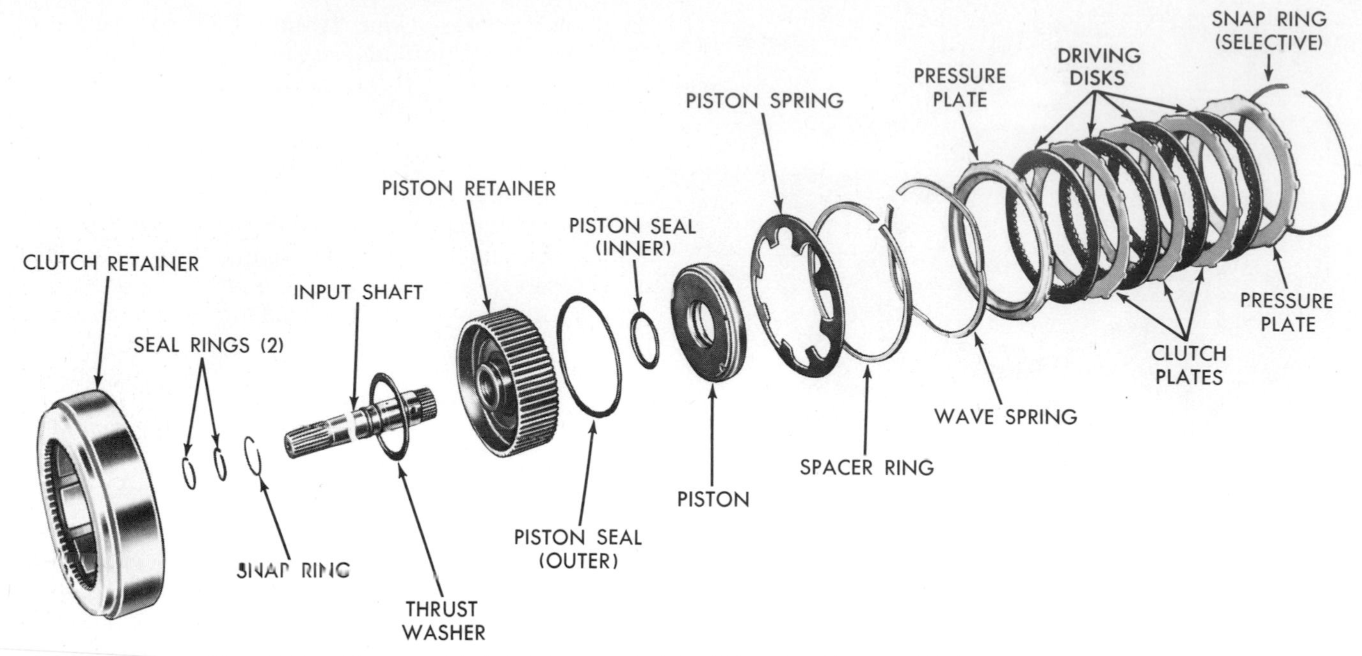

Fig. 18-53. Disassembled A-727 rear clutch. (*Chrysler Corporation*)

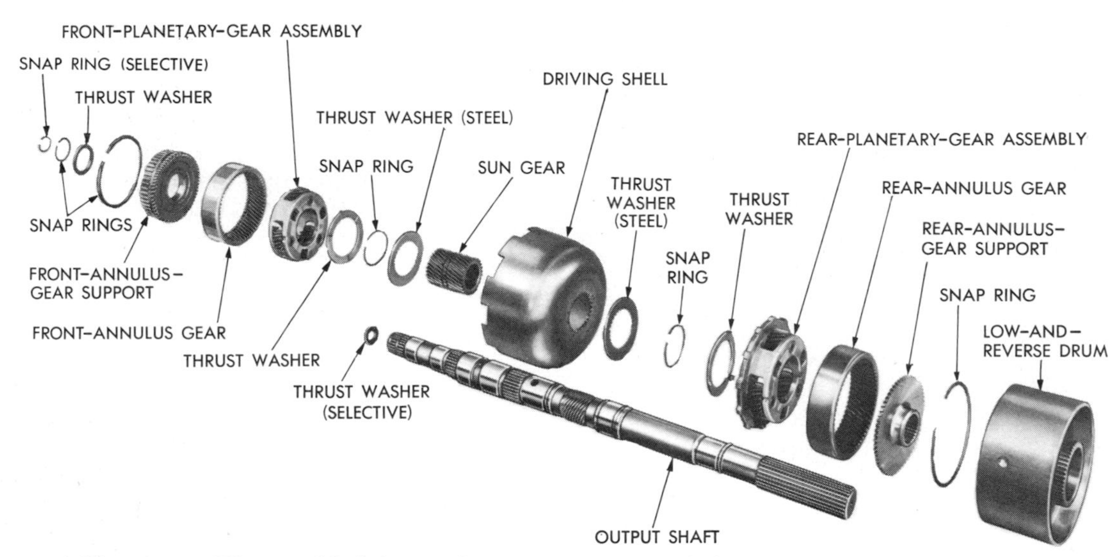

Fig. 18-54. Disassembled A-904 planetary-gear train. (*Chrysler Corporation*)

required, something is not properly installed. Remove parts to find the trouble.

⊘ 18-33 Reassembly Procedure

Reassemble the transmission as follows:

1. Insert the clutch hub and install the overrunning-clutch rollers and springs exactly as shown in Fig. 18-59.
2. Install the low-and-reverse-servo piston in the case with a twisting motion and put the spring, retainer, and snap ring over the piston. See Fig. 18-58. Use a compressor tool, as shown in Fig. 18-33, to compress the spring so that the snap ring can be installed.
3. Put the rear band in the case and install a short strut. Connect a long link and anchor to the band, as shown in Fig. 18-60. Screw in the band adjuster just enough to hold the strut in place. Install the low-and-reverse drum. Make sure the link is installed as shown in Fig. 18-59.
4. Install the kickdown-servo piston with a twisting motion, followed by other parts, as shown in Fig. 18-57. Use a spring compressor, as shown in Fig. 18-33, to compress the spring so that the snap ring can be installed.

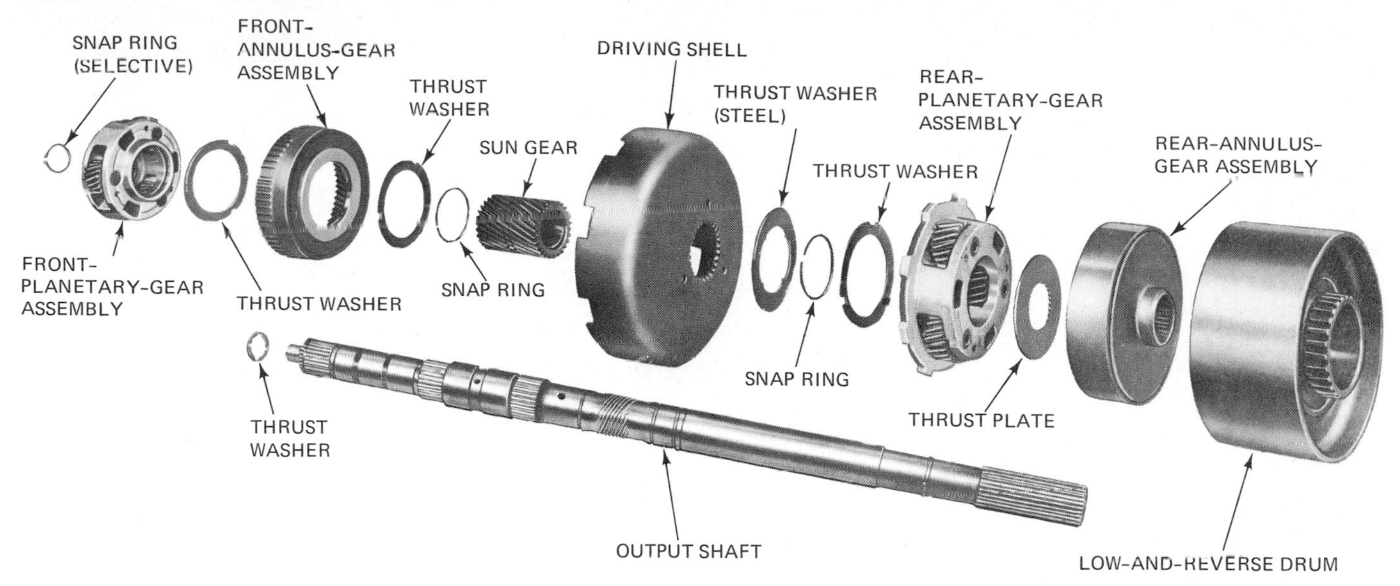

Fig. 18-55. Disassembled A-727 planetary-gear train. (*Chrysler Corporation*)

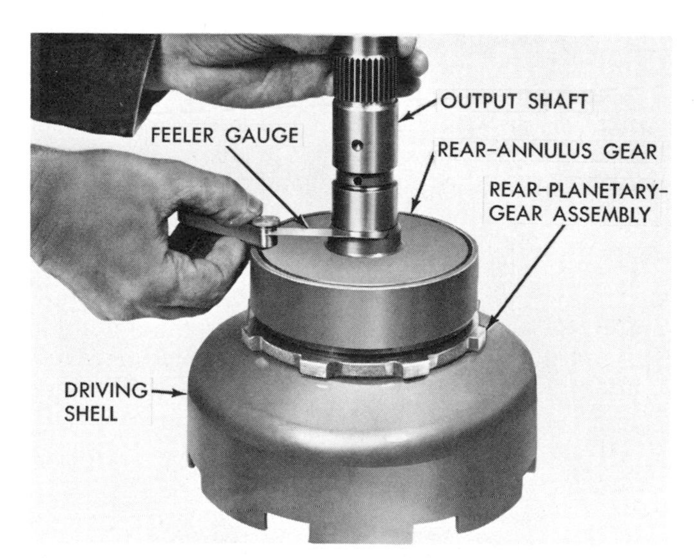

Fig. 18-56. Measuring end play of the planetary-gear assembly. (*Chrysler Corporation*)

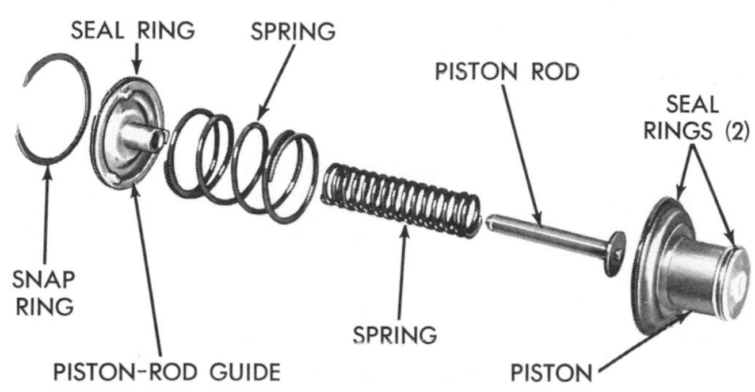

Fig. 18-57. Disassembled view of the kickdown servo. (*Chrysler Corporation*)

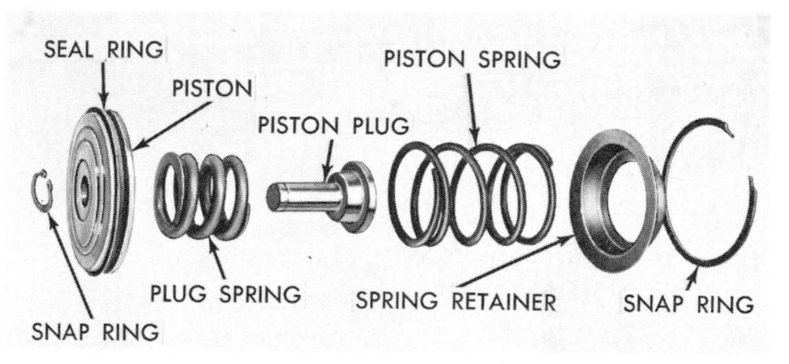

Fig. 18-58. Disassembled view of the low-and-reverse servo. (*Chrysler Corporation*)

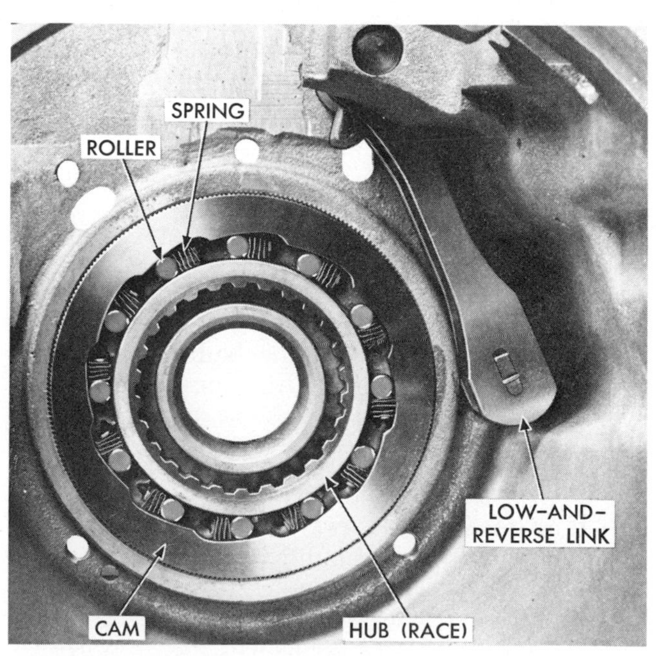

Fig. 18-59. Overruning clutch and low-and-reverse band link in place. (*Chrysler Corporation*)

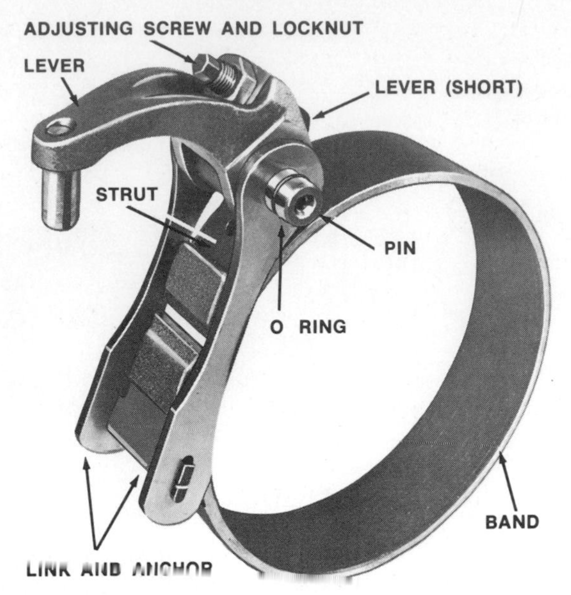

Fig. 18-60. Low-and-reverse brake band and linkage. (*Chrysler Corporation*)

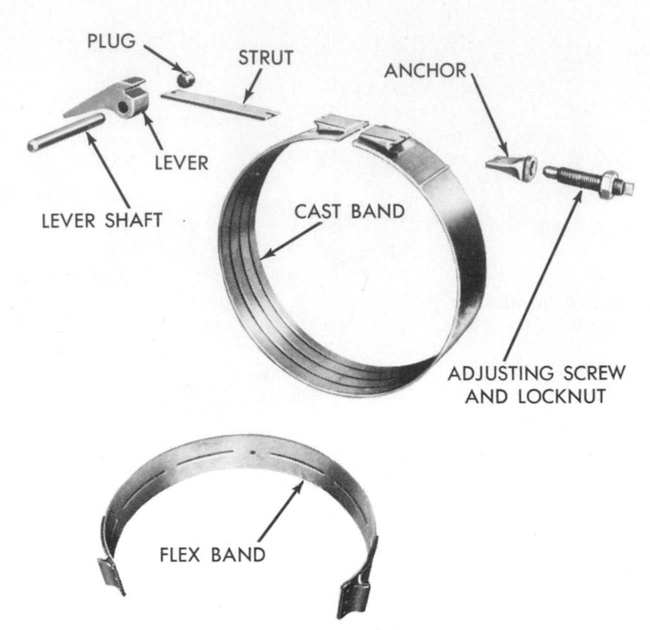

Fig. 18-61. Kickdown band and linkage with flex band for the A-727. (*Chrysler Corporation*)

5. Install the planetary-gear assembly with the sun gear and driving shell, as follows: Support the assembly in the case and insert the output shaft through the rear support. Carefully work the assembly into the case, engaging the rear planetary lugs into the low-and-reverse-drum slots.

CAUTION: Do not damage finished surfaces on the output shaft!

6. The front and rear clutches, front band, oil pump, and reaction-shaft support are more easily installed with the transmission in an upright position. Proceed as follows:

a. On the A-904 transmission, if the end play, shown being checked in Fig. 18-56, is not correct, a selective thrust washer of a different thickness will be required. Stick this thrust washer to the end of the output shaft with a coat of grease. On the A-727, apply a coat of grease on the input-to-output-shaft thrust washer (see Fig. 18-55) to hold it on the end of the output shaft.

b. Align the front-clutch-plate inner splines and put the assembly in position on the rear clutch. Make sure the front-clutch-plate splines are fully engaged on the rear-clutch splines.

c. Align the rear-clutch-plate inner splines, grasp the input shaft, and lower the two clutch assemblies into the transmission case.

d. Carefully work the clutch assemblies in a circular action to engage the rear-clutch splines over the splines of the front annulus gear. Make sure the front-clutch drive lugs are fully engaged in the slots in the driving shell.

7. The front band and associated parts are shown in Fig. 18-61. Slide the band over the front-clutch assembly and install the parts, tightening the adjusting screw tight enough to hold the strut and anchor in place.

8. Install the oil pump by first installing two pilot studs, as shown in Fig. 18-62. Put a new gasket over the studs. Put a new seal ring in the groove on the outer flange of the pump housing. Coat the seal with grease for easy installation. Install the pump in the case, tapping it down with a wood mallet if necessary. Put a deflector over the vent opening and install the pump attaching bolts. Remove two studs and install the other two bolts. Rotate the input and output shafts to make sure nothing is binding. Then tighten the bolts to 175 pound-inches [3.15 kg-mm].

Fig. 18-62. Installing the oil pump in the transmission case. (*Chrysler Corporation*)

9. Install the governor and support, as explained in ⊘ 18-18.
10. Install the extension-housing-and-output-shaft bearing, as explained in ⊘ 18-17.
11. Install the valve-body assembly, as explained in ⊘ 18-20.
12. Install the torque converter as follows:
a. Use a special tool to turn the pump rotors, as shown in Fig. 18-63, until the holes in the tool are vertical. Remove the tool.
b. Make sure converter-impeller-shaft slots are vertical so that they will engage with the oil-pump inner-rotor lugs. Then slide the converter into place.
c. Check for full engagement by placing a straightedge on the face of the case and measuring from the straightedge to one of the front-cover mounting lugs. The distance should be at least ½ inch [12.7 mm].
d. Attach a small C clamp to hold the converter in place.

⊘ 18-34 Transmission Installation To install the transmission on the car, proceed as follows:

1. Check the converter drive plate for distortion or cracks. If it is replaced, torque the bolts attaching it to the crankshaft to 55 pound-feet [7.601 kg-m].
2. Coat the converter-hub hole in the crankshaft with wheel-bearing grease. Put the transmission assembly on a jack and raise it up under the car, aligned for installation. Rotate the converter so that the mark on it will align with the mark on the drive plate.
3. Carefully work the transmission assembly forward so that the converter hub enters the crankshaft hole.

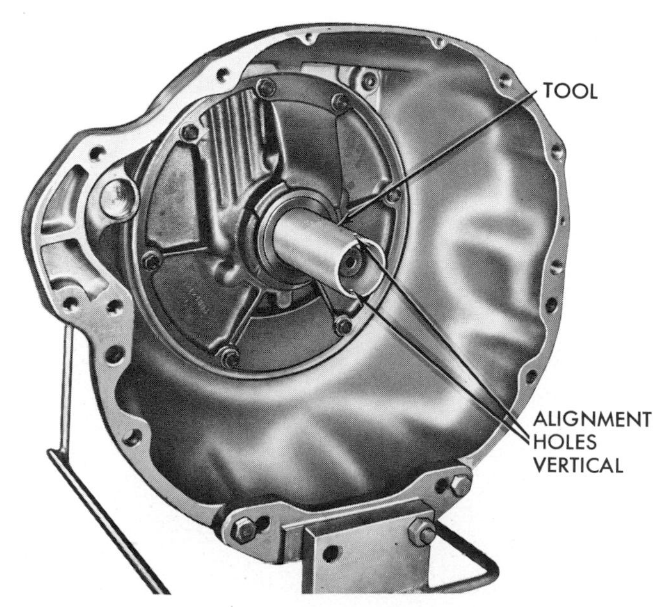

Fig. 18-63. Using a special tool to align the oil-pump rotors. (*Chrysler Corporation*)

4. Install and tighten the converter-housing bolts to 28 pound-feet [3.869 kg-m].
5. Install and tighten the two lower drive-plate-to-converter bolts to 270 pound-inches [4.86 kg-mm].
6. Install the starting motor and connect the battery ground cable. Use the remote-control switch to rotate the converter and install and tighten the other two drive-plate-to-converter bolts to 270 pound-inches [4.86 kg-mm].
7. Install the cross member and also tighten the exhaust system if it was loosened for transmission removal.
8. Check and adjust pressure, linkages, and bands, as described previously. Add fluid as necessary.

CHAPTER 18 CHECKUP

NOTE: Since the following is a chapter review test, you should review the chapter before taking it.

You have come a long way in your studies of automatic transmissions. If you have absorbed the essential information in the past several chapters, you are near your goal of becoming an expert on automatic transmissions. Now test your memory on how well you remember the service details for the TorqueFlite transmission by taking the test below.

Completing the Sentences The sentences that follow are incomplete. After each sentence there are several words or phrases, only one of which will correctly complete the sentence. Write each sentence in your notebook, selecting the proper word or phrase to complete it correctly.

1. To attach the converter to the drive plate during transmission installation, turn the drive plate with: (*a*) a wrench, (*b*) the starting motor, (*c*) the output shaft.
2. For normal service in all but the hemi engine, the recommendation on oil changes is: (*a*) do not change oil, (*b*) change every 12,000 miles [19,312 km], (*c*) change every 24,000 miles [38,624 km].
3. The purpose of the stall test is to check the: (*a*) brake bands, (*b*) oil pressures, (*c*) clutches.
4. The purpose of the air-pressure tests is to find out whether or not the: (*a*) pressures are sufficient, (*b*) clutches and servos work, (*c*) linkages are adjusted.
5. To remove the governor, you first have to remove the: (*a*) valve body, (*b*) extension housing, (*c*) converter.
6. Drive end play is checked by moving the: (*a*) output shaft in and out, (*b*) torque converter back and forth, (*c*) input shaft in and out.
7. A major difference between the front clutches for the A-904 and A-727 transmissions is that the A-727 has a series of small springs while the A-904 has: (*a*) one large spring, (*b*) a flat tension spring, (*c*) no springs.
8. The bands can be adjusted: (*a*) on the car only, (*b*) off the car only, (*c*) either on or off the car.

Troubles and Service Procedures In the following, you should write in your notebook the required trouble causes and service procedures. Do not copy from the book; try to write them in your own words. Writing the trouble causes and procedures in your own words will help you remember them better; you will thereby be greatly benefited when you go into the automotive shop.

1. Explain how to make the stall test.
2. Explain how to test the hydraulic pressures.
3. Explain how to make the two pressure adjustments.
4. Explain how to make the air-pressure tests and their purposes.
5. Refer to a shop manual covering the TorqueFlite and prepare a step-by-step list on how to adjust the gearshift linkage; the throttle linkage.
6. Explain how to make the band adjustments.
7. List the precautions to take when overhauling a transmission.
8. List the procedures required to clean and inspect transmission parts.
9. Explain how to change transmission oil.
10. List the transmission parts in the order in which they are removed from the transmission.
11. Refer to a shop manual covering the TorqueFlite and prepare a transmission-removal story outlining the step-by-step procedure for taking the transmission out of the car.

SUGGESTIONS FOR FURTHER STUDY

There is nothing like practical experience in learning a technical job like transmission overhaul. Thus, you should try to get shop experience on automatic transmissions along with your studies of the book and shop manuals. If you can watch a transmission expert at work, you will learn a great deal. But you will learn still more when you actually check and overhaul transmissions yourself. Of course, the study part is essential, too, because you have to know how the transmission is constructed and how it works before you can really do an intelligent job of servicing it. Write everything of importance you learn in your notebook.

chapter 19

FORD AUTOMATIC TRANSMISSIONS

This chapter describes the construction and operation of the Ford C4 and C6 automatic transmissions used on many Ford Motor Company cars. The two transmissions are similar in construction and operation, the major difference being that the C4 uses a low-and-reverse brake band while the C6 uses a low-and-reverse clutch to perform the same function. Both transmissions are three-speed units and, aside from the low-and-reverse arrangement, use similar clutches, intermediate brake bands, and planetary-gear systems. The hydraulic system and valve arrangements are also very similar.

⊘ **19-1 C4 and C6 Transmissions** Figure 19-1 is a cutaway view of the C6 transmission. It has three multiple-plate clutches, one brake band, and one one-way, or overrunning, clutch. In past chapters, we have covered the operation of servos that operate the brake bands, overrunning clutches, multiple-plate clutches, and planetary-gear systems. In this chapter, we shall explain how the clutches and bands operate to produce the various gear ratios, and point out the special features of the hydraulic system used in the C4 and C6 transmissions.

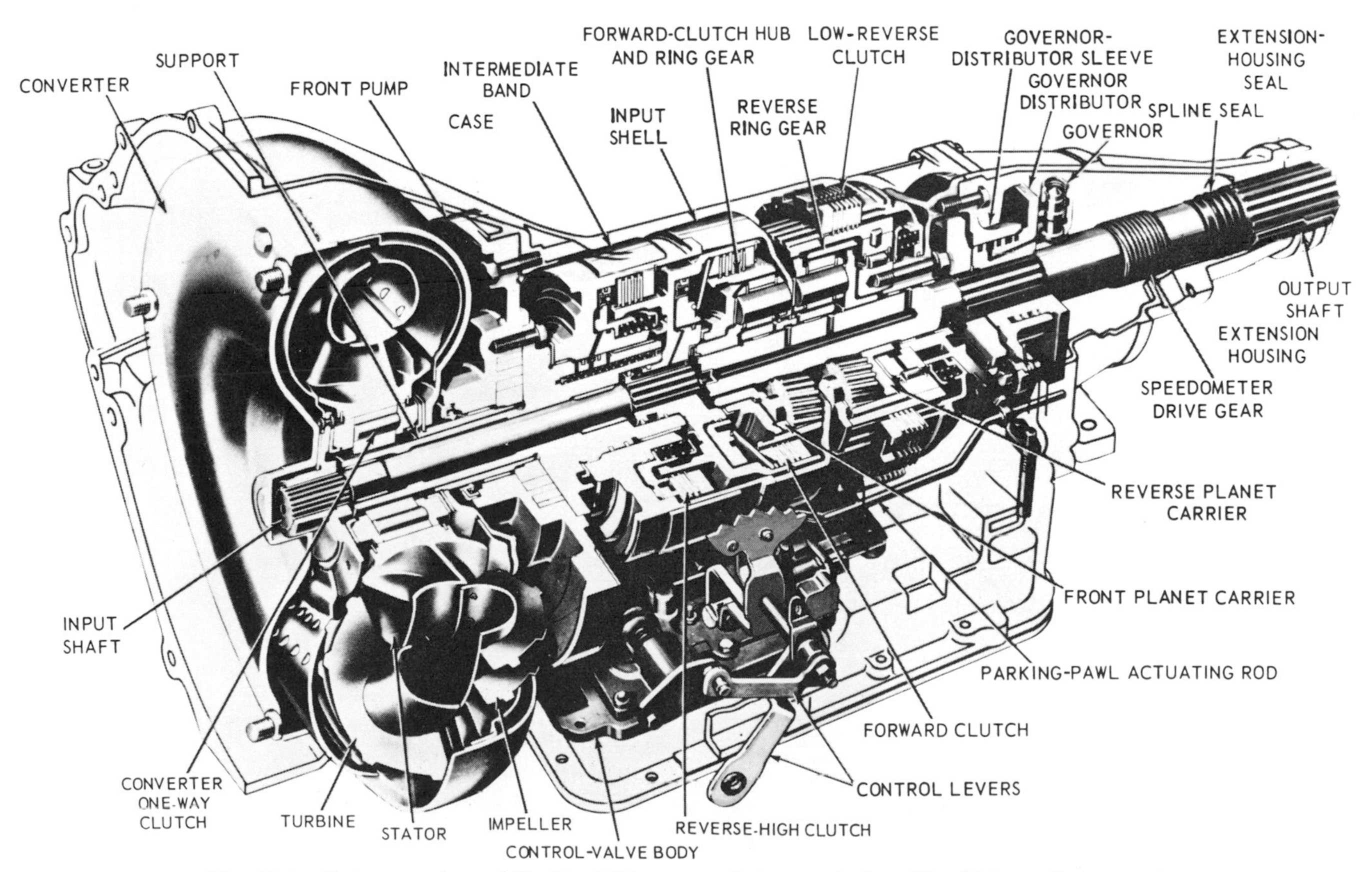

Fig. 19-1. Cutaway view of the Ford C6 automatic transmission. (*Ford Motor Company*)

⊘ 19-2 Building the Transmission Figure 19-2 is a cutaway view of the transmission mechanism for the C4 transmission. Figure 19-3 is a sectional view of the mechanism for the C6 transmission. You will note that the main difference between the two is that the C4 has a low-and-reverse band while the C6 has a low-and-reverse multiple-plate clutch. The effect is the same with both. When the band or clutch is applied, the reverse planet carrier is held stationary.

Let us now build the transmission, part by part. We shall then show the conditions in the transmission in the different gear ratios.

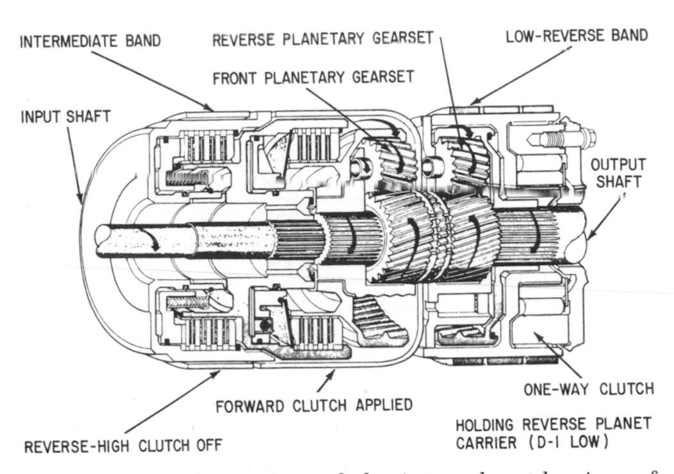

Fig. 19-2. Cutaway view of the internal mechanism of the C4 transmission. (*Ford Motor Company*)

First, we add the input and output shafts to the converter, as shown in Fig. 19-4. Next, we install the forward-clutch cylinder, as shown in Fig. 19-5. It is splined to the input shaft and therefore must turn with it.

Next, we add the sun gear, as shown in Fig. 19-6. The sun gear is installed on the output shaft and has two bushings so that it can turn freely on the shaft.

NOTE: This is not the way the transmission is actually assembled. We are building the transmission by illustrations in this particular fashion just to show how the parts are related and how they work together.

Now we install the forward planetary gear, as shown in Fig. 19-7. It consists of the ring-gear pilot, ring gear (also called the *internal gear*), and planet carrier with planet pinions. Note that there are teeth cut on the outside of the ring gear. These teeth engage with the clutch plates. Thus, the ring gear is also the forward-clutch hub. The planet-pinion carrier is splined to the output shaft.

Next, we install the forward-clutch plates, as shown in Fig. 19-8. The plates are alternately splined to the forward-clutch cylinder and the clutch hub (the outside of the planetary ring gear we installed in Fig. 19-7). Note that when the forward clutch is applied, the forward planetary ring gear is driven at shaft-input speed.

Now, we add the reverse-and-high clutch, as

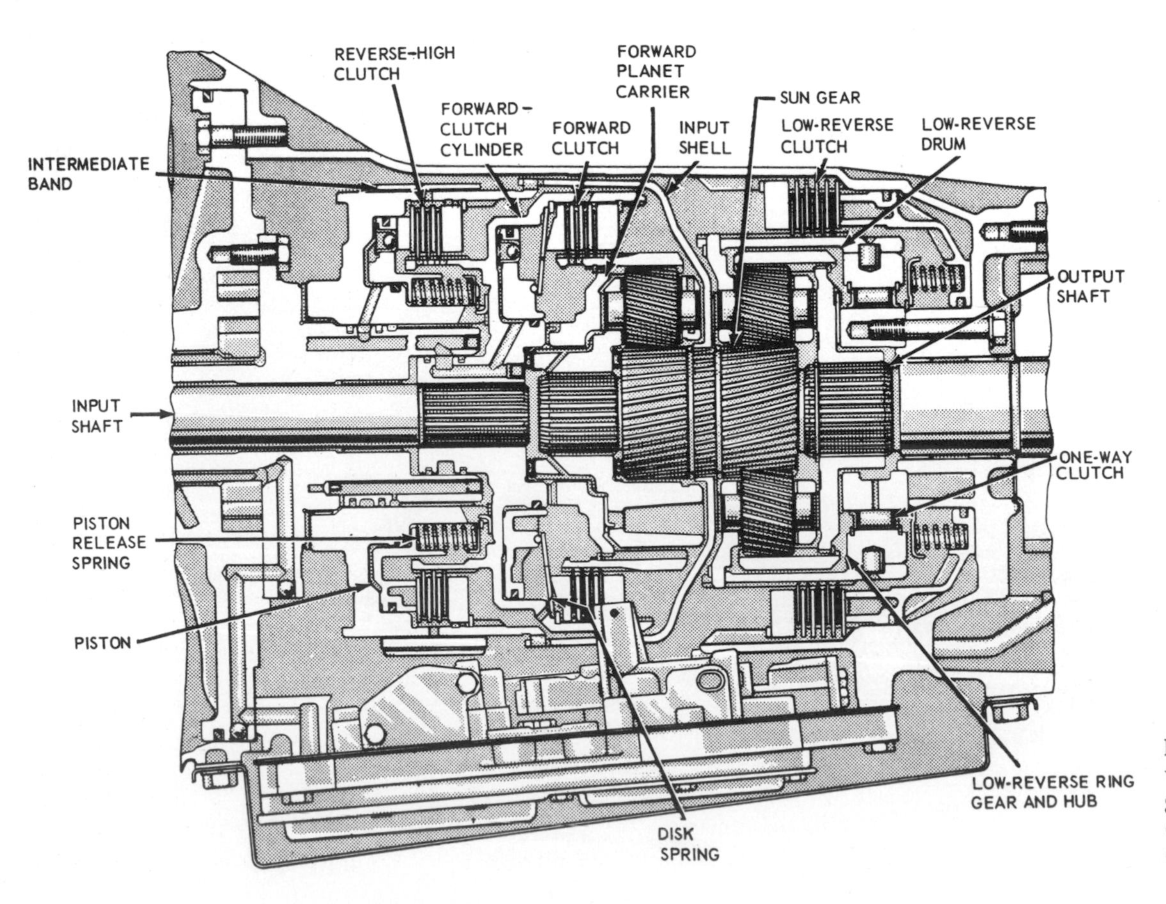

Fig. 19-3. Sectional view of the planetary-gear train and clutches of the C6 transmission. (*Ford Motor Company*)

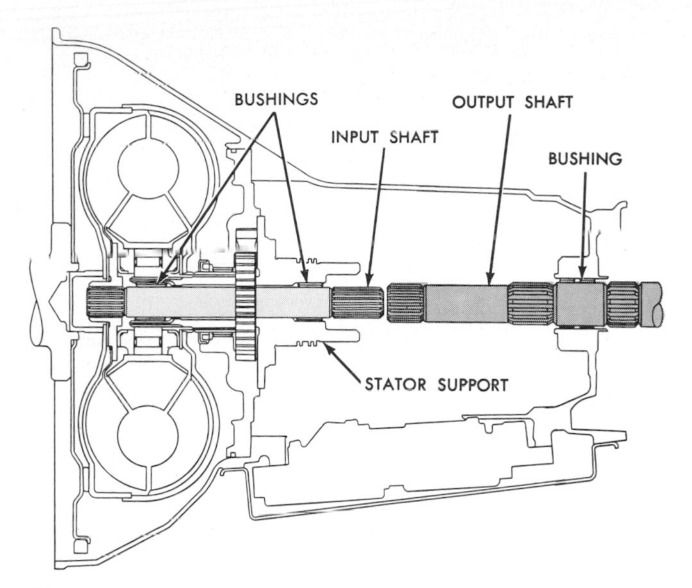

Fig. 19-4. Input and output shafts in place in the transmission. (*Ford Motor Company*)

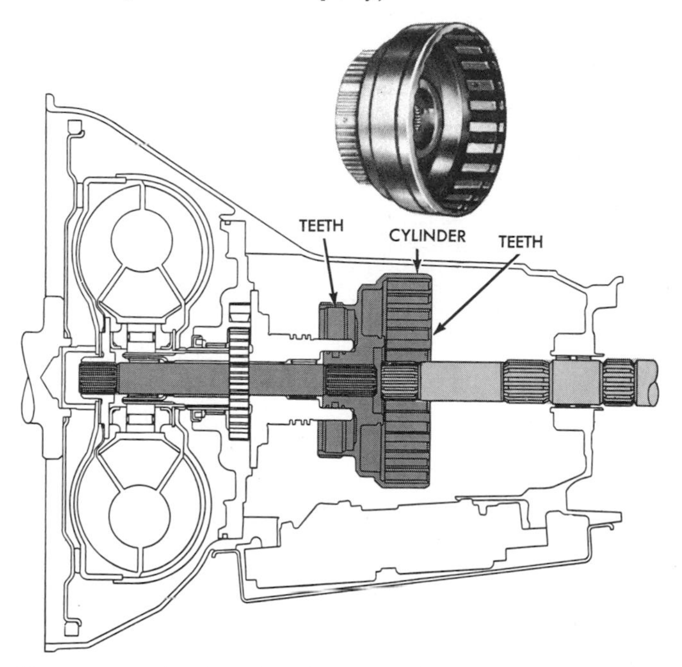

Fig. 19-5. Adding the forward-clutch cylinder. (*Ford Motor Company*)

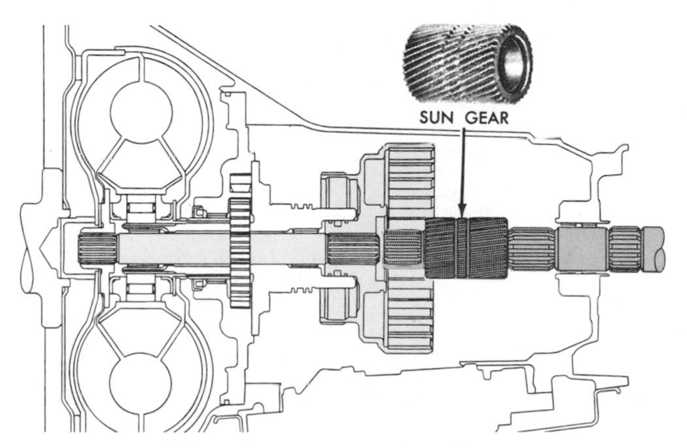

Fig. 19-6. Adding the sun gear. (*Ford Motor Company*)

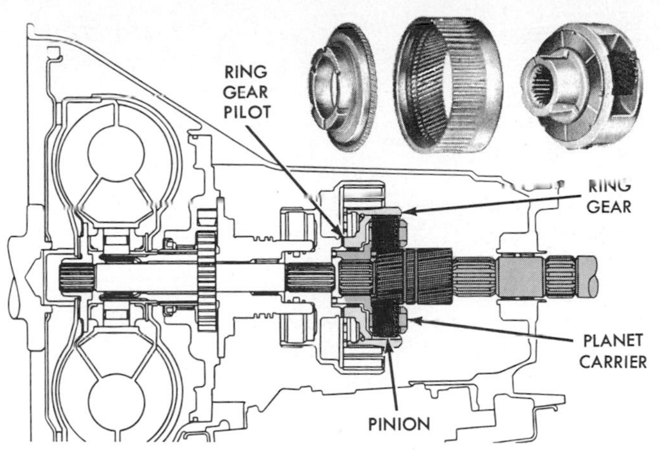

Fig. 19-7. Adding the forward planetary-gear unit. (*Ford Motor Company*)

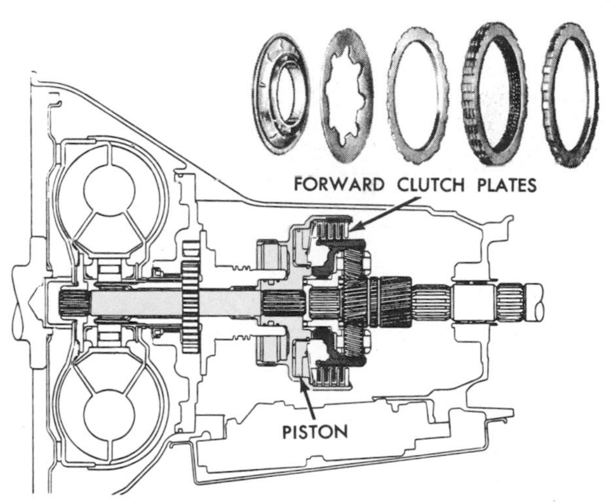

Fig. 19-8. Adding the forward-clutch plates. (*Ford Motor Company*)

shown in Fig. 19-9. Don't be confused by the fact that the reverse-and-high clutch is ahead of the forward clutch in the assembly. Engineers decided this was the most convenient way to arrange the clutches. The reverse-and-high clutch includes the reverse-and-high drum, piston, clutch plates, pressure plate, and input shell, as shown. The input shell is splined to the center of the sun gear. The reverse-and-high drum is free-running on the stationary front-pump hub. The clutch plates are alternately splined to the reverse-and-high-clutch drum and the hub of the forward-clutch cylinder.

Next, we add the intermediate band and servo, as shown in Fig. 19-10. The band is positioned around the outside of the reverse-and-high drum. When hydraulic apply pressure is directed to the servo, it tightens the band on the drum, bringing the drum to a halt. The drum, you will note, is connected to the sun gear through the lugs on the drum that fit the notches in the input shell. The input shell is splined to the center of the sun gear. Thus, the reverse-and-high drum, input shell, and sun gear

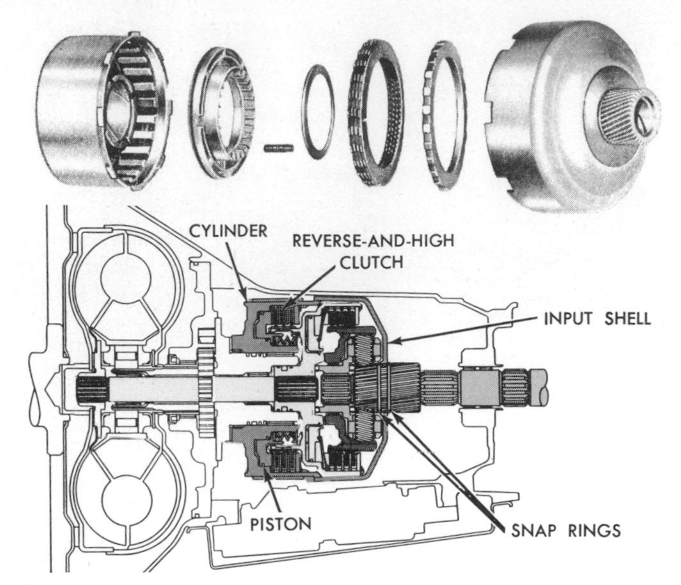

Fig. 19-9. Adding the reverse-and-high clutch. (*Ford Motor Company*)

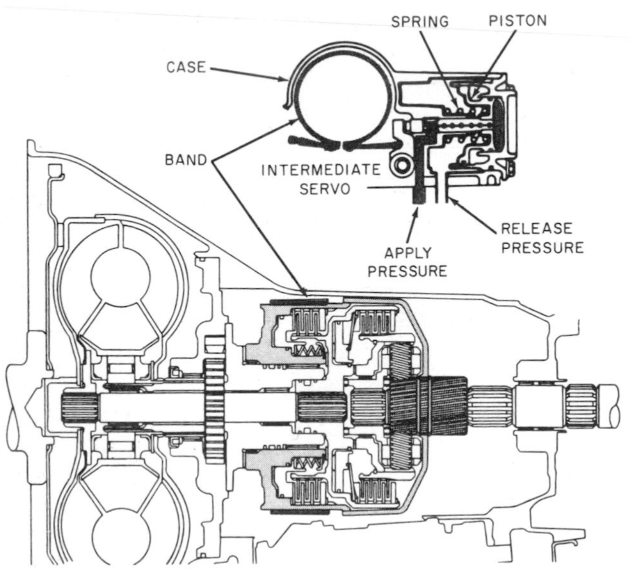

Fig. 19-10. Adding the intermediate band and servo. (*Ford Motor Company*)

always turn or are stationary together. They are, in effect, one part.

Our transmission is almost complete. But first, let us see what gear ratios we can get with what we have put together so far. We can get second gear by applying the band and the forward clutch. This produces the situation shown in Fig. 19-11. The sun gear is held stationary by the band, and the forward planetary ring gear is driven through the forward clutch by the input shaft. The planet pinions are forced to walk around the stationary sun gear, carrying the pinion carrier around with them. The pinion carrier is splined to the output shaft. Thus, the output shaft is turned. But there is gear reduction through the planetary-gear system.

We can also get high gear with our partly built

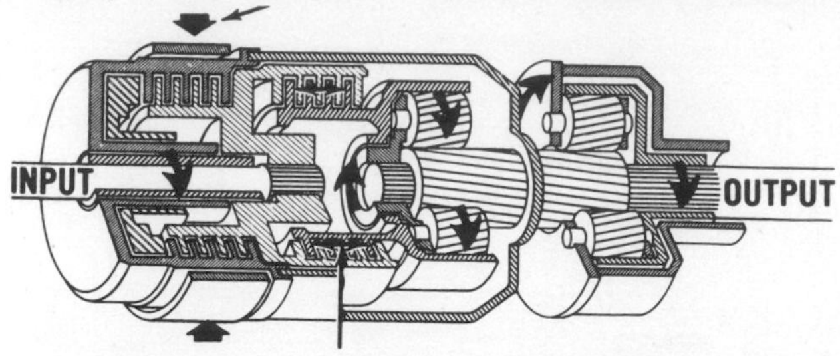

Fig. 19-11. Transmission in second gear, drive range. (*Ford Motor Company*)

transmission, as shown in Fig. 19-12. This will occur when both the reverse-and-high and forward clutches are applied. This action locks the input shaft to the sun gear and the ring gear. Thus, the planetary-gear system is locked up so it turns as a unit. The output shaft turns at the same speed as the input shaft.

To get low gear and reverse, however, we must add more parts to our transmission. First, we add the low-and-reverse ring gear, as shown in Fig. 19-13.

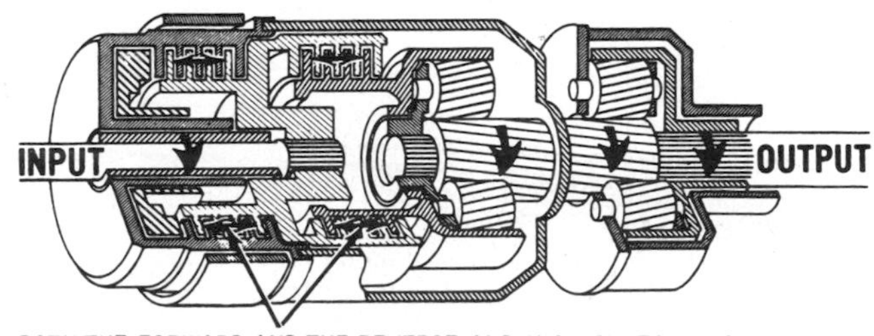

Fig. 19-12. Transmission in high gear, or direct drive. (*Ford Motor Company*)

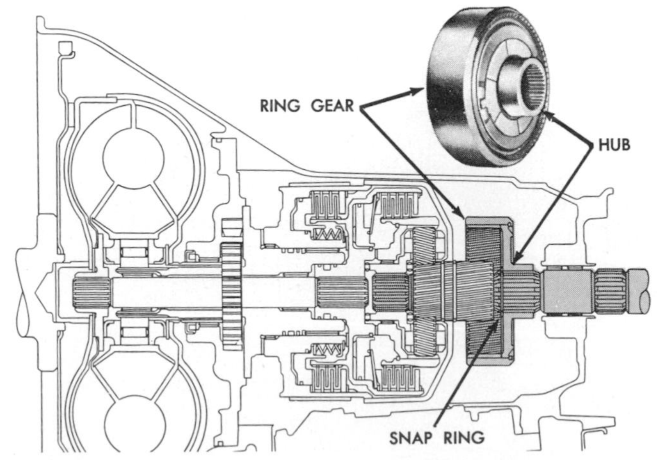

Fig. 19-13. Adding the low-and-reverse ring gear. (*Ford Motor Company*)

This ring gear is splined through a hub to the output shaft. Next, we add the low-and-reverse pinion carrier and clutch, as shown in Fig. 19-14. The clutch plates are alternately splined to the transmission case and to the low-and-reverse-clutch hub. The hub is attached to the carrier by alternating tabs or lugs. When the low-and-reverse clutch is applied, the pinion carrier is locked to the case.

Now we can see how to get reverse. The low-and-reverse clutch is applied to hold the planet-pinion carrier stationary. At the same time, the reverse-and-high clutch is applied so that the input shaft drives the sun gear. This is shown in Fig. 19-15.

With the carrier held and sun gear driven clockwise, the pinions act as idlers and reverse the direction of rotation. Thus, the ring gear is driven counterclockwise at a reduced speed. The ring gear is attached to the output shaft so that it turns in the reverse direction.

Now let us recap the gear positions.

⊘ 19-3 Low Gear in D

⊘ 19-3 Low Gear in D Low, or first, gear is shown in Fig. 19-16. The forward clutch is applied, locking

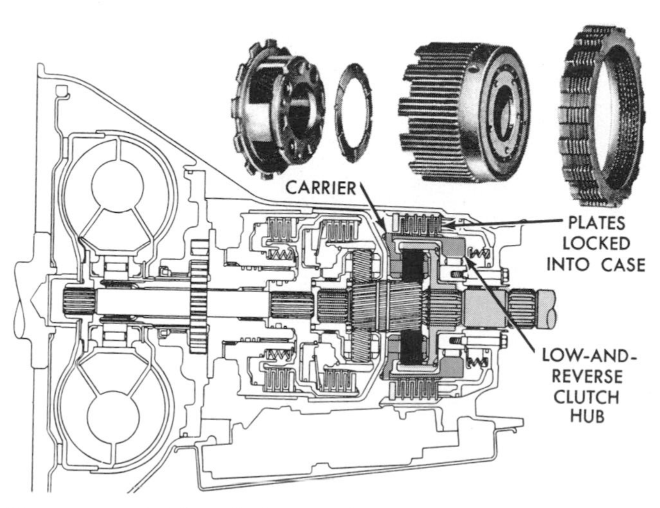

Fig. 19-14. Adding the low-and-reverse pinion carrier and clutch. (*Ford Motor Company*)

THE REVERSE-AND-HIGH CLUTCH IS APPLIED. THE INPUT SHAFT IS LOCKED TO THE REVERSE-AND-HIGH CLUTCH DRUM, THE INPUT SHELL, AND THE SUN GEAR.

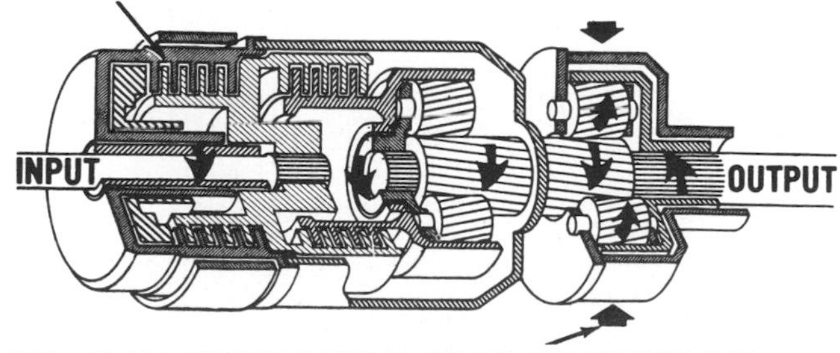

THE LOW-AND-REVERSE CLUTCH IS APPLIED. THE REVERSE-UNIT PLANET CARRIER IS HELD STATIONARY.

REVERSE

Fig. 19-15. Transmission in reverse. (*Ford Motor Company*)

THE FORWARD CLUTCH IS APPLIED. THE FRONT-PLANETARY-UNIT RING GEAR IS LOCKED TO THE INPUT SHAFT.

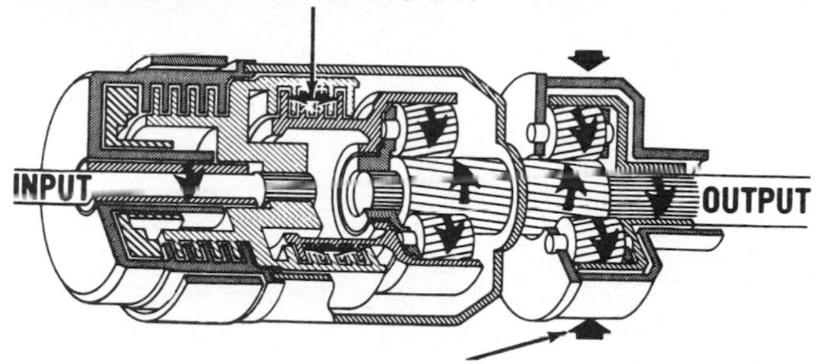

THE LOW-AND-REVERSE CLUTCH (LOW RANGE) OR THE ONE-WAY CLUTCH (DI RANGE) IS HOLDING THE REVERSE-UNIT PLANET CARRIER STATIONARY.

FIRST GEAR

Fig. 19-16. Transmission in low, or first, gear, drive range. (*Ford Motor Company*)

the forward planetary ring gear to the input shaft. This drives the sun gear through the forward planet pinions. The sun gear then drives the low-and-reverse pinions, which drive the low-and-reverse ring gear. The one-way, or overrunning, clutch holds the reverse planet carrier stationary. Since the ring gear is splined to the output shaft, the output shaft turns. There is gear reduction through both planetary gearsets.

Note that the situation might be called a *double reverse.* The forward planetary gearset reverses input rotation and produces gear reduction. The low-and-reverse planetary gearset then re-reverses the rotation to give forward motion at a further gear reduction.

⊘ 19-4 Low Gear in L1 In low gear with the selector lever in L1, the forward and the low-and-reverse clutches are applied. The low-and-reverse pinion carrier is held from turning by the low-and-reverse clutch. Power flow is the same as in low gear in D. The difference is that if the car starts down a hill in L1, the car can drive the engine to produce engine braking of the car. In low gear in D range, if the car tries to drive the engine, the one-way clutch unlocks and the car freewheels.

⊘ 19-5 Second Gear in D We have already described and illustrated the conditions in the transmission with the gears in second. These are shown in Fig. 19-11. The sun gear is held stationary by the band, and the forward planetary ring gear is driven through the forward clutch by the input shaft. The planet pinions walk around the stationary sun gear, carrying the pinion carrier around with them. The pinion carrier is splined to the output shaft so that the output shaft turns, with a gear reduction.

⊘ 19-6 High Gear The condition in high gear is shown in Fig. 19-12. Both the reverse-and-high and forward clutches are applied. This action locks up the planetary gear system so that it turns as a unit and the gear ratio is 1:1.

⊘ 19-7 Reverse Figure 19-15 shows the conditions with the transmission in reverse. The low-and-reverse clutch is applied. The reverse-and-high clutch is also applied. This action holds the carrier and causes the sun gear to be driven clockwise. The pinions act as idlers and drive the ring gear counterclockwise so that the output shaft is driven counterclockwise for reverse.

⊘ 19-8 Hydraulic Control System The complete hydraulic system for the C6 transmission is shown in Fig. 19-17. You will recognize a good many of the valves and other components. The multiple-plate clutches, overrunning clutch, intermediate servo, and band are very similar to those used on other transmissions described in previous chapters, and we shall not discuss them further in this chapter. This also applies to the pump, main oil-pressure regulator, main oil-pressure booster, and governor.

The 1-2 and 2-3 shift valves are similar to the upshift valves used in the 350, 400, and TorqueFlite

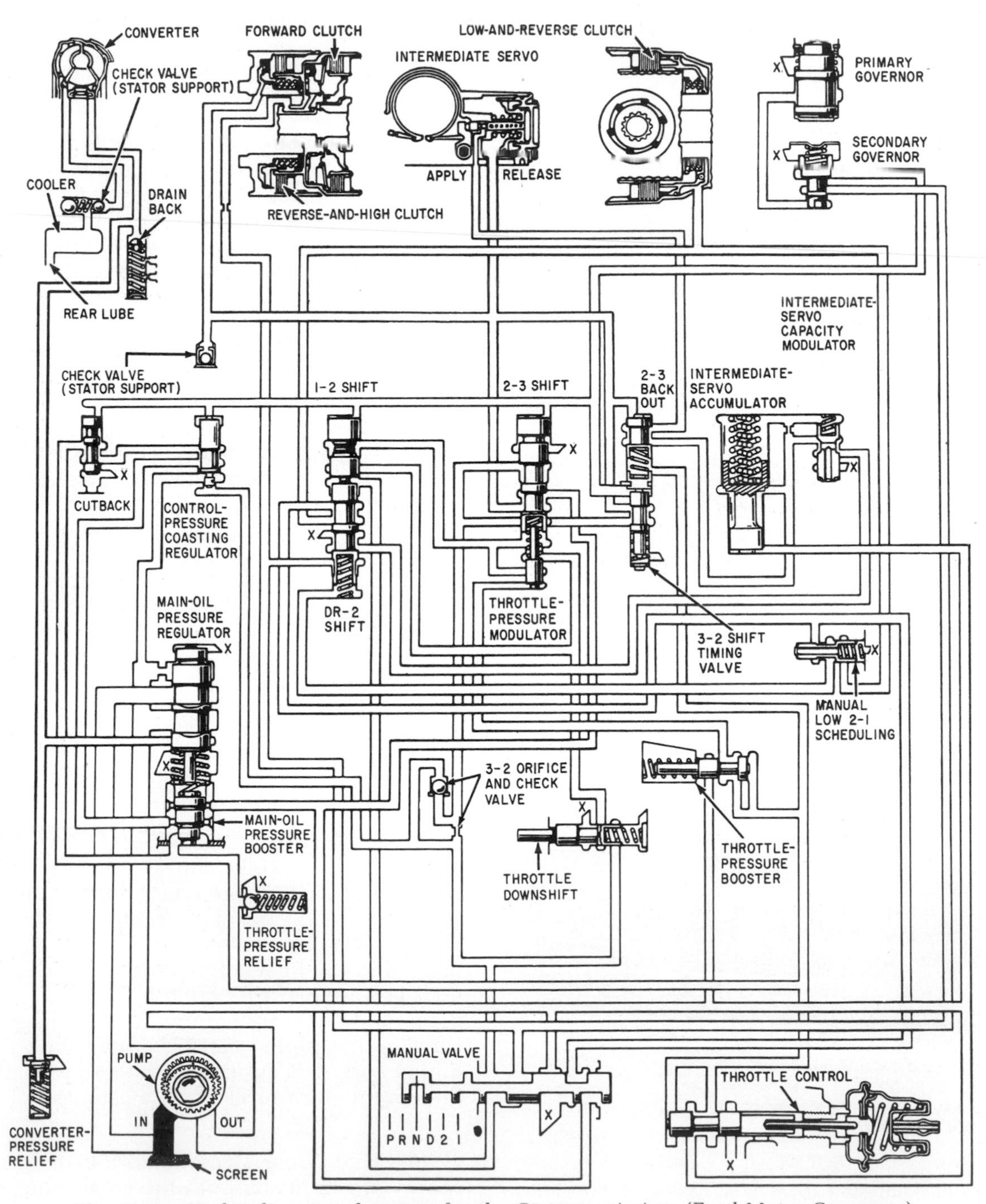

Fig. 19-17. Hydraulic control system for the C6 transmission. (*Ford Motor Company*)

transmissions. They have governor pressure working at one end, and at the other, throttle pressure plus a spring. Now, let us look at these and other valves used in the C6 hydraulic system and find out how they work.

⊘ 19-9 Manual Valve The manual valve, similar to those in other transmissions, has the usual six positions, as shown in Fig. 19-17: P, R, N, D, 2, and 1. Positioning of the manual valve is controlled by movement of the selector lever on the steering mast or on the floor-mounted console.

⊘ 19-10 Throttle Control Valve The throttle control valve produces a varied pressure dependent on intake-manifold vacuum. As the accelerator pedal is pushed down and the throttle is opened, the throttle control valve passes more and more line pressure. This becomes throttle pressure.

The throttle control valve contains an airtight diaphragm against which intake-manifold vacuum is applied. This vacuum causes atmospheric pressure to move the diaphragm and the spool valve connected to it. Opposing this pressure is the pressure of a spring. When the valve is moved, it passes more line pressure, and this becomes throttle pressure.

⊘ 19-11 Main Oil-Pressure Booster This valve boosts line pressure as necessary to provide sufficient holding power at the clutches and band servo. With increased throttle opening and higher engine torques, the throttle pressure increases. This increasing pressure, as it acts on the oil-pressure booster valve, puts additional spring pressure on the main oil-pressure regulator valve. Regulated pressure therefore increases.

During reverse operation, the manual valve directs line pressure to the oil-pressure booster valve so that the booster valve forces the regulated pressure to increase. High holding power is necessary in reverse to prevent band or clutch slippage.

⊘ 19-12 Throttle-Pressure Booster Valve Throttle openings above about 50 mph [80.4 km/h] provide very little additional change in intake-manifold vacuum. The throttle-pressure booster valve boosts throttle pressure to provide the necessary shift delay for higher engine speeds. This is accomplished as follows: When throttle pressure increases above about 65 psi [4.6 kg/cm^2] (which varies with different applications), the higher pressure, acting on the throttle-pressure booster valve, causes it to move against its spring. This movement permits line pressure to enter and take over control of the booster valve. Now, with the higher line pressure at work, the booster valve increases the throttle pressure about 3 psi [0.2 kg/cm^2] for every 1 psi [0.07 kg/cm^2] that the pressure from the throttle control valve increases.

⊘ 19-13 Governor The governor works in the same manner as the governors used with other transmissions. As car speed increases, so does governor pressure. This pressure, working against the 1-2 and 2-3 shift valves, causes upshifts when the right combination of car speed and throttle opening is reached. The governor is a dual unit, with the primary governor regulating at low speeds and the secondary governor regulating at high speeds.

⊘ 19-14 Throttle Downshift Valve The throttle downshift valve is mechanically linked to the throttle so that when the driver opens the throttle wide, the downshift valve is moved. Then, if car speed is below a certain value, the line pressure will actuate the 2-3 shift valve to cause a downshift. Above this speed, the downshift will not occur because the governor pressure will hold the 2-3 shift valve in the third-, or high-, gear position.

⊘ 19-15 Hydraulic System The complete hydraulic system is shown in Plate 31. The conditions shown exist with the selector level in the D range and the transmission in first gear. Study the manual valve at the bottom of Plate 31. An earlier version of the C6 transmission used P, R, N, D2, D1, and L for manual valve positions. The earlier version used the D1 position for the full range of shifts—from low to second to direct or high. The later version conforms to the P R N D S L or P R N D 2 1 arrangement found on other transmissions (see Fig. 19-17). This is about the only significant difference between the earlier and later versions of the C6.

Note also, in Fig. 19-17, the names of the intermediate-servo accumulator and intermediate-servo capacity modulator. In the earlier version these were called the *1-2 accumulator valve* and the *1-2 scheduling valve*. However, the hydraulic system in the later version (see Plate 31) works the same way as in the earlier version. The major difference between the C4 and the C6, as we said previously, is that the C4 has a low-and-reverse band while the C6 has a low-and-reverse clutch.

⊘ 19-16 Hydraulic System in Drive Range, Low Gear The conditions in the hydraulic system during operation in low gear in drive range are shown in Plate 31. With the selector lever and manual valve in drive range, the forward clutch is applied and the power flow is as shown in Fig. 19-16. Governor pressure increases as car speed increases. Governor pressure is applied to the ends of the 1-2 and 2-3 shift valves. When this pressure has increased sufficiently to overcome the spring pressure and the throttle pressure working on the 1-2 shift valve, the shift valve will move, admitting pressure to the intermediate servo. The intermediate servo applies the intermediate band, thus locking the reverse-and-high clutch drum, input shell, and sun gear in a stationary position. Now, the transmission is in second gear, and the power flow is as shown in Fig. 19-11. The 1-2 accumulator valve—or intermediate-servo accumulator, as it is called in Fig. 19-17—cushions the application of the band so that a smooth shift occurs.

⊘ 19-17 Hydraulic System in Drive Range, Direct or High Gear The conditions in the hydraulic system during operation in direct drive are shown in Plate 32. To reach this condition, governor pressure—that is, car speed—must increase enough to overcome the spring pressure and throttle pressure acting on the 2-3 shift valve. When this happens, the 2-3 shift valve moves so that pressure is admitted to the reverse-and-high clutch. At the same time, pressure is admitted to the release side of the intermediate servo so that the band is released. The transmission is in direct drive, and the power flow through the transmission is as shown in Fig. 19-12.

⊘ 19-18 Hydraulic System in Reverse The conditions in the hydraulic system during operation in reverse are as shown in Plate 33. Movement of the manual valve to R permits pressure to flow to the reverse-and-high clutch and the low-and-reverse clutch. Now, the transmission is in reverse, and the power flow is as shown in Fig. 19-15. Note that throttle pressure is applied to the pressure booster valve so that line pressure increases. This ensures the holding power that the clutches need in reverse.

CHAPTER 19 CHECKUP

NOTE: Since the following is a chapter review test, you should review the chapter before taking it.

By this time, you understand almost all there is to know about how automatic transmissions work. Now find out how well you remember the details of the C4 and C6 transmissions you have just studied in this chapter by taking the test that follows.

Completing the Sentences The sentences that follow are incomplete. After each sentence there are several words or phrases, only one of which will correctly complete the sentence. Write each sentence in your notebook, selecting the proper word or phrase to complete it correctly.

1. Counting all the active control members in the transmission, including all clutches and the brake band, there are: (*a*) four, (*b*) five, (*c*) six.
2. The compound planetary-gear system has two sets of planetary pinions, two ring gears, and: (*a*) one sun gear, (*b*) two sun gears, (*c*) three sun gears.
3. The forward-clutch cylinder is splined to the: (*a*) output shaft, (*b*) input shaft, (*c*) input shell.
4. The reverse-and-high clutch and input shell are splined to the: (*a*) input shaft, (*b*) output shaft, (*c*) sun gear.
5. The intermediate band is positioned around the: (*a*) forward-clutch cylinder, (*b*) reverse-and-high clutch cylinder, (*c*) output ring gear.
6. When the low-and-reverse clutch is applied, it locks the reverse-unit pinion carrier to the: (*a*) output shaft, (*b*) input shaft, (*c*) transmission case.
7. In low gear, drive range, the one-way clutch is holding the reverse-unit pinion carrier stationary and: (*a*) the low-and-reverse clutch is applied, (*b*) the forward clutch is applied, (*c*) the band is applied.
8. In second gear, drive range, the forward clutch is applied and: (*a*) the intermediate band is applied, (*b*) the low-and-reverse clutch is applied, (*c*) the one-way clutch is holding.
9. In direct drive, the forward clutch is applied and: (*a*) the reverse-and-high clutch is applied, (*b*) the intermediate band is applied, (*c*) the one-way clutch is holding.
10. In reverse, the reverse-and-high clutch is applied and: (*a*) the intermediate band is applied, (*b*) the forward clutch is applied, (*c*) the low-and-reverse clutch is applied.

Purpose and Operation of Components In the following, you are asked to write the purpose and operation of the various components in the transmission. If you have any difficulty with any of the questions, reread the pages in the chapter that will give you the information you need. Then write your explanation or make your drawing in your notebook. Don't copy; try to tell it in your own words.

1. Make a chart showing the band application and clutch engagements for the different operating conditions.
2. Describe the power flow in drive range, low gear.
3. Describe the power flow in drive range, second gear.
4. Describe the power flow in drive range, direct drive.
5. Describe the power flow in reverse.
6. Explain how a kickdown downshift occurs.
7. Draw the part of the hydraulic system that is active in drive range, low gear.
8. Draw the part of the hydraulic system that is active in drive range, second gear.
9. Draw the part of the hydraulic system that is active in drive range, direct drive.
10. Draw the part of the hydraulic system that is active in reverse.

SUGGESTIONS FOR FURTHER STUDY

As we have stated in previous suggestions for further study, your best bet to learn more about the workings of automatic transmissions is to handle the parts and see how an expert takes a transmission apart and puts it back together again. Be sure to write in your notebook any interesting facts you learn.

chapter 20

FORD C6 TRANSMISSION SERVICE

This chapter describes the maintenance, trouble-diagnosis, and service procedures for the Ford C6 automatic transmission. The service procedures for the Ford C6 automatic transmission are similar to those used for the Ford C4 and FMX automatic transmissions. However, there are differences among the various models. Therefore, you should always refer to the manufacturer's shop manual that covers the specific model you are working on.

Normal Maintenance and Adjustments

Normal maintenance includes periodic checking of fluid level, checking throttle and shift linkages, and possibly band adjustment.

⊘ 20-1 Checking Fluid Level and Changing Fluid Normal maintenance and lubrication requirements do not necessitate periodic changes of the fluid. However, if a major transmission repair is necessary, then the entire transmission, including the converter, cooler, and cooler lines, must be flushed out and fresh fluid added. Checking fluid level and changing fluid are described below.

CAUTION: Use only the transmission fluid approved by Ford. To use another fluid is to risk serious transmission trouble.

1. Check fluid level with the car on level ground. Start the engine and run it at fast idle until the transmission fluid reaches operating temperature. Then operate the engine on slow idle, apply the foot brake firmly, and move the selector lever through all positions. Return to P and, with the engine running, remove the transmission dipstick, wipe it, put it back in, remove it again, and note the fluid level. If the fluid is below the ADD mark, add enough fluid to bring it up between the ADD and FULL marks. Reinstall the dipstick.

CAUTION: Do not overfill. Overfilling can cause aeration and foaming of the fluid, loss of control pressure, and possible damage to the transmission.

If the transmission seems to require frequent additions of fluid, chances are there is a leak and the transmission should be inspected to determine the point of leakage. It may be nothing more serious than a loose oil pan or damaged oil-pan gasket. On the other hand, it could be a bad seal gasket, or a cracked case, which would call for removal of the transmission for major service.

2. When filling a dry transmission and converter, pour in 5 quarts [4.73 l] of fluid. Start the engine and apply the foot brake while moving the selector lever through all positions. This action fills the converter and lines. Then return the selector lever to P and add more fluid, as necessary, to bring the fluid level up between the ADD and FULL marks on the dipstick. The transmission fluid must be hot, and so you may have to run the engine in P for a few minutes before making the final fluid-level check.

⊘ 20-2 Throttle-Linkage Adjustment The linkage between the throttle and the transmission actuates the transmission downshift valve to cause downshift when the throttle is opened wide. Figure 20-1 shows the linkages involved on various cars and the adjustments required. Adjustment is made by holding the downshift rod against the stop and adjusting the downshift screw to provide the proper clearance, as shown. The linkage is then reconnected.

⊘ 20-3 Selector-Lever-Linkage Adjustment The linkage between the selector lever and transmission is different on different models. You should check the shop manual whenever you have an adjustment to make. As a first step, the throttle and downshift linkage should be adjusted as noted in ⊘ 20-2. Figures 20-2 and 20-3 show the adjustments for two column-shift-linkage arrangements. Figure 20-4 shows the linkages for a console or floor shift.

1. COLUMN SHIFT Put the selector lever in D tight against the stop. Loosen the shift-rod adjusting nut at point *A* in Figs. 20-2 and 20-3. Shift the manual

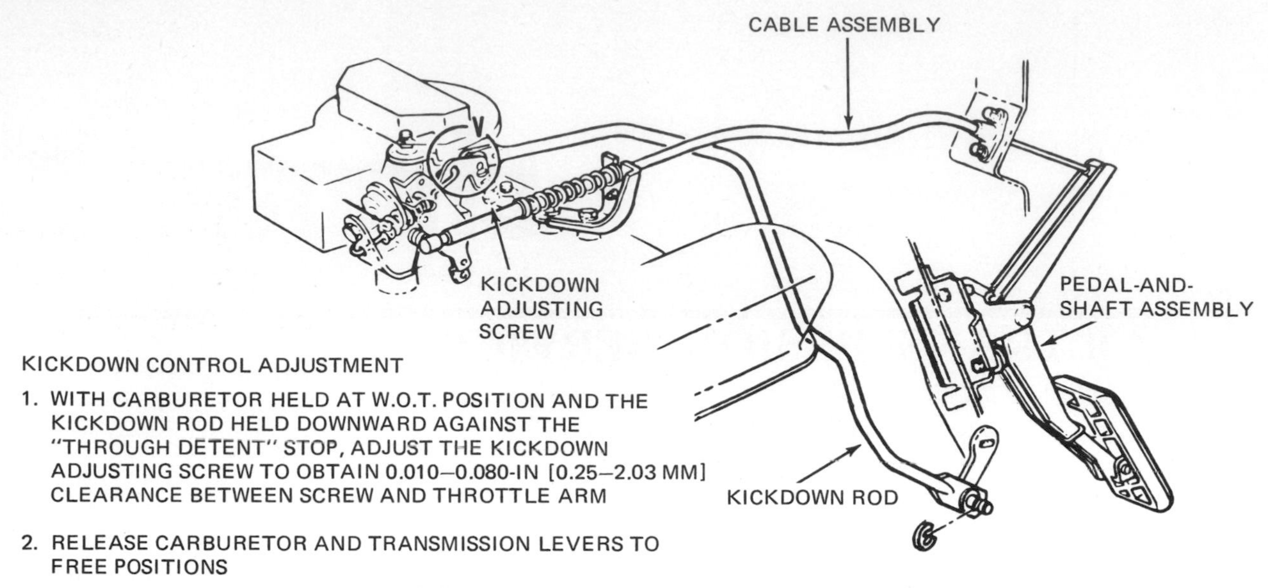

Fig. 20-1. Throttle and downshift linkage. (*Ford Motor Company*)

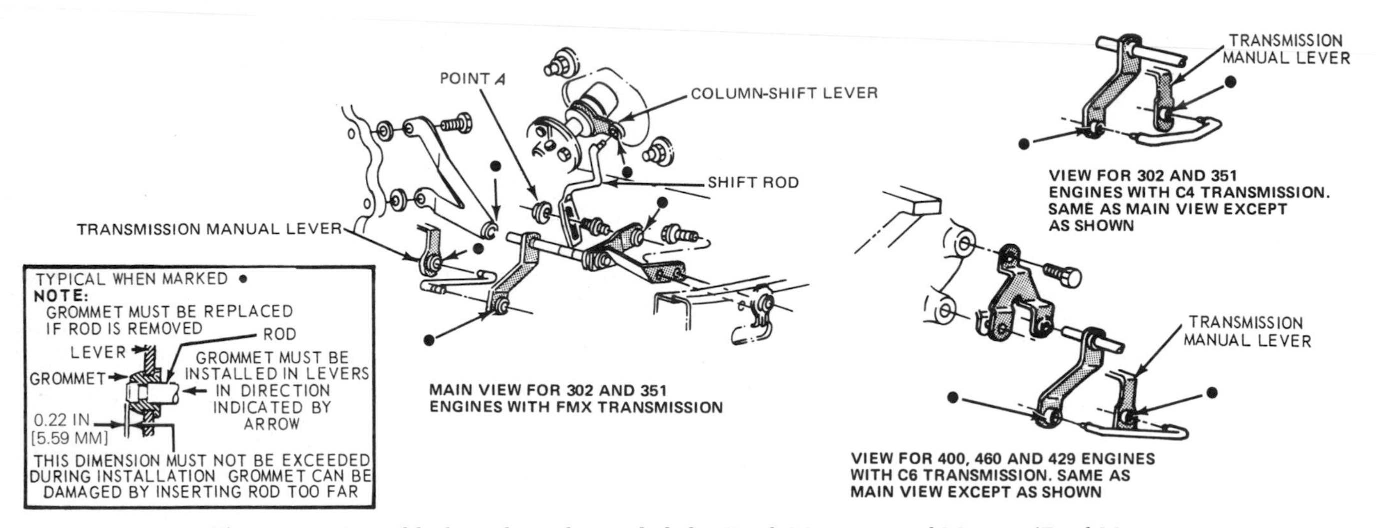

Fig. 20-2. Manual linkage for column shift for Ford, Mercury, and Meteor. (*Ford Motor Company*)

lever at the transmission into the D detent position, third from the rear. Make sure that the selector lever has not moved from the D position. Then tighten the nut at point *A* to 10 to 20 pound-feet [1.382 to 2.764 kg-m]. Check transmission operation for all selector-lever positions.

2. *CONSOLE OR FLOOR SHIFT* Position the selector lever in D. Raise the car on a hoist and loosen the manual-lever-shift-rod retaining nut. Move the transmission manual lever to the D position, third from the rear. Make sure the selector lever has not moved. Tighten the retaining nut to 10 to 20 pound-feet [1.382 to 2.764 kg-m]. Check the transmission in all selector-lever positions.

⊘ 20-4 Neutral-Start-Switch Adjustment The neutral start switch should be closed only with the selector lever in P or N to permit starting only in these positions. If the switch does not do its job, chances are it is out of adjustment. Adjustment is made by loosening the attaching screws and sliding the switch back and forth as necessary to obtain the correct action.

⊘ 20-5 Band Adjustment There is only one band on the C6 transmission: the intermediate band. It is adjusted as follows: Raise the car on a hoist and clean all dirt from around the band-adjustment-screw area. Remove and discard the locknut. Install a new locknut and tighten the adjusting screw to 10 pound-feet [1.382 kg-m] torque, as shown in Fig. 20-5. Then back off the adjusting screw exactly $1\frac{1}{2}$ turns. Hold the adjusting screw and tighten the locknut to specifications. Lower the car.

On the C4 transmission, there is an additional band: the low-reverse band, which is used instead

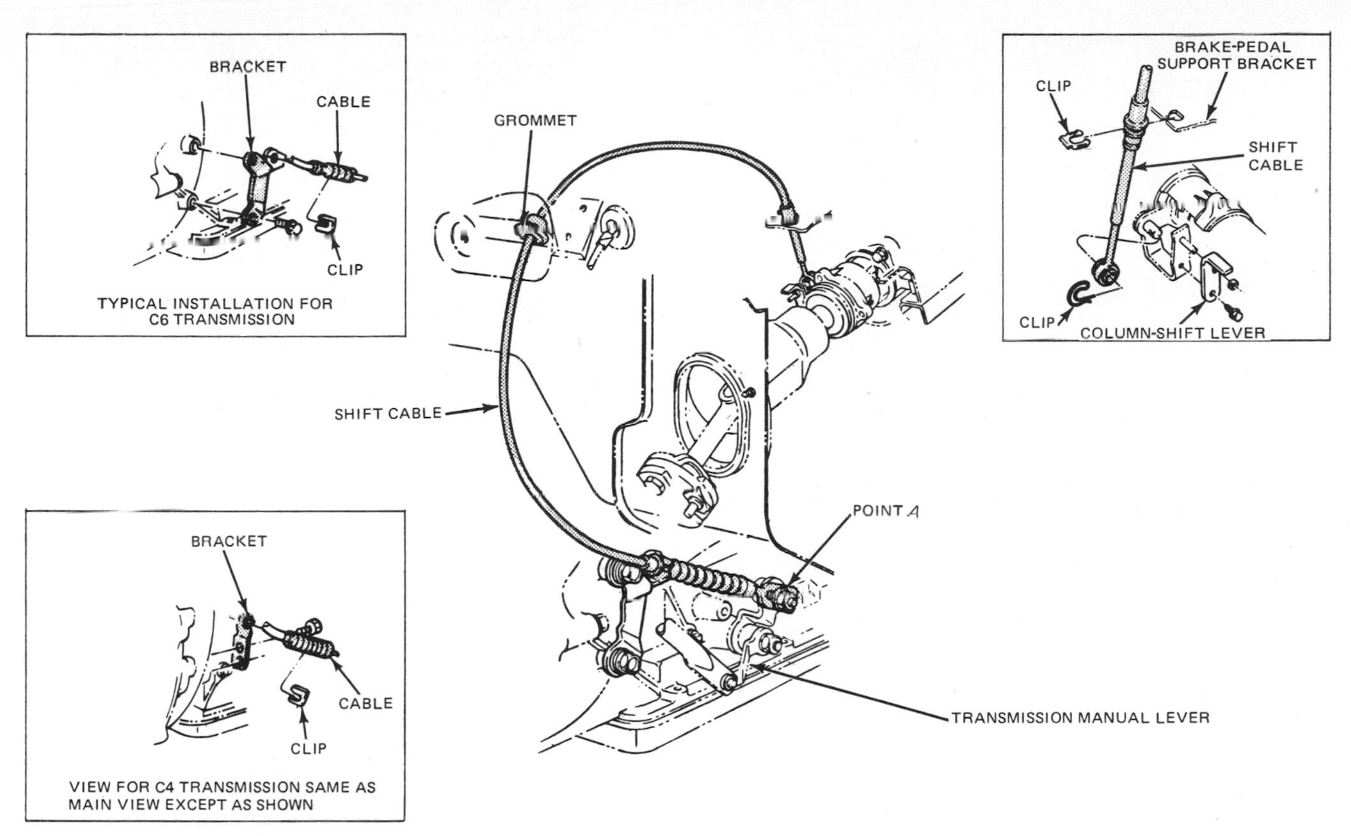

Fig. 20-3. Manual linkage for column shift for Torino, Montego, Thunderbird, and Continental Mark IV. (*Ford Motor Company*)

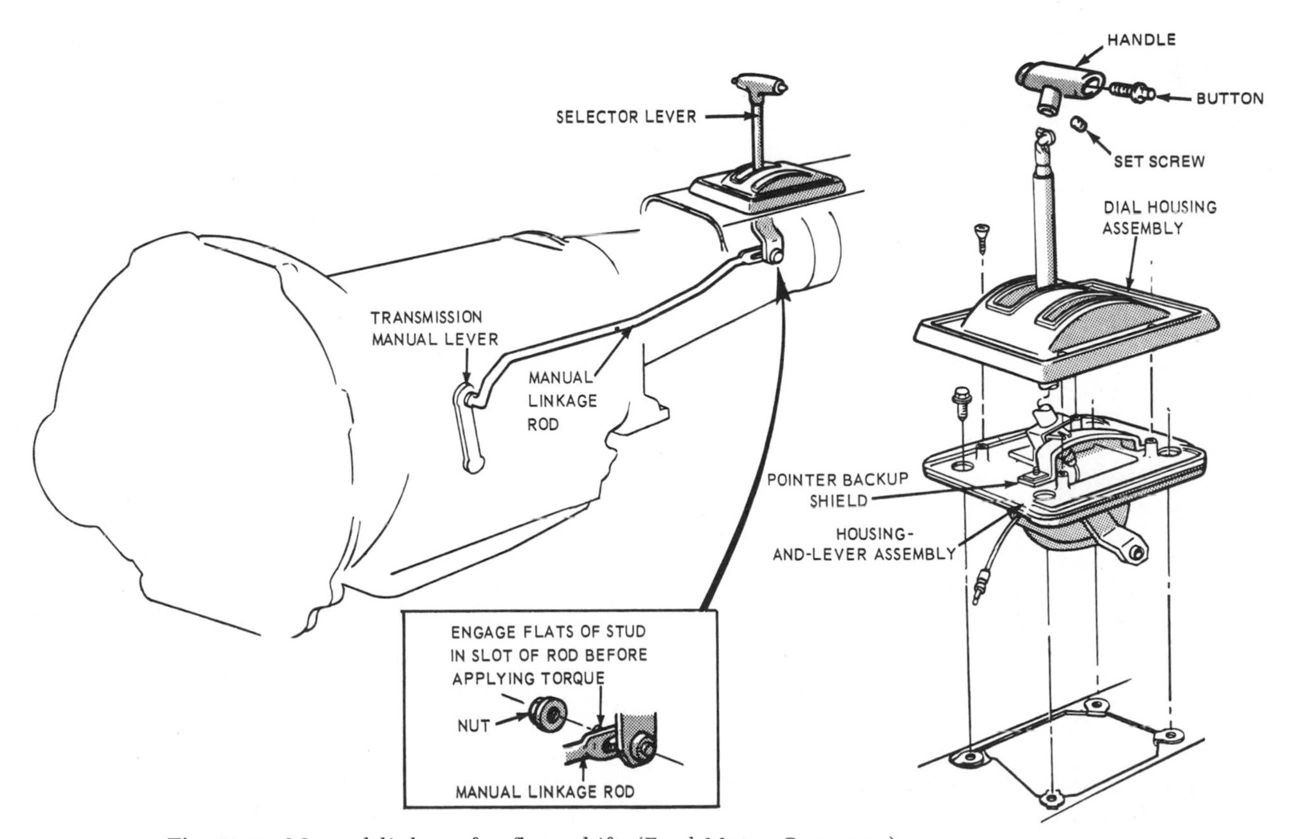

Fig. 20-4. Manual linkage for floor shift. (*Ford Motor Company*)

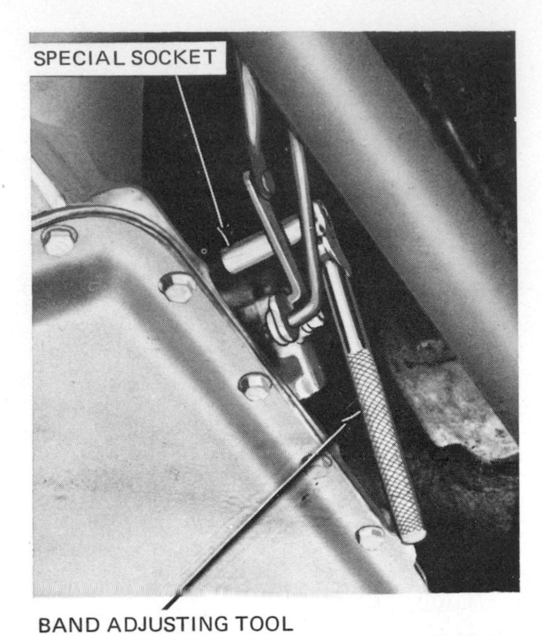

Fig. 20-5. Adjusting the intermediate band on the C6. (*Ford Motor Company*)

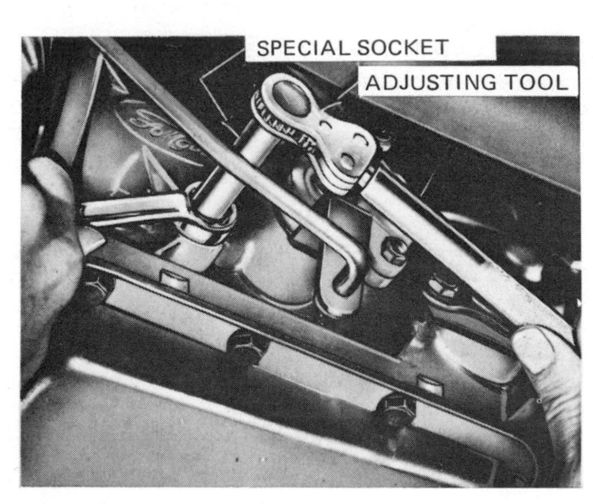

Fig. 20-6. Adjusting the intermediate band on the C4. (*Ford Motor Company*)

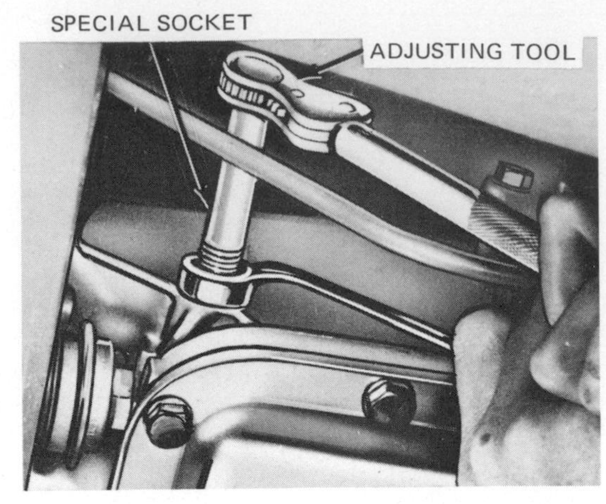

Fig. 20-7. Adjusting the low-reverse band on the C4. (*Ford Motor Company*)

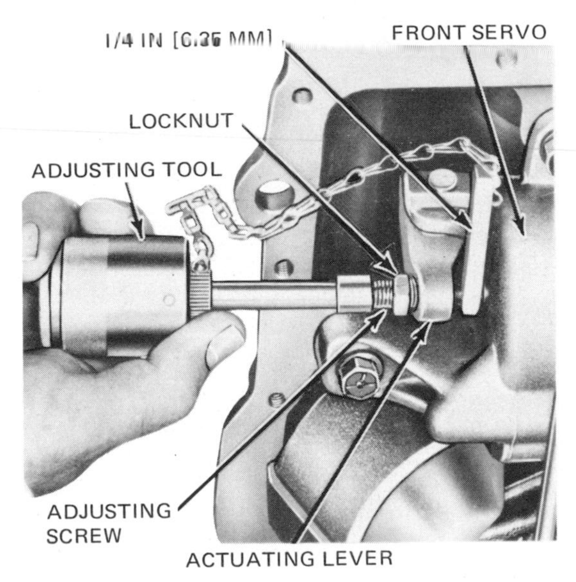

Fig. 20-8. Adjusting the front band on the FMX. (*Ford Motor Company*)

of a clutch as on the C6. On the C4, the intermediate band is adjusted in the same way as in the C6 (see Fig. 20-6). The low-reverse band on the C4 is adjusted in a similar way, but with a special tool. Figure 20-7 shows the special tools required for different cars using the C4. With a new locknut installed in place of the old one, tighten the adjusting screw until the tool handle clicks. It is preset to click at 10 pound-feet [1.382 kg-m] torque. Then back off the adjusting screw 3 turns. Tighten the locknut to the proper tension, holding the screw stationary.

The FMX transmission has both a front and rear band. To adjust the front band, you must remove the oil pan and drain the fluid. Then remove the fluid screen and clip. Next, loosen the locknut, pull back on the actuating rod, and insert a $\frac{1}{4}$-inch [6.35 mm] spacer between the adjusting screw and the servo-piston stem (Fig. 20-8). Tighten the adjusting screw to 10 pound-inches [11.5 kg-cm] torque. Remove the spacer and tighten the screw an additional $\frac{3}{4}$ turn. Hold the adjusting screw stationary and tighten the locknut. Reinstall the fluid screen and clip. Then install the oil pan using a new gasket. Refill the transmission, as already described (⊘ 20-1).

To adjust the rear band on the FMX, remove all dirt from the adjusting-screw threads and oil the threads. Then loosen the locknut. Use the special torque wrench, as shown in Fig. 20-9, and tighten the screw until the tool clicks. Then, back off the screw $1\frac{1}{2}$ turns. Hold the screw and tighten the locknut to specifications.

SPECIAL CAUTION: The transmission can be severely damaged if the adjusting screw is not backed off exactly $1\frac{1}{2}$ turns!

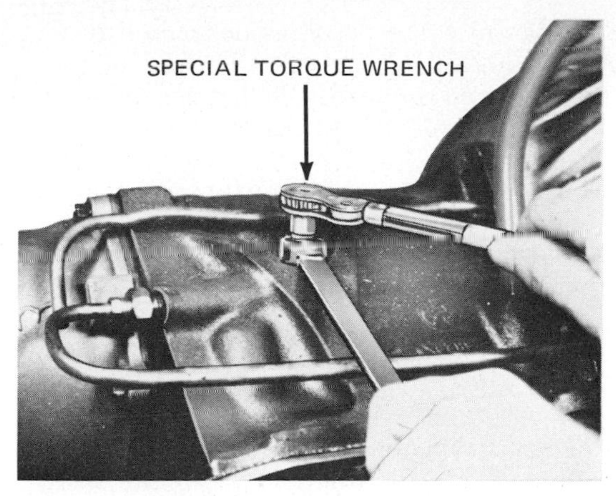

Fig. 20-9. Adjusting the rear band on the FMX. (*Ford Motor Company*)

On-the Car Repairs

A number of C6 transmission parts can be removed from the car without removing the complete transmission. These include the governor and extension housing, the servo, and the valve body.

⊘ 20-6 Governor and Extension-Housing Removal and Replacement First, raise the car on a hoist. Then disconnect the parking-brake cable from the equalizer. On the Lincoln Continental, remove the equalizer. Proceed as follows:

1. Disconnect the drive shaft from the rear-axle flange and remove it from the transmission.
2. Disconnect the speedometer cable from the extension housing.
3. Remove the engine-rear-support-to-extension-housing attaching bolts. On the Lincoln, remove the reinforcement plate from under the transmission oil pan as well.
4. Place a jack under the transmission and raise it just enough to remove the weight from the engine rear support.
5. Remove the bolt that attaches the engine rear support to the cross member and remove the support.
6. Put a drain pan under the transmission.
7. Lower the transmission and remove the extension-housing attaching bolts. Slide the housing off the output shaft and allow the fluid to drain.
8. Remove the governor attaching bolts and slide the governor off the output shaft. See Fig. 20-10.
9. Clean the governor parts with solvent and air-dry them. Check valves and bores for scores. Minor scores can be removed with crocus cloth. Heavier scores or other damage require replacement of the governor assembly. Valves should slide through the bores of their own weight when dry. Fluid passages should be free. Mating surfaces should be smooth and flat.
10. On the C4, the oil screen should be removed from the collector body, cleaned, and air-dried.

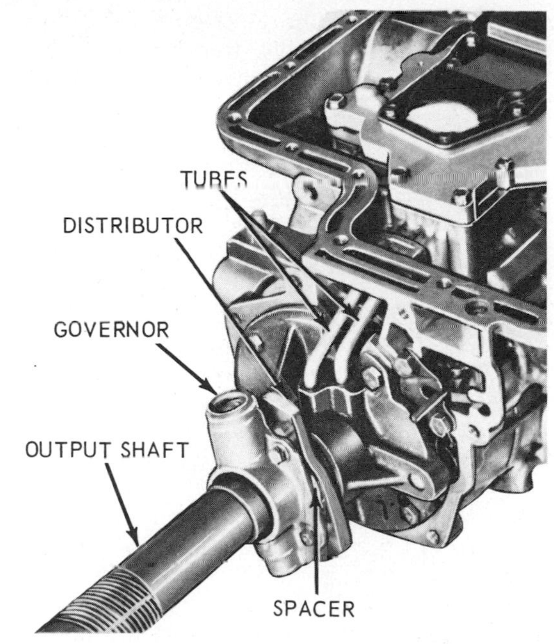

Fig. 20-10. Governor in position in the C6. (*Ford Motor Company*)

11. Check the extension-housing bushing and rear seal. If either needs replacement, use the special tools required to remove the old part and install the new one.
12. To reinstall the governor and extension housing, first attach the governor to the distributor flange and secure it with the attaching bolts, torqued to specifications.
13. Next, make sure the matching surfaces of the extension housing and transmission case are clean. Use a new gasket and install the housing, securing it with the attaching bolts.
14. Raise the transmission high enough to position the engine rear support on the cross member. Secure the support with the attaching bolt and nut, torqued to specifications.
15. Lower the transmission and remove the jack. Install and torque the engine rear-support-to-extension-housing attaching bolts. On the Lincoln, install the reinforcing plate with attaching bolts.
16. Attach the speedometer cable and connect the parking-brake cable to the equalizer. On the Lincoln, connect the parking-brake equalizer. Adjust the parking brakes.
17. Install the drive shaft. Fill the transmission to the correct level with the specified transmission fluid.

⊘ 20-7 Servo Removal and Replacement Remove the servo by raising the car on a hoist. Then:

1. Remove the engine rear-support-to-extension-housing attaching bolts. Raise the transmission high enough to remove the weight from the engine rear support and remove the bolt that secures the engine rear support to the cross member. Remove the support.

2. Lower the transmission and remove the jack. Put a drain pan under the servo. Remove the bolts that attach the servo cover to the transmission case. See Fig. 20-11.
3. Remove the cover, piston, spring, and gasket from the case, screwing the band adjusting screw inward as the piston is removed. This will place enough tension on the band to keep the struts in place when the piston is removed.
4. To change the piston seals, apply air pressure to the port in the servo cover to remove the piston. Then remove the seals from the piston and the seal from the cover. Dip new seals in transmission fluid and install them. Coat two new gaskets with petroleum jelly and install them on the cover. Dip the piston in transmission fluid and install it in the cover.
5. Put the servo spring on the piston rod and insert the piston rod in the case. Secure the cover with the attaching bolts, taking care to back off the band adjusting screw as the cover bolts are tightened. Make sure the vent-tube retaining clip and service-identification tag are in place.
6. Raise the transmission high enough to install the engine rear support. Secure the support to the extension housing with the attaching bolts. Lower the transmission to install the support to the cross member. Torque the attaching bolt to specifications.
7. Remove the jack. Adjust the band as previously explained.
8. Lower the car and replenish the fluid as necessary.

⊘ 20-8 Valve-Body Removal and Replacement Raise the car on a hoist and put a drain pan under the transmission. Loosen the transmission-pan bolts and allow the fluid to drain. Then:

1. Remove the transmission-pan attaching bolts from both sides to complete draining. Remove the rest of the bolts and the pan. Remove and discard the nylon shipping plug from the filler-tube hole. This plug is used to retain transmission fluid during shipping and should be removed when the pan is off.
2. Remove the valve-body attaching bolts and take the valve body off.

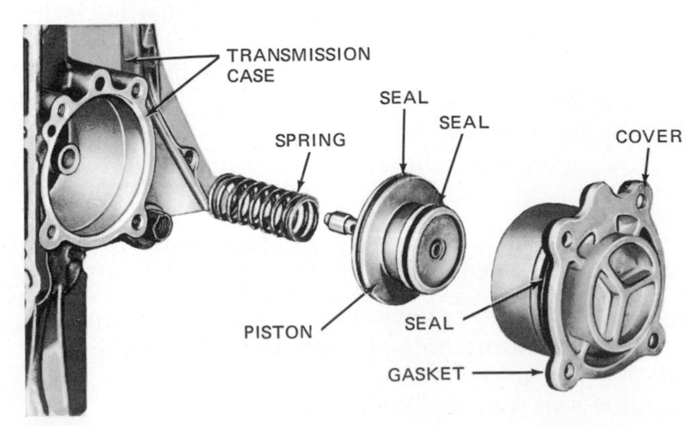

Fig. 20-11. Disassembled view of the servo. (*Ford Motor Company*)

3. Service the valve body, as described in ⊘ 20-21.
4. When reinstalling the valve body, make sure that the selector and downshift levers are engaged. Then install and torque the attaching bolts to specifications.
5. Clean the pan and gasket surfaces thoroughly. Use a new pan gasket and attach the pan with the bolts tightened to specifications.
6. Lower the car and add fluid as necessary.

Trouble Diagnosis

To diagnose transmission troubles, a series of checks are made. These include checks of the shift points and hydraulic control pressures, as well as air-checks of the clutches and servo. In addition, the technician diagnosing the trouble can use one of the diagnosis guides given later in the chapter to help pinpoint the trouble.

⊘ 20-9 Shift-Point Checks The engine should be in good condition, linkages to the transmission should be properly adjusted, and the transmission fluid should be at the proper height. The shift points can be checked on the road or in the shop if a dynamometer or a source of vacuum, such as an ignition-distributor tester, is available. Without the dynamometer, engine loading can be simulated by applying a varying vacuum to the vacuum diaphragm of the transmission throttle control. By varying the vacuum to simulate closed to open throttle conditions and varying engine speed, you can make all the checks you need in the shop. Ford furnishes diagnostic guides, as shown in Fig. 20-12. They have spaces in which to record test data on shift points, stall speed, and pressure tests. Note that there are spaces in which to record the specifications that are to be picked up from the shop manual. The actual test results are then compared with the specifications, and this comparison usually shows up any trouble in the transmission.

1. To make the shift-point test in the shop, raise the car on jacks or a hoist that will allow the wheels to turn. Then disconnect the vacuum line from the transmission and plug it. Connect the vacuum hose from the vacuum source to the transmission. Now start the engine and move the selector lever through all positions. Adjust the vacuum to 18 inches [457.2 mm] to simulate a light throttle. Move the lever to D and gradually increase engine speed. Note the upshift points. You can tell these because the speedometer needle will surge momentarily and you can feel the slight bump indicating the upshift. Slow the engine to idle speed.

NOTE: Do not exceed 60 mph [97 km/h] on the speedometer!

2. Now simulate a high engine load or wide-open throttle by reducing the vacuum to 0 to 2 inches [0

Automatic Transmission Diagnosis Guide
Ford Customer Service Division

General: This form must be completely filled in throughout the steps required to diagnose the condition covering transmission malfunction complaints (e.g., erratic shifting, slippage during shifts, failure to shift, harsh and delayed shifts, noise, etc.). It is not necessary to complete this form on complaints involving external leaks.

Transmission Model ____________ Transmission Date Code/or Serial No. ____________

R.O. No. ____________ Axle Ratio ____________ Tire Size ____________

DIAGNOSIS PROCEDURE

Following steps will provide complete data necessary to perform an accurate diagnosis of transmission difficulties.

1. Check transmission fluid level ☐ Room Temp. ☐ Operating Temp. ☐ OK ☐ Overfilled ☐ Low
2. Engine (CID) and Calibration Number ____________

   Idle RPM in Drive ____________ ____________ ____________

   Specification As Received Set To

   Check EGR System (if so equipped)

   Valve Operation ☐ OK ☐ Other (Explain) ____________

   Restriction ☐ OK ☐ Other (Explain) ____________
3. Check downshift and manual linkage ☐ OK ☐ Other (Explain) ____________
4. Drive the car in each range, and through all shifts, including forced downshifts, observing any irregularities of transmission performance.

Throttle Opening	Range	Shift	Shift Points (MPH)	
			Record Actual	Record Spec.
Minimum (Above 12" Vacuum)	D	1-2		
	D	2-3		
	D	3-1		
	1	2-1		

Throttle Opening	Range	Shift	Shift Points (MPH)	
			Record Actual	Record Spec.
To Detent (Torque Demand)	D	1-2		
	D	2-3		
		3-2		
Thru Detent (WOT)	D	1-2		
	D	2-3		
	D	3-2		
	D	2-1 or 3-1		

5. Control Pressure Test ____________ AND ____________ Stall Speed Data

   Transmission fluid must be normal operating temperatures. DO NOT hold throttle open over five seconds during tests.

   CAUTION: Release throttle immediately if slippage is indicated.

   After each stall test move selector lever to neutral with engine running at 1000 RPM to cool the transmission.

Engine RPM	Manifold Vacuum In-Hg	Throttle	Range	PSI	
				Record Actual	Record Spec.
Idle	Above 12	Closed	P		
			N		
			D		
			2		
			1		
			R		
As Required	10 ①	As Required	D, 2, 1		
As Required	Below 3	Wide Open	D		
			2		
			1		
			R		

Above Specified Engine RPM	
1. Transmission slippage	
2. Clutches or bands not holding	
Below Specified Engine RPM	
1. Poor engine performance, such as need for tune-up	
2. Converter one way clutch slipping or improperly installed	
Specified Engine RPM	**Record Actual Engine RPM**
→	
→	
→	
→	

① On units equipped with a dual area diaphragm, the front port of diaphragm must be vented to atmosphere (hose disconnected and plugged) during this check only.

After the tests, you should know the following items:

- CONTROL PRESSURE – Does the transmission have the CORRECT CONTROL PRESSURE? ☐ Yes ☐ No
- CONTROL VALVES – Beyond the manual valve are all the CONTROL VALVES FUNCTIONING? ☐ Yes ☐ No
- HYDRAULIC CIRCUITS – If the first two items check out good, then check the transmission's internal hydraulic circuits that are beyond the VALVE BODY. These circuits must be checked during transmission disassembly.

6. TORQUE CONVERTER AND OIL COOLER (where applicable)
   - Was torque converter flushed with a mechanical cleaner? ☐ Yes ☐ No
   - Was oil cooler flushed with a mechanical cleaner? ☐ Yes ☐ No
7. The problem was diagnosed to be: ____________

   ____
8. If it was necessary to disassembly the transmission, record the actual problem found: ____________

Fig. 20-12. Automatic-transmission diagnosis guide. (*Ford Motor Company*)

to 50.8 mm]. Accelerate the engine slowly and note the 1-2 upshift point. You don't need to check the 2-3 shift point because you already checked this in the previous test. Do not exceed 60 mph [97 km/h]!
3. Next, test the kickdown system. You do not want to run an unloaded engine at wide-open throttle because this could ruin it. So you need another way to check kickdown. You can do this by leaving the vacuum adjusted to 0 to 2 inches [0 to 50.8 mm] and manually holding the downshift linkage in the wide-open-throttle position. Accelerate the engine just enough to cause a 1-2 upshift. Do not exceed 60 mph [97 km/h]. Note that the wide-open-throttle 1-2 shift should be at a higher mph than the high-engine-load test. This indicates that the kickdown system is working.
4. To make the shift-point tests on the highway, operate the car at various speeds and in the different selector-lever positions to note whether all shifts take place properly.

CAUTION: Drive carefully and obey all traffic laws. Use a chassis dynamometer (⊘ 1-9), if available, to test the transmission.

⊘ 20-10 Control-Pressure and Stall Checks These checks consist in measuring the control pressure in the hydraulic system under varying operating conditions and also seeing if there is transmission slippage. Ford recommends the tests in the shop with a dynamometer or with the transmission tester, shown in Fig. 20-13. The transmission tester has a pressure gauge, vacuum gauge, and tachometer. We shall describe the use of the transmission tester to make the checks.

First, connect the tachometer to the ignition system so that you can measure engine speed. One clip goes on the distributor terminal on the ignition coil, the other clip (the black one) to ground.

Next, connect the vacuum hose from the tester (Fig. 20-13) into the manifold vacuum line by means of a T fitting, as shown in Fig. 20-14. The third connection is the pressure hose to the control-pressure port. The connection to the C6 transmission is shown in Fig. 20-15. The control-pressure port has

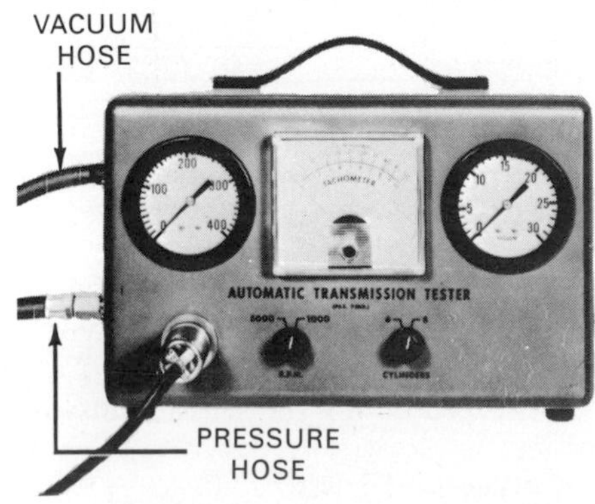

Fig. 20-13. Automatic-transmission tester. (*Ford Motor Company*)

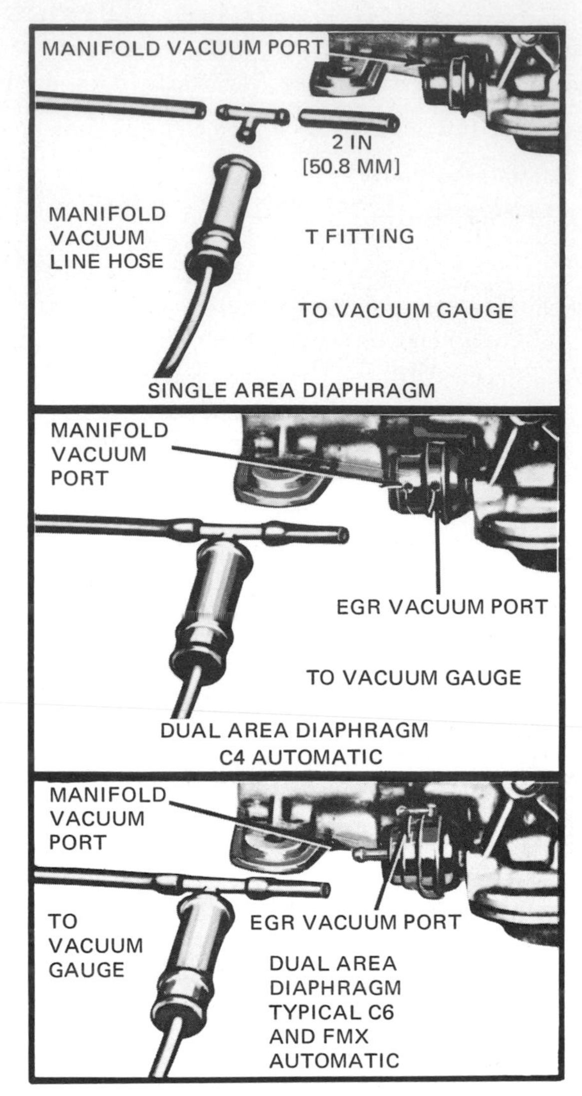

Fig. 20-14. Vacuum-line connections to check control pressure. (*Ford Motor Company*)

different locations in other transmissions; check the shop manual. With the three connections made, proceed as follows:

1. It is assumed that the engine is in satisfactory condition and that the intake-manifold vacuum will read around 18 inches [457.2 mm] at idle.
2. The Stall Speed Data chart in the diagnosis guide of Fig. 20-12 has spaces in which to record the specified engine rpm at which the stall test should be made, as well as spaces in which to record the stall-test results.
3. The Control Pressure Test chart in Fig. 20-12 has spaces in which to record the specified pressures under different conditions, as well as places to record the results of the pressure tests.
4. Prior to the test, you should refer to the shop manual and find the specifications for the transmission and car being tested. These should be recorded on the diagnosis guide.
5. Now, with the transmission at normal operating temperature, apply the foot brake and make the stall

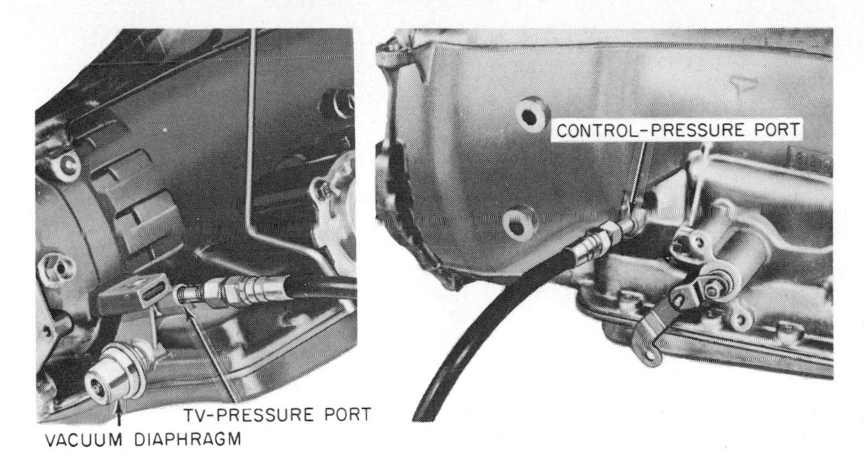

Fig. 20-15. Vacuum-diaphragm and control- and TV-pressure connecting points. (*Ford Motor Company*)

check by increasing the engine speed to the specified value and shifting into the indicated selector-lever position.

CAUTION: Do not hold throttle open more than 5 seconds! After each stall check, return the selector lever to N and idle the engine for a minute or so to cool the transmission.

CAUTION: Release the throttle instantly if the transmission slips, that is, if the speed goes beyond specifications.

6. Excessive speed in any selector position indicates transmission slippage.
7. Now make the pressure check by adjusting the engine idle to the specified speed. With a closed throttle, move the selector lever to all positions and note the pressures. See Fig. 20-12.
8. Next, with a source of vacuum connected to the transmission instead of the intake manifold of the engine, make the pressure tests at 10 and 1 inches [254.0 and 25.4 mm] of vacuum. These tests can be made with the engine idling and only the vacuum changed. In other words, it is not necessary to open the throttle wide to get the vacuums needed for the test. Note that the pressures should be checked in D, D2, and D1 with 10 inches [254 mm] of vacuum and at D, D2, D1, and R with 1 inch [25.4 mm] of vacuum.
9. The chart in Fig. 20-16 tells you what to suspect if the pressures are not within the specifications. As you can see, improper pressures usually require looking into the transmission itself to find the actual unit causing the trouble.

⊘ 20-11 Air-Pressure Checks Further information on the condition of the clutches and servo can be obtained by using air pressure to operate them. This procedure requires removal of the oil pan and control-valve body and draining of the fluid, as explained in ⊘ 20-8. Then air pressure can be directed into the appropriate "apply" hole to see whether or not the brake band or the clutch will apply. Figure 20-17 shows the transmission case for the C6 with the various fluid passages identified. Figure 20-18 shows the FMX transmission case with all fluid passages identified.

⊘ 20-12 Vacuum-Unit Checks and Adjustments The vacuum unit actuates the primary throttle valve, as explained in Chap. 19. The unit should be checked for leakage and for bellows failure. To check for leakage, connect the unit to a source of vacuum, such as a vacuum pump on a distributor tester. Apply 18 inches [457.2 mm] of vacuum to the vacuum unit and see if it will hold this vacuum. If it does not, the diaphragm is leaking.

Next, check the bellows, as shown in Fig. 20-19. Insert a rod in the unit, as shown, and make a reference mark on the rod with the rod bottomed in the hole. Then press down on the unit with the rod resting on a scale, as shown. Increase pressure to 12 pounds [5.443 kg]. If the mark remains visible, the bellows is okay. If the mark disappears before 4 pounds [8.818 kg] is exerted, the bellows is defective and the complete vacuum unit must be replaced.

The vacuum unit can be adjusted by turning the adjusting screw in or out. Adjustment is desirable if the control pressures are high or low and if upshifts or downshifts are high or low. One complete turn of the adjusting screw changes idle control pressure approximately 2 to 3 psi [0.141 to 0.210 kg/cm^2]. If the pressure readings are low, turn the adjusting screw in. If the readings are high, turn the screw out.

⊘ 20-13 Diagnosis Guides The guide in Fig. 20-12 explains how to make various diagnostic tests of the transmission to determine what might be wrong and what is causing the trouble. In addition to these, there are diagnosis guides for specific models, such as shown in Figs. 20-20 and 20-21 (pp. 300 and 301). These guides can help you pinpoint every cause of trouble in the transmission. It is routine for the service technician to use such guides in this checkup work.

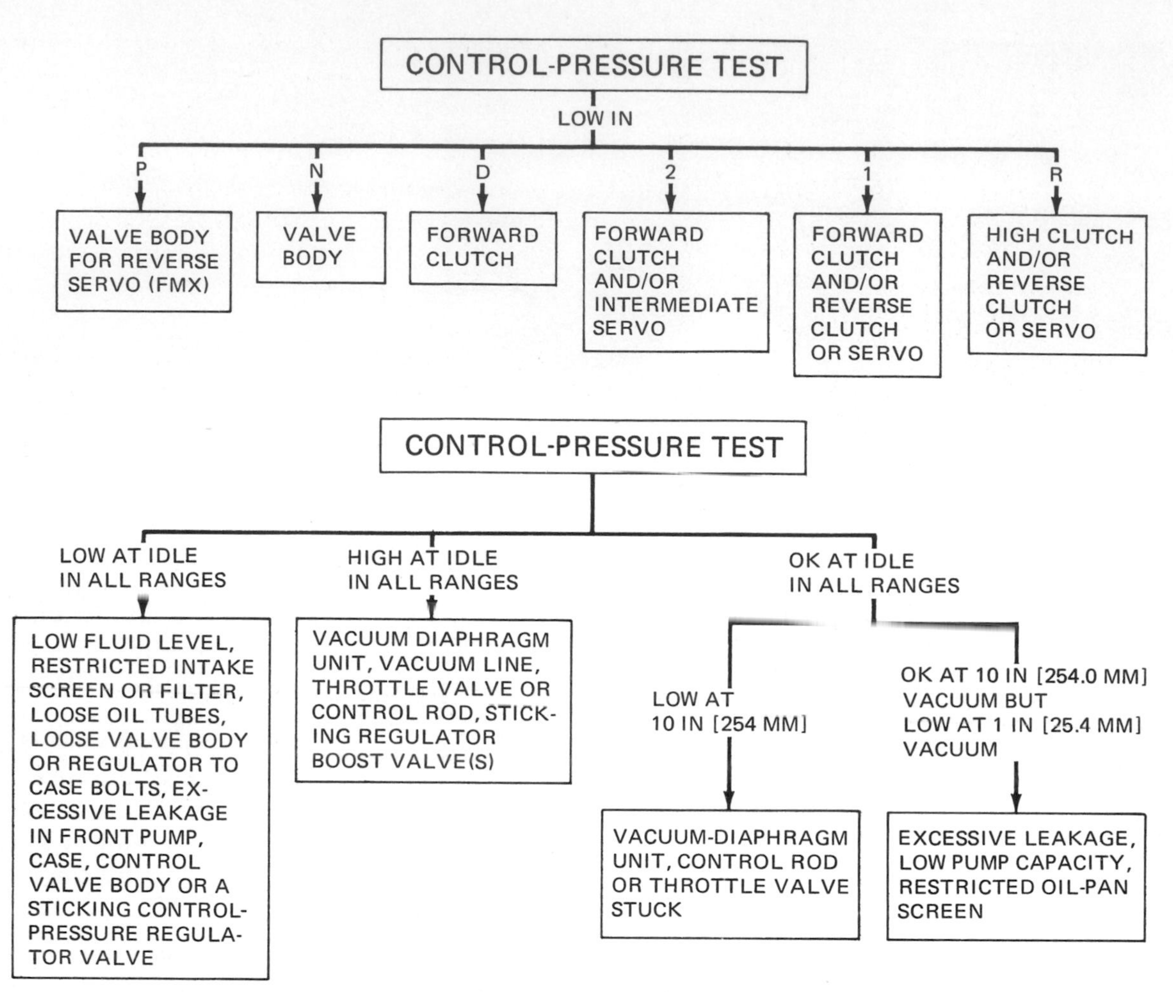

Fig. 20-16. Chart of causes of incorrect pressures. (*Ford Motor Company*)

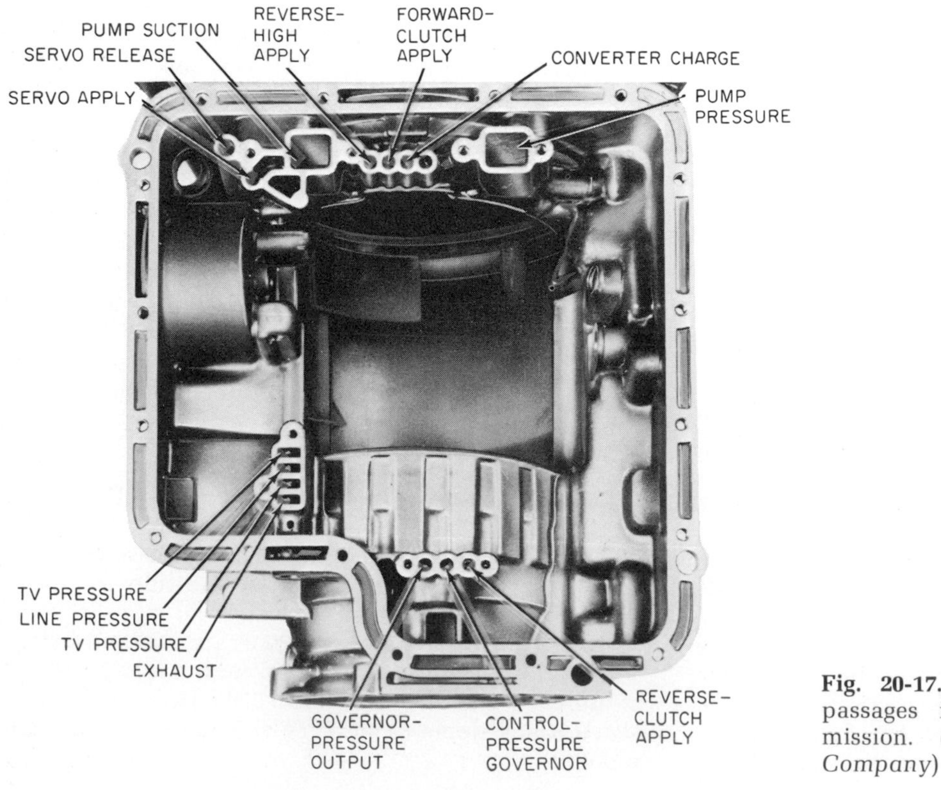

Fig. 20-17. Case fluid passages in C6 transmission. (*Ford Motor Company*)

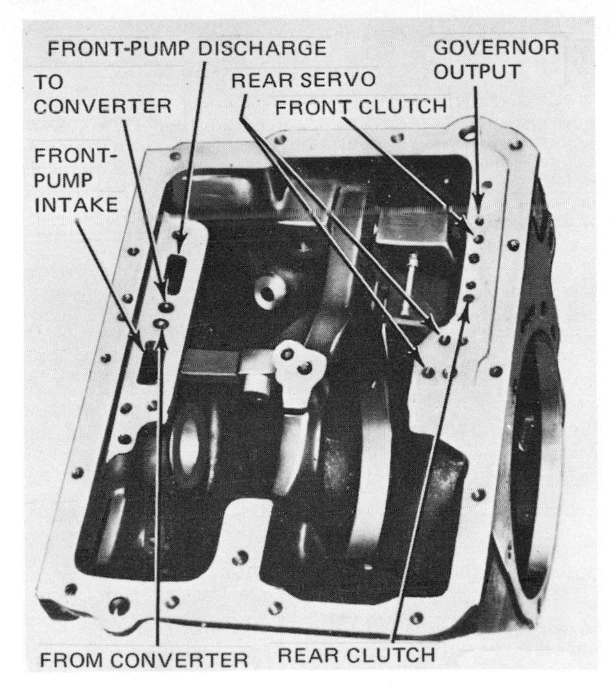

Fig. 20-18. Fluid passage holes in the FMX transmission case. (*Ford Motor Company*)

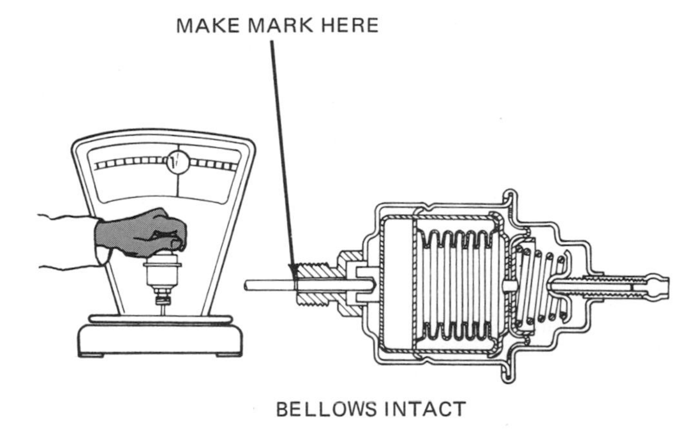

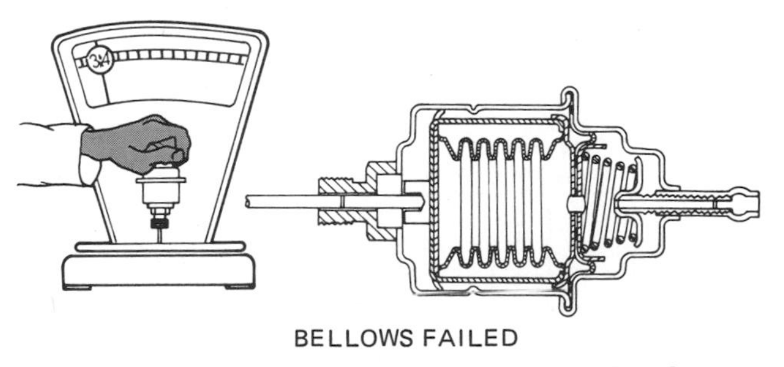

Fig. 20-19. Checking vacuum-unit bellows. (*Ford Motor Company*)

Check Your Progress

Progress Quiz 20-1 By now, you may begin to have the feeling that you've "seen all this before." That is, you begin to realize that all automatic transmissions tend to have similar troubles, rising from similar conditions. Although the various transmissions are different in many details, in methods of adjustment, and in the way they achieve the various gearshifts, they are the same in using planetary gearsets and clutches and bands to control the planetary systems. To find out how well you remember the adjustments and checking procedures for the C4 and C6 transmissions, take the quiz that follows.

Completing the Sentences The sentences that follow are incomplete. After each sentence there are several words or phrases, only one of which will correctly complete the sentence. Write each sentence in your notebook, selecting the proper word or phrase to complete it correctly.

1. The two linkage adjustments are to the: (*a*) throttle linkage and intermediate band linkage, (*b*) throttle linkage and selector-lever linkage, (*c*) selector-lever and band linkage.
2. The band can be adjusted: (*a*) only after removing valve body, (*b*) on the car, (*c*) only after removing the transmission from the car.
3. The three diagnostic checks that can be made on a transmission include shift points, air-checks of the clutches and servo, and: (*a*) hydraulic control pressures, (*b*) stall pressure, (*c*) band pressure.
4. The Ford transmission tester contains three elements: a pressure gauge, a: (*a*) vacuum gauge and an ammeter, (*b*) tachometer and rpm indicator, (*c*) tachometer and a vacuum gauge.
5. If, on the stall test, the engine goes well above the specified rpm, the: (*a*) band or clutches are slipping, (*b*) engine needs a tune-up, (*c*) converter one-way clutch is slipping.
6. If the engine does not reach the specified rpm on the stall test: (*a*) the clutches or band are slipping, (*b*) the engine needs a tune-up, (*c*) line pressure is too low.
7. To make the air-pressure checks of the clutches and servo, you must: (*a*) remove the transmission from the car, (*b*) disassemble the transmission to get to the clutches, (*c*) remove the oil pan and valve body.
8. The vacuum unit can be adjusted by: (*a*) shifting the bracket, (*b*) turning the adjusting screw, (*c*) bending the adjusting lever.

Transmission Overhaul

If there are troubles in the transmission that require its removal and overhaul, proceed as follows.

⊘ 20-14 Transmission Removal and Reinstallation There are many variations in engine-compartment arrangements and methods of transmission support in different car models. Thus, you should always check the shop manual for the car you are working

C6 AUTOMATIC TRANSMISSION
DIAGNOSIS GUIDE

General: This form must be completely filled in throughout the steps required to diagnose the problem and attached to Forms 1863 covering the correction of transmission malfunction complaints (e.g., erratic shifting, slippage during shifts, failure to shift, harsh and delayed shifts, noise, etc.). It is not necessary to complete this form on complaints involving external leaks. Do not use this form on complaints involving the "C4" or "FMX" Transmissions. Instead use Form LM-5185-E-70 for "C4" and LM-5185-F-70 for "FMX".

Transmission Model ______________ Transmission Date Code/or Serial No. ______________

R. O. No. ______________ Axle Ratio ______________ Tester ______________

DIAGNOSIS PROCEDURE

Following steps will provide complete data necessary to perform an accurate diagnosis of transmission difficulties.
(NOTE: *Items 1,2,3 and 5 also should have been performed during pre-delivery.)*

PROCEDURE — **CHECK BOX TO INDICATE FINDINGS**

1. Check transmission fluid level — ☐ Full ☐ Overfilled ☐ Low

SPECIFICATIONS — **FINDINGS**

2. Engine idle: — **AS RECEIVED** — **SET TO**

   Refer to engine idle speed specification in the appropriate section of volume two, 1970 Shop Manual

3. Check kickdown and manual linkage ☐ O.K. ☐ Other (Explain) ______________
4. Perform stall test to check engine performance and for any sign of transmission slippage. ☐ Performed.
   After each stall test move selector lever to neutral with engine running at 1000 R.P.M. for 15 seconds to cool the transmission.
   CAUTION: *Release throttle immediately if slippage is indicated.*

STALL SPEED DATA

Transmission Model	Engine CID	Stall Speed	Record Actual Engine R.P.M.	DIAGNOSIS: Above Specified Engine R.P.M.	DIAGNOSIS: Below Specified Engine R.P.M.
PGA-A4, J4 PGB-G3	390-2V	1600-1860		1. Transmission slippage 2. Clutches or band not holding	1. Poor engine performance, such as need for tune up. 2. Converter one way clutch slipping or improperly installed.
PJA-A, B	429-2V	1760-1960			
PGB-F3	428-4V, P.I.	1800-2020			
PJB-F	429-2V	1800-2020			
PJB-A,B,F,J PJC-E,F	429-4V	1880-2100			
PJD-C,E,F	460-4V	1940-2140			
PGB-AF2	428-4V, C.J.	1840-2060			
PJC-A,B	429-4V, C.J.				
PJB-C,D	429-4V	1860-2080			

NOTE: *Stall test with transmission at operating temperature. DO NOT hold throttle open over five seconds during tests.*

5. Drive the car in each range, and through all shifts, including forced downshifts, observing any irregularities of transmission performance.
   ☐ Road test completed.
   Record malfunctions observed on reverse side of this form. The operating conditions shown on the Performance Chart (**reverse side of this form**) should be checked in the order listed, to avoid performing a major repair where a simple minor repair could correct the malfunction.
6. **Shift Test** – REFER TO PAGES 17-04-34 TO 17-04-37 OF THE 1970 SHOP MANUAL, VOLUME ONE, FOR THE VARIOUS SHIFT SPEED SPECIFICATIONS.
7. CONTROL PRESSURE TEST – Transmission fluid must be at normal operating temperatures.

Engine Speed			Idle						As Required				As Required					
Throttle			Closed						As Required				As Required					
Manifold Vacuum (Inches Hg)			Above 18						10				Below 1.0					
Range			Control Pressure (psi)				TV Pressure (psi)		Control Pressure(psi)		TV Pressure (psi)		Control Pressure (psi)				TV Pressure (psi)	
			P,N,D,2,1		R				D, 2, 1				D, 2, 1		R			
	Barometric Pressure in Inches HG	Nominal Altitude (Feet)	Spec.	Actual	Spec.	Actual	Spec.	Actual	Spec.	Actual	Spec.	Actual	Spec.	Actual	Spec.	Actual	Spec.	Actual
psi Barometric Pressure	29.5	Sea Level	56–62		71–86		7–10		100–115		40–44		160–190		240–300		77–84	
	28.5	1000	49–59		65–80		4–7		99–114		37–41		158–176		233–290		74–80	
	27.5	2000	49–56		60–75		2–5		96–111		35–39		156–174		228–284		72–78	
	26.5	3000	49–56		56–71		0–3		91–106		32–36		151–169		222–277		69–75	
	25.5	4000	49–56		56–65		0		88–103		30–34		146–164		215–269		66–72	
	24.5	5000	49–56		56–65		0		84–98		27–31		143–161		211–264		64–70	
	23.5	6000	49–56		56–65		0		80–95		25–29		138–156		204–256		61–67	

Fig. 20-20. C6 automatic-transmission diagnosis guide, front side. (*Ford Motor Company*)

C6 AUTOMATIC TRANSMISSION DIAGNOSIS GUIDE
SHIFT CONDITIONS AND OPERATING CHARACTERISTICS

NOTE: Under "Components to Check" (see below) for Pressure Check and Valve Body & Position of Valves, the transmission pressure gauge, and the tachometer and engine vacuum gauge will have to be used before and during road test. By driving in all possible ranges and through all shift points it will be possible to determine if the control valves are able to ~~move and~~ can be placed in their correct positions for each gear ratio.

After road test, you should know the following items:

1. CONTROL PRESSURE – Does the transmission have the CORRECT CONTROL PRESSURE? YES ☐ NO ☐
2. CONTROL VALVES – Beyond the manual valve are all the CONTROL VALVES FUNCTIONING? YES ☐ NO ☐
3. HYDRAULIC CIRCUITS – If the first two items check out good, then check the transmission's internal hydraulic circuits that are beyond the VALVE BODY. These circuits must be checked during transmission disassembly.

PERFORMANCE CHART

OPERATING CONDITIONS	(X)	COMPONENTS TO CHECK (In the order indicated) See detailed possible causes below.	
Rough initial Engagement in D or 2		K B W E	a
1-2 or 2-3 Shift Points Incorrect or Erratic		A B L C D W E R	
Rough 1-2 Upshifts		B J G W E	
Rough 2-3 Shifts		B J W G E R	b r
Dragged Out 1-2 Shift		A B J W G E R	c
Engine Overspeeds on 2-3 Shift		C A B J W E G	b r
No 1-2 or 2-3 Shift		C L B D W E G J	b c
No 3-1 Shift in D or 3-2 Shift in 2		D E	
No Forced Downshifts		L W E	
Runaway Engine on Forced 3-2 Downshift		W J G E B	c
Rough 3-2 or 3-1 Shift at Closed Throttle		K B J W E	
Shifts 1-3 in D		G J E D R	
No Engine Braking in 1		C H E D R	c
Creeps Excessively		K	
Slips or Chatters in First Gear, D		A B W E	a c i
Slips or Chatters in Second Gear		A B J G W E R	a c
Slips or Chatters in R		A B C H W E R	b c r t
No Drive in D Only		W E	i
No Drive in 2 Only		A C W J E R G	c
No Drive in 1 Only		A W E R	c
No Drive in R Only		A C H W E R	b c r t
No Drive in D, 2 and 1		C W E R	a c
No Drive in Any Selector Lever Position		A C W E R	c d
Lockup in D Only		E	g c
Lockup in 2 Only		H E	b g c i
Lockup in 1 Only		G J E	b g c
Lockup in R Only		G J E	a g c
Parking Lock Binds or Does Not Hold		C	g
Transmission Overheats		O E W	n s
Maximum Speed too Low, Poor Acceleration		Y Z	n
Transmission Noisy in N and P		A E	d
Transmission Noisy in First, Second, Third or Reverse Gear		A E	h a d i
Fluid Leak		A M N O P Q S U X B J	j m p
Car moves Forward in N		C	a

DETAILED POSSIBLE CAUSES

A. Fluid Level	W. Perform Control Pressure Check
B. Vacuum Diaphragm Unit or Tubes Restricted – Leaking – Adjustment	X. Speedometer Driven Gear Adapter Seal
C. Manual Linkage	Y. Engine Performance
D. Governor	Z. Vehicle Brakes
E. Valve Body	a. Forward Clutch
G. Intermediate Band	b. High Clutch
H. Reverse Clutch	c. Leakage in Hydraulic System
J. Intermediate Servo	d. Front Pump
K. Engine Idle Speed	g. Parking Linkage
L. Downshift Linkage – Including Inner Lever Position	h. Planetary Assembly
M. Converter Drain Plugs	i. Planetary One-Way Clutch
N. Oil Pan Gasket, Filler Tube or Seal	j. Engine Rear Oil Seal
O. Oil Cooler and Connections	m. Front Pump Oil Seal
P. Manual or Downshift Lever Shaft Seal	n. Converter One-Way Clutch
Q. 1/8 Inch Pipe Plug in Side of Case	p. Front Pump to Case Gasket or Seal
R. Perform Air Pressure Check	r. High Clutch Piston Air Bleed Valve
S. Extension Housing to Case Gaskets & Lockwashers	s. Converter Pressure Check Valves
U. Extension Housing Rear Oil Seal	t. Reverse Clutch Piston Air Bleed Valve

TORQUE CONVERTER AND OIL COOLER (Where applicable)

1. Was torque converter turbine and stator end play checked? (Spec. .030" maximum) Actual ________
2. Was torque converter flushed with a mechanical cleaner? YES ☐ NO ☐
3. Was oil cooler flushed with a mechanical cleaner? YES ☐ NO ☐

Fig. 20-21. C6 automatic-transmission diagnosis guide, back side. (*Ford Motor Company*)

on before proceeding with transmission removal. A typical removal procedure follows:

1. Raise the vehicle on a hoist and drain the fluid from the transmission and converter.
2. Disconnect the drive shaft from the rear axle and slide the shaft rearward to remove it from the transmission. Install a seal-installation tool in the extension housing to prevent fluid leakage.
3. Disconnect the ground cable from the battery and remove the starting motor.
4. Remove the four converter-to-flywheel attaching nuts. Do not pry on the converter to turn it. Instead, use a wrench on the crankshaft pulley to turn the crankshaft so that you can gain access to the nuts.
5. Remove the rear-mount-to-cross-member attaching bolt.
6. Remove the two cross-member-to-frame attaching bolts.
7. Remove the two engine-rear-support-to-extension-housing attaching bolts.
8. Disconnect the downshift rod and the manual-linkage rod from the transmission levers. On console models, disconnect the column-lock rod from the transmission.
9. Remove the two bolts securing the bell crank bracket to the converter housing.
10. Raise the transmission with a transmission jack to provide clearance to remove the cross member. Remove the rear mount from the cross member and then remove the cross member.
11. Lower the transmission enough to gain access to the oil-cooler lines. Disconnect the lines from the transmission.
12. Disconnect the vacuum line from the transmission vacuum unit and detach the line from the retaining clip on the transmission.
13. Disconnect the speedometer cable from the extension housing.
14. Remove the bolt that secures the transmission oil filler tube to the cylinder block. Lift the filler tube and dipstick from the transmission.
15. Use a chain to secure the transmission to the jack.
16. Remove the converter-housing-to-engine attaching bolts.
17. Carefully move the transmission away from the engine and lower it.

CAUTION: Don't let the converter slip off!

18. Remove the converter and mount the transmission in a repair fixture, as shown in Fig. 20-22.
19. Reinstallation of the transmission is essentially the reverse of removal. When installing the oil filler tube, use a new O ring. Do not use a wrench on the converter attaching nuts to turn the converter when you are attaching the converter to the flywheel. Instead, use a wrench on the crankshaft-pulley attaching nut. Adjust linkages and refill the transmission with specified fluid, as already noted, when installation is complete. Start the engine and check the transmission in all selector-lever positions.

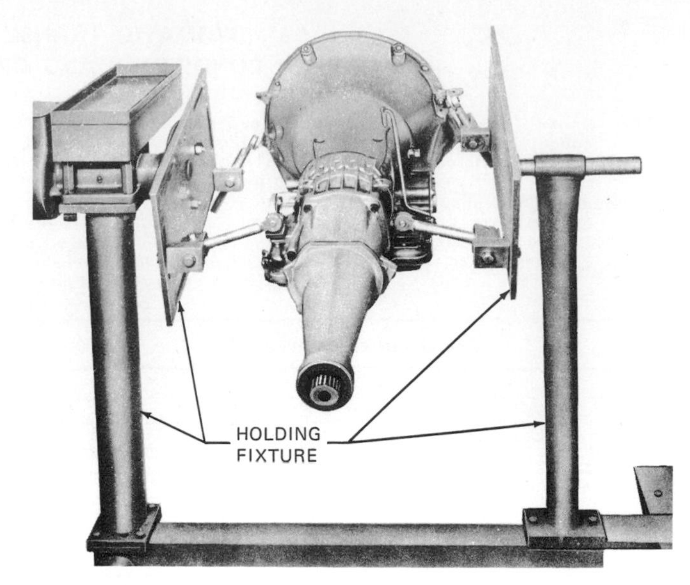

Fig. 20-22. Fixture to hold the transmission for repair. *(Ford Motor Company)*

⊘ 20-15 Transmission-Service Precautions Cleanliness is of the utmost importance in transmission work. The tiniest piece of dirt or lint from a cleaning rag can cause a valve to hang up and prevent normal transmission operation. Under some conditions, this could ruin the transmission. The outside of the transmission must be clean before it is opened. The workbench, tools, your hands, and all parts must be kept clean at all times. There are a number of other precautions you should heed in transmission work. They are discussed in ⊘ 14-10.

⊘ 20-16 Cleaning and Inspecting Transmission Parts During and after disassembly, clean and inspect all parts for wear or other damage. Refer to ⊘ 14-11, for details. If the transmission shows signs of internal damage to clutches or other parts, or if there is evidence that foreign material has been circulating in the oil, then all parts, including the oil cooler and lines and the converter, must be flushed out. A new oil filter should be installed on reassembly.

⊘ 20-17 Transmission Disassembly First, mount the transmission in a holding fixture, as shown in Fig. 20-22. The transmission drive train is shown in disassembled view in Fig. 20-23. Proceed with disassembly as follows:

1. Remove the 17 oil-pan attaching screws and the oil pan.
2. Remove the eight valve-body attaching screws and the valve body.
3. Attach a dial indicator to the front pump, as shown in Fig. 20-24. Install a special tool in the extension housing to center the shaft. Then pry the gear train to the rear of the case, as shown, and press the input shaft inward until it bottoms. Set the dial indicator to zero.

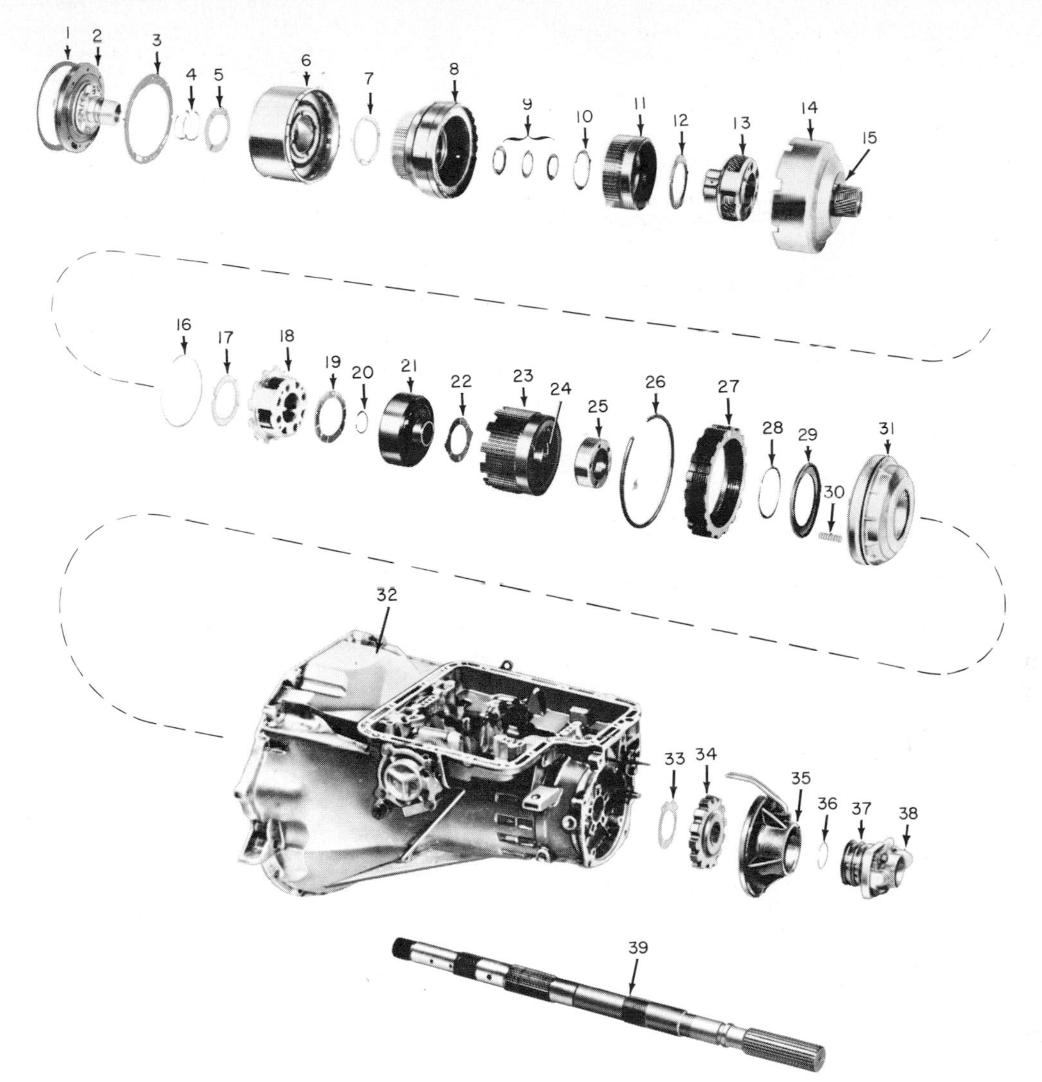

1. Front-pump seal ring
2. Front pump
3. Gasket
4. Seal
5. No. 1 thrust washer (selective)
6. Reverse-high-clutch assembly
7. No. 2 thrust washer
8. Forward clutch assembly
9. No. 3 thrust washer
10. No. 4 thrust washer
11. Forward-clutch-hub assembly
12. No. 5 thrust washer
13. Forward planet assembly
14. Input shell and sun-gear assembly
15. No. 6 thrust washer
16. Snap ring
17. No. 7 thrust washer
18. Reverse planet assembly
19. No. 8 thrust washer
20. Reverse ring gear and hub-retaining ring
21. Reverse ring gear and hub
22. No. 9 thrust washer
23. Low-reverse clutch hub
24. One-way clutch
25. One-way clutch inner race
26. Snap ring
27. Low-reverse clutch
28. Snap ring
29. Low-reverse piston-return-spring retainer
30. Return spring
31. Low-reverse piston
32. Case
33. No. 10 thrust washer
34. Parking gear
35. Governor-distributor sleeve
36. Snap ring
37. Governor distributor
38. Governor
39. Output shaft

Fig. 20-23. Disassembled view of the transmission drive train. (*Ford Motor Company*)

4. Pry the gear train forward and note the amount of gear-train end play as registered on the dial indicator. Record the end play. You will need this figure on reassembly.
5. Remove the vacuum unit, rod, and primary throttle valve. Slip the input shaft out of the pump.
6. Remove the front-pump attaching bolts. Pry the gear train forward, as shown in Fig. 20-25, to remove the pump.
7. Loosen the band adjustment screw and remove the two struts. Rotate the band 90 degrees counterclockwise to align the ends with the slot in the case. Then slide the band off the clutch drum.
8. Remove the forward part of the gear train as an assembly, as shown in Fig. 20-26.
9. Remove the large snap ring that holds the reverse planet carrier in the low-and-reverse clutch hub, as shown in Fig. 20-27. Lift the planet carrier from the drum.
10. Remove the snap ring that secures the reverse

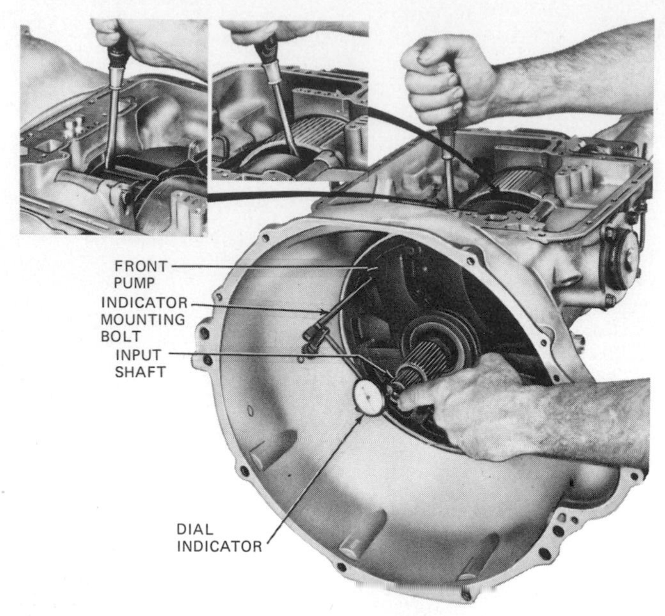

Fig. 20-24. Checking gear-train end play. (*Ford Motor Company*)

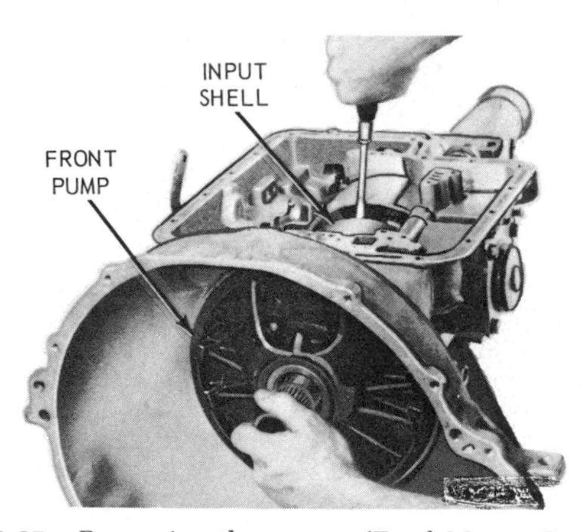

Fig. 20-25. Removing the pump. (*Ford Motor Company*)

ring gear and hub to the output shaft. Slide the ring gear and hub off the shaft.

11. Rotate the low-and-reverse clutch hub in a clockwise direction and at the same time withdraw it from the case.

12. Remove the reverse-clutch snap ring from the case, and then remove the clutch disks, plates, and pressure plate from the case.

13. Remove the extension-housing attaching bolts and vent tube from the case. Remove the extension housing and gasket.

14. Slide the output-shaft assembly from the case.

15. Remove the distributor-sleeve attaching bolts and remove the sleeve, parking-pawl gear, and thrust washer. If the thrust washer is staked in place, use a sharp chisel and cut off the metal from behind the washer. Clean up the rear of the case with air pressure and solvent to remove any metal particles.

16. Compress the reverse-clutch-piston spring with a compressor tool, as shown in Fig. 20-28. Remove the snap ring, tool, and spring retainer.

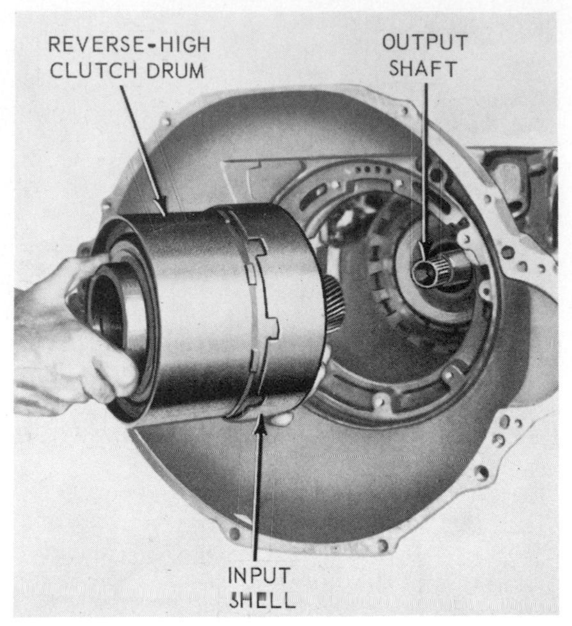

Fig. 20-26. Removing the forward part of the gear train. (*Ford Motor Company*)

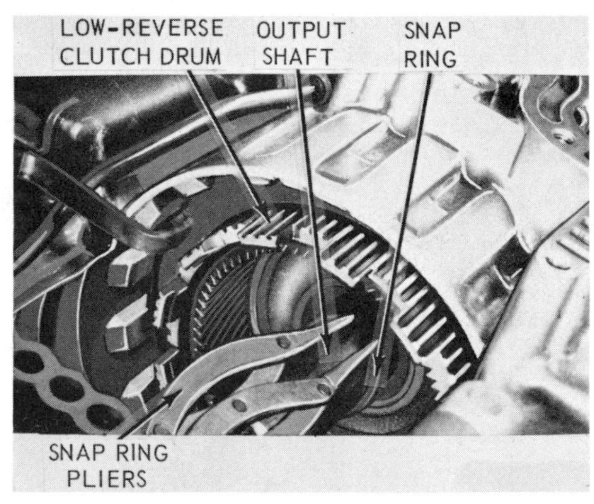

Fig. 20-27. Removing the reverse-ring-gear-hub retaining ring. (*Ford Motor Company*)

17. Remove the one-way-clutch inner-race attaching bolts and the inner race. Then remove the reverse-clutch piston from the case, using air pressure, as shown in Fig. 20-29.

⊘ 20-18 Downshift and Manual Linkages The downshift and manual linkages are shown in the case in Fig. 20-30. Follow this illustration if the parts must be removed. If they are removed, remove the shaft seal from the case and install a new seal by using a special tool to drive it into place.

⊘ 20-19 Parking-Pawl Linkage The parking-pawl linkage is shown in Fig. 20-31. Refer to this illustration if the parts must be removed. Drill a $\frac{1}{8}$-inch

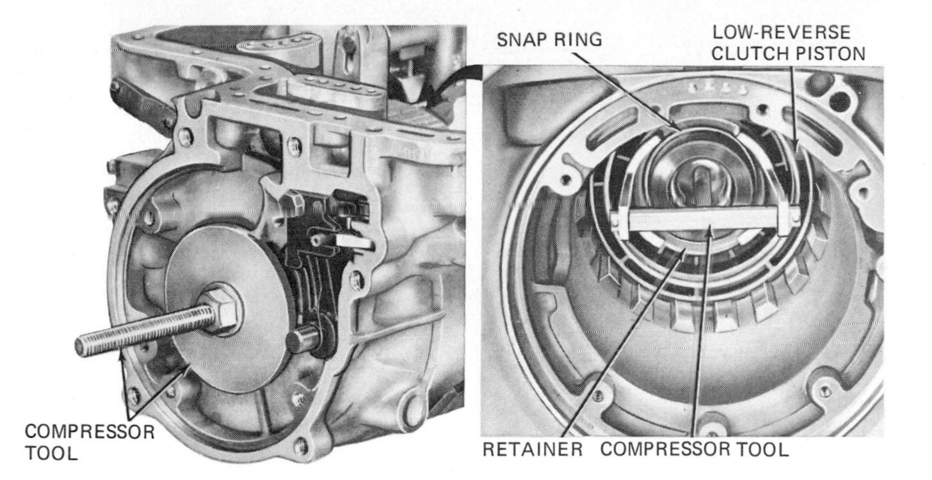

Fig. 20-28. Compressing the low-reverse-clutch springs. (*Ford Motor Company*)

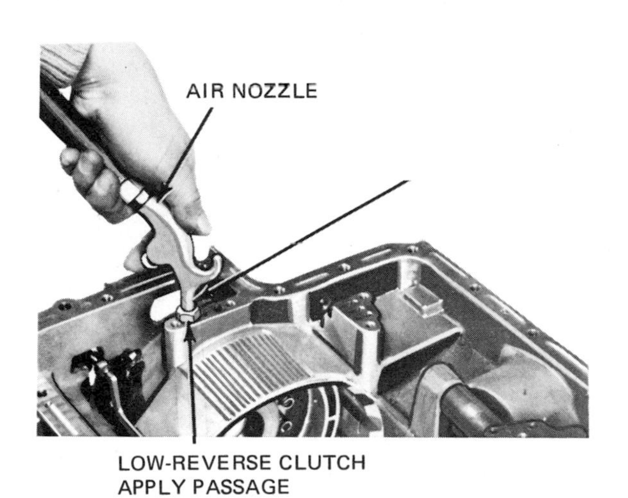

Fig. 20-29. Removing the low-reverse-clutch piston with compressed air. (*Ford Motor Company*)

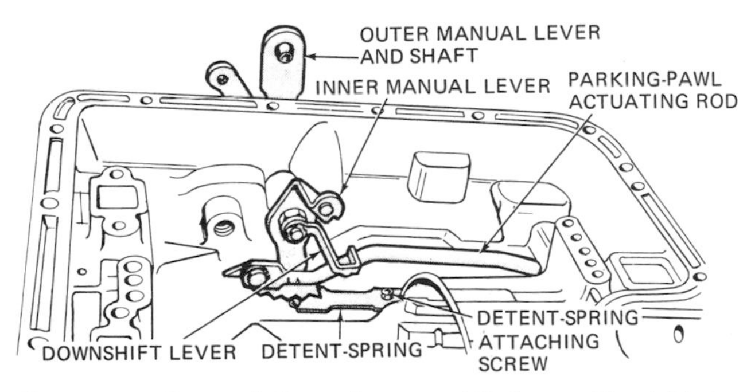

Fig. 20-30. Downshift and manual linkage. (*Ford Motor Company*)

[3.175 mm] hole in the cupped plug and pull it out with a wire hook. Then, after the spring is lifted off the park-plate pin, thread a ¼-20 screw into the shaft and pull it out. You will need a new cupped plug to retain the shaft when you reassemble the parts.

⊘ 20-20 Servo Apply Lever If the servo-apply lever requires replacement, drive the lever shaft from inside the case to drive out the cup plug. Then the shaft can be pulled out. Use a new plug on reinstallation, and coat it with a special sealant, such as Loctite.

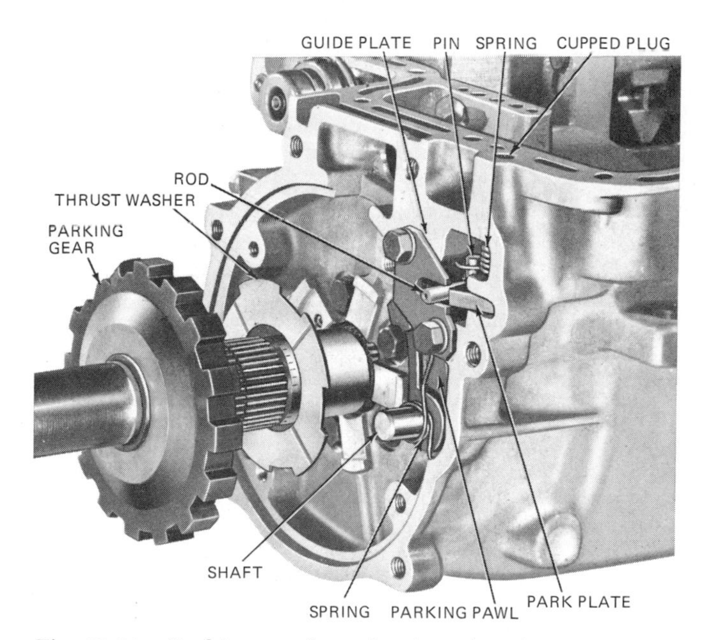

Fig. 20-31. Parking-pawl mechanism. (*Ford Motor Company*)

⊘ 20-21 Valve Body Figure 20-32 shows the upper and lower valve bodies separated. Figure 20-33 shows the upper valve body disassembled. Follow these illustrations to disassemble and reassemble the valve body. To remove the manual valve, you will need a special tool to depress the manual-valve detent spring, as shown in Fig. 20-34. Note, also, the locations of the relief balls and springs, as shown in Fig. 20-35.

⊘ 20-22 Pump Figure 20-36 shows the pump in disassembled view. To disassemble it, remove the selective thrust washer and the two seal rings, the large square-cut seal ring, and the five bolts that

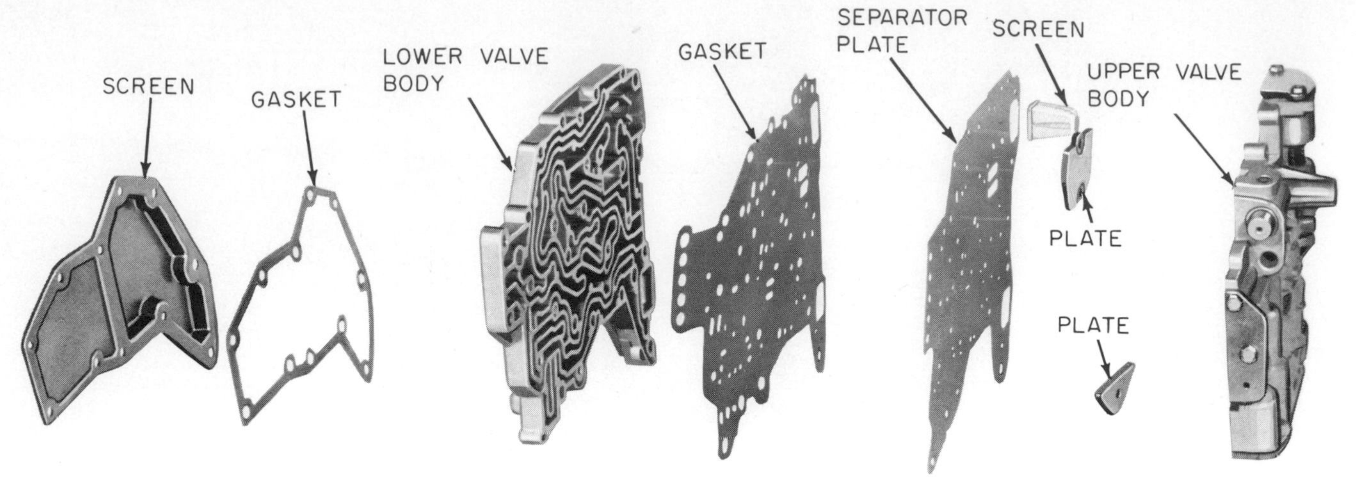

Fig. 20-32. Upper and lower valve bodies separated. (*Ford Motor Company*)

secure the pump support to the pump housing. Now, remove the support and the drive and driven gears from the housing.

When reassembling the pump, make sure that the gears are installed with the identification marks facing the front of the pump housing. Also, make sure that the two locking seal rings on the support are locked. Use new seal rings on reinstallation.

Finally, make sure that the correct thickness of selective thrust washer is used to produce the correct gear-train end play. If the end play is not within the specifications given in the car shop manual, select a thrust washer of a different thickness to give the correct end play.

⊘ 20-23 Reverse-and-High Clutch Separate the drive train into its major component parts, as shown in Fig. 20-23. Then remove the pressure-plate snap ring, as shown in Fig. 20-37, so that the internal parts can be taken out of the clutch drum, as shown in Fig. 20-38. Next, use the special compressor tool, as shown in Fig. 20-39, to compress the springs so that the snap ring can be removed. Then, the tool can be removed along with the spring retainer and springs. Compressed air can then be used to remove the piston, as shown in Fig. 20-40. The piston seals, shown in Fig. 20-38, should be removed from the piston and drum.

If the front bushing is worn, it can be removed by cutting along the bushing seam with a cape chisel. The rear bushing is removed with a removal tool. New bushings can then be pressed into place.

To reassemble the reverse-and-high clutch, dip new seals in transmission fluid and install one on the piston and one in the drum. Put the piston in the drum. Put the piston springs in the piston sockets and place the spring retainer on the springs. Compress the springs with the special tool, as shown in Fig. 20-39, and install the snap ring. Make sure the snap ring is inside the four snap-ring guides on the spring retainer.

Install the clutch plates, starting with a steel drive plate and alternating with the composition plates. Soak the composition plates in transmission fluid for 15 minutes before installing them. Then install the pressure plate and secure it with the pressure-plate snap ring.

Check the clearance between the pressure plate and snap ring with a feeler gauge, as shown in Fig. 20-41. If the clearance is not within specifications, replace the snap ring with another snap ring of the correct thickness to produce the proper clearance. Selective snap rings of three thicknesses are available.

⊘ 20-24 Forward Clutch The forward clutch is shown in disassembled view in Fig. 20-42. It is disassembled in the same way as the reverse-and-high clutch. The snap ring is removed so that the pressure plate and disks can be removed. This clutch also has a forward pressure plate and disk spring, as shown. After these parts are removed, the disk spring must be compressed by a compressor so that the snap ring and spring can be removed. Then the piston can be removed by air pressure applied to the passage in the cylinder.

On reassembly, clearance must be checked between the snap ring and the pressure plate, just as in the forward clutch. Also, there are snap rings of different thicknesses so that the correct one can be installed to provide the proper clearance.

⊘ 20-25 Input Shell and Sun Gear The sun gear is held in place in the input shell by two snap rings and a thrust washer, as shown in Fig. 20-43. Removal of the snap rings permits separation of the parts.

⊘ 20-26 One-way Clutch The one-way clutch is shown disassembled and separated from the low-and-reverse-clutch hub in Fig. 20-44. This clutch is disassembled by removing the snap rings. On

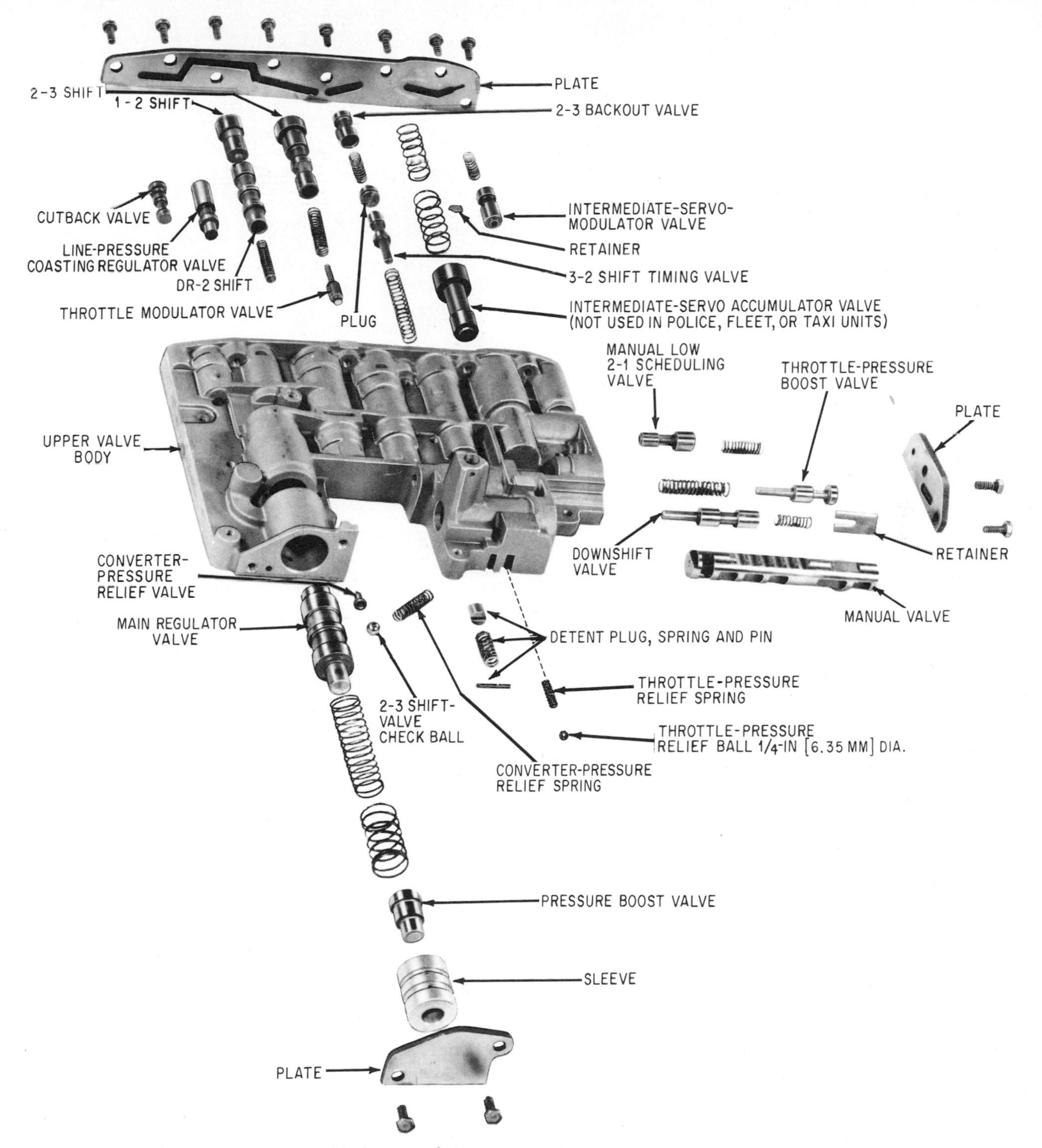

Fig. 20-33. Disassembled view of the upper valve body. (*Ford Motor Company*)

reassembly, first install a snap ring in the forward-snap-ring groove of the clutch hub. Then put the hub forward end down, as shown in Fig. 20-45. Put the forward-clutch bushing into place against the snap ring with the flat side up. Install the spring retainer on top of the bushing. Be sure to install it in the hub so that the springs load the rollers in a counterclockwise direction when one is looking down at the unit. See Fig. 20-45. Install a spring and roller into each of the spring-retainer compartments by slightly compressing each spring and positioning the roller between the spring and spring retainer. Install the rear bushing, flat side down, and install the snap ring to secure the assembly.

⊘ **20-27 Output Shaft** The output shaft contains the governor and governor distributor. The governor is attached by bolts, and the distributor is held on

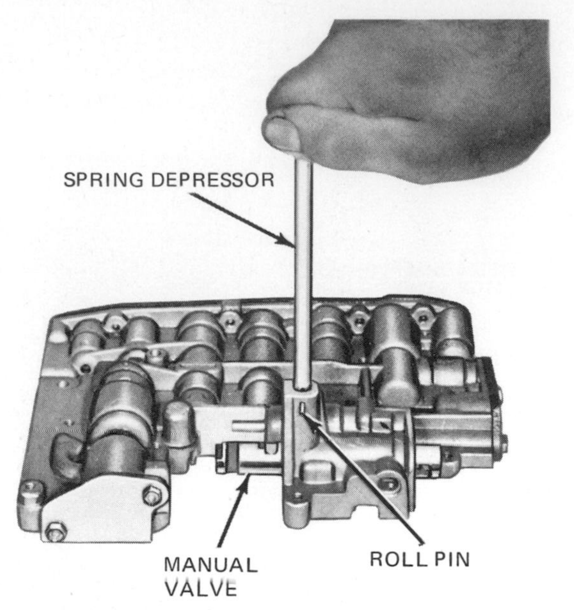

Fig. 20-34. Using a special tool to remove the manual valve. (*Ford Motor Company*)

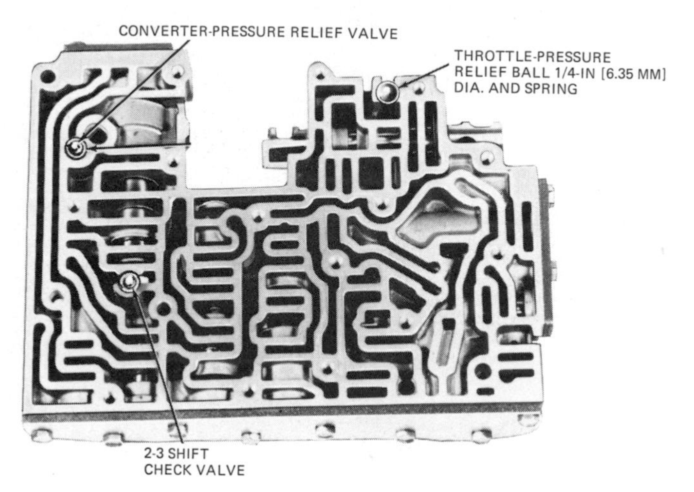

Fig. 20-35. Location of check balls in the valve body. (*Ford Motor Company*)

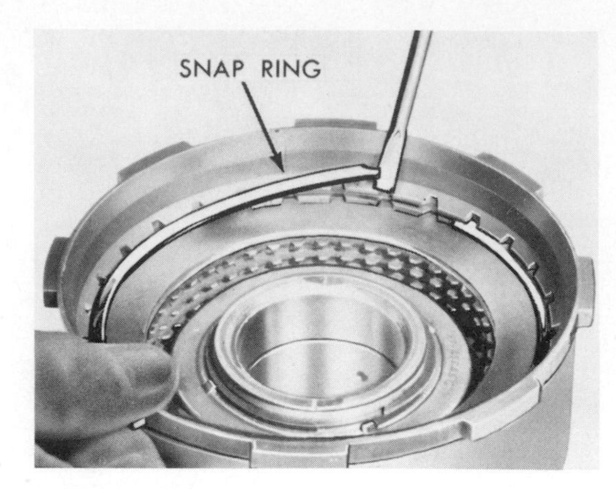

Fig. 20-37. Removing the reverse-high-clutch pressure-plate snap ring. (*Ford Motor Company*)

the shaft by snap rings. Remove bolts and snap rings to remove these parts.

⊘ 20-28 Converter Checks The converter is a welded assembly and cannot be disassembled for service. If it is defective, it must be replaced as a unit. It can be checked for leakage and for end play and stator-clutch action. To check for leakage, the tools shown in Fig. 20-46 are required. With these tools, the converter can be sealed. Then, 20 psi [1.406 kg/cm^2] of air can be applied to the converter through the special tire valve. Finally, the converter can be placed in a tank of water to check it for air leaks.

The end-play check requires a special tool that is inserted into the pump-drive hub and then lifted so that a dial indicator will measure the amount of end play. The stator-clutch check requires another special tool that permits application of torque to determine whether or not the clutch can hold against reverse rotation.

Any defect found in the converter requires converter replacement.

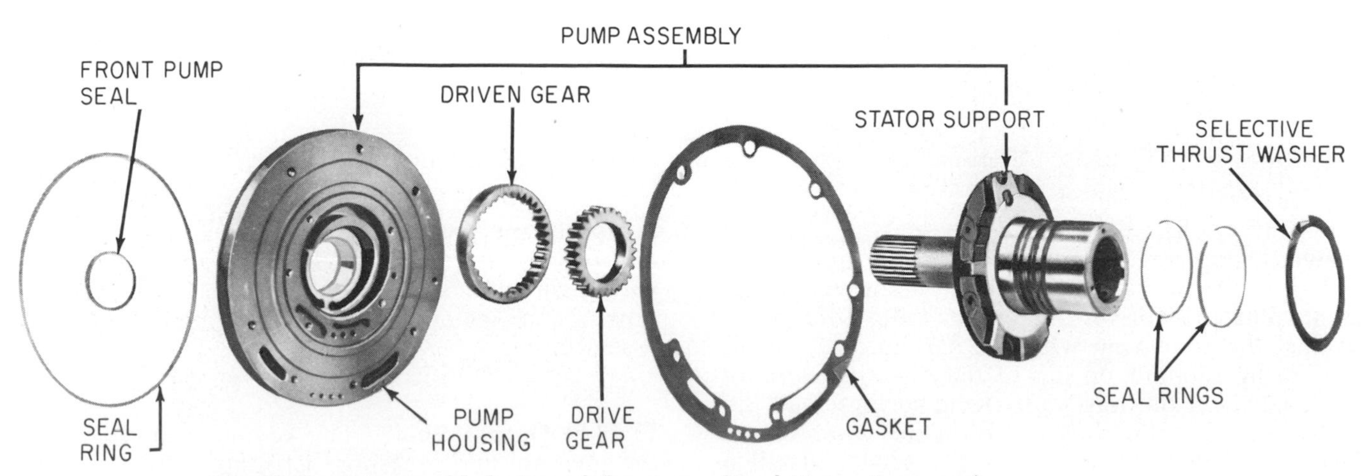

Fig. 20-36. Disassembled view of the pump. (*Ford Motor Company*)

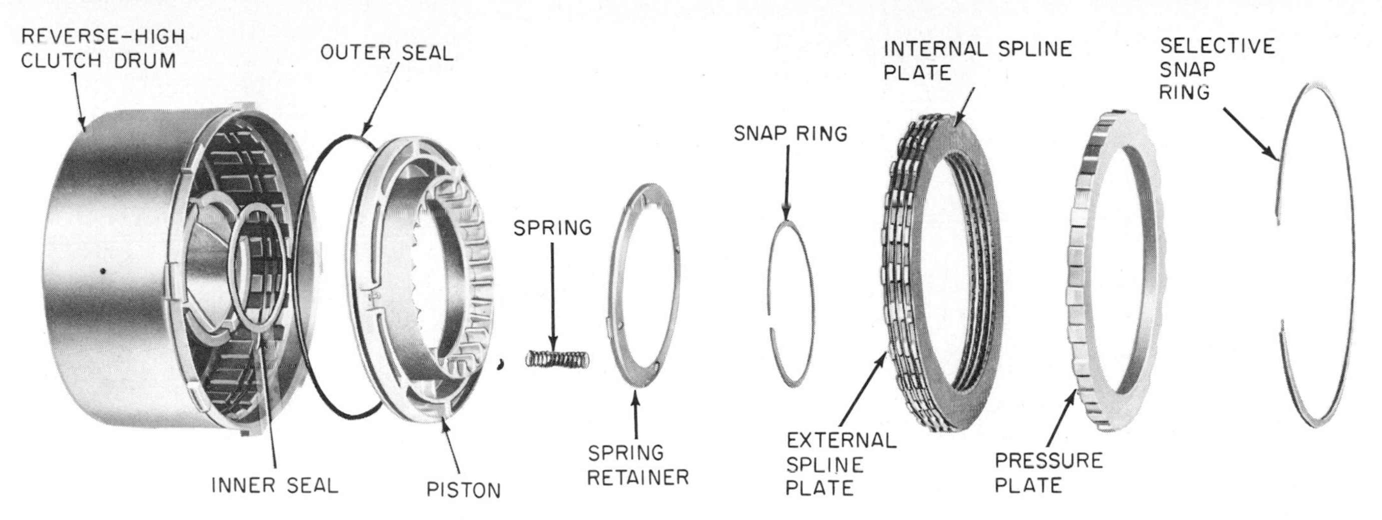

Fig. 20-38. Disassembled view of the reverse-high clutch. (*Ford Motor Company*)

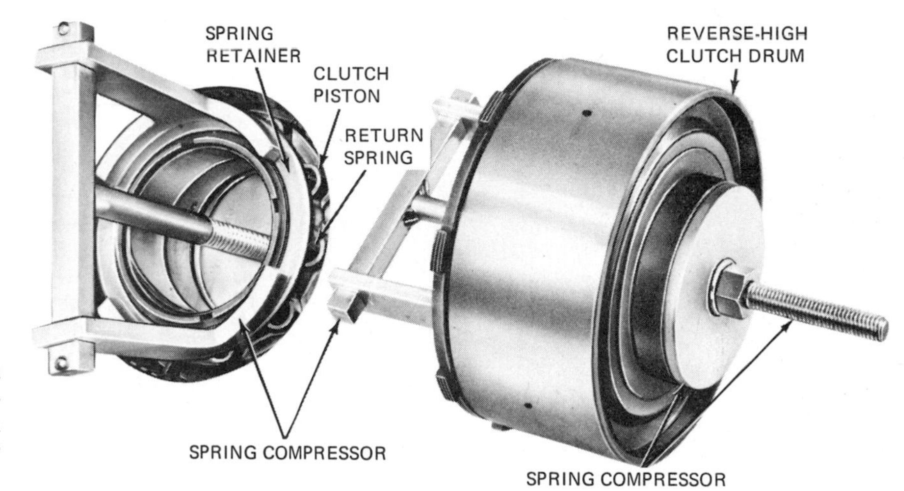

Fig. 20-39. Using a spring compressor to remove the clutch-piston snap ring. (*Ford Motor Company*)

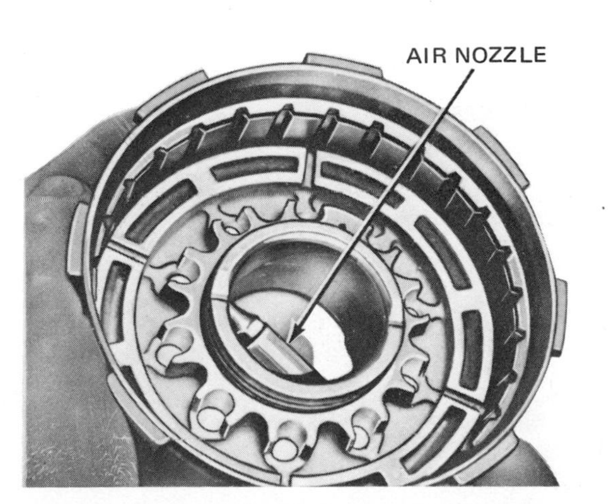

Fig. 20-40. Removing the clutch piston with air pressure. (*Ford Motor Company*)

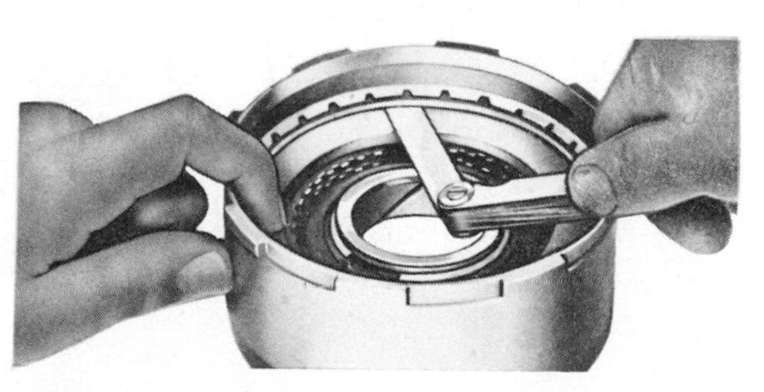

Fig. 20-41. Checking clearance between the snap ring and pressure plate with a feeler gauge. (*Ford Motor Company*)

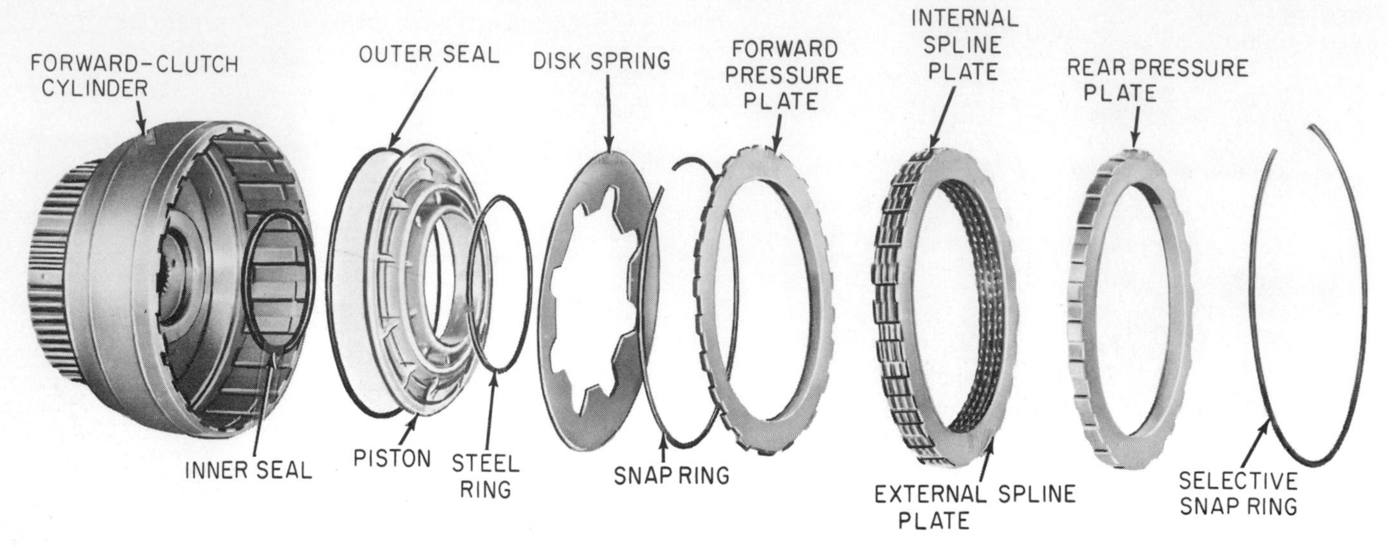

Fig. 20-42 Disassembled view of the forward clutch. (*Ford Motor Company*)

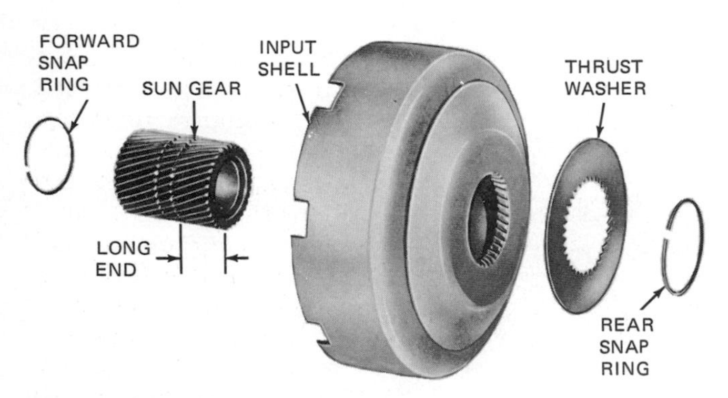

Fig. 20-43. Sun gear separated from the input shell. (*Ford Motor Company*)

⊘ 20-29 Transmission Reassembly

On reassembly, lubricate the parts with transmission fluid. Use petroleum jelly to hold washers in place during reassembly. Proceed as follows, after the case has been installed in a repair fixture:

1. Position the low-and-reverse-clutch piston so that the check ball is at 6 o'clock—at the bottom of the case—and tap it into place with a clean rubber hammer.
2. Hold the one-way-clutch inner race in position and install and torque the attaching bolts to specifications.
3. Install the low-and-reverse-clutch return springs

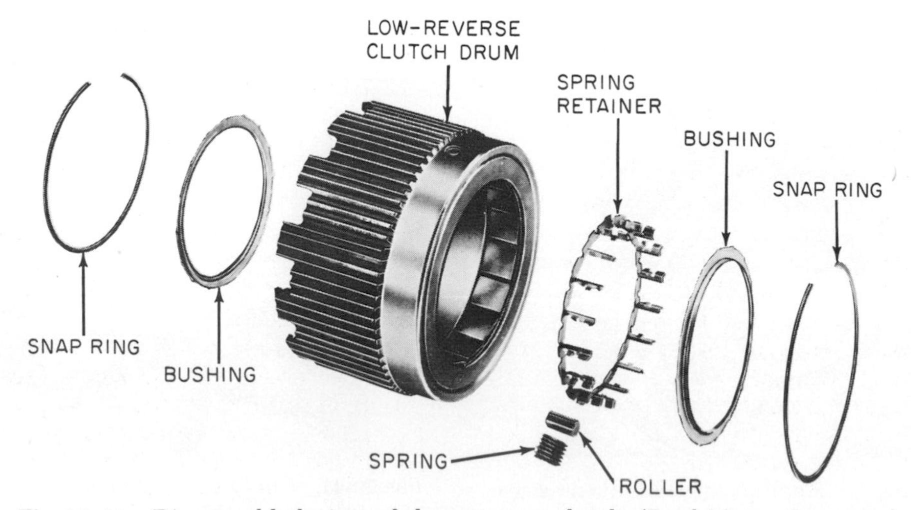

Fig. 20-44. Disassembled view of the one-way clutch. (*Ford Motor Company*)

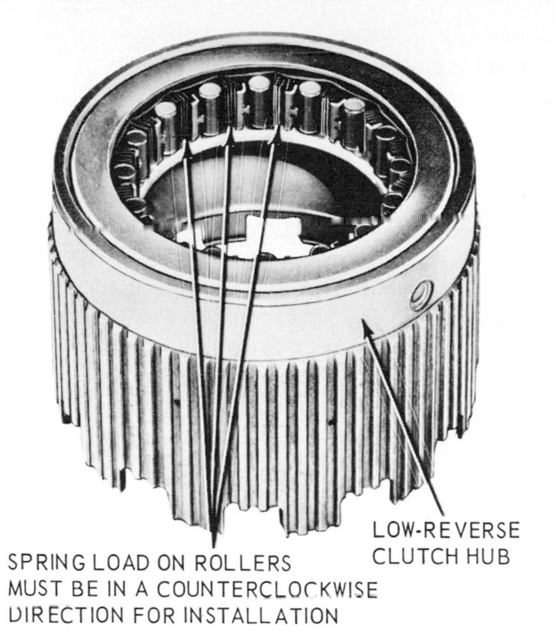

Fig. 20-45. Correct installed positions of the rollers and springs in the one-way clutch. (*Ford Motor Company*)

into the pocket in the piston. Press the springs firmly into place so that they will not fall out.

4. Put the spring retainer over the springs and put the snap ring above it. Use a compressor tool to compress the springs and install the snap ring.

5. Put the case on the bench, front end down.

6. Put the parking-gear thrust washer and gear on the case. Do not stake the washer.

7. Position the oil distributor and tubes on the rear of the case, and torque the attaching bolts to specifications.

8. Install the output shaft and governor.

9. Put a new gasket on the rear of the case and install the extension housing, torquing the bolts to specifications.

10. Put the case back in the repair fixture.

11. Align the low-and-reverse-clutch hub and one-way clutch with the inner race at the rear of the case. Rotate the low-and-reverse-clutch hub clockwise while applying pressure to seat it on the inner race.

12. Install the low-and-reverse-clutch plates, starting with a steel plate and following alternately with composition and steel plates. Retain them with petroleum jelly. New composition plates should be soaked in transmission fluid for 15 minutes before installation. Test the operation of the clutch with compressed air.

13. Install the reverse-planet-ring-gear thrust washer and the ring-gear-and-hub assembly. Insert the snap ring in the groove in the output shaft.

14. Assemble the front and rear thrust washers onto the reverse-planet assembly. Retain them with petroleum jelly. Insert the assembly into the ring gear and secure with the snap ring.

15. Set the reverse-and-high clutch on the bench, front end facing down. Install the thrust washer on the rear end, retaining it with petroleum jelly. Insert

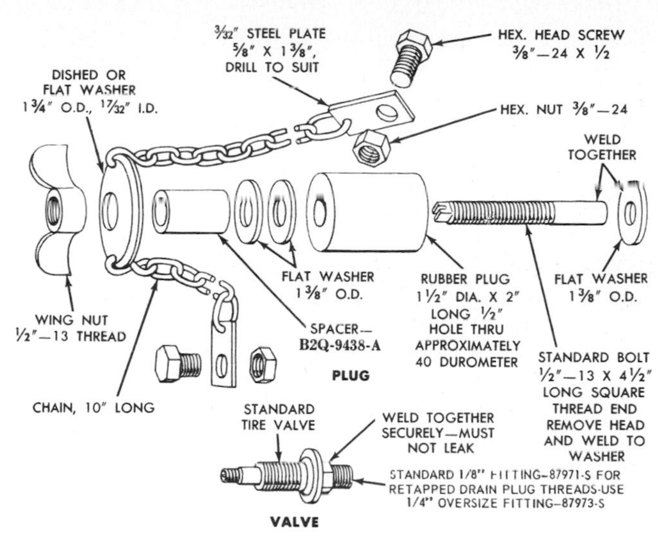

Fig. 20-46. Special tools needed to check the converter for leakage. (*Ford Motor Company*)

the splined end of the forward clutch into the open end of the reverse-and-high clutch so that the splines engage the direct-clutch plates.

16. On the front end of the forward planet ring gear and hub, install the thrust washer and retain it with petroleum jelly. Insert the assembly into the ring gear. Install the input-shell-and-sun-gear assembly.

17. Install the reverse-and-high clutch, forward clutch, forward-planet assembly, and input shell and sun gear as an assembly into the transmission case.

18. Insert the intermediate band into the case around the direct-clutch drum. Install the struts and tighten the band adjusting screw enough to hold the band.

19. Place a selective-thickness washer of the proper thickness as determined by the end-play check during disassembly (see Fig. 20-24) on the shoulder of the stator support. This is the selective thrust washer referred to in ⊘ 20-22 and shown in Fig. 20-36. Retain the washer with petroleum jelly.

20. Lay a new gasket on the rear mounting face of the pump and bring the pump into place. Be careful not to damage the large seal on the outside diameter of the pump housing. Install six of the seven pump mounting screws and torque them to specifications.

21. Adjust the intermediate band, as already noted, and install the input shaft with the long splined end inserted into the forward-clutch assembly.

22. Install the special tool with the dial indicator in the seventh pump bolt hole and recheck the end play. If the end play is correct, remove the tool and install the seventh bolt. If it is incorrect, remove the pump and install a selective thrust washer of the correct thickness.

23. Install the control-valve body in the case, making sure the levers engage the valves properly. Install the primary throttle valve, rod, and vacuum unit.

24. Install a new oil-pan gasket and pan.

25. Install the converter and install the transmission in the car, as previously explained. Attach and adjust linkages.

26. Add transmission fluid, as previously noted, to bring the fluid up to the correct level.
27. Start the engine and check the operation of the transmission in all selector-lever positions.

CHAPTER 20 CHECKUP

NOTE: Since the following is a chapter review test, you should review the chapter before taking it.

Well, you have made it! You have completed all but one of the chapters in the book on automatic transmissions. If you are like most students, you found the earlier chapters rather hard going. But once you got into the subject you found that the later chapters were a little easier because you had learned the fundamentals that apply to all transmissions. Thus, to remove clutch pistons, you need a spring compressor to take out the snap ring. Valves and valve bodies have to be handled in the same careful way regardless of what transmission they come from. And so it is with all transmission parts. There are differences, of course, in the way transmissions are tested and adjusted, and different overhaul procedures are used. Yet they are all much alike. Now test your memory on how well you remember the service details for the C6 transmission by taking the test that follows.

Completing the Sentences The sentences that follow are incomplete. After each sentence there are several words or phrases, only one of which will correctly complete the sentence. Write each sentence in your notebook, selecting the proper word or phrase to complete it correctly.

1. For normal service, the recommendation on oil changes is: (a) change every 12,000 miles [19,312 km], (b) change every 24,000 miles [38,624 km], (c) do not change oil.
2. The band adjustment can be made: (a) on the car, (b) only after the transmission is off the car, (c) only after the valve body is removed.
3. When using the Ford transmission tester to check shift points in the shop, you need: (a) a dynamometer, (b) a source of vacuum, (c) to disconnect the drive shaft.
4. When using the Ford transmission tester in the shop to check shift points, you simulate a high engine load or wide-open throttle by: (a) opening the carburetor throttle, (b) reducing the vacuum to 0 to 2 in [0 to 50.8 mm], (c) opening the kickdown valve.
5. To remove a clutch piston, you must first compress the clutch springs so that you can: (a) remove the clutch plates, (b) loosen the screws, (c) remove the snap ring.
6. Gear-train end play is adjusted by: (a) changing a snap ring, (b) installing different pinion carriers, (c) changing the selective thrust washer.
7. If the transmission slips on the stall-speed test: (a) release the throttle instantly, (b) open the throttle wide to see how high speed will go, (c) shift to D and hold the throttle half open.
8. The pressure checks are made with the engine: (a) running at about 2,500 rpm, (b) idling, (c) not running.

Troubles and Service Procedures In the following, you should write in your notebook the required trouble causes and service procedures. Do not copy from the book; try to write in your own words. Writing the trouble causes and service procedures in your own words will help you remember them better; you will thereby be greatly benefited when you go into the automotive shop.

1. Explain how to make the stall-speed test.
2. Explain how to make the control-pressure test.
3. How do you simulate a wide-open throttle when you make the control-pressure test?
4. What are the cautions to observe when making the stall-speed test?
5. Explain how to check the shift point on kickdown in the shop with the Ford transmission tester.
6. Explain how to check the shift points in the drive range, using the Ford transmission tester.
7. Explain how to change the oil in the transmission.
8. Describe a typical selector-lever-linkage adjustment.
9. Describe a typical throttle-linkage adjustment.
10. Explain how to adjust the band.
11. List those components that can be removed with the transmission in the car.
12. Refer to the diagnosis guide for a specific model of transmission and list the major possible troubles along with their possible causes.
13. Refer to a shop manual covering the C6 transmission and make a list of the transmission parts in the order in which they are removed from the transmission.
14. Refer to a shop manual covering the C6 transmission and prepare a transmission-removal story outlining the step-by-step procedure for removing the transmission from the car.
15. Make a list of the precautions to take when overhauling a transmission.
16. Make a list of the procedures required to clean and inspect transmission parts.

SUGGESTIONS FOR FURTHER STUDY

Now that you have finished all but one of the chapters on automatic transmissions, you are equipped with the theory and fundamentals you need to perform actual service operations on automatics. As we have mentioned previously, you should get all the practical experience possible, too, while studying the book. This combination—learning the fundamentals and working on the job—will make you the transmission expert you want to become. Best of luck to you!

chapter 21

OTHER AUTOMATIC TRANSMISSIONS

This chapter covers the features of other automatic transmissions, which are interesting but not as widely used as those covered in previous chapters. These include the Powerglide for the rear-engine Chevrolet Corvair, the Turbo Hydra-Matic for the front-drive Oldsmobile Toronado and Cadillac Eldorado, the Toyoglide for the Toyota, and the Volkswagen automatic transmission.

⊘ 21-1 Powerglide for Rear Engine Figure 21-1 shows a Powerglide automatic transmission for the Chevrolet rear-engine Corvair. It is very similar in construction and operation to other Powerglide models except that the torque converter and planetary-gear system with the controls have been separated. The differential has been placed between them. Chapter 23 describes differentials.

⊘ 21-2 Turbo Hydra-Matic for Front Drive Figure 21-2 shows the Turbo Hydra-Matic automatic transmission for the front-drive Cadillac and Oldsmobile. It is essentially a Type 400 Turbo Hydra-Matic automatic transmission which has been cut in two just behind the torque converter so that the planetary-gear system can be placed alongside the torque converter, as shown in Fig. 21-3. A pair of sprockets and a drive chain connect the torque converter with the automatic transmission, as shown in Fig. 21-4.

The transmission and torque converter operate essentially in the same manner as the Type 400 Turbo Hydra-Matic automatic transmission described in Chap. 15. The power-flow diagrams, however, do look different because of the side-by-side location of the torque converter and gear system. Figures 21-5 to 21-7 show the power flow under different operating conditions. The end views of the front and rear gearsets, shown in the upper left of the figures, indicate the directions the components of the gearsets rotate under the different operating conditions.

⊘ 21-3 Toyoglide The Toyoglide automatic transmission, used on Toyota cars, is a three-speed unit with a three-member torque converter and compound planetary-gear system controlled by two multiple-disk clutches, two brake bands, and an overrunning clutch. Figure 21-8 is a sectional view of the transmission, and Fig. 21-9 is a partial cutaway view of the planetary-gear system with the controlling clutches and brake bands.

Figure 21-10 is a table showing the conditions in the transmission in the different selector-lever positions. Figures 21-11 to 21-14 show the power flow under different operating conditions. Plate 34 illustrates the hydraulic circuit in the Toyoglide when in direct drive.

⊘ 21-4 Volkswagen Automatic Stick-Shift Transmission This transmission is automatic in the sense that the clutch is operated automatically when the shift lever—the "stick"—is moved from one gear position to another. There is no hydraulic control system to produce shifts from one gear ratio to another, and so all shifting is manual. Figure 21-15 shows the complete transmission. It includes a three-member torque converter similar to others we have discussed in previous chapters. In addition, it has a clutch of the diaphragm-spring, tapering-finger type described in Chap. 2. This clutch is operated automatically by a vacuum-actuated servo cylinder. The gearing in the transmission provides three forward speeds and reverse. The transmission is closely coupled to the differential, as shown in Fig. 21-15.

The gear shifting is similar to gear shifting in other manual transmissions described in previous chapters. Figure 21-16 shows the gearshift lever and five positions: N, L, R, 1, and 2. N is neutral. L is low-speed range, for driving up hills or when the car is heavily loaded. This range gives speeds of 0 to about 34 mph [54.72 km/h]. 1 is for moving off and accelerating and is for use in normal traffic and for rapid acceleration. 2 is for use for normal acceleration and high speed. R is, of course, for reverse.

There is a set of contact points in the lower end of the shift lever which close when the shift lever is moved. This action connects a solenoid in the control valve to the battery. The solenoid is thereby

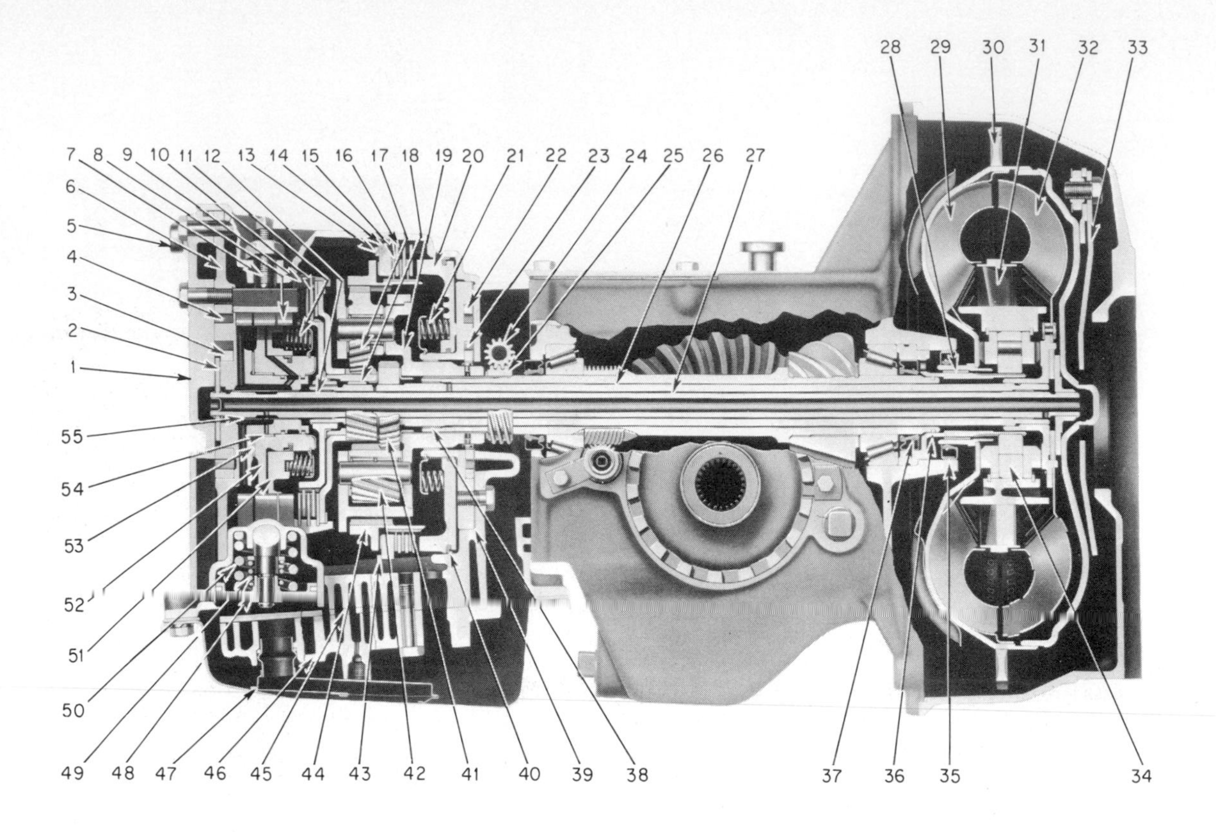

1. Front-pump cover
2. Front-pump-shaft drive hub
3. Front-pump drive gear
4. Front-pump driven gear
5. Transmission vent
6. Front-pump body
7. Low-band adjusting screw and locknut
8. Low band
9. Clutch-drum reaction plate (three used)
10. Clutch-drum faced plate (two used)
11. Clutch piston return spring (15 used)
12. Turbine shaft
13. Reverse-clutch retaining-ring clip
14. Reverse-clutch front reaction plate (thick)
15. Reverse-clutch faced plate (three used)
16. Reverse-clutch reaction plate (three used)
17. Short pinion
18. Low-sun-gear bushing
19. Planet-carrier hub (transmission output)
20. Reverse piston
21. Reverse-piston return spring (17 used)
22. Rear-pump driven gear
23. Rear-pump drive gear
24. Governor drive gear
25. Governor drive gear
26. Turbine shaft
27. Front-pump shaft
28. Converter-hub bushing.
29. Converter pump
30. Starter gear
31. Stator
32. Turbine
33. Engine flex plate
34. Stator cam race
35. Converter-hub seal
36. Stator shaft
37. Pinion-shaft rear oil seal
38. Pinion-shaft bushing
39. Rear-pump wear plate
40. Reverse-piston outer seal
41. Planet-carrier input sun gear
42. Long pinion gear
43. Reverse-clutch-plate retaining ring
44. Ring gear
45. Valve-body ditch plate
46. Valve body
47. Oil pickup pipe
48. Low-servo piston
49. Low-servo piston cushion spring
50. Low-servo-piston return spring
51. Clutch-drum piston
52. Clutch-drum hub
53. Clutch-drum selective thrust washer
54. Clutch-drum bushing
55. Front-pump body bushing

Fig. 21-1. Powerglide used with rear-engine Corvair. (*Chevrolet Motor Division of General Motors Corporation*)

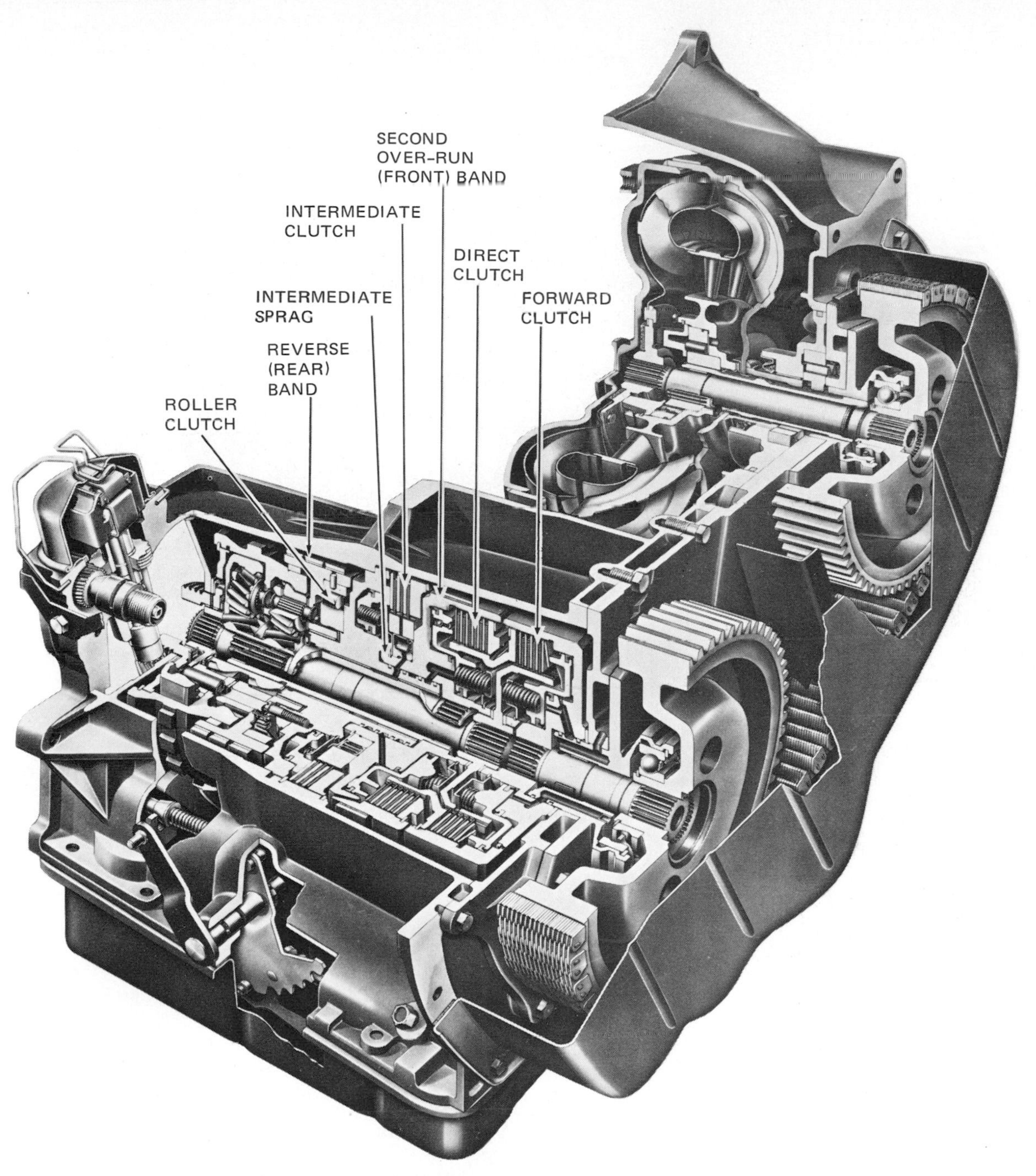

SELECTOR POSITION	PUMP PRESSURE	FORWARD CLUTCH	DIRECT CLUTCH	2ND OVERRUN BAND	INT. CLUTCH	INT. SPRAG	ROLLER CLUTCH	REV. BAND
PARK—NEUT.	60-150	OFF	OFF	OFF	OFF	OFF	OFF	OFF
DRIVE 1 LEFT 2 3	60-150 60-150 60-150	ON ON ON	OFF OFF ON	OFF OFF OFF	OFF ON ON	OFF ON OFF	ON OFF OFF	OFF OFF OFF
DRIVE 1 RIGHT 2	150 150	ON ON	OFF OFF	OFF ON	OFF ON	OFF ON	ON OFF	OFF OFF
LO 1 2	150 150	ON ON	OFF OFF	OFF ON	OFF ON	OFF ON	ON OFF	ON OFF
REV.	95 - 230	OFF	ON	OFF	OFF	OFF	OFF	ON

Fig. 21-2. Turbo Hydra-Matic automatic transmission for a front-drive automobile, partly cut away so that internal construction can be seen. The table shows the internal conditions for different selector positions. (*Cadillac Motor Car Division of General Motors Corporation*)

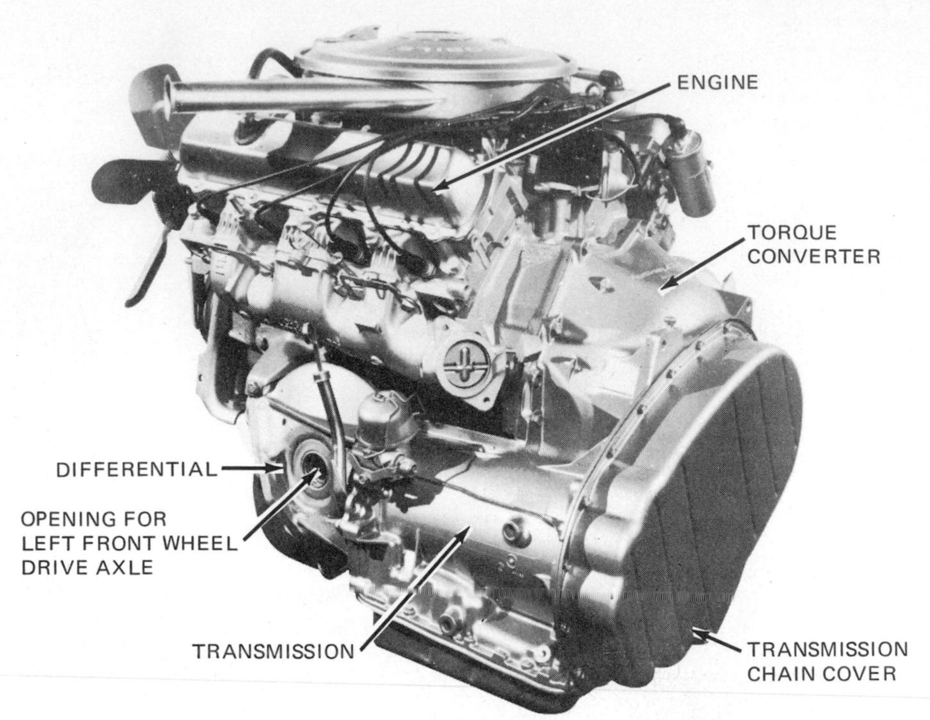

Fig. 21-3. Engine and transmission assembly for a front-wheel-drive car, as seen from the left rear. (*Oldsmobile Division of General Motors Corporation*)

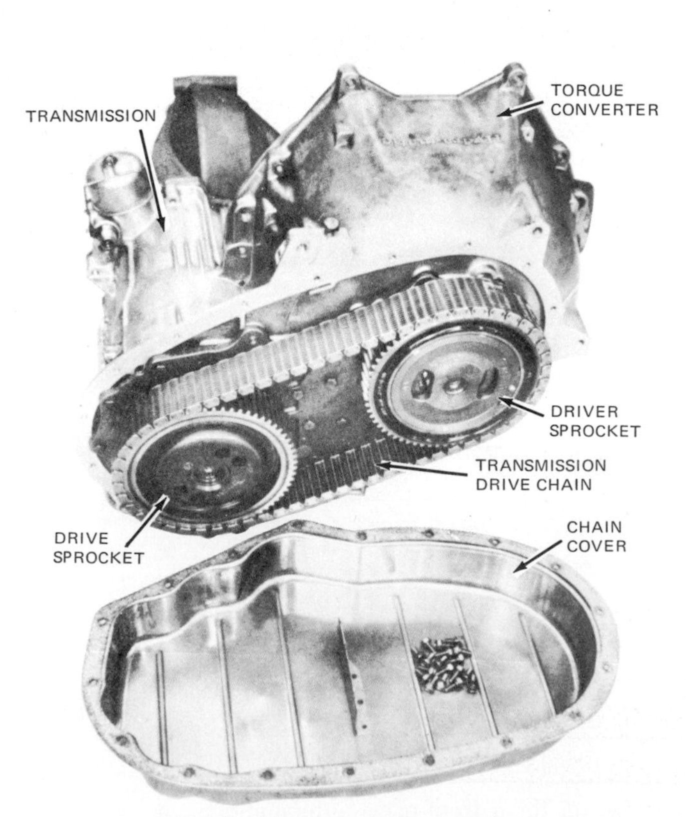

Fig. 21-4. Transmission assembly for a front-wheel-drive car with the transmission-chain cover removed so that the chain and sprockets can be seen. (*Oldsmobile Division of General Motors Corporation*)

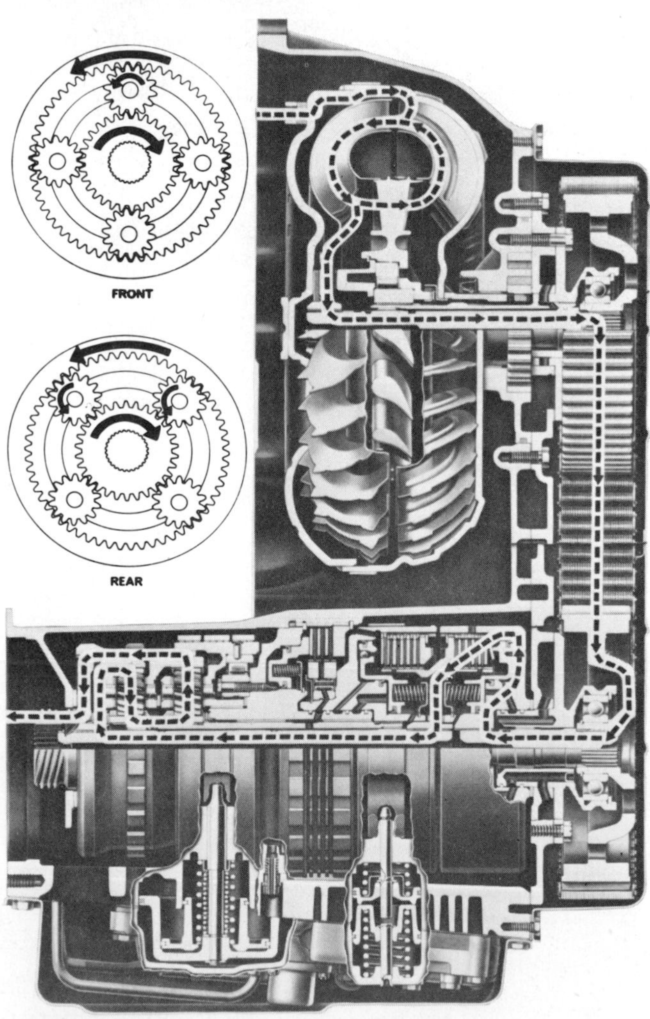

Fig. 21-5. Power flow in drive range, first gear, in the Turbo Hydra-Matic automatic transmission for front-drive cars. (*Cadillac Motor Car Division of General Motors Corporation*)

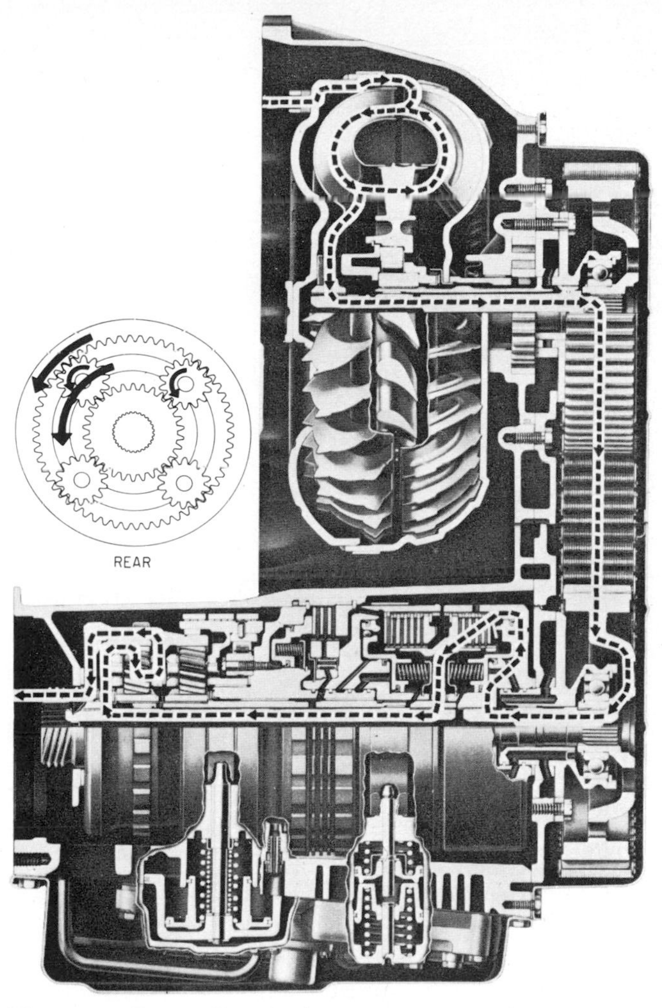

Fig. 21-6. Power flow in drive range, second gear, in the Turbo Hydra-Matic automatic transmission for front-drive cars. (*Cadillac Motor Car Division of General Motors Corporation*)

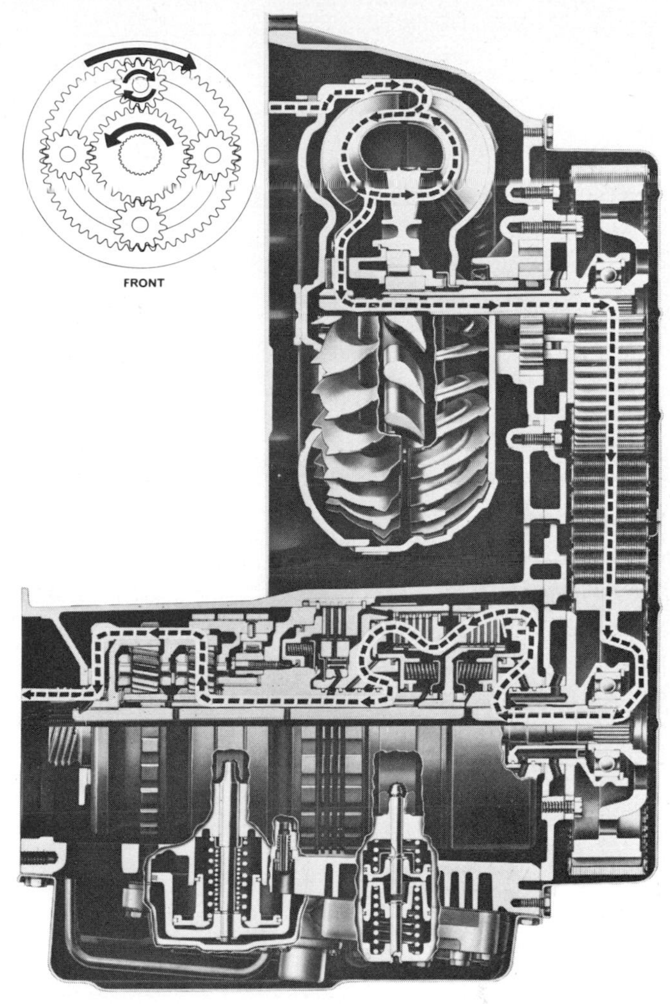

Fig. 21-7. Power flow in reverse in the Turbo Hydra-Matic automatic transmission for front-drive cars. (*Cadillac Motor Car Division of General Motors Corporation*)

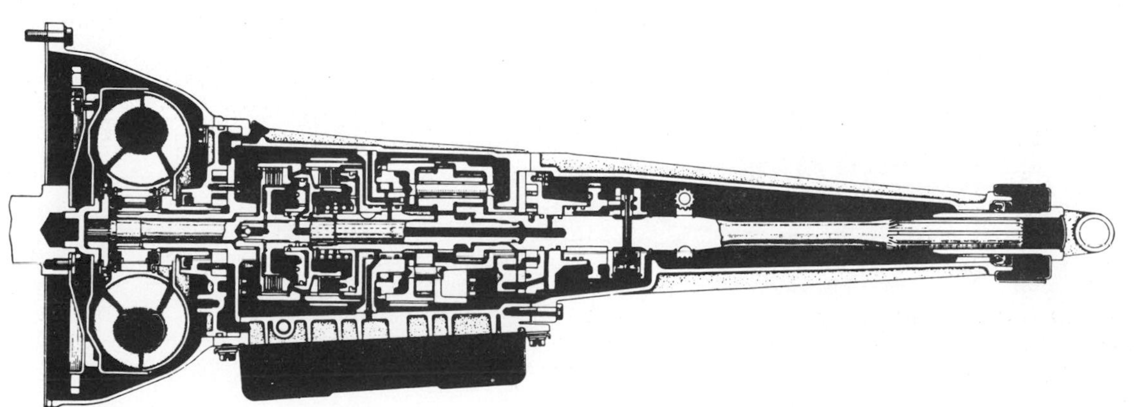

Fig. 21-8. Sectional view of the Toyoglide. (*Toyota Motor Sales Company, Ltd.*)

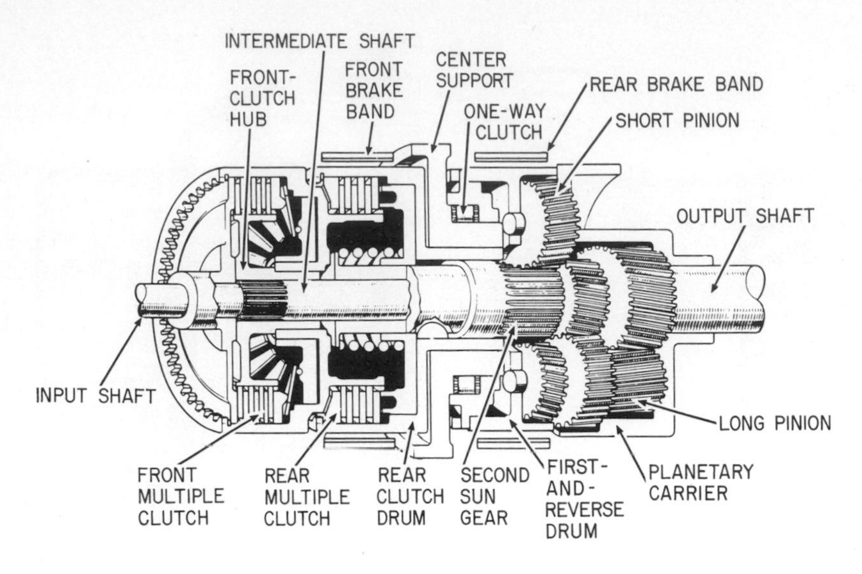

Fig. 21-9. Cutaway view of the planetary-gear system and controls in the Toyoglide. (*Toyota Motor Sales Company, Ltd.*)

SELECTOR-LEVER POSITIONS	GEAR	FRONT CLUTCH	REAR CLUTCH	FRONT BAND	REAR BAND	ONE-WAY CLUTCH	LUBRICATION AND OIL CIRCULATION TO CONVERTER	GEAR RATIO
P	NEUTRAL	X	X	X	O	X	O	......
R	REVERSE	X	O	X	O	X	O	1.920
N	NEUTRAL	X	X	X	X	X	O	......
D OR 2	FIRST	O	X	X	X	HOLDING	O	2.400
D OR 2	SECOND	O	X	O	X	OVER-RUNNING	O	1.479
D	THIRD	O	O	X	X	OVER-RUNNING	O	1.000
L	FIRST	O	X	X	O	X	O	2.400

GEAR RATIOS O = APPLIED X = RELEASED

Fig. 21-10. Table of the conditions in the transmission with the selector lever in different positions. (*Toyota Motor Sales Company, Ltd.*)

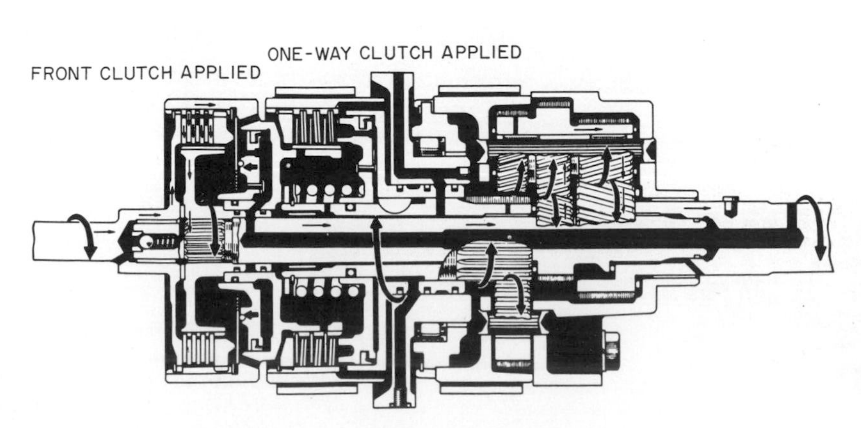

Fig. 21-11. Power flow in the Toyoglide in first gear, drive range. (*Toyota Motor Sales Company, Ltd.*)

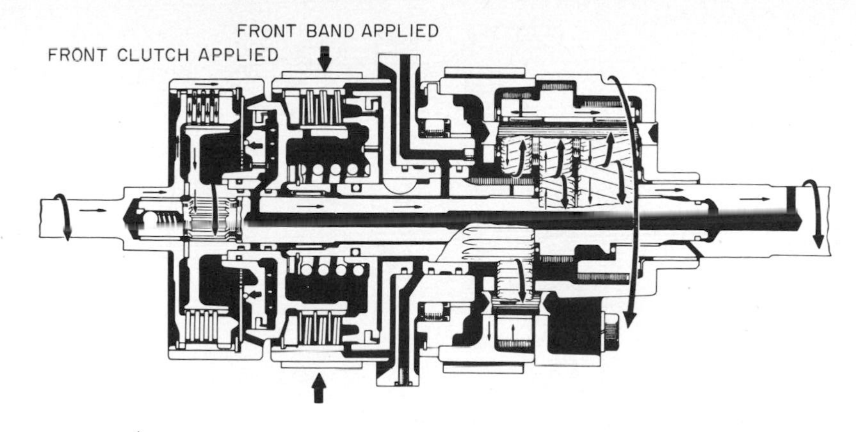

Fig. 21-12. Power flow in the Toyoglide in second gear, drive range. (*Toyota Motor Sales Company, Ltd.*)

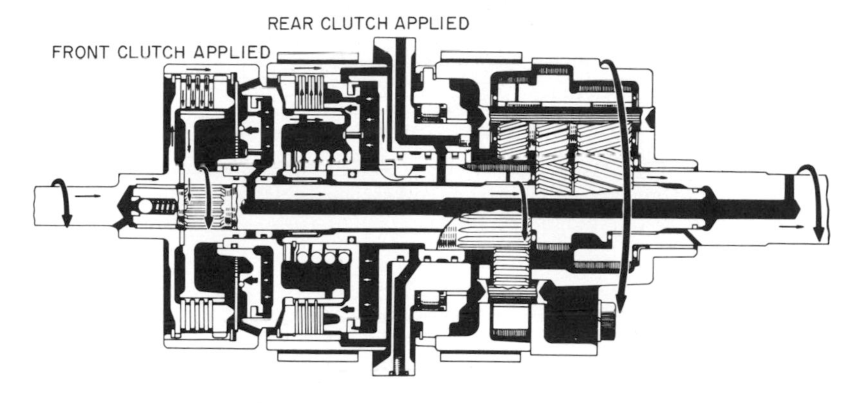

Fig. 21-13. Power flow in the Toyoglide in third gear, drive range. (*Toyota Motor Sales Company, Ltd.*)

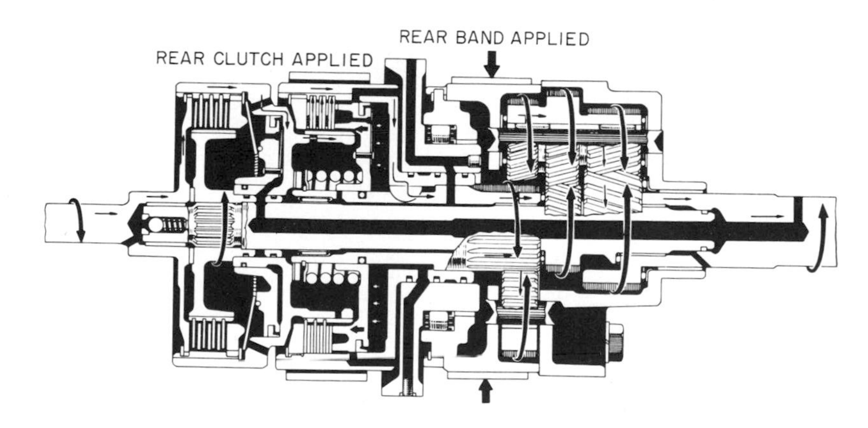

Fig. 21-14. Power flow in the Toyoglide in reverse gear. (*Toyota Motor Sales Company, Ltd.*)

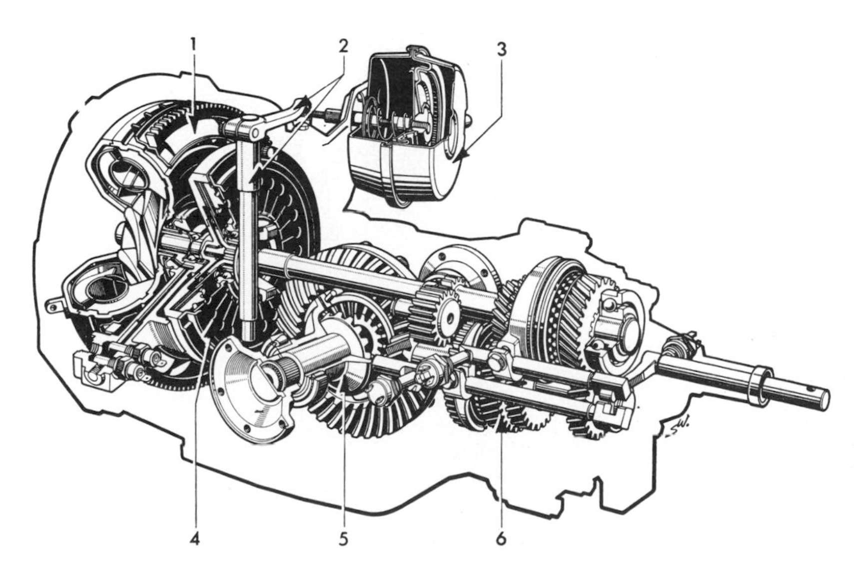

1. Torque converter
2. Clutch shaft
3. Servo cylinder
4. Shift clutch
5. Differential
6. Transmission

Fig. 21-15. Essential operating parts of the Volkswagen automatic stick-shift transmission. (*Volkswagen of America, Inc.*)

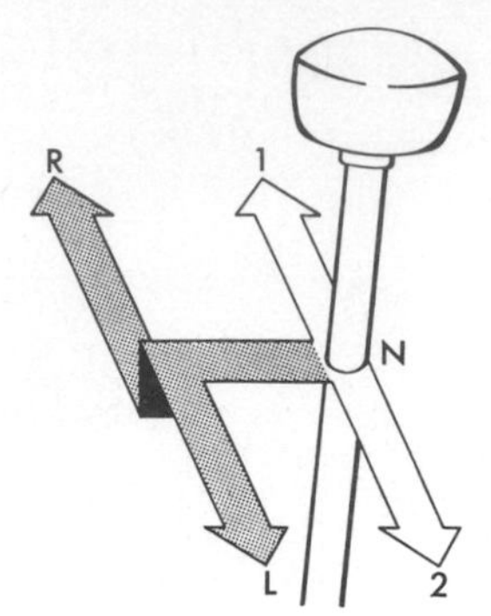

Fig. 21-16. Gear-shifting pattern for the Volkswagen automatic stick-shift transmission. (*Volkswagen of America, Inc.*)

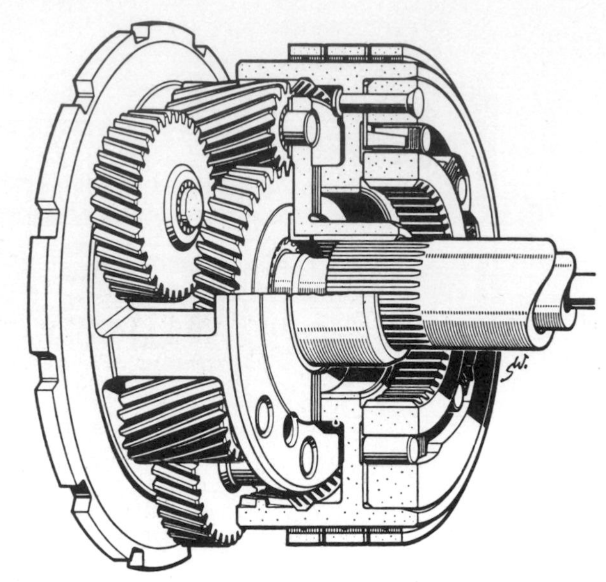

Fig. 21-18. Cutaway view of the compound planetary gearset used in the Volkswagen automatic transmission. (*Volkswagen of America, Inc.*)

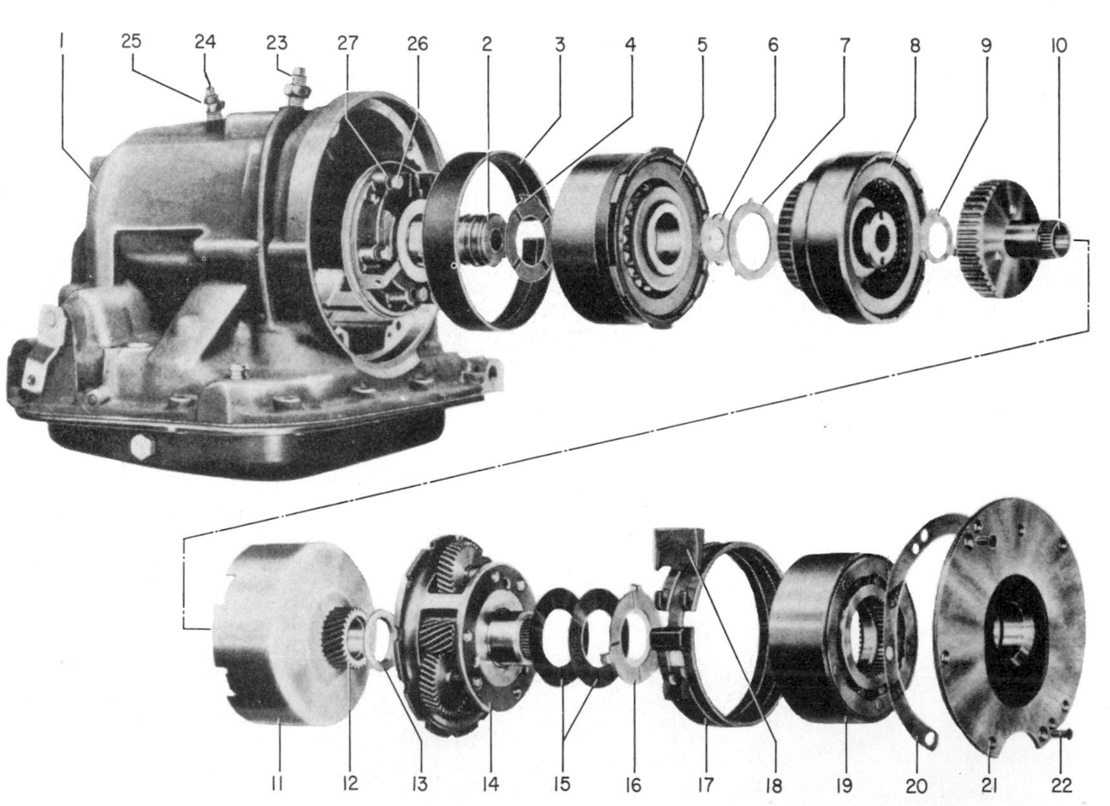

1. Transmission case
2. Fluid pump
3. Second-gear band
4. Thrust washer 1
5. Direct-and-reverse clutch
6. Thrust washer 2
7. Thrust washer 3
8. Forward clutch
9. Thrust washer 4
10. Clutch hub
11. Driving shell
12. Sun gear, small
13. Thrust washer 5
14. Planetary gearset
15. Shim
16. Thrust washer 6
17. First-gear band
18. Support fork
19. Annular gear with one-way clutch
20. Gasket for bearing flange
21. Bearing flange
22. Screw
23. Adjusting screw for first-gear band
24. Adjusting screw for second-gear band
25. Nut for adjusting screw
26. Screw
27. Spring washer B 6

Fig. 21-17. Partial exploded view of the Volkswagen automatic transmission. (*Volkswagen of America, inc.*)

actuated, and it opens the line from the intake manifold to the vacuum-actuated servo cylinder (shown in Fig. 21-15). The vacuum is applied to one side of an airtight diaphragm. The other side of the diaphragm is open to the atmosphere. With the vacuum applied to one side of the diaphragm and atmospheric pressure to the other side, the diaphragm must move. This movement is carried to the throwout bearing in the clutch by linkage and a shaft. The clutch therefore is actuated, and the shift is accomplished smoothly. As soon as the shift lever is released, the contacts separate and the control valve shuts off the line between the servo cylinder and the intake manifold, admitting air to the vacuum side of the diaphragm. Now, with atmospheric pressure on both sides of the diaphragm, the force of a spring

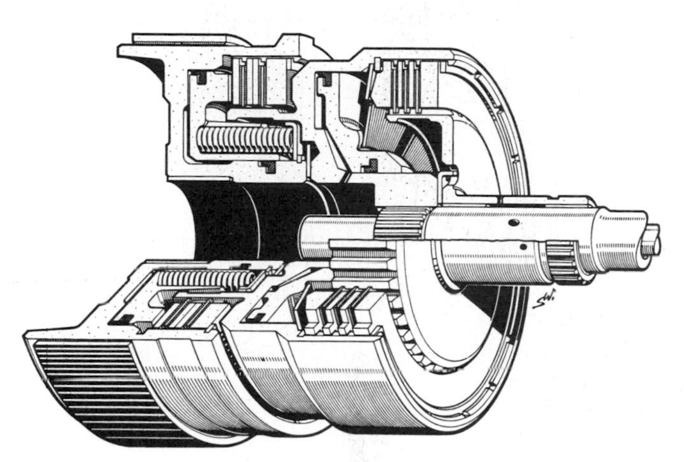

Fig. 21-19. Cutaway view of the forward clutch used in the Volkswagen automatic transmission. (*Volkswagen of America, Inc.*)

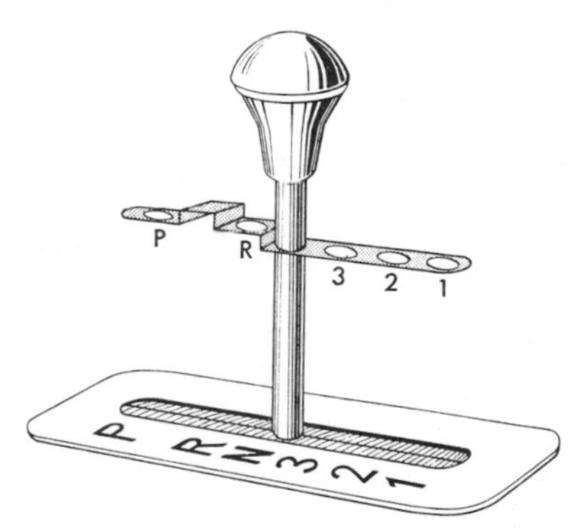

Fig. 21-20. Selector-level positions for the Volkswagen automatic transmission. (*Volkswagen of America, Inc.*)

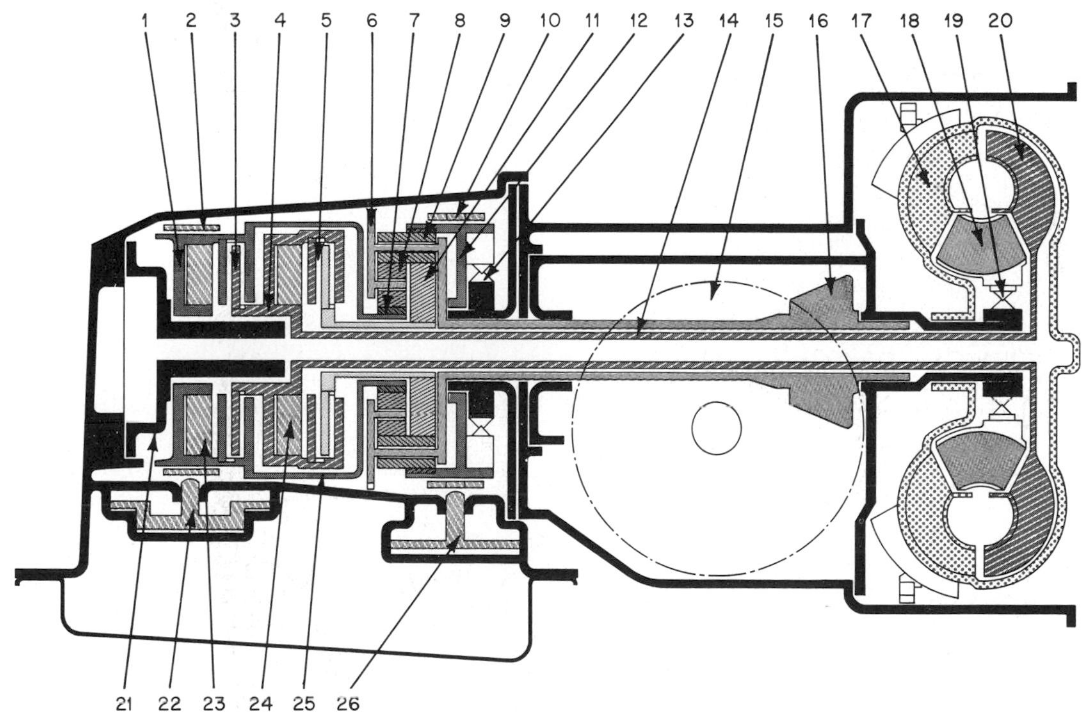

1. Direct-and-reverse-clutch drum
2. Second-gear brake band
3. Direct-and-reverse clutch
4. Forward-clutch drum
5. Forward clutch
6. Planet carrier
7. Small sun gear
8. Large planet pinion
9. Small planet pinion
10. First-gear-and-reverse brake band
11. Large sun gear
12. Annulus gear
13. First-gear one-way clutch
14. Turbine shaft
15. Differential ring gear
16. Drive pinion
17. Impeller
18. Stator
19. One-way clutch for stator
20. Turbine
21. Oil-pump housing
22. Piston for second-gear brake band
23. Piston for direct-and-reverse-gear clutch
24. Piston for forward clutch
25. Driving shell
26. Piston for first-and-reverse-gear brake band

Fig. 21-21. Schematic layout of the essential parts in the Volkswagen automatic transmission. (*Volkswagen of America, Inc.*)

in the servo cylinder moves the diaphragm and allows the clutch to reengage.

The torque converter adds flexibility to the transmission and also smooths the shifts and clutch operation.

⊘ 21-5 Volkswagen Automatic Transmission The Volkswagen automatic transmission provides three forward speeds, automatically selected when the transmission is in the drive range, and reverse. It includes one compound planetary gearset, two multiple-disk clutches, two brake drums, and a one-way, or overrunning, clutch. Figure 21-17 is a partly disassembled view of the transmission, showing all the major components. Figure 21-18 is a cutaway view of the planetary gearset. Figure 21-19 is a cutaway view of the forward clutch. The shifts are controlled, of course, by a hydraulic system similar to those used in other automatic transmissions previously discussed.

Figure 21-20 shows the six selector-lever positions. P for park, R for reverse, and N for neutral are the same as with other transmissions. 3 is the same as D, or drive, in other automatic transmissions and is for all normal driving conditions. 2 is the same as D2 or S in other automatic transmissions and is for intermediate driving conditions. 1 is the same as L, or low, in other automatic transmissions and is used to pull heavy loads or for engine braking when going down a long hill.

Figure 21-21 is a simplified sectional view of the Volkswagen automatic transmission with all essential parts named. Study Fig. 21-17 also to get a good idea of what the various parts look like.

CHAPTER 21 CHECKUP

NOTE: Since the following is a chapter review test, you should review the chapter before taking it.

With the background of knowledge you have acquired in your studies of automatic transmissions, you will have no difficulty understanding how any kind of automatic transmission operates, no matter how different it appears. You know that all automatic transmissions have planetary gears that are controlled by multiple-disk clutches and brake bands. You know that the controlling is done by a hydraulic control system. This system contains valves that operate to apply hydraulic pressure to the clutches or brake-band servos. As these actions take place, the planetary-gear system is controlled so as to produce the gear ratio asked for by the position of the manual valve, engine speed, intake-manifold vacuum, and car speed. Now, take the test that follows to find out how well you understand the different automatic transmissions described in this chapter.

Operation of Automatic Transmissions In the following, you are asked to write explanations of the various constructions and operating conditions described in the chapter. If you have any difficulty, study the illustrations and reread the chapter. You should be able to determine the power flows, for example, by studying the relevant illustrations.

1. Describe the power flow in the Turbo Hydra-Matic automatic transmission for front-drive cars in the drive range, first gear.
2. Describe the power flow in the Turbo Hydra-Matic automatic transmission for front-drive cars in the drive range, second gear.
3. Describe the power flow in the Turbo Hydra-Matic automatic transmission for front-drive cars in reverse.
4. How many clutches does the Turbo Hydra-Matic automatic transmission for front-drive cars have? How many brake bands? How many overrun clutches?
5. How many clutches, brake bands, and overrun clutches does the Toyoglide have?
6. Describe the power flow in the Toyoglide in first gear, drive range.
7. Describe the power flow in the Toyoglide in second gear, drive range.
8. Describe the power flow in the Toyoglide in third gear, drive range.
9. Describe the power flow in the Toyoglide in reverse.
10. Explain how the Volkswagen automatic stick-shift transmission works.
11. How many clutches, brake bands, and overrun clutches does the Volkswagen automatic transmission have?
12. Describe the power flow in the Volkswagen automatic transmission in first gear, drive range.
13. Describe the power flow in the Volkswagen automatic transmission in second gear, drive range.
14. Describe the power flow in the Volkswagen automatic transmission in third gear, drive range.
15. Describe the power flow in the Volkswagen automatic transmission in reverse.

SUGGESTIONS FOR FURTHER STUDY

If you can watch the automatic transmissions described in the chapter being tested, removed from the car, disassembled, overhauled, reassembled, and reinstalled on the car, you will learn many interesting details. There is a great similarity among automatic transmissions, and thus much of the service work is similar. However, each transmission has its own special designs and procedures. Study the different transmissions and the service manuals issued by their manufacturers to learn all you can about them. Write in your notebook any important facts you learn.

chapter 22

DRIVE LINES

This chapter describes the purpose, construction, operation, and service of drive lines. Drive lines, in automobiles, consist of the driving connection between the transmission and the driving wheels that are rotated to move the car (Fig. 22-1). In most automobiles, the drive line consists of a drive, or propeller, shaft, a universal joint or joints, and a slip joint. This combination of mechanisms carries the driving power of the engine from the transmission to the differential at the rear-wheel axles. Some cars do not have a drive shaft—for example, an automobile with the engine at the front and with front-wheel drive. Instead, the transmission is coupled directly to the front-wheel axles by axle drive shafts. Likewise, automobiles with the engines mounted in the rear and with rear-wheel drive have a direct coupling between the transmission and the rear-wheel axles. Note that the two terms "drive shaft" and "propeller shaft" are used interchangeably. Some manufacturers call it a drive shaft; *others call it a* propeller shaft *or* drive line.

⊘ 22-1 Function of the Drive Shaft In automobiles in which the engine is at the front and the rear wheels drive the car, a drive shaft is required to connect the transmission main or output shaft to the differential at the rear axles. Front-wheel-drive cars, which have the engine at the front, and rear-engine cars do not require drive shafts. However, on cars with shafts, rotary motion of the transmission main or output shaft carries through the drive shaft to the differential, causing the rear wheels to rotate.

The drive-shaft design must take into consideration two facts: First, the engine and transmission are more or less rigidly attached to the car frame; second, the rear-axle housing (with wheels and

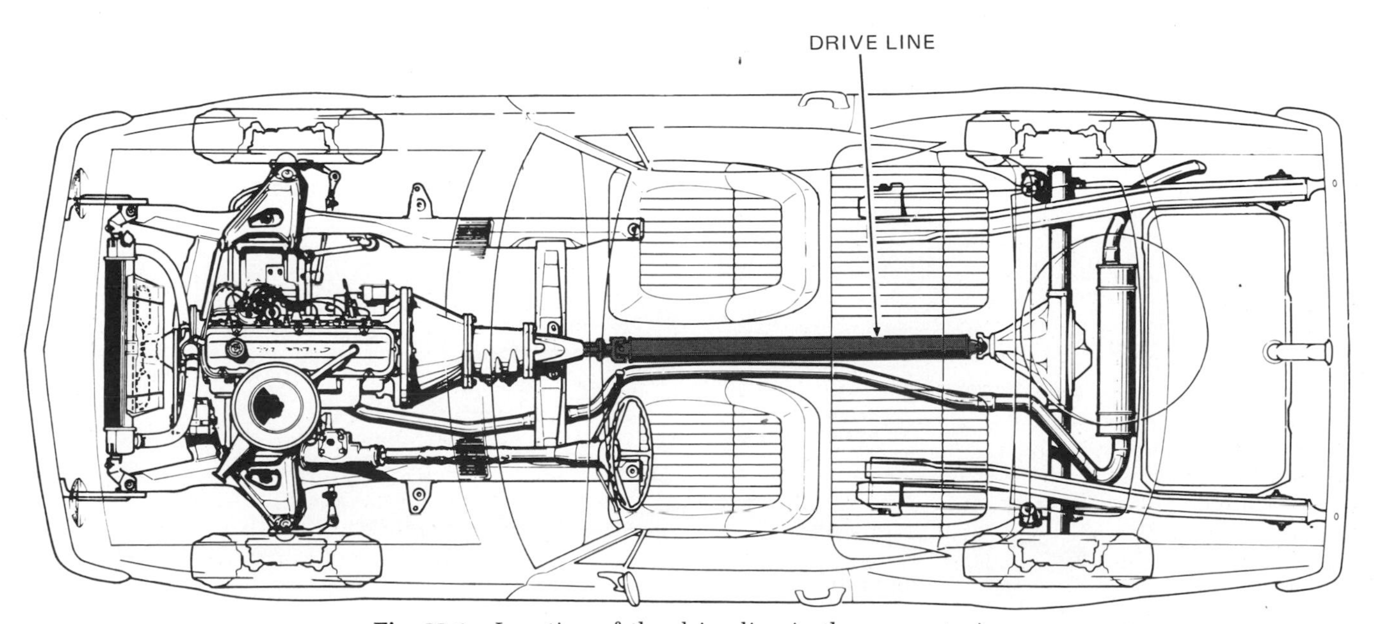

Fig. 22-1. Location of the drive line in the power train.

differential) is attached to the frame by springs. As the rear wheels encounter irregularities in the road, the springs compress or expand. This changes the angle of drive between the drive and transmission shafts. It also changes the distance between the transmission and the differential (see Fig. 22-2). In

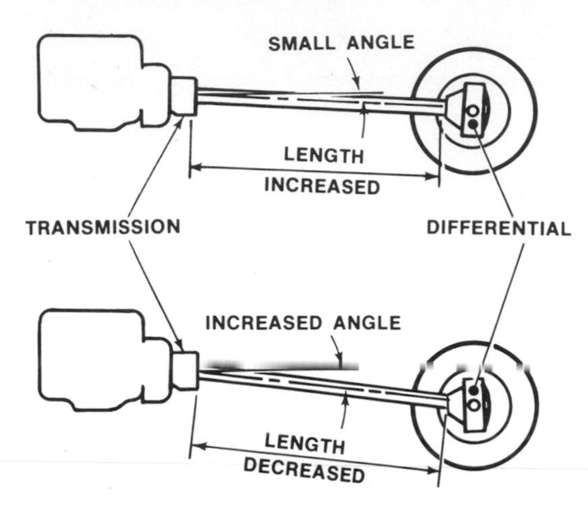

Fig. 22-2. As the rear-axle housing, with differential and wheels, moves up and down, the angle between the transmission output shaft changes and the length of the shaft also changes. The reason the drive shaft shortens as the angle increases is that the rear axle and differential move in a shorter arc than the drive shaft. The center point of the axle-housing arc is the rear spring or control-arm attachment to the frame.

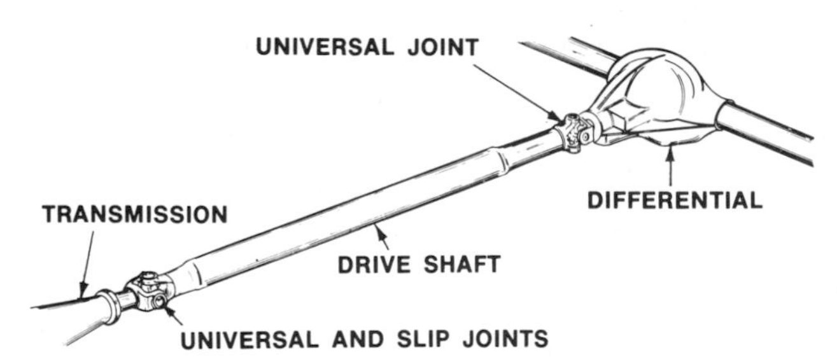

Fig. 22-3. Relationship of drive shaft (propeller shaft) to the transmission, frame, and differential. This is a one-piece drive shaft supported at front and rear by universal joints. (*Pontiac Motor Division of General Motors Corporation*)

order that the drive shaft may take care of these two changes, it must incorporate two separate types of devices: There must be one or more universal joints to permit variations in the angle of drive. There must also be a slip joint that permits the effective length of the drive shaft to change.

The drive, or propeller, shaft is usually a hollow tube with universal joints at front and back and, on some applications, at the center. Many propeller shafts are one piece, as shown in Fig. 22-3. Some have two sections, as shown in Fig. 22-4 to 22-6. The two-section propeller shaft has a center support, as shown in Figs. 22-5 and 22-6.

Typical disassembled views of the two general types of propeller shafts are shown in Figs. 22-7 and 22-8. Note, in Fig. 22-7, the several alternative constructions. The two alternative tube constructions to the lower right show methods of reducing noise and vibration. These include rubber elements in the tube which act as vibration dampers. The two alternative universal-joint constructions to the upper left in Fig. 22-7 show variations in the way the bearings are installed in the universal joint. We shall describe universal joints in ⊘ 22-2.

Figure 22-8 shows a two-piece propeller shaft in disassembled view. This is very similar to the propeller shaft shown in Fig. 22-5.

⊘ 22-2 Universal Joints

⊘ 22-2 Universal Joints A simple universal joint is shown in Fig. 22-9. It is essentially a double-hinged joint consisting of two Y-shaped yokes, one on the driving shaft and the other on the driven shaft, and a cross-shaped member called the *cross*, or *spider*.

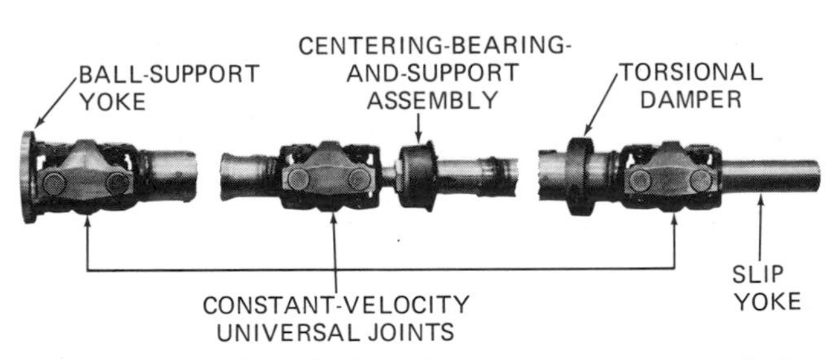

Fig. 22-5. Propeller shaft with center support, two shafts, and three constant-velocity universal joints. (*Cadillac Motor Car Division of General Motors Corporation*)

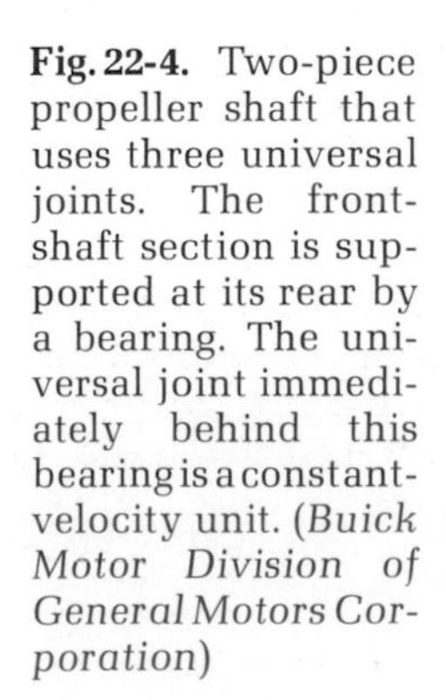

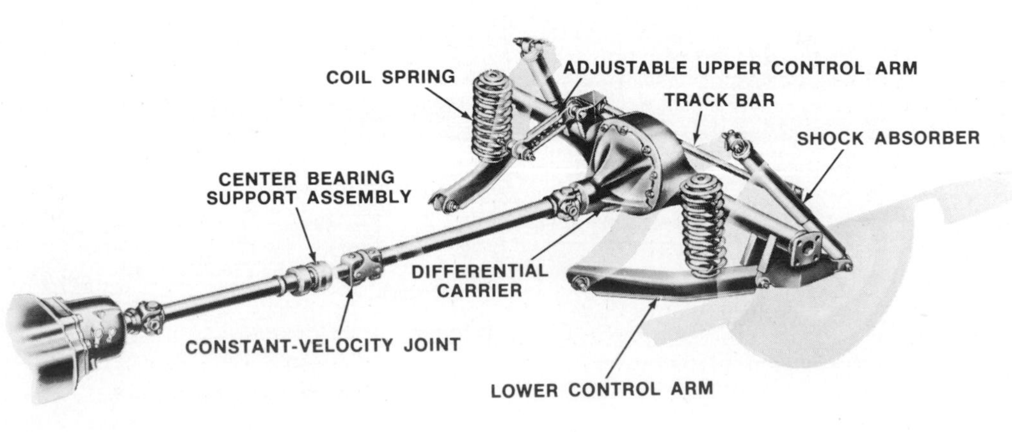

Fig. 22-4. Two-piece propeller shaft that uses three universal joints. The front-shaft section is supported at its rear by a bearing. The universal joint immediately behind this bearing is a constant-velocity unit. (*Buick Motor Division of General Motors Corporation*)

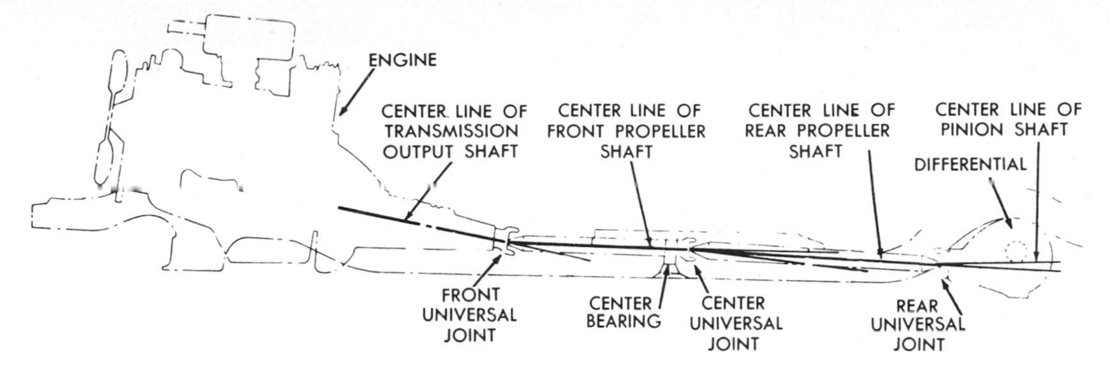

Fig. 22-6. Schematic drawing of the relationship of a two-piece propeller shaft, universal joints, transmission, and differential. (*Chevrolet Motor Division of General Motors Corporation*)

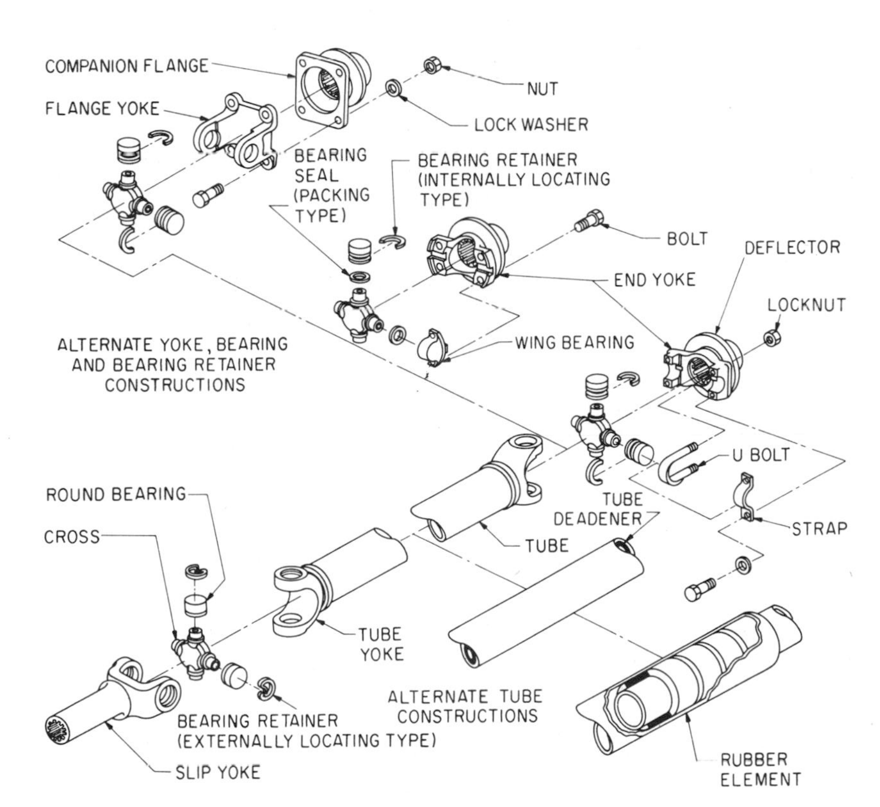

Fig. 22-7. Disassembled view of a one-piece drive shaft showing alternate tube and universal-joint construction. (*Society of Automotive Engineers*)

The four arms of the cross, known as *trunnions,* are assembled into bearings in the ends of the two shaft yokes. The driving shaft causes the cross to rotate, and the other two trunnions of the cross cause the driven shaft to rotate. When the two shafts are at an angle to each other, the bearings in the yokes permit the yokes to swing round on the trunnions with each revolution. A variety of universal joints have been used on automobiles, but the types now in most common use are the cross-and-two-yoke, ball-and-trunnion, and constant-velocity joints.

The cross-and-two-yoke design is essentially the same as the simple universal joint discussed previously except that the bearings are often of the needle type (Figs. 22-7, 22-8, and 22-10). As will be noted, there are four sets of needle bearings, one set for each trunnion of the cross. The bearings are held in place by snap rings that drop into undercuts in the yoke-bearing holes.

The ball-and-trunnion type of universal joint combines the universal and the slip joint in one assembly. A universal joint of this design is shown in Fig. 22-11 in exploded view. The shaft has a pin pressed through it, and around both ends of the pin are placed balls drilled out to accommodate needle bearings. The other member of the universal joint consists of a steel casing or body that has two longitudinal channels into which the balls fit. The body is bolted to a flange on the mating shaft (not shown in Fig. 22-11). The rotary motion is carried through the pin and balls, according to the direction of drive. The balls can move back and forth in the channels of the body to compensate for varying angles of drive. At the same time, they act as a slip joint by slipping in or out of the channels.

The simple universal joint shown in Fig. 22-9, consisting of a cross and two yokes, is not a constant-velocity joint. That is, as the drive shaft rotates,

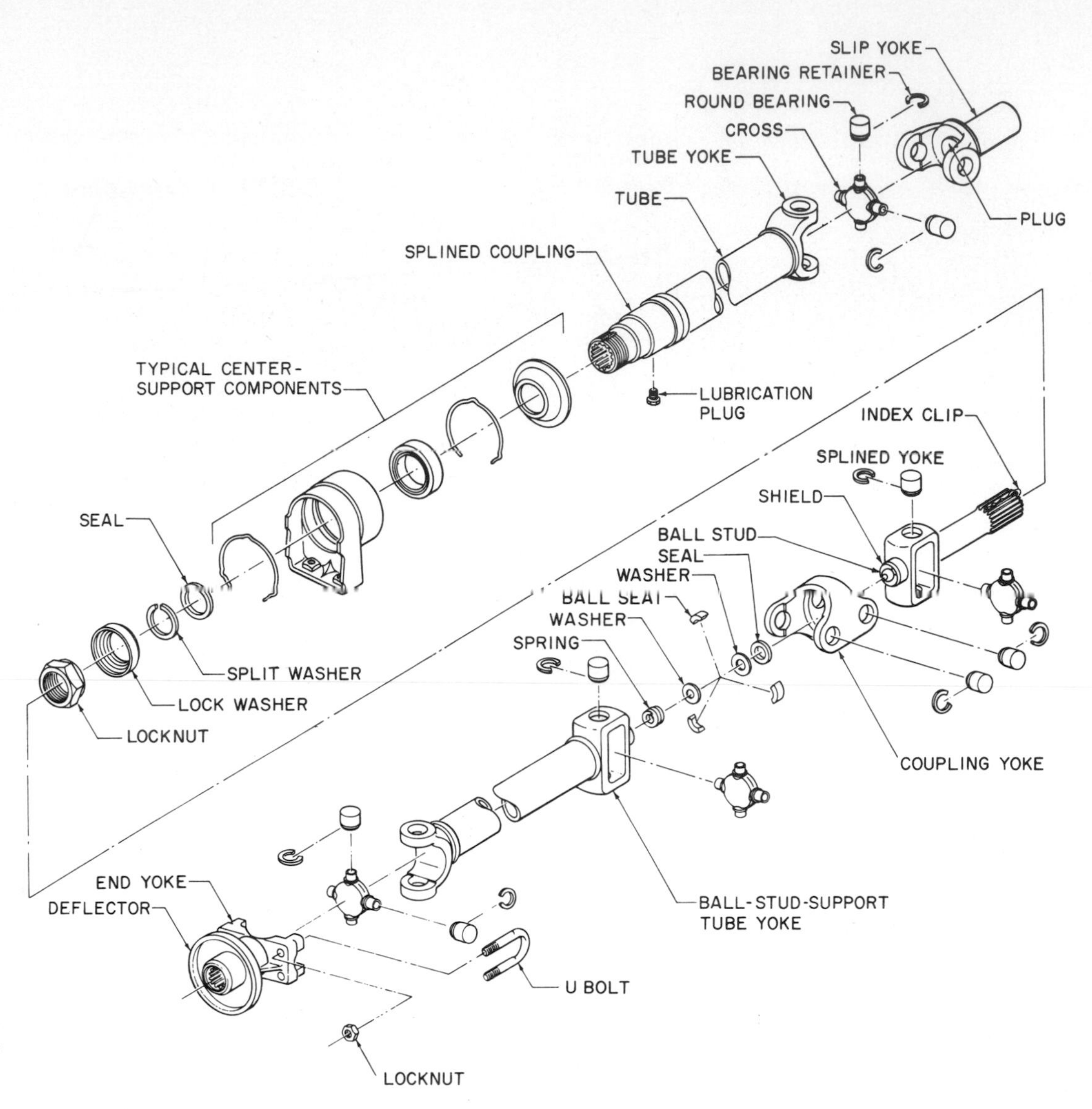

Fig. 22-8. Disassembled view of a two-piece drive shaft. (*Society of Automotive Engineers*)

the driven shaft will be given a variable velocity. The greater the angle between the two, the greater the variation. This is not a healthy situation for the drive line because the loads on the bearings and gears will pulsate. Thus, there will be repeated reductions and increases in the load with every revolution. This condition could cause increased wear of the affected parts. To eliminate this, constant-velocity universal joints are used on many cars.

Constant-velocity universal joints are shown on propeller shafts in Figs. 22-4 and 22-5 and in sectional view in Fig. 22-12. Figure 22-13 is a simplified drawing of a constant-velocity universal joint. This joint, known as a *double-Cardan universal joint,* consists of two individual universal joints linked by a ball and socket. The ball and socket splits the angle of the two propeller shafts between the two universal joints. Because the two joints are operating at the same angle, the normal fluctuations that could result from the use of a single universal joint are canceled out. In other words, the acceleration resulting at any instant from the action of one universal joint is nullified by the deceleration of the other, and vice versa.

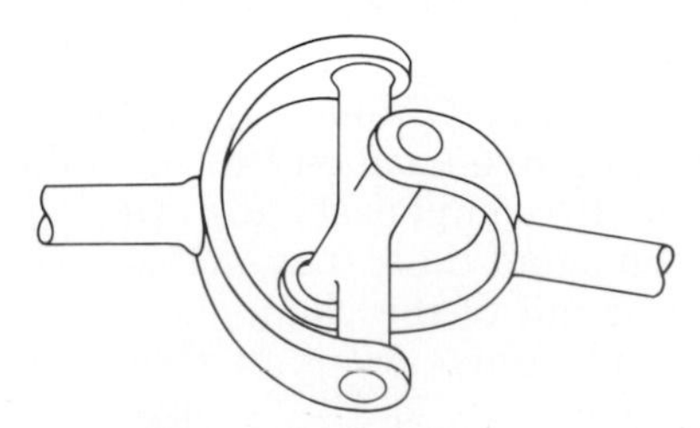

Fig. 22-9. Simple universal joint.

Another type of constant-velocity universal joint is the *Rzeppa joint,* shown in disassembled view in Fig. 22-14. The large balls roll in the curved grooves between the inner housing and the inner race. They are retained in position by the cage. Regardless of the angles between the drive and driven shafts, the velocity imparted is constant because the balls roll between the two races. Note that the inner housing is positioned in the outer housing with a

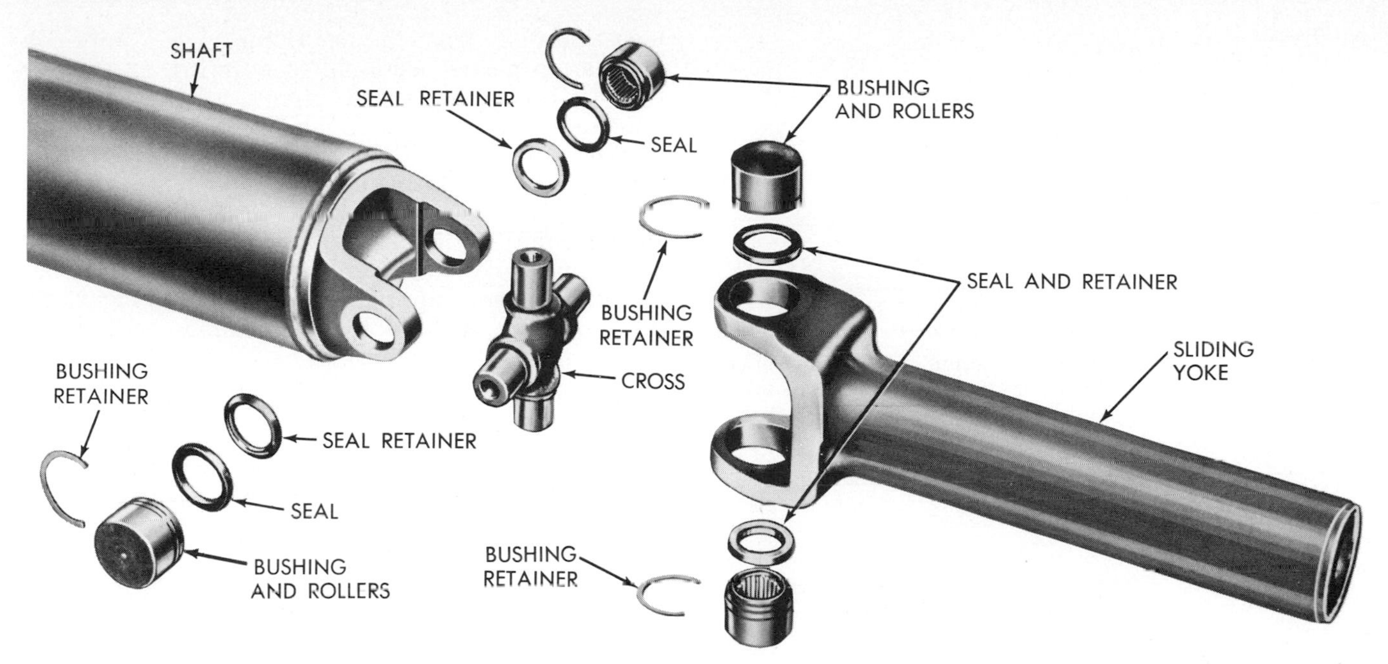

Fig. 22-10. Disassembled view of a universal joint. The cross is also called a *spider* by some manufacturers. (*Chrysler Corporation*)

series of small balls in the straight grooves. This arrangement acts as a slip joint to permit the effective length of the drive shaft to change. Slip joints are described in ⊘ 22-3.

The three-ball-and-trunnion universal joint, shown in Fig. 22-15, is also a constant-velocity universal joint. The balls on the ends of the three trunnions can move back and forth in the grooves in the housing as required by the angle of drive.

In rear-engine rear-drive or front-engine front-drive cars, no drive shaft is required. Instead, the power flow is to a pair of short shafts which deliver the power from the transmission, through universal joints, to the wheels. This is covered in ⊘ 22-4.

⊘ 22-3 Slip Joint A slip joint is shown in Fig. 1-24 and 22-8. As previously explained, the slip joint consists of outside splines on one shaft and matching internal splines on the mating hollow shaft. The splines cause the two shafts to rotate together but permit the two to move endwise with each other. Thus, any effective change of length of the propeller shaft is accommodated as the rear axles move to-

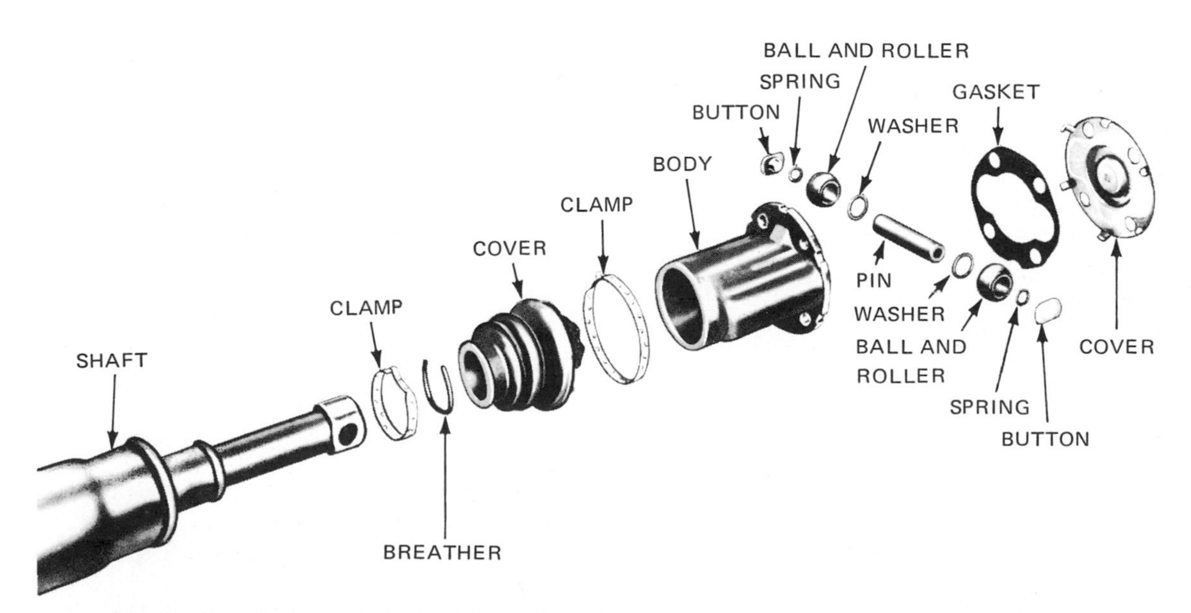

Fig. 22-11. Disassembled view of a ball-and-trunnion universal joint. (*Chrysler Corporation*)

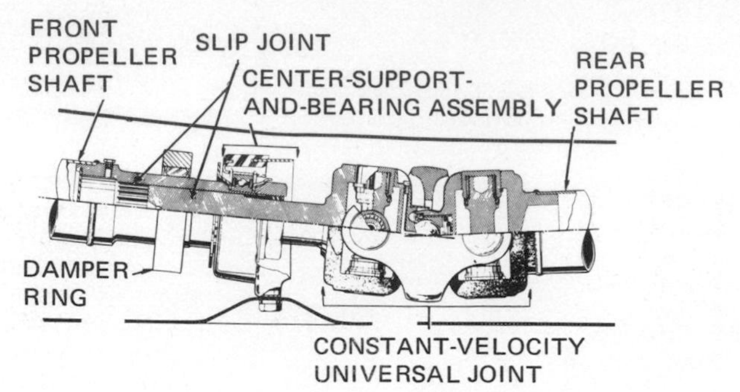

Fig. 22-12. Sectional view of a constant-velocity universal joint. (*Buick Motor Division of General Motors Corporation*)

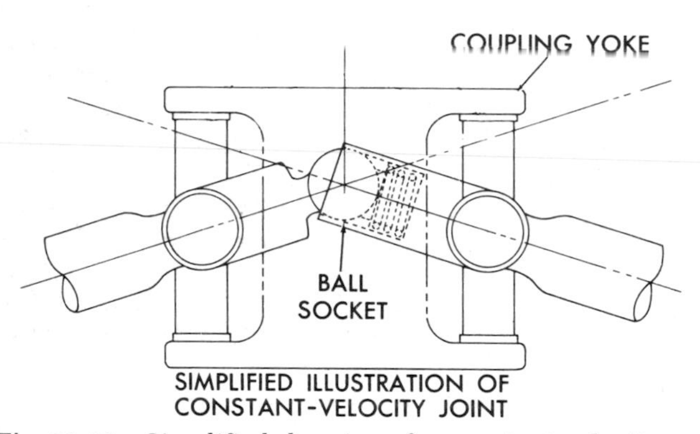

Fig. 22-13. Simplified drawing of a constant-velocity, or double-Cardan, universal joint. (*Chevrolet Motor Division of General Motors Corporation*)

ward or away from the car frame. A slip joint can be seen in partial sectional view in Fig. 22-12.

For certain high-performance or heavy-duty applications, special ball- or needle-bearing slip joints are used. Figure 22-16 shows a slip joint using spring-loaded balls that can roll between grooves cut in the shaft and hollow yoke. This design greatly reduces the frictional force necessary to allow the shaft and yoke to slide back and forth. Another design for special applications is shown in Fig. 22-17. In this design, a series of recirculating needle bearings are used. These bearings can roll back and forth as the yoke and shaft move with respect to each other.

⊘ 22-4 Types of Drives A variety of drive arrangements are used on automobiles. The most common type is the front-engine rear-wheel drive arrangement. The engine is mounted at the front, and the rear wheels drive the car. This arrangement requires a propeller shaft.

Other arrangements include the front-engine front-wheel drive and the rear-engine rear-wheel drive. There is also the rear-engine four-wheel drive, as well as the front-engine four-wheel drive.

1. *FRONT-ENGINE REAR-WHEEL DRIVE* The front-engine rear-wheel drive arrangement requires the use of a propeller shaft. The rotation of the propeller shaft transmits torque through the differential to the rear wheels, causing them to rotate and move the car. The wheels rotate because torque is applied to them. This torque not only rotates the wheels in one direction but also attempts to rotate

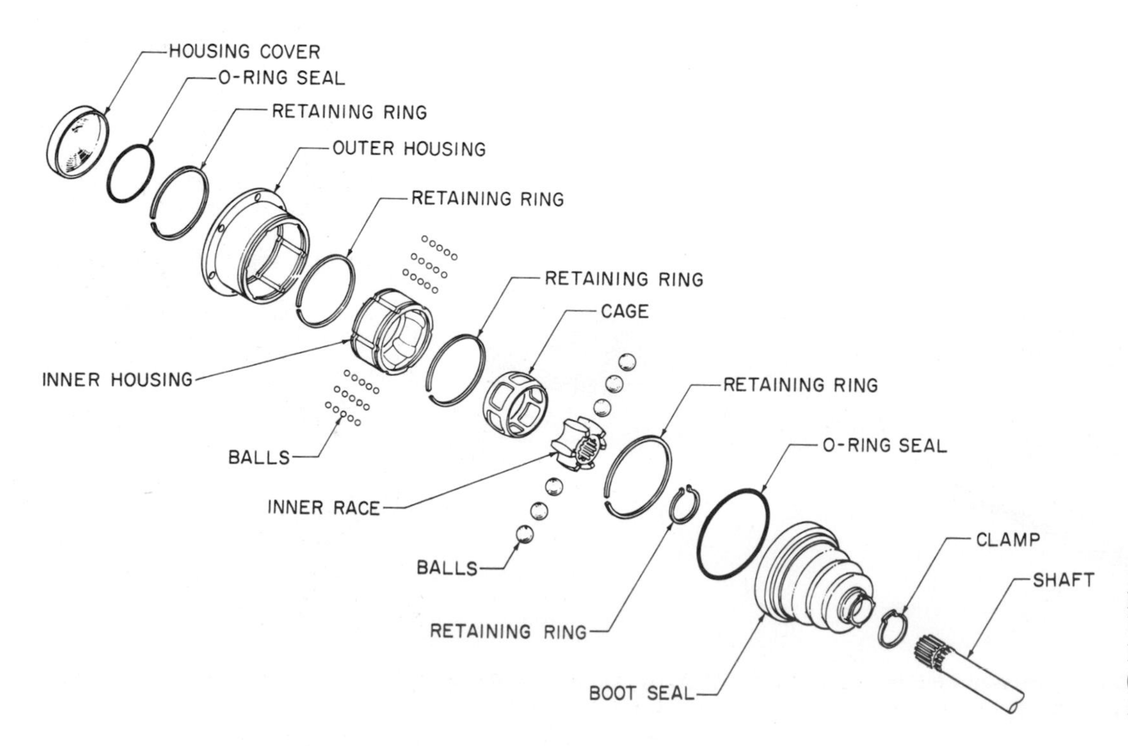

Fig. 22-14. Disassembled view of a ball-spline Rzeppa universal joint. (*Society of Automotive Engineers*)

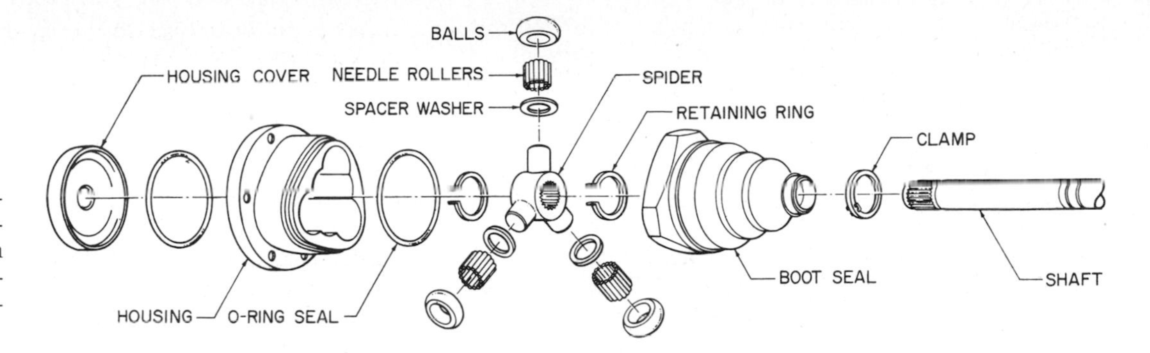

Fig. 22-15. Disassembled view of the three-ball - and - trunnion universal joint. (*Society of Automotive Engineers*)

the differential housing in the opposite direction. To understand why this occurs, it is necessary to review briefly the construction of the differential (Fig. 23-1).

The ring gear is connected through other gears to the rear-wheel axles; the torque applied through the drive pinion forces the ring gear and wheels to rotate. It is the side thrust of the drive-pinion teeth against the ring-gear teeth that makes the ring gear rotate. This side thrust also causes the drive-pinion shaft to push against the shaft bearing. The thrust against the shaft bearing is in a direction opposite to the thrust of the pinion teeth against the ring-gear teeth. Since the drive-pinion bearings are held in the differential housing, the housing tries to rotate in a direction opposite to the ring gear and wheel rotation. This action is termed *rear-end torque,* and to prevent excessive movement of the differential housing from this action, several methods of bracing the housing are used. The two most common types of bracing found on modern automobiles are the torque-tube drive and the Hotchkiss drive (Fig. 22-18). Most cars use the Hotchkiss drive.

In the torque-tube drive, the propeller shaft is encased in a hollow tube. The tube is rigidly bolted to the differential housing at one end and is fastened at the other end of the transmission case through a somewhat flexible joint. On many cars, a pair of truss rods are attached between the rear-axle housings and the transmission end of the torque tube. The torque tube and the truss rods brace the differ-

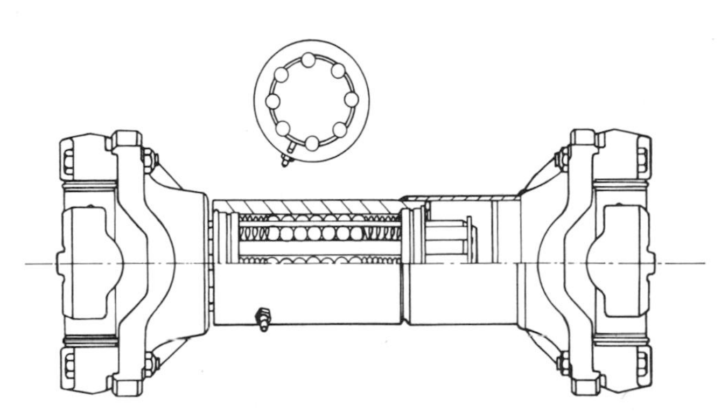

Fig. 22-16. Slip joint using balls that roll in grooves instead of splines.

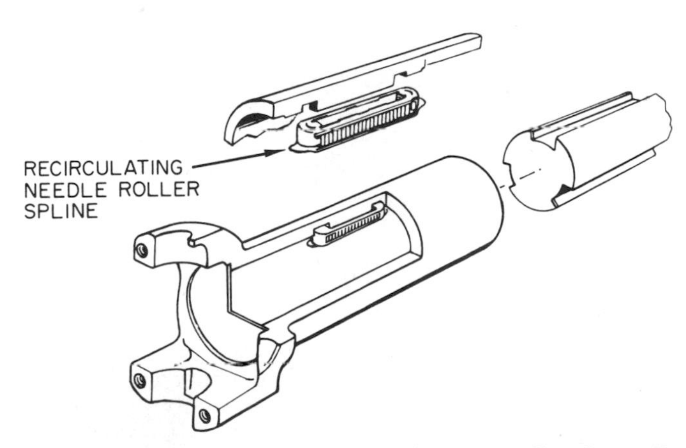

Fig. 22-17. Slip joint using recirculating needles that roll in grooves cut in the shaft instead of splines.

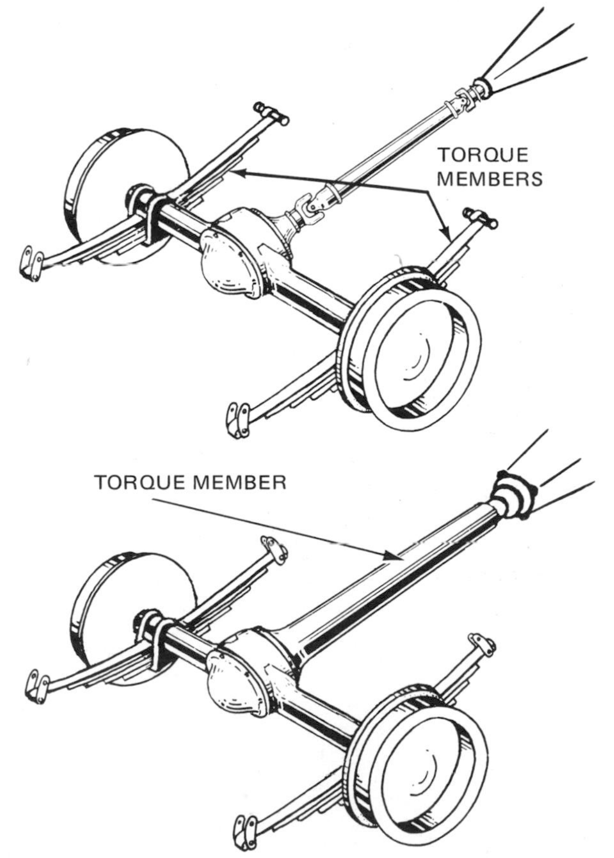

Fig. 22-18. Hotchkiss drive (top) compared with torque-tube drive (bottom).

ential housing to prevent excessive differential-housing movement. In other words, these members absorb the rear-end torque.

In the Hotchkiss drive, the rear-end torque is absorbed by the rear springs and control arms. On the leaf-spring rear suspension, the leaf spring absorbs the rear-end torque. On the coil-spring rear suspension, the control arms absorb the rear-end torque. Real leaf springs are attached to brackets bolted to the rear-axle housings so that the springs themselves act as the torque absorbing members. Thus, when the car moves forward, the rear-end torque causes the front halves of the springs to be compressed while the rear halves of the springs are expanded. Two universal joints are required on the propeller shaft in the Hotchkiss drive, one at each end of the shaft (see Fig. 22-3). The reason for this is obvious; the differential housing does rotate as a result of rear-end torque within the limits imposed by car springs.

A variation on the arrangements shown in Fig. 22-18 is illustrated in Fig. 22-19. Here, the usual axle housing is dispensed with. A transverse leaf spring is used, attached at the center to the axle carrier (differential housing) and at the two ends to the two control arms. Two axle drive shafts connect between the two differential output shafts to the two rear-wheel axles. Each of these drive shafts must have two universal joints, as shown, to take care of the up-and-down motion of the rear wheels. In effect, each of these axle drive shafts is a short propeller shaft.

2. *REAR-ENGINE REAR-WHEEL DRIVE* With this arrangement, the engine mounts at the rear of the car and the rear wheels are driven (Fig. 22-20). You will note that the drive arrangement is very similar in many ways to that shown in Fig. 22-19. Each rear wheel is driven by a short shaft (the axle drive shaft), and each axle drive shaft contains two universal joints to take care of the up-and-down motion of the rear wheels.

3. *FRONT-ENGINE FRONT-WHEEL DRIVE* Some vehicles have front drive. That is, the transmission is connected through driving axles to the front wheels. Figure 22-21 shows the transmission and differential for such an arrangement. Figure 22-22 shows the front suspension system and drive shafts for this same vehicle. Each drive shaft has two uni-

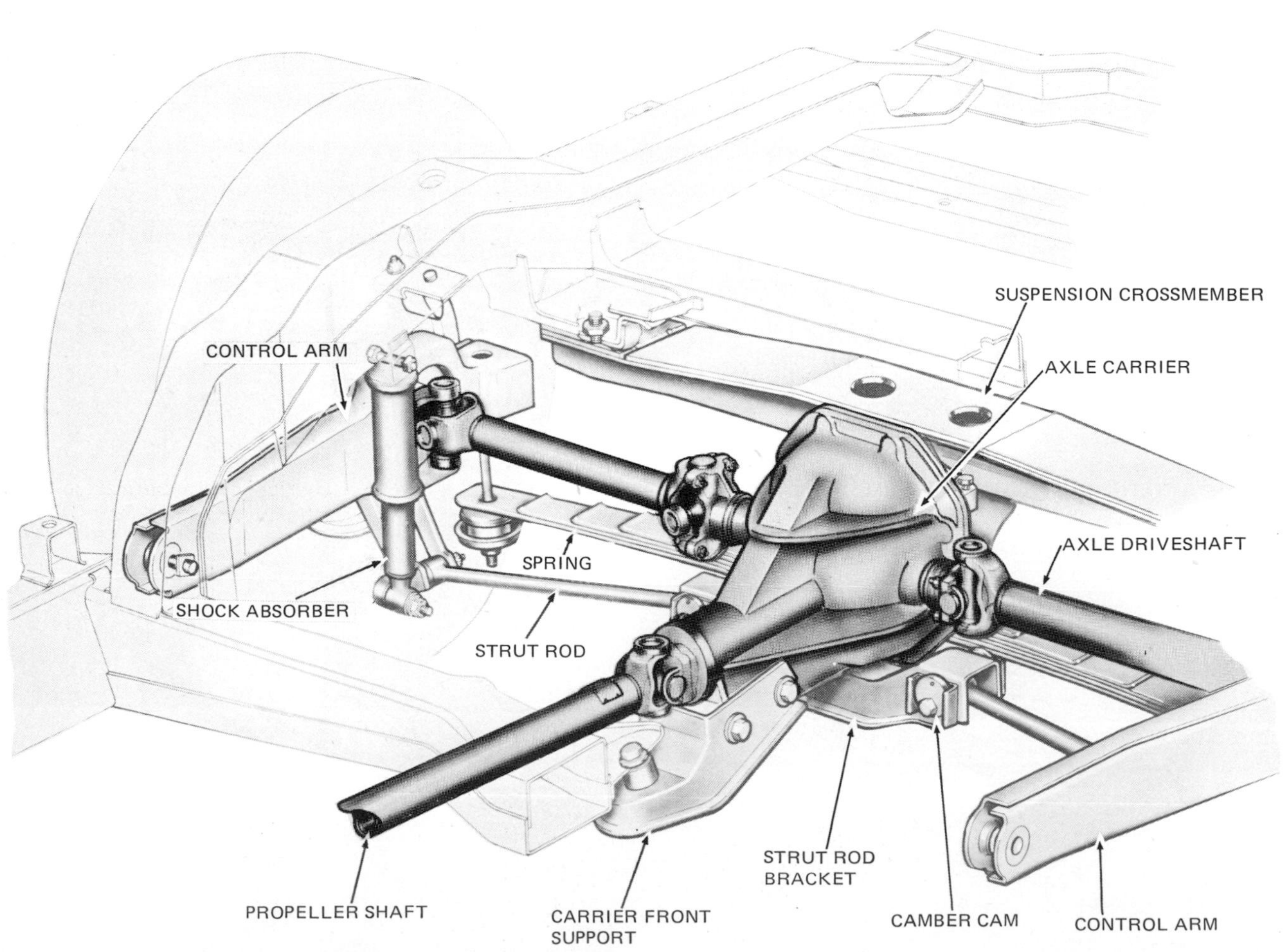

Fig. 22-19. Rear suspension and drive-line components used in the Corvette showing the transverse leaf spring and the axle-drive shafts with their two universal joints each. (*Chevrolet Motor Division of General Motors Corporation*)

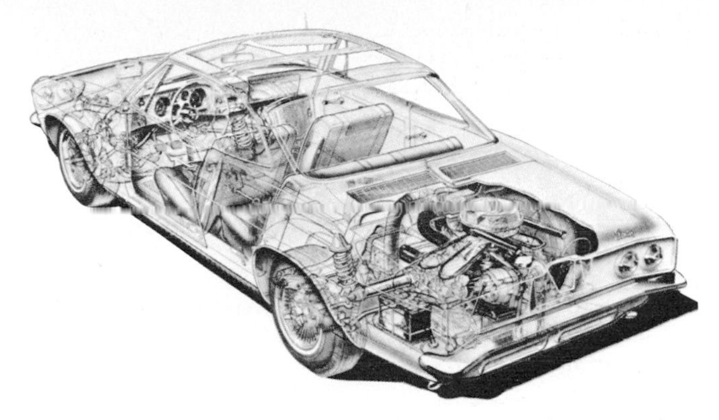

Fig. 22-20. Phantom view of the Corvair showing rear-engine mounting and drive arrangement. (*Chevrolet Motor Division of General Motors Corporation*)

versal joints to permit the front wheels to move up and down and also pivot from one side to the other for steering.

Some models of Cadillac and Oldsmobile are front-drive cars. Figures 21-2 and 21-3 show the transmission and its attachment to the engine. Figure 22-23 shows the transmission and final drive for the right-hand output shaft. The left-hand output shaft is shorter than the right-hand output shaft because the differential is offset to the left. The output shafts maintain their alignment with the differential. They are connected by two short drive shafts to the two front wheels in a manner much like the Simca, shown in Fig. 22-22. Each drive shaft has two universal joints to permit vertical movement of the wheels and pivoting for steering. The right-hand drive axle also has a torsional damper to prevent torsional vibration of the longer drive axle.

4. *REAR-ENGINE FOUR-WHEEL DRIVE* The rear-engine four-wheel drive system is not widely used. Both a front and a rear differential are required, and each of the four axle drive shafts has

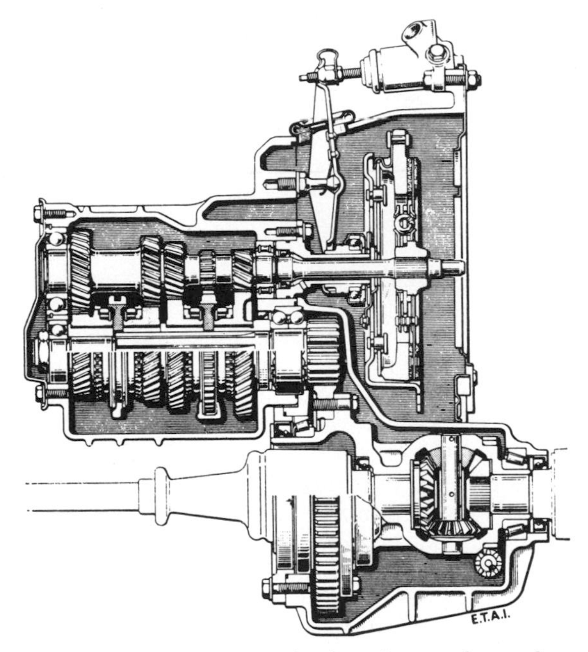

Fig. 22-21. Cutaway view of a four-forward-speed transmission for a cross-mounted engine used on a front-drive car. (*Simca*)

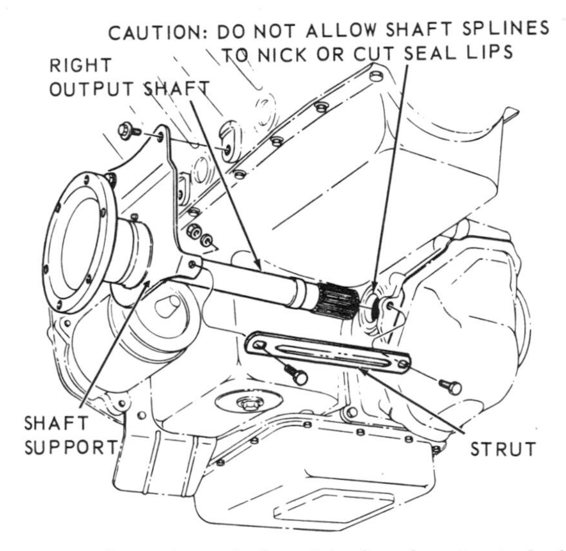

Fig. 22-23. Location of the right-hand output shaft in relation to the final drive. (*Cadillac Motor Car Division of General Motors Corporation*)

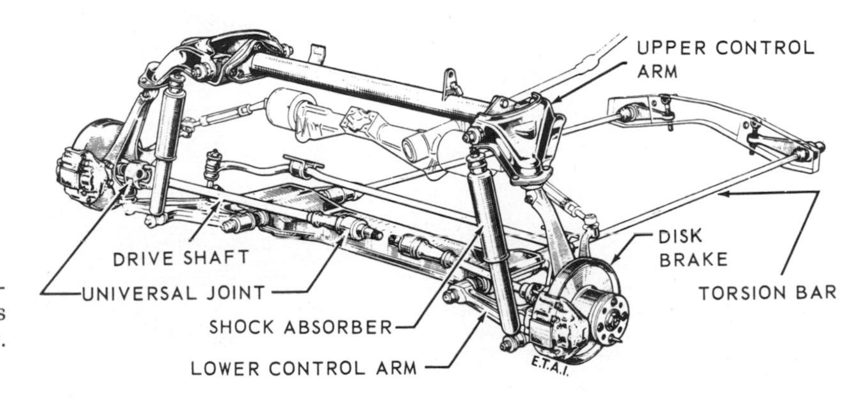

Fig. 22-22. Front suspension and drive shafts for a front-drive car. (*Simca*)

two universal joints. With this arrangement, a propeller shaft is necessary to carry the power from the rear to the front of the car.

5. *FRONT-ENGINE FOUR-WHEEL DRIVE* The front-engine four-wheel drive is a common arrangement. The well-known Jeep has used it for years. The engine mounts at the front, and each of the four wheels is connected to a differential by means of drive shafts containing universal joints. Figure 22-24 is a bottom view of a four-wheel-drive vehicle showing the drive arrangement. Note that the power passes from the transmission to a transfer case which is connected to both front and rear propeller shafts. These shafts, in turn, are connected to the front and rear differentials. The differentials are connected by drive axles or shafts to the front and rear wheels.

⊘ 22-5 Universal-Joint and Drive Line Service Most universal joints require no maintenance. They are lubricated for life and cannot be lubricated on the car. If a universal joint becomes noisy or worn, it must be replaced. Manufacturers supply service kits that include all necessary parts to make the replacement.

The drive shaft is a balanced unit. On some cars, if the shaft is unbalanced enough to cause undesirable vibrations, it can be rebalanced with the addition of hose clamps, as noted later. On other cars, the recommendation is to replace the offending drive shaft. Typical service procedures follow.

NOTE: The propeller shaft and universal joints are carefully balanced during original assembly. To ensure correct relationship so that balance can be maintained, mark all parts before disassembly (if they are not already marked). Then, you can put everything back together in the original positions.

⊘ 22-6 Chevrolet Drive-Line Service To remove the drive-shaft assembly, mark the relationship of the shaft to the companion flange at the differential. Then disconnect the rear universal joint by removing trunnion bearing straps or flange attaching bolts (Figs. 22-25 and 22-26). On the strap-attachment type,

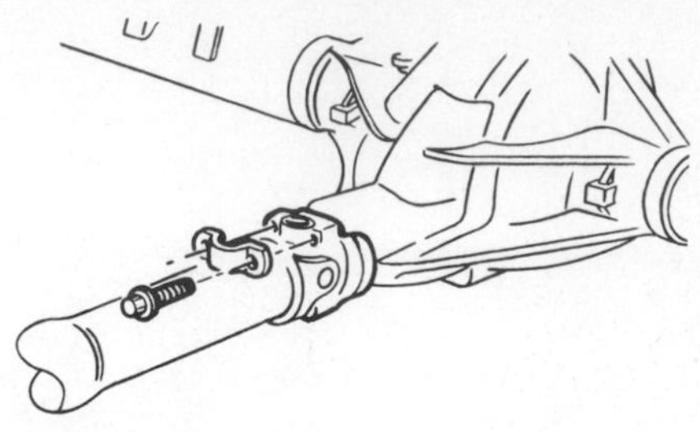

Fig. 22-25. Strap attachment of the rear end of the propeller shaft. (*Chevrolet Motor Division of General Motors Corporation*)

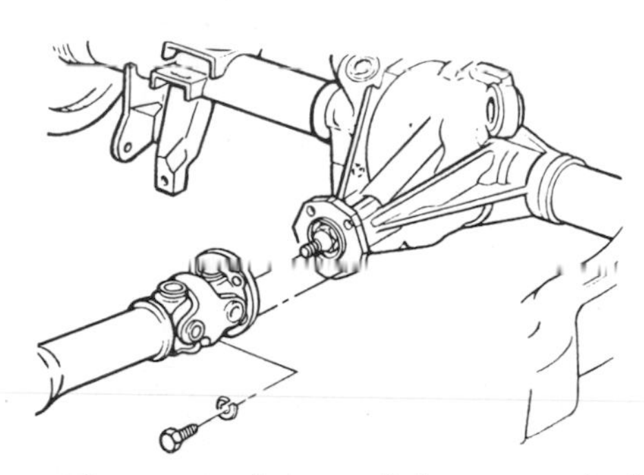

Fig. 22-26. Flange attachment of the rear end of the propeller shaft. (*Chevrolet Motor Division of General Motors Corporation*)

tape the bearing cups to the trunnion to keep bearing rollers from falling out.

Then withdraw the propeller-shaft front yoke from the transmission by moving the shaft to the rear. Pass it under the axle housing. Watch for leakage from the transmission extension housing.

1. *UNIVERSAL-JOINT SERVICE* Three universal-joint designs are used on Chevrolet: the Cleveland, Saginaw, and double-Cardan constant-velocity types. Special service kits are supplied for all these designs (see Figs. 22-27 and 22-28). As examples of service procedure, we shall discuss the Cleveland and constant-velocity universal joints.

a. Cleveland Universal-Joint Disassembly Remove bearing lock rings from the trunnion yoke. Support the trunnion yoke on a piece of $1\frac{1}{4}$ inch [31.75 mm] inside-diameter pipe in an arbor bed (Fig. 22-29).

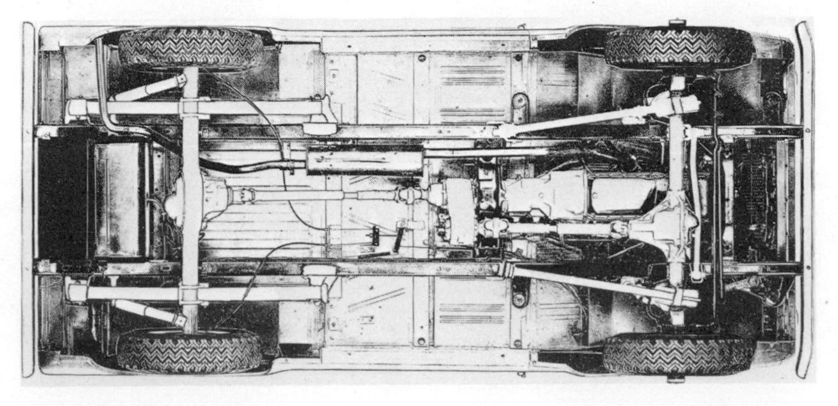

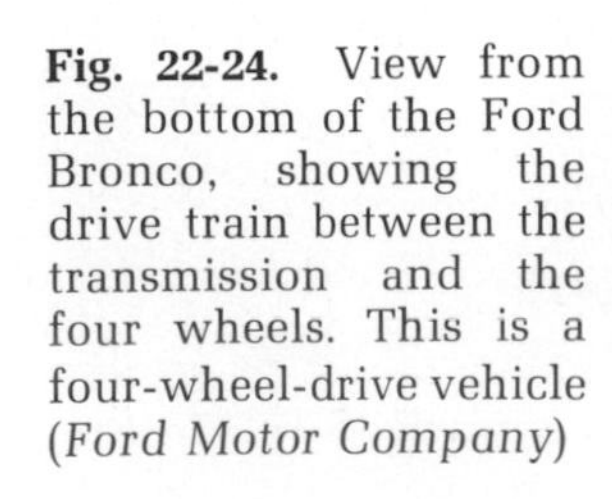

Fig. 22-24. View from the bottom of the Ford Bronco, showing the drive train between the transmission and the four wheels. This is a four-wheel-drive vehicle (*Ford Motor Company*)

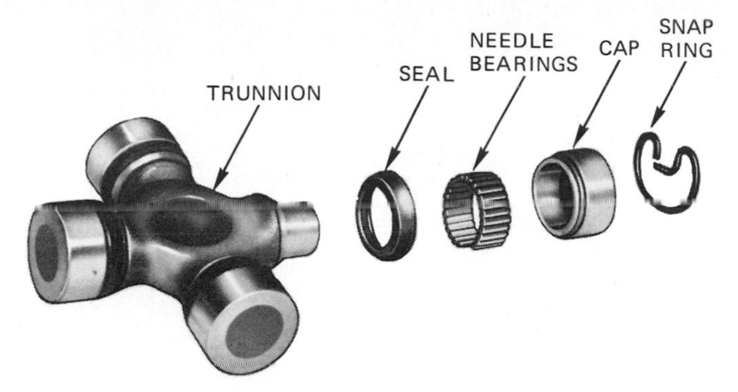

Fig. 22-27. Cleveland-type universal-joint repair kit. (*Chevrolet Motor Division of General Motors Corporation*)

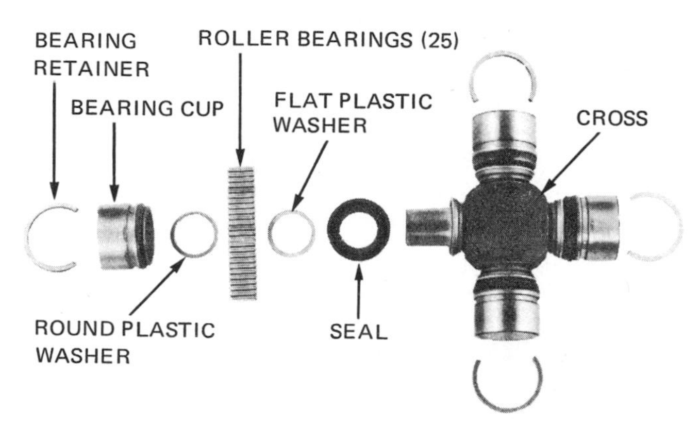

Fig. 22-28. Saginaw-type universal-joint repair kit. (*Chevrolet Motor Division of General Motors Corporation*)

NOTE: A bench vise can also be used to exert the pressure necessary to press the bearing cup loose.

Apply pressure to the trunnion until the bearing cup is almost out. It cannot be pressed all the way out. Grasp the cup in a vise or pliers and work it out of the yoke. Reverse the position of the trunnion and remove the other bearing cup.

Clean and inspect dust seals, bearing rollers, and trunnion. If everything is in order, repack the bearings and lubricate the reservoirs at the ends of trunnions with high-melting-point wheel-bearing lubricant. Use new seals. Then reassemble as explained in *b*. If any part is defective, replace everything with new parts included in the repair kit (Fig. 22-27).

b. Cleveland Universal-Joint Reassembly When packing lubricant into the lubricant reservoirs at the ends of the trunnions, make sure that they are completely filled from the bottom. The use of a squeeze bottle is recommended to prevent air pockets at the bottom.

Use a trunnion seal installer, as shown in Fig. 22-30, to install the seals. Then position trunnion in yoke. Partly install one bearing cup, as shown in Fig. 22-31. Then partly install the other bearing cup.

Fig. 22-29. Removing the bearing cup. (*Chevrolet Motor Division of General Motors Corporation*)

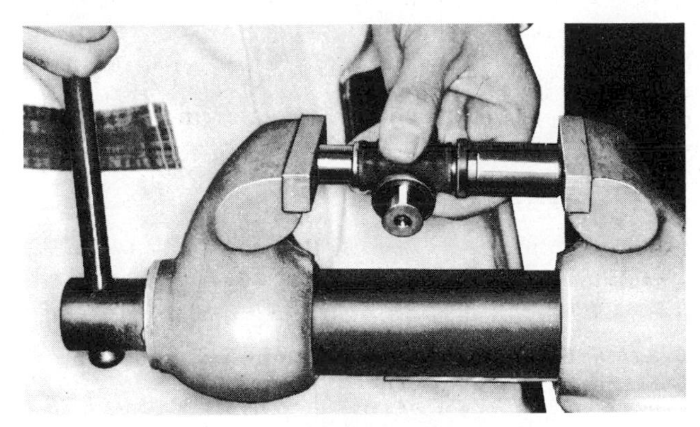

Fig. 22-30. Installing the trunnion seal. (*Chevrolet Motor Division of General Motors Corporation*)

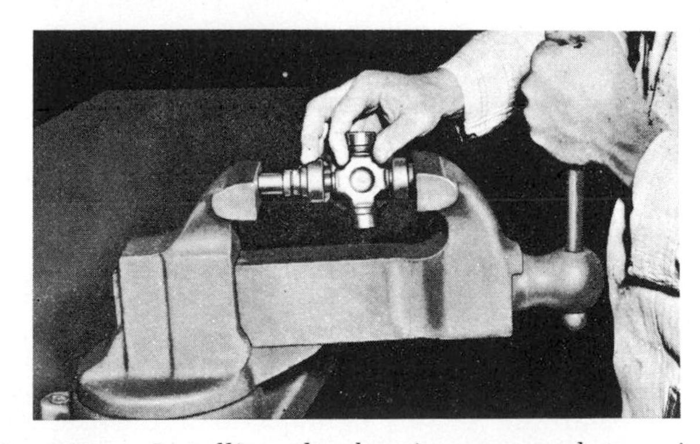

Fig. 22-31. Installing the bearing cap and trunnion. (*Chevrolet Motor Division of General Motors Corporation*)

Align the trunnion into the cups and press the cups into the yoke.

c. Constant-Velocity Universal-Joint Service In Fig. 22-32, the constant-velocity universal joint is shown with the bearing caps numbered in the order in which they should be removed. Figure 22-33 shows the universal joint with alignment punch marks which serve as a guide for reassembly. See also Fig.

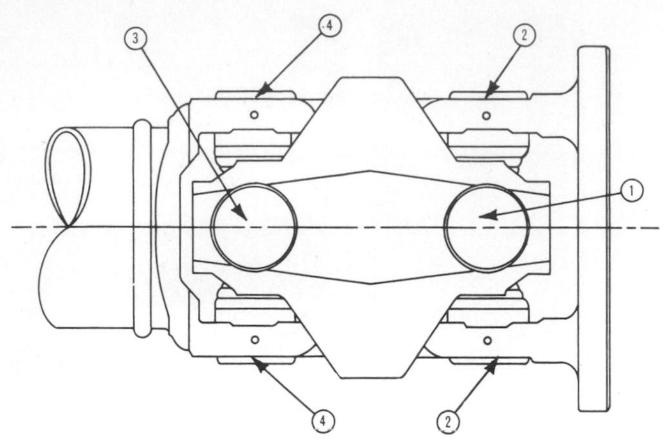

Fig. 22-32. Sequence in which bearing caps should be removed from a constant-velocity universal joint. (*Chevrolet Motor Division of General Motors Corporation*)

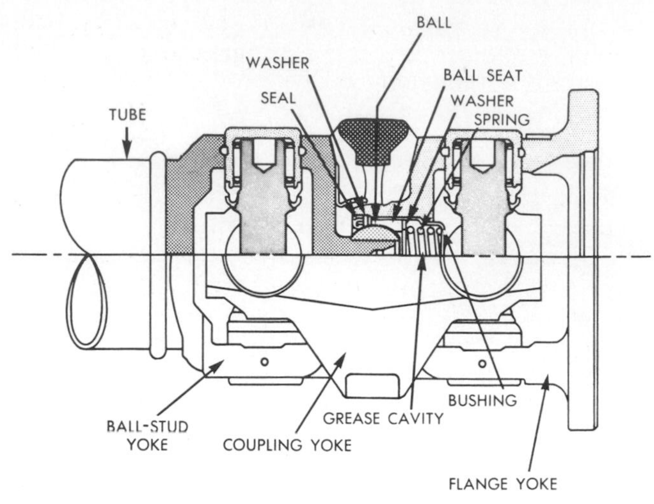

Fig. 22-34. Sectional view of a constant-velocity universal joint. (*Chevrolet Motor Division of General Motors Corporation*)

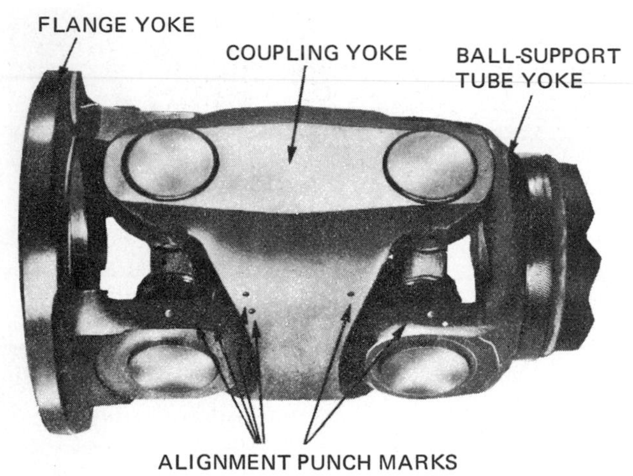

Fig. 22-33. Alignment punch marks on a constant-velocity universal joint which aid in proper reassembly of the joint. (*Chevrolet Motor Division of General Motors Corporation*)

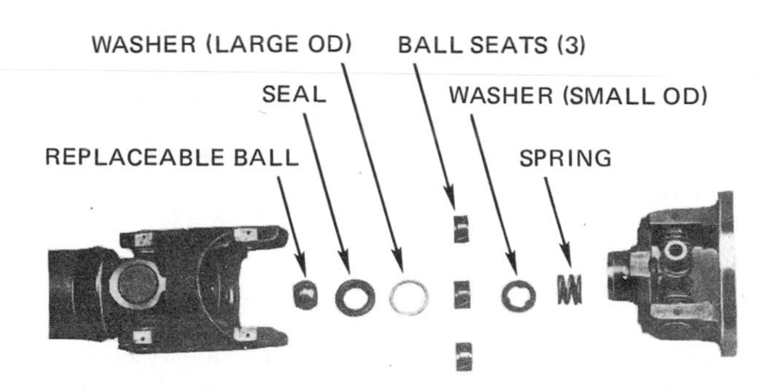

Fig. 22-35. Disassembled view of the centering-ball mechanism in a constant-velocity universal joint. (*Chevrolet Motor Division of General Motors Corporation*)

22-34 for a sectional view and Fig. 22-40 for a disassembled view of this type of joint.

Remove the bearing cups in the same manner as for the Cleveland unit. Then disengage the flange yoke and trunnion from the centering ball. Note, in Fig. 22-35, that the ball socket is part of the flange-yoke assembly. The centering ball is pressed onto a stud and is part of the ball-stud yoke. Pry the seal from the ball socket and remove the washers, spring, and ball seats. Replace everything with a service kit.

Remove all plastic from the groove of the coupling yoke. To replace the centering ball, use the special puller tool, shown in Fig. 22-36. The fingers of the tool are placed under the ball. Then a collar is put on the tool and a nut tightened on the tool screw threads. This pulls the ball off the ball stud. A new ball can then be driven on the stud.

CAUTION: The ball must seat firmly against the shoulder at the base of the stud.

Fig. 22-36. Special tool for removing the ball from the stud. When a collar is placed over the tool and a nut is tightened on the screw threads, the ball is pulled off the stud. (*Chevrolet Motor Division of General Motors Corporation*)

Use the grease furnished with the service kit to lubricate all parts. Install parts into the clean ball seat cavity in this order: spring, washer (small), three ball seats (largest opening outward to receive the ball), large washer, and seal. Press the seal flush with special tool. Fill the cavity with grease provided in the kit. Install the flange yoke to the centering ball, making sure that the alignment marks line up. Install the trunnion and bearing caps in the same manner as for the Cleveland unit.

d. Propeller-Shaft Installation Inspect the yoke seal at the transmission extension. Replace it if necessary. Apply a light coating of transmission oil to the transmission-shaft splines. Insert the front-shaft yoke into the transmission extension, making sure that the output-shaft splines mate with the shaft splines. Align the propeller shaft with companion flange using the reference marks established during removal. Remove the tapes used to retain the bearing caps. Connect the exposed bearing caps to the companion flange by installing the retainer strap and screws or bolts (Figs. 22-25 and 22-26).

e. Correcting Propeller-Shaft Imbalance To check for imbalance, put the car up on a hoist so that the rear wheels are free to rotate. Remove both rear-tire-and-wheel assemblies and brake drums.

CAUTION: Do not apply brakes with the drums removed!

With the transmission in gear and the engine running at the speed at which disturbance was noted, observe the intensity of the disturbance. Stop the engine and check for mud or undercoating on the drive shaft. Remove any found. Check again. If the disturbance is still there, stop engine and disconnect the drive shaft from the companion shaft. Rotate the shaft 180 degrees and reconnect it. Try again. If the problem is not corrected, install a new drive shaft.

NOTE: American Motors Corporation suggests using an electronic wheel balancer to check for unbalance. With rear axles free to spin and rear wheels removed, put the electronic pickup under the axle housing as close to the pinion yoke as possible. Use crayon or chalk and mark four equally spaced horizontal lines on the propeller shaft. Make the lines of different lengths so that you can identify each one. Operate the car and use the wheel balancer to locate the heavy spot as identified by the length of line.

American Motors Corporation further suggests the use of two hose clamps (worm type) as near the rear of the shaft as possible (Fig. 22-37). The heads of the clamps should be 180 degrees from the heavy spot on the shaft. Then start the engine and make the test again. If vibration still exists, move the clamp heads equal distances away from their original spots (Fig. 22-38). Move them only a few degrees and try again. If the problem is not quite solved, move the clamp heads some more. You will find spots at which good balance at the rear can be achieved. Then repeat the procedure at the front of the drive shaft. If the transmission has an aluminum extension, install a steel hose clamp around the transmission so that the magnet will stay put.

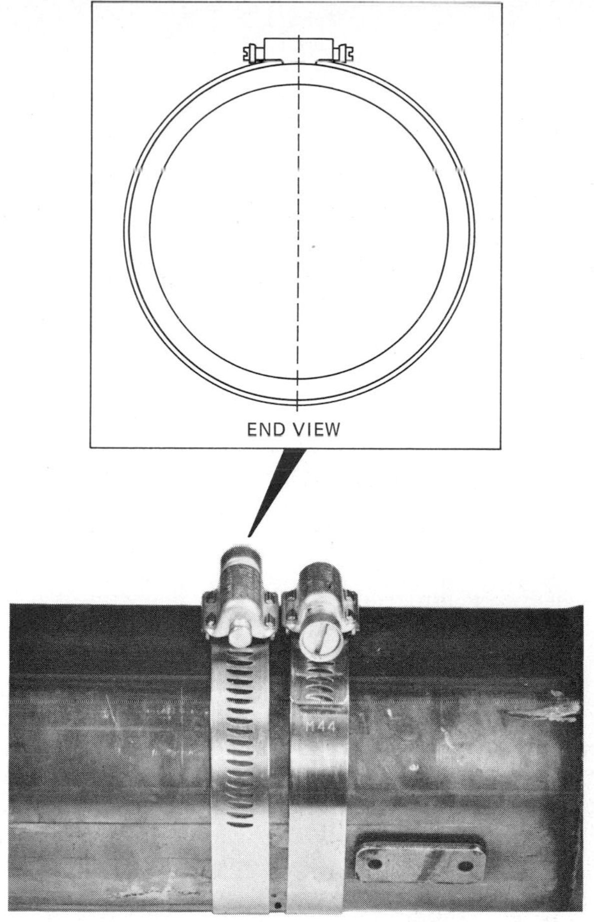

Fig. 22-37. Installation of two worm-type hose clamps on the drive shaft. (*American Motors Corporation*)

⊘ 22-7 Ford Drive-Line Service Ford Motor Company vehicles use two types of universal joints: the single spider and double-Cardan constant-velocity universal joints (Figs. 22-39 and 22-40). The drive shaft is removed and installed in the same manner as for the Chevrolet units, described in ⊘ 22-6.

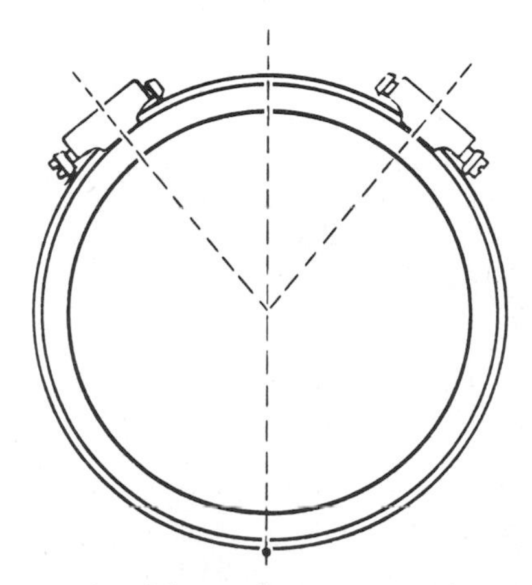

Fig. 22-38. Heads of hose clamps moved equal distance toward the heavy spot to achieve shaft balance. (*American Motors Corporation*)

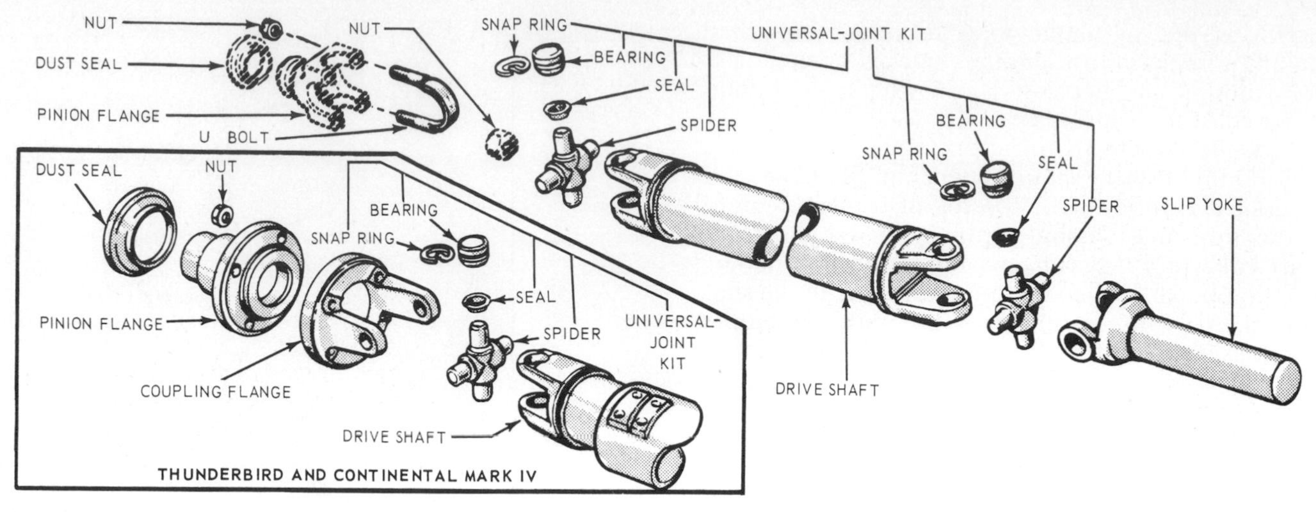

Fig. 22-39. Disassembled view of the drive shaft and universal joints. (*Ford Motor Company*)

Disassembly and reassembly are also simpler, but Ford recommends the use of a special C-clamp type of tool (Figs. 22-41 and 22-42).

1. *SHAFT-BALANCE CHECK* To check drive-shaft balance, bring a crayon or colored pencil up to the rear end of the rotating drive shaft (Fig. 22-43). The car must be on a hoist so that the wheels can spin, and the speedometer should read 40 to 45 mph [64.37 to 72.42 km/h]. If the shaft is unbalanced, the mark on the shaft will be light on one side and heavy on the other.

CAUTION: Keep away from the balance weights on the shaft so that you don't hurt yourself.

Ford recommends the use of two worm-type hose clamps, as already described and recommended by American Motors Corporation. See Figs. 22-44 and 22-45.

⊘ 22-8 Plymouth Drive-Line Service The drive-line service recommended for Chrysler Corporation cars is very similar to that already described for

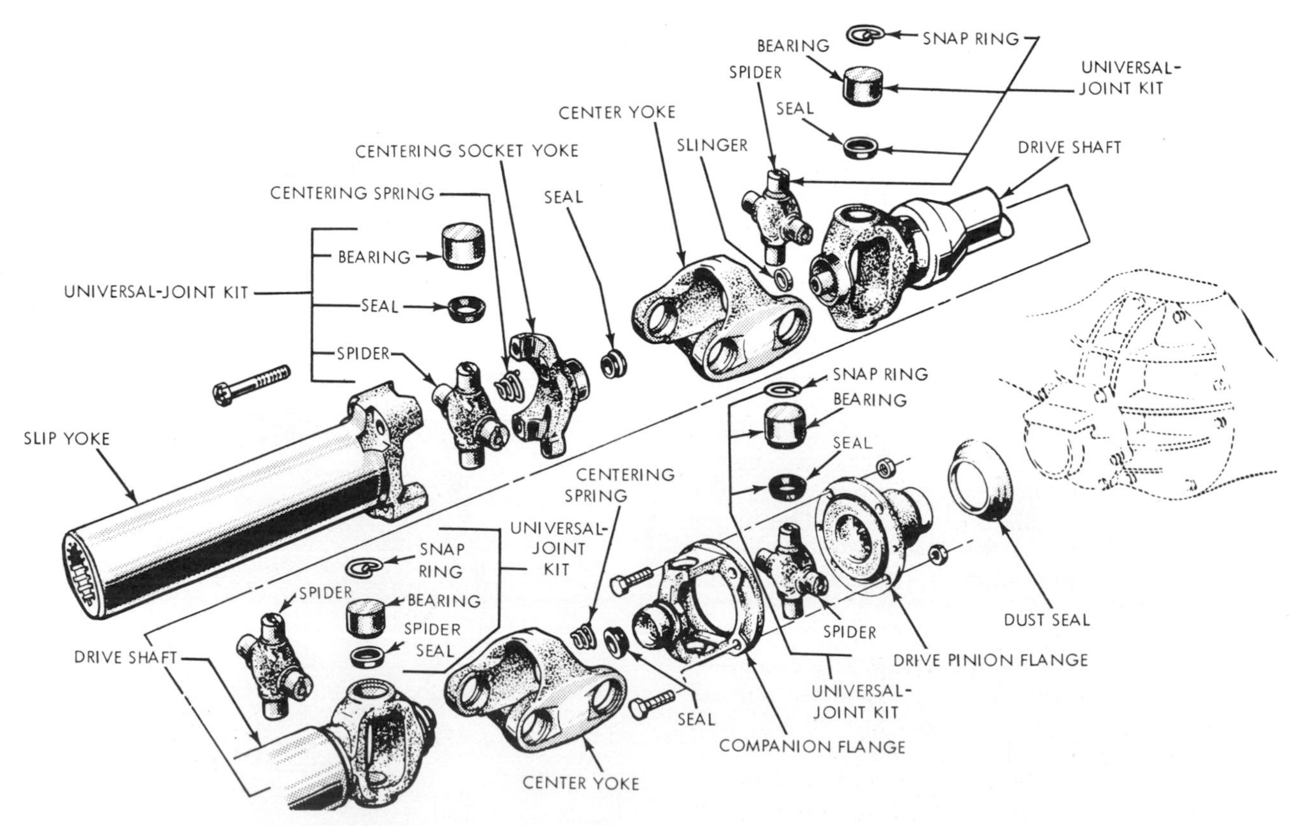

Fig. 22-40. Disassembled view of the drive shaft and constant-velocity universal joint. (*Ford Motor Company*)

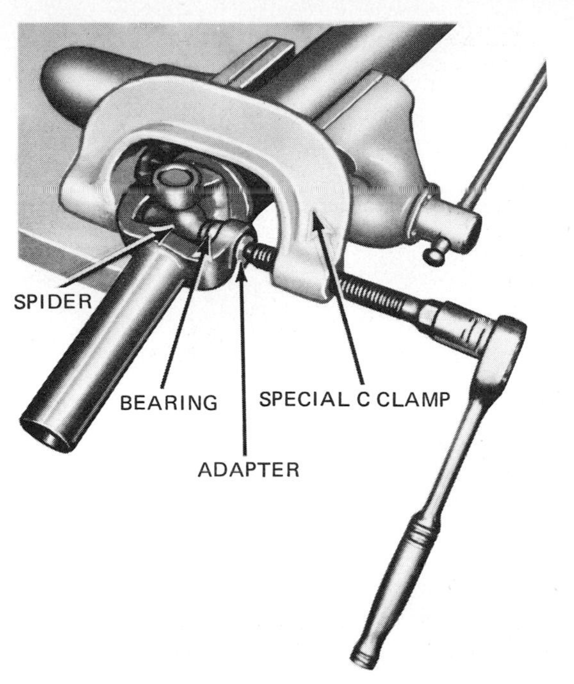

Fig. 22-41. Removing a universal-joint bearing with a special C clamp. (*Ford Motor Company*)

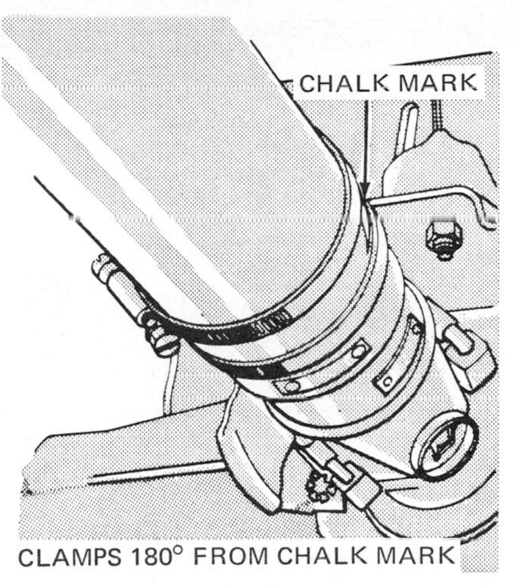

Fig. 22-44. Installation of worm-type hose clamps on a drive shaft. (*Ford Motor Company*)

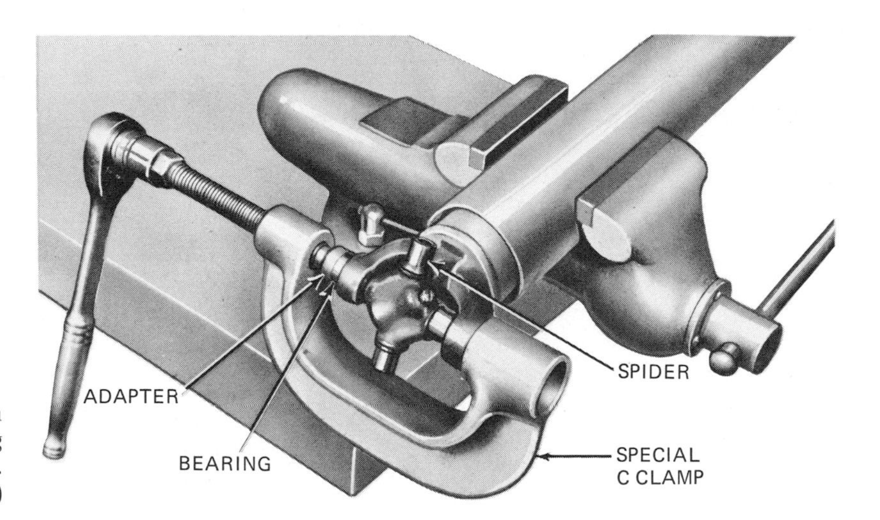

Fig. 22-42. Installing a universal-joint bearing with a special C clamp. (*Ford Motor Company*)

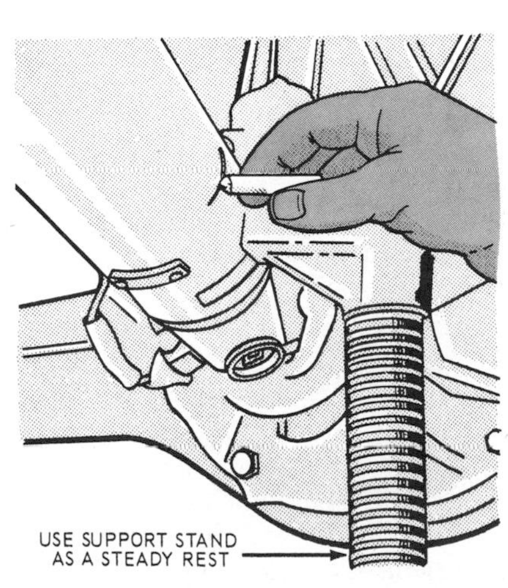

Fig. 22-43. Marking a drive shaft. (*Ford Motor Company*)

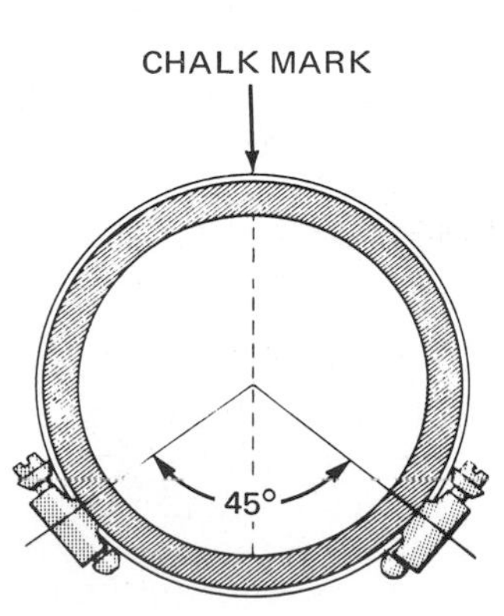

Fig. 22-45. Separating clamps away from light spot to achieve shaft balance. (*Ford Motor Company*)

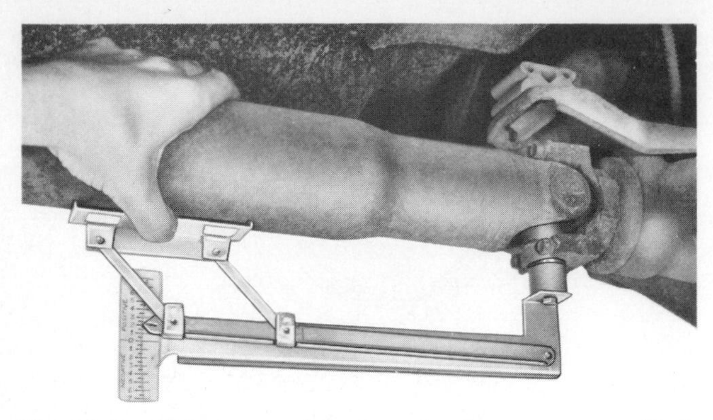

Fig. 22-46. Using a special measuring instrument to check the universal-joint angle at the rear of the drive shaft. (*Chrysler Corporation*)

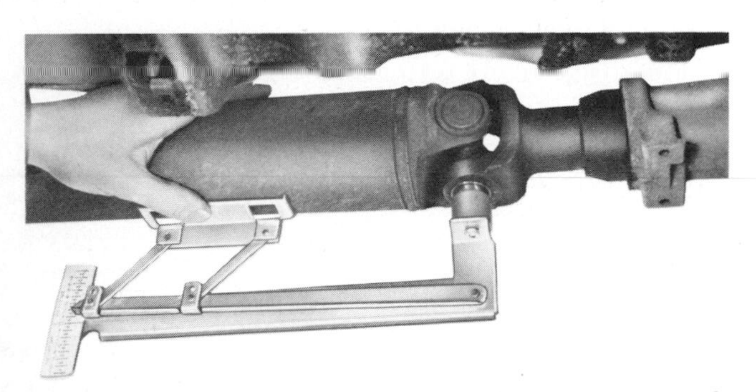

Fig. 22-47. Using a special measuring instrument to check the universal-joint angle at the transmission. (*Chrysler Corporation*)

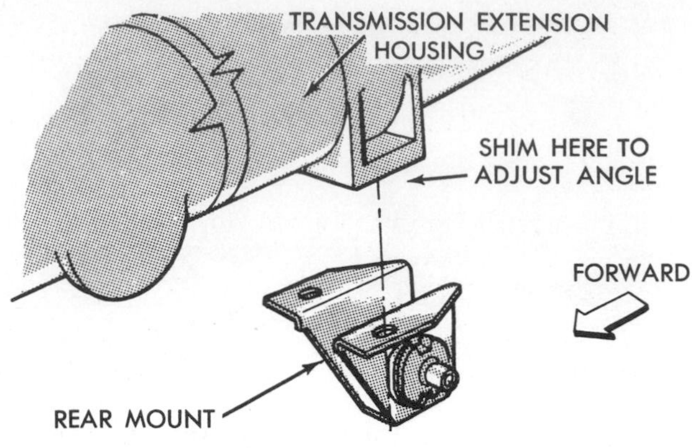

Fig. 22-48. Where shims are installed at the front of the drive shaft to correct universal-joint angle. (*Chrysler Corporation*)

Fig. 22-49. Installing a tapered wedge above the spring to correct the rear universal-joint angle. (*Chrysler Corporation*)

Chevrolet and Ford. However, one additional service is recommended by Chrysler: checking the universal-joint angles. These angles are very important. The greater the angles, the greater the fluctuation of speed through the universal joint (except for the constant-velocity type). A wide difference in speed will cause drive-line vibration.

Figures 22-46 and 22-47 show the use of a special angle-measuring instrument at the rear and front of the drive line. If the front angle requires correction, install shims as shown in Fig. 22-48. To correct the angle at the rear, install tapered shims between the springs and spring seats (Fig. 22-49).

CHAPTER 22 CHECKUP

NOTE: Since the following is a chapter review test, you should review the chapter before taking it.

The drive line, or propeller shaft, with its universal and slip joints, is a small but important part of the car. You should have a good acquaintance with its purpose, construction, operation, and service, as described in this chapter. To check yourself on how well you remember the material you have just read on the drive line, take the following test. Write the answers in your notebook.

Completing the Sentences The sentences that follow are incomplete. After each sentence there are several words or phrases, only one of which will correctly complete the sentence. Write each sentence in your notebook, selecting the proper word or phrase to complete it correctly.

1. The drive line consists of the propeller shaft with: (*a*) drive and universal joints, (*b*) universal and slip joints, (*c*) slip and drive joints.
2. The universal joint permits a change in the: (*a*) length of shaft, (*b*) speed of rotation, (*c*) angle of drive.
3. The slip joint permits a change in the: (*a*) length of shaft, (*b*) speed of rotation, (*c*) angle of drive.
4. The center part of the typical universal joint is called the: (*a*) trunnion, (*b*) joint, (*c*) bearing, (*d*) spider.
5. In the slip joint, slippage occurs between internally and externally mated: (*a*) couplings, (*b*) balls, (*c*) joints, (*d*) splines.
6. The attempt of the rear-axle housing to rotate in

a direction opposite to rear-wheel rotation is due to: (*a*) acceleration, (*b*) rear-end torque, (*c*) the torque tube.

7. In the Hotchkiss drive, the rear-end torque is absorbed by the: (*a*) torque tube, (*b*) rear springs, (*c*) radius rods.

8. In the other type of drive discussed in the chapter, the rear-end torque is absorbed by a: (*a*) torque tube, (*b*) rear spring, (*c*) radius rod.

Service Procedures To help you remember some of the details of drive-line service, write the answers to the following questions in your notebook. Refer to a service manual covering the item mentioned. Don't copy from the manual; try to write the procedures in your own words as you would explain them to a friend.

1. Explain how to service the drive line of a car equipped with a two-piece drive shaft.

2. Explain how to service a Saginaw universal joint.

3. Explain how to check the drive line of a car for balance and how to correct balance.

4. Explain how to service a double-Cardan universal joint.

SUGGESTIONS FOR FURTHER STUDY

Study the various types of drive lines used on different cars. You can see many of these at a service station where cars are put on lifts for lubrication. Your local school automotive shop or a local garage may have universal joints and other parts used in propeller shafts. Examine these if you have a chance. Study car service manuals that describe drive lines. If possible, watch an automotive technician servicing the drive lines of various cars. Write in your notebook any important facts you learn.

chapter 23

REAR AXLES AND DIFFERENTIALS

This chapter describes the purpose, construction, operation, and service of differentials and rear axles (Fig. 23-1). The differential is part of the rear-axle-housing assembly, which includes the differential, rear axles, wheels, and bearings (see Fig. 23-2).

⊘ 23-1 Function of the Differential As explained in ⊘ 1-8, a differential is required to compensate for the difference in distances that the rear wheels travel when the car rounds a curve. If a right-angle turn were made with the inner rear wheel turning on a 20-foot [6.096 m] radius, the inner rear wheel would travel about 31 feet [9.449 m] while the outer rear wheel would travel about 39 feet [11.89 m] (Fig. 23-3). The differential permits application of power to both rear wheels while allowing the wheels to turn different amounts when the car is rounding a curve (see Fig. 1-26).

⊘ 23-2 Construction of the Differential Figure 23-2 is a cutaway view of a differential. Figure 23-4 shows a similar differential in exploded view.

To study differential construction and operation, we shall build up, gear by gear, a simple differential. The two rear wheels are mounted on axles. On the inner ends of the axles are bevel gears, which are called *differential side gears* (see Fig. 23-5). All the teeth are at an angle. When two bevel gears are put together so that their teeth mesh, the driving and driven shafts can be at a 90-degree angle (Fig. 23-6).

Figure 23-7 shows all the essential parts of a differential. The parts are separated so that they can be seen clearly. Keep referring to Fig. 23-7 as we put the parts of the differential together.

First, we add the differential case to the two wheel axles and bevel (differential side) gears, as shown in Fig. 23-8. The differential case has bearings that permit it to rotate on the two axles. Next, we add the two pinion gears and the supporting shaft, as shown in Fig. 23-9. The shaft fits into the differential case. The two pinion gears are meshed with the differential side gears.

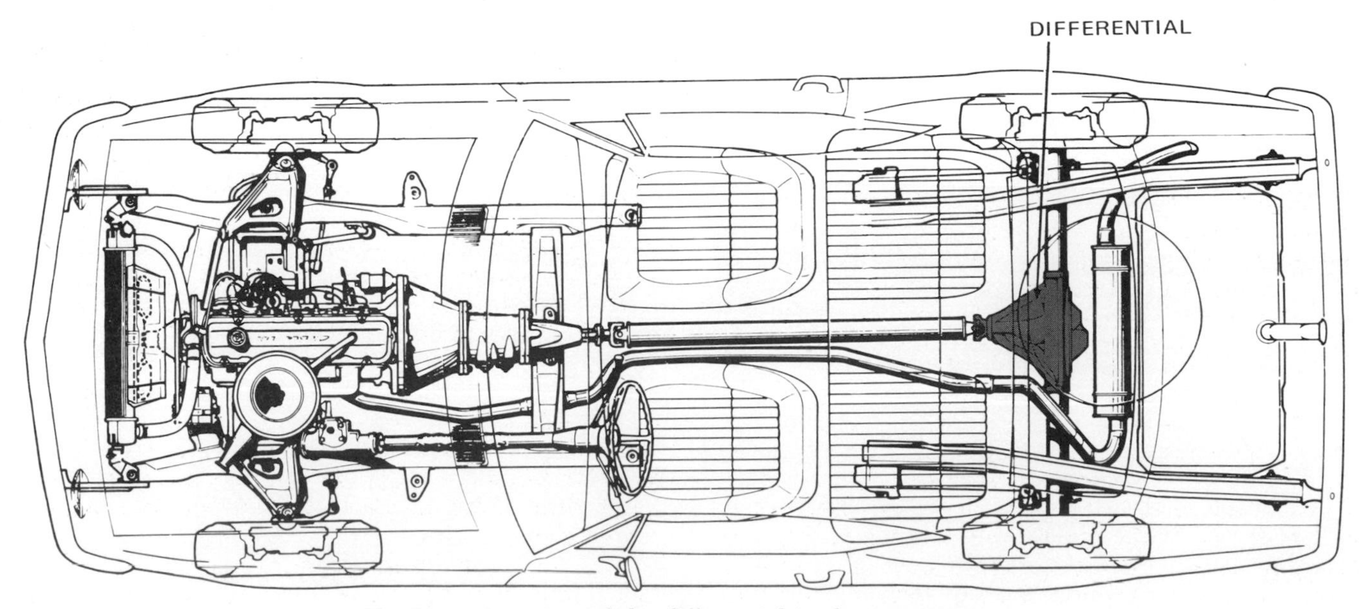

Fig. 23-1. Location of the differential in the power train.

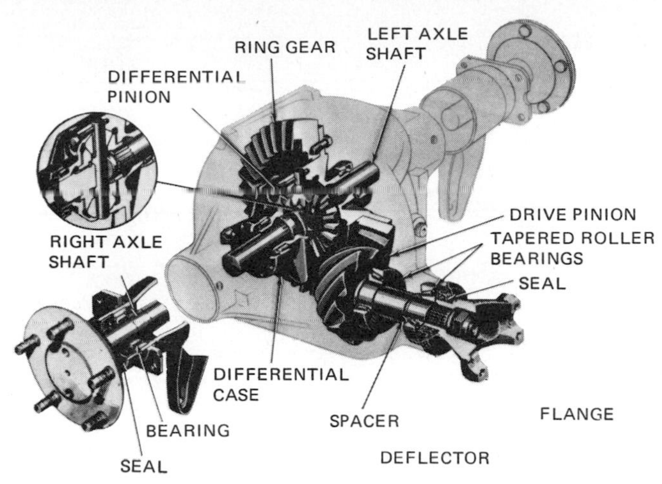

Fig. 23-2. Cutaway view of a differential and rear axle. (*Ford Motor Company*)

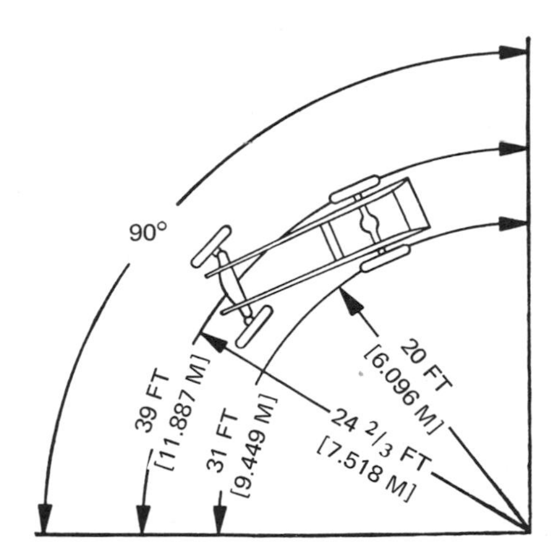

Fig. 23-3. Difference in wheel travel as car makes a 90-degree turn with the inner wheel turning on a 20-foot [6.096 m] radius.

NOTE: Actually, the two pinion gears are also bevel gears, but we call them *pinion gears* so as not to confuse them with the bevel gears (differential side gears) on the ends of the axles.

Now, we add the ring gear, as shown in Fig. 23-10. The ring gear is bolted to the flange on the differential case. Finally, we add the drive pinion, as shown in Fig. 23-11. The drive pinion is at the end of the drive shaft (drive line). When the drive shaft rotates, the drive pinion rotates, which rotates the ring gear.

⊘ 23-3 Operation of the Differential The drive pinion on the end of the drive shaft drives the ring gear. The rotation of the ring gear causes the differential case to rotate. When the differential case rotates, the two pinion gears and their shaft move around in a circle with the differential case. Because the two differential side gears are meshed with the pinion gears, the differential side gears must rotate. This causes the rear axles to rotate. The wheels turn, and the car moves.

Suppose one rear wheel turns slower than the other as the car rounds a curve. As the differential case rotates, the pinion gears must rotate on their shafts. The reason for this is that the pinion gears must walk around the slower-turning differential side gear. Thus, the pinion gears carry additional rotary motion to the faster-turning outer wheel on the turn. The action is shown in the sample (but typical) condition in Fig. 23-12. The differential-case speed is considered to be 100 percent. The rotating action of the pinion gears carries 90 percent of this speed to the slower-rotating inner wheel. It sends 110 percent of the speed to the faster-rotating outer wheel.

You can now see how the differential allows one rear wheel to turn faster than the other. Whenever the car goes around a turn, the outer rear wheel travels a greater distance than the inner rear wheel. The two pinion gears rotate on their shaft and send more rotary motion to the outer wheel.

When the car moves down a straight road, the pinion gears do not rotate on their shaft. They apply equal torque to the bevel gears. Therefore both rear wheels rotate at the same speed.

⊘ 23-4 Differential Gearing Since the ring gear has many more teeth than the drive pinion, a considerable gear reduction is produced in the differential. The gear ratios vary somewhat on different cars, depending on car and engine design. Ratios from 2.75:1 upward to about 5:1 are used on passenger cars. This means that the ring gear has from 2.75 to 5 times as many teeth as the drive pinion, so that the drive pinion must rotate from 2.75 to 5 times (according to gear ratio) in order to cause the ring gear to rotate once. For heavy-duty applications, such as large trucks, ratios of about 9:1 may be used. Such high ratios are secured by use of *double-reduction* gearing (Fig. 23-15), as described in ⊘ 23-6.

The gear ratio in the differential is usually referred to as the *axle ratio,* although it would be more accurate to call it the *differential ratio.*

Early cars used simple spur-gear-type drive pinions and ring gears (Fig. 23-13). In this type of gearing, the lines of the gear teeth are straight and all point toward the center of the gear. The center line of the drive-pinion shaft, if extended, would intersect the center line of the axles.

A later design made use of spiral bevel gears (Fig. 23-13), in which the teeth have a curved, or spiral, shape. This shape permits contact between more than one pair of teeth at a time; more even wear and quieter operation result. Extension of the center line of the drive-pinion shaft would intersect the axle center line.

Modern car design features low bodies, which has created the problem of interference between the drive shaft and the floor of the car body. In order

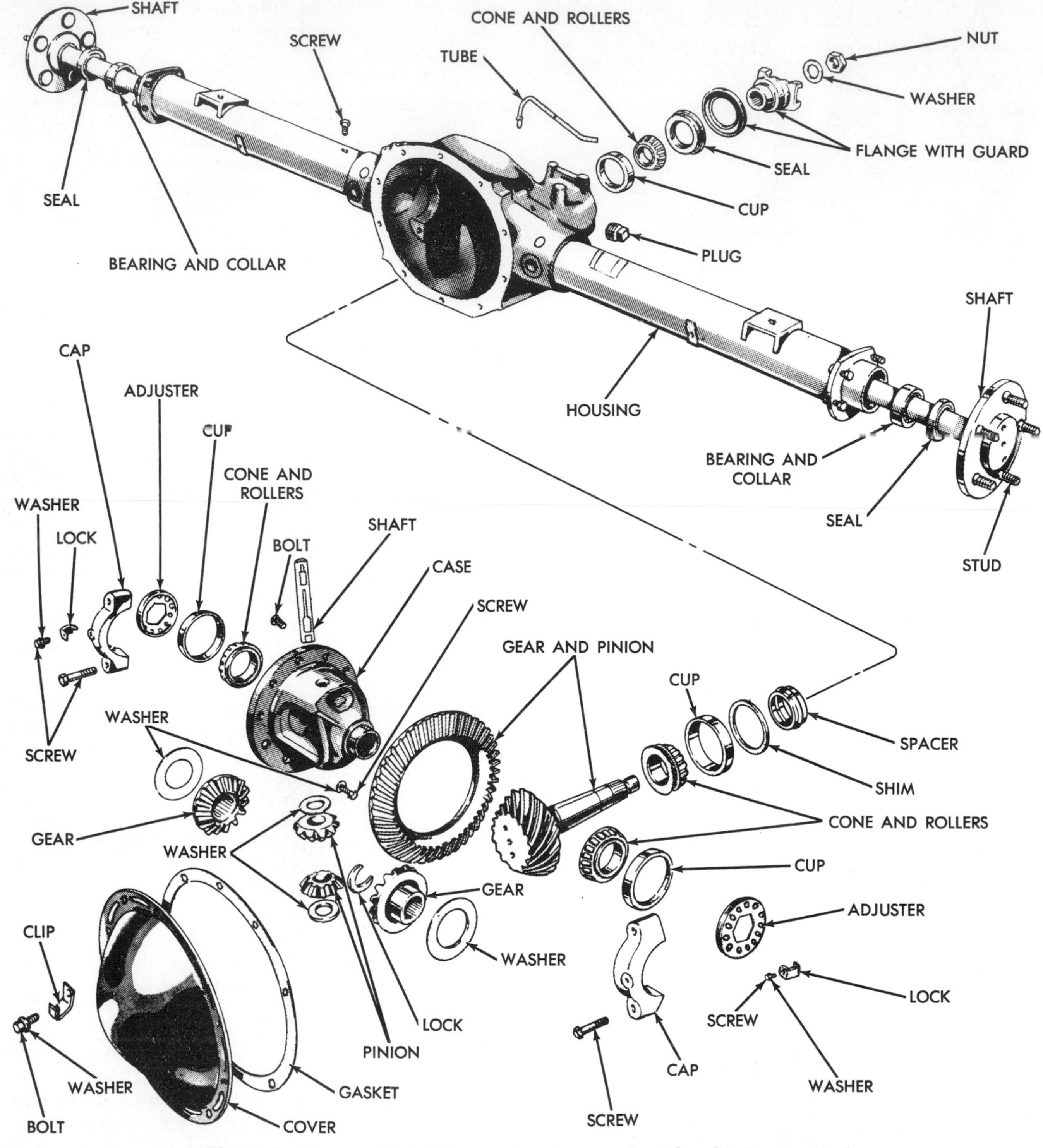

Fig. 23-4. Disassembled differential and rear axle. (*Chrysler Corporation*)

to permit further lowering of the car body without interference with the drive shaft, hypoid differential gears are used (Figs. 23-2 and 23-13). These gears are somewhat similar to the spiral bevel gears except that the tooth formation allows the drive-pinion shaft to be lowered. In this type of gear a wiping action takes place between the teeth as the teeth mesh and unmesh. This wiping action, which is characteristic of hypoid gears, makes the use of special hypoid-gear lubricants necessary.

Figure 23-14 shows gear-tooth nomenclature. The mating teeth to the left illustrate clearance and backlash, and the tooth to the right has the various tooth parts named. *Clearance* is the distance between the top of the tooth of one gear and the valley between adjacent teeth of the mating gear. *Backlash*

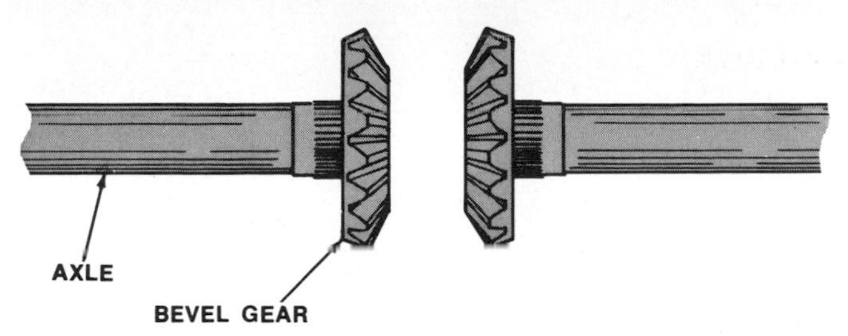

Fig. 23-5. Inner ends of the rear axles with bevel gears (differential side gears) installed on them.

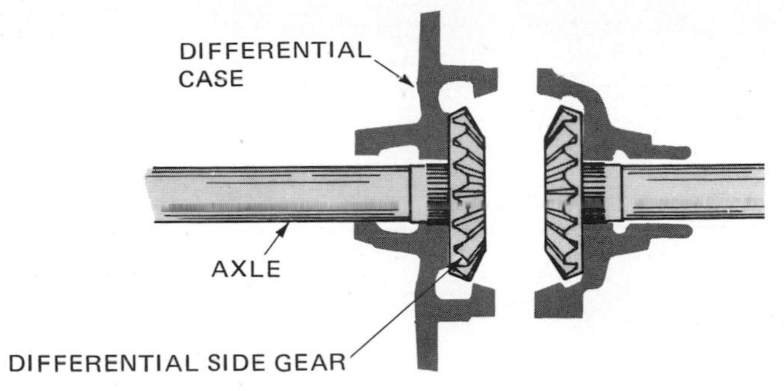

Fig. 23-8. Here, we add the differential case.

Fig. 23-6. Two meshing bevel gears.

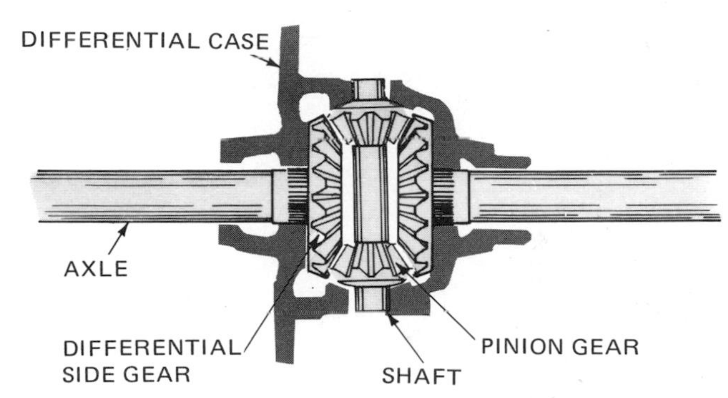

Fig. 23-9. Here, we add the two pinion gears and supporting shaft.

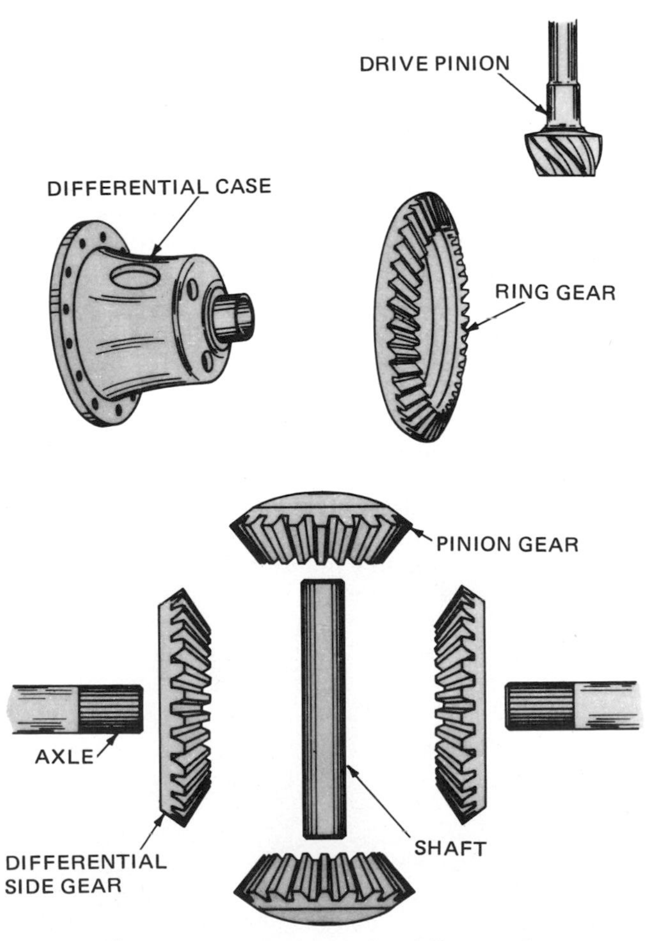

Fig. 23-7. Basic parts of a differential.

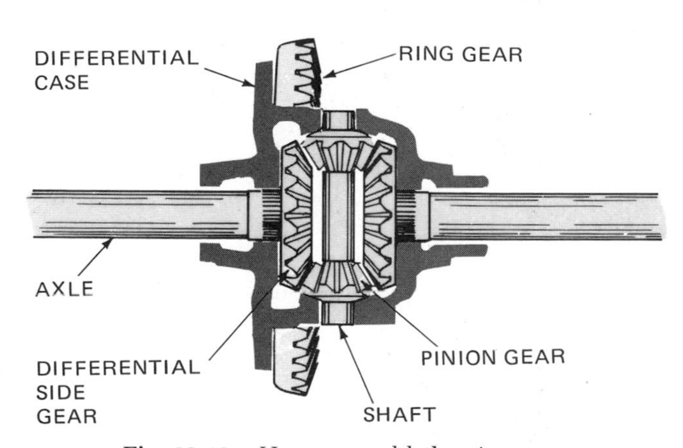

Fig. 23-10. Here, we add the ring gear.

is the distance between adjacent meshing teeth in the driving and driven gears; it is the distance one gear can rotate backward, or backlash, before it will cause the other gear to move. The *toe* is the smaller section of the gear tooth; the *heel* is the larger section of the gear tooth.

⊘ 23-5 Hunting and Nonhunting Gearsets The gearset (consisting of the drive pinion and the drive, or ring, gear) may be of either the *hunting* or the *nonhunting* type. Hunting here refers to the number of drive-gear teeth that each drive-pinion tooth makes contact with. In a hunting-type gearset, any one pinion-gear tooth will come into contact with every drive-gear tooth. In a nonhunting gearset, each pinion-gear tooth will come into contact with only a few drive-gear teeth.

To take a specific example, suppose we have a differential with a drive pinion that has 13 teeth and a drive gear that has 39 teeth. In this gearset, each drive pinion would make contact with only three drive-gear teeth. Every three drive-pinion revolutions, each drive-pinion tooth would meet with the same three drive-gear teeth. This is a nonhunting gearset because the drive-pinion teeth do not "hunt out" different drive-gear teeth; they mesh with the same teeth.

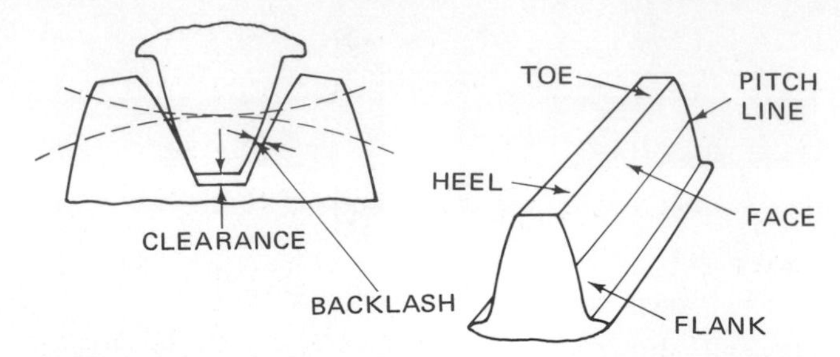

Fig. 23-14. Gear-tooth nomenclature. (*Chrysler Corporation*)

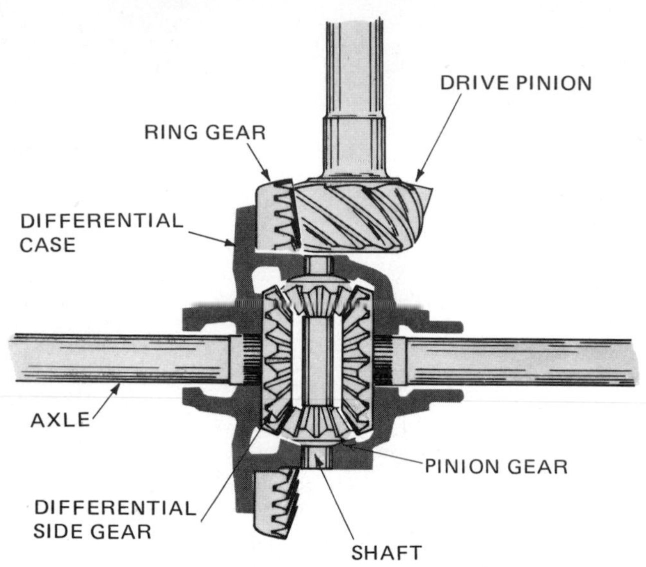

Fig. 23-11. To complete the basic differential we now add the drive pinion. The drive pinion is meshed with the ring gear.

However, suppose that the drive pinion has 9 teeth and the drive gear has 37 teeth. Now, each drive-pinion tooth will mesh with all drive-gear teeth as the gearset revolves. This is a hunting gearset.

When checking the gear-tooth-contact pattern, it is important to know whether the gearset is hunting or nonhunting. The checking procedures and acceptable patterns are different for the two gearsets, as we will explain later.

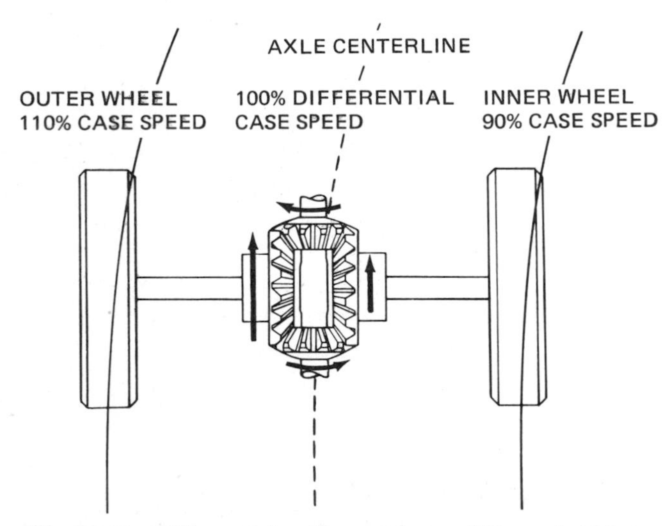

Fig. 23-12. Differential action on turns. (*Chevrolet Motor Division of General Motors Corporation*)

⊘ 23-6 Double-Reduction Differentials In order to secure additional gear reduction through the differential and thus provide a higher gear ratio between the engine and the rear wheels, some heavy-duty applications use *double-reduction differentials* (Fig. 23-15). In this type of differential the drive pinion meshes with a ring gear assembled to a straight shaft on which there is a reduction-drive gearset. The reduction-drive gearset drives a driven gearset that has a greater number of gear teeth. Gear reduction is thus obtained between the drive pinion and the ring gear and also between the two reduction gearsets.

The driven gearset is attached to the differential case, the case being supported by bearings in the differential housing in a manner similar to the differential discussed above. It will be noted that the differential illustrated in Fig. 23-15 has a four-pinion differential instead of the two-pinion differential

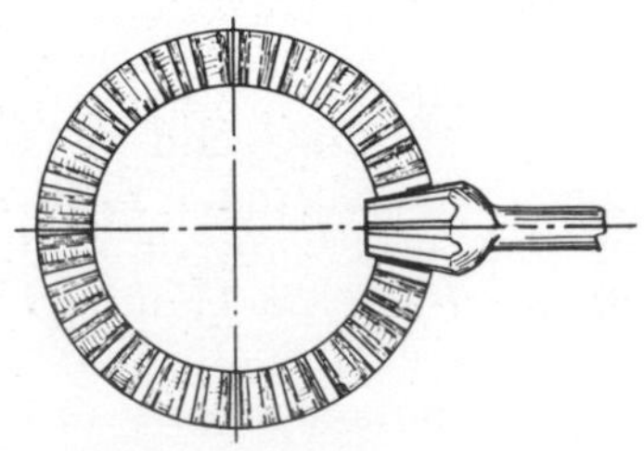
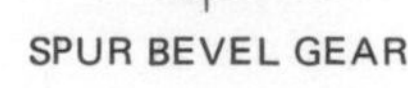

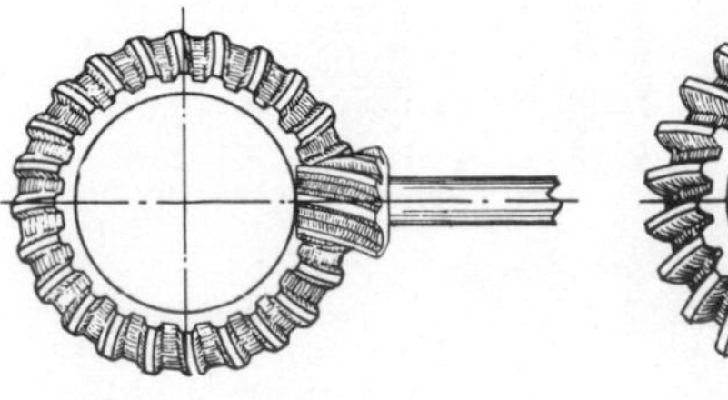

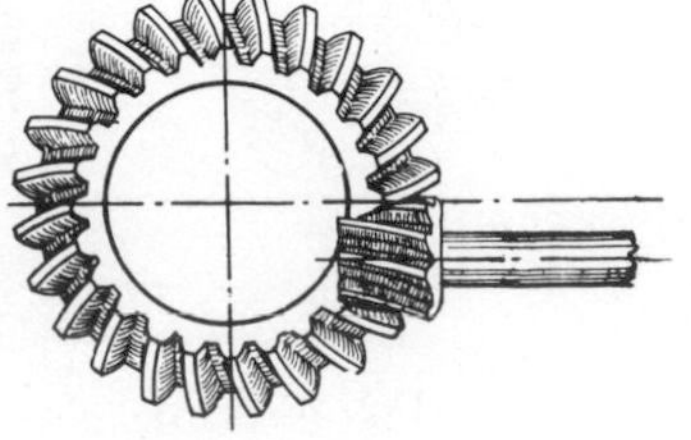

Fig. 23-13. Spur bevel, spiral bevel, and hypoid differential drive pinions and ring gears.

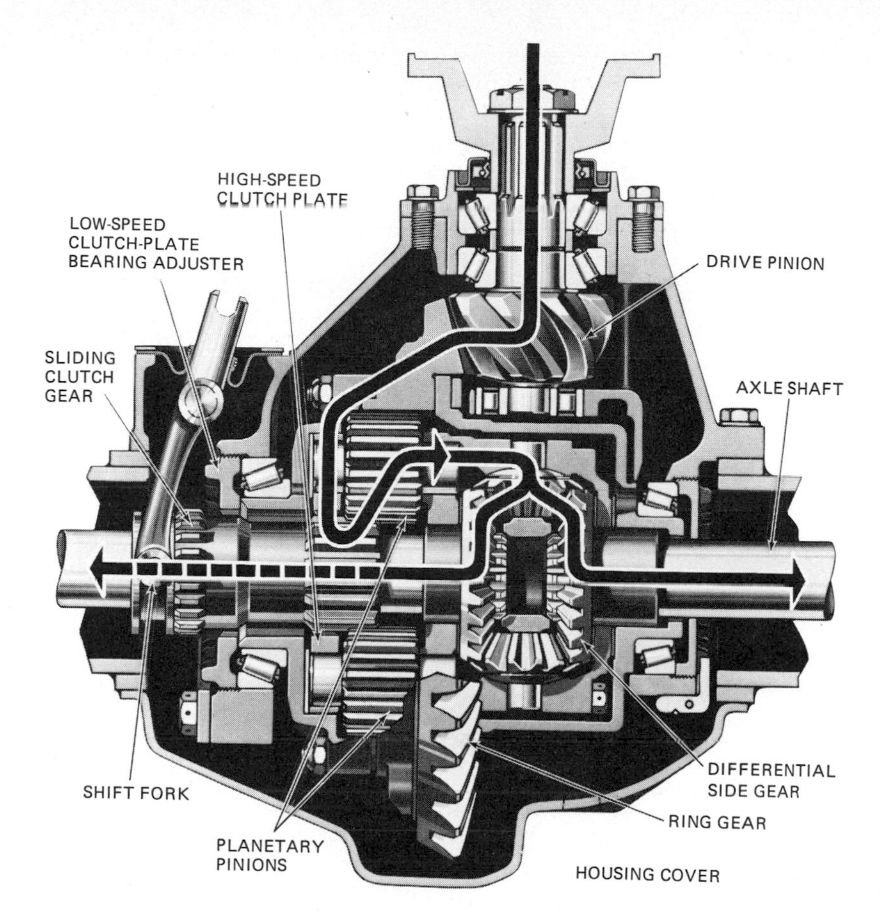

Fig. 23-15. Sectional view of a double-reduction differential. (*Axle Division, Eaton Corporation*)

shown in Figs. 23-2 and 23-4. Otherwise, the construction and principle of operation are the same as on the differential described above.

⊘ 23-7 Nonslip Differentials The standard differential delivers the same amount of torque to each rear wheel when both wheels have equal traction. When one wheel has less traction than the other, for example, when one wheel slips on ice, the other wheel cannot deliver torque. All the turning effort goes to the slipping wheel. To provide good traction even though one wheel is slipping, a *nonslip differential* is used in many cars. It is very similar to the standard unit but has some means of preventing wheel spin and loss of traction.

To sum up, the standard differential delivers maximum torque to the wheel with minimum traction. But the nonslip differential delivers maximum torque to the wheel with maximum traction.

One of the older types of nonslip differentials is shown in Fig. 23-16. It has two sets of clutch plates. Also, the ends of the pinion-gear shafts lie loosely in notches in the two halves of the differential case. Figure 23-17 is a sectional view of the nonslip differential. During normal straight-road driving, the power flow is as shown in Fig. 23-18.

NOTE: In Figs. 23-16 to 23-18, the ring gear is called the *axle drive gear*. The bevel gear is called the *differential side gear*. The pinion gears are called the *differential pinions*.

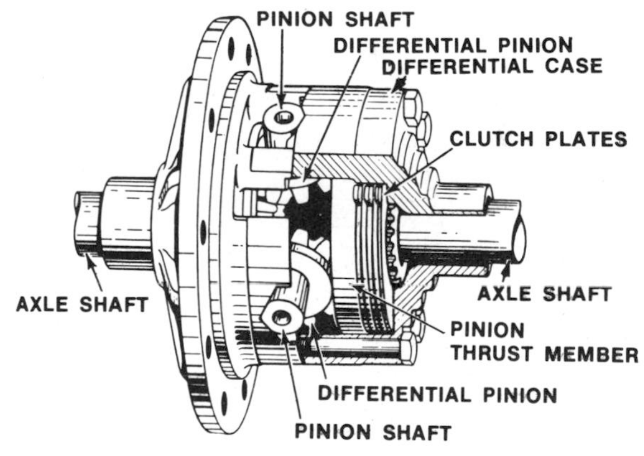

Fig. 23-16. Cutaway view of a nonslip differential. (*Chrysler Corporation*)

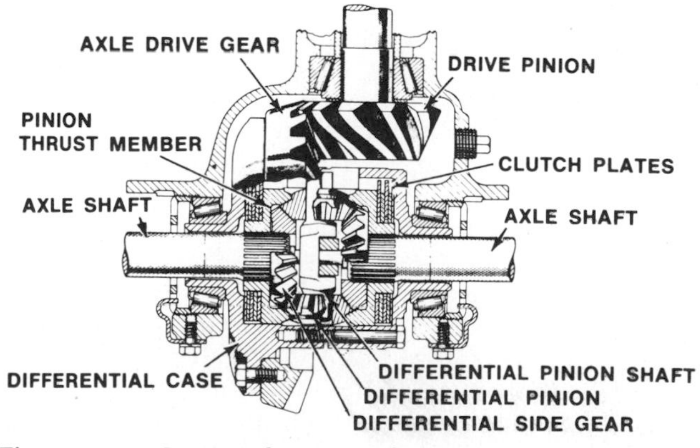

Fig. 23-17. Sectional view of a nonslip differential. (*Chrysler Corporation*)

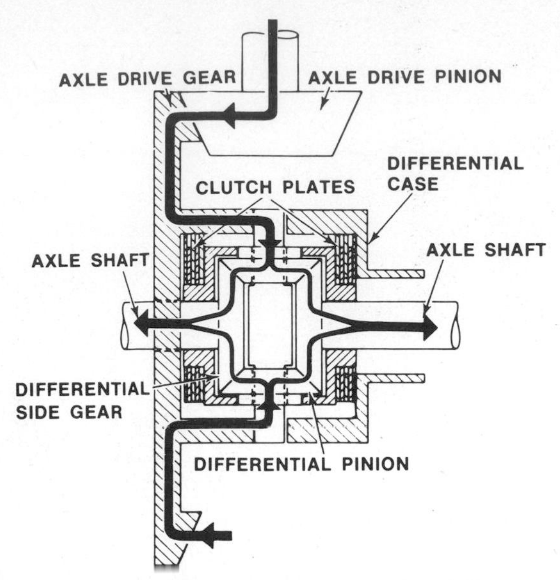

Fig. 23-18. Power flow through a nonslip differential on a straightaway. (*Chrysler Corporation*)

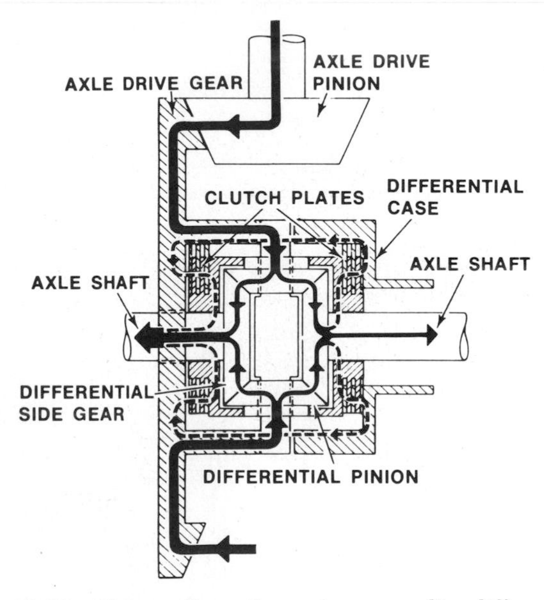

Fig. 23-19. Power flow through a nonslip differential when rounding a turn. Heavy arrows show greater torque to the left axle shaft. (*Chrysler Corporation*)

Note that the rotating differential case carries the pinion-gear shafts around with it. Since there is considerable side thrust, the pinion shafts tend to slide up the sides of the notches in the two halves of the differential case. As the pinion shafts slide up, they are forced outward. This force is carried to the two sets of clutch plates. The clutch plates thus lock the axle shafts to the differential case. Therefore both wheels turn.

Suppose one wheel encounters a patch of ice or snow and loses traction, or tends to slip. Then the pressure is released on the clutch plates feeding power to that wheel. Thus, the torque goes to the other wheel, and the wheel on the ice does not slip.

During normal driving, if the car rounds a curve, pressure is released on the clutch feeding the inner wheel. Just enough pressure is released to permit some slipping. Figure 23-19 shows the action. This release of pressure permits the outer wheel to turn faster than the inner wheel.

Late-model nonslip differentials are of two types: spring-loaded clutch plates and spring-loaded clutch cones. These differentials are shown in disassembled views in Figs. 23-20 and 23-21. The action is the same as in the nonslip differential described above. The essential difference is that the plates or cones are preloaded by springs to give a more positive action.

⊘ 23-8 Rear-Axle and Differential Trouble Diagnosis Most often, it is noise that draws attention to trouble existing in the rear axles or differential. It is not always easy, however, to diagnose the trouble by determining the sort of noise and the operating conditions under which noise is obtained. Such conditions as defective universal joints, rear-wheel bearings, and muffler or tire noises may be improperly diagnosed as differential or rear-axle trouble. Some clue as to the cause of trouble may be gained, however, by noting whether the noise is a hum, growl, or knock; whether it is obtained when the car is operating on a straight road or on turns only; and whether the noise is most noticeable when the engine is driving the car or when the car is coasting.

A humming noise in the differential is often caused by improper drive-pinion or ring-gear adjustment, which prevents normal tooth contact between the gears. This condition produces rapid gear-tooth wear, and so the noise will gradually take on a growling characteristic. Correction should be made before the trouble progresses this far since such wear will require pinion and gear replacement.

If the noise is most prominent when the car accelerates, the probability is that there is heavy heel contact on the gear teeth. The ring gear must be moved near the drive pinion. If the noise is most prominent when the car coasts in gear with the throttle closed, it is probable that there is heavy toe contact on the gear teeth. The ring gear must be moved away from the drive pinion. Following sections described how gear-tooth contact is tested and ring-gear or pinion-drive adjustment is made.

NOTE: Tire noise is sometimes mistaken for differential noise. Since tire noise varies considerably according to the type of pavement while differential noise does not, the car should be driven over various types of pavement to determine whether the noise is resulting from tires or from the differential.

If the noise is present only when the car is rounding a curve, the trouble is due to some condition in the differential-case assembly. Differential pinion gears tight on the pinion shaft, differential side gears tight in the differential case, damaged

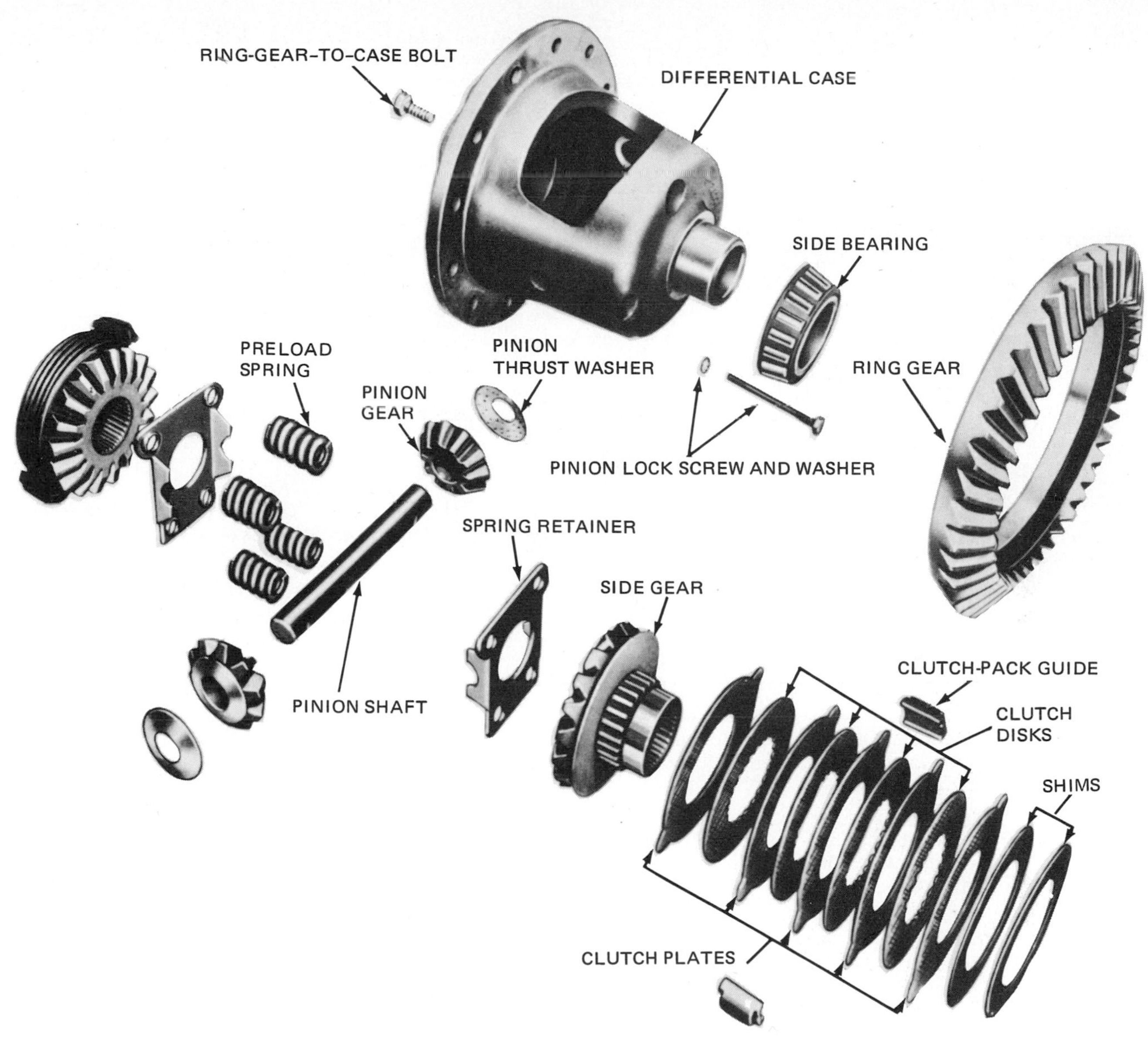

Fig. 23-20. Disassembled view of a nonslip differential with clutch plates. (*Chevrolet Motor Division of General Motors Corporation*)

gears or thrust washers, or excessive backlash between gears could produce noise when the car turns. A knocking noise will result if bearings or gears are damaged or badly worn.

Check Your Progress

Progress Quiz 23-1 Here is your chance to check up on how well you are understanding and remembering what you are reading. Don't be discouraged if any of the following questions stump you. Just review the past few pages and try again.

Completing the Sentences The sentences that follow are incomplete. After each sentence there are several words or phrases, only one of which will correctly complete the sentence. Write each sentence in your notebook, selecting the proper word or phrase to complete it correctly.

1. In the differential, the ring gear is bolted to the: (*a*) differential housing, (*b*) differential case, (*c*) drive pinion, (*d*) axle shaft.
2. The drive pinion is assembled into the: (*a*) carrier, (*b*) differential housing, (*c*) axle housing.
3. In the modern differential, the type of gearing used for the drive pinion and ring gear is: (*a*) spur, (*b*) spiral bevel, (*c*) hypoid.
4. The distance between adjacent meshing teeth of mating gears is called: (*a*) clearance, (*b*) pitch line, (*c*) backlash, (*d*) flank.
5. In a differential with a gear ratio of 4:1, the drive pinion would revolve four times to cause the ring gear to rotate: (*a*) one time, (*b*) two times, (*c*) four times, (*d*) sixteen times.

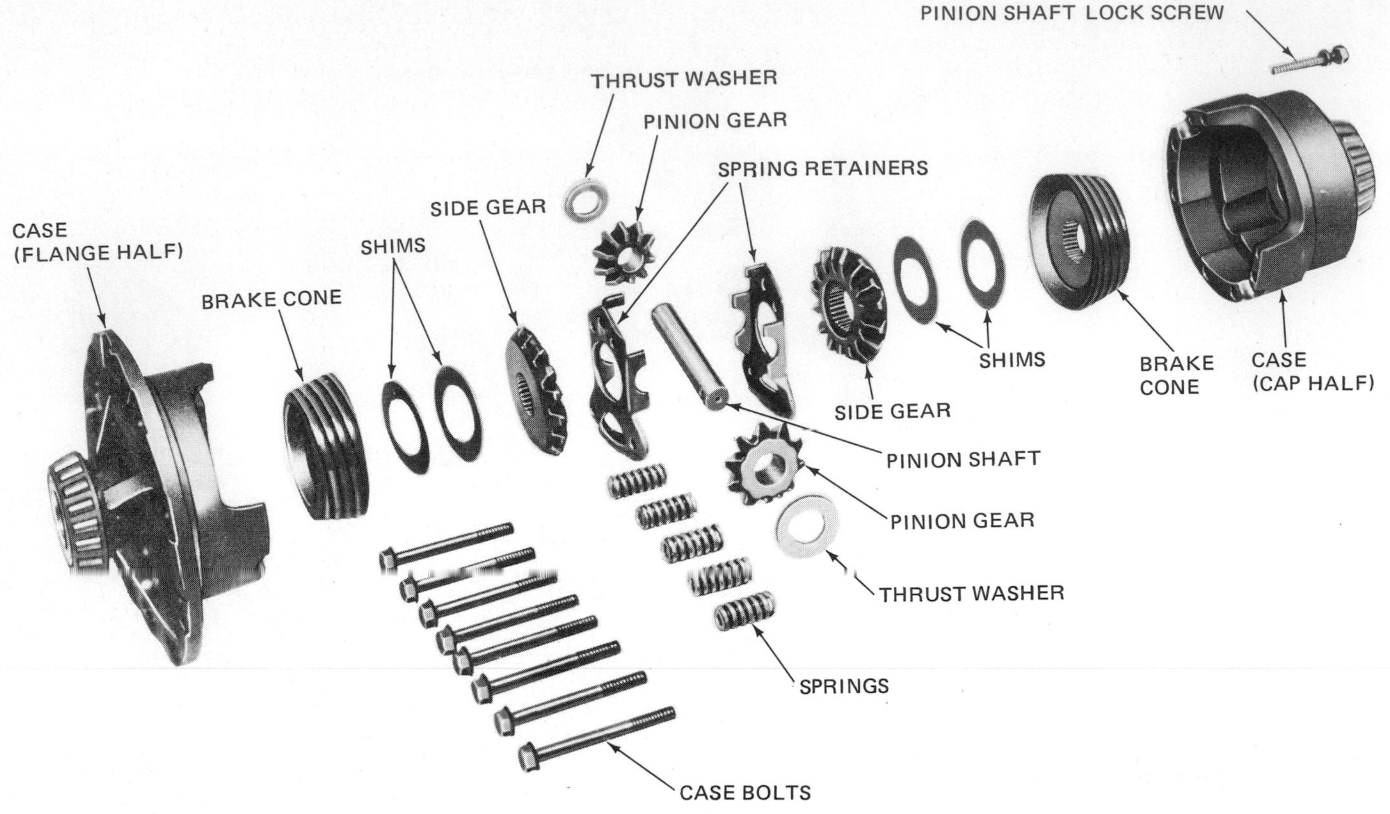

Fig. 23-21. Disassembled view of a nonslip differential with clutch cones. (*Chevrolet Motor Division of General Motors Corporation*)

6. If the drive pinion has 10 teeth and the drive (ring) gear has 30 teeth, the gearset is of the: (*a*) hunting type, (*b*) nonhunting type.
7. If the drive pinion has 10 teeth and the drive (ring) gear has 33 teeth, the gearset is of the: (*a*) hunting type, (*b*) nonhunting type.
8. If a humming noise from the differential is most noticeable when accelerating, chances are there is: (*a*) heavy heel contact on gear teeth, (*b*) heavy toe contact on gear teeth, (*c*) binding in differential case.
9. If a humming noise from the differential is most noticeable when coasting with the car in gear and the throttle closed, chances are there is: (*a*) heavy heel contact on gear teeth, (*b*) heavy toe contact on gear teeth, (*c*) binding in differential case.
10. If the differential noise is present only when the car is rounding a curve, the trouble could be due to: (*a*) heavy heel contact on gear teeth, (*b*) heavy toe contact on gear teeth, (*c*) binding in differential case.

⊘ 23-9 Rear-Axle and Differential Repair A considerable variety of differentials will be found on late-model cars. Plymouth describes five different types in its most recent shop manuals. Ford describes 12, including those used in its heavy-duty trucks; and Chevrolet describes 10, also including those used in its heavy-duty trucks. The repair procedures on these differentials vary, even though the differentials are much alike in basic construction and operation. As examples of service procedures, we include here the procedures for two standard differentials and three nonslip differentials.

CAUTION: Always refer to the applicable shop manual before attempting to service a specific make and model of differential.

⊘ 23-10 Rear-Axle In-Car Repairs (Standard) The rear-axle shafts, wheel bearings, and oil seals can be replaced without removing the differential assembly, as follows:

1. SHAFT, WHEEL BEARING (BALL-BEARING TYPE), AND OIL SEAL Remove the wheel. Detach the brake drum from the axle flange. Work through the hole in the axle-shaft flange to remove the nuts that attach the wheel-bearing-retainer plate. Pull the axle assembly (Fig. 23-22). Do not dislodge the brake-carrier plate. Install one nut to hold it in place after the axle is removed.

To replace the wheel bearing, first loosen the inner retaining ring by nicking it deeply in several places with a cold chisel (Fig. 23-23). This action loosens the ring so that it slide off. Then use the tool shown in Fig. 23-24 to press off the wheel bearing.

Whenever a rear axle is removed, the oil seal must be replaced. The old oil seal is removed with a special tool and slide hammer (Fig. 23-25). The new oil seal must be soaked in SAE10 oil for 30 minutes before installation. It is installed with a special tool

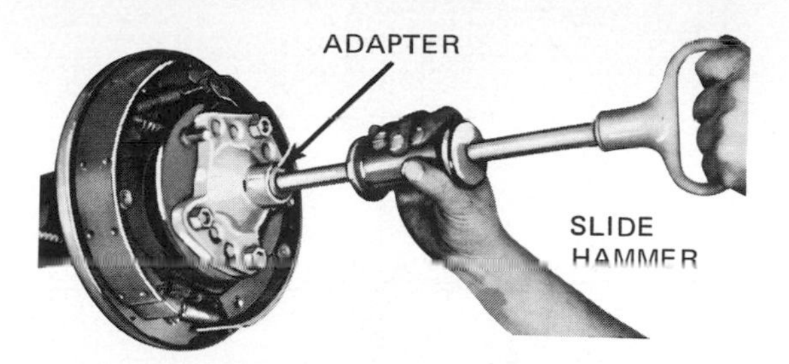

Fig. 23-22. Pulling the axle shaft. (*Ford Motor Company*)

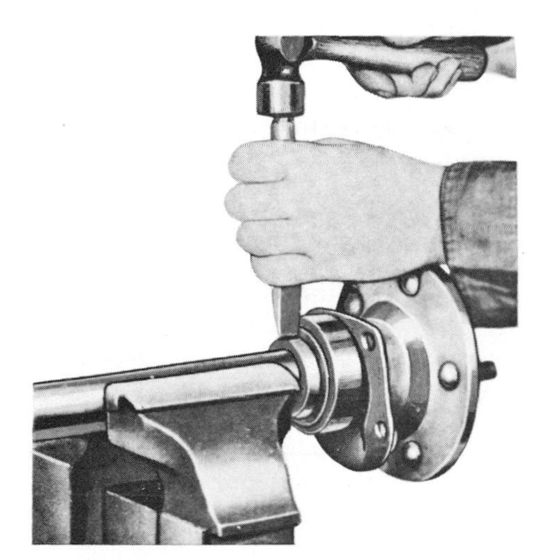

Fig. 23-23. Removing the rear-wheel-bearing retainer ring. (*Ford Motor Company*)

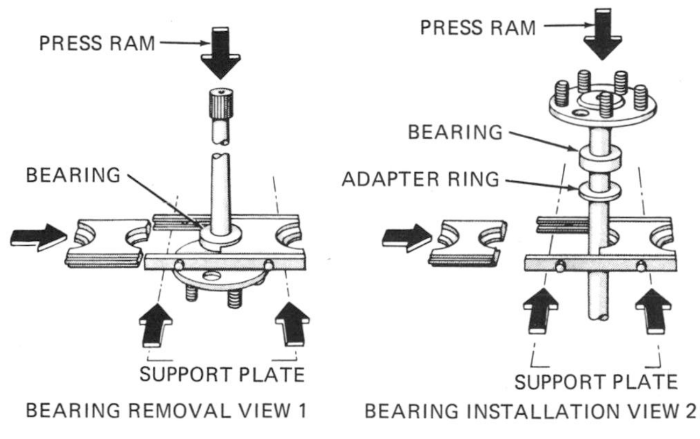

Fig. 23-24. Wheel-bearing removal and replacement. (*Ford Motor Company*)

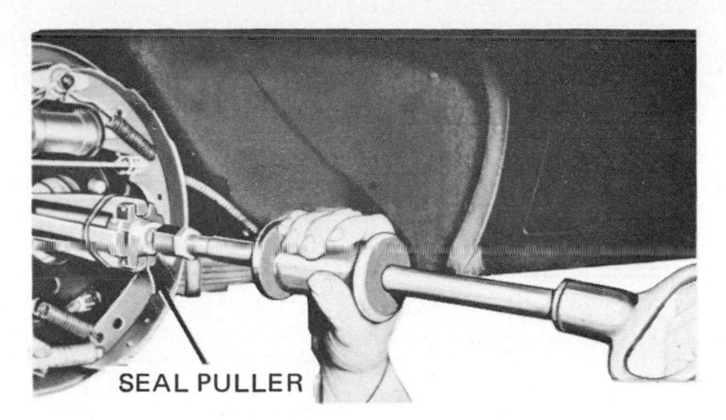

Fig. 23-25. Removing the rear-wheel-bearing oil seal. (*Ford Motor Company*)

Fig. 23-26. Installing the rear-wheel-bearing oil seal. (*Ford Motor Company*)

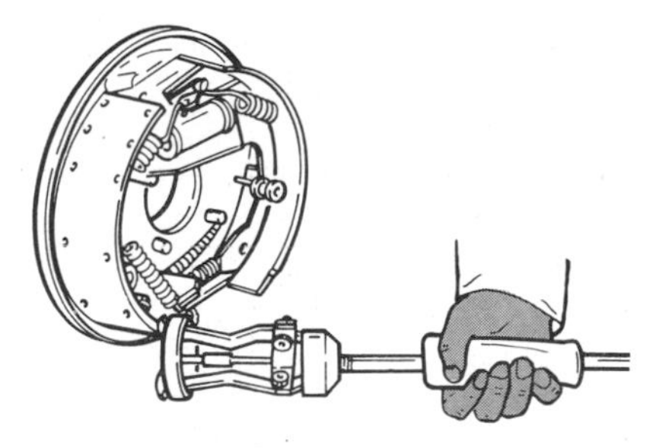

Fig. 23-27. Removing the tapered bearing cup with a slide-hammer puller. (*Ford Motor Company*)

(Fig. 23-26). Coat the outer edge of the oil seal with oil-resistant sealer before installation. Do not get sealer on the sealing lip!

CAUTION: Do not attempt to press the wheel bearing and the bearing-retainer ring on at the same time. Press each on separately (Fig. 23-24).

Put a new gasket on each side of the brake-carrier plate and then slide the axle shaft into the housing, working carefully so that the rough edges on the shaft will not damage the oil seal. Align the axle splines with the side-gear splines, and push the shaft in until the bearing bottoms. Install the bearing-retainer plate and nuts, brake drum and nuts, and wheel.

2. *SHAFT, WHEEL BEARING (TAPERED ROLLER TYPE), AND OIL SEAL* Remove the axle-shaft-retainer nuts and bolts from the housing. The tapered bearing cup will normally stay in the axle housing. A standard slide-hammer puller can be used to remove it (Fig. 23-27). On reinstallation, put the bearing cup on the shaft and install the complete bearing as a unit with the shaft.

To remove the tapered bearing and seal, drill a ¼-inch hole [6.35 mm] in the outside diameter of the inner retainer (Fig. 23-28). Make the hole about three-fourths the thickness of the retainer. Do not drill all the way through because this could damage the axle shaft. Then use a chisel (Fig. 23-29) to break the ring. Use a puller, as shown in Fig. 23-30, to pull the bearing cone. Do not allow the puller jaws to get behind the outer retainer.

NOTE: If the oil seal requires replacement, the entire bearing assembly must also be replaced.

Figure 23-31 shows the installation of a seal and bearing. It is pressed down into place with pressure on the inside race only. The inner cone must seat squarely on the shaft shoulder. Press a new retainer onto the axle shaft so that it seats firmly against the bearing.

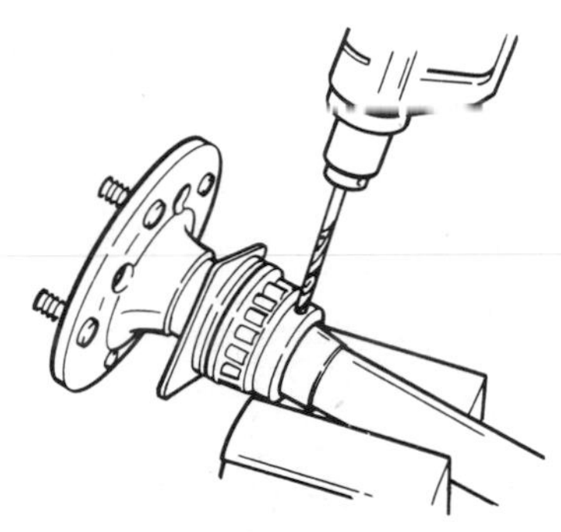

Fig. 23-28. Drilling the retainer ring. (*Ford Motor Company*)

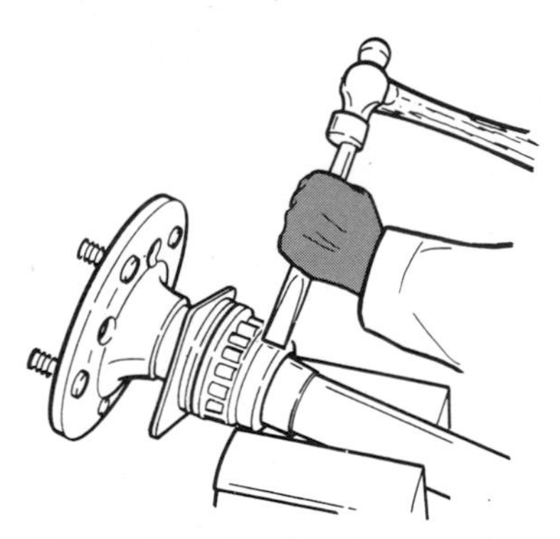

Fig. 23-29. Removing the bearing retainer ring. (*Ford Motor Company*)

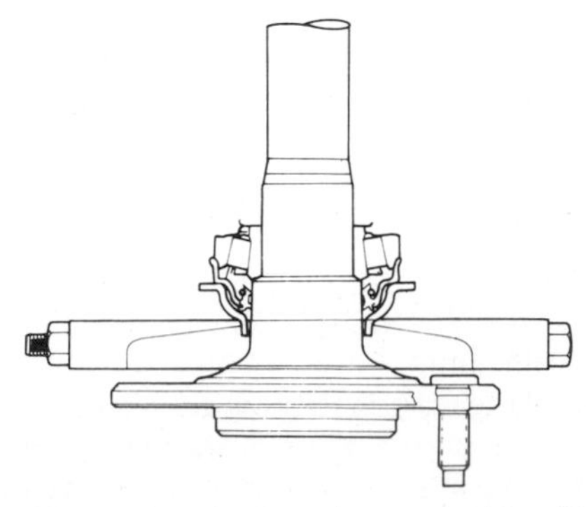

Fig. 23-30. Removing the bearing assembly. (*Ford Motor Company*)

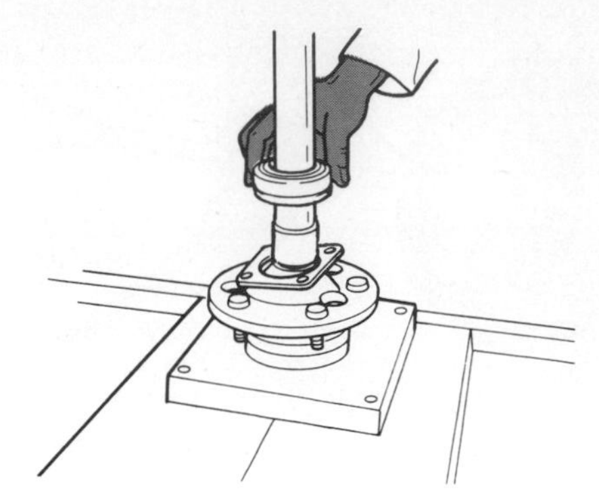

Fig. 23-31. Installing the bearing and seal. (*Ford Motor Company*)

3. *DRIVE-PINION OIL SEAL* The new oil seal should be soaked for 30 minutes in SAE10 oil before installation. Mark scribe lines on the drive-shaft end yoke and the U-joint flange (Fig. 23-32) to ensure proper reassembly. Disconnect the drive shaft, being careful not to drop the loose universal-joint-bearing cups. (See also ⊘ 22-6 on drive-shaft removal.) Detach the drive shaft at the transmission. Install an oil-seal replacer tool in the transmission extension housing to prevent leakage of oil from the transmission.

Make punch marks on the end of the pinion shaft, shaft nut, and U-joint flange inner surface so that parts can be properly aligned on reassembly. Use a special tool to hold the flange and remove the pinion nut and washer (Fig. 23-33).

NOTE: Figures 23-33 to 23-37 show the servicing of a detached differential. However, the special tools shown can also be used on a differential in the car.

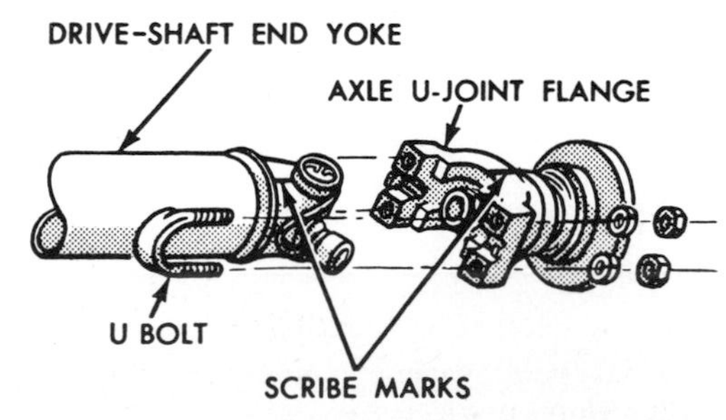

Fig. 23-32. Drive-shaft-to-U-joint connection. (*Ford Motor Company*)

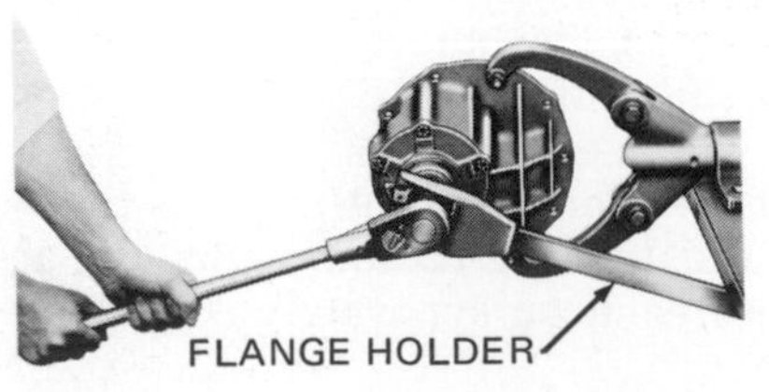

Fig. 23-33. Pinion-shaft-nut removal. (*Ford Motor Company*)

Clean the pinion bearing retainer around the oil seal. Put a drain pan under the seal or raise the front of the car so that the differential oil will not run out. Remove the U-joint flange with a special tool (Fig. 23-34). Then use special tools, as shown in Fig. 23-35, to remove the pinion seal. Clean the oil-seal seat.

Coat the outer edge of the new seal with oil-resistant sealer. Do not get sealer on the sealing lip. Install the seal with a special driver (Fig. 23-36). Then install the U-joint flange with the tool shown in Fig. 23-37. Install the integral retaining nut and washer on the pinion shaft, tightening it until the punch mark on the nut aligns with those on the pinion shaft and inner surface of the U-joint flange.

Tighten the nut an additional $\frac{1}{4}$ turn. Hold the flange with a special tool, as shown in Fig. 23-33. Reinstall the drive shaft, aligning the scribe marks made on the drive yoke and U-joint flange (Fig. 23-27). Check the lubricant level in the differential and add lubricant if necessary.

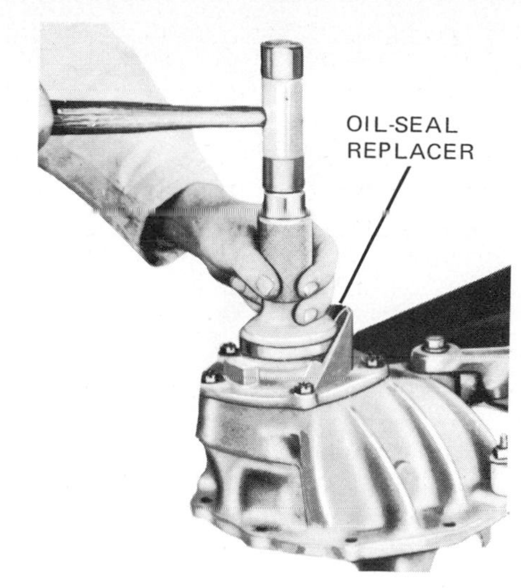

Fig. 23-36. Oil-seal installation. (*Ford Motor Company*)

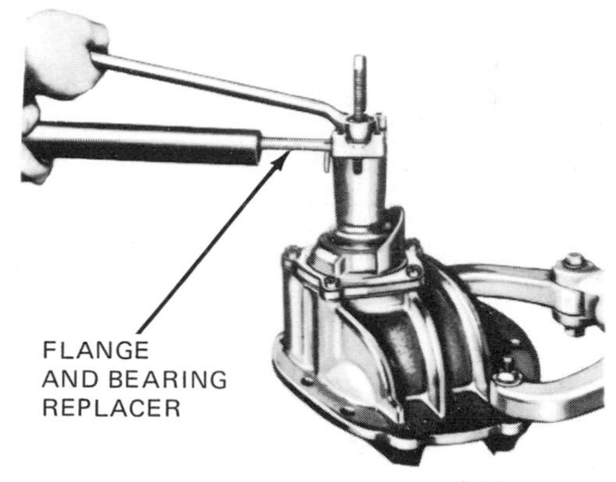

Fig. 23-37. U-joint flange installation. (*Ford Motor Company*)

⊘ 23-11 Differential Removal and Installation

1. *REMOVAL* Raise the rear of the car on a hoist and remove the two rear wheels and axle shafts, as described in ⊘ 22-6 and 23-10. Disconnect the drive shaft, as described in ⊘ 22-6 and 23-10.

Place a drain pan under the differential carrier. Remove the carrier retaining nuts and drain the axle. Remove the carrier assembly.

2. *INSTALLATION* Clean the axle housing and shafts with kerosene and swabs. Do not allow any solvent to run onto the wheel bearings. It could dissolve the lubricant and damage the bearings. Clean the mating surfaces of the axle housing and carrier.

Position the differential carrier on the studs in the axle housing, using a new gasket between the carrier and housing. Install retaining nuts. Then reattach the drive shaft (both ends) and reinstall the axle shafts, brake drums, and rear wheels as described in ⊘ 23-10. Fill the differential to the proper level with differential lubricant.

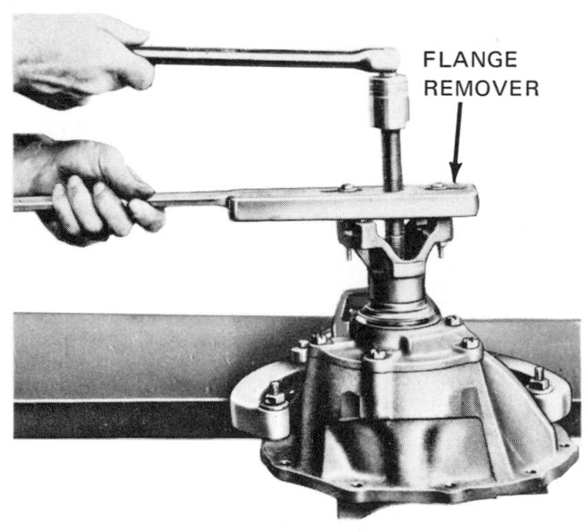

Fig. 23-34. U-joint flange removal. (*Ford Motor Company*)

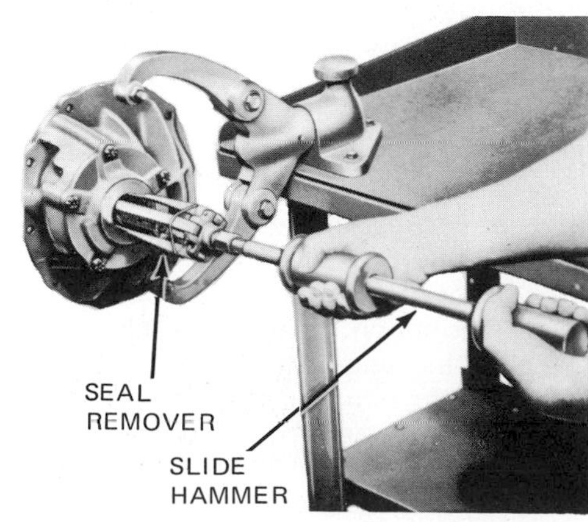

Fig. 23-35. Pinion oil-seal removal. (*Ford Motor Company*)

⊘ 23-12 Axle-Housing Removal and Installation

These procedures vary somewhat with different makes and models of automobile. A typical procedure follows for a car with leaf springs in the rear suspension (Fig. 23-38).

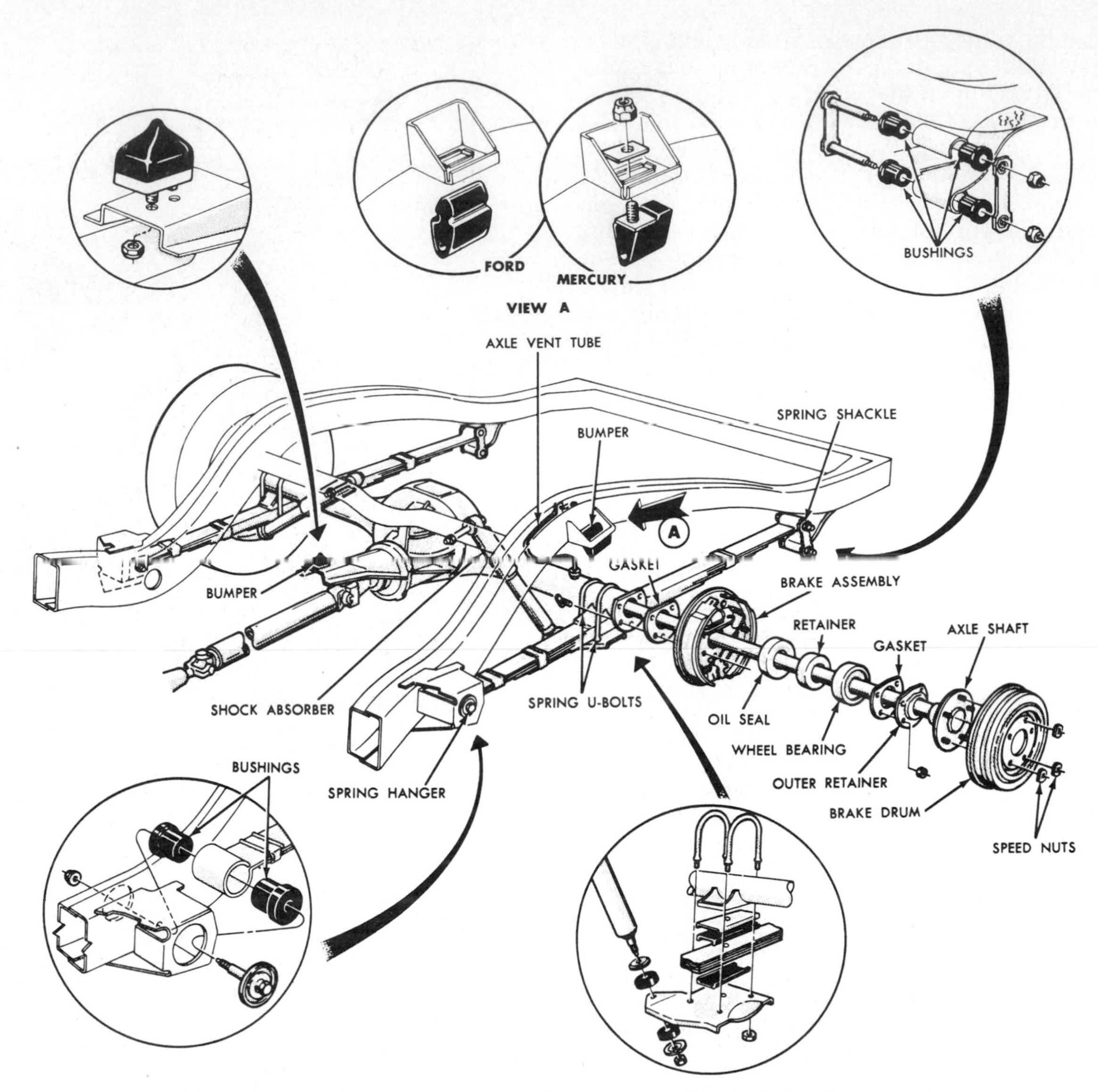

Fig. 23-38. Rear-suspension using leaf-springs. (*Ford Motor Company*)

1. *REMOVAL* Remove the carrier assembly, as described in ⊘ 23-11. Support rear frame members with safety stands. Disconnect the brake line from the clips on the axle housing. Disconnect the vent tube from the housing. After removing the brake-carrier plates from the housing, support them with wire so that the brake lines will not need to be disconnected. Disconnect the shock absorbers and move them out of the way. Lower the rear axle to reduce spring tension and then remove the spring U-bolt nuts, U bolts, and plates. Remove the spring lower insulator and retainer, and remove the axle housing.

2. *INSTALLATION* Install new rear-wheel-bearing oil seals in the ends of the rear-axle housing, as shown in Fig. 23-26 and described in ⊘ 23-10. Position the rear axle on the rear springs with spring upper insulators and retainers between the axle housing and springs with the retainer flange forward (see Fig. 23-38). Install the lower insulators and retainers (flange to rear) and secure with spring U bolts, plates, and nuts. Torque the spring U-bolt nuts evenly, making sure the lower insulator retainer contacts the upper retainer. Connect the vent tube, brake line (with clips), and shock absorbers. Install the carrier assembly and two axle assemblies as already described.

⊘ 23-13 Gear-Tooth-Contact Patterns Before disassembly, the differential should be inspected and the gear-tooth-contact pattern checked. Install the differential in a bench fixture (Fig. 23-39). Wipe lubricant from the gears and check for wear or damage. Rotate the gears to see if there is roughness or signs of scoring or abnormal wear. Set up a dial indicator, as shown in Fig. 23-40, and check for gear backlash at several points around the ring gear.

1. *CHECKING GEAR-TOOTH CONTACT* Paint the gear teeth with a suitable gear-marking com-

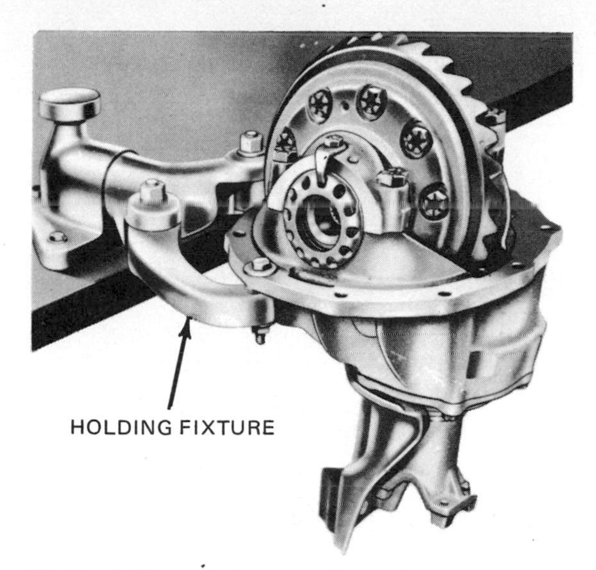

Fig. 23-39. Differential mounted in a bench fixture. (*Ford Motor Company*)

Fig. 23-40. Checking for backlash with a dial indicator. (*Ford motor Company*)

pound, such as red lead mixed with oil, to form a thin paste. Hold the drive-pinion flange with a cloth as a brake and rotate the drive gear back and forth with a box wrench on the drive-gear attachment bolts (Fig. 23-41).

In the nonhunting gearset, only a few rotations are required to obtain the full pattern of gear-tooth contact. In the hunting gearset, quite a few rotations are required. The ideal pattern is shown in Fig. 23-42. However, the pattern can vary somewhat from this ideal and still be satisfactory. Generally speaking, the drive pattern should be fairly well centered on the tooth. The coast pattern should also be fairly well centered, but it can be slightly toward the toe.

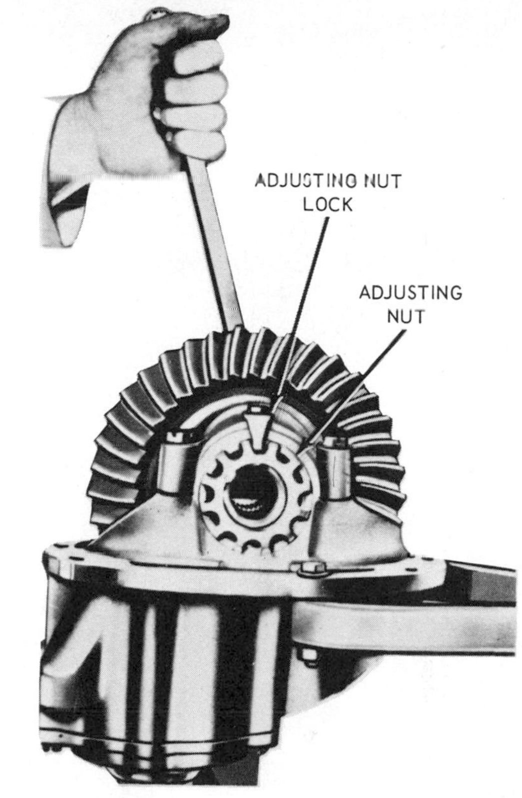

Fig. 23-41. Rotating the drive gear to check gear-tooth contact. (*Ford Motor Company*)

Fig. 23-42. Ideal gear-tooth-contact pattern. (*Ford Motor Company*)

There should be some clearance between the pattern and the top of the tooth, and there should be no hard lines indicating high pressure.

In general, the nonhunting gearset can have a more eccentric pattern than the hunting gearset and still function satisfactorily. For example, Figs. 23-43 and 23-44 show two acceptable patterns for nonhunting gearsets.

If the gear-tooth-contact pattern is not correct, it could be due to drive-gear runout. Check by using a dial indicator, as shown in Fig. 23-45. If drive-gear runout is excessive, the differential must be disassembled so that defective parts can be replaced.

Gear-tooth contact can often be improved if no parts are excessively worn or defective by turning the adjustment nuts or adding or removing shims (Fig. 23-46). For example, see Fig. 23-47, which shows four conditions and the steps required to correct them. In 1, the backlash is correct, as shown by the

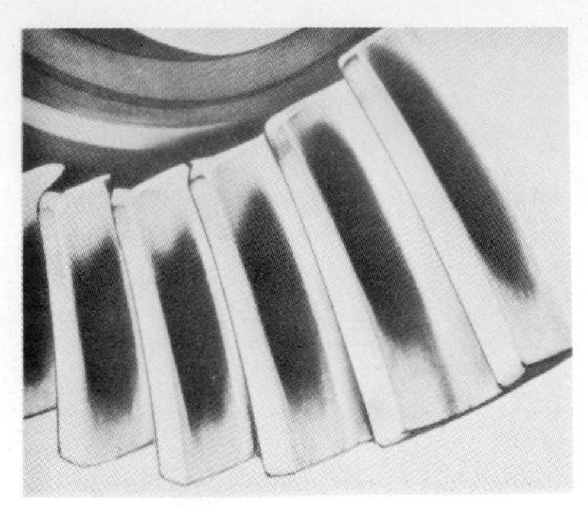

Fig. 23-43. Acceptable pattern for nonhunting gearset showing center-to-toe-to-center eccentricity. (*Ford Motor Company*)

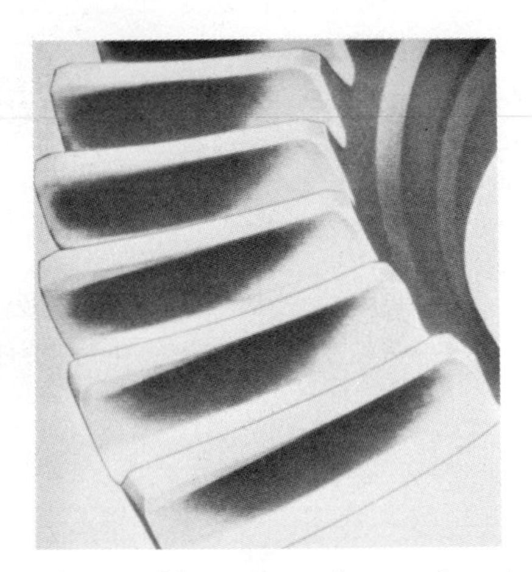

Fig. 23-44. Acceptable pattern for nonhunting gearset showing center-to-heel-to-center eccentricity. (*Ford Motor Company*)

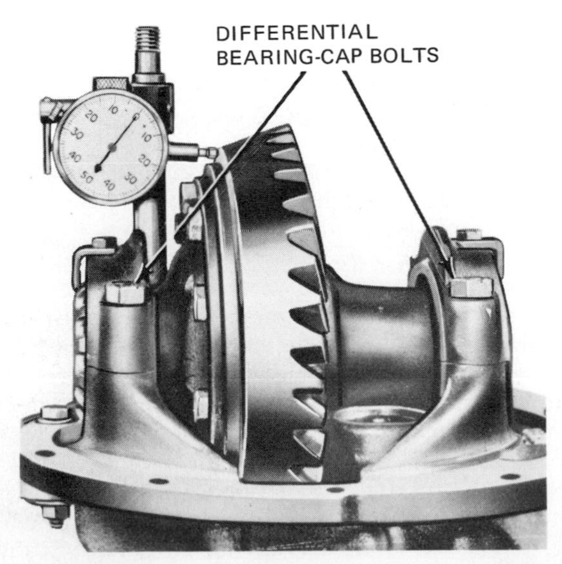

Fig. 23-45. Using a dial indicator to check drive-gear runout. (*Ford Motor Company*)

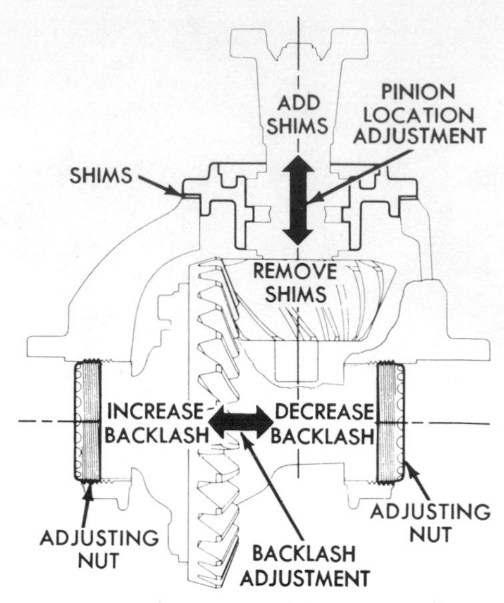

Fig. 23-46. Drive-pinion and drive-gear adjustments to improve gear-tooth-contact patterns requiring correction. (*Ford Motor Company*)

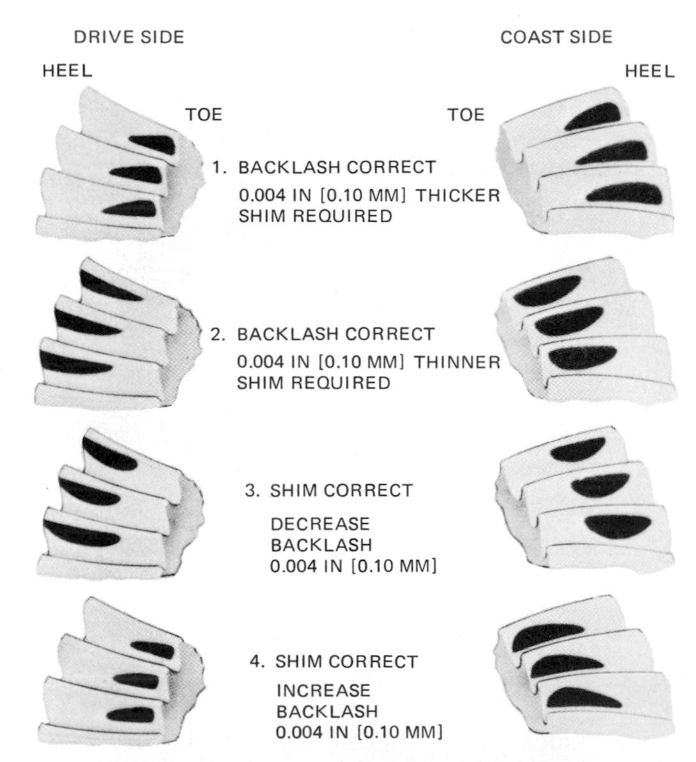

Fig. 23-47. Typical gear-tooth-contact patterns requiring correction. (*Ford Motor Company*)

pattern's being about halfway between the base and the top of the teeth. However, because the pattern is not centered between the heel and toe, a thicker shim is required.

Figure 23-48 illustrates various drive-gear-tooth impressions and corrective adjustments to be made, as recommended by Plymouth for one model of its standard differential. Note that the adjustment procedures are very similar to those shown in Fig. 23-48.

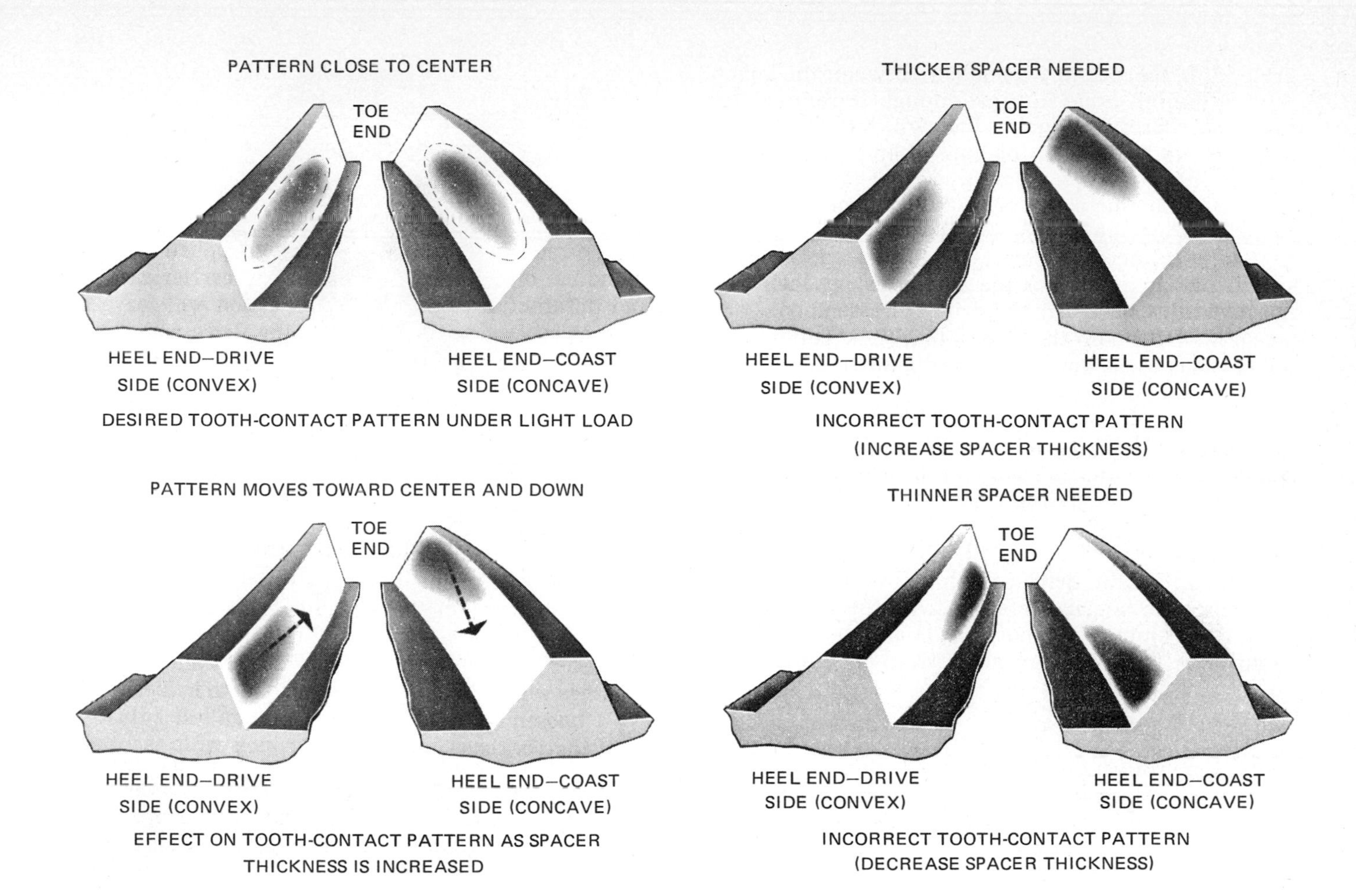

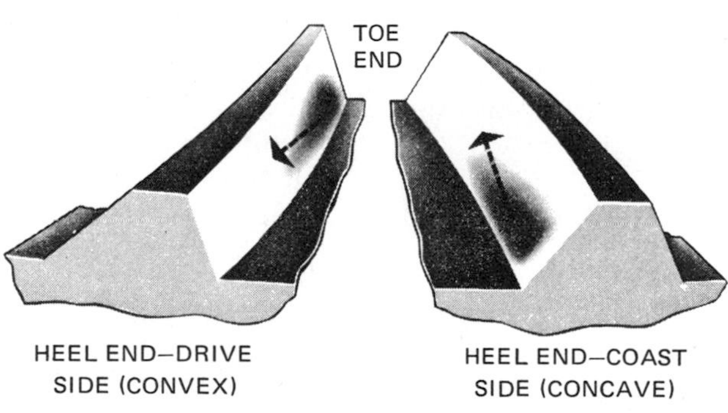

Fig. 23-48. Correct and incorrect gear-tooth-contact patterns and adjustments to correct them. (*Chrysler Corporation*)

Plymouth, you will note, refers to the drive gear as the *ring gear*.

2. *ADJUSTING BACKLASH* To adjust the backlash between the drive pinion and the drive gear, one adjustment nut is tightened while the other is loosened (Fig. 23-46). To make an adjustment, first remove the adjustment-nut locks, loosen the differential bearing-cap bolts, and then torque the bolts to 25 pound-feet [3.455 kg-m].

The left-hand adjustment nut is on the drive-gear side of the carrier, and the right-hand nut is on the pinion side (as shown in Fig. 23-46). As a first step, loosen the right-hand nut until it is away from the cup. Tighten the left-hand nut until the drive gear is just forced into the pinion with no backlash. Recheck the right-hand nut to make sure it is still loose. Now, tighten the right-hand nut 2 notches beyond the position where it first contacts the bearing cup. Rotate the drive gear several revolutions in each direction while the bearings are being loaded so that the bearings will seat. *This is important!*

Again loosen the right-hand nut to release the

preload. If there is any backlash between the gears, tighten the left-hand nut just enough to remove the backlash. Carefully tighten the right-hand nut until it just contacts the bearing cup. Then tighten it further $2\frac{1}{2}$ to 3 notches to apply correct preload. This should force the drive gear away from the drive pinion, giving the proper backlash.

Tighten the differential cap bolts to the correct specifications and check the backlash (Fig. 23-40). If the backlash is uneven, the drive gear is running out (see Fig. 23-45 for the setup to check runout). If backlash is even but incorrect, readjust the adjustment nuts.

CAUTION: Always make the final adjustment in a tightening direction to make sure that the nut is in contact with the bearing cap.

3. ADJUSTING THE DRIVE-PINION LOCATION Adjustment of the drive-pinion location is made by detaching the bearing retainer from the carrier and adding or removing shims (Fig. 23-49). Refer to Figs. 23-46 and 23-47 to determine whether shims must be removed or added.

Be careful not to pinch the O ring during reinstallation. Coat it with axle lubricant, and snap it into the groove; do not roll it. Before installing the bearing retainer on the carrier, determine whether the gearset is of the hunting or the nonhunting type. If of the nonhunting or partial-hunting type, it will be identified by painted timing marks on the gear teeth (Fig. 23-50). Be sure to align these timing marks on reassembly, as shown.

CAUTION: Noise and probability of early failure will be greatly increased if the gears are not properly aligned on nonhunting or partial-hunting gearsets.

On hunting-type gearsets, the drive pinion and drive gear can be reassembled without regard to any special tooth relationship.

Install the retainer-to-carrier mounting bolts and torque to specifications. Adjust the backlash between the drive pinion and the drive gear, as already explained, and recheck the tooth pattern. If the tooth pattern is still not correct, additional changing of the shims will be required.

Fig. 23-50. Gear-timing marks on a nonhunting gearset. (*Ford Motor Company*)

⊘ 23-14 Standard-Differential Disassembly and Reassembly Figure 23-51 is a disassembled view of the differential discussed in this section.

1. DISASSEMBLY Use a punch to mark one bearing cap and mating bearing support, and make scribe marks on one of the bearing adjustment nuts and carrier so that the parts can be reassembled in the proper relationship.

Remove the adjustment-nut locks, bearing caps, and adjustment nuts. Lift the differential assembly from the carrier. Pull the differential bearings with a special tool (Fig. 23-52). Detach the drive gear from the differential case, using a soft-face hammer or an arbor press to loosen it. Use a drift to drive out the differential-pinion-shaft lockpin. Separate the two-piece differential case. Drive out the pinion shaft with a brass drift and remove the gears and thrust washers.

The drive pinion and bearing retainer (Fig. 23-49) can be disassembled by removing the U-joint flange (Figs. 23-33 and 23-34) and pinion seal (Fig. 23-35). Then, special tools are required to press out the bearings. Use a protective sleeve (rubber hose) on the drive-pinion pilot-bearing surface and a fiber block on the end of the shaft to protect finished surfaces.

2. INSPECTION All parts should be carefully inspected. Worn or otherwise defective bearings, gears, or other parts should be discarded.

If the ring-gear runout is excessive, the cause

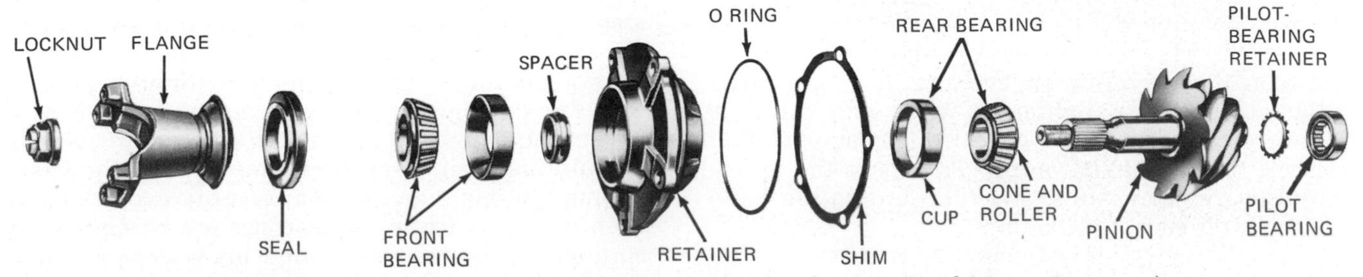

Fig. 23-49. Pinion, bearing retainer, shim, and related parts. (*Ford Motor Company*)

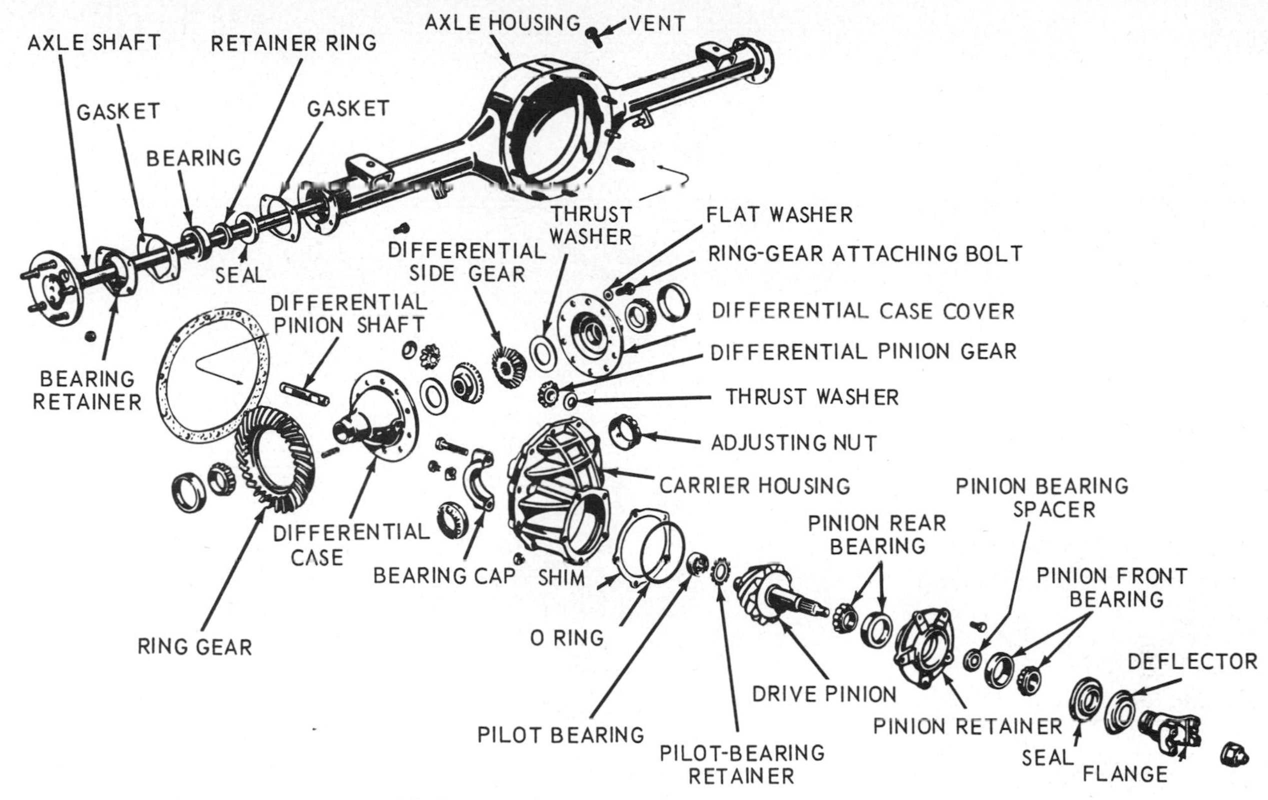

Fig. 23-51. Disassembled view of a standard differential. (*Ford Motor Company*)

may be a warped gear or case or worn differential bearings. To determine the cause, assemble the two halves of the differential case without the drive gear, and press the two differential side bearings on the case hubs. Put the cups on the bearings, and set the differential case in the carrier. Install the bearing caps and adjusting nuts (see item 3) and adjust the bearings, as described in ⊘ 23-13.

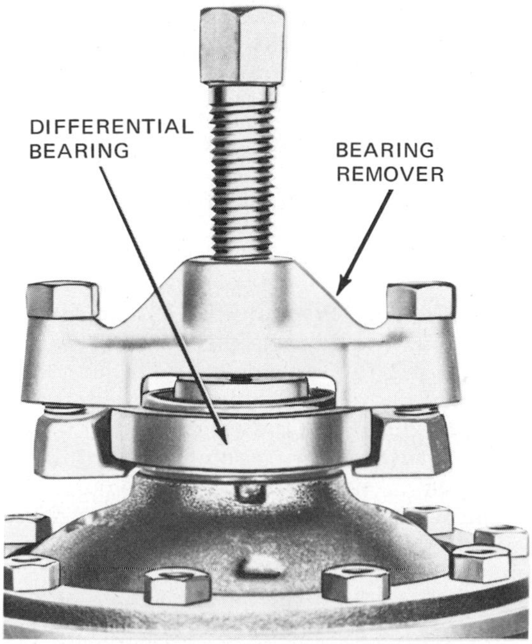

Check the runout of the differential-case flange. If the runout exceeds specifications, the case is defective or the bearings are worn. However, if the runout is within specifications, the trouble is with the drive gear, and a new drive gear is required.

3. REASSEMBLY First reassemble the drive pinion and bearing retainer (Fig. 23-49), using special tools to install the bearings. Install a new oil seal (Fig. 23-36) and the U-joint flange (Fig. 23-37), as previously described.

Refer to Fig. 23-53 for assembly of the differential case. Lubricate all parts liberally during assembly. Drive the pinion shaft into the case only far enough to retain a pinion thrust washer and pinion gear. Put the second pinion gear and thrust washer into position, and drive the pinion shaft into place. Be sure to line up the pinion-shaft lockpin holes.

Put the second side gear and thrust washer into position, and install the cover on the differential case. Insert an axle-shaft spline into a side gear to check for free rotation of the gears.

Insert two $\frac{7}{16}$ N.F. bolts 2 inches [50.8 mm] long through the case flange, and thread them several turns into the drive gear as an aid to aligning the drive gear to the case. Press or tap the gear into place. Install and tighten the gear attachment bolts, using washers. Torque them alternately across the gear to ensure alignment. If the differential bearings have been removed, press them on. Install the

DIFFERENTIAL BEARING
BEARING REMOVER

Fig. 23-52. Removing a differential bearing. (*Ford Motor Company*)

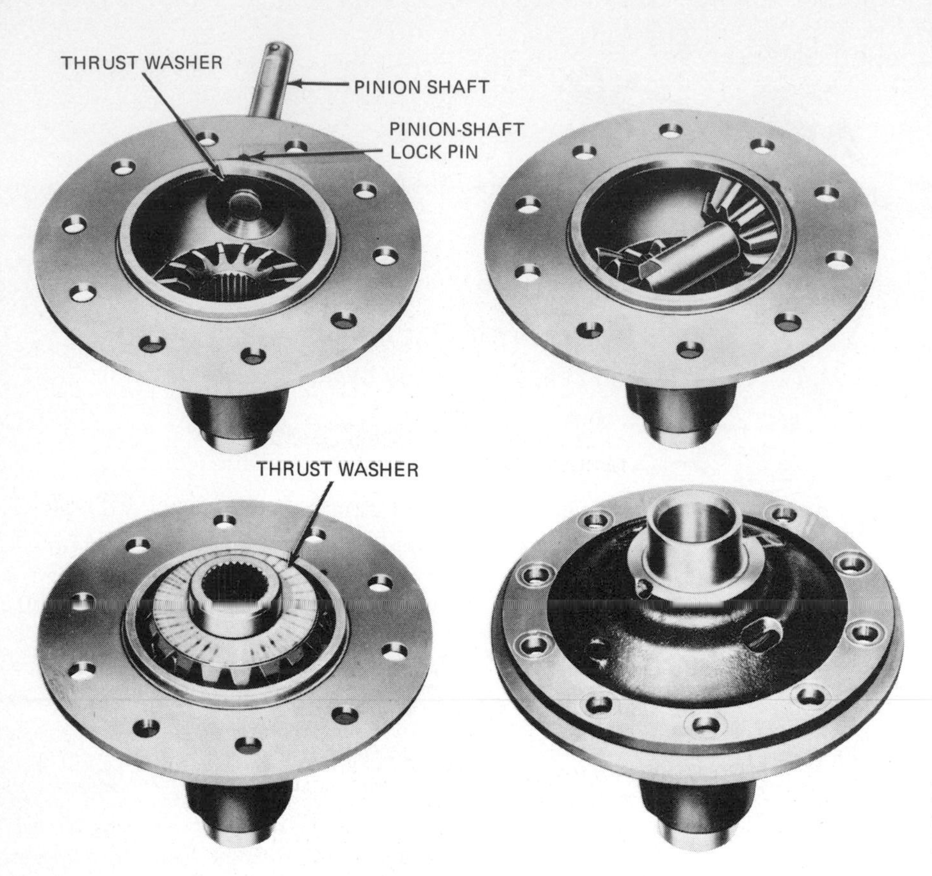

Fig. 23-53. Assembling a differential case. (*Ford Motor Company*)

bearing-retainer-and-drive-pinion assembly, as was described in ⊘ 23-13, making sure that the gearset timing marks are properly aligned, as shown in Fig. 23-50 (nonhunting type).

Adjust backlash and drive-pinion location, as described in ⊘ 23-13, making a final tooth-pattern check before installing the carrier assembly in the axle housing.

⊘ 23-15 Nonslip-Differential Disassembly and Reassembly Figures 23-16 to 23-19 illustrate and ⊘ 23-7 describes the nonslip differential discussed in this section. This nonslip differential is the earlier type, with two sets of clutch plates. The later types, with spring-loaded clutch plates (Fig. 23-20) or spring-loaded cones (Fig. 23-21), are discussed in ⊘ 23-16.

1. *DISASSEMBLY* Remove the axle drive gear (ring gear) and measure the runout of the gear mounting flange on the case. Replace both halves of the case if the runout is excessive. Put scribe marks on the case halves before disassembly (Fig. 23-54). Remove the case-cap attachment bolts, cap, and clutch plates (Figs. 23-55 and 23-56). Then lift off side-gear retainer, side gear, and pinion shafts with pinion gears (Figs. 23-57 and 23-58). Take out the other side gear, side-gear retainer, and remaining clutch plates (Figs. 23-59 and 23-60).

2. *INSPECTION* After cleaning all parts, check them for wear, nicks, burrs, and similar defects. Replace worn or distorted clutch plates. If the case is defective, replace both halves.

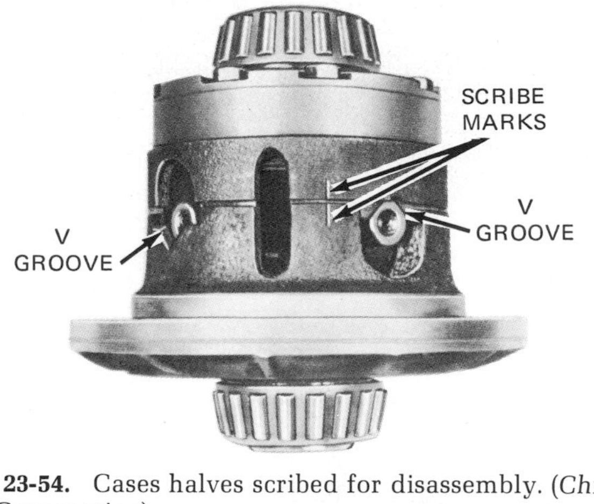

Fig. 23-54. Cases halves scribed for disassembly. (*Chrysler Corporation*)

3. *REASSEMBLY* Position the clutch plates and disks in their proper locations in each half of the case (Fig. 23-61). Put the side gears in their retainers. Insert the splines of the retainers through the splines of the clutch disks. Put the aligning pin through one axle-shaft thrust spacer. Assemble the pinion shafts on the aligning pin. Put the pinion gears on the shafts, and install the assembly on the drive-gear half of the case. Insert the thrust spacer into the pinion shaft (Fig. 23-62).

Slide the cap half of the case over the edge of the bench just far enough to insert one finger up through the assembly to hold it together. Put the assembly on the drive-gear half, matching the scribe

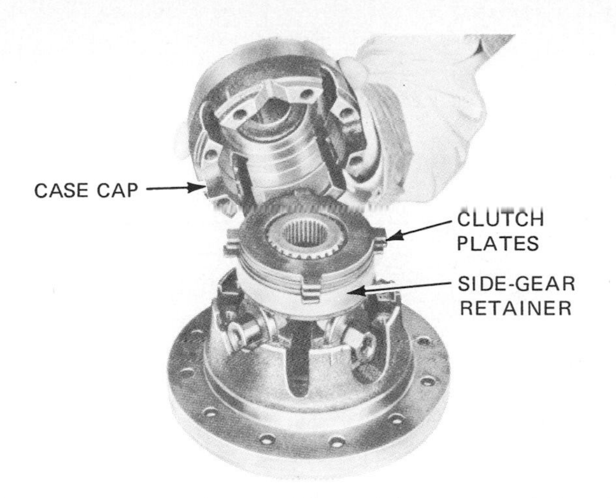

Fig. 23-55. Removing a differential-case cap. (*Chrysler Corporation*)

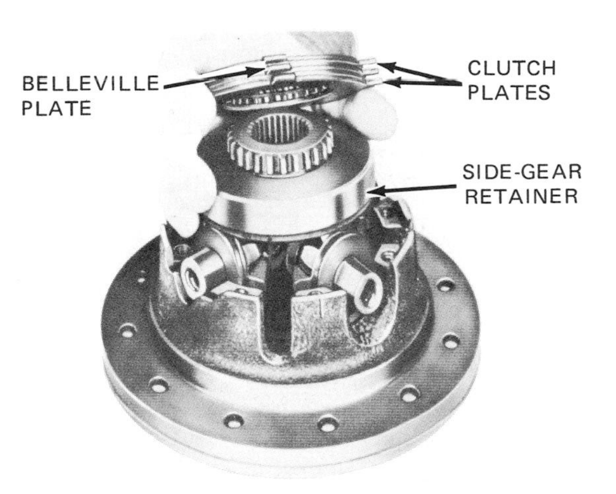

Fig. 23-56. Removing clutch plates. (*Chrysler Corporation*)

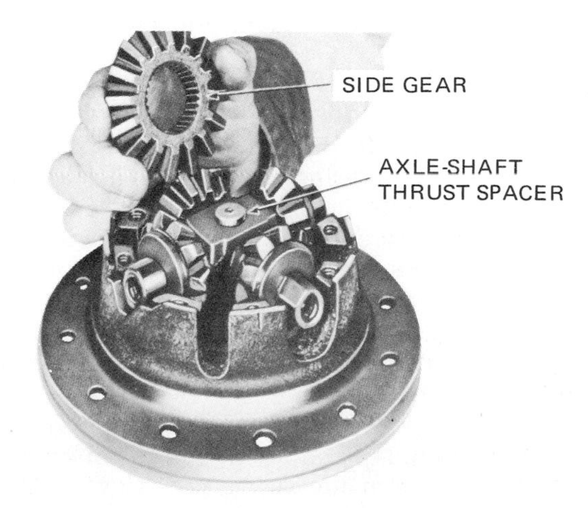

Fig. 23-57. Removing the side gear. (*Chrysler Corporation*)

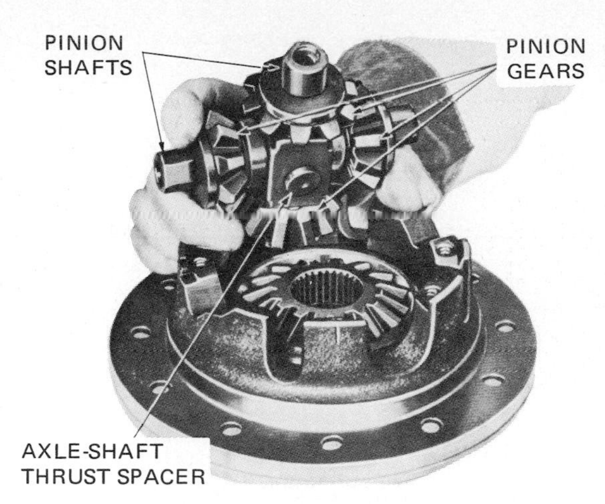

Fig. 23-58. Removing pinion shafts and gears. (*Chrysler Corporation*)

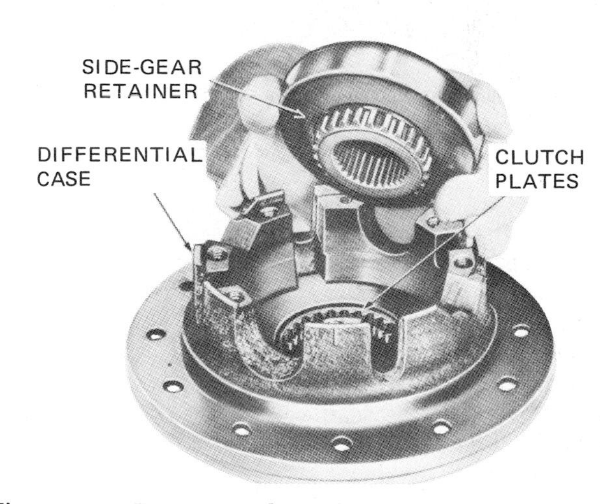

Fig. 23-59. Removing the side-gear retainer. (*Chrysler Corporation*)

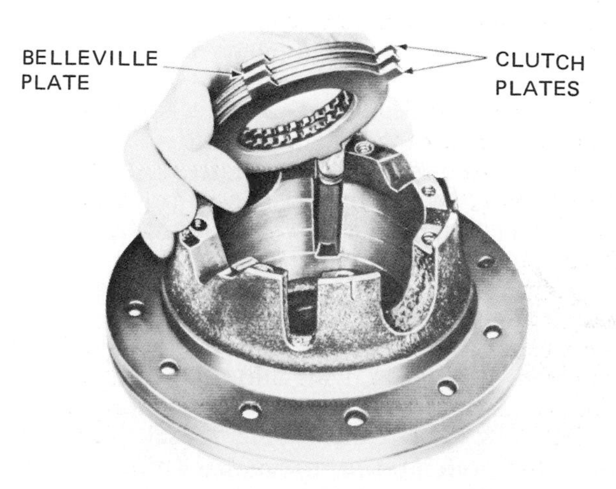

Fig. 23-60. Removing clutch plates. (*Chrysler Corporation*)

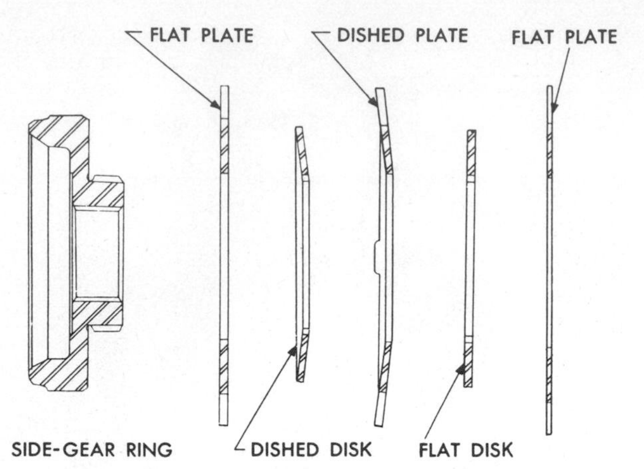

Fig. 23-61. Arrangement of clutch plates and disks. (*Chrysler Corporation*)

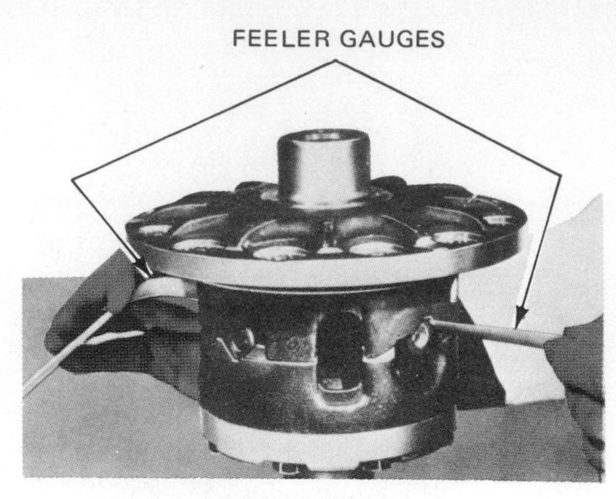

Fig. 23-63. Checking clearance between the pinion shaft and case. (*Chrysler Corporation*)

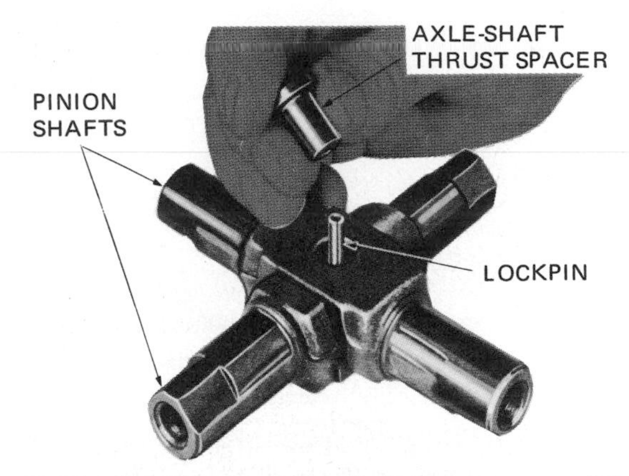

Fig. 23-62. Installing an axle-shaft thrust spacer. (*Chrysler Corporation*)

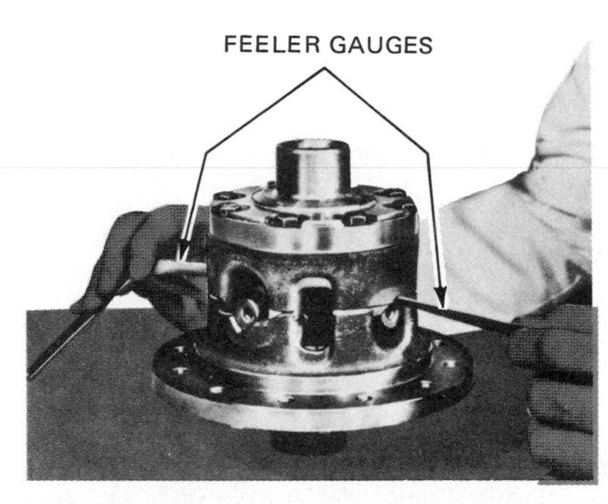

Fig. 23-64. Checking clearance between the pinion shaft and cap. (*Chrysler Corporation*)

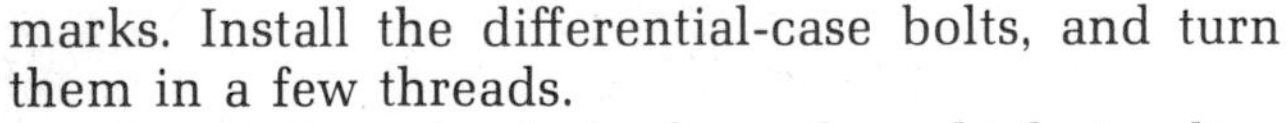

marks. Install the differential-case bolts, and turn them in a few threads.

Insert the axle shafts from the vehicle to align the splines. Make sure the axle shafts engage the side-gear splines as well as the clutch-disk splines.

With the shafts installed, center the cross shafts between the two ramp surfaces in the differential case. Tighten the differential-case bolts evenly by alternately turning the opposite bolts until all are tightened to the specified tension. After reassembly, any slight misalignment of the splines can be corrected by moving the axle shafts back and forth until free. Remove the axle shafts.

With the differential resting on one hub, insert two feeler blades, one over each end of the pinion shaft having ramps above it (Fig. 23-63). Invert the differential and check the opposite pinion shaft in a like manner (Fig. 23-64). Excessive clearance means that the clutch disks are worn and should be replaced. With new disks and plates, the clearance may be as little as 0.002 inch [0.051 mm].

⊘ 23-16 Servicing Spring-Loaded Clutch and Cone Nonslip Differentials Figures 23-20 and 23-21 show the two nonslip differentials described in this section. The type with cones is a Borg-Warner unit (Figs. 23-21 and 23-65) and is not repairable. If it becomes defective, Chevrolet specifies replacement with either the Eaton or Chevrolet unit they supply.

The service procedures for the nonslip differential (Fig. 23-20) will now be described. Overhaul procedures are the same as for the differentials described previously, with the following exceptions:

1. DISASSEMBLING THE EATON NONSLIP DIFFERENTIAL Remove the ring gear and side bearings in the same manner as for other differentials. Remove the pinion shaft. Then remove the preloaded spring retainer and springs by tapping on the spring retainer through the observation hole (Fig. 23-66). Drive the spring retainers out far enough to permit installation of $\frac{1}{4}$-inch [6.35 mm] bolts through the springs. Secure the bolts with nuts to hole the springs in compression (Fig. 23-67). Then tap the spring retainers out enough so that you can put a clamp on the retainers (Fig. 23-68).

Use bar stock, as shown, to avoid damaging the retainers. Tighten the clamp enough to compress the other springs so that the assembly can be removed from the differential case. To disassemble this assembly, clamp it in a vise as shown in Fig. 23-69.

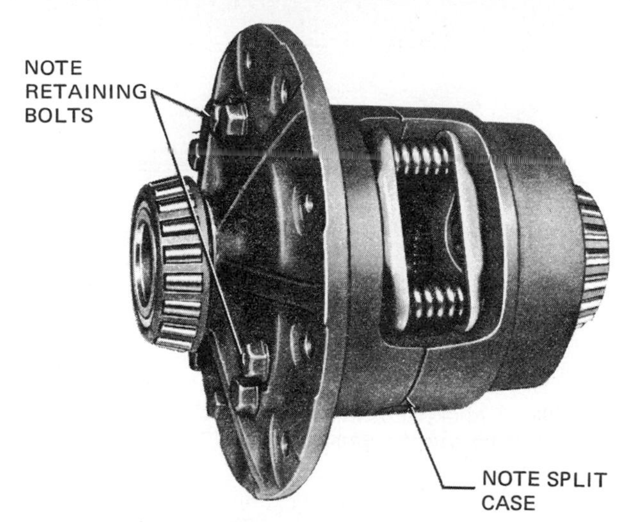

Fig. 23-65. Borg-Warner Positraction nonslip differential. (*Chevrolet Motor Division of General Motors Corporation*)

Fig. 23-68. Using a C clamp and bar stock to remove spring pack. (*Chevrolet Motor Division of General Motors Corporation*)

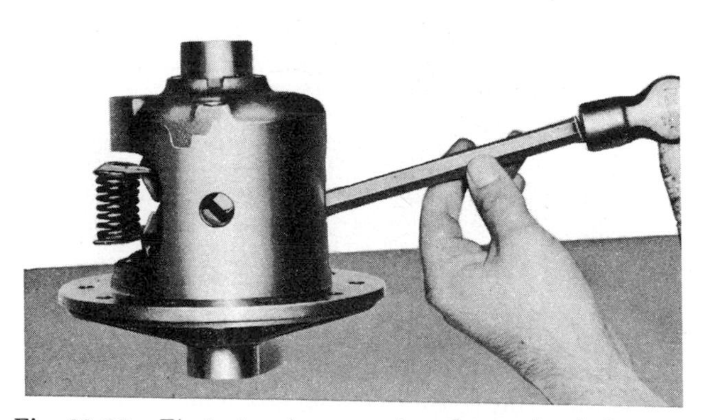

Fig. 23-66. First step in removing the preloaded spring pack. (*Chevrolet Motor Division of General Motors Corporation*)

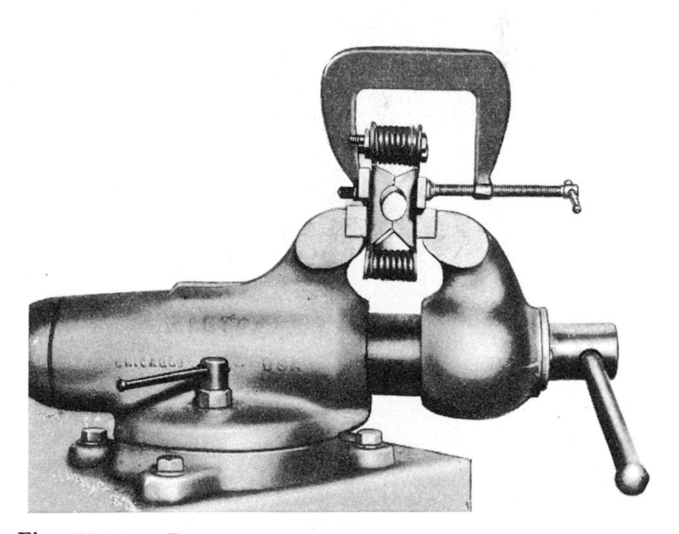

Fig. 23-69. Removing preloaded springs from pack. (*Chevrolet Motor Division of General Motors Corporation*)

Fig. 23-67. Installing bolts to retain springs. (*Chevrolet Motor Division of General Motors Corporation*)

Remove the nuts and bolts. Then alternately loosen the vise and C clamp until spring pressure is relieved.

For truck models, roll out the differential pinions and thrust washers. For passenger-car models, remove the pinion gears by rotating them in one direction only. See Fig. 23-70. Rotate differential case clockwise (viewed from inside) to remove the first gear. Then rotate the case counterclockwise to remove the other gear. It may be necessary to pry on the gear, as shown, to remove the second gear.

Next, remove the side gears, clutch packs, shims, and guides from the case. To remove one side gear, tap on the assembly using a brass drift, as shown in Fig. 23-71. Reverse the case and remove the second side gear.

2. INSPECTING AND ASSEMBLING THE EATON NONSLIP DIFFERENTIAL First, inspect

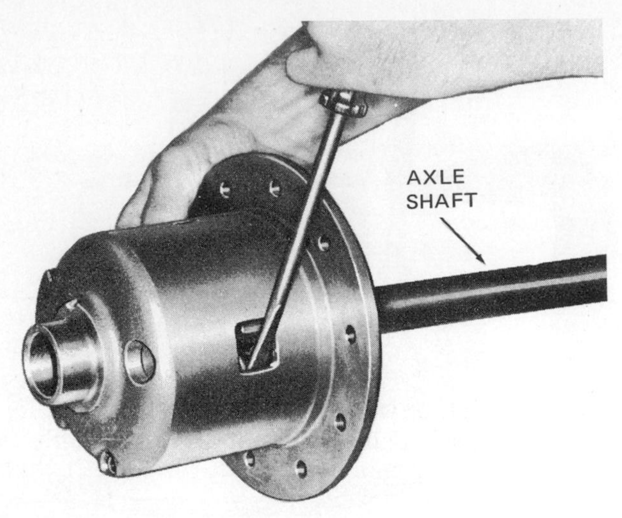

Fig. 23-70. Removing differential pinion gears. (*Chevrolet Motor Division of General Motors Corporation*)

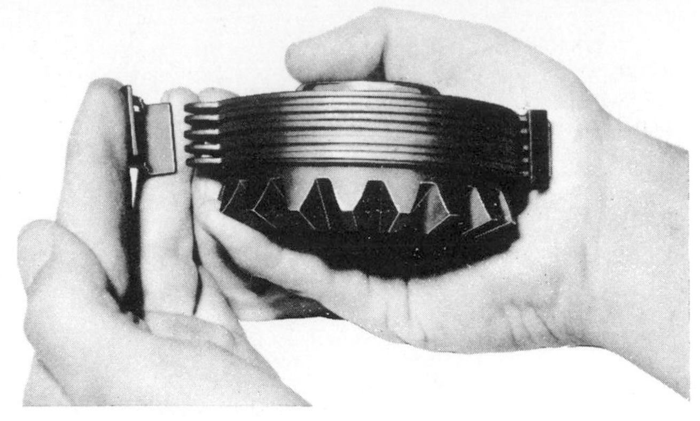

Fig. 23-72. Assembling clutch pack on side gear. (*Chevrolet Motor Division of General Motors Corporation*)

Fig. 23-71. Removing side gear. (*Chevrolet Motor Division of General Motors Corporation*)

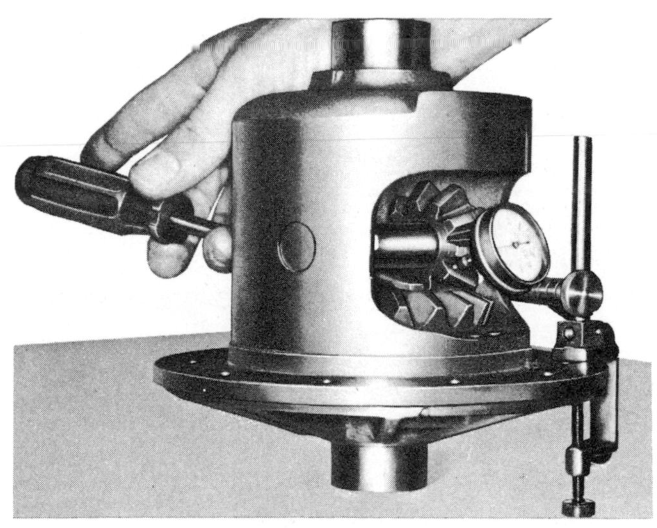

Fig. 23-73. Measuring pinion-gear-tooth clearance. (*Chevrolet Motor Division of General Motors Corporation*)

parts. Check clutch plates and disks for wear and signs of over-heating. If some plates or disks look questionable, install a complete new clutch pack. Inspect springs for weakness or distortion. Make sure the spring retainers are in good alignment and are not worn excessively at the spring seats.

To reassemble, lubricate the plates and disks with special lubricant and assemble the pack on the side gear (Fig. 23-72). Install clutch-pack guides. Select shims of the same thickness as those removed (or use the old shims if they are still in good condition). Install them over the side-gear hub. Lubricate and assemble the opposite clutch pack on the other side-gear hub.

Install one side gear with clutch pack and shims in the case. Position the pinion gears and thrust washers on the side gear. Install the pinion shaft. Position a dial indicator with the contact button pinion gear (Fig. 23-73). Compress the clutch pack with a screwdriver, as shown. Move the pinion gear to check tooth clearance. If required, change shims to get proper clearance. Remove the side gear and repeat the clearance-check procedure for other side gear. Then remove the pinion shaft, gears, and thrust washers.

Install the two side gears with clutch packs and shims. Install the pinion gears and thrust washers. With springs and retainers assembled with C clamp and nuts and bolts, as shown in Fig. 23-68, drive the assembly partly into the case. Then remove the C clamp and bolts to complete the installation. Install the pinion shaft and lock screw.

Check the alignment of spring retainers with side gears. The spring pack may require some adjustment to secure proper alignment. Install the side bearings and ring gear to the case. Then place the differential in the carrier and adjust the bearings in the same manner as for the other differentials previously described.

3. CHECKING OPERATION The operation of the unit can be checked while it is in the car by raising

Fig. 23-74. Measuring nonslip differential rotating torque. (*Chevrolet Motor Division of General Motors Corporation*)

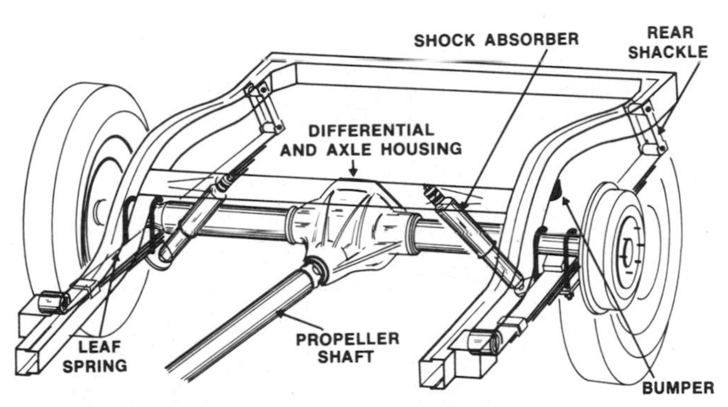

Fig. 23-75. Integral-carrier differential from which the carrier cannot be removed.

the car so that both wheels are off the ground. Then remove one wheel-and-tire assembly. Use the special adapter and torque wrench (Fig. 23-74) to check the torque required to turn the axle. With the other wheel held firmly so that it cannot turn, the torque required should be no less than 40 pound-feet [5.528 kg-m].

⊘ 23-17 Servicing Integral-Carrier Differentials Many late-model cars use a rear-axle assembly from which the differential carrier cannot be removed (Fig. 23-75). Ford calls this type of rear axle an *integral-carrier axle,* while Buick calls it a *salisbury-type axle housing.* This rear-axle assembly is made by pressing and welding tubes (which then become the axle housings) into each side of the differential carrier (Fig. 23-76). A stamped bolt-on cover is used (Fig. 23-4), and the filler plug is threaded into the front part of the carrier casting (Fig. 23-76). Usually, the differential can be serviced without removing the axle housing from the car.

To remove the differential case and bearings for service, first remove the axle shafts (⊘ 23-10). Check the ring-gear-to-pinion backlash. Then mark and remove the differential side-bearing caps. Pull the differential case from the carrier (Fig. 23-77). Some manufacturers recommend the use of a special slide hammer to pull the case.

Remove the companion-flange nut and companion flange. Drive out the pinion gear (Fig. 23-78).

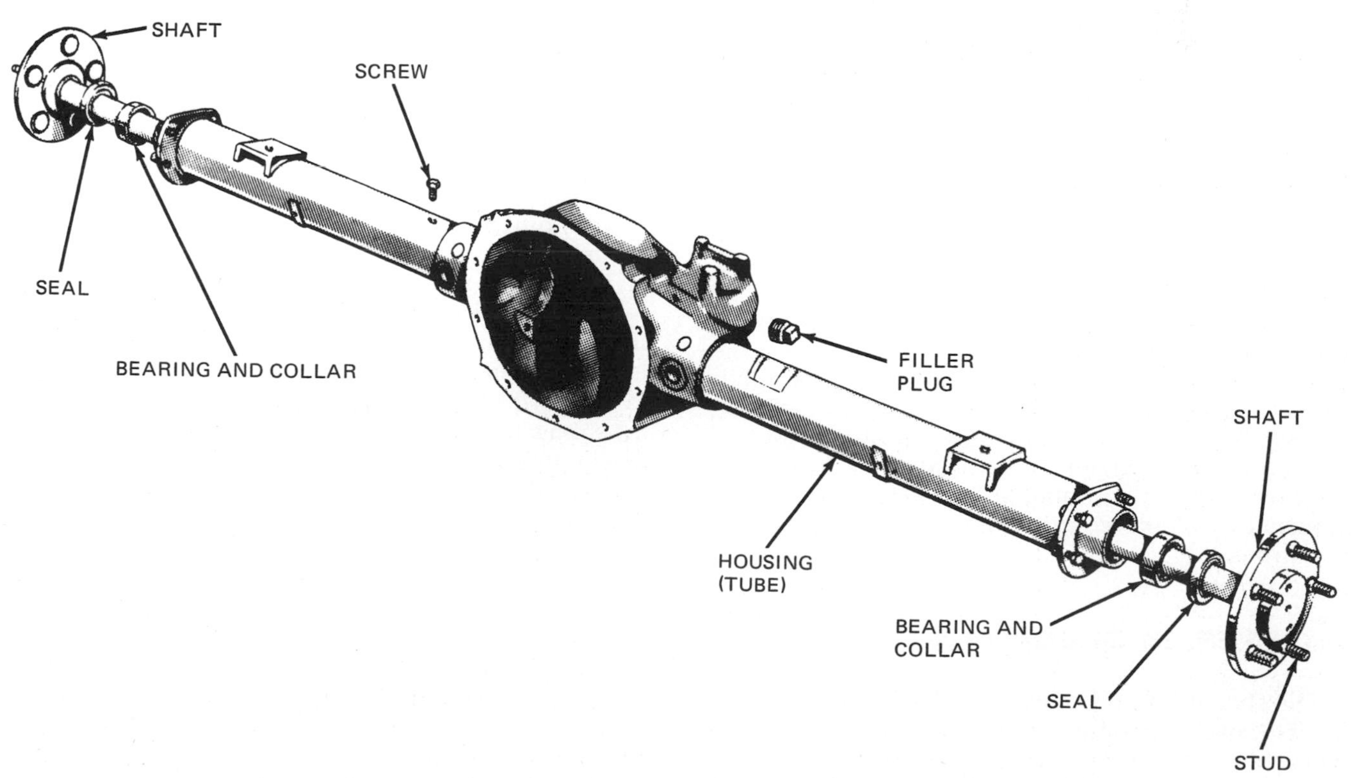

Fig. 23-76. Rear view of an integral-carrier differential. Note the location of the filler plug. (*Chrysler Corporation*)

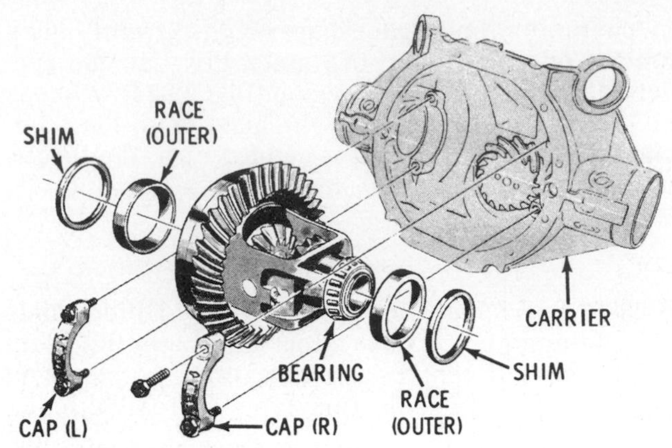

Fig. 23-77. Differential case and bearings. (*Oldsmobile Division of General Motors Corporation*)

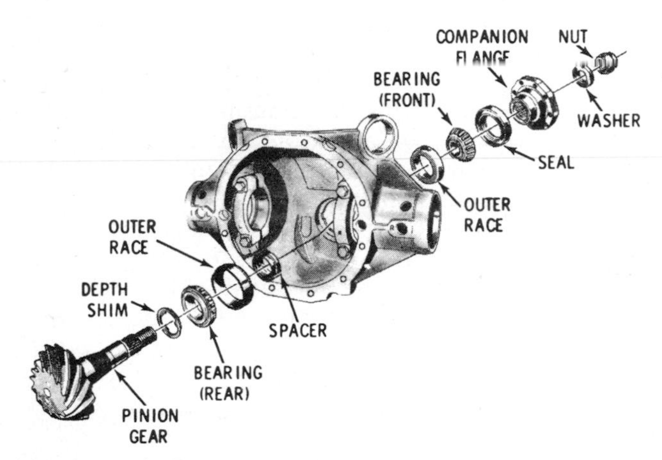

Fig. 23-78. Differential pinion gear and companion flange. (*Oldsmobile Division of General Motors Corporation*)

Cleaning, inspection, and service procedures on the ring gear, pinion gear, and case assembly generally follow those covered in ⊘ 23-13 to 23-16. However, always refer to the applicable shop manual before attempting to service a specific make and model of differential.

To assemble an integral-carrier differential, install the pinion gear, following the procedure in the manufacturer's shop manual for adjusting pinion depth. Then, before installing the pinion in the carrier, install the case and make the differential-side-bearing-preload adjustment. Some integral-carrier differentials use adjusters, as shown in Fig. 23-4. In integral-carrier differentials that do not use adjusters, the differential-side-bearing preload is adjusted by changing the thickness of the right and left shims (Fig. 23-77).

CHAPTER 23 CHECKUP

NOTE: Since the following is a chapter review test, you should review the chapter before taking it.

With this chapter on differentials, you have completed your studies of the power train. Compared with transmissions, which you studied in previous chapters, the differential may seem like a very simple mechanism. However, even though it is relatively uncomplicated, the repair and adjustment procedures are rather complex. Furthermore, they must be followed exactly if proper differential operation and normal differential life are to be obtained. The test that follows is designed to check your memory on how well you remember the various checks and service procedures detailed in the chapter.

Completing the Sentences The sentences that follow are incomplete. After each sentence there are several words or phrases, only one of which will correctly complete the sentence. Write each sentence in your notebook, selecting the proper word or phrase to complete it correctly.

1. Most often, the condition that draws attention to trouble in the differential is: (*a*) rough operation, (*b*) noise, (*c*) power loss.
2. A humming noise in the differential is often caused by improper tooth contact between the: (*a*) drive pinion and drive gear, (*b*) axle and side gear, (*c*) pinion and side gears.
3. If noise is present in the differential only when the car rounds a curve, chances are the trouble is due to some condition in the: (*a*) drive-pinion assembly, (*b*) differential-case assembly, (*c*) wheel bearing.
4. To correct heavy face contact on the drive-gear teeth, move the: (*a*) drive pinion in, (*b*) drive pinion out, (*c*) drive gear out, and then adjust backlash, as necessary.
5. To correct heavy flank contact on the drive-gear teeth, move the: (*a*) drive pinion in, (*b*) drive pinion out, (*c*) drive gear out, and then adjust backlash, as necessary.
6. To correct heavy heel contact on the drive-gear teeth, move the: (*a*) drive pinion in, (*b*) drive gear toward pinion, (*c*) drive gear away from pinion, and adjust backlash.
7. To correct heavy toe contact on the drive-gear teeth, move the: (*a*) drive gear toward pinion, (*b*) drive gear away from pinion, (*c*) drive pinion out, and adjust backlash.
8. On Ford, the drive pinion is adjusted by use of: (*a*) shims, (*b*) adjustment nuts, (*c*) bearing adjusters.
9. The drive gear is adjusted in the differential by use of: (*a*) selective washers of proper thickness, (*b*) adjustment nuts, (*c*) adjustment screws.
10. Bearing adjusters are held in place in the differential carrier by: (*a*) bearing caps, (*b*) selective washers, (*c*) lockpins.

Troubles and Service Procedures In the following, you should write in your notebook the required trouble causes and service procedures. Do not copy from the book; try to write them in your own words. Writing the trouble causes and service procedures in your own words will help you to remember them, which will be of great value to you in the shop.

1. Explain how a differential operates when the car is rounding a turn.
2. List and explain the names used to describe the various parts of gear teeth, including backlash, clearance, heel, toe, face, and flank.
3. List the possible causes of hum in the differential.
4. Explain how to determine axle ratio in a car equipped with Hotchkiss drive.
5. List the main steps in removing and replacing an axle shaft.
6. List the main steps in removing and replacing a differential-carrier assembly.
7. Describe the disassembly and reassembly procedures for one make of differential-carrier assembly.
8. Explain how to adjust the differential in one make of car.
9. Describe the disassembly and reassembly of a nonslip differential.

SUGGESTIONS FOR FURTHER STUDY

Examine the differentials used in various cars. If possible, disassemble and reassemble one, and make the adjustments it requires. If you do have a chance to work on an actual differential, be sure to follow the service instructions as outlined in the applicable automotive shop manual. In addition to studying actual differentials, study the shop manuals also since they contain considerable detailed material applying to particular models of differentials. Be sure you write in your notebook any important facts you learn.

GLOSSARY

This glossary of automotive terms used in the book provides a ready reference for the student. The definitions may differ somewhat from those given in a standard dictionary. They are not intended to be all-inclusive but to refresh the memory on automotive terms. More complete definitions and explanations of the terms are found in the text.

Accelerator The foot-operated pedal linked to the throttle valve in the carburetor.

Accelerator pump In the carburetor, a pump linked to the accelerator, which momentarily enriches the mixture when the accelerator pedal is depressed.

Accessories Devices not considered essential to the operation of the vehicle, for example, the radio, car heater, and electric window lifts.

Accumulator A device used in automatic transmissions to cushion the shock of clutch and servo actions.

Adjustments Necessary or desired changes in clearances, fit, or setting.

Air impact wrench An air-powered hand-held tool that runs nuts and bolts on and off quickly using a series of sharp rapid blows created by pressurized air forced into the wrench.

Air line A hose, pipe, or tube through which air passes.

Air pressure Atmospheric pressure of 14.7 psi (pounds per square inch) [in the metric system, 1.0355 kg/cm^2 (kilograms per square centimeter)] at sea level. The pressure on air produced by a pump, by compression in the engine cylinder, etc.

Allen wrench A type of screwdriver which turns a screw with a matching recessed hex head.

Alternator The device in the electric system that converts mechanical energy into electric energy for charging the battery, etc. Also known as an ac generator, the alternator produces alternating current (ac) which must be changed to direct current (dc) for use in the automobile.

Antifriction bearing Name given to almost any type of ball, roller, or tapered roller bearing.

Arbor press Small hand-operated press used on jobs when light pressure is needed.

Atmospheric pressure See "Air pressure."

Automatic transmission A transmission in which gear ratios are changed automatically.

Axle A crossbar supporting a vehicle on which one or more wheels turn.

Axle ratio The ratio between drive-shaft rpm (revolutions per minute) and rear-wheel rpm; the gear reduction in the differential.

Backlash In gearing, the clearance between meshing teeth of two gears. The amount by which the width of the tooth space exceeds the thickness of the tooth in that space; generally, the amount of free motion, or lash, in a mechanical system.

Balanced valve A type of hydraulic valve that produces pressure changes proportional to movement of mechanical linkage or variations in spring pressure.

Ball-and-trunnion joint A type of universal joint which combines the universal joint and slip joint in one assembly.

Ball bearing An antifriction bearing with an inner and outer race with one or more rows of balls between.

Ball check valve A valve consisting of a ball and seat. Fluid can pass in one direction only; when it attempts to flow the other way, it is checked by the ball seating on the seat.

Band In an automatic transmission, a hydraulically controlled brake band installed around a metal clutch drum, used to stop or permit drum rotation.

Battery An electrochemical device for storing energy in chemical form so that it can be released as electricity. A group of electric cells connected together.

Bearing The part which transmits the load to the support and, in so doing, takes the friction caused by moving parts in contact.

Bearing oil clearance The space purposely provided between the shaft and the bearing through which lubricating oil can flow.

Bevel gear A gear shaped like the lower part of a cone used to transmit motion through an angle.

Body The assembly of sheet-metal sections, together with windows, doors, seats, and other parts, that provides an enclosure for the passengers, engine, etc.

Boiling point The temperature at which a liquid begins to boil.

Bore The diameter of an engine cylinder; the diameter of any hole. Also used to describe the process of enlarging or accurately refinishing a hole, as to bore an engine cylinder.

Brake An energy-conversion device used to slow, stop, or hold a vehicle or mechanism.

Brake drums Metal drums mounted on the car wheels which form the outer shell of the brakes; brake shoes press against the drums to slow or stop drum and wheel rotation for braking.

Brake horsepower (bhp) The power delivered by the engine which is available for driving the vehicle.

Brake lining A high-friction material, usually a form of asbestos, attached by rivets or a bonding process to the brake shoe. The lining takes the wear when the shoe is pressed against the brake drum, or rotor.

Burr A featheredge of metal left on a part being cut with a file or other cutting tool.

Bushing A one-piece sleeve placed in a bore to serve as a bearing surface.

Caliper A measuring tool that can be set to measure the thickness of a block, the diameter of a shaft, or the bore of a hole (inside caliper). In a disk brake, a housing for pistons and brake shoes, connected to the hydraulic system, which holds the brake shoes so that they straddle the disk.

Cam A rotating lobe or eccentric which changes rotary motion to reciprocating motion.

Carburetor The mixing device in the fuel system which meters gasoline into the airstream (vaporizing the gasoline as it does so) in varying proportions to suit engine operating conditions.

Cardan universal joint A universal joint of the ball-and-socket type.

Car lift An air, electrical, or hydraulically operated piece of shop equipment which can lift the entire vehicle, or in some cases, one end of a vehicle.

Celsius In the metric system, a temperature scale on which water boils at 100 degrees and freezes at 0 degrees; equal to a reading on a Fahrenheit thermometer of $\frac{5}{9}$ (°F − 32). Also called *centigrade*.

Centigrade See "Celsius."

Centimeter (cm) In the metric system, a unit of linear measure equal to approximately 0.39 inch.

Centrifugal clutch A clutch that uses centrifugal force to apply a higher pressure against the friction disk as the clutch spins faster.

Chassis The assembly of mechanisms that makes up the major operating part of the vehicle. It is usually assumed to include everything except the car body.

Check valve A valve that opens to permit the passage of air or fluid in one direction only or operates to check, or prevent, excessive pressure rise or other undesirable action.

Clearance The space between two moving parts or between a moving and a stationary part, such as a journal and a bearing. Bearing clearance is considered to be filled with lubricating oil when the mechanism is running.

Clutch In the vehicle, the mechanism in the power train that connects the engine crankshaft to or disconnects it from the transmission and thus the remainder of the power train.

Clutch disk See "Friction disk."

Clutch fork In the clutch, a Y-shaped member, into which is assembled the throw-out bearing.

Clutch gear See "Clutch shaft."

Clutch housing A metal housing that surrounds the flywheel and clutch assembly.

Clutch pedal A pedal in the driver's compartment that operates the clutch.

Clutch shaft The shaft on which the clutch is assembled, with the gear that drives the countershaft in the transmission on one end. It has external splines that can be used by a synchronizer drum to lock the clutch shaft to the main shaft for direct drive.

Coil spring A spring made up of an elastic metal, such as steel, formed into a wire or bar and wound into a coil.

Coil-spring clutch A clutch using coil springs to hold the pressure plate against the friction disk.

Constant-velocity joint Two closely coupled universal joints arranged so that their acceleration-deceleration effects cancel out each other, resulting in an output-drive-shaft speed that is always identical with input-drive-shaft speed.

Coolant The liquid mixture of antifreeze and water used in the cooling system.

Cooling system In the engine, the system that removes heat by the forced circulation of coolant and thereby prevents engine overheating. It includes the water jackets, water pump, radiator, and thermostat.

Core In a radiator, a number of coolant passages surrounded by fins through which air flows to carry away heat.

Corrosion An eating or gradually wearing away of metal, as by the effect of chemical action.

Cotter pin A type of fastener, made in the form of a split pin from soft steel, that can be inserted in drilled holes with the split ends spread for locking.

Countershaft The shaft in the transmission which is driven by the clutch gear; gears on the countershaft drive gears on the main shaft when the latter are shifted "into gear."

Crankshaft The main rotating member, or shaft, of the engine, with cranks to which the connecting rods are attached.

Cubic centimeter (cm^3) A unit of the metric system used to measure volume; equal to approximately 0.061 cubic inch.

Cylinder A round hole or tubular-shaped structure in a block or casting in which a piston reciprocates. In an engine, the circular bore in the block in which the piston moves up and down.

Dashpot A device on the carburetor that prevents excessively sudden closing of the throttle.

Dead axle An axle that simply supports and does not turn or deliver power to the wheel or rotating member.

Degree $\frac{1}{360}$ of the circumference of a circle.

Detent A small depression in a shaft, rail, or rod into which a pawl or ball drops when the shaft, rail, or rod is moved; this provides a locking effect.

Dial indicator A gauge that has a dial face and a needle to register movement; used to measure runout, variations in size, movements too little to be measured conveniently by other means, etc.

Diaphragm A thin rubber sheet used to separate an area

into different compartments; used in fuel pump, modulator valve, vacuum-advance unit, etc.

Diaphragm spring A type of spring which is shaped like a disk with tapering fingers pointed inward or like a wavy disk (crown type).

Diaphragm-spring clutch A type of clutch which uses a diaphragm spring instead of coil springs to apply pressure against the friction disk.

Differential A mechanism between axles that permits one wheel to turn at a different speed than the other while transmitting power from the drive shaft to the wheel axles.

Differential case The metal unit that encases the differential pinions and side gears and to which the ring gear is attached.

Differential side gears The gears driven by the ring gear through the case and pinion gears on the wheel sides of the differential case, which are internally splined to the axle shafts.

Double-Cardan joint A near-constant-velocity universal joint which consists of two Cardan universal joints connected by a coupling yoke.

Double-reduction differential A differential containing an extra set of gears to provide additional gear reduction.

Drive line The driving connection, made up of one or more drive shafts, between the transmission and the differential; consists of the drive shaft with universal and slip joints.

Driven disk Friction disk in a clutch.

Drive pinion A rotating shaft that transmits torque to another gear; used in the differential; also called the *clutch shaft* in the transmission.

Drive shaft An assembly of one or two universal joints connected to a hollow tube and used to transmit torque and motion. A shaft in the power train that extends from the transmission to the differential and transmits power from one to the other.

Dry-disk clutch A clutch in which the friction faces of the friction disk are dry, as opposed to a wet-disk clutch which runs submerged in oil.

Dry friction The friction between two dry solids.

Dynamometer A device for measuring the power output, or brake horsepower, of an engine; may be an engine dynamometer, which measures power output at the flywheel, or a chassis dynamometer, which measures the power output at the drive wheels.

Eccentric A disk or offset section, of a shaft, for example, used to convert rotary to reciprocating motion.

Efficiency The ratio between the effect produced and the power expended to produce the effect; the ratio between the actual and the theoretical.

Electric system In the automobile, the system that electrically cranks the engine for starting, furnishes high-voltage sparks to the engine cylinders to fire the compressed air-fuel charges, lights the lights, operates the heater motor, radio, and so on. Consists, in part, of the starting motor, wiring, battery, alternator, regulator, ignition distributor, and ignition coil.

End play As applied to the crankshaft, the distance that the crankshaft can move forward and back.

Energy The capacity or ability to do work.

Engine A machine that converts heat energy into mechanical energy. The assembly that burns fuel to produce power, sometimes referred to as the *power plant*.

Engine tune-up The procedure of checking and adjusting various engine components so that the engine is restored to top operating condition.

Epoxy A plastic compound that can be used to repair some types of cracks in metal.

Expansion plug A plug that is slightly dished out, used to seal core passages in the cylinder block and cylinder head. When driven into place, it is flattened and expanded to fit tightly.

Extreme-pressure (EP) lubricant A special lubricant for use in hypoid-gear differentials, needed because of the heavy wiping loads imposed on gear teeth.

Fatigue failure A type of metal failure resulting from repeated stress which finally alters the character of the metal so that it cracks. In engine bearings, frequently caused by excessive idling or slow engine idle speed.

Feeler gauge Strips of metal of accurately known thicknesses used to measure clearances.

File A cutting tool with a large number of cutting edges arranged along a surface.

Filter That part in the lubricating or fuel system through which fuel, air, or oil must pass so that dust, dirt, or other contaminants are removed.

Floor jack A small, portable, hydraulically operated lifting device to raise part of a vehicle from the floor for repairs.

Fluid coupling A device in the power train consisting of two rotating members. It transmits power through a fluid from the engine to the remainder of the power train.

Flywheel The rotating metal wheel attached to the crankshaft which helps even out the power surges from the power strokes and also serves as part of the clutch and engine-cranking system.

Four on the floor Slang for a four-speed transmission that has its shift lever mounted on the floor of the driving compartment, frequently as part of a center console.

Four-wheel drive A vehicle that has driving axles that can be engaged on the front and rear so that all four wheels can be driven.

Frame The assembly of metal structural parts and channel sections that supports the engine and body and is supported by the car wheels.

Friction The resistance to motion between two bodies in contact with each other.

Friction bearings Bearings having sliding contact between the moving surfaces. Sleeve bearings, such as those used in connecting rods, are friction bearings.

Friction disk In the clutch, a flat disk faced on both sides with friction material and splined to the clutch shaft. It is positioned between the clutch pressure plate and the engine flywheel. Also called *clutch disk* and *driven plate*.

Front-end drive A vehicle having its drive wheels located on the front axle.

Fuel system In the automobile, the system that delivers

to the engine cylinders the combustible mixture of vaporized fuel and air. It consists of the fuel tank, lines, gauge, carburetor, fuel pump, and intake manifold.

Full throttle Wide-open-throttle position with the accelerator pressed all the way down to the floorboard.

Gasket A flat strip, usually of cork or metal or both, placed between two machined surfaces to provide a tight seal between them.

Gasket cement A liquid adhesive material, or sealer, used to apply gaskets; in some applications the liquid layer of gasket cement is used as the gasket.

Gear lubricant A type of grease or oil designed especially to lubricate gears.

Gear ratio The relative speeds at which two gears (or shafts) turn; the proportional rate of rotation.

Gears Mechanical devices to transmit power, or turning effort, from one shaft to another; gears contain teeth that interlace, or mesh, as the gears turn.

Gearshift A linkage-type mechanism by which the gears in an automobile transmission are engaged.

Gear-type pump A pump using a pair of matching gears that rotate; meshing of the gears forces oil (or other liquid) from between the teeth through the pump outlet.

Generator A device that converts mechanical energy into electric energy; it can produce either ac or dc electricity. In automotive usage, the term applied to a dc generator (such generators are now seldom used).

Governor A device that governs or controls another device, usually in accordance with speed, or rpm. The governor used in certain automatic transmissions is an example; it controls gear shifting in relation to car speed.

Grease Lubricating oil to which thickening agents have been added.

Greasy friction The friction between two solids coated with a thin film of oil.

Grinder A machine for removing metal by means of an abrasive wheel or stone.

Grinding wheel A wheel, made of abrasive material, used for grinding metal objects held against it.

Helical gear A gear in which the teeth are cut or twisted at an angle to the center line of the gear.

Heli-Coil A rethreading device to repair worn or damaged threads. It is installed in a retapped hole to bring the screwthread down to original size.

Hone An abrasive stone that is rotated in a bore or bushing to remove material.

Horsepower (hp) A measure of mechanical power, or the rate at which work is done. One horsepower equals 33,000 foot-pounds of work per minute.

Hotchkiss drive The type of rear suspension in which the springs absorb the rear-end torque.

Hydraulic brakes A brake system that uses hydraulic pressure to force the brake shoes against the brake drums or rotors as the brake pedal is depressed.

Hydraulic clutch A clutch which uses hydraulic pressure to actuate the clutch. Used in heavy-duty equipment and where the engine is away from the driver's compartment so that it would be difficult to use mechanical linkages.

Hydraulic press A piece of shop equipment used to develop a heavy force against an object by use of a hydraulic-piston-and-jack assembly.

Hydraulic pressure Pressure exerted through the medium of a liquid.

Hydraulics A branch of science dealing with the use of liquids under pressure as a means of operation to transfer force or motion, or to increase the force applied.

Hydraulic valve A valve in a hydraulic system that operates on, or controls, hydraulic pressure in the system.

Hypoid gear A type of gear used in the differential (drive pinion and ring gear) cut in a spiral form to permit setting the pinion below the center line of the ring gear. This construction enables the floor of the car to be lower.

Idle port The opening into the carburetor throttle body through which the idle system in the carburetor discharges fuel.

Idle speed The speed, or rpm, at which the engine runs without load when the accelerator pedal is released.

Ignition coil That part of the ignition system which acts as a transformer to step up the battery voltage to many thousands of volts; the high-voltage surge then produces a spark at the spark-plug gap.

Ignition distributor That part of the ignition system which closes and opens the circuit to the ignition coil with correct timing and distributes the resulting high-voltage surges from the ignition coil to the proper spark plugs.

Ignition system In the automobile, the system that furnishes high-voltage sparks to the engine cylinders to fire the compressed air-fuel charges. Consists of the battery, ignition coil, ignition distributor, ignition switch, wiring, and spark plugs.

Impeller The rotating finned disk used in a centrifugal pump, such as the water pump, and in the torque converter.

Inertia Property of objects that causes them to resist any change in speed or direction of travel.

Integral Built into, as part of the whole.

Interchangeability The basis of mass production; manufacture of similar parts so that any of these parts can be assembled into a device and the part will fit and operate properly.

Internal gear A gear with teeth pointing inward toward the hollow center of the gear.

Jack stand Also called a *car stand,* or *safety stand.* A pinned, or locked, type of safety stand placed under a car to support its weight after the car is raised with a floor jack.

Journal The part of a rotating shaft which turns in a bearing.

Key A wedgelike metal piece, usually rectangular or semicircular, inserted in grooves to transmit torque while holding two parts in relative position; the small strip of metal with coded peaks and grooves used to operate a lock, such as in the ignition switch.

Kickdown A system in an automatic transmission which produces a downshift when the accelerator is pushed down to the floorboard.

Kilogram (kg) In the metric system, a unit of weight, or mass, approximately equal to 2.2 pounds.

Kilometer (km) In the metric system, a unit of linear measure equal to 0.621 mile.

Kilowatt (kW) In the metric system, a measure of power. One horsepower equals 0.746 kilowatt.

Kinetic energy The energy of motion; the energy stored in a moving body as developed through its momentum, for example, the kinetic energy stored in a rotating flywheel.

kW See "Kilowatt."

Leaf spring A spring made up of a single leaf, or a series of flat steel plates of graduated length assembled one on top of another, to absorb road shocks by bending, or flexing, in the middle.

Limited-slip differential A differential that supplies, when one wheel is slipping, a major portion of the drive torque to the wheel having the better traction; also called a *nonslip differential.*

Linkage An assembly of rods, or links, to transmit motion.

Liter (l) In the metric system, a measure of volume equal to approximately 0.2642 U.S. gallon.

Live axle An axle, which has the wheels rigidly attached to it, that also drives the attached wheels.

Locknut A second nut turned down on a holding nut to prevent loosening.

Lubricating system The system in the engine that supplies moving engine parts with lubricating oil to prevent actual contact between any moving metal surfaces.

Lugging Low-speed, full-throttle engine operation in which the engine is heavily loaded and overworked; usually caused by failure of the driver to shift to a lower gear.

Machining The process of using a machine to remove metal from a metal part.

Manual low Position of the units in an automatic transmission when the driver moves the shift lever to the low- or first-gear position on the quadrant.

Mechanism A system of interrelated parts that make up a working assembly.

Member Any essential part of a machine or assembly.

Meshing The mating, or engaging, of the teeth of two gears.

Meter (m) In the metric system, a unit of linear measure equal to 39.37 inches. Also, the name given to a test instrument, such as an ammeter that measures as a result of the substance being measured passing through it. Also, any device that measures and controls the discharge of the substance passing through it. For example, a carburetor jet is used to meter fuel flow.

Micrometer A precision measuring device that measures small distances, such as crankshaft or cylinder-bore diameter, or thickness of an object. Also called a *mike.*

Millimeter (mm) In the metric system, a unit of linear measure equal to approximately 0.039 inch.

Mode A term used to designate a particular set of operating characteristics.

Modulator A pressure-regulated, governing device used, for example, in automatic transmissions.

Multiple-disk clutch A clutch which has more than one friction disk; usually there are several driving disks and several driven disks, alternately placed.

Needle bearing Antifriction bearing of the roller type; the rollers are very small in diameter (needle-sized).

Neoprene A synthetic rubber that is not affected by various chemicals harmful to natural rubber.

Neutral In a transmission, the condition that exists when all gears are disengaged and the output shaft is disconnected from the drive wheels.

Neutral start switch A switch wired into the ignition switch which prevents engine cranking unless the transmission shift lever is in neutral.

Nonslip differential See "Limited-slip differential."

Odometer The meter, usually set in the speedometer, that measures the distance the vehicle travels in miles or kilometers.

Oil cooler A small radiator through which the oil flows to lower its temperature.

Oil filter The filter through which the crankcase oil passes to remove any impurities from the oil.

Oil seal A seal placed around a rotating shaft or other moving part to prevent leakage of oil.

Oil strainer A wire-mesh screen placed at the inlet end of the oil-pump-pickup tube to prevent dirt and other large particles from entering the oil pump.

One-way clutch See "Sprag."

Orifice A small opening, or hole, into a cavity.

O ring A type of sealing ring, made of special rubberlike material, which is compressed into grooves to provide the sealing action.

Output shaft The main shaft; the shaft from the transmission that delivers torque to the drive shaft.

Overcenter spring A spring used in some clutch linkages to reduce the foot pressure required to depress the clutch pedal.

Overdrive A device in the power train, usually mounted behind the transmission, which introduces an extra set of gears into the power train. This causes the drive shaft to overdrive, or drive faster than, the engine crankshaft.

Overheat A substance, or material, that is heated to excess; something that gets too hot is said to overheat.

Overrunning-clutch drive A type of clutch drive which transmits rotary motion in one direction only; when rotary motion attempts to pass through in the other direction, the driving member overruns and does not pass the motion to the other member. Widely used as the drive mechanism for starting motors.

Parking brakes Mechanically operated brakes that operate independently of the foot-operated service brakes on the vehicle; they are set when the vehicle is parked.

Passage A small hole or gallery in an assembly or casting through which air, coolant, fuel, or oil flows.

Pawl An arm, pivoted so that its free end can fit into a detent, slot, or groove during certain times to hold a part stationary.

Pilot bearing A small bearing in the center of the flywheel end of the crankshaft which carries the forward end of the clutch shaft.

Pilot shaft A shaft used to align parts which is removed before final installation of the parts to be used; a dummy shaft.

Pinion gear The smaller gear of two meshing gears.

Piston A movable part, fitted in a cylinder, which can receive or transmit motion as a result of pressure changes (fluid, vapor, combustion) in the cylinder.

Planetary-gear system A gearset consisting of a central sun gear surrounded by two or more planet pinions which are, in turn, meshed with a ring (or internal) gear; used in overdrives and automatic transmissions.

Planet carrier In a planetary-gear system, the carrier, or bracket, that contains the shaft upon which the pinion or planet gears turn.

Planet pinions In a planetary-gear system, the pinions that mesh with and revolve about the sun gear. They also mesh with the internal, or ring, gear.

Plastic gasket compound A plastic paste in a tube which can be squeezed out to make a gasket in any shape.

Pivot A pin or short shaft upon which another part rests or turns.

Pour point The lowest temperature at which an oil will flow.

Power plant The engine, or power-producing mechanism, on the vehicle.

Power takeoff An attachment for connecting the engine to devices or other machinery when its use is required.

Power train The group of mechanisms that carry the rotary motion developed in the engine to the car wheels, it includes the clutch, transmission, drive shaft, differential, and axles.

Preload In bearings, the amount of load originally imposed on a bearing before actual operating loads are imposed. This is done by bearing adjustments and ensures alignment and minimum looseness in the system.

Pressure plate That part of the clutch which exerts pressure against the friction disk; it is mounted on and rotates with the flywheel.

Pressure regulator A regulating device which operates to prevent excessive pressure from developing; in the hydraulic systems of certain automatic transmissions, a valve that opens to release oil from a line when the oil pressure attains specified maximum limits.

Pressure-relief valve A valve in the oil line that opens to relieve excessive pressures that the oil pump might develop.

Pressurize To apply more than atmospheric pressure to a gas or liquid.

Propeller shaft See "Drive shaft."

Psi Abbreviation for pounds per square inch; often used to indicate pressure of a liquid or gas.

Puller Generally, a shop tool that permits removal of one closely fitted part from another without damage. Often contains a screw, or screws, which can be turned to apply gradual pressure.

Pump A device that transfers gas or liquid from one plac to another.

Quadrant A term sometimes used to identify the shift-lever selector mounted on the steering column.

Reactor The stator in the torque converter which provides reactive blades against which the fluid can change directions (under certain conditions) as it passes from the turbine to the pump.

Reamer A metal-cutting tool with a series of sharp cutting edges that remove material from a hole when the reamer is turned in it.

Rear-end torque Reaction torque applied to the rear-axle housing as torque is applied to the wheels, which attempts to turn the axle housing in a direction opposite to wheel rotation.

Reassembly Putting the parts of a device back together.

Reciprocating motion Motion of an object between two limiting positions: motion in a straight line, back and forth, or up and down, etc.

Release bearing See "Throw-out bearing."

Release levers In the clutch, the levers that are moved by the throw-out-bearing movement, causing clutch-spring pressure to be relieved so that the clutch is released, or uncoupled from the flywheel.

Relief valve A valve that opens, when a predetermined pressure is reached, to prevent excessive pressure.

Reverse-idler gear In a transmission, an additional gear that must be meshed to obtain reverse gear; a gear used only in reverse, idling when the transmission is in any other position.

Ring gear A large gear carried by the differential case, meshed with and driven by the drive pinion.

Rotor oil pump A type of oil pump using a pair of rotors, one inside the other, to produce the oil pressure required to circulate oil to engine parts.

rpm Abbreviation for revolutions per minute.

Runout Wobble.

Scored Scratched or grooved, as a cylinder wall may be scored by abrasive particles moved up and down by the piston rings.

Screen A fine-mesh screen in the fuel and lubricating systems that prevents large particles from entering the system.

Seal A material, shaped around a shaft, used to close off the operating compartment of the shaft, preventing oil leakage.

Sealer A thick, tacky, compound, usually spread with a brush, which may be used as a gasket, or sealant, to seal small openings or surface irregularities.

Servo A device in a hydraulic system that converts hydraulic pressure into mechanical movement. It consists of a piston which moves in a cylinder as hydraulic pressure acts on it.

Shift lever The lever used to change gears in a transmission; also, the lever on the starting motor which moves the drive pinion into or out of mesh with the flywheel teeth.

Shift valve In an automatic transmission, a valve that moves to produce the shifts from one gear ratio to another.

Shim A slotted strip of metal used as a spacer to adjust front-end alignment on many cars and to make small corrections in the position of body sheet metal and other parts.

Shim stock Sheets of metal of accurately known thickness which can be cut into strips and used to measure or correct clearance.

Shrink fit A tight fit of one part in another, achieved by heating or cooling one part and then assembling it with the other part. If heated, the part shrinks on cooling to provide a shrink fit. If cooled, the part expands on warming to provide the fit.

Slip joint In the power train, a variable-length connection that permits the drive shaft to change effective length.

Solenoid Device which, when connected to an electrical source, such as a battery, produces a mechanical movement. This movement can control a valve or produce other controlling movements.

Speedometer Instrument that indicates vehicle speed, usually driven from the transmission.

Spline Slot or groove cut in a shaft or bore; a splined shaft onto which a hub, wheel, gear, etc., with matching splines in its bore is assembled so that the two must turn together.

Sprag In an automatic transmission, a one-way clutch; a clutch in which power can be transmitted in one direction but not in the other.

Spring An elastic device which yields under stress or pressure but returns to its original state or position when the stress or pressure is removed; the operating component of the automotive suspension system which absorbs the force of road shock by flexing and twisting.

Spur gear A gear in which the teeth parallel the center line of the gear.

Starting motor The electric motor in the electric system that cranks the engine, or turns the crankshaft, for starting.

Stator In the torque converter, a third member (in addition to turbine and pump) which changes direction of fluid under certain operating conditions (when stator is stationary). In the alternator, the assembly that includes the stationary conductors.

Steering-column shift An arrangement in which the transmission shift lever is mounted on the steering column.

Steering system The mechanism that enables the driver to turn the wheels for changing the direction of vehicle movement.

Stud A headless bolt that is threaded on both ends.

Stumble A term related to vehicle drivability; the tendency of an engine to falter and then catch, resulting in a noticeable stumble felt by the driver.

Sun gear In a planetary-gear system, the center gear, which meshes with a series of planet pinions.

Synchronize To make two or more events, operations, or gears, occur at the same time.

Synchronizer A device in the transmission that synchronizes gears about to be meshed so that there will not be any gear clash.

Tachometer A device for measuring engine speed, or rpm.

Taper A shaft or hole that gets gradually smaller toward one end.

Throttle A valve in the carburetor that permits the driver to vary the amount of air-fuel mixture entering the engine, thus controlling the engine speed.

Throttle-return check A device on the carburetor that prevents excessively sudden closing of the throttle; also called a *dashpot*.

Throttle valve The round disk valve in the throttle body of the carburetor that can be turned to admit more or less air, thereby controlling engine speed.

Throw-out bearing In the clutch, the bearing that can be moved in to the release levers by clutch-pedal action so as to cause de-clutching, or a disconnection between the engine crankshaft and power train.

Timing In the engine, refers to timing of valves and timing of ignition and their relation to the piston position in the cylinder.

Timing light A light that is connected to the ignition system to flash each time the No. 1 spark plug fires; used for adjusting the timing of the ignition spark.

Torque Turning or twisting effort, usually measured in pound-feet or kilogram-meters.

Torque converter In an automatic transmission, a fluid coupling which incorporates a stator to permit a torque increase through the torque converter.

TorqueFlite An automatic transmission used on Chrysler-manufactured cars. It has three forward speeds and reverse.

Torque wrench A special wrench that indicates the amount of torque being applied to a nut or bolt.

Transmission The device in the power train that provides different gear ratios between the engine and rear wheels, as well as reverse.

Transmission-controlled spark (TCS) system An exhaust-emission-control system which allows distributor vacuum advance in high gear only; widely used on General Motors vehicles.

Transmission oil cooler A small radiator, mounted separately or part of the engine radiator, to cool the transmission lubricating oil.

Transmission-regulated spark (TRS) system Name used by Ford to identify an exhaust-emission-control system; similar to the General Motors transmission-controlled-spark system; an exhaust-emission-control system which allows distributor-vacuum advance in high gear only.

Trouble diagnosis The detective work necessary to run down the cause of a trouble. Also implies the correction of the trouble by elimination of the cause.

Turbine A device that produces rotary motion as a result of gas, vapor, or hydraulic pressure. In the torque converter, the driven member.

Turbo Hydra-Matic An automatic transmission used on many models of General Motors car. It has three forward speeds and reverse.

Two-disk clutch A clutch having two friction disks for

additional holding power. Used in heavy-duty equipment.

Universal joint In the power train, a jointed connection in the drive shaft that permits a change of driving angle.
Upshift To shift a transmission into a higher gear.

Vacuum An absence of air or other substance.
Vacuum gauge In automotive-engine service, a device that measures intake-manifold vacuum and thereby indicates actions of engine components.
Vacuum modulator A device in automatic transmissions that modulates, or changes, the mainline hydraulic pressure to meet changing engine loads.
Valve A device that can be opened or closed to allow or stop the flow of a liquid, gas, or vapor from one place to another.
Vanes Blades attached to a rotating shaft or drum.
Vent An opening from an enclosed chamber through which air can pass.
Vibration A complete rapid motion back and forth; oscillation.
Viscosity The resistance of a liquid to flow. A thick oil has greater viscosity than a thin oil.
Viscous Thick, tending to resist flowing.
Viscous friction Friction between layers of a liquid.
Vise A gripping device for holding a piece while it is being worked on.

Welding The process of joining pieces of metal by fusing them together with heat.
Wet-disk clutch A clutch in which the friction disk (or disks) is operated in a bath of oil.
Wire feeler gauge A round wire gauge used for checking clearances between electrical contacts, such as distributor points and spark plug electrodes.
Work The changing of the position of a body against an opposing force, measured in foot-pounds or meter-kilograms. Product of force times the distance through which it acts.

INDEX

A

B

C

D

F

G

H

L

M

N

O

U

V

ANSWERS TO QUESTIONS

The answers to the questions in the progress quizzes and chapter checkups are given here. In chapter checkups, you may be asked to list parts in components, describe the purpose and operation of components, define certain terms, and so on. Obviously, no answers to these could be given here since that would mean repeating substantially the entire book. Therefore, you are asked to refer to the book to check your answers.

If you want to figure your grade on any quiz, divide the number of questions in the quiz into 100. This gives you the value of each question. For instance, suppose there are 8 questions: 8 goes into a hundred 12.5 times. Each correct answer therefore gives you 12.5 points. If you answer 6 correct out of the 8, then your grade would be 75 (6 × 12.5).

If you are not satisfied with the grade you make on a quiz or checkup, restudy the chapter or section and retake the test. This review will help you remember the important facts.

Remember, when you take a course in school, you can pass and graduate even though you make a grade of less than 100. But in the automotive shop, you must score 100 percent all the time. If you make 1 error out of 100 service jobs, for example, your average would be 99. In school that is a fine average. But in the automotive shop that one job you erred on could cause such serious trouble (a ruined engine or a wrecked car) that it would outweigh all the good jobs you performed. Therefore, always proceed carefully in performing any service job and make sure you know exactly what you are supposed to do and how you are to do it.

CHAPTER 1

Progress Quiz 1-1

Correcting Parts Lists 1. hoist 2. steering knuckle 3. rear axle 4. 16:1

Completing the Sentences 1. (b) 2. (d) 3. (c) 4. (d) 5. (b) 6. (c) 7. (b) 8. (b)

Progress Quiz 1-2

Correcting Parts Lists 1. motor 2. crankshaft 3. knee joint 4. countershaft bevel gear

Completing the Sentences 1. (c) 2. (a) 3. (c) 4. (b) 5. (d) 6. (a) 7. (c) 8. (b)

Chapter 1 Checkup

Completing the Sentences 1. (d) 2. (c) 3. (d) 4. (b) 5. (c) 6. (c) 7. (c) 8. (c)

CHAPTER 2

Chapter 2 Checkup

Completing the Sentences 1. (c) 2. (c) 3. (a) 4. (c) 5. (b) 6. (c) 7. (d) 8. (c) 9. (a) 10. (c)

CHAPTER 3

Progress Quiz 3-1

Correcting Troubles Lists 1. pressure springs weak 2. high-octane gas 3. differential out of line 4. gears worn 5. drive shaft out of line 6. spring weak 7. loose floating axle. 8. loose friction-disk facings

Chapter 3 Checkup

Completing the Sentences 1. (c) 2. (c) 3. (c) 4. (c) 5. (c) 6. (b) 7. (a) 8. (c)

CHAPTER 4

Progress Quiz 4-1
Completing the Sentences 1. (b) 2. (a) 3. (c) 4. (b) 5. (a) 6. (b)

Chapter 4 Checkup
Completing the Sentences 1. (a) 2. (c) 3. (b) 4. (b) 5. (b) 6. (c) 7. (c) 8. (b)

CHAPTER 5

Chapter 5 Checkup
Completing the Sentences 1. (b) 2. (c) 3. (c) 4. (c) 5. (b) 6. (b) 7. (b) 8. (b) 9. (a) 10. (a)

CHAPTER 6

Progress Quiz 6-1
Correcting Troubles Lists 1. crankshaft worn 2. distributor jammed 3. clutch not releasing 4. idler gear loose 5. clutch slipping 6. drive shaft bent 7. synchromesh cones worn 8. steering gear worn 9. overrunning clutch slipping 10. clutch slipping

Chapter 6 Checkup
Completing the Sentences 1. (a) 2. (b) 3. (b) 4. (c) 5. (b) 6. (b) 7. (b) 8. (b)

CHAPTER 7

Progress Quiz 7-1
Completing the Sentences 1. (b) 2. (a) 3. (a) 4. (c) 5. (d) 6. (a) 7. (a) 8. (c) 9. (c) 10. (b)

Chapter 7 Checkup
Completing the Sentences 1. (c) 2. (b) 3. (c) 4. (b) 5. (d) 6. (c) 7. (c) 8. (a) 9. (b) 10. (b)

CHAPTER 8

Chapter 8 Checkup
Completing the Sentences 1. (c) 2. (c) 3. (a) 4. (b) 5. (c) 6. (a) 7. (b) 8. (c) 9. (a) 10. (b)

CHAPTER 9

Chapter 9 Checkup
Completing the Sentences 1. (b) 2. (c) 3. (a) 4. (c) 5. (c) 6. (b) 7. (a) 8. (b) 9. (b) 10. (c)

CHAPTER 10

Progress Quiz 10-1
Completing the Sentences 1. (b) 2. (d) 3. (b) 4. (c)

Correcting Troubles Lists 1. forward clutch locked 2. clutch not releasing 3. band not holding 4. brake band not applying 5. forward clutch not applying

Chapter 10 Checkup
Completing the Sentences 1. (b) 2. (a) 3. (c) 4. (c) 5. (c) 6. (b) 7. (c) 8. (c)

CHAPTER 11

Chapter 11 Checkup
Completing the Sentences 1. (b) 2. (a) 3. (c) 4. (a) 5. (b) 6. (b) 7. (a) 8. (b) 9. (a) 10. (b) 11. (c) 12. (c) 13. (a) 14. (b) 15. (b)

CHAPTER 12

Chapter 12 Checkup
Completing the Sentences 1. (b) 2. (c) 3. (b) 4. (c) 5. (a) 6. (c) 7. (a) 8. (b)

CHAPTER 13

Progress Quiz 13-1
Completing the Sentences 1. (c) 2. (b) 3. (a) 4. (c) 5. (c) 6. (a) 7. (a) 8. (b) 9. (b) 10. (c)

Chapter 13 Checkup
Completing the Sentences 1. (c) 2. (b) 3. (c) 4. (b) 5. (a) 6. (b) 7. (c) 8. (a) 9. (b) 10. (c)

CHAPTER 14

Chapter 14 Checkup
Completing the Sentences 1. (c) 2. (c) 3. (a) 4. (c) 5. (b) 6. (a) 7. (b) 8. (b) 9. (a) 10. (b)

CHAPTER 15

Chapter 15 Checkup
Completing the Sentences 1. (b) 2. (c) 3. (a) 4. (b) 5. (c) 6. (a) 7. (b) 8. (c)

CHAPTER 16

Progress Quiz 16-1
Correcting Troubles Lists 1. forward clutch not releasing 2. forward clutch not applying 3. intermediate clutch not applying 4. governor valve stuck 5. front band not applying 6. governor valve stuck 7. direct clutch not releasing 8. bands not applying 9. front band not applying 10. direct clutch not applying

Chapter 16 Checkup
Completing the Sentences 1. (b) 2. (a) 3. (c) 4. (b) 5. (b)

CHAPTER 17

Chapter 17 Checkup
Completing the Sentences 1. (b) 2. (c) 3. (c) 4. (b) 5. (c) 6. (c) 7. (a) 8. (c) 9. (b) 10. (b)

CHAPTER 18

Progress Quiz 18-1
Completing the Sentences 1. (c) 2. (c) 3. (b) 4. (a) 5. (a) 6. (b) 7. (b) 8. (a) 9. (c) 10. (b)

Chapter 18 Checkup
Completing the Sentences 1. (b) 2. (a) 3. (c) 4. (b) 5. (b) 6. (c) 7. (a) 8. (c)

CHAPTER 19

Chapter 19 Checkup
Completing the Sentences 1. (b) 2. (a) 3. (b) 4. (c) 5. (b) 6. (c) 7. (b) 8. (a) 9. (a) 10. (c)

CHAPTER 20

Progress Quiz 20-1
Completing the Sentences 1. (b) 2. (b) 3. (a) 4. (c) 5. (a) 6. (b) 7. (c) 8. (b)

Chapter 20 Checkup
Completing the Sentences 1. (c) 2. (a) 3. (b) 4. (b) 5. (c) 6. (c) 7. (a) 8. (b)

CHAPTER 22

Chapter 22 Checkup
Completing the Sentences 1. (b) 2. (c) 3. (a) 4. (d) 5. (d) 6. (b) 7. (b) 8. (a)

CHAPTER 23

Progress Quiz 23-1
Completing the Sentences 1. (b) 2. (a) 3. (c) 4. (c) 5. (a) 6. (b) 7. (a) 8. (a) 9. (b) 10. (c)

Chapter 23 Checkup
Completing the Sentences 1. (b) 2. (a) 3. (b) 4. (a) 5. (b) 6. (b) 7. (b) 8. (a) 9. (b) 10. (c)